EXPLAINING
PSYCHOLOGICAL
STATISTICS

THIRD EDITION

LIBRARY

Learning
Resource Centre

EXPLAINING PSYCHOLOGICAL STATISTICS

THIRD EDITION

Barry H. Cohen
New York University

John Wiley & Sons, Inc.

For Leona

PART **Two**
ONE- AND TWO-SAMPLE HYPOTHESIS TESTS   **123**

Chapter 5 ――――――――――――――――
INTRODUCTION TO HYPOTHESIS TESTING:
THE ONE-SAMPLE *z* TEST   **123**

Chapter 6
INTERVAL ESTIMATION AND THE *t* DISTRIBUTION   156

Chapter 10
LINEAR REGRESSION   286

Chapter 11

THE MATCHED t TEST   317

PART Five
ANALYSIS OF VARIANCE WITH REPEATED MEASURES

Chapter 15
REPEATED MEASURES ANOVA

Chapter 16 _____
TWO-WAY MIXED DESIGN ANOVA   528

PART Six
MULTIPLE REGRESSION AND ITS CONNECTION TO ANOVA   571

Chapter 17
MULTIPLE REGRESSION   571

Chapter 18

THE REGRESSION APPROACH TO ANOVA   627

Chapter 21
STATISTICAL TESTS FOR ORDINAL DATA   728

Appendix A
STATISTICAL TABLES   **757**

Appendix B
ANSWERS TO SELECTED EXERCISES   **777**

The second edition of this text raised its level from that of an advanced course for undergraduates to a graduate-level course primarily for students in a master's program in psychology or a related field. This edition raises the level a bit further, making the text more suitable for doctoral programs in psychology. However, this revision retains the modular nature of the text, making it relatively easy to ignore the more advanced material, allowing for its use in master's level, and even advanced undergraduate courses.

The major additions to this edition begin in Chapter 5, wherein I have included a spam-filter analogy in order to explain Bayes' Theorem in a concrete form, and demonstrated how Bayes' Theorem can be used to gain a deeper understanding of null hypothesis testing (NHT). I return to the spam-filter analogy in Chapter 8 as part of my explanation of power analysis and the practical effects of NHT in serving as a screen against small effects in the population. In Chapter 6, I added material on the calculation of robust statistics—especially as based on the 20% trimmed mean recommended by Wilcox (2003). I return to the use of trimmed means in the context of two-sample t tests in Chapter 7.

The largest additions come from a chapter that I posted on the Web as a supplement to the second edition, and which was labeled "Chapter 22." The three-way (independent-samples) ANOVA has been added to the end of Chapter 14, and MANOVA has been added to the end of Chapter 18. I moved the topic of Trend Analysis up from Chapter 18 to Chapter 13, which allowed me to include the analysis of trend components in Chapters 14, 15 and 16 (and made room in Chapter 18 for MANOVA). Finally, I have expanded my coverage of effect-size estimates (including measures of proportion of variance accounted for) in Chapters 12, 14, 15, 16, and 21. The additions just described principally involve the C sections of the chapters (as explained next), though in many cases I moved to Section B material that had been in Section C of that chapter in the second edition. To make room for some of these additions I deleted some topics that seemed less central to the theme of this text (e.g., calculating percentiles by use of a formula, the sampling distribution of the variance, the intraclass correlation).

ABC Format

As adopters of previous editions will recall, each chapter is divided into three sections, labeled A, B, and C. Section A provides the "Conceptual Foundation" for the chapter. In this section I focus on the simplest case of the procedure dealt with in that chapter (e.g., one-way ANOVA with equal-sized groups; 2×2 balanced ANOVA), and explain the definitional formulas so that students can gain some insight into why and how statistical formulas work (e.g., the sample size is in the denominator of the denominator of the one-sample test, so it is effectively in the numerator—making the sample size larger tends to increase the size of z or t). In my experience, students learn statistics more easily not when statistical formulas are presented as arbitrary strings of characters to be memorized or looked up when needed but rather when the connections between seemingly diverse formulas are elucidated, and the structure of the formulas are made clear (e.g., the SS components of a repeated-measures ANOVA are calculated exactly as a corresponding two-way ANOVA with "subjects" as the second factor; the error term of the RM ANOVA is explained in terms of the interaction between subject "levels" and the treatment levels). Some instructors may prefer an approach in which concepts are explained first without reference to statistical formulas at all. I don't feel I could do that well, nor is it my goal in this text. I believe that all of the formulas in this text can be made understandable to the average graduate student in psychology (or advanced undergraduate), and I believe that it is worth the effort to do so.

Section A has its own exercises to help ensure that students grasp the basic concepts and definitional formulas before moving on to the complications of Section B. Section B, "Basic Statistical Procedures," presents the more general case of that chapter's procedure and includes computational formulas, significance tests, and comments on research design so that students will be equipped to analyze real data and interpret their results. Section B also includes supplementary statistical procedures (e.g., confidence intervals) and information on how to report such statistical results in the latest APA format, usually illustrated with an excerpt from a published journal article. Section B has a variety of exercises so that students can practice the basic computations. Moreover, these exercises usually refer to previous exercises, sometimes in previous chapters, to make instructive comparisons (e.g., that a one-way RM ANOVA can be calculated on the same data that had been used to illustrate the computation of a matched t test).

Section C presents "Optional Material" that is usually more conceptually advanced and less central to students' needs than the topics covered in Sections A and B, and can be omitted without any loss in continuity. (Note that the Section Cs of later chapters often refer to material presented in the Section C of a previous chapter.) Students who have both read and successfully performed the exercises for Sections A and B should be ready to tackle Section C, which contains its own exercises. Even if an instructor decides not to assign Section C for a particular chapter, its presence in that chapter means that students who want to go beyond their course will have the opportunity to do so (perhaps after the course has ended, and the student is using the text as a reference for analyzing data). One option is that instructors can use some of the C Sections as extra-credit assignments.

Each of the three sections ends with a detailed summary, and the chapter ends with an annotated list of all the major formulas in that chapter.

The Organization of the Text

This edition retains the basic organization of the previous two editions, including my personal (and sometimes idiosyncratic) preferences for the ordering of the chapters. Fortunately, adopters of the previous editions had no difficulty teaching some of the chapters in a different order than they appear in this text. The main organizational choices, and the rationale for each, are as follows. At the end of Part One (Descriptive Statistics), I describe probability in terms of smooth mathematical distributions only (mainly the normal curve), and postpone any discussion of probability in terms of discrete events until Part Seven (Nonparametric Statistics). In my experience, a presentation of discrete mathematics (e.g., combinatorics) at this point would interrupt the smooth flow from the explanation of areas under the normal distribution to the use of p values in inferential parametric statistics.

I also postpone Correlation and Linear Regression until after completing the basic explanation of (univariate) inferential statistics in terms of the t test for two independent samples. I have never understood the inclusion of correlation as part of descriptive statistics, mainly because I have never seen correlations used for purely descriptive purposes. More controversial is my decision to separate the matched (or repeated-measures) t test from the one- and two-sample tests, and to present the matched test only after the chapters on correlation and regression. My feeling is that the conceptual importance of explaining the increased power of the matched t test in terms of correlation outweighs the advantage of the computational similarity of the matched *t* test to the one-sample test. However, the students' basic familiarity with the concept of correlation makes it reasonable to teach Chapter 11 (the matched *t* test) directly after Chapter 7 (the two-sample *t* test), or even after Chapter 6 (the one-sample *t* test). In sum, the unifying theme of Part Two is the basics of univariate inference, whereas Part Three deals with the different (bivariate) methods that can be applied when each participant is measured twice (or participants are matched in pairs).

Part Four of the text is devoted to the basics of analysis of variance without the added complications of repeated measures. Moreover, by detailing the analysis of the two-way and three-way between-groups ANOVA before introducing repeated measures, I am able to describe the analysis of the one-way and two-way RM ANOVA in terms of those two former analyses, respectively. Part Five includes the two-way mixed design ANOVA, after the simple RM ANOVA is explained. Part Six introduces the basic concepts of multiple regression and then draws several connections between multiple regression and ANOVA, including the significance test for eta squared, the analysis of unbalanced factorial designs, the analysis of covariance, and multivariate analysis of variance.

Finally, Part Seven of the text begins with a demonstration of how the basics of probability with discrete events can be used to construct the binomial distribution and draw inferences from it. More complex inferential statistics for categorical variables are then described in terms of the chi-square test. The discussion of nonparametric statistics concludes with a chapter that presents the most popular methods for analyzing ordinal data, and explains their use as alternatives to the parametric procedures described in previous chapters of the text.

Tailoring the Use of This Text for Different Courses

This edition retains all of the foundational material needed for an introductory course in statistics, although it is hard to imagine any course covering all 21 chapters in a single semester. Parts One, Two, and Three would easily fill the first half of a two-semester sequence, leaving Parts Four through Seven for the second semester (with the possible deletion of Chapter 18 and 21). For those instructors who want to move back and forth between parametric tests and their nonparametric counterparts, it would make sense to teach Chapter 19 directly after Chapter 4 or Chapter 5.

A typical one-semester master's course could begin with Chapter 5, requiring students to refresh Part One on their own, perhaps before the first day of class. Chapter 18 (and even Chapter 17) could easily be left out before proceeding to Chapters 19 and 20 (and, if there is time, Chapter 21). A one-semester doctoral course might leave Part One to the students as needed, move fairly quickly through Parts Two and Three in class, and devote the bulk of the semester to Parts Four, Five, and Six. Doctoral instructors should also see my stats web page (http://www.psych.nyu.edu/cohen/statstext.html) for additional advanced material (e.g., ANOVA designs with nested and/or random effects, three-way mixed designs) that should be posted by the time this edition is published.

Another way to tailor this text is by the inclusion or exclusion of each chapter's C section. A less-advanced statistics course can be created by assigning few if any of the C sections. Note, however, that most of these sections contain more than one topic, so an instructor may want to retain selected parts of the various C sections. Also, note that C sections in later chapters will often build on material presented in the C sections of earlier chapters, although I have been careful to make sure that none of the A or B section material in later chapters requires students to read any part of Section C from an earlier chapter. A beginning master's (or advanced undergraduate) course might skip most of the C section material, while doctoral courses might spend a good deal of time on the C sections.

A Note about the Use of Calculators

Beginning with Chapter 14, this text assumes that the student is using a scientific or statistical calculator that has both the biased and unbiased standard deviations as built-in functions. This is a reasonable assumption in that, as of the publication of this edition, of new calculators, only the simplest four-function calculators and some specialized ones lack these functions. Any student using a calculator that is too old or basic to include these functions should be urged to buy a new one. The new scientific calculators are so inexpensive and will save so much tedious calculation effort that they are certainly worth the purchase. If a student has a calculator with these built-in functions but does not know how to access them and has lost the instruction booklet, it is worth some effort to try to figure out how to put the calculator in statistics mode (this is not always the default), which key to press after each score to enter it, and which keys access the desired functions, but it is not worth a great deal of effort given the affordability of a new calculator.

Appendixes and Supplements

Appendix A contains all the statistical tables that are required by the analyses described in this text. Appendix B contains answers to selected exercises (those marked with asterisks in the text) or, in some cases, selected parts of selected exercises. Note that answers are almost always given for exercises from earlier chapters that are referred to in later chapters. If an exercise you would like to assign refers to a previous exercise, it is preferable to assign the previous exercise as well, but students can usually use Appendix B to obtain the answer to the previous exercise for comparison purposes. There are also several useful supplements that are not in print, but are included without charge on our student and instructor web sites.

Web Site for Instructors

Complete solutions to all the exercises, including those not marked by asterisks, are contained in the Instructor's Manual, which can be found on the publisher's web page for instructors using this text : http://www.wiley.com/college/psyc/cohen007184/instructor.html. Also in the Instructor's Manual are explanations of the concepts each exercise was designed to illustrate. Another useful publication on this site is a battery of test items in multiple-choice format that covers the material in every section of every chapter. Test items cover a range of difficulty, and each item can easily be converted to an open-ended question. These items can also be given to students as additional homework exercises or practice for your own exams if you don't want to use them as test items. This web site can only be accessed by instructors who have been given a password. Supplements that do not need password protection wil be available from a separate location on the publisher's web site labeled "Student Resources."

Web Site for Students

Both instructors and students will want to check out the supplements available on the publisher's web page, without charge or the need for a password, for student users of this text: http://www.wiley.com/college/psyc/cohen 007184/student.html. First, there is a two-part Basic Math Review, which provides practice with the basic arithmetic and algebraic operations that will be required of students to calculate the results from statistical formulas and solve exercises in this text. Students should be encouraged to take the diagnostic quiz at the beginning of the math review to determine how much refreshing of basic math they will be likely to need at the start of their stats course. Basic Math Review contains plenty of practice exercises with answers, as well as a second quiz at the end, so that students can gauge their own progress.

Second, there is a Study Guide, with additional worked-out computational examples from each chapter of the text, accompanied by tips about common errors to avoid, and tricks to make calculations less painful and/or error prone. Each chapter of the Study Guide also includes a glossary of important terms and other useful, supplemental material. To enable more rapid updating of important supplemental material, especially with respect to the use of computer programs for data analysis, other supplements will be made available from my New York University web page.

My Own Statistics Web Pages

I am planning to post two major text supplements, available in pdf format, from links on my main stats web page: http://www.psych.nyu.edu/cohen/statstext.html.

The first consists of a Section D for each of the 21 chapters in my text. Each of these 21 D sections will explain how to use the Statistical Package for the Social Sciences (SPSS; in some cases via its Syntax Window) to perform the statistical procedures presented in the corresponding text chapter, and how to relate the terms and symbols used by SPSS in its output to those used in my text. I will include tips and tricks I have learned over the years, but I will also be able to update these sections for each new, major version of SPSS that is released (while keeping material relevant only to users of earlier versions in brackets). Each D section will conclude with exercises that can be performed with any major statistical software package, along with an answer key. All of these computer exercises will be based on the same large data set, available from my web page as an Excel spreadsheet, an SPSS save file, or a simple rich-text file (rtf).

The second major supplement that will be available from my web page is an additional chapter (a new Chapter 22 to replace the one that was previously on the Web) entitled "Some Advanced ANOVA Designs," which will contain sections on nested designs, designs with random effects, and three-way mixed designs (the latter comes from Section B of the old Chapter 22). My web pages will also include up-to-date errata tables, and additional examples and supplemental study materials from my other texts published by John Wiley and Sons. Instructors and students should check my web site, perhaps monthly, for new additions or corrections.

Finally, I hope that students and instructors alike will feel free to send me corrections, suggestions for future editions, requests for clarification, or just general feedback at barry.cohen@nyu.edu. I look forward to hearing from you, and will try to publicly acknowledge all substantive contributions.

The first edition of this text, and therefore all subsequent editions, owes its existence to the encouragement and able assistance of my former teaching assistant, Dr. R. Brooke Lea, who is now tenured on the faculty of Macalester College. I also remain indebted to the former acquisition editors of Brooks/Cole, whose encouragement and guidance were instrumental to the publication of the first edition: Philip Curson, Marianne Taflinger, and James Brace-Thompson. However, this book seemed destined to be a victim of the mergers and acquisitions in the world of textbook publishing when it was rescued by Jennifer Simon of the Professional and Trade division of John Wiley & Sons. I am grateful to Ms. Simon for seeing the potential for a second edition of my text, geared primarily for courses on the master's level. Similarly, I am grateful to Ms. Simon's wily successor, Patricia Rossi, for her wise counsel and support in the preparation of the present edition. It is no less a pleasure to acknowledge all of the help and constant support I have received from Ms. Simon's, and then Ms. Rossi's, editorial assistant, Isabel Pratt, now herself an editor at Wiley. My gratitude also extends to Judi Knott, marketing manager for psychology, who has promoted *Explaining Psychological Statistics* with such intelligence and persistence, and kept me in touch with its users. The attractive look of the present edition is due in large part to the efforts of Linda Witzling (associate managing editor, John Wiley & Sons), and Susan Dodson (full service coordinator, Graphic Composition, Inc., Georgia).

The content of this text has been improved by the helpful comments and corrections of reviewers and adopters of this and previous editions. The remaining errors and faults are mine alone. Specifically, I would like to acknowledge the contributions of the following statistics instructors: Ronan Bernas, PhD., Eastern Illinois University; Mark Brady, PhD., North Dakota State University; Lawrence DeCarlo, PhD., Columbia University; George Domino, PhD., University of Arizona; Michael McGuire, Washburn University; John Pittenger, PhD., University of Arkansas at Little Rock; and, especially, Karen Caplowitz Barrett, PhD. Since the second edition of this text was published several of my statistics students at New York University have contributed helpful comments and/or pointed out mistakes; they include Natasha Burke, Ihno Lee, Valentyna Moskvina, Suzanne Riela, Linda Ruzickova, and Tina Wyman. Several doctoral students serving as teaching assistants for my statistics courses have also been extremely helpful; they are Alison Ledgerwood, Jennifer Thorpe, and Aaron Wallen.

Finally, I want to acknowledge my colleagues at New York University, several of whom have directly influenced and inspired my teaching of statistics over the past 20 years. In particular, I would like to thank Doris Aaronson, Elizabeth Bauer, Patrick Shrout, and Gay Snodgrass, and, in memoriam, Jacob Cohen and Joan Welkowitz. Of course, I cannot conclude without thanking my friends and family for their understanding support while I was preoccupied with revising this text—especially my wife, Leona, who once again did much of the additional typing required.

Barry H. Cohen
New York University
July 2007

EXPLAINING PSYCHOLOGICAL STATISTICS

THIRD EDITION

INTRODUCTION TO PSYCHOLOGICAL STATISTICS

Chapter

A

CONCEPTUAL FOUNDATION

If you have not already read the Preface, please do so now. Many readers have developed the habit of skipping the Preface because it is often used by the author as a soapbox, or as an opportunity to give his or her autobiography and to thank many people the reader has never heard of. The preface of this text is different and plays a particularly important role. You may have noticed that this book uses a unique form of organization (each chapter is broken into A, B, and C sections). The preface explains the rationale for this unique format and explains how you can derive the most benefit from it.

What Is (Are) Statistics?

An obvious way to begin a text about statistics is to pose the rhetorical question, "What *is* statistics?" However, it is also proper to pose the question "What *are* statistics?"—because the term *statistics* can be used in at least two different ways. In one sense *statistics* refers to a collection of numerical facts, such as a set of performance measures for a baseball team (e.g., batting averages of the players) or the results of the latest U.S. census (e.g., the average size of households in each state of the United States). So the answer is that statistics are observations organized into numerical form.

In a second sense, *statistics* refers to a branch of mathematics that is concerned with methods for understanding and summarizing collections of numbers. So the answer to "What is statistics?" is that it is a set of methods for dealing with numerical facts. Psychologists, like other scientists, refer to numerical facts as *data*. The word *data* is a plural noun and always takes a plural verb, as in "the data *were* analyzed." (The singular form, datum, is rarely used.) Actually, there is a third meaning for the term *statistics*, which distinguishes a statistic from a parameter. To explain this distinction, I have to contrast samples with populations, which I will do at the end of this section.

As a part of mathematics, statistics has a theoretical side that can get very abstract. This text, however, deals only with *applied statistics*. It describes methods for data analysis that have been worked out by statisticians, but does not show how these methods were derived from more fundamental mathematical principles. For that part of the story, you would need to read a text on *theoretical* or *mathematical statistics* (e.g., Hogg & Craig, 1995).

The title of this text uses the phrase "psychological statistics." This could mean a collection of numerical facts about psychology (e.g., how large a percentage of the population claims to be happy), but as you have probably guessed, it actually refers to those statistical methods that are commonly applied to the analysis of psychological data. Indeed, just about

1

every kind of statistical method has been used at one time or another to analyze some set of psychological data. The methods presented in this text are the ones usually taught in an intermediate (advanced undergraduate or graduate level) statistics course for psychology students, and they have been chosen because they are not only commonly used but are also simple to explain. Unfortunately, some methods that are now used frequently in psychological research (e.g., structural equation modeling) are too complex to be covered adequately at this level.

One part of applied statistics is concerned only with summarizing the set of data that a researcher has collected; this is called *descriptive statistics*. If all sixth graders in the United States take the same standardized exam, and you want a system for describing each student's standing with respect to the others, you need descriptive statistics. However, most psychological research involves relatively small groups of people from which inferences are drawn about the larger population; this branch of statistics is called *inferential statistics*. If you have a random sample of 100 patients who have been taking a new antidepressant drug, and you want to make a general statement about the drug's possible effectiveness in the entire population, you need inferential statistics. This text begins with a presentation of several procedures that are commonly used to create descriptive statistics. Although such methods can be used just to describe data, it is quite common to use these descriptive statistics as the basis for inferential procedures. The bulk of the text is devoted to some of the most common procedures of inferential statistics.

Statistics and Research

The reason a course in statistics is nearly universally required for psychology students is that statistical methods play a critical role in most types of psychological research. However, not all forms of research rely on statistics. For instance, it was once believed that only humans make and use tools. Then chimpanzees were observed stripping leaves from branches before inserting the branches into holes in logs to "fish" for termites to eat (van Lawick-Goodall, 1971). Certainly such an observation has to be replicated by different scientists in different settings before becoming widely accepted as evidence of toolmaking among chimpanzees, but statistical analysis is not necessary.

On the other hand, suppose you wanted to know whether a glass of warm milk at bedtime will help insomniacs get to sleep faster. In this case, the results are not likely to be obvious. You don't expect the warm milk to knock out any of the subjects, or even to help every one of them. The effect of the milk is likely to be small and noticeable only after averaging the time it takes a number of participants to fall asleep (the sleep latency) and comparing that to the average for a (control) group that does not get the milk. Descriptive statistics is required to demonstrate that there is a difference between the two groups, and inferential statistics is needed to show that if the experiment were repeated, it would be likely that the difference would be in the same direction. (If warm milk really has *no* effect on sleep latency, the next experiment would be just as likely to show that warm milk slightly increases sleep latency as to show that it slightly decreases it.)

Variables and Constants

A key concept in the above example is that the time it takes to fall asleep varies from one insomniac to another and also varies after a person drinks warm milk. Because sleep latency varies, it is called a *variable*. If sleep latency were the same for everyone, it would be a *constant*, and you really wouldn't need statistics to evaluate your research. It would be obvious after testing a few participants whether the milk was having an effect. But, because sleep latency varies from person to person and from night to night, it would not be obvious whether a particular case of shortened sleep latency was due to warm milk or just to the usual variability. Rather than focusing on any one instance of sleep latency, you would probably use statistics to compare a whole set of sleep latencies of people who drank warm milk with another whole set of people who did not.

In the field of physics there are many important constants (e.g., the speed of light, the mass of a proton), but most human characteristics vary a great deal from person to person. The number of chambers in the heart is a constant for humans (four), but resting heart rate is a variable. Many human variables (e.g., beauty, charisma) are easy to observe but hard to measure precisely or reliably. Because the types of statistical procedures that can be used to analyze the data from a research study depend in part on the way the variables involved were measured, we turn to this topic next.

Scales of Measurement

Measurement is a system for assigning values to observations in a consistent and reproducible way. When most people think of measurement, they think first of physical measurement, in which numbers and measurement units (e.g., minutes and seconds for sleep latency) are used in a precise way. However, in a broad sense, measurement need not involve numbers at all.

Nominal Scales

Facial expressions can be classified by the emotions they express (e.g., anger, happiness, surprise). The different emotions can be considered values on a *nominal scale*; the term *nominal* refers to the fact that the values are simply named, rather than assigned numbers. (Some emotions can be identified quite reliably, even across diverse cultures and geographical locations; see Ekman, 1982.) If numbers are assigned to the values of a nominal scale, they are assigned arbitrarily and therefore cannot be used for mathematical operations. For example, the *Diagnostic and Statistical Manual* of the American Psychiatric Association (the latest version is *DSM-IV*) assigns a number as well as a name to each psychiatric diagnosis (e.g., the number 300.3 designates obsessive-compulsive disorder). However, it makes no sense to use these numbers mathematically; for instance, you cannot average the numerical diagnoses of all the members in a family to find out the average mental illness of the family. Even the order of the assigned numbers is arbitrary; the higher *DSM-IV* numbers do not indicate more severe diagnoses.

Many variables that are important to psychology (e.g., gender, type of psychotherapy) can be measured only on a nominal scale, so we will be dealing with this level of measurement throughout the text. Nominal scales are often referred to as *categorical scales* because the different levels of the scale represent distinct categories; each object measured is assigned to one and only one category. A nominal scale is also referred to as a *qualitative* level of measurement because each level has a different quality and therefore cannot be compared with other levels with respect to quantity.

Ordinal Scales

A quantitative level of measurement is being used when the different values of a scale can be placed in order. For instance, an elementary school teacher may rate the handwriting of each student in a class as excellent, good, fair, or poor. Unlike the categories of a nominal scale, these designations have a meaningful order and therefore constitute an *ordinal scale*. One can add the percentage of students rated excellent to the percentage of students rated good, for instance, and then make the statement that a certain percentage of the students have handwriting that is "better than fair."

Often the levels of an ordinal scale are given numbers, as when a coach rank-orders the gymnasts on a team based on ability. These numbers are not arbitrary like the numbers that may be assigned to the categories of a nominal scale; the gymnast ranked number 2 *is* better than the gymnast ranked number 4, and gymnast number 3 is somewhere between. However, the rankings cannot be treated as real numbers; that is, it cannot be assumed that the third-ranked gymnast is midway between the second and the fourth. In fact, it could be the case that the number 2 gymnast is much better than either number 3 or 4, and that number 3 is only slightly better than number 4 (as shown in Figure 1.1). Although the average of the numbers 2 and 4 is 3, the average of the abilities of the number 2 and 4 gymnasts is not equivalent to the abilities of gymnast number 3.

A typical example of the use of an ordinal scale in psychology is when photographs of human faces are rank-ordered for attractiveness. A less obvious example is the measurement of anxiety by means of a self-rated questionnaire (on which subjects indicate the frequency of various anxiety symptoms in their lives using numbers corresponding to never, sometimes, often, etc.). Higher scores can generally be thought of as indicating greater amounts of anxiety, but it is not likely that the anxiety difference between subjects scoring 20 and 30 is going to be exactly the same as the anxiety difference between subjects scoring 40 and 50. Nonetheless, scores from anxiety questionnaires and similar psychological measures are usually dealt with mathematically by researchers as though they were certain the scores were equally spaced throughout the scale.

The topic of measurement is a complex one to which entire textbooks have been devoted, so we will not delve further into measurement

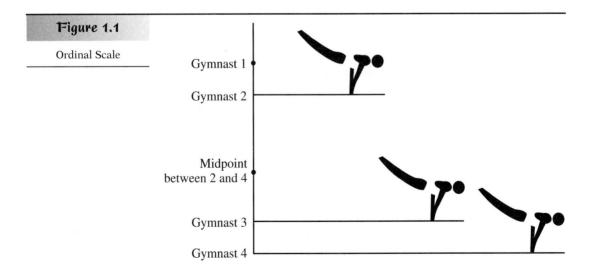

Figure 1.1

Ordinal Scale

Gymnast 1

Gymnast 2

Midpoint between 2 and 4

Gymnast 3

Gymnast 4

controversies here. For our purposes, you should be aware that when dealing with an ordinal scale (when you are sure of the order of the levels but not sure that the levels are equally spaced), you should use statistical procedures that have been devised specifically for use with ordinal data. The descriptive statistics that apply to ordinal data will be discussed in the next two chapters. The use of inferential statistics with ordinal data will not be presented until Chapter 21. (Although it can be argued that inferential ordinal statistics should be used more frequently, such procedures are not used very often in psychological research.)

Interval and Ratio Scales

In general, physical measurements have a level of precision that goes beyond the ordinal property previously described. We are confident that the inch marks on a ruler are equally spaced; we know that considerable effort goes into making sure of this. Because we know that the space, or interval, between 2 and 3 inches is the same as that between 3 and 4 inches, we can say that this measurement scale possesses the *interval property* (see Figure 1.2a). Such scales are based on *units* of measurement (e.g., the inch); a unit at one part of the scale is always the same size as a unit at any other part of the scale. It is therefore permissible to treat the numbers on this kind of scale as actual numbers and to assume that a measurement of three units is exactly halfway between two and four units.

In addition, most physical measurements possess what is called the *ratio property*. This means that when your measurement scale tells you that you now have twice as many units of the variable as before, you really *do* have twice as much of the variable. Measurements of sleep latency in minutes and seconds have this property. When a subject's sleep latency is 20 minutes, it has taken that person twice as long to fall asleep as a subject with a sleep latency of 10 minutes. Measuring the lengths of objects with a ruler also involves the ratio property. Scales that have the ratio property in addition to the interval property are called *ratio scales* (see Figure 1.2b).

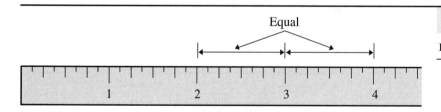

a. Interval scale

Figure 1.2

Interval and Ratio Scales

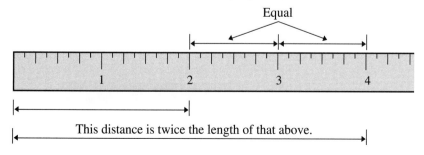

This distance is twice the length of that above.

b. Ratio scale

Whereas all ratio scales have the interval property, there are some scales that have the interval property but not the ratio property. These scales are called *interval scales*. Such scales are relatively rare in the realm of physical measurement; perhaps the best known examples are the Celsius (also known as centigrade) and Fahrenheit temperature scales. The degrees are equally spaced, according to the interval property, but one cannot say that something that has a temperature of 40 degrees is twice as hot as something that has a temperature of 20 degrees. The reason these two temperature scales lack the ratio property is that the zero point for each is arbitrary. Both scales have different zero points ($0\,°C = 32\,°F$), but in neither case does zero indicate a total lack of heat. (Heat comes from the motion of particles within a substance, and as long as there is some motion, there is some heat.) In contrast, the Kelvin scale of temperature is a true ratio scale because its zero point represents *absolute* zero temperature—a total lack of heat. (Theoretically, the motion of internal particles has stopped completely.)

Although interval scales may be rare when dealing with physical measurement, they are not uncommon in psychological research. If we grant that IQ scores have the interval property (which is open to debate), we still would not consider IQ a ratio scale. It doesn't make sense to say that someone who scores a zero on a particular IQ test has no intelligence at all, unless intelligence is defined very narrowly. And does it make sense to say that someone with an IQ of 150 is exactly twice as intelligent as someone who scores 75?

Continuous versus Discrete Variables

One distinction among variables that affects the way they are measured is that some variables vary continuously, whereas others have only a finite (or countable) number of levels with no intermediate values possible. The latter variables are said to be discrete (see Figure 1.3). A simple example of a *continuous variable* is height; no matter how close two people are in height, it is possible to find someone whose height is midway between those two people. (Quantum physics has shown that there are limitations to the precision of measurement, and it may be meaningless to talk of continuous variables at the quantum level, but these concepts have no practical implications for psychological research.)

Figure 1.3

Discrete and Continuous Variables

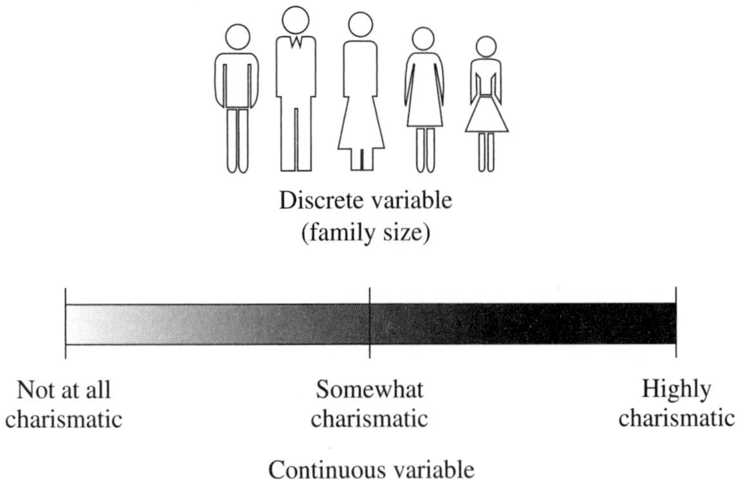

Discrete variable
(family size)

Not at all   Somewhat   Highly
charismatic   charismatic   charismatic

Continuous variable

An example of a *discrete variable* is the size of a family. This variable can be measured on a ratio scale by simply counting the family members, but it does not vary continuously—a family can have two or three children, but there is no meaningful value in between. The size of a family will always be a whole number and never involve a fraction. The distinction between discrete and continuous variables affects some of the procedures for displaying and describing data, as you will see in the next chapter. Fortunately, however, the inferential statistics discussed in Parts II through VI of this text are *not* affected by whether the variable measured is discrete or continuous.

Scales versus Variables

It is important not to confuse variables with the scales with which they are measured. For instance, the temperature of the air outside can be measured on an ordinal scale (e.g., the hottest day of the year, the third hottest day), an interval scale (degrees Celsius or Fahrenheit), or a ratio scale (degrees Kelvin); these three scales are measuring the same physical quantity but yield very different measurements. In many cases, a variable that varies continuously, such as charisma, can only be measured crudely, with relatively few levels (e.g., highly charismatic, somewhat charismatic, not at all charismatic). On the other hand, a continuous variable such as generosity can be measured rather precisely by the exact amount of money donated to charity in a year (which is at least one aspect of generosity). Although in an ultimate sense all scales are discrete, scales with very many levels relative to the quantities measured are treated as continuous for display purposes, whereas scales with relatively few levels are usually treated as discrete (see Chapter 2). Of course, the scale used to measure a discrete variable is always treated as discrete.

Parametric versus Nonparametric Statistics

Because the kinds of statistical procedures described in Parts II through VI of this text are just as valid for interval scales as they are for ratio scales, it is customary to lump these two types of scales together and refer to *interval/ratio scales* or interval/ratio data when some statement applies equally to both types of scales. The data from interval/ratio scales can be described in terms of smooth distributions, which will be explained in greater detail in the next few chapters. These data distributions sometimes resemble well-known mathematical distributions, which can be described by just a few values called parameters. (I will expand on this point in Section C.) Statistical procedures based on distributions and their parameters are called *parametric statistics*. With interval/ratio data it is often (but not always) appropriate to use parametric statistics. Conversely, parametric statistics are truly valid only when you are dealing with interval/ratio data. The bulk of this text (i.e., Parts II through VI) is devoted to parametric statistics. If all of your variables have been measured on nominal or ordinal scales, or your interval/ratio data do not meet the distributional assumptions of parametric statistics (which will be explained at the appropriate time), you should be using *nonparametric statistics* (described in Part VII).

Independent versus Dependent Variables

Returning to the experiment in which one group of insomniacs gets warm milk before bedtime and the other does not, note that there are actually *two* variables involved in this experiment. One of these, sleep latency, has already been discussed; it is being measured on a ratio scale. The other variable is less obvious; it is group membership. That is, subjects *vary* as to which experimental condition they are in—some receive milk, and some do not. This variable, which in this case has only two levels, is called the *independent variable*. A subject's level on this variable—that is, which group a subject is placed in—is determined at random by the experimenter and is independent of anything that happens during the experiment. The other variable, sleep latency, is called the *dependent variable* because its value depends (it is hoped) at least partially on the value of the independent variable. That is, sleep latency is expected to depend in part on whether the subject drinks milk before bedtime. Notice that the independent variable is measured on a nominal scale (the two categories are "milk" and "no milk"). However, because the dependent variable is being measured on a ratio scale, parametric statistical analysis can be performed. If neither of the variables were measured on an interval or ratio scale (for example, if sleep latency were categorized as simply less than or greater than 10 minutes), a nonparametric statistical procedure would be needed (see Part VII). If the independent variable were also being measured on an interval/ratio scale (e.g., amount of milk given) you would still use parametric statistics, but of a different type (see Chapter 9). I will discuss different experimental designs as they become relevant to the statistical procedures I am describing. For now, I will simply point out that parametric statistics can be used to analyze the data from an experiment, even if the independent variable is measured on a nominal scale.

Experimental versus Correlational Research

It is important to realize that not all research involves experiments; much of the research in some areas of psychology involves measuring differences between groups that were not created by the researcher. For instance, insomniacs can be compared to normal sleepers on variables such as anxiety. If inferential statistics shows that insomniacs, in general, differ from normal sleepers in daily anxiety, it is interesting, but we still do not know whether the greater anxiety causes the insomnia, the insomnia causes the greater anxiety, or some third variable (e.g., increased muscle tension) causes both. We cannot make causal conclusions because we are not in control of who is an insomniac and who is not. Nonetheless, such *correlational* studies can produce useful insights and sometimes suggest confirmatory experiments.

To continue this example: If a comparison of insomniacs and normal sleepers reveals a statistically reliable difference in the amount of sugar consumed daily, these results suggest that sugar consumption may be interfering with sleep. In this case, correlational research has led to an interesting hypothesis that can be tested more conclusively by means of an experiment. A researcher randomly selects two groups of sugar-eating insomniacs; one group is restricted from eating sugar and the other is not. If the sugar-restricted insomniacs sleep better, that evidence supports the notion that sugar consumption interferes with sleep. If there is no sleep difference between the groups, the causal connection may be in the opposite direction (i.e., lack of sleep may produce an increased craving for sugar), or the insomnia may be due to some as yet unidentified third variable (e.g., maybe anxiety produces both insomnia *and* a craving for sugar). The statistical analysis is generally the same for both experimental and correlational research; it is the causal conclusions that differ.

Populations versus Samples

In psychological research, measurements are often performed on some aspect of a person. The psychologist may want to know about people's ability to remember faces, solve anagrams, or experience happiness. The collection of all people who could be measured, or in whom the psychologist is interested, is called the *population*. However, it is not always people who are the subjects of measurement in psychological research. A population can consist of laboratory rats, mental hospitals, married couples, small towns, and so forth. Indeed, as far as theoretical statisticians are concerned, a population is just a set (ideally one that is infinitely large) of numbers. The statistical procedures used to analyze data are the same regardless of where the numbers come from (as long as certain assumptions are met, as subsequent chapters will make clear). In fact, the statistical methods you will be studying in this text were originally devised to solve problems in agriculture, beer manufacturing, human genetics, and other diverse areas.

If you had measurements for an entire population, you would have so many numbers that you would surely want to use descriptive statistics to summarize your results. This would also enable you to compare any individual to the rest of the population, compare two different variables measured on the same population, or even to compare two different populations measured on the same variable. More often, practical limitations will prevent you from gathering all of the measurements that you might want. In such cases you would obtain measurements for some subset of the population. This subset is called a *sample* (see Figure 1.4).

Sampling is something we all do in daily life. If you have tried two or three items from the menu of a nearby restaurant and have not liked any of them, you do not have to try everything on the menu before deciding not to dine at that restaurant anymore. When you are conducting research, you follow a more formal sampling procedure. If you have obtained measurements on a sample, you would probably begin by using descriptive statistics to summarize the data in your sample. But it is not likely that you would stop there. Usually, you would then use the procedures of inferential statistics to draw some conclusions about the entire population from which you obtained your sample. Strictly speaking, these conclusions would be valid only if your sample was a *random sample*. In reality, truly random samples are virtually impossible to obtain, and most research is conducted on *samples of convenience* (e.g., students in an introductory psychology class who must either "volunteer" for some experiments or complete some alternative assignment). To the extent that one's sample is not truly random, it may be difficult to generalize one's results to the larger population. The role of sampling in inferential statistics will be discussed at greater length in Part II.

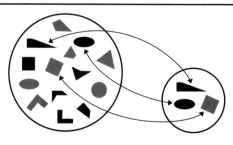

Population   Sample

Figure 1.4

A Population and
a Sample

Now we come to the third definition for the term *statistic*. A *statistic* is a value derived from the data in a sample rather than a population. It could be a value derived from all of the data in the sample, such as the mean, or it could be just one measurement in the sample, such as the maximum value. If the same mathematical operation used to derive a statistic from a sample is performed on the entire population from which you selected the sample, the result is called a population *parameter* rather than a sample statistic. As you will see, sample statistics are often used to make estimates of, or draw inferences about, corresponding population parameters.

Statistical Formulas

Many descriptive statistics, as well as sample statistics that are used for inference, are found by means of statistical formulas. Often these formulas are applied to all of the measurements that have been collected, so a notational system is needed for referring to many data points at once. It is also frequently necessary to add many measurements together, so a symbol is needed to represent this operation. Throughout the text, Section B will be reserved for a presentation of the nuts and bolts of statistical analysis. The first Section B will present the building blocks of all statistical formulas: subscripted variables and summation signs.

SUMMARY

1. Descriptive statistics is concerned with summarizing a given set of measurements, whereas inferential statistics is concerned with generalizing beyond the given data to some larger potential set of measurements.
2. The type of descriptive or inferential statistics that can be applied to a set of data depends, in part, on the type of measurement scale that was used to obtain the data.
3. If the different levels of a variable can be named, but not placed in any specific order, a *nominal scale* is being used. The categories in a nominal scale can be numbered, but the numbers cannot be used in any mathematical way—even the ordering of the numbers would be arbitrary.
4. If the levels of a scale can be ordered, but the intervals between adjacent levels are not guaranteed to be the same size, you are dealing with an *ordinal scale*. The levels can be assigned numbers, as when subjects or items are rank-ordered along some dimension, but these numbers cannot be used for arithmetical operations (e.g., we cannot be sure that the average of ranks 1 and 3, for instance, equals rank 2).
5. If the intervals corresponding to the units of measurement on a scale are always equal (e.g., the difference between 2 and 3 units is the same as between 4 and 5 units), the scale has the interval property. Scales that have equal intervals but do not have a true zero point are called *interval scales*.
6. If an interval scale has a true zero point (i.e., zero on the scale indicates a total absence of the variable being measured), the ratio between two measurements will be meaningful (a fish that is 30 inches long is twice as long as one that is 15 inches long). A scale that has both the interval and the ratio properties is called a *ratio scale*.
7. A variable that has countable levels with no values possible between any two adjacent levels is called a *discrete variable*. A variable that can be measured with infinite precision (i.e., intermediate measurements are always possible), at least in theory, is called a *continuous variable*.

In practice, most physical measurements are treated as continuous even though they are not infinitely precise.

8. The entire set of measurements about which one is concerned is referred to as a *population*. The measurements that comprise a population can be from individual people, families, animals, hospitals, cities, and so forth. A subset of a population is called a *sample*, especially if the subset is considerably smaller than the population and is chosen at random.

9. Values that are derived from and in some way summarize samples are called *statistics*, whereas values that describe a population are called *parameters*.

10. If at least one of your variables has been measured on an interval or ratio scale, and certain additional assumptions have been met, it may be appropriate to use *parametric statistics* to draw inferences about population parameters from sample statistics. If all of your variables have been measured on ordinal or nominal scales, or the assumptions of parametric statistics have not been met, it may be necessary to use *nonparametric statistics*.

EXERCISES

1. Give two examples of each of the following:
   a. Nominal scale
   b. Ordinal scale
   c. Interval scale
   d. Ratio scale
   e. Continuous variable
   f. Discrete variable

*2. What type of scale is being used for each of the following measurements?
   a. Number of arithmetic problems correctly solved
   b. Class standing (i.e., one's rank in the graduating class)
   c. Type of phobia
   d. Body temperature (in °F)
   e. Self-esteem, as measured by self-report questionnaire
   f. Annual income in dollars
   g. Theoretical orientation toward psychotherapy
   h. Place in a dog show
   i. Heart rate in beats per minute

*3. Which of the following variables are discrete and which are continuous?
   a. The number of people in one's social network
   b. Intelligence
   c. Size of vocabulary
   d. Blood pressure
   e. Need for achievement

4. a. Give two examples of a population that does not consist of individual people.
   b. For each population described in part a, indicate how you might obtain a sample.

*5. A psychologist records how many words participants recall from a list under three different conditions: large reward for each word recalled, small reward for each word recalled, and no reward.
   a. What is the independent variable?
   b. What is the dependent variable?
   c. What kind of scale is being used to measure the dependent variable?

6. Patients are randomly assigned to one of four types of psychotherapy. The progress of each subject is rated at the end of 6 months.
   a. What is the independent variable?
   b. What is the dependent variable?
   c. What kind of scale is formed by the levels of the independent variable?
   d. Describe one type of scale that might be used to measure the dependent variable.

*7. Which of the following studies are experimental and which are correlational?
   a. Comparing pet owners with those who don't own pets on an empathy measure
   b. Comparing men and women with respect to performance on a video game that simulates landing a space shuttle
   c. Comparing participants run by a male experimenter with participants run by a female experimenter with respect to the number of tasks completed in 1 hour

d. Comparing the solution times of participants given a hint with those not given a hint

8. Which of the following would be called a statistic and which a parameter?
   a. The average income for 100 U.S. citizens selected at random from various telephone books
   b. The average income of citizens in the United States
   c. The highest age among respondents to a sex survey in a popular magazine

Throughout the text, asterisks (*) will be used in each section of exercises to indicate exercises whose answers appear in the Appendix.

B

BASIC STATISTICAL PROCEDURES

Variables with Subscripts

Recognizing that statistics is a branch of mathematics, you should not be surprised that its procedures are usually expressed in terms of mathematical notation. For instance, you probably recall from high school math that a variable whose value is unknown is most commonly represented by the letter X. This is also the way a statistician would represent a random variable. However, to describe statistical manipulations with samples, we need to refer to collections of random variables. Because this concept is rather abstract, I will use a very concrete example.

When describing the characteristics of a city to people who are considering living there, a realtor typically gives a number of facts such as the average income and the size of the population. Another common statistic is the average temperature in July. The usual way to find the average temperature for the entire month of July is to take the average temperature for each day in July and then average these averages. To express this procedure symbolically it would be helpful to find a way to represent the average temperature for any particular day in July. It should be obvious that it would be awkward to use the same letter, X, for each day of the month if we then want to write a formula that tells us how to combine these 31 different averages into a single average. On the other hand, we certainly cannot use a different letter of the alphabet for each day. The solution is to use subscripts. The average temperature for July 1 can be written X_1, for July 2, X_2, and so on up to X_{31}. We now have a compact way of referring to 31 different variables. If we wanted to indicate a different type of variable, such as high or low temperature for each day, we would need to use a different letter (e.g., Y_1, Y_2, up to Y_{31}). If we want to make some general statement about the average temperature for any day in July without specifying which particular day, we can write X_i. The letter i used as a subscript stands for the word *index* and can take the place of any numerical subscript.

The Summation Sign

To get the average temperature for the month of July, we must add up the average temperatures for each day in July and then divide by 31. Using the subscripts introduced above, the average temperature for July can be expressed as $(X_1 + X_2 + X_3 + \cdots + X_{31})/31$. (Note that because it would take up a lot of room to write out all of the 31 variables, dots are used to indicate that variables are left out.) Fortunately, there is a neater way of indicating that all the variables from X_1 to X_{31} should be added. The mathematical symbol that indicates that a string of variables is to be added is called the summation sign, and it is symbolized by the uppercase Greek letter sigma (Σ). The summation sign works in conjunction with the subscripts on the variables in the following manner. First, you write $i = 1$ under the summation sign to indicate that the summing should start with the variable that has the subscript 1. (You could write $i = 2$ to indicate that you want to begin with the second variable, but it is rare to start with any subscript other than 1.) On top of the summation sign you indicate the subscript of the last variable to be added. Finally, next to the summation sign you write the letter that stands for the collection of variables to be added, using the subscript i. So the sum of the average temperatures for each day in July can be symbolized as follows:

$$\sum_{i=1}^{31} X_i$$

This expression is a neat, compact way of telling you to perform the following:

1. Take X_i and replace i with the number indicated under the summation sign (in this case, X_1).
2. Put a plus sign to the right of the previous expression (X_1+).
3. Write X_i again, this time replacing i with the next integer, and add another plus sign ($X_1 + X_2+$).
4. Continue the above process until i has been replaced by the number on top of the summation sign ($X_1 + X_2 + X_3 + \cdots + X_{31}$).

If you wanted to write a general expression for the sum of the average temperatures on all the days of any month, you could not use the number 31 on top of the summation sign (e.g., June has only 30 days). To be more general, you could use the letter N to stand for the number of days in any month, which leads to the following expression:

$$\sum_{i=1}^{N} X_i$$

To find the average temperature for the month in question, we would divide the above sum by N (the number of days in that month). The whole topic of finding averages will be dealt with in detail in Chapter 3. For now we will concentrate on the mathematics of finding sums.

Summation notation can easily be applied to samples from a population, where N represents the sample size. For instance, if N is the number of people that are allowed by law on a particular elevator, and X_i is the weight of any one particular person, the previous expression represents the total weight of the people on some elevator that is full to its legal capacity. When statisticians use summation signs in statistical formulas, $i = 1$ almost always appears under the summation sign and N appears above it. Therefore, most introductory statistics texts leave out these indexes and simply write the summation sign by itself, expecting the reader to assume that the summation goes from $i = 1$ to N. Although mathematical statisticians dislike this lack of precision, I will, for the sake of simplicity, go along with the practice of leaving off the indexes from summation signs, and usually from variables, as well.

The summation sign plays a role in most of the statistical formulas in this text. To understand those formulas fully it is helpful to know several interesting mathematical properties involved with the use of the summation sign. The most important of those properties will be presented in the remainder of this section.

Properties of the Summation Sign

The first property we will discuss concerns the addition of two collections of variables. Returning to our example about the temperature in July, suppose that you are interested in a temperature-humidity index (THI), which is a better indicator of comfort than temperature alone. Assume that the average THI for any day is just equal to the average temperature of that day (X_i) plus the average humidity of that day (Y_i) (although this is not the index that is usually used). Thus we can express the THI for any day as $X_i + Y_i$. If you wanted to add the THI for all the days in the month, you could use the following general expression: $\Sigma(X_i + Y_i)$. This expression produces the same result as adding the Xs and Ys separately. This leads to a first rule for dealing with summation signs.

Summation Rule 1A

$$\sum(X_i + Y_i) = \sum X_i + \sum Y_i$$

The rule works in exactly the same way for subtraction.

Summation Rule 1B

$$\sum(X_i - Y_i) = \sum X_i - \sum Y_i$$

Rule 1A works because if all you're doing is adding, it doesn't matter what order you use. Note that $\Sigma(X_i + Y_i)$ can be written as:

$$(X_1 + Y_1) + (X_2 + Y_2) + (X_3 + Y_3) + \cdots + (X_N + Y_N)$$

If you remove the parentheses and change the order, as follows,

$$X_1 + X_2 + X_3 + \cdots + X_N + Y_1 + Y_2 + Y_3 + \cdots + Y_N$$

you can see that the above expression is equal to $\Sigma X_i + \Sigma Y_i$. The proof for Rule 1B is exactly parallel.

Sometimes the summation sign is applied to a constant: ΣC_i. In this case, we could write $C_1 + C_2 + C_3 + \cdots + C_N$, but all of these terms are just equal to C, the value of the constant. The fact that the number of Cs being added is equal to N leads to the following rule.

Summation Rule 2

$$\sum C = NC$$

In the equation above, the subscript on the letter C was left off because it is unnecessary and is not normally used.

Quite often a variable is multiplied or divided by a constant before the summation sign is applied: $\Sigma C X_i$. This expression can be simplified without changing its value by placing the constant in front of the summation sign. This leads to the next summation rule.

Summation Rule 3

$$\sum C X_i = C \sum X_i$$

The advantage of this rule is that it reduces computational effort. Instead of multiplying every value of the variable by the constant before adding, we can first add up all the values and then multiply the sum by the constant. You can see why Rule 3 works by writing out the expression and rearranging the terms:

$$\sum CX_i = CX_1 + CX_2 + CX_3 + \cdots + CX_N$$

The constant C can be factored out of each term, and the rest can be placed in parentheses, as follows: $C(X_1 + X_2 + X_3 + \cdots + X_N)$. The part in parentheses is equal to $\sum X_i$, so the entire expression equals $C\sum X_i$.

The last rule presents a simplification that is *not* allowed. Because $\sum(X_i + Y_i) = \sum X_i + \sum Y_i$, it is tempting to assume that $\sum X_iY_i$ equals $(\sum X_i)(\sum Y_i)$ but unfortunately this is *not* true. In the case of Rule 1A, only addition is involved, so the order of operations does not matter (the same is true with a mixture of subtraction and addition). But when multiplication and addition are mixed together, the order of operations cannot be changed without affecting the value of the expression. This leads to the fourth rule.

Summation Rule 4

$$\sum (X_iY_i) \neq \left(\sum X_i\right)\left(\sum Y_i\right)$$

This inequality can be demonstrated with a simple numerical example. Assume that:

$$X_1 = 1 \quad X_2 = 2 \quad X_3 = 3 \quad Y_1 = 4 \quad Y_2 = 5 \quad Y_3 = 6.$$

$$\sum (X_iY_i) = 1 \cdot 4 + 2 \cdot 5 = 3 \cdot 6 = 4 + 10 + 18 = 32$$

$$\left(\sum X_i\right)\left(\sum Y_i\right) = (1 + 2 + 3)(4 + 5 + 6) = (6)(15) = 90$$

As you can see, the two sides of the above inequality do not yield the same numerical value.

An important application of Rule 4 involves the case in which X and Y are equal, so we have $\sum(X_iX_i) \neq (\sum X_i)(\sum X_i)$. Because X_iX_i equals X_i^2 and $(\sum X_i)(\sum X_i) = (\sum X_i)^2$, a consequence of Rule 4 is that:

$$\sum X_i^2 \neq \left(\sum X_i\right)^2$$

This is an important property to remember because both terms play an important role in statistical formulas, and in some cases both terms appear in the same formula. The term on the left, $\sum X_i^2$, says that each X value should be squared *before the values are added*. If $X_1 = 1$, $X_2 = 2$, and $X_3 = 3$, $\sum X_i^2 = 1^2 + 2^2 + 3^2 = 1 + 4 + 9 = 14$. On the other hand, the term on the right $(\sum X_i)^2$ says that all of the X values should be added *before the total is squared*. Using the same X values as above, $(\sum X_i)^2 = (1 + 2 + 3)^2 = 6^2 = 36$. Notice that 36 is larger than 14. When all the values are positive, $(\sum X_i)^2$ will always be larger than $\sum X_i^2$.

In this text, I will use only one summation sign at a time in the main formulas. Summation signs can be doubled or tripled to create more complex formulas (see Section C). There's no limit to the number of summation signs that can be combined, but matters soon become difficult to keep track of, so I will use other notational tricks to avoid such complications.

Rounding Off Numbers

Whereas discrete variables can be measured exactly, the measurement of continuous variables always involves some rounding off. If you are using an interval or ratio scale, the precision of your measurement will depend on the unit you are using. If you are measuring height with a ruler in which the inches are divided into tenths, you must round off to the nearest tenth of an inch. When you report someone's height as 65.3 inches, it really means that the person's height was somewhere between 65.25 inches (half a unit below the reported measurement) and 65.35 inches (half a unit above). You can choose to round off to the nearest inch, of course, but you cannot be more precise than the nearest tenth of an inch.

Rounding off also occurs when calculating statistics, even if the data come from a discrete variable. If three families contain a total of eight people, the average family size is 8/3. To express this fraction in terms of decimals requires rounding off because this is a number with repeating digits past the decimal point (i.e., 2.666 . . . and so on infinitely). When the original data come in the form of whole numbers, it is common to express calculations based on those numbers to two decimal places (i.e., two digits to the right of the decimal point). In the case of 8/3, 2.666 . . . is rounded off to 2.67. The rule is simple: When rounding to two decimal places, look at the digit in the third decimal place (e.g., 2.666). If this digit is 5 or more, the digit to its left is raised by 1 and the rest of the digits are dropped (e.g., 2.666 becomes 2.67 and 4.5251 is rounded off to 4.53). If the digit in the third decimal place is less than 5, it is just dropped, along with any digits to its right (e.g., 7/3 = 2.333 . . . is rounded to 2.33, 4.5209 is rounded to 4.52).

The only exception to this simple rule occurs when the digit in the third decimal place is 5 and the remaining digits are all zero (e.g., 3/8 = .375). In this case, add 1 to the digit in the second decimal place if it is odd, and drop the remaining digits (.375 is rounded to .38); if the digit in the second decimal place is even, simply drop the digits to its right (.425 is rounded to .42). This convention is arbitrary, but it is useful in that about half the numbers will have an odd digit in the second decimal place and will be rounded up and the other half will be rounded down. Of course, these rules can be applied no matter how many digits to the right of the decimal point you want to keep. For instance, if you want to keep five such digits, you look at the sixth one to make your decision.

Extra care must be taken when rounding off numbers that will be used in further calculations (e.g., the mean family size may be used to calculate other statistics, such as a measure of variability). If you are using a calculator, you may want to jot down all the digits that are displayed. When this is not convenient, a good strategy is to hold on to two more decimal places than you want to have in your final answer. If you are using whole numbers and want to express your final answer to two decimal places, your intermediate calculations should be rounded off to not less than four decimal places (e.g., 2.66666 would be rounded to 2.6667).

The amount of *round-off error* that is tolerable depends on what your results are being used for. When it comes to homework exercises or exam questions, your instructor should give you some idea of what he or she considers a tolerable degree of error due to rounding. Fortunately, with the use of computers for statistical analysis, rounding error is rapidly disappearing as a problem in psychological research.

1. Summation Rules
Summation Rule 1A

$$\sum(X_i + Y_i) = \sum X_i + \sum Y_i$$

This rule says that when summing the sums of two variables (e.g., summing the combined weights of male and female members of tennis teams), you can get the same answer by summing each variable separately (sum the weights of all of the men and then the weights of all of the women) and then adding these two sums together.

Summation Rule 1B

$$\sum(X_i - Y_i) = \sum X_i - \sum Y_i$$

This rule says that when summing the differences of two variables (e.g., summing the height differences of male-female couples), you can get the same answer by summing each variable separately (sum the heights of all of the men and then the heights of all of the women) and then subtracting the two sums at the end.

Summation Rule 2

$$\sum C = NC$$

For instance, if everyone working for some company earns the same annual salary, C, and there are N of these workers, the total wages paid in a given year, $\sum C$, is equal to NC.

Summation Rule 3

$$\sum CX_i = C \sum X_i$$

For instance, in a company where the workers earn different annual salaries (X_i), if each worker's salary were multiplied by some constant, C, the total wages paid during a given year $(\sum X_i)$ would be multiplied by the same constant. Because the constant can be some fraction, there is no need to have a separate rule for dividing by a constant.

Summation Rule 4

$$\sum(X_iY_i) \neq \left(\sum X_i\right)\left(\sum Y_i\right)$$

An important corollary of this rule is that $\sum X_i^2 \neq (\sum X_i)^2$.

2. Rules for Rounding Numbers
If you want to round to N decimal places, look at the digit in the $N + 1$ place.
    a. If it is less than 5, do not change the digit in the Nth place.
    b. If it is 5 or more, increase the digit in the Nth place by 1.
    c. If it is 5 and there are no more digits to the right (or they are all zero), raise the digit in the Nth place by 1 only if it is an odd number. Leave the Nth digit as is if it is an even number.
    In all cases, the last step is to drop the digit in the $N + 1$ place and any other digits to its right.

The first two exercises are based on the following values for two variables: $X_1 = 2$, $X_2 = 4$, $X_3 = 6$, $X_4 = 8$, $X_5 = 10$; $Y_1 = 3$, $Y_2 = 5$, $Y_3 = 7$, $Y_4 = 9$, $Y_5 = 11$.

*1. Find the value of each of the following expressions:

a. $\sum_{i=2}^{5} X_i$ b. $\sum_{i=1}^{4} Y_i$ c. $\sum 5X_i$

d. $\sum 3Y_i$ e. $\sum X_i^2$ f. $\left(\sum 5X_i\right)^2$

g. $\sum Y_i^2$ h. $\left(\sum Y_i\right)^2$

*2. Find the value of each of the following expressions:

a. $\sum (X + Y)$ b. $\sum XY$ c. $\left(\sum X\right)\left(\sum Y\right)$

d. $\sum (X^2 + Y^2)$ e. $\sum (X - Y)$ f. $\sum (X + Y)^2$

g. $\sum (X + 7)$ h. $\sum (Y - 2)$

3. Make up your own set of at least five numbers and demonstrate that $\sum X_i^2 \neq (\sum X_i)^2$.

*4. Use the appropriate summation rule(s) to simplify each of the following expressions (assume all letters represent variables rather than constants):

a. $\sum (9)$ b. $\sum (A - B)$ c. $\sum (3D)$

d. $\sum (5G + 8H)$ e. $\sum (Z^2 + 4)$

5. Using the appropriate summation rules, show that, as a general rule, $\sum (X_i + C) = \sum X_i + NC$.

*6. Round off the following numbers to two decimal places (assume digits to the right of those shown are zero):

a. 144.0135   b. 67.245 c. 99.707 d. 13.345
e. 7.3451   f. 5.9817 g. 5.997

7. Round off the following numbers to four decimal places (assume digits to the right of those shown are zero):

a. .76995   b. 3.141627   c. 2.7182818
d. 6.89996   e. 1.000819   f. 22.55555

*8. Round off the following numbers to one decimal place (assume digits to the right of those shown are zero):

a. 55.555   b. 267.1919   c. 98.951
d. 99.95   e. 1.444   f. 22.14999

OPTIONAL MATERIAL

Double Summations

In the example in Section B, we summed the average temperatures of each day in the month of July. It is easy to extend this example to sum the average temperatures for each of the 12 months of a particular year. However, if we continue using the notation of Section B, how do we indicate at any point the month with which we are dealing? Were we to use different letters, it would look like we were referring to different types of variables. The solution is to add a second subscript, so the first subscript can represent the month (1 to 12), and the second subscript can represent the day. For instance, $X_{1,21}$ would stand for the average temperature for January 21, whereas $X_{12,1}$ would stand for December 1. (You can see the need for the comma separating the two subscripts.)

To represent any possible day, it is customary to write X_{ij}. (If a third subscript were needed, the letter k would be used, and so on.) The ij notation is called a double subscript, and it indicates a collection of variables that can be categorized along two different dimensions (e.g., days in a month and months in a year). According to this notation, the sum of temperatures for July would be written as:

$$\sum_{j=1}^{31} X_{ij}$$

The sum of temperatures for the entire year requires a double summation sign, as follows:

$$\sum_{i=1}^{M} \sum_{j=1}^{N} X_{ij}$$

where M equals the number of months in the year, and N equals the number of days in the given month. (I could have written 12 on top of the first summation sign, but I wanted to show what the double summation looks like in the general case. Also, strictly speaking I should have written N_i, because the number of days is different for different months, but that would make the expression look too complicated.)

The double summation sign indicates that the first step is to set i and j to 1. Then, keep increasing j by 1 and summing until you get to N. (For this example, $N = 31$; when N has been reached, January has been summed.) Next, i is increased to 2, and the second summation sign is once again incremented from 1 to N (now $N = 28$ or 29, and when N has been reached, February has been summed and added to the sum for January.) This process continues until the first summation sign (using index i) has been incremented from 1 to M.

If you wanted to represent the preceding summation for a series of several years, a triple summation could be used:

$$\sum_{k=1}^{O} \sum_{i=1}^{M} \sum_{j=1}^{N} X_{kij}$$

In this case subscript k changes for each year, and O represents the total number of years to be summed.

Double and triple summation signs are often used to express statistical formulas in more mathematically oriented statistics texts, which is why I wanted to expose you to this notation. The good news is that I will use various computational tricks in this text that make the use of double and triple summations unnecessary.

Random Variables and Mathematical Distributions

The relatively simple rules of statistical inference that I will be explaining in this text were not derived for sampling finite populations of actual people and then measuring them for some characteristic of interest. Mathematical statisticians think in terms of *random variables* that may take on an infinite range of values, but are more likely to yield some values than others. If the random variable is continuous, and the population is infinitely large, the curve that indicates the relative likelihood of the values is called a *probability density function*. When the probability density function of some random variable corresponds to some well-understood mathematical distribution, it becomes relatively easy to predict what will happen when random samples of some fixed size are repeatedly drawn.

In mathematical statistics, a parameter is a value that helps describe the form of the mathematical distribution that describes a population. (The parameter is usually a part of the mathematical equation that generates the distribution.) The methods for estimating or drawing inferences about population parameters from sample statistics are therefore called inferential parametric statistics; the methods of Parts II through VI fall into this category. If you are dealing with discrete random variables that have only a few possible values, or if you have no basis for guessing what kind of distribution exists for your variable in your population, it may be appropriate to use one of the procedures known as distribution-free, or nonparametric, statistics. The methods of Part VII fall in this category.

Much of the controversy surrounding the use of parametric statistics to evaluate psychological research arises because the distributions of many psychological variables, measured on actual people, do not match the theoretical mathematical distributions on which the common methods are based. Often the researcher has collected so few data points that the empirical distribution (i.e., the distribution of the data collected) gives no clear basis for determining which theoretical distribution would best represent the population. Moreover, using any theoretical distribution to represent a finite population of psychological measurements involves some degree of approximation.

Fortunately, the procedures described in this text are applicable to a wide range of psychological variables, and computer simulation studies have shown that the approximations involved usually do not produce errors large enough to be of practical significance. You can rest assured that I will not have much more to say about the theoretical basis for the applied statistics presented in this text, except to explain, where appropriate, the assumptions underlying the use of mathematical statistics to analyze the data from psychological research.

SUMMARY

1. Double subscripts can be used to indicate that a collection of variables differs along two dimensions. For instance, a double subscript can be used to represent test scores of school pupils, with one subscript indicating a student's seat number within a class and the other subscript indicating the number of the classroom (e.g., $X_{12,3}$ can represent a score for the twelfth student in room 3).

2. When a double summation sign appears in front of a variable with a double subscript, both subscripts are set to the first value, and then the second summation is completed. Next, the first subscript is incremented by 1, and the second summation is completed again, and so on.

3. The procedures of mathematical statistics are derived from the study of random variables and theoretical mathematical distributions. Applying these procedures to psychological variables measured on finite populations of people involves some approximation, but if some reasonable assumptions can be made (to be discussed in later chapters), the error involved is not a practical problem.

EXERCISES

1. a. Describe an example for which it would be useful to have a variable with a double subscript.
   b. Describe an example for which a triple subscript would be appropriate.

*2. Imagine that there are only nine baseball players on a team and 12 teams in a particular league. If X_{ij} represents the number of home runs hit by the ith player on the jth team, write an expression using a double summation (including the indexes) to indicate the sum of all the home runs in the entire league.

3. Referring to exercise 2, assume Y_{ij} represents the number of hits other than home runs for each player.
   a. Write an expression to indicate the sum of all hits (including home runs) in the league.

   b. Assume Z_{ij} represents the number of times at bat for each player. Write an expression to indicate the *average* of all the batting averages in the league. (Batting average = number of hits/number of times at bat.)

*4. A die is a cube that has a different number of spots (1 to 6) on each of its sides. Imagine that you have N dice in all and that X_{ij} represents the number of spots on the ith side of the jth die. Write an expression using a double summation (including the indexes) to indicate the sum of all the spots on all of the dice.

5. Perform the summation indicated by the expression you wrote in exercise 4 for $N = 10$.

FREQUENCY TABLES, GRAPHS, AND DISTRIBUTIONS

You will need to use the following from the previous chapter:

Symbols
Σ: Summation sign

Concepts
Continuous versus discrete scales

I used to give a diagnostic quiz during the first session of my course in statistics. The quiz consisted of 10 multiple-choice questions requiring simple
algebraic manipulations, designed to show whether students had the basic
mathematical tools to handle a course in statistics. (I have since stopped
giving the quiz because the results were frequently misleading, especially
when very bright students panicked and produced deceptively low scores.)
Most students were curious to see the results of the quiz and to know how
their performance compared to those of their classmates. To show how the
class did on the quiz, about the cruelest thing I could have done would have
been to put all of the raw scores, in no particular order, on the blackboard.
This is the way the data would first appear after I graded the quizzes. For a
class of 25 students, the scores would typically look like those in Table 2.1.

A

**CONCEPTUAL
FOUNDATION**

8	6	10	9	6	6	8	7	
4	9	6	2	8	6	10	4	
5	6	8	4	7	8	4	7	6

Table 2.1

You can see that there are a lot of 6s and 8s and not a lot of 10s or
scores below 4, but this is not the best way to get a sense of how the class
performed. A very simple and logical step makes it easier to understand the
scores: put them in order. A string of scores arranged in numerical order
(customarily starting with the highest value) is often called an *array*.
Putting the scores from Table 2.1 into an array produces Table 2.2 (read
left to right starting with the top row).

10	10	9	9	8	8	8	8	
8	7	7	7	6	6	6	6	
6	6	6	5	4	4	4	4	2

Table 2.2

Frequency Distributions

The array in Table 2.2 is certainly an improvement, but the table could be
made more compact. If the class contained 100 students, an array would be
quite difficult to look at. A more informative way to display these data is in
a *simple frequency distribution*, which is a table consisting of two columns.
The first column lists all of the possible scores, beginning with the highest
score in the array and going down to the lowest score. The second column
lists the *frequency* of each score—that is, how many times that score is
repeated in the array. You don't have to actually write out the array before
constructing a simple frequency distribution, but doing so makes the task

Table 2.3	
X	**f**
10	2
9	2
8	5
7	3
6	7
5	1
4	4
3	0
2	1
$\Sigma f =$	25

easier. Table 2.3 is a simple frequency distribution of the data in Table 2.2. X stands for any score, and f stands for the frequency of that score. Notice that the score of 3 is included in the table even though it has a frequency of zero (i.e., there are no 3s in the data array). The rule is to list all the possible scores from the highest to the lowest, whether a particular score actually appears in the data or not. To check whether you have included all your scores in the frequency distribution, add up all of the frequencies (i.e., Σf), and make sure that the total is equal to the number of scores in the array (i.e., check that $\Sigma f = N$).

A simple frequency distribution is very helpful when the number of different values listed is not very high (nine in the case of Table 2.3), but imagine 25 scores on a midterm exam graded from 0 to 100. The scores might range from 64 to 98, requiring 35 different scores to be listed, at least 10 of which would have zero frequencies. In that case a simple frequency distribution would not be much more informative than a data array. A better solution would be to group the scores into equal-sized intervals (e.g., 80–84, 85–89, etc.) and construct a *grouped frequency distribution*. Because the mechanics of dealing with such distributions are a bit more complicated, I will save this topic for Section B.

The Mode of a Distribution

The score that occurs most frequently in a distribution is called the *mode* of the distribution. For the preceding distribution, the mode is 6 because that score occurs seven times in the distribution—more often than any other score. Complicating things is the fact that a distribution can have more than one mode (e.g., if there were seven instead of only five 8s in Table 2.3, the distribution would have two modes: 6 and 8). The mode will be discussed further in the next chapter, when I deal with ways for summarizing a distribution with just one number.

The Cumulative Frequency Distribution

To evaluate his or her own performance in a class, a student will frequently ask, "How many students in the class had lower scores than mine?" To answer this question for any particular student you need only sum the frequencies for scores below that student's own score. However, you can perform a procedure that will answer that question for any student in the class: You can construct a *cumulative frequency distribution*. The X and f columns of such a distribution are the same as in the simple frequency distribution, but each entry in the cumulative frequencies (cf) column contains a sum of the frequencies for the corresponding score and all scores below it. Table 2.4 shows the cumulative frequencies for the data in Table 2.3.

Table 2.4		
X	**f**	**cf**
10	2	25
9	2	23
8	5	21
7	3	16
6	7	13
5	1	6
4	4	5
3	0	1
2	1	1

If a student attained a score of 7 on the quiz, we can look at the entry in the cf column for a score of 6 to see that this student performed better than 13 other students. The cf entry corresponding to a score of 7 (i.e., 16) answers the question, How many scores are either lower than or tied with a score of 7?

The mechanics of creating the cf column are easy enough. The cf entry for the lowest score is just the same as the frequency of that score. The cf for the next highest score is the frequency of that score plus the frequency of the score below. Each cf entry equals the frequency of that score plus the cf for the score below. For example, the cf for a score of 7 is the frequency of 7, which is 3, plus the cf for 6, which is 13: cf for 7 = 3 + 13 = 16. The entry at the top of the cf column should equal the total number of scores, N, which also equals Σf.

The Relative Frequency and Cumulative Relative Frequency Distributions

Although it may be satisfying to know that you scored better than many other students, what usually matters in terms of getting good grades is what *fraction* of the class scored below you. Outscoring 15 students in a class of 25 is pretty good because you beat 3/5 of the class. Having 15 students below you in a class of 100 is not very good because in that case you have outperformed only 3/20, or .15, of the class. The kind of table that can tell you what fraction of the scores are lower than yours is called a *cumulative relative frequency distribution*. There are two different ways to arrive at this distribution.

As a first step, you can create a relative frequency distribution by dividing each entry in the f column of a simple frequency distribution by N. The resulting fraction is the relative frequency (rf), and it tells you what proportion of the group attained each score. Notice that in Table 2.5, each rf entry was created by dividing the corresponding f by 25. The cumulative relative frequencies (crf) are then found by accumulating the rf's starting from the bottom, just as we did with the f column to obtain the cf entries. Alternatively, you can convert each entry in the cf column into a proportion by dividing it by N. (For example, the crf of .64 for a score of 7 can be found either by dividing 16 by 25 or by adding .12 to the crf of .52 for the score below.) Either way you get the crf column, as shown in Table 2.5. Note that the crf for the top score in the table must be 1.0—if it isn't, you made some kind of mistake (perhaps too much rounding off in lower entries).

X	f	cf	rf	crf		
10	2	25	.08	1.00		**Table 2.5**
9	2	23	.08	.92		
8	5	21	.20	.84		
7	3	16	.12	.64		
6	7	13	.28	.52		
5	1	6	.04	.24		
4	4	5	.16	.20		
3	0	1	0	.04		
2	1	1	.04	.04		

The Cumulative Percentage Distribution

Let us again focus on a quiz score of 7. I pointed out earlier that by looking at the cf entry for 6 you can see that 13 students scored below 7. Now we can look at the crf entry for 6 to see that a score of 7 beats .52, or a little more than half, of the class ($\frac{13}{25} = 52$). A score of 6, however, beats only .24 (the crf entry for 5), or about one fourth, of the class. Sometimes people find it more convenient to work with percentages. If you want a *cumulative percentage frequency* (cpf) column, you need only multiply each crf entry by 100. A score of 7 is better than the scores of 52% of the class; a 6 beats only 24% of the scores. Because the cpf column is especially useful for describing scores in a group, let's look at Table 2.6 and focus only on that column. The entries in the cpf column have a special name: They are called *percentile ranks* (PR). By convention, a percentile rank is defined as the percentage of the group that is at or below a given score. To find the PR of a

	Table 2.6	
X	f	cpf
10	2	100%
9	2	92
8	5	84
7	3	64
6	7	52
5	1	24
4	4	20
3	0	4
2	1	4

particular score we look straight across at the cpf entry, rather than looking at the score below. Thus, the PR of a score of 7 is 64; 64% of the group scored 7 or below. Similarly, the PR for 6 is 52. The way percentile ranks are found changes a bit when dealing with a continuous scale or when dealing with grouped frequency distributions, but the concept is the same, as you will see in Section B.

Percentiles

Instead of being concerned with the percentage at or below a particular score, sometimes you may want to focus on a particular percentage and find the score that has that percentile rank. For instance, before seeing the results of the diagnostic quiz, a professor might decide that the bottom 20% of the class must receive some remedial training on algebra, regardless of their actual scores on the quiz. That is, whether the whole class does well or poorly, whoever is in the bottom 20% will have to get help. In this case, we want to find the score in the distribution that has a PR of 20. You can see from Table 2.6 that a score of 4 has a PR of 20, so that is the score we are interested in. This score is called the twentieth *percentile*. Anyone with this score or a lower score will have to get algebra help.

A percentile can be defined as a score that has a given PR—the 25th percentile is a score whose PR is 25. In other words, a percentile is the score at or below which a given percentage of the group falls. The most interesting percentiles are either *quartiles* (i.e., 25%, 50%, or 75%) or *deciles* (i.e., 10%, 20%, etc.). Unfortunately, these convenient percentiles rarely appear as entries in a cpf column. In Table 2.6, the only convenient percentile is the twentieth. The score of 6 comes close to the 50th percentile (52%), and the score of 5 is a good approximation for the 25th percentile. On the other hand, the 75th percentile is almost exactly midway between 7 and 8. Later in this section, I will show how you can use a graph to more precisely estimate percentiles (and PRs) that do not appear as entries in a table.

Graphs

The information in a frequency distribution table can usually be presented more clearly and dramatically in the form of a graph. A typical graph is made with two perpendicular lines, one horizontal and the other vertical. The values of some variable (X) are marked off along the horizontal line, which is also called the *horizontal axis* (or X axis). A second variable, labeled Y, is marked off along the *vertical axis* (or Y axis). When graphing a frequency distribution, the variable of interest (e.g., quiz scores) is placed along the X axis, and distance (i.e., height) along the Y axis represents the frequency count for each variable.

The Bar Graph

Probably the simplest type of graph is the *bar graph*, in which a rectangle, or bar, is erected above each value of X. The higher the frequency of that value, the greater the height of the bar. The bar graph is appropriate when the values of X come from a discrete rather than a continuous scale (as defined in Chapter 1). A good example of a variable that always produces discrete values is family size. Whereas quiz scores can sometimes be measured with fractions, family size is *always* a whole number. The appropriate way to display a frequency distribution of family size is with a bar graph.

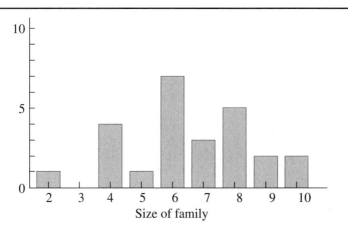

Figure 2.1

Bar Graph

Imagine that the X values in Table 2.3 are not quiz scores but the sizes (number of parents plus number of children) of 25 randomly selected families in which the parents have been taking fertility drugs. The bar graph for these data is shown in Figure 2.1. Notice that the bars do not touch; we wouldn't want to give the impression that the values come from a continuous scale—that a family can be, for instance, between 3 and 4 in size.

The advantage of a bar graph as compared to a table should be clear from Figure 2.1; the bar graph shows at a glance how the family sizes are distributed among the 25 families. Bar graphs are also appropriate when the variable in question has been measured on a nominal or ordinal scale. For instance, if the 25 members of a statistics class were sorted according to eye color, the values along the X axis would be blue, brown, green, and so forth, and the heights of the bars would indicate how many students had each eye color.

The Histogram

A slightly different type of graph is more appropriate if the variable is measured on a continuous scale. A good example of a variable that is almost always measured on a continuous scale is height. Unlike family size, height varies continuously, and it is often represented in terms of fractional values. By convention, however, in the United States height is most commonly reported to the nearest inch. If you ask someone how tall she is, she might say, for example, 5 feet 5 inches, but you know she is rounding off a bit. It is not likely that she is *exactly* 5 feet 5 inches tall. You know that her height could be anywhere between 5 feet $4\frac{1}{2}$ inches and 5 feet $5\frac{1}{2}$ inches. Because height is being measured on a continuous scale, a value like 5 feet 5 inches generally stands for an interval that goes from 5 feet $4\frac{1}{2}$ inches (the lower *real limit*) to 5 feet $5\frac{1}{2}$ inches (the upper real limit). When constructing a bar graph that involves a continuous scale, the bar for each value is drawn wide enough so that it goes from the lower real limit to the upper real limit. Therefore, adjacent bars touch each other. A bar graph based on a continuous scale, in which the bars touch, is called a *frequency histogram*. The data from Table 2.3 can be displayed in a histogram if we assume that the X values represent inches above 5 feet for a group of 25 women whose heights have been measured. (That is, a value of 2 represents 5 feet 2 inches, or 62 inches; 3 is 5 feet 3 inches, or 63 inches; etc.) The histogram is shown in Figure 2.2. As with the bar graph, the heights of the bars represent the

Figure 2.2

Frequency Histogram

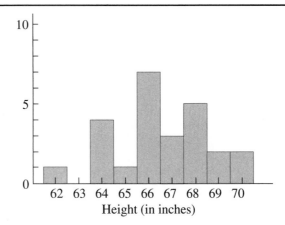

Height (in inches)

frequency count for each value. A glance at this figure shows you how the women are distributed in terms of height.

The Frequency Polygon

For some purposes researchers find the bars of a histogram to be distracting and prefer an alternative format, the *frequency polygon*. An easy way to think of a frequency polygon is to imagine placing a dot in the middle of the top of each bar in a histogram and connecting the dots with straight lines (and then getting rid of the bars), as shown in Figure 2.3a. Of course, normally a frequency polygon is drawn without first constructing the histogram, as shown in Figure 2.3b. Notice that the frequency polygon is connected to the horizontal axis at the high end by a straight line from the bar representing the frequency count of the highest value, and is similarly connected at the low end. Thus, the area enclosed by the polygon is clearly defined and can be used in ways to be described later. A frequency polygon is particularly useful when comparing two overlapping distributions on the same graph. The bars of a histogram would only get in the way in that case.

Just as a simple frequency distribution can be displayed as a histogram or as a polygon, so too can the other distributions we have discussed: the relative frequency distribution, the cumulative frequency distribution, and so forth. It should be obvious, however, that a histogram or polygon based on a relative frequency distribution will have exactly the same shape as the corresponding graph of a simple frequency distribution—only the scale of the Y axis will change (because all of the frequency counts are divided by

Figure 2.3

Frequency Polygon

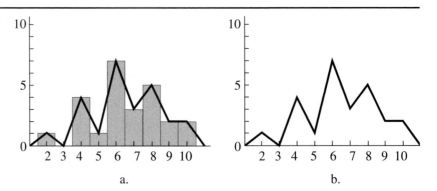

a.

b.

the same number, $N = \Sigma\, f$). Whether it is more informative to display actual frequencies or relative frequencies depends on the situation. If the group from which the data have been taken is very large, relative frequencies will probably make more sense.

Whether your polygon is based on simple or relative frequencies, it is easy to find the mode of your distribution (defined earlier in this section) from looking at the polygon. The mode is the score on the X axis that is directly under the highest point of the polygon. Because the height of the polygon at each point represents the frequency of the score below it, the score at which the polygon is highest is the most popular score in the distribution, and therefore it is the mode. However, as mentioned before, there can be more than one mode in a distribution (e.g., the polygon can look a bit like a camel with two humps). Even if one mode is actually a bit higher than the other (in which case, technically, there is really only one mode), if the polygon rises to one distinct peak, decreases, and then rises again to another distinct peak, it is common to say that the distribution has two modes. The role of the mode in describing distributions will be discussed further in the next chapter.

The Cumulative Frequency Polygon

A *cumulative frequency polygon* (also called an *ogive*) has a very different shape than a simple frequency polygon does. For one thing, the cf polygon never slopes downward as you move to the right in the graph, as you can see in Figure 2.4 (which was drawn using the same data as in all the examples above). That is because the cumulative frequency can never decrease. It can stay the same, if the next value has a zero frequency, but there are no negative frequency counts, so a cumulative frequency can never go down as the number of values increases. This is a case for which the polygon is definitely easier to look at and interpret than the corresponding histogram. Notice that in the cumulative frequency polygon the dots of the graph are not centered above the values being counted, but rather are above the *upper real limit* of each value (e.g., 5 feet $4\frac{1}{2}$ inches, instead of 5 feet 4 inches). The rationale is that to make sure you have accumulated, for instance, all of the heights labeled 5 feet 4 inches, you have to include all measurements up to 5 feet $4\frac{1}{2}$ inches.

The ogive can be quite useful when the percentile, or PR, in which you are interested falls between two of the entries in a table. In these common

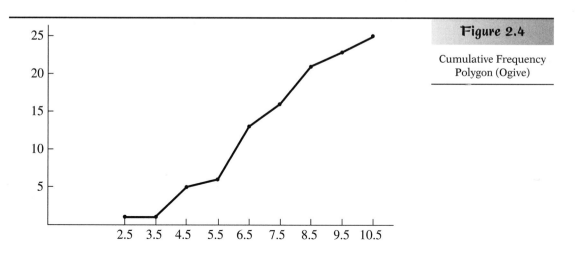

Figure 2.4

Cumulative Frequency
Polygon (Ogive)

cases, expressing the Y axis of the ogive in terms of cumulative percentages can help you to estimate the intermediate value that you are seeking. In the case of Figure 2.4, you would need only to multiply the frequencies on the Y axis by 4 (i.e., $100/N$) to create a cumulative percentage polygon. Then, to find the percentile rank of any score you first find the score on the X axis of the polygon, draw a vertical line from that score up to intersect the cumulative polygon, and finally draw a horizontal line from the point of intersection to the Y axis. The percentage at the point where the horizontal line intersects the Y axis is the PR of the score in question. For example, if you start with a score of 6.0 on the horizontal axis of Figure 2.4, move up until you hit the curve, and then move to the left, you will hit the vertical axis near the frequency of 10, which corresponds to 40%. So the PR of a score of 6.0 is about 40.

Naturally, the procedure for finding percentiles is exactly the opposite of the one just described. For instance, to find the score at the 70th percentile, start at this percentage on the Y axis of Figure 2.4 (midway between the frequencies of 15 and 20, which correspond to 60 and 80%, respectively), and move to the right on a horizontal line until you hit the ogive. From the point of intersection, go straight down to the horizontal axis, and you should hit a score of about 7.8 on the X axis. So the 70th percentile is about 7.8.

Of course, the accuracy of these graphical procedures depends on how carefully the lines are drawn. Drawing the graph to a larger scale tends to increase accuracy. Also, note that the cumulative percentage polygon consists of straight *lines*; therefore, these approximations are a form of *linear interpolation*. The procedure can be made more accurate by fitting a curve to the points of the cumulative polygon, but how the curve is drawn depends inevitably on assumptions about how the distribution would look if it were smooth (i.e., if you had infinitely precise measurements of the variable on an infinitely large population). These days you are usually better off just letting your computer draw the graphs and/or find the percentiles and PRs in which you are interested (see Section D).

I will discuss the preceding graphs again when I apply these graphing techniques to grouped frequency distributions in Section B. For now, I will compare the relative frequency polygon with the concept of a *theoretical frequency distribution*.

Real versus Theoretical Distributions

Frequency polygons make it easy to see the distribution of values in your data set. For instance, if you measured the anxiety level of each new student entering a particular college and made a frequency polygon out of the distribution, you could see which levels of anxiety were common and which were not. If you are a dean at the college and you see a high and wide peak over some pretty high anxiety levels, you would be concerned about the students and would consider various interventions to make the students more at ease. If new anxiety measurements were taken after some interventions, you would hope to see the frequency polygon change so that the line is high over the low anxiety values and gets quite low for high anxiety levels.

Unfortunately, the frequency polygon is harder to look at and interpret simply when it is based on a small number of scores. For instance, the frequency polygon in Figure 2.3b is based on only 25 height measurements (rounded to the nearest inch), and therefore it is not at all smooth; it consists of straight lines and sharp angles, which at no point resemble a curve. However, if height were measured to the nearest tenth of an inch and many

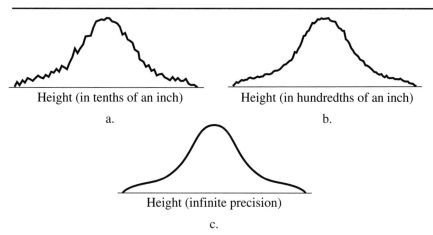

Figure 2.5

Frequency Polygons with
Precise Measurements

Height (in tenths of an inch)   Height (in hundredths of an inch)

a.   b.

Height (infinite precision)

c.

more people were included in the distribution, the polygon would consist of many more lines, which would be shorter, and many more angles, which would be less sharp (see Figure 2.5a). If height were measured to the nearest hundredth of an inch on a large population, the frequency polygon would consist of very many tiny lines, and it would begin to look fairly smooth (see Figure 2.5b). If we kept increasing the precision of the height measurements and the number of people measured, eventually the frequency polygon will not be distinguishable from a smooth curve (see Figure 2.5c). Smooth curves are easier to summarize and to interpret in a simple way.

The frequency polygons that psychological researchers create from their own data are usually far from smooth due to relatively few measurements and, often, an imprecise scale (that is one reason why psychologists are not likely to publish such displays, using them instead as tools for inspecting their data). On the other hand, a mathematical (or theoretical) distribution is determined by an equation and usually appears as a perfectly smooth curve. The best-known mathematical distribution is the *normal distribution*, which looks something like a bell viewed from the side (as in Figure 2.5c). With a precise scale, and enough people of one gender in a distribution of height (the distribution gets more complicated when the heights of both genders are included, as you will see in the next chapter), the frequency polygon for height will look a lot like the normal curve (except that the true normal curve actually never ends, extending to infinity in both directions before touching the horizontal axis). This resemblance is important because many advanced statistical procedures become quite easy if you assume that the variable of interest follows a normal distribution. I will have much more to say about the normal distribution in the next few chapters, and about other theoretical distributions in later chapters.

1. Often the first step in understanding a group of scores is to put them in order, thus forming an *array*.
2. If the number of different values in the group is not too large, a *simple frequency distribution* may make it easy to see where the various scores lie. To create a simple frequency distribution, write down all the possible scores in a column with the highest score in the group on top and the lowest on the bottom, even if some of the intermediate possible scores do not appear in the group. Add a second column in which you record the frequency of occurrence in the group for each value in the first column. The score with the highest frequency is the *mode* of the distribution.

SUMMARY

3. It is easy to create a cumulative frequency (cf) distribution from a simple frequency distribution: The cf entry for each score is equal to the frequency for that score plus the frequencies for all lower scores. (This is the same as saying that the cf for a given score is the frequency for that score, plus the cf for the next lower score.) The cf entry for the highest score must equal $\Sigma f = N$ (the total number of scores in the group).

4. To convert a simple or cumulative frequency distribution to a relative or cumulative relative distribution, divide each entry by N. The relative distribution tells you the proportion of scores at each value, and the cumulative relative distribution tells you what proportion of the scores is at or below each value.

5. Multiplying each entry of a cumulative relative frequency distribution by 100 gives a cumulative percentage distribution. The entries of the latter distribution are *percentile ranks* (PRs); each entry tells you the percentage of the distribution that is at or below the corresponding score. A percentile, on the other hand, is the score corresponding to a particular cumulative percentage. For example, the 40th percentile is the score that has a PR of 40. If the percentile or PR of interest does not appear in the table, it can be estimated with the appropriate graph (see point 9).

6. If the scores in a distribution represent a discrete variable (e.g., number of children in a family), and you want to display the frequency distribution as a graph, a *bar graph* should be used. In a bar graph, the heights of the bars represent the frequency counts, and adjacent bars do not touch. A bar graph is also appropriate for distributions involving nominal or ordinal scales (e.g., the frequency of different eye colors in the population).

7. When dealing with a continuous scale (e.g., height measured in inches), the distribution can be graphed as a *histogram,* which is a bar graph in which adjacent bars *do* touch. In a histogram, the width of the bar that represents a particular value goes from the *lower* to the *upper real limit* of that value.

8. An alternative to the histogram is the frequency polygon, in which a point is drawn above each value. The height of the point above the value on the X axis represents the frequency of that value. These points are then connected by straight lines, and the polygon is connected to the X axis at either end to form a closed figure. It is usually easier to compare two polygons on the same graph (e.g., separate distributions for males and females) than two histograms.

9. A cumulative frequency distribution can be graphed as a cumulative frequency polygon, called an *ogive,* in the same manner as the ordinary frequency polygon—just place the dot representing the cf over the upper real limit of each corresponding score. If you convert the cumulative frequencies to cumulative percentages, the ogive can be used to estimate percentiles and PRs not in your original table. Move straight up from a score until you hit the curve and then horizontally to the left until you hit the Y axis to find the PR of the score. Reversing this procedure allows you to estimate percentiles.

10. A frequency polygon can let you see at a glance which scores are popular in a distribution and which are not. As the number of people in the distribution and the precision of the measurements increase, the polygon begins to look fairly smooth. Ideally, the frequency polygon can somewhat resemble a perfectly smooth mathematical distribution, such as the normal curve.

EXERCISES

*1. A psychotherapist has rated all 20 of her patients in terms of their progress in therapy, using a 7-point scale. The results are shown in the following table:

	f
Greatly improved	5
Moderately improved	4
Slightly improved	6
Unchanged	2
Slightly worse	2
Moderately worse	1
Greatly worse	0

a. Draw a bar graph to represent the above results. To answer the following questions, create relative frequency (rf), cumulative frequency (cf), cumulative relative frequency (crf), and cumulative percentage frequency (cpf) columns for the table.
b. What proportion of the patients was greatly improved?
c. How many patients did not improve (i.e., were unchanged or became worse)? What proportion of the patients did not improve?
d. What is the percentile rank of a patient who improved slightly? Of a patient who became slightly worse?
e. Which category of improvement corresponds to the third quartile (i.e., 75th percentile)? To the first quartile?

*2. A cognitive psychologist is training volunteers to use efficient strategies for memorizing lists of words. After the training period, 25 participants are each tested on the same list of 30 words. The numbers of words correctly recalled by the participants are as follows: 25, 23, 26, 24, 19, 25, 24, 28, 26, 21, 24, 24, 29, 23, 19, 24, 23, 24, 25, 23, 24, 25, 26, 28, 25. Create a simple frequency distribution to display these data, and then add columns for rf, cf, crf, and cpf.
a. What proportion of the participants recalled exactly 24 words?
b. How many participants recalled no more than 23 words? What proportion of the total does this represent?
c. What is the percentile rank of a participant who scored 25? Of a participant who scored 27?

d. Which score is close to being at the first quartile? The third quartile?
e. Draw a histogram to represent the data.

3. A boot camp sergeant recorded the number of attempts each of 20 soldiers required to complete an obstacle course. The results were 2, 5, 3, 1, 2, 7, 1, 4, 2, 4, 8, 1, 3, 2, 6, 5, 2, 4, 3, 1. Create a simple frequency table to display these data. Add columns for cf, rf, crf, and cpf. (*Note*: Because lower numbers reflect better scores, you may want to put the lowest number on top of the table.)
a. What proportion of the soldiers could complete the course on the first attempt?
b. What proportion of them needed four or more attempts?
c. What is the percentile rank of someone who needed five attempts?
d. What score is closest to being the third quartile?
e. Draw a frequency polygon to represent the data.

4. An ethnographer surveyed 25 homes to determine the number of people per household. She found the following household sizes: 2, 1, 3, 5, 1, 4, 3, 2, 2, 6, 3, 4, 5, 1, 2, 4, 2, 7, 4, 6, 5, 5, 6, 6, 5. Construct a simple frequency table to display these results. Add columns for cf, rf, crf, and cpf.
a. What percentage of households have three or fewer members?
b. What household size corresponds to the 80th percentile?
c. How many households have only one member? To what proportion does that correspond?
d. What proportion of households have five or more members?
e. Draw a bar graph to represent the data.

*5. A physics professor gave a quiz with 10 questions to a class of 20 students. The scores were 10, 3, 8, 7, 1, 6, 5, 9, 8, 4, 2, 7, 7, 10, 9, 6, 8, 3, 8, 5. Create a simple frequency table to display these results. Add columns for cf, rf, crf, and cpf.
a. How many students obtained a perfect score? What proportion does that represent?
b. What score is closest to the 50th percentile?
c. What is the percentile rank of a student who scored a 5? Of a student who scored a 9?

d. What proportion of the students scored 9 or more?

e. Draw a frequency polygon to represent the data.

6. Draw a cumulative percentage polygon (ogive) to represent the data in Exercise 3. Use your graph to answer the following questions (approximate your answer to the nearest tenth of a point):

a. What score is at the 30th percentile?

b. What score is at the 50th percentile?

c. What is the percentile rank that corresponds to a score of 3.5?

d. What is the percentile rank that corresponds to a score of 6.5?

*7. Draw a cumulative percentage polygon (ogive) to represent the data in Exercise 5. Use your graph to answer the following questions (approximate your answer to the nearest tenth of a point):

a. What score is at the 50th percentile?

b. What score is at the 75th percentile?

c. What is the percentile rank that corresponds to a score of 4?

d. What is the percentile rank that corresponds to a score of 7?

8. The following data represent the scores of 50 students on a difficult 20-question quiz: 17, 12, 6, 13, 9, 15, 11, 16, 4, 15, 12, 13, 10, 13, 2, 11, 13, 10, 20, 14, 12, 17, 10, 15, 12, 17, 9, 14, 11, 15, 11, 16, 9, 13, 18, 10, 13, 0, 11, 16, 9, 8, 12, 13, 12, 17, 8, 16, 12, 15. Create a simple frequency table for these data, add columns for cf and cpf, and then graph the cumulative percentage polygon in order to answer the following questions.

a. Find the (approximate) values for the three quartiles of this distribution.

b. Find the (approximate) values for the first and ninth deciles of this distribution.

c. What is the (approximate) percentile rank of a student who scored an 8 on the quiz?

d. What is the (approximate) percentile rank of a student who scored an 18 on the quiz?

B

BASIC STATISTICAL PROCEDURES

Grouped Frequency Distributions

Constructing a simple frequency distribution is, as the name implies, simple. Unfortunately, measurements on an interval/ratio scale usually result in too many different values for a simple frequency distribution to be helpful. The example of quiz scores was particularly convenient because there were only eight different values. However, suppose the example involved 25 scores on a midterm exam graded from 0 to 100. Hypothetical scores are listed in the form of an array in Table 2.7, as defined in Section A.

Table 2.7

98	96	93	92	92	89	89	88	86	86	86	85	85
84	83	81	81	81	81	79	75	75	72	68	64	

To put these scores in a simple frequency distribution, we would have to include all of the values from 98 down to 64, which means that many potential scores would have a frequency of zero (e.g., 97, 95, 94).

The simple frequency distribution obviously would not be very helpful in this case. In fact, it seems little better than merely placing the scores in order in an array. The problem, of course, is that the simple frequency distribution has too many different values. The solution is to group the possible score values into equal-sized ranges, called *class intervals*. A table that shows the frequency for each class interval is called a *grouped frequency distribution*. The data from Table 2.7 were used to form the grouped frequency distribution in Table 2.8. Notice how much more informative the frequency distribution becomes when scores are grouped in this way.

Table 2.8

Class Interval X	f	Class Interval X	f
95–99	2	75–79	3
90–94	3	70–74	1
85–89	8	65–69	1
80–84	6	60–64	1

Apparent versus Real Limits

To describe the construction of a grouped frequency distribution, I will begin by focusing on just one class interval from Table 2.8—for example,

80–84. The interval is defined by its *apparent limits*. A score of 80 is the *lower apparent limit* of this class interval, and 84 is the *upper apparent limit*. If the variable is thought of as continuous, however, the apparent limits are not the *real* limits of the class interval. For instance, if the score values from 64 to 98 represented the heights of 1-year-old infants in centimeters, any fractional value would be possible. In particular, any height above 79.5 cm would be rounded to 80 cm and included in the interval 80–84. Similarly, any height below 84.5 cm would be rounded to 84 and also included in the interval 80–84. Therefore, the *real limits* of the class interval are 79.5 (lower real limit) and 84.5 (upper real limit).

In general, the real limits are just half a unit above or below the apparent limits—whatever the unit of measurement happens to be. In the example of infant heights, the unit is centimeters. If, however, you were measuring the lengths of people's index fingers to the nearest tenth of an inch, you might have an interval (in inches) from 2.0 to 2.4, in which case the real limits would be 1.95 to 2.45. In this case, half a unit of measurement is half of a tenth of an inch, which is one twentieth of an inch, or .05. To find the width of a class interval (usually symbolized by i), we use the real limits rather than the apparent limits. The width of the interval from 2.0 to 2.4 inches would be 2.45 − 1.95 = .5 inch. In the case of the 80–84 interval we have been discussing, the width is 84.5 − 79.5 = 5 cm (if the values are thought of as the heights of infants), not the 4 cm that the apparent limits would suggest. If the values are thought of as midterm grades, they will not include any fractional values (exams graded from 0 to 100 rarely involve fractions). Nevertheless, the ability being measured by the midterm is viewed as a continuous variable.

Constructing Class Intervals

Notice that the different class intervals in Table 2.8 do not overlap. Consider, for example, the interval 80–84 and the next highest one, 85–89. It is impossible for a score to be in both intervals simultaneously. This is important because it would become very confusing if a single score contributed to the frequency count in more than one interval. It is also important that there is no gap between the two intervals; otherwise, a score could fall between the cracks and not get counted at all. Bear in mind that even though there appears to be a gap when you look at the apparent limits (80–84, 85–89), the gap disappears when you look at the real limits (79.5–84.5, 84.5–89.5) and yet there is still no overlap. Perhaps you are wondering what happens if a score turns out to be exactly 84.5. First, when dealing with a continuous scale, the probability of any particular *exact* value (e.g., 84.500) arising is con-sidered too small to worry about. In reality, however, measurements are not so precise, and such values do arise. In that case, a simple rule can be adopted, such as any value ending in exactly .5 being placed in the higher interval if the number before the .5 is even.

Choosing the Class Interval Width

Before you can create a grouped frequency distribution, you must first decide how wide to make the class intervals. This is an important decision. If you make the class interval too large, there will be too few intervals to give you much detail about the distribution. For instance, suppose we chose to put the data from Table 2.7 into a grouped frequency distribution in which i (the interval width) equals 10. The result would be as shown in Table 2.9. Such a grouping could be useful if these class intervals actually

Table 2.9

Class Interval X	f
90–99	5
80–89	14
70–79	4
60–69	2

corresponded with some external criterion; for instance, the class intervals could correspond to the letter grades A, B, C, and D. However, in the absence of some external criterion for grouping, it is preferable to have at least 10 class intervals to get a detailed picture of the distribution. On the other hand, if you make the class intervals too narrow, you may have so many intervals that you are not much better off than with the simple frequency distribution. In general, more than 20 intervals is considered too many to get a good picture of the distribution.

You may have noticed that Table 2.8, with only eight intervals, does not follow the recommendation of 10 to 20 intervals. There is, however, at least one other guideline to consider in selecting a class interval width: multiples of 5 are particularly easy to work with. To have a number of class intervals between 10 and 20, the data from Table 2.7 would have to be grouped into intervals with $i = 3$ or $i = 2$. The distribution with $i = 2$ is too similar to the simple frequency distribution (i.e., $i = 1$) to be of much value, but the distribution with $i = 3$ is informative, as shown in Table 2.10.

Finally, note that it is a good idea to make all of the intervals the same size. Although there can be reasons to vary the size of the intervals within the same distribution, it is rarely done, and this text will not discuss such cases.

Table 2.10			
Class Interval X	f	Class Interval X	f
96–98	2	78–80	1
93–95	1	75–77	2
90–92	2	72–74	1
87–89	3	69–71	0
84–86	6	66–68	1
81–83	5	63–65	1

Finding the Number of Intervals Corresponding to a Particular Class Width

Whether Table 2.10 is really an improvement over Table 2.8 depends on your purposes and preferences. In trying to decide which size class interval to use, you can use a quick way to determine how many intervals you will wind up with for a particular interval width. First, find the *range* of your scores by taking the highest score in the array and subtracting the lowest score. (Actually, you have to start with the *upper real limit* of the highest score and subtract the *lower real limit* of the lowest score. If you prefer, instead of dealing with real limits, you can usually just subtract the lowest from the highest score and add 1.) For the midterm scores, the range is $98.5 - 63.5 = 35$. Second, divide the range by a convenient interval width, and round up if there is any fraction at all. This gives you the number of intervals. For example, using $i = 3$ with the midterm scores, we get $35/3 = 11.67$, which rounds up to 12, which is the number of intervals in Table 2.10. Note that if the range of your values is less than 20 to start with, it is reasonable to stick with the simple frequency distribution, although you may want to use $i = 2$ if the number of scores in your array is small (which would result in many zero frequencies). To avoid having too many intervals with low or zero frequency, it has been suggested that the number of classes not be much more than the square root of the sample size (e.g., if $N = 25$, this rule suggests the use of $\sqrt{25} = 5$ classes; this rule would argue in favor of Table 2.9, but Table 2.8 would still be considered a reasonable choice).

Choosing the Limits of the Lowest Interval

Having chosen the width of your class interval, you must decide on the apparent limits of the lowest interval; the rest of the intervals will then be determined. Naturally, the lowest class interval must contain the lowest score in the array, but that still leaves room for some choice. A useful guideline is to make sure that either the lower apparent limit or the upper apparent limit of the lowest interval is a multiple of i. (If the lower limit of one interval is a multiple of i, all the lower limits will be multiples of i.)

This is true in Table 2.10: The lower limit of the lowest interval (63) is a multiple of i, which is 3. It also would have been reasonable to start with 64–66 as the lowest interval because then the upper limit (66) would have been a multiple of i. Choosing the limits of the lowest interval is a matter of convenience, and a judgment can be made after seeing the alternatives.

Relative and Cumulative Frequency Distributions

Once a grouped frequency distribution has been constructed, it is easy to add columns for cumulative, relative, and cumulative relative frequencies, as described in Section A. These columns have been added to the grouped frequency distribution in Table 2.8 to create Table 2.11.

Interval	f	cf	rf	crf	
95–99	2	25	.08	1.00	**Table 2.11**
90–94	3	23	.12	.92	
85–89	8	20	.32	.80	
80–84	6	12	.24	.48	
75–79	3	6	.12	.24	
70–74	1	3	.04	.12	
65–69	1	2	.04	.08	
60–64	1	1	.04	.04	

Cumulative Percentage Distribution

Perhaps the most useful table of all is one that shows cumulative percent frequencies because (as noted in Section A) such a table allows you to find percentile ranks (PRs) and percentiles. The cumulative percent frequencies for the midterm scores are shown in Table 2.12. It is important to note that the cumulative percentage entry (as with any cumulative entry) for a particular interval corresponds to the *upper real limit* of that interval. For example, across from the interval 85–89 is the cpf% entry of 80. This means that a score of 89.5 is the 80th percentile (that is why the table includes a separate column for the upper real limit, labeled url). To score better than 80% of those in the class, a student must have a score that beats not only all the scores below the 85–89 interval but all the scores *in* the 85–89 interval. And the only way a student can be sure of beating all the scores in the 85–89 interval is to score at the top of that interval: 89.5.

Interval	f	pf%	url	cf	cpf%	
95–99	2	8	99.5	25	100	**Table 2.12**
90–94	3	12	94.5	23	92	
85–89	8	32	89.5	20	80	
80–84	6	24	84.5	12	48	
75–79	3	12	79.5	6	24	
70–74	1	4	74.5	3	12	
65–69	1	4	69.5	2	8	
60–64	1	4	64.5	1	4	

On the other hand, if you wanted to know what your percentile rank would be if you scored 79.5 on the midterm, you would look at the cumulative percent frequency entry for the 75–79 interval, which tells you that your PR is 24 (i.e., you beat 24% of the group). If you wanted to know the PR for a score of 67 or 81, or you wanted to know what score was at the

40th percentile, you could not find that information directly in Table 2.12. However, you could use a graph to help you estimate these answers, as demonstrated in Section A, or you could use linear interpolation more directly, as described next.

Estimating Percentiles and Percentile Ranks by Linear Interpolation

If you are dealing with a grouped distribution, and therefore know how many scores are in each interval but not where within each interval those scores lie (i.e., I am assuming that you don't have access to the raw data from which the frequency table was constructed), you can use *linear interpolation* to estimate both percentiles and percentile ranks. The key assumption behind linear interpolation is that the scores are spread evenly (i.e., linearly) throughout the interval.

Estimating Percentile Ranks

Consider the interval 85–89 in Table 2.12, for which the frequency is 8. We assume that the eight scores are spread evenly from 84.5 to 89.5 so that, for instance, four of the scores are between 84.5 and 87 (the *midpoint* of the interval), and the other four are between 87 and 89.5. This reasoning also applies to the percentages. The cumulative percentage at 84.5 is 48 (the cpf entry for 80–84), and at 89.5 it is 80 (the cpf entry for 85–89), as shown in Figure 2.6. On the basis of our assumption of linearity, we can say that the midpoint of the interval, 87, should correspond to a cumulative percentage midway between 48 and 80, which is 64% [(48 + 80)/2 = 128/2 = 64)]. Thus, the PR for a score of 87 is 64.

A more complicated question to ask is: What is the PR for a score of 86 in Table 2.12? Because 86 is not right in the middle of an interval, we need to know how far across the interval it is. Then we can use linear interpolation to find the cumulative percentage that corresponds to that score. To go from the lower real limit, 84.5, to 86 we have to go 1.5 score points. To go across the entire interval requires five points (the width of the

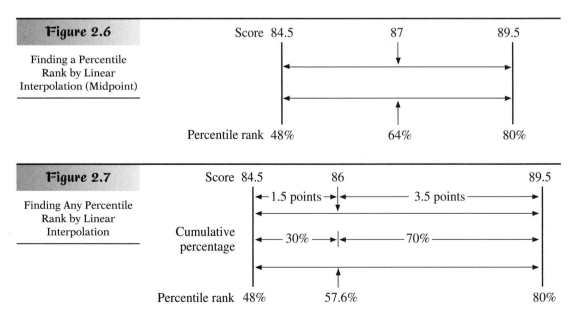

Figure 2.6

Finding a Percentile Rank by Linear Interpolation (Midpoint)

Figure 2.7

Finding Any Percentile Rank by Linear Interpolation

interval). So 86 is 1.5 out of 5 points across the interval; 1.5 out of 5 = 1.5/5 = .3, or 30%. A score of 86 is 30% of the way across the interval. That means that to find the cumulative percentage for 86, we must go 30% of the way from 48 to 80, as shown in Figure 2.7. From 48 to 80 there are 32 percentage points, and 30% of 32 is (.3)(32) = 9.6. So we have to add 9.6 percentage points to 48 to get 57.6, which is the PR for a score of 86. In sum, 86 is 30% of the way from 84.5 to 89.5, so we go 30% of the way from 48 to 80, which is 57.6.

Bear in mind that it is not terribly important to be exact about estimating a percentile rank from a grouped frequency distribution. First, the estimate is based on the assumption that the scores are spread evenly throughout the interval, which may not be true. Second, the estimate may be considerably different if the class interval width or the starting score of the lowest interval changes, Now that I have described how to estimate a PR corresponding to any score in a grouped distribution, it will be easy to describe the reverese process of estimating the score that corresponds to a given percentile rank.

Estimating Percentiles

Suppose you want to find the sixtieth percentile (i.e., the score for which the PR is 60) for the midterm exam scores. First, you can see from Table 2.12 that 60% lands between the entries 48% (correspoding to 84.5) and 80% (corresponding to 89.5). Because 60 is somewhat closer to 48 than it is to 80, you know that the 60th percentile should be somewhat closer to 84.5 than to 89.5—i.e., in the neighborhood of 86. More exactly, the proportion of the way from 48 to 80 you have to go to get to 60 (which is the same proportion you will have to go from 84.5 to 89.5) is $(60 - 48)/32 = 12/32 = .375$. Adding .375 of 5 (the width of the class interval) to 84.5 yields $84.5 + (.375) \cdot 5 = 84.5 + 1.875 = 86.375$. It would be reasonable to round off in this case, and say that the 60th percentile is 86.4.

Graphing a Grouped Frequency Distribution

A grouped frequency distribution can be displayed as a histogram, like the one used to represent the simple frequency distribution in Section A (see Figure 2.2). In a graph of a grouped distribution, however, the width of each bar extends from the lower real limit to the upper real limit of the class interval that the bar represents. As before, the height of the bar indicates the frequency of the interval. (This is only true when all the class intervals have the same width, but because this is the simplest and most common arrangement, we will consider only this case.) A histogram for a grouped frequency distribution is shown in Figure 2.8, which is a graph of the data in Table 2.8.

If you prefer to use a frequency polygon, place a dot at the top of each bar of the histogram at the midpoint of the class interval. (A quick way to calculate the midpoint is to add the upper and lower apparent limits and divide by 2—this also works with the real limits.) Place dots on the horizontal axis (to represent zero frequency) on either side of the distribution—that is, at the midpoint of the next interval below the lowest and above the highest, as shown in Figure 2.9. Connecting the polygon to these additional dots on either side closes the polygon, with the horizontal axis serving as one of the sides. Thus, the frequency polygon encloses a particular amount of area, which represents the total number of scores in the distribution. A third of that area, for example, would represent a third of the scores. I will have a lot more to say about the areas enclosed by frequency polygons and smooth dis-

Figure 2.8

Frequency Histogram for
a Grouped Distribution

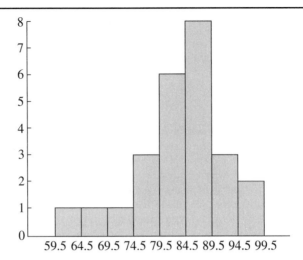

Figure 2.9

Frequency Polygon for a
Grouped Distribution

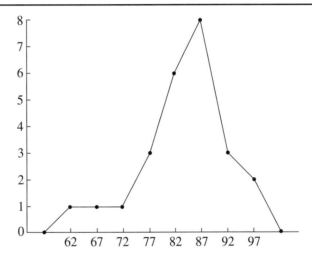

tributions in Chapter 4. Of course, you can also create a cumulative frequency or percentage polygon (an ogive) as described in Section A. Just place the dot representing the cumulative frequency or percentage over the upper real limit of the interval to which it corresponds. Then, you can use the ogive you plotted to find percentiles and PRs, also as described in Section A.

Guidelines for Drawing Graphs of
Frequency Distributions

Graphs of frequency distributions are not often published in psychological journals, but there are general guidelines for creating any kind of line graph that should be followed to make the graphs easier to interpret. (These guidelines appear in most statistics texts; here, I adapt them for use with frequency distributions.) The first guideline is that you should make the X axis longer than the Y axis by about 50% so that the height of the graph is only about two thirds of the width. (Some researchers suggest that the height be closer to three quarters of the width. The exact ratio is not critical, but a proportion in this vicinity is considered easiest to interpret visually.) The second guideline is that the scores or measurement values should

Figure 2.10

Frequency Histograms:
Continuous Scale and
Broken Scale

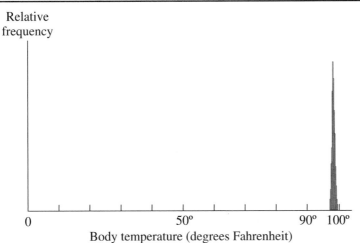

Relative
frequency

Body temperature (degrees Fahrenheit)

a.

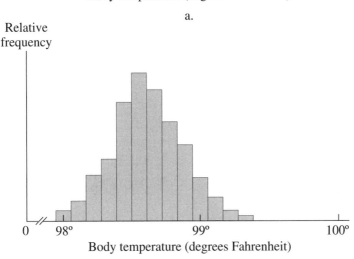

Relative
frequency

Body temperature (degrees Fahrenheit)

b.

be placed along the horizontal axis and the frequency for each value indicated on the vertical axis. This creates a profile, like the skyline of a big city in the distance, that is easy to grasp. A third guideline is obvious: The units should be equally spaced on both axes (e.g., a single frequency count should be represented by the same distance anywhere along the Y axis). The fourth guideline is that the intersection of the X and Y axes should be the zero point for both axes, with numbers getting larger (i.e., more positive) as you move up or to the right. The fifth guideline is that you should choose a measurement unit and a scale (i.e., how much distance on the graph equals one unit) so that the histogram or polygon fills up nearly all of the graph without at any point going beyond the axes of the graph.

Sometimes it is difficult to satisfy the last two guidelines simultaneously. Suppose you want to graph a distribution of normal body temperatures (measured to the nearest tenth of a degree Fahrenheit) for a large group of people. You would like to mark off the X axis in units of .1 degree, but if you start with zero on the left and mark off equal intervals, each representing .1 degree, you will have to mark off 1,000 intervals to get to 100 degrees. Assuming that your distribution extends from about 97° F to about

100° F (98.6° being average body temperature), you will be using only a tiny portion of the *X* axis, as indicated in Figure 2.10a. The solution to this dilemma is to increase the scale so that .1 degree takes more distance along the *X* axis and not to mark off units continuously from 0 to 97°. Instead, you can indicate a break along the *X* axis, as shown in Figure 2.10b so that the zero point can still be included but the distribution can fill the graph. Similarly, a break can be used on the *Y* axis if all the frequency counts are high but do not differ greatly.

The sixth and last guideline is that both axes should be clearly labeled. In the case of a frequency distribution the *X* axis should be labeled with the name of the variable and the unit in which it was measured.

B

SUMMARY

1. When a distribution contains too many possible values to fit conveniently in a regular frequency distribution, class intervals may be created (usually all of which are the same size) such that no two intervals overlap and that there are no gaps between intervals. If the variable is measured on a continuous scale, the upper and lower real limits of the interval are half of a measurement unit above and below the upper and lower apparent limits, respectively.

2. One way to help you decide on a class width to use is to first find the range of scores by subtracting the lowest score in your distribution from the highest and adding 1. Then divide the range by a convenient class width (a multiple of 5, or a number less than 5, if appropriate), and round up if there is any fraction, to find the number of intervals that would result. If the number of intervals is between 10 and 20, the width is probably reasonable; otherwise, you can try another convenient value for *i*. However, you may want to use fewer intervals if there are fewer than 100 scores in your distribution.

3. The lowest class interval must contain the lowest score in the distribution. In addition, it is highly desirable for the lower or upper limit to be a multiple of the chosen interval width.

4. In a grouped cumulative percentage distribution, the entry corresponding to a particular class interval is the percentile rank of the upper real limit of that interval. To find the PR of a score that is not at one of the upper real limits in your table, you can use linear interpolation. If the score is *X*% of the interval width above the lower limit of some interval, look at the PR for the upper and lower limits of that interval, and add *X*% of the difference of the two PRs to the lower one.

5. To find a percentile that does not appear as an entry in your table, first locate the two table entries for cumulative percentage that it is between—that will determine the interval that the percentile is in. You can then interpolate within that interval to estimate the percentile.

6. In a histogram for a grouped frequency distribution, the bars for each class interval extend from its lower to its upper real limit, and therefore neighboring bars touch each other. To create a polygon for a grouped distribution, place the dot over the midpoint of the class interval, and for a cumulative polygon (ogive), the dot is placed over the upper real limit of the interval.

7. The guidelines for graphs of frequency distributions that follow apply, for the most part, to other types of line graphs published in psychological journals.
   a. The *Y* axis should be only about two-thirds as long as the *X* axis.
   b. For frequency distributions, the variable of interest is placed along the *X* axis and the frequency counts (or relative frequency) are represented along the *Y* axis.

c. The measurement units are equally spaced along the entire length of both axes.

d. The intersection of the X and Y axes is the zero point for both dimensions.

e. Choose a scale to represent the measurement units on the graph so that the histogram or polygon fills the space of the graph as much as possible. Indicating a break in the scale on one or both axes may be necessary to achieve this goal.

f. Both axes should be clearly labeled, and the X axis should include the name of the variable and the unit of measurement.

EXERCISES

*1. The following are the IQ scores for the 50 sixth-grade students in Happy Valley Elementary school: 104, 111, 98, 132, 128, 106, 126, 99, 111, 120, 125, 106, 99, 112, 145, 136, 124, 130, 129, 114, 103, 121, 109, 101, 117, 119, 122, 115, 103, 130, 120, 115, 108, 113, 116, 109, 135, 121, 114, 118, 110, 136, 112, 105, 119, 111, 123, 115, 113, 117.

a. Construct the appropriate grouped frequency distribution, and add crf and cpf columns (treat IQ as a continuous scale).

b. Draw a frequency histogram to represent the above data.

c. Estimate the first and third quartiles.

d. Estimate the 40th and 60th percentiles.

e. Estimate the percentile rank of a student whose IQ is 125.

f. Estimate the percentile rank of a student whose IQ is 108.

*2. An industrial psychologist has devised an aptitude test for selecting employees to work as cashiers using a new computerized cash register. The aptitude test, on which scores can range from 0 to 100, has been given to 60 new applicants, whose scores were as follows: 83, 76, 80, 81, 74, 68, 92, 64, 95, 96, 55, 70, 78, 86, 85, 94, 76, 77, 82, 85, 81, 71, 72, 99, 63, 75, 76, 83, 92, 79, 82, 69, 91, 84, 87, 90, 80, 65, 84, 87, 97, 61, 73, 75, 77, 86, 89, 92, 79, 80, 85, 87, 82, 94, 90, 89, 85, 84, 86, 56.

a. Construct a grouped frequency distribution table for the above data.

b. Draw a frequency polygon to display the distribution of these applicants.

c. Suppose the psychologist is willing to hire only those applicants who scored at the 80th percentile or higher (i.e., the top 20%). Estimate the appropriate cutoff score.

d. Estimate the 75th and 60th percentiles.

e. If the psychologist wants to use a score of 88 as the cutoff for hiring, what percentage of the new applicants will qualify?

f. Estimate the percentile rank for a score of 81.

3. A telephone company is interested in the number of long-distance calls its customers make. Company statisticians randomly selected 40 customers and recorded the number of long-distance calls they made the previous month. They found the following results: 17, 0, 52, 35, 2, 8, 12, 28, 9, 43, 53, 39, 4, 21, 17, 47, 19, 13, 7, 32, 6, 2, 0, 45, 4, 29, 5, 10, 8, 57, 9, 41, 22, 1, 31, 6, 30, 12, 11, 20.

a. Construct a grouped frequency distribution for the data.

b. Draw a cumulative percentage polygon for these data.

c. What percentage of customers made fewer than 10 long-distance calls?

d. What is the percentile rank of a customer who made 50 calls?

e. What percentage of customers made 30 or more calls?

*4. A state trooper, interested in finding out the proportion of drivers exceeding the posted speed limit of 55 mph, measured the speed of 25 cars in an hour. Their speeds in miles per hour were as follows: 65, 57, 49, 75, 82, 60, 52, 63, 49, 75, 58, 66, 54, 59, 72, 63, 85, 69, 74, 48, 79, 55, 45, 58, 51.

a. Create a grouped frequency distribution table for these data. Add columns for cf and cpf.

b. Approximately what percentage of the drivers were exceeding the speed limit?

c. Suppose the state trooper only gave tickets to those exceeding the speed limit by 10 mph or more. Approximately what

proportion of these drivers would have received a ticket?
d. Estimate the 40th percentile.
e. Estimatethe first and third quartiles.
f. What is the percentile rank of a driver going 62 mph?

*5. A psychologist is interested in the number of dreams people remember. She asked 40 participants to write down the number of dreams they remember over the course of a month and found the following results: 21, 15, 36, 24, 18, 4, 13, 31, 26, 28, 16, 12, 38, 26, 0, 13, 8, 37, 22, 32, 23, 0, 11, 33, 19, 11, 1, 24, 38, 27, 7, 14, 0, 13, 23, 20, 25, 3, 23, 26.
a. Create a grouped frequency distribution for these data with a class interval width of 5. Add columns for cf and cpf. (*Note*: Treat the number of dreams as a continuous variable.)
b. Draw a frequency histogram to display the distribution of the number of dreams remembered.
c. Suppose that the psychologist would like to select participants who remembered 30 or more dreams for further study. How many participants would she select? What proportion does this represent? What percentile rank does that correspond to?
d. Approximately what number of dreams corresponds to the 90th percentile?
e. What is the percentile rank of a participant who recalled 10 dreams?
f. What is the percentile rank of a participant who recalled 20 dreams?

6. Estimate all three quartiles for the data in the following table. (*Hint*: Each value for X can be assumed to represent a class that ranges from a half unit below to a half unit above the value shown; for example, $X = 16$ represents the range from 15.5 to 16.5.)

X	f	X	f
18	1	9	1
17	0	8	3
16	2	7	5
15	0	6	5
14	1	5	7
13	0	4	5
12	0	3	4
11	1	2	2
10	2	1	1

*7. Construct a grouped frequency distribution (width = 2) for the data in Exercise 2A8.
a. Add a cpf column and graph the cumulative percentage polygon.
b. Find the (approximate) values for all three quartiles.
c. Find the (approximate) values for the first and ninth deciles.
d. What is the (approximate) PR of a student who scored an 8 on the quiz?
e. What is the (approximate) PR of a student who scored an 18 on the quiz?

8. Redo Exercise 5 using a class interval width of 3. Discuss the similarities and differences between your answers to this exercise and your answers to Exercise 5. Describe the relative advantages and disadvantages of using a class interval of 3 for these data as compared to a width of 5.

Note: Some chapters will refer to exercises from previous sections of the chapter or from earlier chapters for purposes of comparison. A shorthand notation, consisting of the chapter number and section letter followed by the problem number, will be used to refer to exercises. For example, Exercise 3B2a refers to Chapter 3, Section B, Exercise 2, part a.

OPTIONAL MATERIAL

Stem-and-Leaf Displays

In this chapter I have been describing how a set of quiz scores or exam scores can be displayed so that it is easy to see how a group of people performed and how one particular individual performed in comparison to the group. Sometimes a group of scores represents the performance of a sample of the population, as described in the previous chapter (e.g., a group of allergy sufferers that has been given a new drug, and the scores are number of sneezes per hour). In this case, you would be less interested in individual scores than in some summary statistic you can use to extrapolate to the population of all allergy sufferers. However, the use of summary statistics depends on certain assumptions, including an assumption about the shape of the distribution for the entire population. Exploring the shape of the distribution for the sample can give you a clue about the shape of the population distribution and which types of summary statistics would be most valid. Thus, the graphs and tables described in this chapter can be useful not only for describing a particular group but also for exploring a sample and the degree to which that sample may be representative of a population.

When dealing with a sample, however, it is tempting to jump quickly to the summary statistics and then to conclusions about populations. J.W. Tukey (1977), the well-known statistician, recommended a more extensive inspection of sample data before moving on to inferential statistics. He called this preliminary inspection *exploratory data analysis*, and he proposed a number of simple but very useful alternatives to the procedures described so far in this chapter. One of these, the *stem-and-leaf display*, will be described in this section. Another technique will be presented in Section C of Chapter 3.

One disadvantage of using a grouped frequency distribution to find percentile ranks is that you don't know where the individual scores are within each interval. Of course, if you still have the raw data from which you constructed the grouped distribution, this is no problem. But if you do not have access to the raw data (perhaps because you are working from someone else's grouped distribution or because you lost your own raw data), the grouped distribution represents a considerable loss of information. An alternative that has the basic advantage of a grouped frequency distribution without the loss of the raw data is a stem-and-leaf display (also called a *stem plot*). The simplest stem-and-leaf display resembles a grouped distribution with $i = 10$. I will use the data from Table 2.7 to construct this kind of display.

First, construct the *stem*. To do this you must notice that all of the scores in Table 2.7 are two-digit numbers. The first (or leftmost) of the two digits in each score is referred to as the "leading" digit, and the second (or rightmost) digit is the "trailing" digit. The stem is made from all of the leading digits by writing these digits in a vertical column starting with the lowest at the top, as follows:

$$6$$
$$7$$
$$8$$
$$9$$

The 6 stands for scores in the sixties, the 7 for scores in the seventies, and so forth. Note that as with the simple frequency distribution, the 7, for example, is included even if there were no scores in the seventies. However, 5 is not included because there are no scores in the fifties or below 50.

Now you are ready to add the leaves to the stem. The leaves are the trailing digits, and they are written to the right of the corresponding leading digit, in order with higher numbers to the right, even if there are duplicates, as shown in Table 2.13.

The 2559 next to the 7 means that the following scores have 7 as their first digit: 72, 75, 75, 79. You can check this in Table 2.7. Compare the display in Table 2.13 with the grouped distribution in Table 2.9. The grouped distribution tells you only how many scores are in each interval. The stem-and-leaf display gives you the same information, plus you can see just

Stem	Leaf											
6	4	8										
7	2	5	5	9								
8	1	1	1	1	3	4	5	5	6	6	6	8	9	9
9	2	2	3	6	8							

Table 2.13

where the scores are within each interval. Moreover, the stem-and-leaf display functions like a histogram on its side, with the horizontal rows of numbers (the leaves) serving as the bars. As long as the leaf numbers are all equally spaced, the length of any row will be proportional to the frequency count corresponding to the adjacent leading digit. (For example, there are twice as many scores in the seventies as in the sixties, so the row next to the 7 should be twice as long as the row next to the 6.)

The stem-and-leaf display in Table 2.13 has a serious drawback, as you have probably noticed. It has the same problem pointed out with respect to Table 2.9: too few categories and therefore too little detail about the shape of the distribution. Obviously, a class width of 10 will not always be appropriate. The solution suggested by Tukey (1977) for this situation is to divide each stem in half, essentially using $i = 5$. The stem is constructed by listing each leading digit twice, once followed by an asterisk (trailing digits 0 to 4 are included in this category) and once followed by a period (for digits 5 to 9). The leaves are then added to the appropriate parts of the stem, as shown in Table 2.14.

Table 2.14

Stem	Leaf							
6*	4							
6.	8							
7*	2							
7.	5	5	9					
8*	1	1	1	1	3	4		
8.	5	5	6	6	6	8	9	9
9*	2	2	3					
9.	6	8						

Compare Table 2.14 to Table 2.8. Table 2.8 gives you the same frequency count for each interval but does not tell you where the raw scores are and does not automatically show you the shape of the distribution.

Of course, your data will not always range conveniently between 10 and 100. Suppose you have recorded the weight (in pounds) of every young man who tried out for your college's football team last year. The numbers might range from about 140 to about 220. Using just the leading digits for your stem would give only the numbers 1 and 2 written vertically. (Even if you used the asterisk and period to double the length of your stem, as described above, your stem would still be too short.) Also, the leaves would consist of two digits each, which is not desirable. The solution is to use the first two digits for the stem (14, 15, etc.) and the third digit for the leaf. On the other hand, if you are dealing with SAT scores, which usually range from 200 to 800, you can use the first digit as the stem, the second digit as the leaf, and simply ignore the third digit, which is always a zero anyway. Table 2.15 shows a stem-and-leaf display of hypothetical verbal SAT scores for 30 students in an introductory psychology class. The SAT scores range from a low of 370 (stem = 3, leaf = 7) to a high of 740 (stem = 7, leaf = 4).

Table 2.15

Stem	Leaf										
3	7	8									
4	0	2	2	5	7	8	9	9			
5	0	0	1	3	4	4	5	5	5	8	8
6	1	1	2	5	6	9					
7	0	2	4								

Yet another advantage of the stem-and-leaf display is that a second distribution can easily be compared to the first as part of the same display. Suppose you wanted to compare the math SAT scores to the verbal SAT scores for the students in the hypothetical introductory psychology class. You can use the same stem but place the leaves for the second distribution to the left, as shown in Table 2.16. You can see from Table 2.16 that the math scores are somewhat lower than the verbal scores. If the data were real, one might speculate that psychology attracts students whose verbal skills are relatively better than their math skills. In subsequent chapters you will learn how to test such hypotheses. Before you are ready for hypothesis testing, however, you need to learn more about descriptive statistics in the next few chapters. The topic of exploratory data analysis, in particular, will be pursued further in Section C of Chapter 3.

Math	Stem	Verbal	
9 9 6 4	3	7 8	**Table 2.16**
9 8 8 7 5 5 3 2 2 1	4	0 2 2 5 7 8 9 9	
8 6 5 3 3 2 1 1 0	5	0 0 1 3 4 4 5 5 5 8 8	
6 3 3 1 0	6	1 1 2 5 6 9	
2 1	7	0 2 4	

C

SUMMARY

1. *Exploratory data analysis* can be an important step before computing sample statistics and drawing inferences about populations. One of the exploratory techniques devised by Tukey (1977) is called a *stem-and-leaf display*, which is similar to a simple frequency distribution.
2. When dealing with double-digit numbers, the possible first (i.e., leading) digits are written in a column, with the lowest on top, to form the stem. The leaves are the second (i.e., trailing) digits that actually appear in the group of scores. All of the trailing digits (even repetitions) are written in a row to the right of the leading digit to which they correspond.
3. A stem-and-leaf display has the advantages of preserving all of the raw score information, as well as serving as a histogram on its side (the length of the row of trailing digits shows how often the leading digit has occurred).
4. If the stem is too short, each leading digit can be used twice—once for low trailing digits (i.e., 0 to 4) and once for high trailing digits (i.e., 5 to 9). If the scores are in the three-digit range, it is possible to use the first two digits for the stem or, depending on the range of the numbers, to use the first digits for the stem, the second digits for the leaves, and to ignore the third digits entirely.

EXERCISES

1. Construct a stem-and-leaf display for the IQ data in Exercise 2B1. Compare it to the frequency histogram you drew for the same data.
*2. Construct a stem-and-leaf display for the aptitude test scores in Exercise 2B2.
3. Construct a stem-and-leaf display for the speed measurements in Exercise 2B4.
*4. Construct a stem-and-leaf display for the numbers of dreams in Exercise 2B5. Compare it to the frequency histogram you drew for the same data.

MEASURES OF CENTRAL TENDENCY AND VARIABILITY

Chapter 3

You will need to use the following from previous chapters:

Symbols
Σ: Summation sign

Concepts
Scales of measurement
Frequency histograms and polygons

Procedures
Rules for using the summation sign

A

**CONCEPTUAL
FOUNDATION**

In Chapter 2, I began with an example in which I wanted to tell a class of 25 students how well the class had performed on a diagnostic quiz and make it possible for each student to evaluate how his or her score compared to the rest of the class. As I demonstrated, a simple frequency distribution, especially when graphed as a histogram or a polygon, displays the scores at a glance, and a cumulative percentage distribution makes it easy to find the percentile rank for each possible quiz score. However, the first question a student is likely to ask about the class performance is, "What is the average for the class?" And it is certainly a question worth asking. Although the techniques described in Chapter 2 provide much more information than does a simple average for a set of scores, the average is usually a good summary of that information. In trying to find the average for a group of scores, we are looking for one spot that seems to be the center of the distribution. Thus, we say that we are seeking the *central tendency* of the distribution. But, as the expression "central tendency" implies, there may not be a single spot that is clearly and precisely at the center of the distribution. In fact, there are several procedures for finding the central tendency of a group of scores, and which procedure is optimal can depend on the shape of the distribution involved. I will begin this chapter by describing the common ways that central tendency can be measured and the reasons for choosing one measure over another. Then, I will explain how central tendency measures can be used as a basis from which to quantify the variability of a distribution. Finally, I will consider some advanced measures for assessing the shape of a distribution.

Measures of Central Tendency

The Arithmetic Mean

When most students ask about the average on an exam, they have in mind the value that is obtained when all of the scores are added and then divided by the total number of scores. Statisticians call this value the *arithmetic mean*, and it is symbolized by the Greek letter μ (mu, pronounced "myoo") when it refers to the mean of a population. Later in this chapter, we will also be interested in the mean for a sample, in which case the mean is symbolized either by a bar over the letter representing the variable (e.g., $\overline{X}$, called "X bar") or by the capital letter M, for mean. There are other types of means, such as the harmonic mean (which will be introduced in Chapter 8) and the geometric mean, but the arithmetic mean is by far the most commonly used. Therefore, when I use the terms *mean* or *average* without further specification, it is the arithmetic mean to which I am referring. The arithmetic mean, when applicable, is undoubtedly the most useful measure of central tendency. However, before we consider the many statistical properties of the mean, we need to consider two lesser known, but nonetheless useful, measures of central tendency.

The Mode

Often the main purpose in trying to find a measure of central tendency is to characterize a large group of scores by one value that could be considered the most typical of the group. If you want to know how smart a class of students is (perhaps because you have to prepare to teach them), you would like to know how smart the typical student in that class is. If you want to know how rich a country is, you might want to know the annual income of a typical family. The simplest and crudest way to define the most typical score in a group is in terms of which score occurs with the highest frequency. That score is called the *mode* of the distribution.

The mode is easy to find once you have constructed a frequency distribution; it is the score that has the highest frequency. It is perhaps even easier to identify the mode when a frequency distribution has been displayed as a histogram or a graph. Simply look for the highest bar in the histogram or the highest point in the polygon—the score that is directly below that highest bar or point is the mode. The mode is defined in the same way for a grouped distribution as for a simple distribution, except with a grouped distribution the mode is the most frequently occurring *interval* (or the midpoint of that interval) rather than a single score. One potential drawback of using the mode with grouped distributions is that the mode depends a good deal on the way the scores are grouped (i.e., on your choice for the lowest interval and your choice for the width of the interval). However, even with a simple frequency distribution the mode has its problems, as I will show next.

Advantages and Disadvantages of the Mode A major disadvantage of the mode is that it is not a very reliable measure of central tendency. Consider the simple frequency distribution on the left side of Table 3.1. The mode of that distribution is 7 because that score has the highest frequency (6). However, if just one of the students who scored a 7 was later found really to have scored a 4, that frequency distribution would change to the one on the right side of Table 3.1, and the mode would consequently move from a score of 7 to a score of 4. Naturally, we would like to see more stability in our measures of central tendency. Moreover, if the score with a frequency of 6 in either distribution in Table 3.1 had a frequency of 5, there would be a whole range of scores at the mode, which would make the mode a rather imprecise measure.

Imagine that in either of the distributions in Table 3.1, the scores of 4 and 7 *both* have a frequency of 6. Now the distribution has more than one mode. If a distribution has many modes, finding these modes is not likely to be useful. However, a distribution that contains two distinct subgroups (e.g., men and women measured on the amount of weight they can lift over their heads) may have two meaningful modes (one for each subgroup), as shown in Figure 3.1. Such a distribution is described as *bimodal*. If a distribution has two or three distinct modes (it is hard to imagine a realistic situation with more modes), finding these modes can be useful indeed, and the modes would provide information not available from the more commonly used mean or median. The most common shape for a smooth, or nearly smooth, distribution involves having only one mode. Such distributions are

		Table 3.1		
X	f	X	f
10	1	10	1
9	0	9	0
8	3	8	3
7	6	7	5
6	5	6	5
5	5	5	5
4	5	4	6
3	2	3	2
2	1	2	1
1	1	1	1
0	1	0	1

Figure 3.1

A Bimodal Distribution

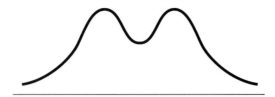

described as *unimodal* and are the only types of distributions that we will be dealing with in this text.

When dealing with interval/ratio scales, it seems that the main advantage of the mode as a measure of central tendency is in terms of distinguishing multimodal from unimodal distributions. The ease with which the mode can be found used to be its main advantage, but in the age of high-speed computers, this is no longer a significant factor. However, the mode has the unique advantage that it can be found for any kind of measurement scale. In fact, when dealing with nominal scales, other measures of central tendency (such as the mean) cannot be calculated; the mode is the *only* measure of central tendency in this case. For instance, suppose you are in charge of a psychiatric emergency room and you want to know the most typical diagnosis of a patient coming for emergency treatment. You cannot take the average of 20 schizophrenics, 15 depressives, and so forth. All you can do to assess central tendency is to find the most frequent diagnosis (e.g., schizophrenia may be the *modal* diagnosis in the psychiatric emergency room).

The Median

If you are looking for one score that is in the middle of a distribution, a logical score to focus on is the score that is at the 50th percentile (i.e., a score whose PR is 50). This score is called the *median*. The median is a very useful measure of central tendency, as you will see, and it is very easy to find. If the scores in a distribution are arranged in an array (i.e., in numerical order), and there are an *odd* number of scores, the median is literally the score in the middle. If there are an *even* number of scores, as in the distribution on the left side of Table 3.1 ($N = \Sigma f = 30$), the median is the average of the two middle scores (as though the scores were measured on an interval/ratio scale). For the left distribution in Table 3.1, the median is the average of 5 and 6, which equals 5.5.

The Median for Ordinal Data   Unlike the mode, the median cannot be found for a nominal scale because the values (e.g., different psychiatric diagnoses) do not have any inherent order (e.g., we cannot say which diagnoses are "above" bipolar disorder and which "below"). However, if the values can be placed in a meaningful order, you are then dealing with an ordinal scale, and the median *can* be found for ordinal scales. For example, suppose that the coach of a debating team has rated the effectiveness of the 25 members of the team on a scale from 1 to 10. The data in Table 2.3 (reproduced here) could represent those ratings.

Once the scores have been placed in order, the median is the middle score. (Unfortunately, if there are two middle scores and you are dealing with ordinal data, it is not proper to average the two scores, although this is sometimes done anyway as an approximation.) Even though the ratings from 1 to 10 cannot be considered equally spaced, we can assume, for

Table 2.3

X	f
10	2
9	2
8	5
7	3
6	7
5	1
4	4
3	0
2	1

example, that the debaters rated between 1 and 5 are all considered less effective than one who is rated 6. Thus, we can find a ranking or rating such that half the group is below it and half above, except for those who are tied with the middle score (or one of the two middle scores). The median is more informative if there are not many ties. In general, having many tied scores diminishes the usefulness of an ordinal scale, as will be made clear in Part VII of this text.

Dealing with Undeterminable Scores and Open-Ended Categories
One situation that is particularly appropriate for the use of the median occurs when the scores for some subjects cannot be determined exactly, but we know on which end of the scale these scores fall. For instance, in a typical study involving reaction time (RT), an experimenter will not wait forever for a subject to respond. Usually some arbitrary limit is imposed on the high end—for example, if the subject does not respond after 10 seconds, record 10 s as the RT and go on to the next trial. Calculating the mean would be misleading, however, if any of these 10-second responses were included. First, these 10-second responses are really *undeterminable scores*—the researcher doesn't know how long it would have taken for the subject to respond. Second, averaging in a few 10-second responses with the rest of the responses, which may be less than 1 second, can produce a mean that misrepresents the results. On the other hand, the median will not change if the response is recorded as 10 or 100 s (assuming that the median is less than 10 s to begin with). Thus, when some of the scores are undeterminable, the median has a strong advantage over the mean as a descriptive statistic.

Sometimes when data are collected for a study, some of the categories are deliberately left *open ended*. For instance, in a study of AIDS awareness, subjects might be asked how many sexual partners they have had in the last 6 months, with the highest category being 10 or more. Once a subject has had at least 10 different partners in the given period, it may be considered relatively unimportant to the study to determine exactly how many more than 10 partners were involved. (Perhaps the researchers fear that the accuracy of numbers greater than 10 could be questioned.) However, this presents the same problem for calculating the mean as an undeterminable score. It would be misleading to average in the number 10 when the subject reported having 10 *or more* partners. Again, this is not a problem for finding the median; all of the subjects reporting 10 or more partners would simply be tied for the highest position in the distribution. (A problem in determining the median would arise only if as many as half the subjects reported 10 or more partners.)

The Median and the Area of a Distribution   As mentioned, the mode is particularly easy to find from a frequency polygon—it is the score that corresponds to the highest point. The median also bears a simple relationship to the frequency polygon. If a vertical line is drawn at the median on a frequency polygon so that it extends from the horizontal axis until it meets the top of the frequency polygon, the area of the polygon will be divided in half. This is because the median divides the total number of scores in half, and the area of the polygon is proportional to the number of scores.

To better understand the relation between the frequency of scores and the area of a frequency polygon, take another look at a frequency histogram (see Figure 3.2). The height of each bar in the histogram is proportional to the frequency of the score or interval that the bar represents. (This is true

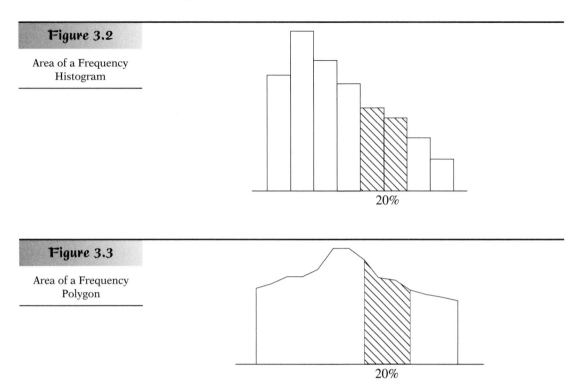

Figure 3.2

Area of a Frequency
Histogram

20%

Figure 3.3

Area of a Frequency
Polygon

20%

for the simplest type of histogram, which is the only type we will consider.) Because the bars all have the same width, the area of each bar is also proportional to the frequency. You can imagine that each bar is a building and that the taller the building, the more people live in it. The entire histogram can be thought of as the skyline of a city; you can see at a glance where (in terms of scores on the X axis) the bulk of the people live. All the bars together contain all the scores in the distribution. If two of the bars, for instance, take up an area that is 20% of the total, you know that 20% of the scores fall in the intervals represented by those two bars.

A relationship similar to the one between scores and areas of the histogram bars can be observed in a frequency polygon. The polygon encloses an area that represents the total number of scores. If you draw two vertical lines within the polygon, at two different values on the X axis, you enclose a smaller area, as shown in Figure 3.3. Whatever proportion of the total area is enclosed between the two values (.20 in Figure 3.3) is the proportion of the scores in the distribution that fall between those two values. We will use this principle to solve problems in the next chapter. At this point I just wanted to give you a feeling for why a vertical line drawn at the median divides the distribution into two equal areas.

Measures of Variability

Finding the right measure of central tendency for a distribution is certainly important, and I will have more to say about this process with respect to the shape of the distribution, but there is another very important aspect of describing a set of data that I do not want to postpone any longer. The following hypothetical situation will highlight the importance of this other dimension.

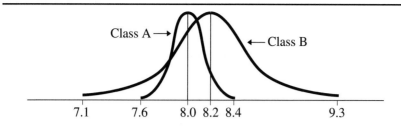

Figure 3.4

Mean Reading Levels in
Two Eighth-Grade
Classes

Suppose you're an eighth-grade English teacher entering a new school, and the principal is giving you a choice of teaching either class A or class B. Having read this chapter thus far, you inquire about the mean reading level of each class. (To simplify matters you can assume that the distributions of both classes are unimodal.) The principal tells you that class A has a mean reading level of 8.0, whereas class B has a mean of 8.2. All else being equal, you are inclined to take the slightly more advanced class. But all is not equal. Look at the two distributions in Figure 3.4.

What the principal neglected to mention is that reading levels in class B are much more spread out. It should be obvious that class A would be easier to teach. If you geared your lessons toward the 8.0 reader, no one in class A is so much below that level that he or she would be lost, nor is anyone so far above that level that he or she would be completely bored. On the other hand, teaching class B at the 8.2 level could leave many students either lost or bored.

The fact is that no measure of central tendency is very representative of the scores, if the distribution contains a great deal of variability. The principal could have shown you both distributions to help you make your decision; the difference in variability (also called the *dispersion*) is so obvious that if you had seen the distributions you could have made your decision instantly. For less obvious cases, and for the purposes of advanced statistical techniques, it would be useful to measure the width of each distribution. However, there is more than one way to measure the spread of a distribution. The rest of this section is about the different ways of measuring variability.

The Range

The simplest and most obvious way to measure the width of a distribution is to subtract the lowest score from the highest score. The resulting number is called the *range* of the distribution. For instance, judging Figure 3.4 by eye, in class A the lowest reading score appears to be about 7.6 and the highest about 8.4. Subtracting these two scores we obtain 8.4 − 7.6 = .8. However, if these scores are considered to be measured on a continuous scale, we should subtract the lower real limit of 7.6 (i.e., 7.55) from the upper real limit of 8.4 (i.e., 8.45) to obtain 8.45 − 7.55 = .9. For class B, the lowest and highest scores appear to be 7.1 and 9.3, respectively, so the range would be 9.35 − 7.05 = 2.3—considerably larger than the range for class A.

The major drawback to the range as a measure of variability is that, like the mode, it can be quite unreliable. The range can be changed drastically by moving only one score in the distribution, if that score happens to be either the highest or the lowest. For instance, adding just one excellent reader to class A can make the range of class A as large as the range of class B. But the range of class A would then be very misleading as a descriptor of

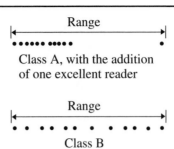

Figure 3.5

The Ranges of Two
Different Distributions

the variability of the bulk of the distribution (see Figure 3.5). In general, the range will tend to be misleading whenever a distribution includes a few extreme scores (i.e., outliers). Another drawback to the range is that it cannot be determined for a distribution that contains undeterminable scores at one end or the other.

On the positive side, the range not only is the easiest measure of variability to find, it also has the advantage of capturing the entire distribution without exception. For instance, in designing handcuffs for use by police departments, a manufacturer would want to know the entire range of wrist sizes in the adult population so that the handcuffs could be made to adjust over this range. It would be important to make the handcuffs large enough so that no wrist would be too large to fit but able to become small enough so that no adult could wriggle out and get free.

The Semi-Interquartile Range

There is one measure of variability that can be used with open-ended distributions and is virtually unaffected by extreme scores because, like the median, it is based on percentiles. It is called the *interquartile (IQ) range*, and it is found by subtracting the 25th percentile from the 75th percentile. The 25th percentile is often called the first quartile and symbolized as $Q1$; similarly, the 75th percentile is known as the third quartile ($Q3$). Thus the interquartile range (IQ) can be symbolized as $Q3 - Q1$. The IQ range gives the width of the middle half of the distribution, therefore avoiding any problems caused by outliers or undeterminable scores at either end of the distribution. A more popular variation of the IQ range is the *semi-interquartile (SIQ) range*, which is simply half of the interquartile range, as shown in Formula 3.1:

$$\text{SIQ range} = \frac{Q3 - Q1}{2} \qquad \textbf{Formula 3.1}$$

The SIQ range is preferred because it gives the distance of a typical score from the median; that is, roughly half the scores in the distribution will be closer to the median than the length of the SIQ range, and about half will be further away. The SIQ range is often used in the same situations for which the median is preferred to the mean as a measure of central tendency, and it can be very useful for descriptive purposes. However, the SIQ range's chief advantage—its unresponsiveness to extreme scores—can also be its chief disadvantage. Quite a few scores on both ends of a distribution can be moved much further from the center without affecting the SIQ range. Thus the SIQ range does not always give an accurate indication of the width of the entire distribution (see Figure 3.6). Moreover, the SIQ

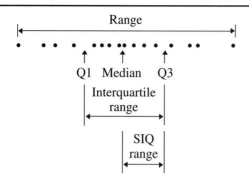

Figure 3.6

The Interquartile and
Semi-interquartile
Ranges

range shares with the median the disadvantage of not fitting easily into advanced statistical procedures.

The Mean Deviation

The SIQ range can be said to indicate the typical distance of a score from the median. This is a very useful way to describe the variability of a distribution. For instance, if you were teaching an English class and were aiming your lessons at the middle of the distribution, it would be helpful to know how far off your teaching level would be, on the average. However, the SIQ range does not take into account the distances of *all* the scores from the center. A more straightforward approach would be to find the distance of every score from the middle of the distribution and then average those distances. Let us look at the mathematics involved in creating such a measure of variability.

First, we have to decide on a measure of central tendency from which to calculate the distance of each score. The median would be a reasonable choice, but because we are developing a measure to use in advanced statistical procedures, the mean is preferable. The distance of any score from the mean $(X_i - \mu)$ is called a *deviation score*. (A deviation score is sometimes symbolized by a lowercase x; but in my opinion that notation is too confusing, so it will not be used in this text.) The average of these deviation scores would be given by $\Sigma(X_i - \mu)/N$. Unfortunately, there is a problem with using this expression. According to one of the properties of the mean (these properties will be explained more fully in Section B), $\Sigma(X_i - \mu)$ will always equal zero, which means that the average of the deviation scores will also always equal zero (about half the deviations will be above the mean and about half will be below). This problem disappears when you realize that it is the distances we want to average, regardless of their direction (i.e., sign). What we really want to do is take the *absolute values* of the deviation scores before averaging to find the typical amount by which scores deviate from the mean. (Taking the absolute values turns the minus signs into plus signs and leaves the plus signs alone; in symbols, $|X|$ means take the absolute value of X.) This measure is called the *mean deviation*, or more accurately, the mean absolute deviation (MAD), and it is found using Formula 3.2:

$$\text{Mean deviation} = \frac{\Sigma |X_i - \mu|}{N}$$ **Formula 3.2**

To clarify the use of Formula 3.2, I will find the mean deviation of the following three numbers: 1, 3, 8. The mean of these numbers is 4. Applying Formula 3.2 yields:

$$\frac{|1 - 4| + |3 - 4| + |8 - 4|}{3} = \frac{|-3| + |-1| + |+4|}{3} = \frac{3 + 1 + 4}{3} = \frac{8}{3} = 2.67$$

The mean deviation makes a lot of sense, and it should be easy to understand; it is literally the average amount by which scores deviate from the mean. It is too bad that the mean deviation does not fit in well with more advanced statistical procedures. Fortunately, there is a measure that is closely related to the mean deviation that does fit well with the statistical procedures that are commonly used. I will get to this measure soon. First, another intermediate statistic must be described.

The Variance

If you square all the deviations from the mean, instead of taking the absolute values, and sum all of these squared deviations together, you get a quantity called the *sum of squares* (SS), which is less for deviations around the mean than for deviations around any other point in the distribution. (Note that the squaring eliminates all the minus signs, just as taking the absolute values did.) Formula 3.3 for SS is:

$$SS = \sum (X_i - \mu)^2 \qquad \qquad \textbf{Formula 3.3}$$

If you divide SS by the total number of scores (N), you are finding the mean of the squared deviations, which can be used as a measure of variability. The mean of the squared deviations is most often called the *population variance*, and it is symbolized by the lowercase Greek letter sigma squared (σ^2; the uppercase sigma, Σ, is used as the summation sign). Formula 3.4A for the variance is as follows:

$$\sigma^2 = \frac{\sum (X_i - \mu)^2}{N} \qquad \qquad \textbf{Formula 3.4A}$$

Because the variance is literally the mean of the squared deviations from the mean, it is sometimes referred to as a mean square, or *MS* for short. This notation is commonly used in the context of the analysis of variance procedure, as you will see in Part IV of this text. Recall that the numerator of the variance formula is often referred to as SS; the relationship between MS and SS is expressed in Formula 3.5A:

$$\sigma^2 = MS = \frac{SS}{N} \qquad \qquad \textbf{Formula 3.5A}$$

It is certainly worth the effort to understand the variance because this measure plays an important role in advanced statistical procedures, especially those included in this text. However, it is easy to see that the variance does not provide a good descriptive measure of the spread of a distribution. As an example, consider the variance of the numbers 1, 3, and 8:

$$\sigma^2 = \frac{(1 - 4)^2 + (3 - 4)^2 + (8 - 4)^2}{3}$$

$$= \frac{3^2 + 1^2 + 4^2}{3} = \frac{9 + 1 + 16}{3} + \frac{26}{3} = 8.67$$

The variance (8.67) is larger than the range of the numbers. This is because the variance is based on *squared* deviations. The obvious remedy to this problem is to take the square root of the variance, which leads to our final measure of dispersion.

The Standard Deviation

Taking the square root of the variance produces a measure that provides a good description of the variability of a distribution and one that plays a role in advanced statistical procedures as well. The square root of the population variance is called the *population standard deviation (SD)*, and it is symbolized by the lowercase Greek letter sigma (σ). (Notice that the symbol is *not* squared—squaring the standard deviation gives the variance.) The basic definitional formula for the standard deviation is:

$$\sigma = \sqrt{\frac{\sum (X_i - \mu)^2}{N}}$$ **Formula 3.4B**

An alternative way to express this relationship is:

$$\sigma = \sqrt{MS} = \sqrt{\frac{SS}{N}}$$ **Formula 3.5B**

To remind you that each formula for the standard deviation will be the square root of a variance formula, I will use the same number for both formulas, adding "A" for the variance formula and "B" for the corresponding *SD* formula. Because σ is the square root of *MS*, it is sometimes referred to as the *root-mean-square* (RMS) of the deviations from the mean.

At this point, you may be wondering why you would bother squaring all the deviations if after averaging you plan to take the square root. First, we need to make it clear that squaring, averaging, and then taking the square root of the deviations is not the same as just averaging the absolute values of the deviations. If the two procedures were equivalent, the standard deviation would always equal the mean deviation. An example will show that this is not the case. The standard deviation of the numbers 1, 3, and 8 is equal to the square root of their variance, which was found earlier to be 8.67. So, $\sigma = \sqrt{8.67} = 2.94$, which is clearly larger than the mean deviation (2.67) for the same set of numbers.

The process of squaring and averaging gives extra weight to large scores, which is not removed by taking the square root. Thus the standard deviation is never smaller than the mean deviation, although the two measures can be equal. In fact, the standard deviation will be equal to the mean deviation whenever there are only two numbers in the set. In this case, both measures of variability will equal half the distance between the two numbers. I mentioned previously that the standard deviation gives more weight to large scores than does the mean deviation. This is true because squaring a large deviation has a great effect on the variance. This sensitivity to large scores can be a problem if there are a few very extreme scores in a distribution, which result in a misleadingly large standard deviation. If you are dealing with a distribution that contains a few extreme scores (whether low, high, or some of each), you may want to consider an alternative to the standard deviation, such as the mean deviation, which is less affected by extreme scores, or the semi-interquartile range, which may not be affected at all. On the other hand, you could consider a method for eliminating outliers or transforming the data, such as those outlined in Section C.

The Variance of a Sample

Thus far the discussion of the variance and standard deviation has confined itself to the situation in which you are describing the variability of an entire population of scores (i.e., your interests do not extend beyond describing the set of scores at hand). Later chapters, however, will consider the case in which you have only a sample of scores from a larger population, and you want to use your description of the sample to extrapolate to that population. Anticipating that need, I will now consider the case in which you want to describe the variability of a sample.

To find the variance of a sample, you can use the procedure expressed in Formula 3.4A, but it will be appropriate to change some of the notation. First, I will use s^2 to symbolize the sample variance, according to the custom of using Roman letters for sample statistics. Along these lines, the mean subtracted from each score will be symbolized as $\overline{X}$ instead of μ, because it is the mean of a sample. Thus Formula 3.4A becomes:

$$s^2 = \frac{\sum (X_i - \overline{X})^2}{N}$$

The Biased and Unbiased Sample Variances   The preceding formula represents a perfectly reasonable way to describe the variability in a sample, but a problem arises when the variance thus calculated is used to estimate the variance of the larger population. The problem is that the variance of the sample tends to underestimate the variance of the population. Of course, the variance of every sample will be a little different, even if all of the samples are the same size and they are from the same population. Some sample variances will be a little larger than the population variance and some a little smaller, but unfortunately the average of infinitely many sample variances (when calculated by the formula above) will be *less* than the population variance. This tendency of a sample statistic to consistently underestimate (or overestimate) a population parameter is called *bias*. The sample variance as defined by the (unnumbered) formula above is therefore called a *biased estimator*.

Fortunately, the underestimation just described is so well understood that it can be corrected easily by making a slight change in the formula for calculating the sample variance. To calculate an *unbiased sample variance*, you can use Formula 3.6A:

$$s^2 = \frac{\sum (X_i - \overline{X})^2}{N - 1} \qquad \textbf{Formula 3.6A}$$

If infinitely many sample variances are calculated with Formula 3.6A, the average of these sample variances *will* equal the population variance σ^2.

Notation for the Variance and the Standard Deviation

You've seen that there are two different versions of the variance of a sample: biased and unbiased. Some texts use different symbols to indicate the two types of sample variances, such as an uppercase S for biased and a lowercase s for unbiased, or a plain s for biased and $\hat{s}$ (pronounced "s hat") for unbiased. I will adopt the simplest notation by assuming that the variance of a sample will always be calculated using Formula 3.6A (or its algebraic equivalent). Therefore, the symbol s^2 for the sample variance will always (in my text, at least) refer to the *unbiased* sample variance. Whenever the biased formula is used (i.e., the formula with N rather than $N - 1$ in the denominator), you can assume that the set of numbers at hand is being treated like a population, and therefore the variance will be identified by σ^2. When you are finding the variance of a population, you are never interested in extrapolating to a larger group, so there would be no reason to calculate an unbiased variance. Thus when you see σ^2, you know that it was obtained by Formula 3.4A (or its equivalent), and when you see s^2, you know that Formula 3.6A (or its equivalent) was used.

As you might guess from the preceding discussion, using Formula 3.4B to find the standard deviation of a sample produces a biased estimate of the population standard deviation. The solution to this problem would seem to be to use the square root of the unbiased sample variance whenever you are finding the standard deviation of a sample. This produces a new formula for the standard deviation:

$$s = \sqrt{\frac{\sum (X_i - \overline{X})^2}{N - 1}}$$ **Formula 3.6B**

Surprisingly, this formula does not entirely correct the bias in the standard deviation, but fortunately the bias that remains is small enough to be ignored (at least that is what researchers in psychology do). Therefore, I will refer to s (defined by Formula 3.6B) as the *unbiased sample standard deviation*, and I will use σ (defined by Formula 3.4B) as the symbol for the standard deviation of a population.

Degrees of Freedom

The adjustment in the variance formula that made the sample variance an unbiased estimator of the population variance was quite simple: $N - 1$ was substituted for N in the denominator. Explaining why this simple adjustment corrects the bias described previously is not so simple, but I can give you some feeling for why $N - 1$ makes sense in the formula. Return to the example of finding the variance of the numbers 1, 3, and 8. As you saw before, the three deviations from the mean are -3, -1, and 4, which add up to zero (as will always be the case). The fact that these three deviations must add up to zero implies that knowing only two of the deviations automatically tells you what the third deviation will be. That is, if you know that two of the deviations are -1 and -3, you know that the third deviation must be $+4$ so that the deviations will sum to zero. Thus, only two of the three deviations are free to vary (i.e., $N - 1$) from the mean of the three numbers; once two deviations have been fixed, the third is determined. The number of deviations that are free to vary is called the number of *degrees of freedom* (df). Generally, when there are N scores in a sample, df = $N - 1$.

Another way to think about degrees of freedom is as the number of separate pieces of information that you have about variability. If you are trying to find out about the body temperatures of a newly discovered race of humans native to Antarctica and you sample just one person, you have one piece of information about the population mean, but no ($N - 1 = 1 - 1 = 0$) information about variability. If you sample two people, you have just one piece of information about variability ($2 - 1 = 1$)—the difference between the two people. Note, however, that the number of pieces of information about variability would be N rather than $N - 1$ if you knew the population mean before doing any sampling. If you knew that the Antarcticans must have 98.6 degrees Fahrenheit as their population mean for body temperature, but that they could have more or less variability than other people, a single Antarctican would give you one piece of information about variability. If that one Antarctican had a normal body temperature of 96, more variability for Antarcticans would be suggested than if he or she had a temperature of 98.2. It is when you do not know the population mean that variability must be calculated from the mean of your sample, and that entails losing one degree of freedom.

Once the deviation scores have been squared and summed (i.e., SS) for a sample, dividing by the number of degrees of freedom is necessary to produce an unbiased estimate of the population variance. This new notation can be used to create shorthand formulas for the sample variance and standard deviation, as follows:

$$s^2 = \frac{SS}{N} - 1 = \frac{SS}{df}$$ **Formula 3.7A**

$$s = \sqrt{\frac{SS}{N-1}} = \sqrt{\frac{SS}{df}}$$ **Formula 3.7B**

In applying these formulas to the sample of three numbers (1, 3, 8), you do not have to recalculate SS, which was the numerator when we found σ^2 by Formula 3.5A. Given that $SS = 26$, Formula 3.7A tells you that $s^2 = SS/(N-1) = 26/2 = 13$, which is considerably larger than σ^2 (8.67). The increase from σ^2 to s^2 is necessary to correct the underestimation created by Formula 3.5A when estimating the true variance of the larger population. Formula 3.7B shows that $\sigma = \sqrt{13} = 3.61$, which, of course, is considerably larger than σ (2.94). The large differences between the biased and unbiased versions of the variance and standard deviation are caused by our unusually tiny sample ($N = 3$). As N becomes larger, the difference between N and $N - 1$ diminishes, as does the difference between σ^2 and s^2 (or σ and s). When N is very large (e.g., over 100), the distinction between the biased and unbiased formulas is so small that for some purposes, it can be ignored.

Skewed Distributions

There are many ways in which the shapes of two unimodal distributions can differ, but one aspect of shape that is particularly relevant to psychological variables and plays an important role in choosing measures of central tendency and variability is *skewness*. A distribution is *skewed* if the bulk of the scores are concentrated on one side of the scale, with relatively few scores on the other side. When graphed as a frequency polygon, a skewed distribution will look something like those in Figure 3.7. The distribution in Figure 3.7a is said to be *negatively skewed*, whereas the one in Figure 3.7b is called *positively skewed*. To remember which shape involves a negative skew and which a positive skew, think of the *tail of the distribution* as a long, thin skewer. If the skewer points to the left (in the direction in which the numbers eventually become negative), the distribution is negatively "skewered" (i.e., negatively skewed); if the skewer points to the right (the direction in which the numbers become positive), the distribution is positively skewed.

Figure 3.7

Skewed Distributions

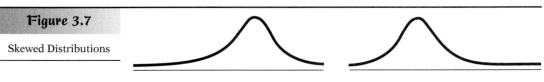

a. Negatively skewed distribution   b. Positively skewed distribution

Recalling the description of the relation between the area of a polygon and the proportion of scores can help you understand the skewed distribution. A section of the tail with a particular width (i.e., range along the horizontal axis) will have a relatively small area (and therefore relatively few scores) as compared to a section with the same width in the thick part of the distribution (the "hump"). The latter section will have a lot more area and thus a lot more scores.

The Central Tendency of a Skewed Distribution

When a unimodal distribution is strongly skewed, it can be difficult to decide whether to use the median or the mean to represent the central tendency of the distribution (the mode would never be the best of the three in this situation). On the other hand, for a symmetrical unimodal distribution, as depicted in Figure 3.8a, the mean and the median are both exactly in the center, right at the mode. Because the distribution is symmetrical, there is the same amount of area on each side of the mode. Now let's see what happens when we turn this distribution into a positively skewed distribution by adding a few high scores, as shown in Figure 3.8b. Adding a small number of scores on the right increases the area on the right slightly. To have the same area on both sides, the median must move to the right a bit. Notice, however, that the median does not have to move very far along the *X* axis. Because the median is in the thick part of the distribution, moving only slightly to the right shifts enough area to compensate for the few high scores that were added. (See how the shaded area on the right end of the graph in Figure 3.8b equals the shaded area between the median and the mode.) Thus, the median is not strongly affected by the skewing of a distri-

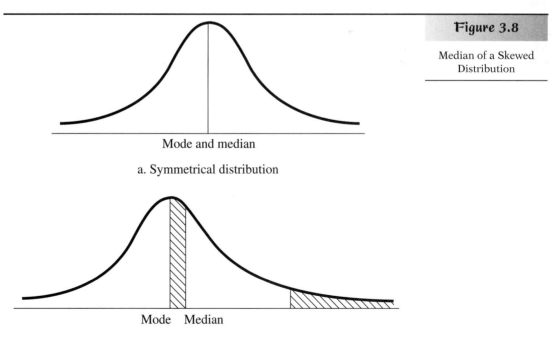

Figure 3.8

Median of a Skewed Distribution

Mode and median

a. Symmetrical distribution

Mode   Median

b. Positively skewed distribution

Figure 3.9

Mean of a Skewed
Distribution

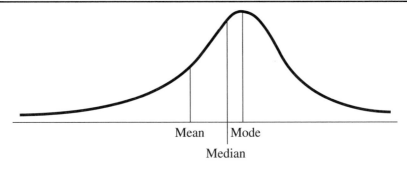

a. Negatively skewed distribution

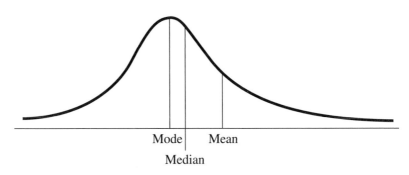

b. Positively skewed distribution

bution, and that can be an advantage in describing the central tendency of
a distribution.

In fact, once you have found the median of a distribution, you can
take a score on one side of the distribution and move it much further away
from the median. As long as the score stays on the same side of the
median, you can move it out as far as you want—the median will not
change its location. This is *not* true for the mean. The mean is affected by
the numerical value of every score in the distribution. Consequently the
mean will be pulled in the direction of the skew, sometimes quite a bit, as
illustrated in Figure 3.9. When the distribution is negatively skewed (Figure 3.9a), the mean will be to the left of (i.e., more negative than) the
median, whereas the reverse will be true for a positively skewed distribution (Figure 3.9b). Conversely, if you find both the mean and the median
for a distribution, and the median is higher (i.e., more positive), the distribution has a negative skew; if the mean is higher, the skew is positive. In a
positively skewed distribution, more than half of the scores will be below
the mean, whereas the opposite is true when dealing with a negative skew.
If the mean and median are the same, the distribution is probably symmetric around its center.

Choosing between the Mean and Median

Let us consider an example of a skewed distribution for which choosing a
measure of central tendency has practical consequences. There has been
much publicity in recent years about the astronomical salaries of a few
superstar athletes. However, the bulk of the professional athletes in any
particular sport (we'll use baseball as our example) are paid a more reason-

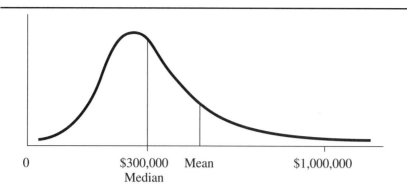

Figure 3.10

Annual Salaries of Major
League Baseball Players

able salary. Therefore, the distribution of salaries for major league baseball players is positively skewed, as shown in Figure 3.10. When the Players' Association is negotiating with management, guess which measure of central tendency each side prefers to use? Of course, management points out that the average (i.e., mean) salary is already quite high. The players can point out, however, that the high mean is caused by the salaries of relatively few superstars, and that the mean salary is not very representative of the majority of players. The argument of the Players' Association would be that the median provides a better representation of the salaries of the majority of players. In this case, it seems that the players have a good point. However, the mean has some very useful properties, which will be explored in detail in Section B.

Floor and Ceiling Effects

Positively skewed distributions are likely whenever there is a limit on values of the variable at the low end but not the high end, or when the bulk of the values are clustered near the lower limit rather than the upper limit. This kind of one-sided limitation is called a *floor effect*. One of the most common examples in psychological research is reaction time (RT). In a typical RT experiment, the subject waits for a signal before hitting a response button; the time between the onset of the signal and the depression of the button is recorded as the reaction time. There is a physiological limit to how quickly a subject can respond to a stimulus, although this limit is somewhat longer if the subject must make some complex choice before responding. After the subjects have had some practice, most of their responses will cluster just above an approximate lower limit, with relatively few responses taking considerably longer. The occasional long RTs may reflect momentary fatigue or inattention, and they create the positive skew (the RT distribution would have a shape similar to the distributions shown in Figures 3.7b and 3.10). Another example of a floor effect involves measurements of clinical depression in a large random group of college students. A third example is scores on a test that is too difficult for the group being tested; many scores would be near zero and there would be only a few high scores.

The opposite of a floor effect is, not surprisingly, a *ceiling effect*, which occurs when the scores in a distribution approach an upper limit but are not near any lower limit. A rather easy exam will show a ceiling effect, with most scores near the maximum and relatively few scores (e.g., only those of students who didn't study or didn't do their homework) near the low end.

For example, certain tests are given to patients with brain damage or chronic schizophrenia to assess their orientation to the environment, knowledge of current events, and so forth. Giving such a test to a random group of adults will produce a negatively skewed distribution (such as the one shown in Figure 3.7a). For descriptive purposes, the median is often preferred to the mean whenever either a floor or a ceiling effect is exerting a strong influence on the distribution.

Variability of Skewed Distributions

The standard deviation (*SD*) is a very useful measure of variability, but if you take a score at one end of your distribution and move it much further away from the center, it will have a considerable effect on the *SD*, even though the spread of the bulk of your scores has not changed at all. The mean deviation (*MD*) is somewhat less affected because the extreme score is not being squared, but like the *SD*, the *MD* also becomes misleading when your distribution is very skewed. Of course, the ordinary range is even more misleading in such cases. The only well-known measure of variability that is not affected by extreme scores, and therefore gives a good description of the spread of the main part of your distribution, even if it is very skewed, is the SIQ range. A measure closely related to the SIQ range will be described in Section C as part of a system for exploring your data, especially with respect to identifying extreme scores.

SUMMARY

1. The mode of a distribution, the most frequent score, is the only descriptive statistic that must correspond to an actual score in the distribution. It is also the only statistic that can be used with all four measurement scales and the only statistic that can take on more than one value in the same distribution (this can be useful, for instance, when a distribution is distinctly bimodal). Unfortunately, the mode is too unreliable for many statistical purposes.

2. The median of a distribution, the 50th percentile, is a particularly good descriptive statistic when the distribution is strongly skewed. Also, it is the point that minimizes the magnitude (i.e., absolute value) of the sum of the (unsquared) deviations. However, the median lacks many of the convenient properties of the mean.

3. The arithmetic mean, the simple average of all the scores, is the most convenient measure of central tendency for use with inferential statistics.

4. The simplest measure of the variability (or *dispersion*) of a distribution is the *range*, the difference between the highest and lowest scores in the distribution. The range is the only measure of variability that tells you the total extent of the distribution, but unfortunately, it tends to be too unreliable for most statistical purposes.

5. The *semi-interquartile range*, half the distance between the first and third quartiles, is a particularly good descriptive measure when dealing with strongly skewed distributions and outliers, but it does not play a role in advanced statistical procedures.

6. The *mean deviation* (*MD*), the average distance of the scores from the mean, is a good description of the variability in a distribution and is easy to understand conceptually, but is rarely used in advanced statistics.

7. The *variance*, the average of the *squared* deviations from the mean, plays an important role in advanced statistics, but it does not provide a convenient description of the spread of a distribution.

8. The *standard deviation*, the square root of the variance, serves as a good description of the variability in a distribution (except when there are very extreme scores), and it also lends itself to use in advanced statistics.

9. Some additional properties of the measures discussed in this section are as follows: the mode, median, range, and SIQ range all require a minimal amount of calculation, and all can be used with ordinal scales; the mode, median, and SIQ range can be used even when there are undeterminable or open-ended scores, and they are virtually unaffected by outliers; the mean, mean deviation, variance, and standard deviation can be used only with an interval or ratio scale, and each of these measures is based on (and is affected by) all of the scores in a distribution.

10. The population variance formula, when applied to data from a sample, tends to underestimate the variance of the population. To correct this *bias*, the sample variance (s^2) is calculated by dividing the sum of squared deviations (SS) by $N - 1$, instead of by N. The symbol σ^2 will be reserved for any calculation of variance in which N, rather than $N - 1$, is used in the denominator.

11. The denominator of the formula for the unbiased sample variance, $N - 1$, is known as the *degrees of freedom* (df) associated with the variance, because once you know the mean, df is the number of deviations from the mean that are free to vary. Although the sample standard deviation ($\sqrt{s^2} = s$) is not a perfectly unbiased estimation of the standard deviation of the population, the bias is so small that s is referred to as the unbiased sample standard deviation.

12. A *floor effect* occurs when the scores in a distribution come up against a lower limit but are not near any upper limit. This often results in a positively skewed distribution, such that the scores are mostly bunched up on the left side of the distribution with relatively few scores that form a *tail* of the distribution pointing to the right. On the other hand, a *ceiling effect* occurs when scores come close to an upper limit, in which case a negatively skewed distribution (tail pointing to the left) is likely.

13. In a positively skewed distribution, the mean will be pulled toward the right more (and therefore be larger) than the median. The reverse will occur for a negatively skewed distribution.

14. In a strongly skewed distribution, the median is usually the better descriptive measure of central tendency because it is closer to the bulk of the scores than the mean. The mean deviation is less affected by the skewing than the standard deviation, but the SIQ range is less affected still, making it the best descriptive measure of the spread of the bulk of the scores.

EXERCISES

*1. Select the measure of central tendency (mean, median, or mode) that would be most appropriate for describing each of the following hypothetical sets of data:
   a. Religious preferences of delegates to the United Nations
   b. Heart rates for a group of women before they start their first aerobics class
   c. Types of phobias exhibited by patients attending a phobia clinic
   d. Amounts of time participants spend solving a classic cognitive problem, with some of the participants unable to solve it
   e. Height in inches for a group of boys in the first grade.

2. Describe a realistic situation in which you would expect to obtain each of the following:
   a. A negatively skewed distribution
   b. A positively skewed distribution
   c. A bimodal distribution

*3. A midterm exam was given in a large introductory psychology class. The median score was 85, the mean was 81, and the mode was 87. What kind of distribution would you expect from these exam scores?

4. A veterinarian is interested in the life span of golden retrievers. She recorded the age at death (in years) of the retrievers treated in her clinic. The ages were 12, 9, 11, 10, 8, 14, 12, 1, 9, 12.
   a. Calculate the mean, median, and mode for age at death.
   b. After examining her records, the veterinarian determined that the dog that had died at 1 year was killed by a car. Recalculate the mean, median, and mode without that dog's data.
   c. Which measure of central tendency in part b changed the most, compared to the values originally calculated in part a?

5. Which of the three most popular measures of variability (range, SIQ range, standard deviation) would you choose in each of the following situations?
   a. The distribution is badly skewed with a few extreme outliers in one direction.
   b. You are planning to perform advanced statistical procedures (e.g., draw inferences about population parameters).
   c. You need to know the maximum width taken up by the distribution.
   d. You need a statistic that takes into account every score in the population.
   e. The highest score in the distribution is "more than 10."

*6. a. Calculate the mean, SS, and variance (i.e., σ^2) for the following set of scores: 11, 17, 14, 10, 13, 8, 7, 14.
   b. Calculate the mean deviation and the standard deviation (i.e., σ) for the set of scores in part a.

*7. How many degrees of freedom are contained in the set of scores in Exercise 6? Calculate the unbiased sample variance (i.e., s^2) and standard deviation (i.e., s) for that set of scores. Compare your answers to σ^2 and σ, which you found in Exercise 6.

8. Eliminate the score of 17 from the data in Exercise 6, and recalculate both MD and σ. Compared to the values calculated in Exercise 6b, which of these two statistics changed more? What does this tell you about these two statistical measures?

*9. Calculate the mean, mode, median, range, SIQ range, mean deviation, and standard deviation (s) for the following set of scores: 17, 19, 22, 23, 26, 26, 26, 27, 28, 28, 29, 30, 32, 35, 35, 36.

10. a. Calculate the range, SIQ range, mean deviation, and standard deviation (s) for the following set of scores: 3, 8, 13, 23, 26, 26, 26, 27, 28, 28, 29, 30, 32, 41, 49, 56.
    b. How would you describe the relationship between the set of data above and the set of data in Exercise 9?
    c. Compared to the values calculated in Exercise 9, which measures of variability have changed the most, which the least, and which not at all?

B

BASIC STATISTICAL PROCEDURES

Formulas for the Mean

In Section A the arithmetic mean was defined informally as the sum of all of the scores divided by the number of scores added. It is more useful to express the mean as a formula in terms of the summation notation that was presented in the first chapter. The formula for the *population mean* is:

$$\mu = \frac{\sum_{i=1}^{N} X_i}{N} = \frac{1}{N} \sum_{i=1}^{N} X_i$$

which tells you to sum all the X's from X_1 to X_N before dividing by N. If you simplify the summation notation by leaving off the indexes (as I promised I would in Chapter 1), you end up with Formula 3.8:

$$\mu = \frac{\sum X}{N}$$

Formula 3.8

The procedure for finding the mean of a sample is exactly the same as the procedure for finding the mean of a population, as shown by Formula 3.9 for the sample mean:

$$\overline{X} = \frac{\sum X}{N}$$ **Formula 3.9**

Suppose that the following set of scores represents measurements of clinical depression in seven normal college students: 0, 3, 5, 6, 8, 8, 9. I will use Formula 3.8 to find the mean: $\mu = 39/7 = 5.57$. (If I had considered this set of scores a sample, I would have used Formula 3.9 and of course obtained the same answer, which would have been referred to by $\overline{X}$, the symbol for the sample mean.) To appreciate the sensitivity of the mean to extreme scores, imagine that all of the students have been measured again and all have attained the same rating as before, except for the student who had scored 9. This student has become clinically depressed and therefore receives a new rating of 40. Thus, the new set of scores is 0, 3, 5, 6, 8, 8, 40. The new mean is $\mu = 70/7 = 10$. Note that although the mean has changed a good deal, the median is 6, in both cases.

The Weighted Mean

The statistical procedure for finding the *weighted mean*, better known as the *weighted average*, has many applications in statistics as well as in real life. I will begin this explanation with the simplest possible example. Suppose a professor who is teaching two sections of statistics has given a diagnostic quiz at the first meeting of each section. One class has 30 students who score an average of 7 on the quiz, whereas the other class has only 20 students who average an 8. The professor wants to know the average quiz score for all of the students taking statistics (i.e., both sections combined). The naive approach would be to take the average of the two section means (i.e., 7.5), but as you have probably guessed, this would give you the wrong answer. The correct thing to do is to take the *weighted* average of the two section means. It's not fair to count the class of 30 equally with the class of 20 (imagine giving equal weights to a class of 10 and a class of 100). Instead, the larger class should be given more *weight* in finding the average of the two classes. The amount of weight should depend on the class size, as it does in Formula 3.10. Note that Formula 3.10 could be used to average together any number of class sections or other groups, where N_i is the number of scores in one of the groups and $\overline{X}_i$ is the mean of that group. The formula uses the symbol for the sample mean because weighted averages are often applied to samples to make better guesses about populations.

$$\overline{X}_w = \frac{\sum N_i X_i}{\sum N_i} = \frac{N_1 X_1 + N_2 X_2 + \cdots}{N_1 + N_2 + \cdots}$$ **Formula 3.10**

We can apply Formula 3.10 to the means of the two statistics sections:

$$\overline{X}_w = \frac{(30)(7) + (20)(8)}{30 + 20} = \frac{210 + 160}{50} = \frac{370}{50} = 7.4$$

Notice that the weighted mean (7.4) is a little closer to the mean of the larger class (7) than to the mean of the smaller class. The weighted average of two groups will always be between the two group means and closer to the mean of the larger group. For more than two groups, the weighted average will be somewhere between the smallest and the largest of the group means.

Let us look more closely at how the weighted average formula works. In the case of the two sections of the statistics course, the weighted average indicates what the mean would be if the two sections were combined into one large class of 50 students. To find the mean of the combined class directly, you would need to know the sum of scores for the combined class and then to divide it by 50. To find ΣX for the combined class, you would need to know the sum for each section. You already know the mean and N for each section, so it is easy to find the sum for each section. First, take another look at Formula 3.9 for the sample mean:

$$\bar{X} = \frac{\Sigma X}{N}$$

If you multiply both sides of the equation by N, you get $\Sigma X = N\bar{X}$. (Note that it is also true that $\Sigma X = N\mu$; we will use this equation in the next subsection.) You can use this new equation to find the sum for each statistics section. For the first class, $\Sigma X_1 = (30)(7) = 210$, and for the second class, $\Sigma X_2 = (20)(8) = 160$. Thus, the total for the combined class is $210 + 160 = 370$, which divided by 50 is 7.4. Of course, this is the same answer we obtained with the weighted average formula. What the weighted average formula is actually doing is finding the sum for each group, adding all the group sums to find the total sum, and then dividing by the total number of scores from all the groups.

Computational Formulas for the Variance and Standard Deviation

The statistical procedure for finding SS, which in turn forms the basis for calculating the variance and standard deviation, can be tedious, particularly if you are using the definitional Formula 3.3, as reproduced below:

$$SS = \sum(X_i - \mu)^2$$

Formula 3.3 is also called the *deviational formula* because it is based directly on deviation scores. The reason that using this formula is tedious is that each score must be subtracted from the mean, usually resulting in fractions even when all the scores are integers, and then each of these differences must be squared. Compare this process to the *computational formula* for SS:

$$SS = \sum X^2 - N\mu^2 \qquad \textbf{Formula 3.11}$$

Note that according to this formula all the X^2 values must be summed, and then the term $N\mu^2$ is subtracted only once, after ΣX^2 has been found. (It may seem unlikely to you that Formula 3.11 yields exactly the same value as the more tedious Formula 3.3—except that the latter is likely to produce more error due to rounding off at intermediate stages—but it takes just a few steps of algebra to transform one formula into the other.)

Some statisticians might point out that if you want a "raw-score" formula for SS, Formula 3.11 does not qualify because it requires that the mean be computed first. I think that anyone would want to find the mean before assessing variability—but if you want to find SS more directly from the data, you can use Formula 3.12:

$$SS + \sum X^2 - \frac{(\sum X)^2}{N} \qquad \textbf{Formula 3.12}$$

As I pointed out in Chapter 1, ΣX^2 and $(\Sigma X)^2$ are very different values; the parentheses in the latter term instruct you to add up all the X values *before* squaring (i.e., you square only once at the end), whereas in the former term you square each X before adding.

All you need to do to create a computational formula for the population variance (σ^2) is to divide any formula for SS by N. For example, if you divide Formula 3.11 by N, you get Formula 3.13A for the population variance:

$$\sigma^2 = \frac{\sum X^2}{N} - \mu^2 \qquad \textbf{Formula 3.13A}$$

There is an easy way to remember this formula. The term $\sum X^2/N$ is the mean of the squared scores, whereas the term μ^2 is the square of the mean score. So the variance, which is the mean of the squared deviation scores, is equal to the mean of the squared scores minus the square of the mean score. A raw-score formula for the population variance, which does not require you to compute μ first, is found by dividing Formula 3.12 by N, as follows:

$$\sigma^2 = \frac{1}{N}\left[\sum X^2 - \frac{(\sum X)^2}{N}\right] \qquad \textbf{Formula 3.14A}$$

The formula above may look a bit awkward, but it lends itself to an easy comparison with a similar formula for the unbiased sample variance, which I will present shortly.

As usual, formulas for the population standard deviation (σ) are created simply by taking the square root of the variance formulas. (Note that I am continuing to use "A" for the variance formula and "B" for the standard deviation.)

$$\sigma = \sqrt{\frac{\sum X^2}{N} - \mu^2} \qquad \textbf{Formula 3.13B}$$

$$\sigma = \sqrt{\frac{1}{N}\left[\sum X^2 - \frac{(\sum X)^2}{N}\right]} \qquad \textbf{Formula 3.14B}$$

To illustrate the use of these computational formulas I will find SS, using Formula 3.11, for the three numbers (1, 3, 8) that I used as an example in Section A. The first step is to find that the mean of the three numbers is 4. Next, $\sum X^2 = 1^2 + 3^2 + 8^2 = 1 + 9 + 64 = 74$. Then, $N\mu^2 = 3 \times 4^2 = 3 \times 16 = 48$. Finally, $SS = \sum X^2 - N\mu^2 = 74 - 48 = 26$. Of course, all you have to do is divide 26 by N, which is 3 in this case, to get the population variance, but because it is common to use one of the variance formulas directly, without stopping to calculate SS first, I will next illustrate the use of Formulas 3.13A and 3.14A for the numbers 1, 3, 8:

$$\sigma^2 = \frac{\sum X^2}{N} - \mu^2 = \frac{74}{3} - 4^2 = 24.67 - 16 = 8.67$$

$$\sigma^2 = \frac{1}{N}\left[\sum X^2 - \frac{(\sum X)^2}{N}\right] = \frac{1}{3}\left[74 - \frac{12^2}{3}\right] = \frac{1}{3}(74 - 48) = \frac{1}{3}(26) = 8.67$$

Finding the population standard deviation entails nothing more than taking the square root of the population variance, so I will not bother to illustrate the use of the standard deviation formulas at this point.

Unbiased Computational Formulas

When calculating the variance of a set of numbers that is considered a sample of a larger population, it is usually desirable to use a variance formula that yields an unbiased estimate of the population variance. An unbiased sample variance (s^2) can be calculated by dividing SS by $N - 1$, instead of by N. A computational formula for s^2 can therefore be derived by taking any computational formula for SS and dividing by $N - 1$. For instance, dividing Formula 3.12 by $N - 1$ produces Formula 3.15A:

$$s^2 = \frac{1}{N-1}\left[\sum X^2 - \frac{(\sum X)^2}{N}\right] \qquad \textbf{Formula 3.15A}$$

You should recognize the portion of the above formula in brackets as Formula 3.12. Also, note the similarity between Formulas 3.14A and 3.15A—the latter being the unbiased version of the former.

The square root of the unbiased sample variance is used as an unbiased estimate of the population standard deviation, even though, as I pointed out before, it is not strictly unbiased. Taking the square root of Formula 3.15A yields Formula 3.15B for the standard deviation of a sample (s):

$$s = \sqrt{\frac{1}{N-1}\left[\sum X^2 - \frac{(\sum X)^2}{N}\right]} \qquad \textbf{Formula 3.15B}$$

Obtaining the Standard Deviation Directly from Your Calculator

As I am writing this edition of the text, scientific calculators that provide standard deviation as a built-in function have become very common and very inexpensive. These calculators have a statistics mode; once the calculator is in that mode, there is a special key that must be pressed after each score in your data set to enter that number. When all your numbers have been entered, a variety of statistics are available by pressing the appropriate keys. Usually the key for the biased standard deviation is labelled σ_N; the subscript N is used to remind you that N rather than $N - 1$ is being used to calculate this standard deviation. Unfortunately, the symbol for the *unbiased* standard deviation is often σ_{N-1}, which is not consistent with my use of s for the sample statistic, but at least the $N - 1$ is there to remind you that this standard deviation is calculated with the unbiased formula. To get either type of variance on most of these calculators, you must square the corresponding standard deviation, and to get SS, you must multiply the variance by N or $N - 1$, depending on which standard deviation you started with.

Converting Biased to Unbiased Variance and Vice Versa

If your calculator has only the biased or unbiased standard deviation built in (but not both), it is easy to obtain the other one with only a little additional calculation. The procedure I'm about to describe could also be used if you see one type of standard deviation published in an article and would like to determine the other one. If you are starting with the biased standard deviation, square it and then multiply it by N to find SS. Then, to obtain s you divide the SS you just found by $N - 1$ and take its square root. Fortunately, there is an even shorter way to do this, as shown in Formula 3.16A:

$$s = \sigma\sqrt{\frac{N}{N-1}} \qquad \textbf{Formula 3.16A}$$

For the numbers 1, 3, and 8, I have already calculated the biased variance (8.67), and therefore the biased standard deviation is $\sqrt{8.67} = 2.94$. To find the unbiased standard deviation, you can use Formula 3.16A:

$$s = 2.94\sqrt{\frac{3}{2}} = 2.94(1.225) = 3.60$$

This result agrees, within rounding error, with the unbiased standard deviation I found for these numbers more directly at the end of Section A.

If you are starting out with the unbiased standard deviation, you can use Formula 3.16A with N and $N - 1$ reversed, as follows:

$$\sigma = s\sqrt{\frac{N-1}{N}} \qquad\qquad \textbf{Formula 3.16B}$$

If you are dealing with variances instead of standard deviations, you can use the preceding formulas by removing the square root signs and squaring both s and σ.

Properties of the Mean

The mean and standard deviation are often used together to describe a set of numbers. Both of these measures have a number of mathematical properties that make them desirable not only for descriptive purposes but also for various inferential purposes, many of which will be discussed in later chapters. I will describe some of the most important and useful properties for both of these measures beginning with the mean:

1. *If a constant is added (or subtracted) to every score in a distribution, the mean is increased (or decreased) by that constant.* For instance, if the mean of a midterm exam is only 70, and the professor decides to add 10 points to every student's score, the new mean will be $70 + 10 = 80$ (i.e., $\mu_{new} = \mu_{old} + C$).The rules of summation presented in Chapter 1 prove that if you find the mean after adding a constant to every score (i.e., $\Sigma(X + C)/N$), the new mean will equal $\mu + C$. First, note that $\Sigma(X + C) = \Sigma X + \Sigma C$ (according to Summation Rule 1A). Next, note that $\Sigma C = NC$ (according to Summation Rule 3). So,

$$\frac{\Sigma(X + C)}{N} = \frac{\Sigma X + \Sigma C}{N} = \frac{\Sigma X + NC}{N} = \frac{\Sigma X}{N} + \frac{NC}{N} = \frac{\Sigma X}{N} + C = \mu + C$$

(A separate proof for subtracting a constant is not necessary; the constant being added could be negative without changing the proof.)

2. *If every score is multiplied (or divided) by a constant, the mean will be multiplied (or divided) by that constant.* For instance, suppose that the average for a statistics quiz is 7.4 (out of 10), but later the professor wants the quiz to count as one of the exams in the course. To put the scores on a scale from 0 to 100, the professor multiplies each student's quiz score by 10. The mean is also multiplied by 10, so the mean of the new exam scores is $7.4 \times 10 = 74$.

   We can prove that this property holds for any constant. The mean of the scores after multiplication by a constant is $(\Sigma CX)/N$. By Summation Rule 2A, you know that $\Sigma CX = C \Sigma X$, so

$$\frac{\Sigma CX}{N} = \frac{C \Sigma X}{N} = C \frac{\Sigma X}{N} = C\mu$$

There is no need to prove that this property also holds for dividing by a constant because the constant in the above proof could be less than 1.0 without changing the proof.

3. *The sum of the deviations from the mean will always equal zero.* To make this idea concrete, imagine that a group of waiters has agreed to share all of their tips. At the end of the evening, each waiter puts his or her tips in a big bowl; the money is counted and then divided equally among the waiters. Because the sum of all the tips is being divided by the number of waiters, each waiter is actually getting the mean amount of the tips. Any waiter who had pulled in more than the average tip would lose something in this deal, whereas any waiter whose tips for the evening were initially below the average would gain. These gains or losses can be expressed symbolically as deviations from the mean, $X_i - \mu$, where X_i is the amount of tips collected by the ith waiter and μ is the mean of the tips for all the waiters. The property above can be stated in symbols as $\Sigma(X_i - \mu) = 0$.

In terms of the waiters, this property says that the sum of the gains must equal the sum of the losses. This makes sense—the gains of the waiters who come out ahead in this system come entirely from the losses of the waiters who come out behind. Note, however, that the *number* of gains does not have to equal the *number* of losses. For instance, suppose that ten waiters decided to share tips and that nine waiters receive $10 each and a tenth waiter gets $100. The sum will be $(9 \times 10) + 100 = \$90 + \$100 = \$190$, so the mean is $190/10 = \$19$. The nine waiters who pulled in $10 each will each gain $9, and one waiter will lose $81 ($100 − $19). But although there are nine gains and one loss, the total amount of gain $(9 \times \$9 = \$81)$ equals the total amount of loss $(1 \times \$81 = \$81)$. Note that in this kind of distribution, the majority of scores can be below the mean (the distribution is positively skewed, as described in the previous section).

The property above can also be proven to be generally true. First, note that $\Sigma(X_i - \mu) = \Sigma X_i - \Sigma \mu$, according to Summation Rule 1B. Because μ is a constant, $\Sigma\mu = N\mu$ (Summation Rule 3), so $\Sigma(X_i - \mu) = \Sigma X - N\mu$. Multiplying both sides of the equation for the mean by N, we get: $\Sigma X_i = N\mu$, so $\Sigma(X_i - \mu) = N\mu - N\mu = 0$.

4. *The sum of the squared deviations from the mean will be less than the sum of squared deviations around any other point in the distribution.* To make matters simple, I will use my usual example: 1, 3, 8. The mean of these numbers is 4. The deviations from the mean (i.e., $X_i - 4$) are −3, −1, and +4. (Note that these sum to zero, as required by property 3 above.) The squared deviations are 9, 1, and 16, the sum of which is 26. If you take any number other than the mean (4), the sum of the squared deviations from that number will be more than 26. For example, the deviations from 3 (which happens to be the median) are −2, 0, +5; note that these do *not* sum to zero. The squared deviations are 4, 0, and 25, which sum to more than 26. Also note, however, that the absolute values of the deviations from the median add up to 7, which is less than the sum of the absolute deviations from the mean (8). It is the median that minimizes the sum of absolute deviations, whereas the mean minimizes the sum of *squared* deviations. Proving that the latter property is always true is a bit tricky, but the interested reader can find such a proof in some advanced texts (e.g., Hays, 1994). This property, often called the *least-squares property*, is a very important one and will be mentioned in the context of several statistical procedures later in this text.

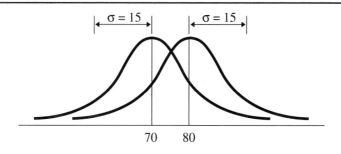

Figure 3.11

Adding a Constant to a
Distribution

Properties of the Standard Deviation

Note: These properties apply equally to the biased and unbiased formulas.

1. *If a constant is added (or subtracted) from every score in a distribution, the standard deviation will not be affected.* To illustrate a property of the mean, I used the example of an exam on which the mean score was 70. The professor decided to add 10 points to each student's score, which caused the mean to rise from 70 to 80. Had the standard deviation been 15 points for the original exam scores, the standard deviation would still be 15 points after 10 points were added to each student's exam score. Because the mean moves with the scores, and the scores stay in the same relative positions with respect to each other, shifting the location of the distribution (by adding or subtracting a constant) does not alter its spread (see Figure 3.11). This can be shown to be true in general by using simple algebra. The standard deviation of a set of scores after a constant has been added to each one is:

$$\sigma_{new} = \sqrt{\frac{\sum(X + C - \mu_{new})^2}{N}}$$

According to the first property of the mean just described, $\mu_{new} = \mu_{old} + C$. Therefore,

$$\sigma_{new} = \sqrt{\frac{\sum[(X + C) - (\mu_{old} + C)]^2}{N}}$$

Rearranging the order of terms gives the following expression:

$$\sigma_{new} = \sqrt{\frac{\sum(X - \mu_{old} + C - C)^2}{N}} = \sqrt{\frac{\sum(X - \mu_{old})^2}{N}} = \sigma_{old}$$

The above proof works the same way if you are subtracting, rather than adding, a constant.

2. *If every score is multiplied (or divided) by a constant, the standard deviation will be multiplied (or divided) by that constant.* In describing a corresponding property of the mean, I used an example of a quiz with a mean of 7.4; each student's score was multiplied by 10, resulting in an exam with a mean of 74. Had the standard deviation of the quiz been 1.8, the standard deviation after scores were multiplied by 10 would have been 18. Whereas adding a constant does not increase the spread of the distribution, multiplying by a constant does (see Figure 3.12). For example, quiz scores of 4 and 7 are spread by only 3 points, but after they are multiplied by 10 the scores are 40 and 70, which are 30

Figure 3.12

Multiplying a
Distribution by a
Constant

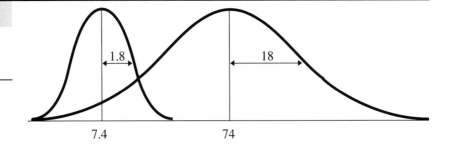

points apart. Once again we can show that this property is true in general by using some algebra and the rules of summation. The standard deviation of a set of scores after multiplication by a constant is:

$$\sigma_{new} = \sqrt{\frac{\sum (CX_i - \mu_{new})^2}{N}}$$

According to the second property of the mean described above, $\mu_{new} = C\mu_{old}$. Therefore:

$$\sigma_{new} = \sqrt{\frac{\sum (CX_i - C\mu_{old})^2}{N}} = \sqrt{\frac{\sum [C(X_i - \mu_{old})]^2}{N}} = \sqrt{\frac{\sum C^2(X_i - \mu_{old})^2}{N}}$$

The term C^2 is a constant, so according to Summation Rule 2, we can move this term in front of the summation sign. Then a little bit of algebraic manipulation proves the preceding property:

$$\sigma_{new} = \sqrt{\frac{C^2 \sum (X_i - \mu_{old})^2}{N}} = \sqrt{C^2} \sqrt{\frac{\sum (X_i - \mu_{old})^2}{N}} = C\sigma_{old}$$

3. *The standard deviation from the mean will be smaller than the standard deviation from any other point in the distribution.* This property follows from property 4 of the mean, as described previously. If SS is minimized by taking deviations from the mean rather than from any other location, it makes sense that σ, which is $\sqrt{SS/N}$, will also be minimized. Proving this requires some algebra and the rules of summation; the proof can be found in some advanced texts (e.g., Hays, 1994, p. 188).

$\mathcal{B}$

SUMMARY

1. If several groups are to be combined into a larger group, the mean of the larger group will be the weighted average of the means of the smaller groups, where the weights are the sizes of the groups. Finding the weighted average, in this case, can be accomplished by finding the sum of each group (which equals its size times its mean), adding all the sums together, and then dividing by the size of the combined group (i.e., the sum of the sizes of the groups being combined).

2. Convenient computational formulas for the variance and SD can be created by starting with a computational formula for SS (e.g., the sum of the squared scores minus N times the square of the mean) and then dividing by N for the biased variance, or $N - 1$ for the unbiased variance. The computational formula for the biased or unbiased SD is just the square root of the corresponding variance formula.

3. The standard deviation can be found directly using virtually any inexpensive scientific or statistical calculator (the calculator must be in statistics mode, and a special key must be used to enter each score in your data set). The variance is then found by squaring the *SD*, and the *SS* can be found by multiplying the variance by N (if the variance is biased), or $N - 1$ (if the variance is unbiased).

4. A biased *SD* can be converted to an unbiased *SD* by multiplying it by the square root of the ratio of N over $N - 1$, a factor that is very slightly larger than 1.0 for large samples. To convert from unbiased to biased, the ratio is flipped over.

5. Properties of the Mean
   a. If a constant is added (or subtracted) to every score in a distribution, the mean of the distribution will be increased (or decreased) by that constant (i.e., $\mu_{new} = \mu_{old} \pm C$).
   b. If every score in a distribution is multiplied (or divided) by a constant, the mean of the distribution will be multiplied (or divided) by that constant (i.e., $\mu_{new} = C\mu_{old}$).
   c. The sum of the deviations from the mean will always equal zero [i.e., $\Sigma(X_i - \mu) = 0$].
   d. The sum of the squared deviations from the mean will be less than the sum of squared deviations from any other point in the distribution (i.e., $\Sigma(X_i - \mu)^2 < \Sigma(X_i - C)^2$, where C represents some location in the distribution other than the mean).

6. Properties of the Standard Deviation
   a. If a constant is added (or subtracted) from every score in a distribution, the standard deviation will remain the same (i.e., $\sigma_{new} = \sigma_{old}$).
   b. If every score is multiplied (or divided) by a constant, the standard deviation will be multiplied (or divided) by that constant (i.e., $\sigma_{new} = C\sigma_{old}$).
   c. The standard deviation around the mean will be smaller than it would be around any other point in the distribution.

EXERCISES

*1. There are three fourth-grade classes at Happy Valley Elementary School. The mean IQ for the 10 pupils in the gifted class is 119. For the 20 pupils in the regular class, the mean IQ is 106. Finally, the five pupils in the special class have a mean IQ of 88. Calculate the mean IQ for all 35 fourth-grade pupils.

2. A student has earned 64 credits so far, of which 12 credits are As, 36 credits are Bs, and 16 credits are Cs. If A = 4, B = 3, and C = 2, what is this student's grade point average?

*3. A fifth-grade teacher calculated the mean of the spelling tests for his 12 students; it was 8. Unfortunately, now that the teacher is ready to record the grades, one test seems to be missing. The 11 available scores are 10, 7, 10, 10, 6, 5, 9, 10, 8, 6, 9. Find the missing score. (*Hint*: You can use property 3 of the mean.)

4. A psychology teacher has given an exam on which the highest possible score is 200 points. The mean score for the 30 students who took the exam was 156, and the standard deviation was 24. Because there was one question that every student answered incorrectly, the teacher decides to give each student 10 extra points and then divide each score by 2, so the total possible score is 100. What will the mean and standard deviation of the scores be after this transformation?

5. The IQ scores for 10 sixth-graders are 111, 103, 100, 107, 114, 101, 107, 102, 112, 109.

a. Calculate σ for the IQ scores using the definitional formula.

b. Calculate σ for the IQ scores using the computational formula.

c. Describe one condition under which it is easier to use the definitional than the computational formula.

d. How could you transform the scores above to make it easier to use the computational formula?

*6. Use the appropriate computational formulas to calculate both the biased and unbiased standard deviations for the following set of numbers: 21, 21, 24, 24, 27, 30, 33, 39.

*7. a. Calculate s for the following set of numbers: 7, 7, 10, 10, 13, 16, 19, 25. (*Note*: This set of numbers was created by subtracting 14 from each of the numbers in the previous exercise.) Compare your answer to this exercise with your answer to Exercise 6. What general principle is being illustrated?

b. Calculate s for the following set of numbers: 7, 7, 8, 8, 9, 10, 11, 13. (*Note*: This set of numbers was created by dividing each of the numbers in Exercise 6 by 3.) Compare your answer to this exercise with your answer to Exercise 6. What general principle is being illustrated?

8. a. For the data in Exercise 6 use the definitional formula to calculate s around the *median* instead of the mean.

b. What happens to s? What general principle is being illustrated?

*9. If σ for a set of data equals 4.5, what is the corresponding value for s

a. When $N = 5$?

b. When $N = 20$?

c. When $N = 100$?

10. If s for a set of data equals 12.2, what is the corresponding value for σ

a. When $N = 10$?

b. When $N = 200$?

**OPTIONAL
MATERIAL**

In Section C of the previous chapter, I introduced a technique for displaying data called the stem-and-leaf plot. I mentioned that that technique is commonly associated with a general approach to dealing with data called Exploratory Data Analysis (EDA). I will soon describe another technique associated with EDA, but first I want to describe EDA more generally. EDA can be viewed as a collection of graphical methods for displaying all of your data at a glance and for highlighting important characteristics of your data that relate to the assumptions of the statistical methods usually used to analyze data. However, EDA can also be considered a "movement," or philosophy concerning the optimal way to deal with data, especially in psychological research. The father of this movement is the statistician John W. Tukey, who devised many of the graphic methods associated with EDA, including the stem-and-leaf plot described in the previous chapter and the method described later in this section (he also invented the HSD statistic presented in Chapter 13). A chief tenet of the movement's philosophy involves approaching a newly collected set of data with an open mind, like a detective who has no preconceived notions about what happened but wants to explore every clue and get to the truth (Behrens, 1997).

One apparent aim of EDA is to counter the tendencies of psychological reserchers to summarize their data before looking at all of the scores carefully and to apply inferential statistics before being certain that their data are appropriate for those methods. Whether or not you use any of the specific methods associated with EDA, the main message is too important to ignore. Before you summarize your data with just a few statistics and/or perform inferential tests (such as those in Parts II through VI of this text), you should look at all of your data with techniques that help you to see important patterns. Your data may have surprises in store for you and suggest analyses you hadn't planned.

A very simple example of the need for EDA is a set of data with a few extreme scores (i.e., outliers). If the researcher automatically calculates the mean, because that is what he or she had planned to do, and uses that mean to describe the data and perform further tests, it is quite possible that the results will be misleading and that very different results will be found upon replication of the original study. Graphing the distribution of the data in your sample by the methods in Section B of the previous chapter would make it possible to see the outlier, but there are various EDA methods that not only make it easier to see outliers but also make it easier to determine whether a point is really extreme enough to be considered an outlier. The stem-and-leaf plot is good for digesting relatively small data sets, but if you need to deal with a relatively large data set, and want to be able to identify outliers quickly, the method I'll describe next is what you want.

Box-and-Whisker Plots

Proponents of EDA prefer to base their descriptions of data sets on statistical measures that are called "resistant" (or *robust*), because they are resistant to the misleading effects of outliers. The median and SIQ range are two examples of resistant measures. If you want a procedure that will describe a distribution at a glance, and help you identify outliers, it makes sense to base your procedure on resistant measures. That is the case for the graphic method known as the *box-and-whisker plot*, or *boxplot*, for short. The boxplot uses the median as its center, and the sides of the "box," which are called the *hinges*, are basically the 25th and 75th percentiles. (Technically, the sides of the box are defined as the median of the scores above the 50th percentile and the median of the scores below the 50th percentile, and Tukey's method for calculating these medians depends on whether there are an odd or even total number of scores, but with a fairly large sample size, there is very little error involved in just using the usual calculation of the 25th and 75th percentiles.) The median will be somewhere between the hinges, but if the distribution is not symmetric, it won't be exactly midway. In fact, the position of the median with respect to the two hinges shows the skewing of the distribution at a glance.

The lengths of the "whiskers" depend, to some extent, on the length of the box; the length of the box is called the *H-spread* because it is the separation of the two hinges (because the hinges are approximately the 25th and 75th percentiles, the H-spread is essentially the same as the IQ range).

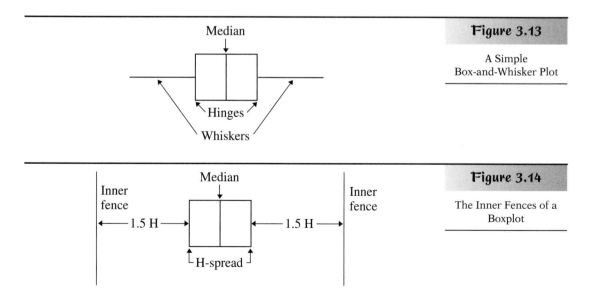

Figure 3.13

A Simple
Box-and-Whisker Plot

Figure 3.14

The Inner Fences of a
Boxplot

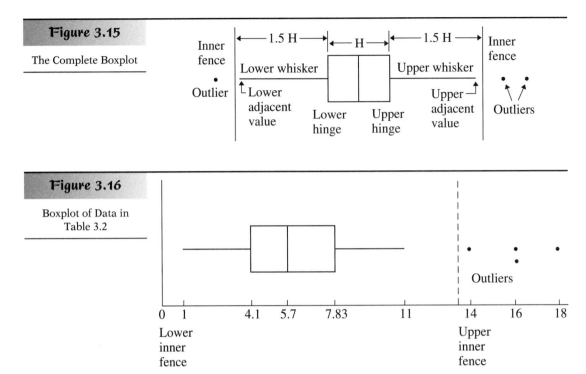

Figure 3.15

The Complete Boxplot

Figure 3.16

Boxplot of Data in
Table 3.2

Figure 3.13 illustrates the basic parts of a box-and-whisker plot. To draw
the whiskers, it is first necessary to mark the outermost possible limits of
the whiskers on either side of the box. These outermost limits are called the
inner fences. To find each inner fence, you start at one of the hinges and
then move away from the box a distance equal to 1.5 times the H-spread, as
shown in Figure 3.14. (There are variations in defining the locations of the
inner fences, but I am following the most common procedure, which is
based on the suggestions of Tukey, 1977.) However, the whiskers generally
do *not* extend all the way to the inner fences. The end of the upper whisker
is drawn at the highest value in the distribution that is not higher than the
upper inner fence; this value, called an *adjacent value*, could turn out to be
not even close to the inner fence. Similarly, the lower whisker ends at the
lowest (i.e., leftmost) value that is not past the lower inner fence. Finally,
any scores in the distribution that fall beyond the ends of the whiskers are
indicated as points in the plot; these points are called *outliers*. Figure 3.15
depicts all of the parts and dimensions of a box-and-whisker plot.

　　One of the main purposes of a boxplot is to provide a systematic procedure for identifying outliers. Because outliers can so strongly affect the
location of the mean in a distribution, it is important to investigate these
points. Sometimes an outlier is found to be the result of a gross error in
recording or transcribing the data or some procedural mishap, such as a
subject misreading the instructions or becoming ill. Once such outliers are
detected, perhaps with the help of a boxplot, they can be removed from the
data at once, and any statistics that had been already calculated can be
recalculated. Other outliers may represent legitimate data that just happen
to be quite unusual (e.g., the score of a subject with a photographic
memory in a picture recall experiment). If a data point is very unusual, but
no error or mishap can be found to account for it, the value should *not* be

removed unless you have chosen a procedure for trimming your data (see the following). The unusual data point is part of the phenomenon being studied. However, you should make an effort to find the underlying causes for extreme values so that you can study such data separately, or simply avoid collecting such data in future experiments.

The boxplot also lets you see the spread of a distribution, and the symmetry of that spread, at a glance. I will illustrate the construction of a boxplot for data from a hypothetical problem-solving experiment in which the number of minutes taken to solve the problem is recorded for each subject. The data for 40 subjects are presented in the form of a simple frequency distribution in Table 3.2 (this is the table from Exercise 2B6). (Although it is instructive to see how the boxplot is constructed, bear in mind that popular statistical packages (e.g., SPSS) now create a boxplot for your data, if you request it from the appropriate menu.)

The first step is to find the median. This was already found to be 5.7 in Exercise 2B6. The two hinges were also calculated for that exercise: the 25th percentile was found to be 4.1 and the 75th was found to be 7.83. Now that we have the two hinges, we can find the H-spread (or H, for short): H = 7.83 − 4.1 = 3.73. Next, we multiply H by 1.5 to get 5.6; this value is added to the upper hinge (7.83) to give the upper inner fence, which is therefore equal to 7.83 + 5.6 = 13.43. Normally, we would also subtract 5.6 from the lower hinge (4.1), but because this would give a negative number, and time cannot be negative, we will take 0 to be the lower inner fence.

Finally, we have to find the upper and lower adjacent values. The upper adjacent value is the highest score in the distribution that is not higher than 13.43 (i.e., the upper inner fence). From Table 3.2 we can see that 11 is the upper adjacent value. The lower adjacent value, in this case, is simply the lowest score in the distribution, which is 1. The outliers are the points that are higher than 11: 14, 16, 16, and 18; there is no room for outliers on the low side. The boxplot is shown in Figure 3.16 and the skew of the distribution can be seen easily from this figure.

Table 3.2			
X	f	X	f
18	1	9	1
17	0	8	3
16	2	7	5
15	0	6	5
14	1	5	7
13	0	4	5
12	0	3	4
11	1	2	2
10	2	1	1

Dealing with Outliers

Trimming

Now that we have identified four outliers, what do we do with them? Of course, any outlier that is clearly the result of some kind of mistake or mishap is deleted immediately. However, it is quite possible to have outliers that represent values that are legitimate but relatively rare. If a certain percentage of outliers can be expected in your experiment, there are methods that can be used to reduce their impact. One method is *trimming.* Suppose previous research has determined that the most extreme 20% of the data points from a certain type of experiment are likely to be misleading—that is, not representative of the phenomenon being studied but attributable to extraneous factors, such as poor motivation, fatigue, cheating, and so forth. A 10% *trimmed* sample is what you want; it is defined as what remains after eliminating *both* the top and bottom 10% of the data points (i.e., the most extreme 20%). The mean of the trimmed sample is called a *trimmed mean.*

A related method for dealing with outliers is called *Winsorizing.* A percentage is chosen, as with trimming, but instead of being eliminated, the extreme values are replaced by the most extreme value that is not being thrown away; this is done separately for each tail unless the distribution is so skewed that there is no tail on one side. For instance, if in Table 3.2 the

18, 16, 16, and 14 were in the percentage to be trimmed, they would all be replaced by the value 11—the highest value not being trimmed. The *Winsorized sample* then has five 11s as its highest values, followed by two 10s, one 9, and so forth, as in the rest of Table 3.2. The trimmed mean is one of several measures of central tendency that are known as "robust estimators of location" because they are virtually unaffected by outliers. However, trimming or Winsorizing your samples will have some effects on the inferential procedures that should be applied to your data; I will introduce the topic of robust inferential statistics in Section C of Chapter 6.

Data Transformation

A more sophisticated method for lessening the impact of extreme scores or marked skewing involves performing a data transformation. For instance, if your data contain a few very large scores, taking the square root of each of the values will reduce the influence of the large scores. Suppose you have recorded the number of minutes each individual takes to solve a problem, and the data for five participants are 4, 9, 9, 16, and 100 (you are a very patient researcher). The mean is 27.6, because of the influence of the one extremely high score. If you take the square root of each score, you obtain 2, 3, 3, 4, 10. Now, the mean is 4.4, and the skewing is much less extreme. Even if you square the new mean to return it to the original units, the mean will be $4.4^2 = 19.4$, which is still less influenced than the original mean of 27.6.

Transforming the data to make the distribution more symmetrical may seem like cheating to you. And how can it make sense to deal with, for instance, the square root of your scores? First, it can be pointed out that data transformations are not always arbitrary. Imagine a study in which children draw coins from memory, in an attempt to see if a child's economic background is related to the size of the coin he or she draws. If the area of the coin that is drawn is used as the dependent variable, suppose the data for five subjects drawing the same type of coin are 4, 9, 9, 16, 100. To reduce the skew of the data it would be perfectly reasonable to measure the radius of each coin drawn instead of its area. But the radius of each coin would be the square root of each number above (i.e., 2, 3, 3, 4, 10) divided by the square root of the constant known as π (dividing by a constant does *not* change the shape of the distribution in any way). Until we know more about the process underlying the drawing of the coin from memory, the radius would be just as reasonable a measure of size as the area.

When dealing with reaction time data, the skew can be quite strong, so a logarithmic transformation is usually used; taking the "log" of each score has an even greater impact on large scores than taking the square root. For instance, the logs (using 10 as the base) of 4, 9, 9, 16, 100 are .60, .95, .95, 1.2, 2 (the $\log_{10}$ of 100 is 2 because you have to raise 10 to that power to get 100). Another common transformation is the reciprocal (the reciprocal of a number C is $1/C$). For instance, if you give a participant a series of similar tasks to perform, you can measure how long it takes that individual to complete all of the tasks (perhaps, as a function of reward level). However, if you measure his or her speed in terms of tasks completed per minute, the speed will be some constant (e.g., the total number of tasks) times the reciprocal of the time for completion. Supporters of the EDA approach prefer to say that the data is being reexpressed, rather than transformed, in part to emphasize that there may be several equally legitimate ways to express the same data.

The goal of a data transformation is usually to reduce the impact of outliers and/or make the distribution more symmetric. In fact, a transformation that makes your distribution more closely resemble the normal distribution is often considered very desirable, for reasons which will be made clear in later chapters. It is considered acceptable to try a number of different transformations (e.g., square root, cube root, and so forth) until you find the one that comes closest to making the distribution of your sample data look like the normal distribution. It is, of course, not acceptable to try different transformations until your data yield the summary statistics or inferences that you would like to see. As you can imagine, it used to be quite tedious to perform several transformations on a large set of data, but now statistical software makes it easy to try a wide selection of functions to see which works best.

At present, data transformations are more commonly applied than procedures for trimming data, but the results they produce can be more difficult to interpret. There are rules of thumb to help you choose the best transformation (e.g., if among several samples, the mean is highly correlated with the standard deviation of the sample, a log transformation is suggested; if the mean is highly correlated with the variance, a square root transformation will probably work better), but the final choice of a transformation should be based on an understanding of the processes that underlie the dependent variable you are measuring and a familiarity with previous results in the literature, as well as the characteristics of your data after the transformation.

Measuring Skewness

Skewness can be detected informally by inspecting a frequency polygon, a stem-and-leaf plot, or a boxplot of your data. However, quantifying skewness can be useful in deciding when the skewing is so extreme that you ought to transform the distribution or use different types of statistics. For this reason most statistical packages provide a measure of skewness when a full set of descriptive statistics is requested. Whereas the variance is based on the average of squared deviations from the mean, skewness is based on the average of *cubed* deviations from the mean:

$$\text{Average cubed deviation} = \frac{\sum (X_i - \mu)^3}{N}$$

Recall that when you square a number, the result will be positive whether the original number was negative or positive. However, the cube (or third power) of a number has the same sign as the original number. If the number is negative, the cube will be negative ($-2^3 = -2 \times -2 \times -2 = -8$), and if the number is positive, the cube will be positive (e.g., $+2^3 = +2 \times +2 \times +2 = +8$). Deviations below the mean will still be negative after being cubed, and positive deviations will remain positive after being cubed. Thus skewness will be the average of a mixture of positive and negative numbers, which will balance out to zero *only* if the distribution is symmetric. (Note that the deviations will always average to zero before being cubed, but after being cubed they need not.) Any negative skew will cause the skewness measure to be negative, and any positive skew will produce a positive skewness measure. Unfortunately, like the variance, the measure of skewness does not provide a good description of a distribution because it is in cubed units. Rather than taking the cube root of the preceding formula, you can derive a more useful measure of the skewness of a population distribution by dividing that formula by σ^3 (the cube of the standard deviation calculated as for a population) to produce Formula 3.17:

$$\text{Skewness} = \frac{\sum (X_i - \mu)^3}{N\sigma^3} \qquad \textbf{Formula 3.17}$$

Formula 3.17 has the very useful property of being dimensionless (cubed units are being divided, and thus canceled out, by cubed units); it is a pure measure of the shape of the distribution. Not only is this measure of skewness unaffected by adding or subtracting constants (as is the variance), it is also unaffected by multiplying or dividing by constants. For instance, if you take a large group of people and measure each person's weight in pounds, the distribution is likely to have a positive skew that will be reflected in the measure obtained from Formula 3.17. Then, if you convert each person's weight to kilograms, the *shape* of the distribution will remain the same (although the variance will be multiplied by a constant), and fortunately the skewness measure will also remain the same. The only drawback to Formula 3.17 is that if you use it to measure the skewness of a sample, the result will be a biased estimate of the population skewness. This is only a problem if you plan to test your measure of skewness with inferential methods, but this is rarely done. However, a measure of skewness can be used as a quick check on the success of a data transformation.

To illustrate the use of Formula 3.17, I will calculate the skewness of four numbers: 2, 3, 5, 10. First, using Formula 3.4B you can verify that $\sigma = 3.082$, so $\sigma^3 = 29.28$. Next, $\Sigma(X - \mu)^3 = (2 - 5)^3 + (3 - 5)^3 + (5 - 5)^3 + (10 - 5)^3 = -3^3 + -2^3 + 0^3 + 5^3 = -27 + (-8) + 0 + 125 = 90$. (Note how important it is to keep track of the sign of each number.) Now we can plug these values into Formula 3.17:

$$\text{Skewness} = \frac{90}{4(29.28)} = \frac{90}{117.1} = .768$$

As you can see, the skewness is positive. Although the total amount of deviation below the mean is the same as the amount of deviation above, one larger deviation (i.e., +5) counts more than two smaller ones (i.e., −2 and −3).

Measuring Kurtosis

Two distributions can both be symmetric (i.e., skewness equals zero), unimodal, and bell-shaped and yet not be identical in shape. (Bell-shaped is a crude designation—many variations are possible.) Moreover, these two distributions can even have the same mean and variance and still differ fundamentally in shape. The simplest way that such distributions can differ is in the degree of flatness that characterizes the curve. If a distribution tends to have relatively thick or heavy tails and then bends sharply so as to have a relatively greater concentration near its center (more "peakedness"), that distribution is called a *leptokurtic distribution*. Compared to the normal distribution, the leptokurtic distribution lacks scores in the "shoulders" of the distribution (the areas on each side of the distribution that are between the tails and the middle of the distribution). On the other hand, a distribution that tends to be flat (i.e., it has no shortage of scores in the shoulder area and therefore does not bend sharply in that area), with relatively thin tails and less peakedness, is called a *platykurtic distribution*. (The Greek prefixes platy- and lepto- describe the middle portion of the distribution, lepto-

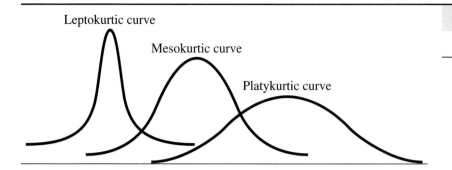

Figure 3.17

Degrees of Kurtosis

meaning "slim," and platy- meaning "wide.") These two different shapes are illustrated in Figure 3.17, along with a distribution that is midway between in its degree of *kurtosis—a mesokurtic distribution*. The normal distribution is used as the basis for comparison in determining kurtosis, so it is mesokurtic, by definition. [Because a distribution can be leptokurtic due to very heavy tails *or* extreme peakedness, there are debates about the relative importance of these two factors in determining kurtosis. This debate goes well beyond the scope of this text, but you can read more about it in an article by DeCarlo (1997).]

Just as the measure of skewness is based on cubed deviations from the mean, the measure of kurtosis is based on deviations from the mean raised to the fourth power. This measure of kurtosis must then be divided by the standard deviation raised to the fourth power, to create a dimensionless measure of distribution shape that will not change if you add, subtract, multiply, or divide all the data by a constant. Formula 3.18 for the kurtosis of a population is as follows:

$$\text{Kurtosis} = \frac{\sum (X_i - \mu)^4}{N\sigma^4} - 3 \qquad \textbf{Formula 3.18}$$

Besides the change from the third to the fourth power, you will notice another difference between Formulas 3.18 and 3.17 for skewness: the subtraction of 3 in the kurtosis formula. The subtraction appears in most kurtosis formulas to facilitate comparison with the normal distribution. Subtracting 3 ensures that the kurtosis of the normal distribution will come out to zero. Thus a distribution that has relatively fatter tails than the normal distribution (and greater peakedness) will have a positive kurtosis (i.e., it will be leptokurtic), whereas a relatively thin-tailed, less peaked distribution will have a negative kurtosis (it will be platykurtic). In fact, unless you subtract 3, the population kurtosis will never be less than $+1$. After you subtract 3, kurtosis can range from -2 to positive infinity.

I will illustrate the calculation of kurtosis for the following four numbers: 1, 5, 7, 11. First, we find that the population variance of these numbers is 13. To find σ^4, we need only square the biased variance: $13^2 = 169$. (Note that in general, $(x^2)^2 = x^4$.) Next, we find $\Sigma(X - \mu)^4 = (1 - 6)^4 + (5 - 6)^4 + (7 - 6)^4 + (11 - 6)^4 = -5^4 + (-1)^4 + 1^4 + 5^4 = 625 + 1 + 1 + 625 = 1252$. Now we are ready to use Formula 3.18:

$$\text{Kurtosis} = \frac{1252}{4(169)} - 3 = 1.85 - 3 = -1.15$$

The calculated value of –1.15 suggests that the population from which the four numbers were drawn has negative kurtosis (i.e., somewhat lighter tails than the normal distribution). In practice, however, one would never draw any conclusions about kurtosis when dealing with only four numbers.

The most common reason for calculating the skewness and kurtosis of a set of data is to help you decide whether your sample comes from a population that is normally distributed. All of the statistical procedures in Parts II through VI of this text are based on the assumption that the variable being measured has a normal distribution in the population. If you use a data transformation to normalize your data, you can use skewness and/or kurtosis measures to determine the degree of your success. If the distribution of your data is very strange, and your sample size is quite small, it may be a good idea to give up on parametric statistics entirely and use one of the alternative statistical procedures described in Part VII of this text.

SUMMARY

1. One technique of exploratory data analysis that is useful in displaying the spread and symmetry of a distribution, and especially in identifying outliers, is the *box-and-whisker plot*, sometimes called a *boxplot*, for short.

2. The left and right sides of the box, called the *hinges*, are closely approximated by the 25th and 75th percentiles, respectively. The median is drawn wherever it falls *inside* the box. The distance between the two hinges is called the *H-spread*.

3. The farthest the whiskers can possibly reach on either side is determined by the *inner fences*. Each inner fence is 1.5 times the H-spread away from its hinge, so the separation of the inner fences is 4 times the H-spread (1.5 H on either side of the box equals 3 H, plus 1 H for the box itself).

4. The *upper adjacent value* is defined as the highest score in the distribution that is not higher than the upper inner fence. The *lower adjacent value* is defined analogously. The *upper whisker* is drawn from the upper hinge to the upper adjacent value; the *lower whisker* is drawn in a corresponding fashion.

5. Any scores in the distribution that are higher than the upper inner fence or lower than the lower inner fence are indicated as points in the boxplot. These values are referred to as *outliers*.

6. If an outlier is found to be the result of an error in measurement or data handling, or is clearly caused by some factor extraneous to the experiment, it can be legitimately removed from the data set. If the outlier is merely unusual, it should not be automatically thrown out; investigating the cause of the extreme response may uncover some previously unnoticed variable that is relevant to the response being measured.

7. The influence of outliers or extreme skewing can be reduced by *trimming* the sample (deleting a fixed percentage of the most extreme scores) or *Winsorizing* the sample (replacing a fixed percentage of extreme scores with the most extreme value, in the same direction, that is not deleted). Finally, a *data transformation*, such as taking the square root or logarithm of each data value, can be used to reduce the influence of outliers.

8. *Skewness* can be measured by cubing (i.e., raising to the third power) the deviations of scores from the mean of a distribution, taking their average, and then dividing by the cube of the population standard deviation. The measure of skewness will be a negative number for a negatively skewed distribution, a positive number for a positively skewed distribution, and zero if the distribution is perfectly symmetric around its mean.

9. *Kurtosis* can be measured by raising deviations from the mean to the fourth power, taking their average, and then dividing by the square of the population variance. If the kurtosis measure is set to zero for the normal distribution, positive kurtosis indicates relatively fat tails and more peakedness in the middle of the distribution (a leptokurtic distribution), whereas negative kurtosis indicates relatively thin tails and a lesser peakedness in the middle (a platykurtic distribution).

EXERCISES

*1. Create a box-and-whisker plot for the data in Exercise 2B1. Be sure to identify any outliers.

2. Create a box-and-whisker plot for the data in Exercise 2B3. Be sure to identify any outliers.

*3. Create a box-and-whisker plot for the data in Exercise 2B4. Be sure to identify any outliers.

*4. Calculate the population standard deviation and skewness for the following set of data: 2, 4, 4, 10, 10, 12, 14, 16, 36.

5. a. Calculate the population standard deviation and skewness for the following set of data: 1, 2, 2, 5, 5, 6, 7, 8, 18. (This set was formed by halving each number in Exercise 4.)

b. How does each value calculated in part a compare to its counterpart calculated in Exercise 4? What general principles are being illustrated?

6. a. Calculate the population standard deviation and skewness for the following set of data: 1, 2, 2, 5, 5, 6, 7, 8. (This set was formed by dropping the highest number from the set in Exercise 5.)

b. Using your answer to Exercise 5 as a basis for comparison, what can you say about the effect of one extreme score on variability and skewness?

*7. Take the square root of each of the scores in Exercise 4, and recalculate σ and the skewness. What effect does this transformation have on these measures?

8. a. Calculate the skewness and kurtosis for the data in Exercise 3A4.

b. Are the data in Exercise 3A4 positively or negatively skewed? Are they platykurtic or leptokurtic?

*9. Calculate the kurtosis for the following set of data: 3, 9, 10, 11, 12, 13, 19.

10. a. Calculate the kurtosis for the following set of data: 9, 10, 11, 12, 13.

b. Compare your answer to your answer for Exercise 9. What is the effect on kurtosis when you remove extreme scores from both sides of a distribution?

The semi-interquartile range after the 25th (Q1) and 75th (Q3) percentiles have been determined:

$$\text{SIQ range} = \frac{Q3 - Q1}{2}$$

Formula 3.1

The mean deviation (after the mean of the distribution has been found):

$$\text{Mean deviation} = \frac{\sum |X_i - \mu|}{N}$$

Formula 3.2

The sum of squares, definitional formula (requires that the mean of the distribution be found first):

$$SS = \sum (X_i - \mu)^2$$

Formula 3.3

The population variance, definitional formula (requires that the mean of the distribution be found first):

$$\sigma^2 = \frac{\sum (X_i - \mu)^2}{N}$$

Formula 3.4A

The population standard deviation, definitional formula (requires that the mean of the distribution be found first):

$$\sigma = \sqrt{\frac{\sum (X_i - \mu)^2}{N}}$$

Formula 3.4B

The population variance (after SS has already been calculated):

$$\sigma^2 = MS = \frac{SS}{N}$$

Formula 3.5A

The population standard deviation (after SS has been calculated):

$$\sigma = \sqrt{MS} = \sqrt{\frac{SS}{N}}$$

Formula 3.5B

The unbiased sample variance, definitional formula:

$$s^2 = \frac{\sum (X_i - \overline{X})^2}{N - 1}$$

Formula 3.6A

The unbiased sample standard deviation, definitional formula:

$$s = \sqrt{\frac{\sum (X_i - \overline{X})^2}{N - 1}}$$

Formula 3.6B

The unbiased sample variance (after SS has been calculated):

$$s^2 = \frac{SS}{N} - 1 = \frac{SS}{df}$$

Formula 3.7A

The unbiased sample standard deviation (after SS has been calculated):

$$s = \sqrt{\frac{SS}{N - 1}} = \sqrt{\frac{SS}{df}}$$

Formula 3.7B

The arithmetic mean of a population:

$$\mu = \frac{\sum X}{N}$$

Formula 3.8

The arithmetic mean of a sample:

$$\overline{X} = \frac{\sum X}{N}$$

Formula 3.9

The weighted mean of two or more samples:

$$\overline{X}_w = \frac{\sum N_i \overline{X}_i}{\sum N_i} = \frac{N_1 \overline{X}_1 + N_2 \overline{X}_2 + \cdots}{N_1 + N_2 + \cdots}$$

Formula 3.10

The sum of squares, computational formula (requires that the mean has been calculated):

$$SS = \sum X^2 - N\mu^2$$

Formula 3.11

The sum of squares, computational formula (direct from raw data):

$$SS + \sum X^2 - \frac{(\sum X)^2}{N}$$ **Formula 3.12**

The population variance, computational formula (requires that the mean has been calculated):

$$\sigma^2 = \frac{\sum X^2}{N} - \mu^2$$ **Formula 3.13A**

The population standard deviation, computational formula (requires that the mean has been calculated):

$$\sigma = \sqrt{\frac{\sum X^2}{N} - \mu^2}$$ **Formula 3.13B**

The population variance, computational formula (direct from raw data):

$$\sigma^2 = \frac{1}{N}\left[\sum X^2 - \frac{(\sum X)^2}{N}\right]$$ **Formula 3.14A**

The population standard deviation, computational formula (direct from raw data):

$$\sigma = \sqrt{\frac{1}{N}\left[\sum X^2 - \frac{(\sum X)^2}{N}\right]}$$ **Formula 3.14B**

The unbiased sample variance, computational formula (direct from raw data):

$$s^2 = \frac{1}{N-1}\left[\sum X^2 - \frac{(\sum X)^2}{N}\right]$$ **Formula 3.15A**

The unbiased sample standard deviation, computational formula (direct from raw data):

$$s = \sqrt{\frac{1}{N-1}\left[\sum X^2 - \frac{(\sum X)^2}{N}\right]}$$ **Formula 3.15B**

The unbiased standard deviation (if the biased formula has already been used):

$$s = \sigma\sqrt{\frac{N}{N-1}}$$ **Formula 3.16A**

The biased standard deviation (if the unbiased formula has already been used):

$$\sigma = s\sqrt{\frac{N-1}{N}}$$ **Formula 3.16B**

Skewness of a population in dimensionless units:

$$\text{Skewness} = \frac{\sum (X_i - \mu)^3}{N\sigma^3}$$

Formula 3.17

Kurtosis of a population in dimensionless units, adjusted so that the normal distribution has zero kurtosis:

$$\text{Kurtosis} = \frac{\sum (X_i - \mu)^4}{N\sigma^4} - 3$$

Formula 3.18

STANDARDIZED SCORES AND THE NORMAL DISTRIBUTION

You will need to use the following from previous chapters:

Symbols
Σ: Summation sign
μ: Population mean
σ: Population standard deviation
σ^2: Population variance

Concepts
Percentile ranks
Mathematical distributions
Properties of the mean and standard deviation

4

Chapter

A

CONCEPTUAL FOUNDATION

A friend meets you on campus and says, "Congratulate me! I just got a 70 on my physics test." At first, it may be hard to generate much enthusiasm about this grade. You ask, "That's out of 100, right?" and your friend proudly says, "Yes." You may recall that a 70 was not a very good grade in high school, even in physics. But if you know how low exam grades often are in college physics, you might be a bit more impressed. The next question you would probably want to ask your friend is, "What was the average for the class?" Let's suppose your friend says 60. If your friend has long been afraid to take this physics class and expected to do poorly, you should offer congratulations. Scoring 10 points above the mean isn't bad.

On the other hand, if your friend expected to do well in physics and is doing a bit of bragging, you would need more information to know if your friend has something to brag about. Was 70 the highest grade in the class? If not, you need to locate your friend more precisely within the class distribution to know just how impressed you should be. Of course, it is not important in this case to be precise about your level of enthusiasm, but if you were the teacher trying to decide whether your friend should get a B+ or an A−, more precision would be helpful.

z Scores

To see just how different a score of 70 can be in two different classes, even if both of the classes have a mean of 60, take a look at the two class distributions in Figure 4.1. (To simplify the comparison, I am assuming that the classes are large enough to produce smooth distributions.) As you can see, a score of 70 in class A is excellent, being near the top of the class, whereas the same score is not so impressive in class B, being near the middle of the distribution. The difference between the two class distributions is visually obvious—class B is much more spread out than class A. Having read the previous chapter, you should have an idea of how to quantify this difference in variability. The most useful way to quantify the variability is to calculate the standard deviation (σ). The way the distributions are drawn in Figure 4.1, σ would be about 5 points for class A and about 20 for class B.

An added bonus from calculating σ is that, in conjunction with the mean (μ), σ provides us with an easy and precise way of locating scores in a distribution. In both classes a score of 70 is 10 points above the mean, but in class A those 10 points represent two standard deviations, whereas 10 points in class B is only half of a standard deviation. Telling someone how

Figure 4.1

Distributions of Scores
on a Physics Test

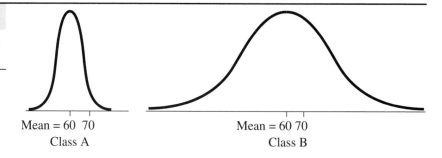

Mean = 60 70   Mean = 60 70
      Class A   Class B

many standard deviations your score is above or below the mean is more informative than telling your actual (raw) score. This is the concept behind the *z score*. In any distribution for which μ and σ can be found, any raw score can be expressed as a *z* score by using Formula 4.1:

$$z = \frac{X - \mu}{\sigma}$$ **Formula 4.1**

Let us apply this formula to the score of 70 in class A:

$$z = \frac{70 - 60}{5} = \frac{10}{5} = +2$$

and in class B:

$$z = \frac{70 - 60}{20} = \frac{10}{20} = +.5$$

In a compact way, the *z* scores tell us that your friend's exam score is more impressive if your friend is in class A ($z = +2$) rather than class B ($z = +.5$). Note that the plus sign in these *z* scores is very important because it tells you that the scores are above rather than below the mean. If your friend had scored a 45 in class B, her *z* score would have been:

$$z = \frac{45 - 60}{20} = \frac{-15}{20} = -.75$$

The minus sign in this *z* score informs us that in this case your friend was three quarters of a standard deviation *below* the mean. The *sign* of the *z* score tells you whether the raw score is above or below the mean; the *magnitude* of the *z* score tells you the raw score's distance from the mean in terms of standard deviations.

z scores are called standardized scores because they are not associated with any particular unit of measurement. The numerator of the *z* score is associated with some unit of measurement (e.g., the difference of someone's height from the mean height could be in inches), and the denominator is associated with the same unit of measurement (e.g., the standard deviation for height might be 3 inches), but when you divide the two, the result is dimensionless. A major advantage of standardized scores is that they provide a neutral way to compare raw scores from different distributions. To continue the previous example, suppose that your friend scores a 70 on the physics exam in class B and a 60 on a math exam in a class where $\mu = 50$ and $\sigma = 10$. In which class was your friend further above the mean? We have already found that your friend's *z* score for the physics exam in class B was $+.5$. The *z* score for the math exam would be:

$$z = \frac{60 - 50}{10} = \frac{10}{10} = +1$$

Thus, your friend's z score on the math exam is higher than her z score on the physics exam, so she seems to be performing better (in terms of class standing on the last exam) in math than in physics.

Finding a Raw Score from a z Score

As you will see in the next section, sometimes you want to find the raw score that corresponds to a particular z score. As long as you know μ and σ for the distribution, this is easy. You can use Formula 4.1 by filling in the given z score and solving for the value of X. For example, if you are dealing with class A (as shown in Figure 4.1) and you want to know the raw score for which the z score would be -3, you can use Formula 4.1 as follows: $-3 = (X - 60)/5$, so $-15 = X - 60$, so $X = -15 + 60 = 45$. To make the calculation of such problems easier, Formula 4.1 can be rearranged in a new form that I will designate Formula 4.2:

$$X = z\sigma + \mu \qquad\qquad \textbf{Formula 4.2}$$

Now if you want to know, for instance, the raw score of someone in class A who obtained a z score of -2, you can use Formula 4.2, as follows:

$$X = z\sigma + \mu = -2(5) + 60 = -10 + 60 = 50$$

Note that you must be careful to retain the minus sign on a negative z score when working with a formula, or you will come up with the wrong raw score. (In the previous example, $z = +2$ would correspond to a raw score of 70, as compared to a raw score of 50 for $z = -2$.) Some people find negative z scores a bit confusing, probably because most measurements in real life (e.g., height, IQ) cannot be negative. It may also be hard to remember that a z score of zero is not bad; it is just average (i.e., if $z = 0$, the raw score $= \mu$). Formula 4.2 will come in handy for some of the procedures outlined in Section B. The structure of this formula also bears a strong resemblance to the formula for a confidence interval, for reasons that will be made clear when confidence intervals are defined in Chapter 6.

Sets of z Scores

It is interesting to see what happens when you take a group of raw scores (e.g., exam scores for a class) and convert all of them to z scores. To keep matters simple, we will work with a set of only four raw scores: 30, 40, 60, and 70. First, we need to find the mean and standard deviation for these numbers. The mean equals $(30 + 40 + 60 + 70)/4 = 200/4 = 50$. The standard deviation can be found by Formula 3.13B, after first calculating ΣX^2: $30^2 + 40^2 + 60^2 + 70^2 = 900 + 1600 + 3600 + 4900 = 11,000$. The standard deviation is found as follows:

$$\sigma = \sqrt{\frac{\Sigma X^2}{N} - \mu^2} = \sqrt{\frac{11,000}{4} - 50^2} = \sqrt{2750 - 2500} = \sqrt{250} = 15.81$$

Each raw score can now be transformed into a z score using Formula 4.1:

$$z = \frac{30 - 50}{15.81} = \frac{-20}{15.81} = -1.265$$

$$z = \frac{40 - 50}{15.81} = \frac{-10}{15.81} = -.6325$$

$$z = \frac{60 - 50}{15.81} = \frac{+10}{15.81} = +.6325$$

$$z = \frac{70 - 50}{15.81} = \frac{+20}{15.81} = +1.265$$

By looking at these four z scores, it is easy to see that they add up to zero, which tells us that the mean of the z scores will also be zero. This is not a coincidence. The mean for any complete set of z scores will be zero. This follows from Property 1 of the mean, as discussed in the previous chapter: If you subtract a constant from every score, the mean is decreased by the same constant. To form z scores, you subtract a constant (namely, μ) from all the scores before dividing. Therefore, this constant must also be subtracted from the mean. But because the constant being subtracted *is* the mean, the new mean is μ (the old mean) minus μ (the constant), or zero.

It is not obvious what the standard deviation of the four z scores will be, but it will be instructive to find out. We will use Formula 3.13B again, substituting z for X:

$$\sigma_z = \sqrt{\frac{\sum z^2}{N} - \mu_z^2}$$

The term that is subtracted is the mean of the z scores squared. But as you have just seen, the mean of the z scores is always zero, so this term drops out. Therefore, $\sigma_z = (\sum z^2/N)$. The term $\sum z^2$ equals $(-1.265)^2 + (-.6325)^2 + (.6325)^2 + (1.265)^2 = 1.60 + .40 + .40 + 1.60 = 4.0$. Therefore, σ equals $\sqrt{z^2/N} = \sqrt{(4/4)} = \sqrt{1} = 1$. As you have probably guessed, this is also no coincidence. The standard deviation for a complete set of z scores will always be 1. This follows from two of the properties of the standard deviation described in the last chapter. Property 1 implies that subtracting the mean (or any constant) from all the raw scores will not change the standard deviation. Then, according to Property 2, dividing all the scores by a constant will result in the standard deviation being divided by the same constant. The constant used for division when creating z scores *is* the standard deviation, so the new standard deviation is σ (the old standard deviation) divided by σ (the constant divisor), which always equals 1.

Properties of z Scores

I have just derived two important properties that apply to any set of z scores: (1) the mean will be zero, and (2) the standard deviation will be 1. Now we must consider an important limitation of z scores. I mentioned that z scores can be useful in comparing scores from two different distributions (in the example discussed earlier in the chapter, your friend performed relatively better in math than in physics). However, the comparison is reasonable only if the two distributions are similar in shape. Consider the distributions for classes D and E, shown in Figure 4.2. In the negatively skewed distribution of class D, a z score of $+2$ would put you very near the top of the distribution. In class E, however, the positive skewing implies that although there may not be a large percentage of scores above $z = +2$, there are some scores that are much higher.

Another property of z scores is relevant to the previous discussion. Converting a set of raw scores into z scores will not change the shape of the original distribution. For instance, if all the scores in class E were transformed into z scores, the distribution of z scores would have a mean of zero, a standard deviation of 1, and exactly the same positive skew as the original distribution. We can illustrate this property with a simple example,

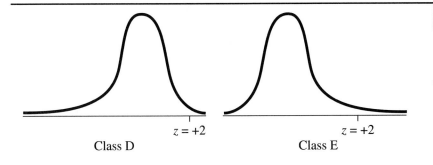

Figure 4.2

Comparing *z* Scores in
Differently Skewed
Distributions

again involving only four scores: 3, 4, 5, 100. The mean of these scores is
28, and the standard deviation is 41.58 (you should calculate these yourself
for practice). Therefore, the corresponding *z* scores (using Formula 4.1) are
−.6, −.58, −.55, and +1.73. First, note the resemblance between the distribution of these four *z* scores and the distribution of the four raw scores: In
both cases there are three numbers close together with a fourth number
much higher. You can also see that the *z* scores add up to zero, which
implies that their mean is also zero. Finally, you can calculate $\sigma = \sqrt{(\Sigma z^2/N)}$
to see that $\sigma = 1$.

SAT, *T*, and IQ Scores

For descriptive purposes, standardized scores that have a mean of zero and
a standard deviation of 1.0 may not be optimal. For one thing, about half
the *z* scores will be negative (even more than half if the distribution has a
positive skew), and minus signs can be cumbersome to deal with; leaving
off a minus sign by accident can lead to a gross error. For another thing,
most of the scores will be between 0 and 2, requiring two places to the right
of the decimal point to have a reasonable amount of accuracy. Like minus
signs, decimals can be cumbersome to deal with. For these reasons, it can
be more desirable to standardize scores so that the mean is 500 and the
standard deviation is 100. Because this scale is used by the Educational
Testing Service (Princeton, New Jersey) to report the results of the Scholastic Assessment Test, standardized scores with $\mu = 500$ and $\sigma = 100$ are
often called *SAT scores*. (The same scale is used to report the results of the
Graduate Record Examination and several other standardized tests.)

Probably the easiest way to convert a set of raw scores into SAT scores
is to first find the *z* scores with Formula 4.1 and then use Formula 4.3 to
transform each *z* score into an SAT score:

SAT = 100*z* + 500   **Formula 4.3**

Thus a *z* score of −3 will correspond to an SAT score of 100(−3) +
500 = −300 + 500 = 200. If *z* = +3, the SAT = 100(+3) + 500 = 300 +
500 = 800. (Notice how important it is to keep track of the sign of the
z score.) For any distribution of raw scores that is not extremely skewed,
nearly all of the *z* scores will fall between −3 and +3; this means (as shown
previously) that nearly all the SAT scores will be between 200 and 800.
There are so few scores that would lead to an SAT score below 200 or above
800 that generally these are the most extreme scores given; thus, we don't
have to deal with any negative SAT scores. (Moreover, from a psychological
point of view, it must feel better to score 500 or 400 on the SAT than to be
presented with a zero or negative *z* score.) Because *z* scores are rarely

expressed to more than two places beyond the decimal point, multiplying by 100 also ensures that the SAT scores will not require decimal points at all. Less familiar to students, but commonly employed for reporting the results of psychological tests, is the *T score*. The *T* score is very similar to the SAT score, as you can see from Formula 4.4:

$$T = 10z + 50$$ **Formula 4.4**

A full set of *T* scores will have a mean of 50 and a standard deviation of 10. If *z* scores are expressed to only one place past the decimal point, the corresponding *T* scores will not require decimal points.

The choice of which standardized score to use is usually a matter of convenience and tradition. The current convention regarding intelligence quotient (IQ) scores is to use a formula that creates a mean of 100. The Stanford-Binet test uses the formula $16z + 100$, resulting in a standard deviation of 16, whereas the Wechsler test uses $15z + 100$, resulting in a standard deviation of 15.

The Normal Distribution

It would be nice if all variables measured by psychologists had identically shaped distributions because then the *z* scores would always fall in the same relative locations, regardless of the variable under study. Although this is unfortunately not the case, it is useful that the distributions for many variables somewhat resemble one or another of the well-known mathematical distributions. Perhaps the best understood distribution with the most convenient mathematical properties is the normal distribution (mentioned in Chapter 2). Actually, you can think of the normal distribution as a family of distributions. There are two ways that members of this family can differ. Two normal distributions can differ either by having different means (e.g., heights of men and heights of women) and/or by having different standard deviations (e.g., heights of adults and IQs of adults). What all normal distributions have in common is the same shape—and not just any bell-like shape, but rather a very precise shape that follows an exact mathematical equation (see Section C).

Because all normal distributions have the same shape, a particular *z* score will fall in the same relative location on any normal distribution. Probably the most useful way to define relative location is to state what proportion of the distribution is above (i.e., to the right of) the *z* score and what proportion is below (to the left of) the *z* score. For instance, if $z = 0$, .5 of the distribution (i.e., 50%) will be above that *z* score and .5 will be below it. (Because of the symmetry of the normal distribution, the mean and the median fall at the same location, which is also the mode.) A statistician can find the proportions above and below any *z* score. In fact, these proportions have been found for all *z* scores expressed to two decimal places (e.g., 0.63, 2.17, etc.) up to some limit, beyond which the proportion on one side is too small to deal with easily. These proportions have been put into tables of the normal distribution, such as Table A.1 in Appendix A of this text.

The Standard Normal Distribution

Tables that give the proportion of the normal distribution below and/or above different *z* scores are called tables of the standard normal distribu-

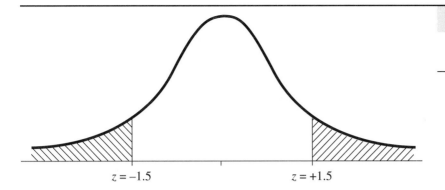

Figure 4.3

Areas Beyond
$z = \pm 1.5$

$z = -1.5$ $z = +1.5$

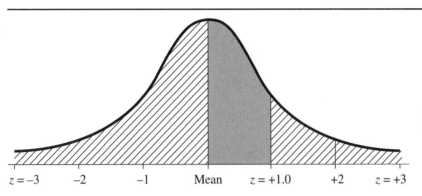

Figure 4.4

Proportion of the Normal
Distribution between the
Mean and $z = +1.0$

$z = -3$ -2 -1 Mean $z = +1.0$ $+2$ $z = +3$

tion; the standard normal distribution is just a normal distribution for
which $\mu = 0$ and $\sigma = 1$. It is the distribution you get when you transform all
of the scores from any normal distribution into z scores. Of course, you
could work out a table for any particular normal distribution. For example,
a table for a normal distribution with $\mu = 60$ and $\sigma = 20$ would show that
the proportion of scores above 60 is .5, and there would be entries for 61,
62, and so forth. However, it should be obvious that it would be impractical
to have a table for every possible normal distribution. Fortunately, it is easy
enough to convert scores to z scores (or vice versa) when necessary and use
a table of the standard normal distribution (see Table A.1 in the Appendix).
It is also unnecessary to include negative z scores in the table. Because of
the symmetry of the normal distribution, the proportion of scores above a
particular positive z score is the same as the proportion below the corresponding negative z score (e.g., the proportion above $z = +1.5$ equals the
proportion below $z = -1.5$; see Figure 4.3).

Suppose you want to know the proportion of the normal distribution
that falls between the mean and one standard deviation above the mean
(i.e., between $z = 0$ and $z = +1$). This portion of the distribution corresponds to the shaded area in Figure 4.4. Assume that all of the scores in the
distribution fall between $z = -3$ and $z = +3$. (The fraction of scores not
included in this region of the normal distribution is so tiny that you can
ignore it without fear of making a noticeable error.) Thus for the moment,
assume that the shaded area plus the cross-hatched areas of Figure 4.4 represent 100% of the distribution, or 1.0 in terms of proportions. The question about proportions can now be translated into areas of the normal

distribution. If you knew what proportion of the area of Figure 4.4 is shaded, you would know what proportion of the scores in the entire distribution were between $z = 0$ and $z = +1$. (Recall from Chapter 2 that the size of an "area under the curve" represents a proportion of the scores.) The shaded area looks like it is about one third of the entire distribution, so you can guess that in any normal distribution about one third of the scores will fall between the mean and $z = +1$.

Table of the Standard Normal Distribution

Fortunately, you do not have to guess about the relative size of the shaded area in Figure 4.4; Table A.1 can tell you the exact proportion. A small section of that table has been reproduced in Table 4.1. The column labeled "Mean to z" tells you what you need to know. First, go down the column labeled z until you get to 1.00. The column next to it contains the entry .3413, which is the proportion of the normal distribution enclosed between the mean and z when $z = 1.00$. This tells you that the shaded area in Figure 4.4 contains a bit more than one third of the scores in the distribution—it contains .3413. The proportion between the mean and $z = -1.00$ is the same: .3413. Thus about 68% (a little over two thirds) of any normal distribution is within one standard deviation on either side of the mean. Section B will show you how to use all three columns of Table A.1 to solve various practical problems.

Table 4.1		
z	Mean to z	Beyond z
.98	.3365	.1635
.99	.3389	.1611
1.00	.3413	.1587
1.01	.3438	.1562
1.02	.3461	.1539

Introducing Probability: Smooth Distributions versus Discrete Events

The main reason that finding areas for different portions of the normal distribution is so important to the psychological researcher is that these areas can be translated into statements about probability. Researchers are often interested in knowing the probability that a totally ineffective treatment can accidentally produce results as promising as the results they have just obtained in their own experiment. The next two chapters will show how the normal distribution can be used to answer such abstract questions about probability rather easily.

Before we can get to that point, it is important to lay out some of the basic rules of probability. These rules can be applied either to discrete events or to a smooth, mathematical distribution. An example of a discrete event is picking one card from a deck of 52 playing cards. Predicting which five cards might be in an ordinary poker hand is a more complex event, but it is composed of simple, discrete events (i.e., the probability of each card). Applying the rules of probability to discrete events is useful in figuring out the likely outcomes of games of chance (e.g., playing cards, rolling dice, etc.) and in dealing with certain types of nonparametric statistics. I will postpone any discussion of discrete probability until Part VII, in which nonparametric statistics are introduced. Until Part VII, I will be dealing with parametric statistics, which are based on measurements that lead to smooth distributions. Therefore, at this point, I will describe the rules of probability only as they apply to smooth, continuous distributions, such as the normal curve.

A good example of a smooth distribution that resembles the normal curve is the distribution of height for adult females in a large population. Finding probabilities involving a continuous variable like height is very different from dealing with discrete events like selecting playing cards. With a deck of cards there are only 52 distinct possibilities. On the other hand, how many different measurements for height are there? It depends on how precisely height is measured. With enough precision, everyone in the population can be determined to have a slightly different height from everyone else. With an infinite population (which is assumed when dealing with the true normal distribution), there are infinitely many different height measurements. Therefore, instead of trying to determine the probability of any particular height being selected from the population, it is only feasible to consider the probability associated with a range of heights (e.g., 60 to 68 inches or 61.5 to 62.5 inches).

Probability as Area under the Curve

In the context of a continuous variable measured on an infinite population, an "event" can be defined as selecting a value within a particular range of a distribution (e.g., picking an adult female whose height is between 60 and 68 inches). Having defined an event in this way, we can next define the probability that a particular event will occur if we select one adult female at random. The probability of some event can be defined as the proportion of times this event occurs out of an infinite number of random selections from the distribution. This proportion is equal to the area of the distribution under the curve that is enclosed by the range in which the event occurs. This brings us back to finding areas of the distribution. If you want to know the probability that the height of the next adult female selected at random will be between one standard deviation below the mean and one standard deviation above the mean, you must find the proportion of the normal distribution enclosed by $z = -1$ and $z = +1$. I have already pointed out that this proportion is equal to about .68, so the probability is .68 (roughly two chances out of three) that the height of the next woman selected at random will fall in this range. If you wanted to know whether the next randomly selected adult female would be between 60 and 68 inches tall, you would need to convert both of these heights to z scores so that you could use Table A.1 to find the enclosed area according to procedures described in Section B.

Real Distributions versus the Normal Distribution

It is important to remember that this text is dealing with methods of *applied* statistics. We are taking theorems and laws worked out by mathematicians for ideal cases and applying them to situations involving humans or animals or even abstract entities (e.g., hospitals, cities, etc.). To a mathematical statistician, a population has nothing to do with people; it is simply an infinite set of numbers that follow some distribution. Usually some numbers are more popular in the set than others, so the curve of the distribution is higher over those numbers than others (although you can have a uniform distribution, in which all of the numbers are equally popular). These distributions are determined by mathematical equations. On the other hand, the distribution that a psychologist is dealing with (or speculating about) is a set of numbers that is not infinite. The numbers would come from measuring each individual in some very large, but finite, population (e.g., adults in the United States) on some variable of interest (e.g., need for achievement, ability to recall words from a list). Thus, we can be sure that such a population of numbers will not follow some simple mathematical distribution exactly. This leads to some warnings about using Table A.1, which is based on a perfect, theoretical distribution: the normal distribution.

First, we can be sure that even if the human population were infinite, none of the variables studied by psychologists would produce a perfect normal distribution. I can state this with confidence because the normal distribution never ends. No matter what the mean and standard deviation, the normal distribution extends infinitely in both directions. On the other hand, the measurements psychologists deal with have limits. For instance, if a psychophysiologist is studying the resting heart rates of humans, the distribution will have a lowest and a highest value and therefore will differ from a true normal distribution. This means that the proportions found in Table A.1 will not apply exactly to the variables and populations in the problems of Section B. However, for many real-life distributions the deviations from Table A.1 tend to be small, and the approximation involved can be a very useful tool. More importantly, when you group scores together before finding the distribution, the distribution tends to look like the normal distribution, even if the distribution of individual scores does not. Because experiments are usually done with groups rather than individuals, the normal distribution plays a pervasive role in evaluating the results of experiments. This is the topic I will turn to next. But first, one more warning.

Because z scores are usually used only when a normal distribution can be assumed, some students get the false impression that converting to z scores somehow makes any distribution more like the normal distribution. In fact, as I pointed out earlier, converting to z scores does not change the shape of a distribution at all. Certain transformations *will* change the shape of a distribution (as described in Section C of Chapter 3) and in some cases will normalize the distribution, but converting to z scores is not one of them. (The z score is a linear transformation, and linear transformations don't change the shape of the distribution. These kinds of transformations will be discussed further in Chapter 9.)

z Scores as a Research Tool

You can use z scores to locate an individual within a normal distribution and to see how likely it is to encounter scores randomly in a particular range. However, of interest to psychological research is the fact that determining how unusual a score is can have more general implications. Suppose you know that heart rate at rest is approximately normally distributed, with a mean of 72 beats per minute (bpm) and a standard deviation of 10 bpm. You also know that a friend of yours, who drinks an unusual amount of coffee every day—five cups—has a resting heart rate of 95 bpm. Naturally, you suspect that the coffee is related to the high heart rate, but then you realize that some people in the ordinary population must have resting heart rates just as high as your friend's. Coffee isn't necessary as an explanation of the high heart rate because there is plenty of variability within the population based on genetic and other factors. Still, it may seem like quite a coincidence that your friend drinks so much coffee *and* has such a high heart rate. How much of a coincidence this really is depends in part on just how unusual your friend's heart rate is. If a fairly large proportion of the population has heart rates as high as your friend's, it would be reasonable to suppose that your friend was just one of the many with high heart rates that have nothing to do with coffee consumption. On the other hand, if a very small segment of the population has heart rates as high as your friend's, you must believe either that your friend happens to be one of those rare individuals who naturally have a high heart rate or that the coffee is elevating his heart rate. The more unusual your friend's heart rate, the harder it is to believe that the coffee is not to blame.

You can use your knowledge of the normal distribution to determine just how unusual your friend's heart rate is. Calculating your friend's z score (Formula 4.1), we find:

$$z = \frac{X - \mu}{\sigma} = \frac{95 - 72}{10} = \frac{23}{10} = 2.3$$

From Table A.1, the area beyond a z score of 2.3 is only about .011, so this is quite an unusual heart rate; only a little more than 1% of the population has heart rates that are as high or higher. The fact that your friend drinks a lot of coffee could be just a coincidence, but it also suggests that there may be a connection between drinking coffee and having a high heart rate (such a finding may not seem terribly shocking or interesting, but what if you found an unusual association between coffee drinking and some serious medical condition?).

The above example suggests an application for z scores in psychological research. However, a researcher would not be interested in finding out whether coffee has raised the heart rate of one particular individual. The more important question is whether coffee raises the heart rates of humans in general. One way to answer this question is to look at a random series of individuals who are heavy coffee drinkers and, in each case, find out how unusually high the heart rate is. Somehow all of these individual probabilities would have to be combined to decide whether these heart rates are just too unusual to believe that the coffee is uninvolved. There is a simpler way to attack the problem. Instead of focusing on one individual at a time, psychological researchers usually look at a group of subjects as a whole. This is certainly not the only way to conduct research, but because of its simplicity and widespread use, the group approach is the basis of statistics in introductory texts, including this one.

Sampling Distribution of the Mean

It is at this point in the text that I will begin to shift the focus from individuals to groups of individuals. Instead of the heart rate of an individual, we can talk about the heart rate of a group. To do so we have to find a single heart rate to characterize an entire group. Chapter 3 showed that the mean, median, and mode are all possible ways to describe the central tendency of a group, but the mean has the most convenient mathematical properties and leads to the simplest statistical procedures. Therefore, for most of this text, the mean will be used to characterize a group; that is, when I want to refer to a group by a single number, I will use the mean.

A researcher who wanted to explore the effects of coffee on resting heart rate would assemble a group of heavy coffee drinkers and find the mean of their heart rates. Then the researcher could see if the group mean was unusual or not. However, to evaluate how unusual a group mean is, you cannot compare the group mean to a distribution of individuals. You need, instead, a distribution of groups (all the same size). This is a more abstract concept than a population distribution that consists of individuals, but it is a critical concept for understanding the statistical procedures in the remainder of this text.

If we know that heart rate has a nearly normal distribution with a mean of 72 and a standard deviation of 10, what can we expect for the mean heart rate of a small group? There is a very concrete way to approach this question. First, you have to decide on the size of the groups you want to deal with—this makes quite a difference, as you will soon see. For our first example, let us say that we are interested in studying groups that have

25 participants each. So we take 25 people at random from the general population and find the mean heart rate for that group. Then we do this again and again, each time recording the mean heart rate. If we do this many times, the mean heart rates will start to pile up into a distribution. As we approach an infinite number of group means, the distribution becomes smooth and continuous. One convenient property of this distribution of means is that it will be a normal distribution, provided that the variable has a normal distribution for the individuals in the population.

Because the groups that we have been hypothetically gathering are supposed to be random samples of the population, the group means are called *sample means* and are symbolized by $\overline{X}$ (read as "X bar"). The distribution of group means is called a *sampling distribution*. More specifically, it is called the *sampling distribution of the mean*. (Had we been taking the median of each group of 25 and piling up these medians into a distribution, that would be called the sampling distribution of the *median*.) Just as the population distribution gives us a picture of how the individuals are spread out on a particular variable, the sampling distribution shows us how the sample means (or medians or whatever is being used to summarize each sample) would be spread out if we grouped the population into very many samples. To make things simple, I will assume for the moment that we are always dealing with variables that have a normal distribution in the population. Therefore, the sampling distribution of the mean will always be a normal distribution, which implies that we need only know its mean and standard deviation to know everything about it.

First, consider the mean of the sampling distribution of the mean. This term may sound confusing, but it really is very simple. The mean of all the group means will always be the same as the mean of the individuals (i.e., the population mean, μ). It should make sense that if you have very many random samples from a population, there is no reason for the sample means to be more often above or below the population mean. For instance, if you are looking at the average heights for groups of men, why should the average heights of the groups be any different from the average height of individual men? However, finding the standard deviation of the sampling distribution is a more complicated matter. Whereas the standard deviation of the individuals within each sample should be roughly the same as the standard deviation of the individuals within the population as a whole, the standard deviation of the sample means is a very different kind of thing.

Standard Error of the Mean

The means of samples do not vary as much as the individuals in the population. To make this concrete, consider again a very familiar variable: the height of adult men. It is obvious that if you were to pick a man off the street at random, it is somewhat unlikely that the man would be over 6 feet tall, but not very unlikely (in some countries, the chance would be better than .1). On the other hand, imagine selecting a group of 25 men *at random* and finding their average height. The probability that the 25 men would average over 6 feet in height is extremely small. Remember that the group was selected at random. It is not difficult to find 25 men whose average height is over 6 feet tall (you might start at the nearest basketball court), but if the selection is truly random, men below 5 feet 6 inches will be just as likely to be picked as men over 6 feet tall. The larger the group, the smaller the chance that the group mean will be far from the population mean (in this case, about 5 feet 9 inches). Imagine finding the average height of men in each of the 50 states of the United States. Could the average height of men in Wisconsin be much different from the average height of men in Pennsylvania or Alabama? Such extremely large groups will not vary much from each other or from the population mean. That sample means vary less from each other than do individuals is a critical concept for understanding the statistical procedures in most of this book. The concept is critical because we will be judging whether groups are unusual, and the fact that groups vary less than individuals do implies that it takes a smaller deviation for a group to be unusual than for an individual to be unusual. Fortunately, there is a simple formula that can be used to find out just how much groups tend to vary.

Because sample means do not vary as much as individuals, the standard deviation for the sampling distribution will be less than the standard deviation for a population. As the samples get larger, the sample means are clustered more closely, and the standard deviation of the sample means therefore gets smaller. This characteristic can be expressed by a simple formula, but first I will introduce a new term. The standard deviation of the sampling distribution of the mean is called the *standard error of the mean* and is symbolized as $\sigma_{\bar{X}}$. For any particular sampling distribution all of the samples must be the same size, symbolized by N (for the number of observations in each sample). How many different random samples do you have to select to make a sampling distribution? The question is irrelevant because nobody really creates a sampling distribution this way. The kinds of sampling distributions that I will be discussing are mathematical ideals based on drawing an infinite number of samples all of the same size. (This approach creates a sampling distribution that is analogous to the population distribution, which is based on an infinite number of individuals.)

Now I can show how the standard error of the mean decreases as the size of the samples increases. This relationship is expressed as Formula 4.5:

$$\sigma_{\bar{X}} = \frac{\sigma}{\sqrt{N}}$$

Formula 4.5

To find the standard error, you start with the standard deviation of the population and then divide by the square root of the sample size. This means that for any given sample size, the more the individuals vary, the more the groups will vary (i.e., if σ gets larger, $\sigma_{\bar{X}}$ gets larger). On the other hand, the larger the sample size, the *less* the sample means will vary (i.e., as

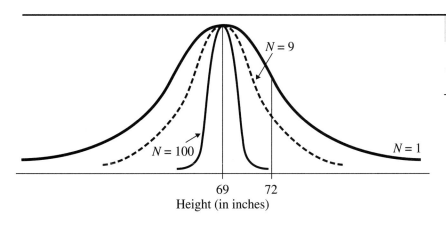

Figure 4.5

Sampling Distributions of the Mean for Different Sample Sizes

N increases, $\sigma_{\bar{X}}$ decreases). For example, if you make the sample size 4 times larger, the standard error is cut in half (e.g., σ is divided by 5 for a sample size of 25, but it is divided by 10 if the sample size is increased to 100).

Sampling Distribution versus Population Distribution

In Figure 4.5, you can see how the sampling distribution of the mean compares with the population distribution for a specific case. We begin with the population distribution for the heights of adult men. It is a nearly normal distribution with μ = 69 inches and σ = 3 inches. For *N* = 9, the sampling distribution of the mean is also approximately normal. We know this because there is a statistical law that states that if the population distribution is normal, the sampling distribution of the mean will also be normal. Moreover, there is a theorem that states that when the population distribution is not normal, the sampling distribution of the mean will be closer to the normal distribution than the population distribution (I'm referring to the Central Limit Theorem, which will be discussed further in the next two sections). So, if the population distribution is close to normal to begin with (as in the case of height for adults of the same gender), we can be sure that the sampling distribution of the mean for this variable will be very similar to the normal distribution.

Compared to the population distribution, the sampling distribution of the mean will have the same mean but a smaller standard deviation (i.e., standard error). The standard error for height when *N* is 9 is 1 inch ($\sigma_{\bar{X}} = \sigma/\sqrt{N} = 3/\sqrt{9} = 3/3 = 1$). For *N* = 100, the sampling distribution becomes even narrower; the standard error equals 0.3 inch. Notice that the means of groups tend to vary less from the population mean than do the individuals and that large groups vary less than small groups.

Referring to Figure 4.5, you can see that it is not very unusual to pick a man at random who is about 72 inches, or 6 feet, tall. This is just one standard deviation above the mean, so nearly one in six men is 6 feet tall or taller. On the other hand, to find a group of nine randomly selected men whose average height is over 6 feet is quite unusual; such a group would be three standard errors above the mean. This corresponds to a *z* score of 3, and the area beyond this *z* score is only about .0013. And to find a group of 100 randomly selected men who averaged 6 feet or more in height would be extremely rare indeed; the area beyond *z* = 10 is too small to appear in most standard tables of the normal distribution. Section B will illustrate the various uses of *z* scores when dealing with both individuals and groups.

SUMMARY

1. To localize a score within a distribution or compare scores from different distributions, *standardized scores* can be used. The most common standardized score is the *z* score. The *z* score expresses a raw score in terms of the mean and standard deviation of the distribution of raw scores. The magnitude of the *z* score tells you how many standard deviations away from the mean the raw score is, and the sign of the *z* score tells you whether the raw score is above (+) or below (−) the mean.

2. If you take a set of raw scores and convert each one to a *z* score, the mean of the *z* scores will be zero and the standard deviation will be 1. The shape of the distribution of *z* scores, however, will be exactly the same as the shape of the distribution of raw scores.

3. z scores can be converted to *SAT scores* by multiplying by 100 and then adding 500. SAT scores have the advantages of not requiring minus signs or decimals to be sufficiently accurate. *T scores* are similar but involve multiplication by 10 and the addition of 50.

4. The *normal distribution* is a symmetrical, bell-shaped mathematical distribution whose shape is precisely determined by an equation (given in Section C). The normal distribution is actually a family of distributions, the members of which differ according to their means and/or standard deviations.

5. If all the scores in a normal distribution are converted to z scores, the resulting distribution of z scores is called the *standard normal distribution*, which is a normal distribution that has a mean of zero and a standard deviation of 1.

6. The proportion of the scores in a normal distribution that falls between a particular z score and the mean is equal to the amount of area under the curve between the mean and z, divided by the total area of the distribution (defined as 1.0). This proportion is the probability that one random selection from the normal distribution will have a value between the mean and that z score. The areas between the mean and z and the areas beyond z (into the tail of the distribution) are given in Table A.1.

7. Distributions based on real variables measured in populations of real subjects (whether people or not) can be similar to, but not exactly the same as, the normal distribution. This is because the true normal distribution extends infinitely in both the negative and positive directions.

8. Just as it is sometimes useful to determine if an individual is unusual with respect to a population, it can also be useful to determine how unusual a group is compared to other groups that could be randomly selected. The group mean (more often called the *sample mean*) is usually used to summarize the group with a single number. To find out how unusual a sample is, the sample mean ($\overline{X}$) must be compared to a distribution of sample means, called, appropriately, the *sampling distribution of the mean*.

9. This sampling distribution could be found by taking very many samples from a population and gathering the sample means into a distribution, but there are statistical laws that tell you just what the sampling distribution of the mean will look like if certain conditions are met. If the population distribution is normal, and the samples are *independent random samples*, all of the same size, the sampling distribution of the mean will be a normal distribution with a mean of μ (the same mean as the population) and a standard deviation called the *standard error of the mean*.

10. The larger the sample size, N, the smaller the standard error of the mean, which is equal to the population standard deviation divided by the square root of N.

EXERCISES

*1. Assume that the mean height for adult women (μ) is 65 inches, and that the standard deviation (σ) is 3 inches.

a. What is the z score for a woman who is exactly 5 feet tall? Who is 5 feet 5 inches tall?

b. What is the z score for a woman who is 70 inches tall? Who is 75 inches tall? Who is 64 inches tall?

c. How tall is a woman whose z score for height is -3? -1.33? -0.3? -2.1?

d. How tall is a woman whose z score for height is $+3$? $+2.33$? $+1.7$? $+.9$?

2. a. Calculate μ and σ for the following set of scores and then convert each score to a z score: 64, 45, 58, 51, 53, 60, 52, 49.
   b. Calculate the mean and standard deviation of these z scores. Did you obtain the values you expected? Explain.
*3. What is the SAT score corresponding to
   a. $z = -0.2$?
   b. $z = +1.3$?
   c. $z = -3.1$?
   d. $z = +1.9$?
4. What is the z score that corresponds to an SAT score of
   a. 520?
   b. 680?
   c. 250?
   d. 410?
*5. Suppose that the verbal part of the SAT contains 30 questions and that $\mu = 18$ correct responses, with $\sigma = 3$. What SAT score corresponds to
   a. 15 correct?
   b. 10 correct?
   c. 20 correct?
   d. 27 correct?
6. Suppose the mean for a psychological test is 24 with $\sigma = 6$. What is the T score that corresponds to a raw score of
   a. 0?
   b. 14?
   c. 24?
   d. 35?
*7. Use Table A.1 to find the area of the normal distribution between the mean and z, when z equals
   a. .18
   b. .50
   c. .88
   d. 1.25
   e. 2.11

8. Use Table A.1 to find the area of the normal distribution beyond z, when z equals
   a. .09
   b. .75
   c. 1.05
   d. 1.96
   e. 2.57
9. Assuming that IQ is normally distributed with a mean of 100 and a standard deviation of 15, describe completely the sampling distribution of the mean for a sample size (N) equal to 20.
*10. If the population standard deviation (σ) for some variable equals 17.5, what is the value of the standard error of the mean when
   a. $N = 5$?
   b. $N = 25$?
   c. $N = 125$?
   d. $N = 625$?
   If the sample size is cut in half, what happens to the standard error of the mean for a particular variable?
11. a. In one college, freshman English classes always contain exactly 20 students. An English teacher wonders how much these classes are likely to vary in terms of their verbal scores on the SAT. What would you expect for the standard deviation (i.e., standard error) of class means on the verbal SAT?
   b. Suppose that a crew for the space shuttle consists of seven people, and we are interested in the average weights of all possible shuttle crews. If the standard deviation for weight is 30 pounds, what is the standard deviation for the mean weights of shuttle crews (i.e., the standard error of the mean)?
*12. If for a particular sampling distribution of the mean we know that the standard error is 4.6, and we also know that $\sigma = 32.2$, what is the sample size (N)?

$\mathcal{B}$

BASIC STATISTICAL PROCEDURES

As you have seen, z scores can be used for descriptive purposes to locate a score in a distribution. Later in this section, I will show that z scores can also be used to describe groups, although when we are dealing with groups, we usually have some purpose in mind beyond pure description. For now, I want to expand on the descriptive power of z scores when dealing with a population of individuals and some variable that follows the normal distribution in that population. As mentioned in Chapter 2, one of the most informative ways of locating a score in a distribution is by finding the percentile rank (PR) of that score (i.e., the percentage of the distribution that is below that score). To find the PR of a score within a small set of scores, the techniques described in Chapter 2 are appropriate. However, if you want to find the PR of a score with respect to a very large group of scores whose distribution resembles the normal distribution (and you know both the mean and standard deviation of this reference group), you can use the following procedure.

Finding Percentile Ranks

I'll begin with the procedure for finding the PR of a score that is above the mean of a normal distribution. The variable we will use for the examples in this section is the IQ of adults, which has a fairly normal distribution and is usually expressed as a standardized score with $\mu = 100$ and (for the Stanford-Binet test) $\sigma = 16$. To use Table A.1, however, IQ scores will have to be converted back to z scores. I will illustrate this procedure by finding the PR for an IQ score of 116. First find z using Formula 4.1:

$$z = \frac{116 - 100}{16} = \frac{16}{16} = +1.0$$

Next, draw a picture of the normal distribution, always placing a vertical line at the mean ($z = 0$), and at the z score in question ($z = +1$, for this example). The area of interest, as shown by the crosshatching in Figure 4.6, is the portion of the normal distribution to the left of $z = +1.0$. The entire crosshatched area does not appear as an entry in Table A.1 (although some standard normal tables include a column that would correspond to the shaded area). Notice that the crosshatched area is divided in two portions by the mean of the distribution. The area to the left of the mean is always half of the normal distribution and therefore corresponds to a proportion of .5. The area between the mean and $z = +1.0$ can be found in Table A.1 (under "Mean to z"), as demonstrated in Section A. This proportion is .3413. Adding .5 to .3413, we get .8413, which is the proportion represented by the crosshatched area in Figure 4.6. To convert a proportion to a percentage, we need only multiply by 100. Thus the proportion .8413 corresponds to a PR of 84.13%. Now we know that 84.13% of the population have IQ scores lower than 116. I emphasize the importance of drawing a picture of the normal distribution to solve these problems. In the problem above, it would have been easy to forget the .5 area to the left of the mean without a picture to refer to.

It is even easier to find the PR of a score below the mean if you use the correct column of Table A.1. Suppose you want to find the PR for an IQ of 84. Begin by finding z:

$$z = \frac{84 - 100}{16} = \frac{-16}{16} = -1.0$$

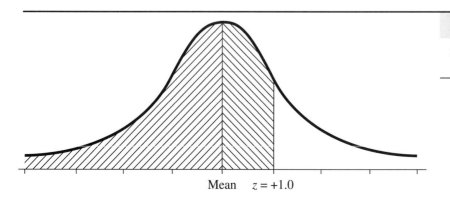

Figure 4.6

Percentile Rank: Area below $z = +1.0$

Mean $z = +1.0$

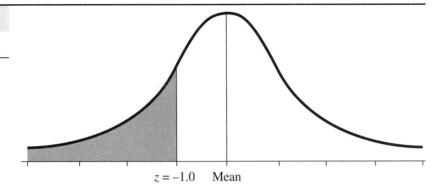

Figure 4.7

Area beyond $z = -1.0$

$z = -1.0$ Mean

Next, draw a picture and shade the area to the left of $z = -1.0$ (see Figure 4.7). Unlike the previous problem, the shaded area this time consists of only one section, which *does* correspond to an entry in Table A.1. First, you must temporarily ignore the minus sign of the z score and find 1.00 in the first column of Table A.1. Then look at the corresponding entry in the column labeled "Beyond z," which is .1587. This is the proportion represented by the shaded area in Figure 4.7 (i.e., the area to the left of $z = -1.0$). The PR of 84 = .1587 × 100 = 15.87%; only about 16% of the population have IQ scores less than 84.

The area referred to as Beyond z (in the third column of Table A.1) is the area that begins at z and extends *away* from the mean in the direction of the closest tail. In Figure 4.7, the area between the mean and $z = -1.0$ is .3413 (the same as between the mean and $z = +1.0$), and the area beyond $z = -1$ is .1587. Notice that these two areas add up to .5000. In fact, for any particular z score, the entries for Mean to z and Beyond z will add up to .5000. You can see why by looking at Figure 4.7. The z score divides one half of the distribution into two sections; together those two sections add up to half the distribution, which equals .5.

Finding the Area between Two z Scores

Now we are ready to tackle more complex problems involving two different z scores. I'll start with two z scores on opposite sides of the mean (i.e., one z is positive and the other is negative). Suppose you have devised a teaching technique that is not accessible to someone with an IQ below 76 and would be too boring for someone with an IQ over 132. To find the proportion of the population for whom your technique would be appropriate, you must first find the two z scores and locate them in a drawing.

$$z = \frac{76 - 100}{16} = \frac{-24}{16} = -1.5$$

$$z = \frac{132 - 100}{16} = \frac{32}{16} = +2.0$$

From Figure 4.8 you can see that you must find two areas of the normal distribution, both of which can be found under the column "Mean to z." For $z = -1.5$ you ignore the minus sign and find that the area from the mean to z is .4332. The corresponding area for $z = +2.0$ is .4772. Adding these two areas together gives a total proportion of .9104. Your teaching technique would be appropriate for 91.04% of the population.

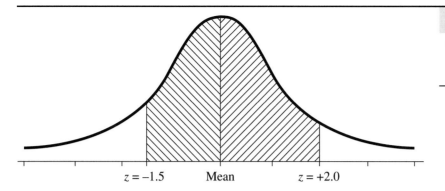

Figure 4.8

The Area between Two
z Scores on Opposite
Sides of the Mean

$z = -1.5$ Mean $z = +2.0$

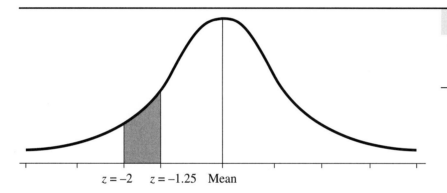

Figure 4.9

The Area between Two
z Scores on the Same
Side of the Mean

$z = -2$ $z = -1.25$ Mean

Finding the area enclosed between two z scores becomes a bit trickier when both of the z scores are on the same side of the mean (i.e., both are positive or both are negative). Suppose that you have designed a remedial teaching program that is only appropriate for those whose IQs are below 80 but would be useless for someone with an IQ below 68. As in the problem above, you can find the proportion of people for whom your remedial program is appropriate by first finding the two z scores and locating them in your drawing.

$$z = \frac{80 - 100}{16} = \frac{-20}{16} = -1.25$$

$$z = \frac{68 - 100}{16} = \frac{-32}{16} = -2.0$$

The shaded area in Figure 4.9 is the proportion you are looking for, but it does not correspond to any entry in Table A.1. The trick is to notice that if you take the area from $z = -2$ to the mean and remove the section from $z = -1.25$ to the mean, you are left with the shaded area. (You could also find the area beyond $z = -1.25$ and then remove the area beyond $z = -2.0$.) The area between $z = 2$ and the mean was found in the previous problem to be .4772. From this we subtract the area between $z = 1.25$ and the mean, which is .3944. The proportion we want is .4772 − .3944 = .0828. Thus the remedial teaching program is suitable for use with 8.28% of the population. Note that you cannot subtract the two z scores and then find an area corresponding to the difference of the two z scores; z scores just don't work that

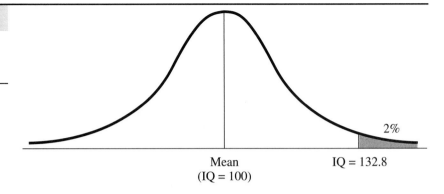

Figure 4.10

Score Cutting Off the
Top 2% of the Normal
Distribution

way (e.g., the area between z scores of 1 and 2 is much larger than the area between z scores of 2 and 3).

Finding the Raw Scores Corresponding to a Given Area

Often a problem involving the normal distribution will be presented in terms of a given proportion, and it is necessary to find the range of raw scores that represents that proportion. For instance, a national organization called MENSA is a club for people with high IQs. Only people in the top 2% of the IQ distribution are allowed to join. If you were interested in joining and you knew your own IQ, you would want to know the minimum IQ score required for membership. Using the IQ distribution from the problems above, this is an easy question to answer (even if you are not qualified for MENSA). However, because you are starting with an area and trying to find a raw score, the procedure is reversed. You begin by drawing a picture of the distribution and shading in the area of interest, as in Figure 4.10. (Notice that the score that cuts off the upper 2% is also the score that lands at the 98th percentile; that is, you are looking for the score whose PR is 98.) Given a particular area (2% corresponds to a proportion of .0200), you cannot find the corresponding IQ score directly, but you can find the z score using Table A.1. Instead of looking down the z column, you look for the area of interest (in this case, .0200) first and then see which z score corresponds to it. From Figure 4.10, it should be clear that the shaded area is the *area beyond* some as yet unknown z score, so you look in the "Beyond z" column for .0200. You will not be able to find this exact entry, as is often the case, so look at the closest entry, which is .0202. The z score corresponding to this entry is 2.05, so $z = +2.05$ is the z score that cuts off (about) the top 2% of the distribution. To find the raw score that corresponds to $z = +2.05$, you can use Formula 4.2:

$$X = z\sigma + \mu = +2.05(16) + 100 = 32.8 + 100 = 132.8$$

Rounding off, you get an IQ of 133—so if your IQ is 133 or above, you are eligible to join MENSA.

Areas in the Middle of a Distribution

One of the most important types of problems involving normal distributions is to locate a given proportion in the middle of a distribution. Imagine an organization called MEZZA, which is designed for people in the middle range of intelligence. In particular, this organization will only accept those

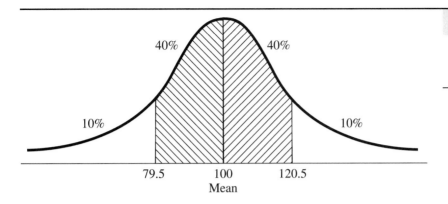

Figure 4.11

Scores Enclosing the
Middle 80% of the
Normal Distribution

in the middle 80% of the distribution—those in the upper or lower 10% are
not eligible. What is the range of IQ scores within which your IQ must fall if
you are to be eligible to join MEZZA? The appropriate drawing is shown
in Figure 4.11. From the drawing you can see that you must look for .1000
in the column labeled "Beyond z". The closest entry is .1003, which corresponds to $z = 1.28$. Therefore, $z = +1.28$ cuts off (about) the upper 10%,
and $z = -1.28$ the lower 10% of the distribution. Finally, both of these z
scores must be transformed into raw scores, using Formula 4.2:

$$X = -1.28(16) + 100 = -20.48 + 100 = 79.52$$
$$X = +1.28(16) + 100 = +20.48 + 100 = 120.48$$

Thus (rounding off) the range of IQ scores that contain the middle 80% of
the distribution extends from 80 to 120.

From Score to Proportion and Proportion to Score

The above procedures relate raw scores to areas under the curve, and vice
versa, by using z scores as the intermediate step, as follows:

Raw score $\leftrightarrow$ (Formula) $\leftrightarrow$ z score $\leftrightarrow$ (Table A.1) $\leftrightarrow$ Area

When you are given a raw score to start with, and you are looking for a proportion or percentage, you move from left to right in the preceding diagram. A raw score can be converted to a z score using Formula 4.1. Then an
area (or proportion) can be associated with that z score by looking down
the appropriate column of Table A.1. Drawing a picture will make it clear
which column is needed. When given a proportion or percentage to start
with, you move from right to left. First, use Table A.1 backwards (look up
the area in the appropriate column to find the z score), and then use Formula 4.2 to transform the z score into a corresponding raw score.

Describing Groups

You can use z scores to find the location of one group with respect to all
other groups of the same size, but to do that you must work with the sampling distribution of the mean. The z score has to be modified only slightly
for this purpose, as you will see. As an example of a situation in which you
may want to compare groups, imagine the following. Suppose there is a

university that encourages women's basketball by assigning all of the female students to one basketball team or another at random. Assume that each team has exactly the required five players. Imagine that a particular woman wants to know the probability that the team she is assigned to will have an average height of 67 inches or more. I will show how the sampling distribution of the mean can be used to answer that question.

We begin by assuming that the heights of women at this large university form a normal distribution with a mean of 65 inches and a standard deviation of 3 inches. Next, we need to know what the distribution would look like if it were composed of the means from an infinite number of basketball teams, each with five players. In other words, we need to find the sampling distribution of the mean for $N = 5$. First, we can say that the sampling distribution will be a normal one because we are assuming that the population distribution is (nearly) normal. Given that the sampling distribution is normal, we need only specify its mean and standard deviation.

The z Score for Groups

The mean of the sampling distribution is the same as the mean of the population, that is, μ. For this example, $\mu = 65$ inches. The standard deviation of the sampling distribution of the mean, called the standard error of the mean, is given by Formula 4.5. For this example the standard error, $\sigma_{\bar{X}}$, equals:

$$\sigma_{\bar{X}} = \frac{\sigma}{\sqrt{N}} = \frac{3}{\sqrt{5}} = \frac{3}{2.24} = 1.34$$

Now that we know the parameters of the distribution of the means of groups of five (e.g., the basketball teams), we are prepared to answer questions about any particular group, such as the team that includes the inquisitive woman in our example. Because the sampling distribution is normal, we can use the standard normal table to determine, for example, the probability of a particular team having an average height greater than 67 inches. However, we first have to convert the particular group mean of interest to a z score—in particular, a z score with respect to the sampling distribution of the mean, or more informally, a z score for groups. The z score for groups closely resembles the z score for individuals and is given by Formula 4.6:

$$z = \frac{\bar{X} - \mu}{\sigma_{\bar{X}}}$$ **Formula 4.6**

in which $\sigma_{\bar{X}}$ is a value found using Formula 4.5.

To show the parallel structures of the z score for individuals and the z score for groups, Formula 4.1 for the individual z score follows:

$$z = \frac{X - \mu}{\sigma}$$

Comparing Formula 4.1 with Formula 4.6, you can see that in both cases we start with the score of a particular individual (or mean of a particular sample), subtract the mean of those scores (or sample means), and then divide by the standard deviation of those scores (or sample means). Note that we could put a subscript X on the σ in Formula 4.1 to make it clear that that formula is dealing with individual scores, but unless we are dealing with more than one variable at a time (as in Chapter 9), it is common to leave off the subscript for the sake of simplicity.

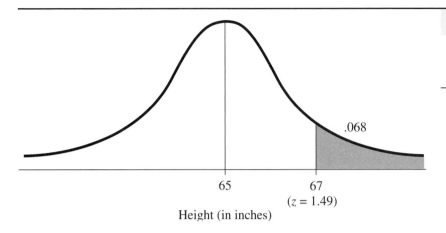

Figure 4.12

Area of the Sampling
Distribution above a
z Score for Groups

In the present example, if we want to find the probability that a randomly selected basketball team will have an average height over 67 inches, it is necessary to convert 67 inches to a z score for groups, as follows:

$$z = \frac{67 - 65}{1.34} = 1.49$$

The final step is to find the area beyond $z = 1.49$ in Table A.1; this area is approximately .068. As Figure 4.12 shows, most of the basketball teams have mean heights that are less than 67 inches; an area of .068 corresponds to fewer than 7 chances out of 100 (or about 1 out of 15) that the woman in our example will be on a team whose average height is at least 67 inches.

Using the z score for groups, you can answer a variety of questions about how common or unusual a particular group is. For the present example, because the standard error is 1.34 and the mean is 65, we know immediately that a little more than two thirds of the basketball teams will average between 63.66 inches (i.e., $65 - 1.34$) and 66.34 inches (i.e., $65 + 1.34$) in height. Teams with average heights in this range would be considered fairly common, whereas teams averaging more than 67 inches or less than 63 inches in height would be relatively uncommon.

The most common application for determining the probability of selecting a random sample whose mean is unusually small or large is to test a research hypothesis, as you will see in the next chapter. For instance, you could gather a group of heavy coffee drinkers, find the average heart rate for that group, and use the preceding procedures to determine how unusual it would be to find a random group (the same size) with an average heart rate just as high. The more unusual the heart rate of the coffee-drinking group turns out to be, the more inclined you would be to suspect a link between coffee consumption and heart rate. (Remember, however, that the observation that heavy coffee drinkers do indeed have higher heart rates does not imply that drinking coffee *causes* an increase in heart rate; there are alternative explanations that can only be ruled out by a *true* experiment, in which the experimenter decides at random who will be drinking coffee and who will not.) However, the p value that we look up for a particular z score is not very accurate unless we can make a few important assumptions about how the sample was selected, and about the population it was selected from.

Assumptions

The use of the sampling distribution of the mean in the example of the basketball teams was based on assuming that the sampling distribution would be a normal distribution, and that its standard deviation would be given by Formula 4.5. These assumptions are based in turn on the following conditions.

The Variable Follows a Normal Distribution in the Population
Strictly speaking, the variable must be normally distributed in the population before you can be certain that the sampling distribution will also be normal. However, as discussed in Section A, the variables that psychological researchers measure do not follow perfectly normal distributions. In fact, many variables that are important to psychology, such as reaction time, often form distributions that are quite skewed or are otherwise considerably different from the normal distribution. Fortunately there is a statistical law, known as the *Central Limit Theorem*, that states that for nonnormal population distributions, the sampling distribution of the mean will be closer to the normal distribution than the original population distribution. Moreover, as the samples get larger, the sampling distribution more closely approaches the shape of the normal distribution.

If the variable is close to being normally distributed in the population, the sampling distribution can be assumed to be normal, with a negligible amount of error, even for rather small samples. On the other hand, if the population distribution is extremely skewed or is very different from the normal distribution in some other way, the sampling distribution won't be very close to normal unless the samples are fairly large. It is, therefore, important *not* to assume that the sampling distribution of the mean is normal, if you know that the population distribution is far from normal *and* you are using small samples. Fortunately, sample sizes of 30 or 40 tend to be adequate even if the population distribution is not very close to being normal. The Central Limit Theorem and its implications for inferential statistics will be discussed in greater detail in Section C.

The Samples are Selected Randomly Many statistical laws, such as the Central Limit Theorem, that lead to the kinds of simple statistical procedures described in this text assume that groups are being created by *simple random sampling*. The easiest way to define this method of sampling is to stipulate that if the sample size is equal to N, every possible sample of size N that could be formed from the population must have the same probability of being selected. It is important to understand that this rule goes beyond stating that every individual in the population must have an equal chance of being selected. In addition, simple random sampling also demands that each selection must be independent of all the others.

For instance, imagine that you are collecting a sample from your neighborhood, but you refuse to include both members of a married couple; if one member of a couple has already been selected, and that person's spouse comes up at random as a later selection, the spouse is excluded and another selection is drawn. Adopting such a rule means that your groups will not be simple random samples. Even though everyone in the population has an equal chance at the outset, the probabilities change as you go along. After a particular married individual has been selected, the spouse's probability of being selected drops to zero. The selections are *not* independent of each other.

To state the above condition more formally: For all samples of the same size to have the same chance of being selected, two conditions must be satisfied. (1) All individuals in the population must have the same probability of being selected, and (2) each selection must be independent of all others. Because of the second condition, this type of sampling is sometimes called *independent random sampling*. Stipulating that both members of a married couple cannot be included is only one way to violate the independence rule. Another is to deliberately include friends and/or family members of someone who had been randomly selected.

Technically, the sampling I am describing is assumed to occur *with replacement*, which means that after a subject is selected, his or her name is put back in the population so that the same subject could be selected more than once for the same group. Of course, in experiments involving real people sampling is always done *without* replacement. Fortunately, when dealing with large populations, the chance of randomly selecting the same subject twice for the same group is so tiny that the error introduced by not sampling with replacement is negligible. There are statistical laws that apply to sampling from relatively small populations, which lead to modifications of the statistical formulas in this text (see Hays, 1994). However, even though psychologists almost never draw random samples from the general population of interest, and typically use "samples of convenience" instead, they almost always use statistical procedures based on the random sampling of infinite populations without modification. In Chapter 7 I will explain why this possibility for statistical inaccuracy leads to little concern among psychological researchers.

B

SUMMARY

1. If a variable is normally distributed and you know both the mean and standard deviation of the population, it is easy to find the proportion of the distribution that falls above or below any raw score or between any two raw scores. Conversely, for a given proportion at the top, bottom, or middle of the distribution, you can find the raw score or scores that form the boundary of that proportion.

2. It is easier to work with Table A.1 if you begin by drawing a picture of the normal distribution and then draw a vertical line in the middle and one that corresponds roughly to the z score or area with which you are working.

3. To find the proportion below a given raw score first convert the raw score to a z score with Formula 4.1. If the z score is negative, the area below (i.e., to the left of) the z score is the area in the "Beyond z" column. If the z score is positive, the area below the z score is the area in the "Mean to z" column *plus* .5. Multiply the proportion by 100 to get the percentile rank for that score. To find the proportion above a particular raw score, you can alter the procedure just described appropriately or find the proportion below the score and subtract from 1.0.

4. To find the proportion between two raw scores,
   a. If the corresponding z scores are opposite in sign, you must find two areas—the area between the mean and z for each of the two z scores—and add them.
   b. If both z scores are on the same side of the mean, you can find the area between the mean and z for each and then *subtract* the smaller from the larger area (alternatively, you can find the area beyond z for both and subtract). Reminder: You cannot subtract the two z scores first and then look for the area.

5. To find the raw score corresponding to a given proportion, first draw the picture and shade in the appropriate area. If the area is the top $X\%$, convert to a proportion by dividing by 100. Then find that proportion (or the closest one to it) in the "Beyond z" column. The corresponding z score can then be converted to a raw score using Formula 4.2. If the area is the bottom $X\%$ (i.e., left side of distribution), the procedure is the same, but when you are using Formula 4.2 don't forget that the z score has a negative sign.

6. To find the raw scores enclosing the middle $X\%$ of the distribution, first subtract $X\%$ from 100% and then divide by 2 to find the percentage in each tail of the distribution. Convert to a proportion by dividing by 100 and then look for this proportion in the "Beyond z" column of Table A.1 (e.g., if you are trying to locate the middle 90%, you would be looking for a proportion of .05). Transform the corresponding z score to a raw score to find the upper (i.e., right) boundary of the enclosed area. Then, put a minus sign in front of the same z score and transform it to a raw score again to find the lower boundary.

7. If you are working with the mean of a sample instead of an individual score, and you want to know the proportion of random samples of the same size that would have a larger (or smaller) mean, you have to convert your sample mean to a z score for groups. Subtract the population mean from your sample mean, and then divide by the standard error of the mean, which is the population standard deviation divided by the square root of the sample size. The area above or below this z score can be found from Table A.1 just as though the z score were from an individual score in a normal distribution.

8. To use Table A.1 for a z score from a sample mean, you must assume that the sampling distribution of the mean is a normal distribution. This is a reasonable assumption under the following conditions. It is assumed that all of the samples are the same size, N, and that an infinite number of samples are drawn. We also assume that the samples are obtained by a process known as *simple*, or *independent*, random sampling, in which each individual in the population has the same chance of being selected *and* each selection is *independent* of all others. Furthermore, we assume that the population distribution is normal. However, if the sample size is large, this last assumption is not critical. Unless the population distribution is virtually the opposite of the normal distribution, a sample size of at least about 30 guarantees that the error involved in assuming the sampling distribution to be normal will be quite small.

EXERCISES

*1. Suppose that a large Introduction to Psychology class has taken a midterm exam, and the scores are normally distributed (approximately) with $\mu = 75$ and $\sigma = 9$. What is the percentile rank (PR) for a student
   a. Who scores 90?
   b. Who scores 70?
   c. Who scores 60?
   d. Who scores 94?

2. Find the area between

   a. $z = -0.5$ and $z = +1.0$
   b. $z = -1.5$ and $z = +0.75$
   c. $z = +0.75$ and $z = +1.5$
   d. $z = -0.5$ and $z = -1.5$

*3. Assume that the resting heart rate in humans is normally distributed with $\mu = 72$ bpm (i.e., beats per minute) and $\sigma = 8$ bpm.
   a. What proportion of the population has resting heart rates above 82 bpm? Above 70 bpm?

b. What proportion of the population has resting heart rates below 75 bpm? Below 50 bpm?

c. What proportion of the population has resting heart rates between 80 and 85 bpm? Between 60 and 70 bpm? Between 55 and 75 bpm?

4. Refer again to the population of heart rates described in Exercise 3:

a. Above what heart rate do you find the upper 25% of the people? (That is, what heart rate is at the 75th percentile, or third quartile?)

b. Below what heart rate do you find the lowest 15% of the people? (That is, what heart rate is at the 15th percentile?)

c. Between which two heart rates do you find the middle 75% of the people?

*5. A new preparation course for the math SAT is open to those who have already taken the test once and scored in the middle 90% of the population. In what range must a test-taker's previous score have fallen for the test-taker to be eligible for the new course?

6. A teacher thinks her class has an unusually high IQ, because her 36 students have an average IQ ($\overline{X}$) of 108. If the population mean is 100, and $\sigma = 15$,

a. What is the z score for this class?

b. What percentage of classes ($N = 36$, randomly selected) would be even higher on IQ?

*7. An aerobics instructor thinks that his class has an unusually low resting heart rate. If $\mu = 72$ bpm and $\sigma = 8$ bpm, and his class of 14 pupils has a mean heart rate ($\overline{X}$) of 66,

a. What is the z score for the aerobics class?

b. What is the probability of randomly selecting a group of 14 people with a mean resting heart rate lower than the mean for the aerobics class?

8. Imagine that a test for spatial ability produces scores that are normally distributed in the population with $\mu = 60$ and $\sigma = 20$.

a. Between which two scores will you find the middle 80% of the people?

b. Considering the means of groups, all of which have 25 participants, between what two scores will the middle 80% of these means be?

*9. Suppose that the average person sleeps 8 hours each night and that $\sigma = .7$ hour.

a. If a group of 50 joggers is found to sleep an average of 7.6 hours per night, what is the z score for this group?

b. If a group of 200 joggers also has a mean of 7.6, what is the z score for this larger group?

c. Comparing your answers to parts a and b, can you determine the mathematical relation between sample size and z (when $\overline{X}$ remains constant)?

10. Referring to the information in Exercise 7, if an aerobics class had a mean heart rate ($\overline{X}$) of 62, and this resulted in a group z score of −7.1, how large must the class have been?

*11. Suppose that the mean height for a group of 40 women who had been breastfed for at least the first 6 months of life was 66.8 inches.

a. If $\mu = 65.5$ inches and $\sigma = 2.6$ inches, what is the z score for this group?

b. If height had been measured in centimeters, what would the z score be? (*Hint:* Multiply $\overline{X}$, μ, and σ by 2.54 to convert inches to centimeters.)

c. Comparing your answers to parts a and b, what can you say about the effect on z scores of changing units? Can you explain the significance of this principle?

12. Suppose that the mean weight of adults (μ) is 150 pounds with $\sigma = 30$ pounds. Consider the mean weights of all possible space shuttle crews ($N = 7$). If the space shuttle cannot carry a crew that weighs more than a total of 1190 pounds, what is the probability that a randomly selected crew will be too heavy? (Assume that the sampling distribution of the mean would be approximately normal.)

The Mathematics of the Normal Distribution

The true normal curve is determined by a mathematical equation, just as a straight line or a perfect circle is determined by an equation. The equation for the normal curve is a mathematical function into which you can insert any *X* value (usually represented on the horizontal axis) to find one corre-

**OPTIONAL
MATERIAL**

sponding Y value (usually plotted along the vertical axis). Because Y, the height of the curve, is a function of X, Y can be symbolized as $f(X)$. The equation for the normal curve can be stated as follows:

$$f(x) = \frac{1}{\sqrt{2\pi\sigma^2}}\, e^{-(X-\mu)^2/2\sigma^2}$$

where π is a familiar mathematical constant, and so is e ($e = 2.7183$, approximately). The symbols μ and σ^2 are called the *parameters* of the normal distribution, and they stand for the ordinary mean and variance. These two parameters determine which normal distribution is being graphed.

The preceding equation is a fairly complex one, but it can be simplified by expressing it in terms of z scores. This gives us the equation for the standard normal distribution and shows the intimate connection between the normal distribution and z scores:

$$f(z) = \frac{1}{\sqrt{2\pi}}\, e^{-z^2/2}$$

Because the variance of the standard normal distribution equals 1.0, $2\pi\sigma^2 = 2\pi$, and the power that e is raised to is just $-\frac{1}{2}$ times the z score squared. The fact that z is being squared tells us that the curve is symmetric around zero. For instance, the height for $z = -2$ is the same as the height for $z = +2$ because in both cases, $z^2 = +4$. Because the exponent of e has a minus sign, the function is largest (i.e., the curve is highest) when z is smallest, namely, zero. Thus, the mode of the normal curve (along with the mean and median) occurs in the center, at $z = 0$. Note that the height of the curve is never zero; that is, the curve never touches the X axis. Instead, the curve extends infinitely in both directions, always getting closer to the X axis. In mathematical terms, the X axis is the *asymptote* of the function, and the function touches the asymptote only at infinity.

The height of the curve ($f(X)$, or Y) is called the density of the function, so the preceding equation is often referred to as a *probability density function*. In more concrete terms, the height of the curve can be thought of as representing the relative likelihood of each X value; the higher the curve, the more likely is the X value at that point. However, as I pointed out in Section A, the probability of any *exact* value occurring is infinitely small, so when one talks about the probability of some X value being selected, it is common to talk in terms of a range of X values (i.e., an interval along the horizontal axis). The probability that the next random selection will come from that interval is equal to the proportion of the total distribution that is contained in that interval. This is the area under the curve corresponding to the given interval, and this area can be found mathematically by *integrating* the function over the interval using the calculus. Fortunately, the areas between the mean and various z scores have already been calculated and entered into tables (see Table A.1), and other areas not in a table can be found by using the techniques described in Section B.

The Central Limit Theorem

Strictly speaking, the sampling distribution of the mean will follow the normal distribution perfectly only if the population does, and we know that real-world populations do not perfectly follow a normal distribution, no matter what variable is measured. Fortunately, the Central Limit Theorem (CLT) says that even for population distributions that are very far from normal, the sampling distribution of the mean can be assumed to be normal,

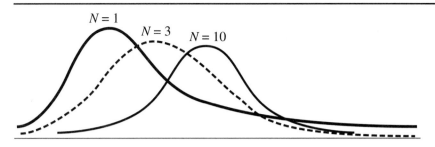

Figure 4.13

Sampling Distribution of the Mean for Different Sample Sizes (Population Distribution Is Positively Skewed)

provided that the sample size is large enough. It is because of the Central Limit Theorem that psychologists do not worry much about the shape of the population distributions of the variables they use, and the process of drawing statistical inferences is greatly simplified. Recognizing the importance of the Central Limit Theorem (which is central to the statistics in this text), I will state it formally and then discuss its implications.

> *Central Limit Theorem:* For any population that has mean μ and a finite variance σ², the distribution of sample means (each based on *N* independent observations) will approach a normal distribution with mean μ and variance σ²/*N*, as *N* approaches infinity.

The statement about variance translates easily into a statement about standard deviation; if the population has standard deviation σ, the standard deviation of the sampling distribution (called the standard error of the mean) will be σ/$\sqrt{N}$. (This principle was presented as Formula 4.5.) However, the statement about *N* approaching infinity may sound strange. If each sample must be nearly infinite in size to guarantee that the sampling distribution will have a normal shape, why is the CLT so helpful? It is fortunate that the sampling distribution becomes very nearly normal long before the sample size becomes infinitely large. How quickly the *sampling distribution* becomes nearly normal as sample size increases depends on how close to normal the *population distribution* is to begin with. If the population distribution is normal to begin with, the sampling distribution will be normal for any sample size. If the population distribution is fairly close to being normal, the sampling distribution will be very nearly normal even for small sample sizes. On the other hand, if the population distribution is virtually the opposite of normal, the sample size must be fairly large before the sampling distribution becomes nearly normal. The CLT guarantees us that no matter how bizarre the population distribution (provided that its variance is finite), the sample size can always be made large enough so that the sampling distribution approximates the normal distribution. The condition of having a finite population variance is never a problem in psychological research; none of the variables psychologists use would ever have an infinite variance.

I mentioned previously that fairly large sample sizes are needed if the population distribution is very far from normal. It has been generally found for a wide range of population distribution shapes that when the sample size exceeds about 30, the sampling distribution closely approximates a normal distribution. Figure 4.13 depicts the type of sampling distribution of the mean that is obtained for different sample sizes, when the population distribution is strongly skewed. Notice that in each case the sampling distribution is narrower, and more symmetrical, than the population distribution (*N* = 1), and that the sampling distribution becomes increasingly narrower and more symmetrical (and, of course, more normal) as the sample size increases.

Psychologists often deal with variables that are based on new measures for which little is known about the population distribution. Nonetheless, thanks to the Central Limit Theorem, they can assume that the sampling distribution of the mean will be nearly normal, as long as sufficiently large samples (N of at least 30 or 40) are used. If, however, a researcher has good reason to suspect that the population distribution of the measured variable will be extremely far from normal, it may be safer to use sample sizes of at least 100. Finally, if practical considerations prevent a researcher from obtaining a sample that he or she feels is sufficiently large to ensure a nearly normal sampling distribution, the use of nonparametric, or distribution-free, statistical tests should be considered (see Part VII).

Probability

As I pointed out in Section A, statements about areas under the normal curve translate directly to statements about probability. For instance, if you select one person at random, the probability that that person will have an IQ between 76 and 132 is about .91, because that is the amount of area enclosed between those two IQ scores, as we found in Section B (see Figure 4.8). To give you a more complete understanding of probability and its relation to problems involving distributions, I will lay out some specific rules. To represent the probability of an event symbolically, I will write $p(A)$, where A stands for some event. For example, $p(IQ > 110)$ stands for the probability of selecting someone with an IQ greater than 110.

Rule 1

Probabilities range from 0 (the event is certain *not* to occur) to 1 (the event is *certain* to occur) or from 0 to 100 if probability is expressed as a percentage instead of a proportion. As an example of $p = 0$, consider the case of adult height. The distribution ends somewhere around $z = -15$ on the low end and $z = +15$ on the high end. So for height, the probability of selecting someone for whom z is greater than $+20$ (or less than -20) is truly zero. An example of $p = 1$ is the probability that a person's height will be between $z = -20$ and $z = +20$ [i.e, $p(-20 < z < +20) = 1$].

Rule 2: The Addition Rule

If two events are *mutually exclusive*, the probability that either one event *or* the other will occur is equal to the sum of the two individual probabilities. Stated as Formula 4.7, the addition rule for mutually exclusive events is:

$$p(A \text{ or } B) = p(A) + p(B) \hspace{2cm} \textbf{Formula 4.7}$$

Two events are mutually exclusive if the occurrence of one rules out the occurrence of the other. For instance, if we select one individual from the IQ distribution, this person cannot have an IQ that is both above 120.5 and also below 79.5—these are mutually exclusive events. As I demonstrated in the discussion of the hypothetical MEZZA organization in Section B, the probability of each of these events is .10. We can now ask: What is the probability that a randomly selected individual will have an IQ above

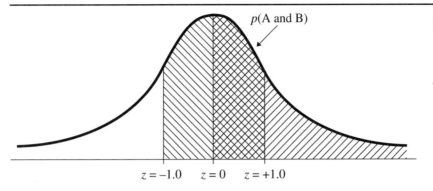

Figure 4.14

The Area Corresponding
to Two Overlapping
Events

120.5 *or* below 79.5? Using Formula 4.7, we simply add the two individual probabilities: .1 + .1 = .2. In terms of a single distribution, two mutually exclusive events are represented by two areas under the curve that do *not* overlap. (In contrast, the area from $z = -1$ to $z = +1$ and the area above $z = 0$ are *not* mutually exclusive because they *do* overlap.) If the areas do not overlap, we can simply add the two areas to find the probability that an event will be in one area *or* the other. The addition rule can be extended easily to any number of events, if all of the events are mutually exclusive (i.e., no event overlaps with any other). For a set of mutually exclusive events the probability that one of them will occur (p[A or B or C, etc.]) is the sum of the probabilities for each event (i.e., p[A] + p[B] + p[C], etc.).

The Addition Rule for Overlapping Events

The addition rule must be modified if events are not mutually exclusive. If there is some overlap between two events, the overlap must be subtracted after the two probabilities have been added. Stated as Formula 4.8, the addition rule for two events that are *not* mutually exclusive is:

$$p(\text{A or B}) = p(\text{A}) + p(\text{B}) - p(\text{A and B}) \qquad \text{\textbf{Formula 4.8}}$$

where p(A and B) represents the overlap (the region where A and B are both true simultaneously). For example, what is the probability that a single selection from the normal distribution will be either within one standard deviation of the mean *or* above the mean? The probability of the first event is the area between $z = -1$ and $z = +1$, which is about .68. The probability of the second event is the area above $z = 0$, which is .5. Adding these we get .68 + .5 = 1.18, which is more than 1.0 and therefore impossible. However, as you can see from Figure 4.14, these events are not mutually exclusive; the area of overlap corresponds to the interval from $z = 0$ to $z = +1$. The area of overlap (i.e., p(A and B)) equals about .34, and because it is actually being added in twice (once for each event), it must be subtracted once from the total. Using Formula 4.8, we find that p(A or B) = .68 + .5 − .34 = 1.18 − .34 = .84 (rounded off).

A Note about Exhaustive Events

Besides being mutually exclusive, two events can also be *exhaustive* if one or the other *must* occur (together they exhaust all the possible events). For

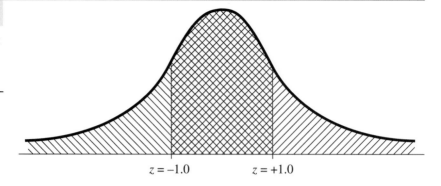

Figure 4.15

The Area Corresponding to Two Exhaustive, but not Mutually Exclusive, Events

$z = -1.0$ $z = +1.0$

instance, consider the event of being above the mean and the event of being below the mean; these two events are not only mutually exclusive, they are exhaustive as well. The same is true of being within one standard deviation from the mean and being at least one standard deviation away from the mean. When two events are both mutually exclusive and exhaustive, one event is considered the *complement* of the other, and the probabilities of the two events must add up to 1.0. If the events A and B are mutually exclusive and exhaustive, we can state that $p(B) = 1.0 - p(A)$.

Just as two events can be mutually exclusive but not exhaustive (e.g., $z > +1.0$ and $z < -1.0$), two events can be exhaustive without being mutually exclusive. For example, the two events $z > -1.0$ and $z < +1.0$ are exhaustive (there is no location in the normal distribution that is not covered by one event or the other), but they are not mutually exclusive; the area of overlap is shown in Figure 4.15. Therefore, the two areas represented by these events will not add up to 1.0, but rather somewhat more than 1.0.

Rule 3: The Multiplication Rule

If two events are *independent*, the probability that both will occur (i.e., A *and* B) is equal to the two individual probabilities multiplied together. Stated as Formula 4.9, the multiplication rule for independent events is:

$$p(A \text{ and } B) = p(A)p(B) \hspace{2cm} \textbf{Formula 4.9}$$

Two events are said to be independent if the occurrence of one in no way affects the probability of the other. The most common example of independent events is two flips of a coin. As long as the first flip does not damage or change the coin in some way—and it's hard to imagine how flipping a coin could change it—the second flip will have the same probability of coming up heads as the first flip. (If the coin is unbiased, $p(H) = p(T) = .5$.) Even if you have flipped a fair coin and have gotten 10 heads in a row, the coin will not be altered by the flipping; the chance of getting a head on the eleventh flip is still .5. It may seem that after 10 heads, a tail would become more likely than usual, so as to even out the total number of heads and tails. This belief is a version of the *gambler's fallacy;* in reality, the coin has no memory—it doesn't keep track of the previous 10 heads in a row. The multiplication rule can be extended easily to any number of events that are all independent of each other ($p(A \text{ and } B \text{ and } C, \text{ etc.}) = p(A)p(B)p(C), \text{ etc.}$).

Consider two independent selections from the IQ distribution. What is the probability that if we choose two people at random, their IQs will both be within one standard deviation of the mean? In this case, the probability of both individual events is the same, about .68 (assuming we replace the first person in the pool of possible choices before selecting the second; see the next section). Formula 4.9 tells us that the probability of both events occurring jointly (i.e., p[A and B]) equals (.68)(.68) = .46. When the two events are *not* independent, the multiplication rule must be modified. If the probability of an event changes because of the occurrence of another event, we are dealing with a *conditional probability*.

Conditional Probability

A common example of events that are *not* independent are those that involve successive samplings from a finite population *without replacement*. Let us take the simplest possible case: A bag contains three marbles; two are white and one is black. If you grab marbles from the bag without looking, what is the probability of picking two white marbles in a row? The answer depends on whether you select marbles *with replacement* or *without replacement*. In selecting with replacement, you take out a marble, look at it, and then replace it in the bag before picking a second marble. In this case, the two selections are independent; the probability of picking a white marble is the same for both selections: 2/3. The multiplication rule tells us that when the two events are independent (e.g., sampling *with* replacement), we can multiply their probabilities. Therefore, the probability of picking two white marbles in a row with replacement is (2/3)(2/3) = 4/9, or about .44.

On the other hand, if you are sampling *without* replacement, the two events will *not* be independent because the first selection will alter the probabilities for the second. The probability of selecting a white marble on the first pick is still 2/3, but if the white marble is *not* replaced in the bag, the probability of selecting a white marble on the second pick is only 1/2 (there is one white marble and one black marble left in the bag). Thus the conditional probability of selecting a white marble, *given that* a white marble has already been selected and not replaced (i.e., p[W|W]) is 1/2 (the vertical bar between the W's represents the word "given"). To find the probability of selecting two white marbles in a row when not sampling with replacement, we need to use Formula 4.10 (the multiplication rule for dependent events):

$$p(\text{A and B}) = p(\text{A})p(\text{B} \mid \text{A}) \hspace{3cm} \textbf{Formula 4.10}$$

In this case, both A and B can be symbolized by W (picking a white marble): $p(\text{W})p(\text{W}|\text{W})$ = (2/3) (1/2) =1/3, or .33 (less than the probability of picking two white marbles when sampling with replacement). The larger the population, the less difference it will make whether you sample with replacement or not. (With an infinite population, the difference is infinitesimally small.) For the remainder of this text I will assume that the population from which a sample is taken is so large that sampling without replacement will not change the probabilities enough to have any practical consequences. Conditional probability will have a large role to play, however, in the logical structure of null hypothesis testing, as described in the next chapter.

SUMMARY

1. The equation of the normal distribution depends on two parameters: the mean and the variance of the population. If the equation is expressed in terms of z scores, the result is the standard normal distribution. The highest point of the curve (i.e., the mode) is at $z = 0$, which is also the mean and the median, because the normal curve is perfectly symmetrical around $z = 0$.

2. When the population distribution is normal, the sampling distribution of the mean is also normal, regardless of sample size. If the population is *not* normally distributed (but the population variance is finite), the *Central Limit Theorem* states that the sampling distribution of the mean will become more like the normal distribution as sample size (N) increases—becoming normal when N reaches infinity. For sample sizes of 30 to 40 or more, the sampling distribution of the mean will be approximately normal, regardless of the shape of the population distribution (except, perhaps, for the most extreme possible cases).

3. The Rules of Probability

   Rule 1: The probability of an event (e.g., $p[A]$) is usually expressed as a proportion, in which case $p(A)$ can range from zero (the event A is certain *not* to occur) to 1.0 (the event A *is* certain to occur).

   Rule 2: The addition rule for mutually exclusive events states that if the occurrence of event A precludes the occurrence of event B, the probability of either A *or* B occurring, $p(A \text{ or } B)$, equals $p(A) + p(B)$. The addition rule must be modified as follows if the events are *not* mutually exclusive: $p(A \text{ or } B) = p(A) + p(B) - p(A \text{ and } B)$. Also, if two events are both mutually exclusive and *exhaustive* (one of the two events must occur), $p(A) + p(B) = 1.0$, and therefore, $p(B) = 1.0 - p(A)$.

   Rule 3: The multiplication rule for independent events states that the probability that two independent events will both occur, $p(A \text{ and } B)$, equals $p(A)p(B)$. Two events are *not* independent if the occurrence of one event changes the probability of the other. The probability of one event, given that another has occurred, is called a *conditional probability*.

4. The probability of two *dependent* events both occurring is given by a modified multiplication rule: The probability of one event is multiplied by the conditional probability of the other event, *given that the first event has occurred*. When you are sampling from a finite population without replacement, successive selections will not be independent. However, if the population is very large, *sampling without replacement* is barely distinguishable from *sampling with replacement* (any individual in the population has an exceedingly tiny probability of being selected twice), so successive selections can be considered independent even without replacement.

EXERCISES

*1. If you convert each score in a set of scores to a z score, which of the following will be true about the resulting set of z scores?
   a. The mean will equal 1.
   b. The variance will equal 1.
   c. The distribution will be normal in shape.
   d. All of the above.
   e. None of the above.

2. The distribution of body weights for adults is somewhat positively skewed—there is much more room for people to be above average than below. If you take the mean weights for random groups of 10 adults each and form a new distribution, how will this new distribution compare to the distribution of individuals?

a. The new distribution will be more symmetrical than the distribution of individuals.
b. The new distribution will more closely resemble the normal distribution.
c. The new distribution will be narrower (i.e., have a smaller standard deviation) than the distribution of individuals.
d. All of the above.
e. None of the above.

*3. The sampling distribution of the mean will tend toward the normal distribution with increasing *N* only if
a. The samples are randomly drawn.
b. The samples are all the same size.
c. The variance is finite.
d. All of the above.
e. None of the above.

*4. Consider a normally distributed population of resting heart rates with $\mu = 72$ bpm and $\sigma = 8$ bpm:
a. What is the probability of randomly selecting someone whose heart rate is either below 58 or above 82 bpm?
b. What is the probability of randomly selecting someone whose heart rate is either between 67 and 75 bpm, above 80 bpm or below 60 bpm?
c. What is the probability of randomly selecting someone whose heart rate is either between 66 and 77 bpm, or above 74 bpm?

5. Refer again to the population of heart rates described in the previous exercise:
a. What is the probability of randomly selecting two people in a row whose resting heart rates are both above 78 bpm?
b. What is the probability of randomly selecting *three* people in a row whose resting heart rates are all below 68 bpm?

c. What is the probability of randomly selecting two people, one of whom has a resting heart rate below 70 bpm, while the other has a resting heart rate above 76 bpm?

*6. What is the probability of selecting each of the following at random from the population (assume $\sigma = 15$):
a. One person whose IQ is either above 110 or below 95?
b. One person whose IQ is either between 95 and 110 or above 105?
c. Two people with IQs above 90?
d. One person with an IQ below 90 and one person with an IQ above 115?

*7. An ordinary deck of playing cards consists of 52 different cards, 13 in each of four suits (hearts, diamonds, clubs, and spades).
a. What is the probability of randomly drawing two hearts in a row if you replace the first card before picking the second?
b. What is the probability of randomly drawing two hearts in a row if you draw *without* replacement?
c. What is the probability of randomly drawing one heart and then one spade in two picks *without* replacement?

8. A third-grade class contains 12 boys and 8 girls. Two blackboard monitors are chosen at random. What is the probability that
a. Both will be girls?
b. Both will be boys?
c. There will be one of each gender?
d. Sum the probabilities found in parts a, b, and c. Can you say the events described in parts a, b, and c are all mutually exclusive and exhaustive?

The *z* score corresponding to a raw score:

$$z = \frac{X - \mu}{\sigma}$$

Formula 4.1

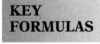
KEY FORMULAS

The raw score that corresponds to a given *z* score:

$$X = z\sigma + \mu$$

Formula 4.2

The SAT score corresponding to a raw score, if the *z* score has already been calculated:

$$SAT = 100z + 500$$

Formula 4.3

The T score corresponding to a raw score, if the z score has already been calculated:

$$T = 10z + 50$$

Formula 4.4

The standard error of the mean:

$$\sigma_{\bar{X}} = \frac{\sigma}{\sqrt{N}}$$

Formula 4.5

The z score for groups (the standard error of the mean must be found first):

$$z = \frac{\bar{X} - \mu}{\sigma_{\bar{X}}}$$

Formula 4.6

The addition rule for mutually exclusive events:

$$p(A \text{ or } B) = p(A) + p(B)$$

Formula 4.7

The addition rule for events that are *not* mutually exclusive:

$$p(A \text{ or } B) = p(A) + p(B) - p(A \text{ and } B)$$

Formula 4.8

The multiplication rule for independent events:

$$p(A \text{ and } B) = p(A)p(B)$$

Formula 4.9

The multiplication rule for events that are *not* independent (based on *conditional probability*):

$$p(A \text{ and } B) = p(A)p(B|A)$$

Formula 4.10

INTRODUCTION TO HYPOTHESIS TESTING: THE ONE-SAMPLE z TEST

You will need to use the following from previous chapters:

Symbols
μ: Mean of a population
$\overline{X}$: Mean of a sample
σ: Standard deviation of a population
$\sigma_{\overline{X}}$: Standard error of the mean
N: Number of subjects (or observations) in a sample

Formulas
Formula 4.5: The standard error of the mean
Formula 4.6: The z score for groups

Concepts
Sampling distribution of the mean

The purpose of this chapter is to explain the concept of hypothesis testing in the simple case in which only one group of subjects is being compared to a population and we know the mean and standard deviation for that population. This situation is not common in psychological research. However, a description of the one-sample hypothesis test (with a known standard deviation) is useful as a simple context in which to introduce the logic of hypothesis testing and, more specifically, as a stepping stone for understanding the very common two-sample t test (see Chapter 7). In this overview I employ a rather fanciful example to help simplify the explanations. In Section B, I consider a more realistic example and suggest general situations in which the one-sample hypothesis test may serve a useful function.

Consider the following situation. You have met someone who tells you he can determine whether any person has a good aptitude for math just by looking at that person. (He gets "vibrations" and sees "auras," which are what clue him in to someone's math aptitude.) Suppose you are inclined to test the claim of this self-proclaimed psychic—maybe you want to show him he's wrong (or maybe you're excited by the possibility that he's right). Your first impulse might be to find a person whose math aptitude you know and ask the psychic to make a judgment. If the psychic were way off, you might decide at once that his powers could not be very impressive. If he were close, you'd probably repeat this test several times, and only if the psychic came close each time would you be impressed.

Selecting a Group of Subjects

Would you be interested if the psychic had only a very slight ability to pick people with high math aptitude? Perhaps not. There wouldn't be much practical application of the psychic's ability in that case. But the possibility of even a slight degree of extrasensory perception is of such great theoretical interest that we should consider a more sensitive way to test our psychic friend. The approach that a psychological researcher might take is to have the psychic select from the population at large a group of people he believed to have a high mathematical aptitude. Then math SAT scores would be obtained for each person selected and the mean of these scores would be calculated. Assume that 500 is the mean math SAT score for the

CONCEPTUAL FOUNDATION

entire population. Note that for the mean of the sample group to be above average it is not necessary for each selected person to be above average. The mean is useful because it permits the selection of one person with a very high math SAT score to make up for picking several people slightly below average.

It is important to make clear that this experiment would not be fair if the psychic were roaming around the physics department of a university. If we are checking to see if the psychic can pick people above the average for a particular population (e.g., adults in the United States), theoretically, all of the people in that population should have an equal chance of being selected. Furthermore, each selection should be independent of all the other selections. For instance, it would not be fair if the psychic picked a particular woman and then decided it would be a good idea to include her sister as well.

The Need for Hypothesis Testing

Suppose we calculate the mean math SAT score of the people the psychic has selected and find that mean to be slightly higher than the mean of the population. Should we be convinced the psychic has some extrasensory ability—even though it appears to be only a slight ability? One reason for being cautious is that it is extremely unlikely that the mean math SAT score for the selected group of people will be *exactly* equal to the population mean. The group mean almost always will be slightly above or below the population mean even if the psychic has no special powers at all. And that, of course, implies that even if the psychic has zero amount of psychic ability, he will pick a group that is at least slightly above the population mean in approximately half of the experiments that you do. Knowing how easily chance fluctuations can produce positive results in cases in which there really is no reason for them, researchers are understandably cautious. This caution seems to underlie the creation of a statistical procedure known as *null hypothesis testing*. Null hypothesis testing involves what is called an indirect proof. You take the worst-case scenario—that is, you take the opposite of what you want to prove—and assume that it is true. Then you try to show that this assumption leads to ridiculous consequences and that therefore you can reject the assumption. Because the assumption was the opposite of what you wanted to prove, you are happy to reject the assumption.

It would not be surprising if the above explanation sounds a bit confusing. Cognitive psychologists have found that logical arguments are sometimes more understandable when they are cast in concrete and familiar terms (Griggs and Cox, 1982; Johnson-Laird, Legrenzi, and Legrenzi, 1972)—so I will take a concrete approach here. The indirect proof described above can be thought of as a "devil's advocate" procedure: You argue from the position of your opponent (the devil, so to speak) and try to show that your opponent's position has unreasonable implications. The idea of the devil's advocate inspired me to create an imaginary character who personifies scientific skepticism: a permanent opponent who is always skeptical of all the types of experiments I will describe in this text. I call this character Dr. Null, for reasons that will soon become obvious. My intention in creating this character is not to be cute or amusing (although I will not be appalled if this is a by-product) and certainly not to be condescending. I'm simply trying to make the logic easier to comprehend by making it concrete. (If you are comfortable with the logic of indirect proofs, feel free to ignore my comments about Dr. Null.)

The Logic of Null Hypothesis Testing

The character of Dr. Null can be used in several ways. Imagine that whenever you run an experiment, like the psychic test just described, and write up the results and send them to a journal to be published, the first person to see your article (no matter which journal you send it to) is Dr. Null. Dr. Null has the same skeptical reaction to any results that are submitted. No matter how plausible the results may seem, Dr. Null insists that the experimental treatment was completely ineffective. In the case of the psychic tests, he would claim the psychic had no ability to pick people with high math aptitude at all. How does Dr. Null explain that the group selected by the psychic has a slightly higher mean than the overall population? Dr. Null is quick to point out the example of flipping a perfectly fair coin 10 times. You don't always get five heads and five tails. You can easily get six of one and four of the other—which doesn't mean there's anything biased or "loaded" about the coin. In fact, in about 1 in a 1,000 sets of 10 tosses, a perfectly fair coin will give you 10 heads in a row. The same principle applies when sampling from a population, Dr. Null would explain. Sometimes, by chance, you'll get a few high scores and the mean will be above the population average; or, it's just as likely that some low scores will put you below the population mean. So regardless of how good your results look, Dr. Null offers the same simple hypothesis: Your experimental results merely come from sampling the original population *at random*; that is, your experiment does not involve a special selection process or a different population (or subpopulation). This hypothesis, about which I will have much more to say in a little while, is generally formalized into what is known as the *null hypothesis* (from which Dr. Null gets his name).

Imagine further that Dr. Null offers you a challenge. Dr. Null insists that before you proclaim that your experiment really worked, he will go out and try to duplicate (or even surpass) your experimental results *without* adding the crucial experimental ingredient, just by taking perfectly random samples of the population. In the case of the psychic test, Dr. Null would select the same number of people but would make sure to do so randomly, to avoid any possibility of psychic ability entering into the process. Then if the mean of Dr. Null's group was as high or even higher than the mean of the psychic's group, Dr. Null would take the opportunity to ridicule your experiment. He might say, "How good could your psychic be if I just did even better (or just as well) without using any psychic ability at all?" You might want to counter by saying that Dr. Null got his results by chance and you got your results through the psychic's ability, but how could you be sure? And how could you convince the scientific community to ignore Dr. Null's arguments? Dr. Null's beating the psychic's results (or even just matching them) is an embarrassment to be avoided if possible. Your only chance to avoid such an embarrassment is to know your enemy, Dr. Null, as well as possible.

The Null Hypothesis Distribution

Is there ever a situation when experimental results are so good (i.e., so far from the population mean) that Dr. Null has no chance of beating them at all? Theoretically, the answer is No, but in some cases the chances of Dr. Null's beating your results are so ridiculously small that nobody would worry about it. The important thing is to know what the probability is that Dr. Null will beat your results so that you can make an informed decision and take a calculated risk (or decide not to take the risk).

How can you know Dr. Null's chances of beating your results? That's relatively easy. Dr. Null is only taking random samples, and the laws of statistics can tell us precisely what happens when you select random samples of a particular size from a particular population. These statistical laws are simple to work with only if we make certain assumptions (to be discussed in the next section), but fortunately these assumptions are reasonable to make in many of the situations that psychological researchers face. Most of the inferential statistics that psychologists use can be informally thought of as ways for researchers to figure out the probability of Dr. Null's beating their results, and arbitrary rules to help them decide whether or not to risk embarrassment.

When you are comparing the mean of one sample to the mean of a population, the statistical laws mentioned in the preceding lead to a very simple formula (given some assumptions). To map out what Dr. Null can do in the one-sample case, imagine that he keeps drawing random samples of the same size, from the same population, and measuring the mean for each sample. If Dr. Null did this sampling many, many times (ideally, an infinite number of times), the sample means he would get would form a smooth distribution. This distribution is called the *null hypothesis distribution* because it shows what can happen (what is likely and not so likely) when the null hypothesis is true.

The Null Hypothesis Distribution for the One-Sample Case

You should recall from Chapter 4 that when you take many groups of the same size and form a distribution from the group means, you get the sampling distribution of the mean. If the population has a normal distribution for the variable of interest (in this case, math SAT scores), the sampling distribution will be normal and will have the same mean as the population from which the samples are being drawn (500 in the case of math SAT scores). But the sampling distribution will be narrower than the population distribution. As you saw in Chapter 4, the standard deviation for the sample means would equal the standard deviation of the population divided by the square root of the sample size. The null hypothesis distribution does not usually work out this simply, but in the simplest possible case—the one-sample test with a known population standard deviation—the null hypothesis distribution is just the sampling distribution of the mean, which you learned about in Chapter 4.

It should now be clear why I introduced the sampling distribution of the mean in the first place. Generally, experiments are conducted not by looking at one individual at a time but by selecting a group and looking at the mean of the group. To get our map of what Dr. Null can do in the one-sample case, we have to know what happens when many groups are selected and their means are calculated. So in this case our map of what Dr. Null can do is just the well-known sampling distribution of the mean (see Figure 5.1). We know the mean of Dr. Null's distribution because we know the mean of the population from which he is taking samples. Similarly, we can easily calculate the standard error of Dr. Null's distribution because we know the standard deviation of the population and the sample size. Because the sampling distribution of the mean tends to be a normal distribution, it can extend infinitely in each direction. This implies that, theoretically, there is no limit to what Dr. Null can do by random sampling. However, the tails of the distribution get very thin, which tells us that Dr. Null has little chance of drawing a sample whose mean is far from the

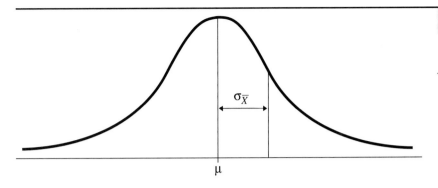

Figure 5.1

One-Sample Null
Hypothesis Distribution

population mean. If we assume that the sampling distribution *is* normal, we can use the standard normal table to answer questions about group means, as described in Chapter 4.

z Scores and the Null Hypothesis Distribution

If we want to know the probability of Dr. Null's surpassing our psychic's results, there is a simple procedure for finding that out. We just have to see where our experimental results fall on Dr. Null's distribution. Then we look at how much area of the curve is beyond that point, because that represents the proportion of times Dr. Null will get a result equal to or better than ours. To find where our experimental result falls on Dr. Null's distribution, all we have to do is convert our result to a *z* score—a *z* score for groups, of course. We can use Formula 4.6 (which is presented below as Formula 5.1). Once we know the *z* score we can find the area beyond that *z* score, using the tables for the standard normal distribution as we did in Chapter 4. To do a one-sample hypothesis test, you begin by simply figuring out the *z* score of your results on the null hypothesis distribution and finding the area beyond that *z* score.

I will illustrate this procedure using actual numbers from a specific example. Suppose the psychic selects a group of 25 people and the mean for that group is 530. This is clearly above the population mean of 500, but remember that Dr. Null claims he can do just as well or better using only random sampling. To find the probability of Dr. Null's beating us, we must first find the *z* score of our experimental result (i.e., a mean of 530) with respect to the null hypothesis distribution, using Formula 5.1:

$$z = \frac{\overline{X} - \mu}{\sigma_{\overline{X}}}$$

Formula 5.1

where $\overline{X}$ is our experimental result, μ is the population mean, and $\sigma_{\overline{X}}$ is the standard deviation of the sampling distribution.

To get the standard deviation of the sampling distribution (i.e., the standard error of the mean) we need to know the standard deviation for the population. This is not something a researcher usually knows, but some variables (such as math SAT scores and IQ) have been standardized on such large numbers of people that we can say we know the standard deviation for the population. The standard deviation for math SAT scores is 100. Using Formula 4.5 we find that the standard error for groups of 25 is:

$$\sigma_{\overline{X}} = \frac{\sigma}{\sqrt{N}} = \frac{100}{\sqrt{25}} = \frac{100}{5} = 20$$

Figure 5.2

Null Hypothesis
Distribution for Math
SAT Scores

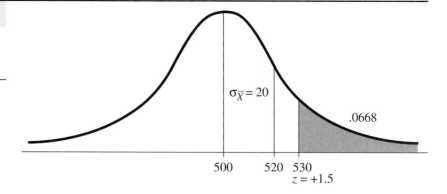

The *z* score for the psychic's group therefore is:

$$z = \frac{530 - 500}{20} = \frac{30}{20} = 1.5$$

How often can Dr. Null do better than a *z* score of +1.5? We need to find the area to the right of (beyond) *z* = +1.5; that is, the area shaded in Figure 5.2. Table A.1 tells us that the area we are looking for is .0668. This means that if Dr. Null performs his fake psychic experiment many times (i.e., drawing random samples of 25 people each, using no psychic ability at all), he will beat us about 67 times out of 1,000, or nearly 7 times out of 100. The probability of Dr. Null's beating us is usually called the *p value* associated with our experimental results (*p* for *probability*). In this case the *p* value is .0668.

Statistical Decisions

Now that we know the probability of Dr. Null embarrassing us by getting even better results than we did without using our experimental manipulation, we can make an informed decision about what we can conclude from our results. But we still must decide how much risk to take. If Dr. Null's chance of beating us is less than 1 in 10, should we take the risk of declaring our results statistically significant? The amount of risk you decide to take is called the *alpha* (α) *level*, and this should be decided *before* you do the experiment, or at least before you see the results. Alpha is the chance you're willing to take that Dr. Null will beat you. When the chance of Dr. Null beating you is low enough—that is, when it is less than alpha—you can say that you *reject* the null hypothesis. This is the same as saying your results have *statistical significance*. What you are telling your fellow scientists is that the chance of Dr. Null beating you is too low to worry about and that it is safe to ignore this possibility.

Does each researcher decide on his or her own alpha level? Certainly not. Among psychologists in particular, an alpha level of 1 chance in 20 (a probability corresponding to .05) is generally considered the largest amount of risk worth taking. In other words, by convention, other psychologists agree to ignore Dr. Null's claims and protests concerning your experiment as long as you can show that Dr. Null has less than a .05 chance of beating you (i.e., the *p* value for your experiment is less than .05). If you are using an alpha level of .05 and your *p* value is less than .05, you can reject the null hypothesis at the .05 level of significance (see Figure 5.3).

But what if your *p* value is greater than .05? What must you conclude then? Unless you have declared an alpha level greater than .05 (which is not

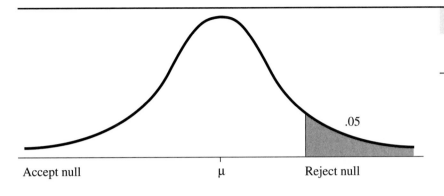

Figure 5.3

One-Tailed Hypothesis
Test, Alpha = .05

likely to be accepted by your fellow psychologists without some special justification, which I will discuss later), you must *accept*, or *retain*, the null hypothesis. Does this mean you are actually agreeing with the infamous Dr. Null? Are you asserting that your experiment has absolutely no effect and that your experimental results are attributable entirely to chance (the luck involved in random sampling)? Generally, no. In fact, many psychologists and statisticians prefer to say not that they accept the null hypothesis in such cases, but that they *fail to reject it*—that they have insufficient evidence for rejecting it. When the *p* value for your experiment is greater than .05, you may still feel that your experiment really worked and that your results are not just due to luck. However, convention dictates that in this situation you must be cautious. Dr. Null's claims, although somewhat farfetched if your *p* value is only a little over .05, are not ridiculous or totally unreasonable. The null hypothesis in such a case must be considered a reasonable possibility for explaining your data, so you cannot declare your results statistically significant.

Although Dr. Null is a fictional character, there is a real person responsible for the emphasis on null hypothesis testing that is so prevalent in most fields of psychological research. That person was Sir Ronald A. Fisher (1890–1962), an English mathematician, whose *Statistical Methods for Research Workers* was first published in 1925; that seminal work led to many subsequent editions over the course of Fisher's lifetime and had an enormous impact on the way data are analyzed, especially in most fields of psychology. Fisher may not have invented the .05 level, but his formalization of this criterion in his early work seems largely responsible for its widespread acceptance as the default alpha level (Cowles, 1989). I will have more to say about Fisher's contributions to statistics in this and subsequent chapters.

The *z* Score as Test Statistic

In describing math SAT scores earlier, I stated that the null hypothesis distribution was normal, with a mean of 500 and a standard error of 20. But there is no table for areas under this particular distribution, so we had to use *z* scores. When you convert to *z* scores, the *z* scores follow a standard normal distribution (mean = 0 and standard deviation = 1), for which there is a convenient table. Used this way, the *z* score can be called a *test statistic*: It is based on one or more sample statistics (in this case just the sample mean), and it follows a well-known distribution (in this case, the standard normal distribution). For convenience, it is the distribution of the test statistic that is viewed as the null hypothesis distribution. In later chapters, I will deal with more complicated test statistics and other well-known distributions (e.g., t and F), but the basic principle for making statistical decisions will remain the same.

The larger the z score, the lower the p value, so in general you would like the calculated z to be as large as possible. But how large can the z score in a one-sample test get? If you look at the standard normal table (Table A.1), you might be misled into thinking that the largest possible z score is 4.0 because that is where this particular table ends. For most research purposes it is not important to distinguish among probabilities that are less than .00001, so I did not include them in the table. If your calculated z comes out to be 22, you can just say that p is less than .0001 and leave it at that. Even though statistical programs can find the area beyond even a very large z score, the p values printed in the output are usually rounded off to three or four decimal places, so for large z scores you may see $p = .0000$. This can be misleading. There is, of course, always *some* area beyond any z score, no matter how large, until you get to infinity (e.g., $p = .0000$ can represent any p that is less than .00005).

Sometimes people are confused when they see a very large z score as the result of a one-sample test because they think of such a z score as almost impossibly rare. They are forgetting that a large z score for a particular experiment is very rare only if the null hypothesis happens to be true for that experiment. The whole point of the one-sample test is to obtain a large z score, if possible, to show that we shouldn't believe the null hypothesis—precisely because believing that the null hypothesis is true leads to a very implausible conclusion, namely, that our results are a very rare fluke produced entirely by sampling error. Admittedly, alpha is normally set to .05, which does not represent a very rare event (i.e., it is one in twenty). Why psychologists seem to be comfortable rejecting the null hypothesis at that probability will be discussed further in this section, but in much greater detail in Section C.

You should also keep in mind that the z score for groups can get very large even if the sample mean does not deviate much from the population mean (as compared to the population standard deviation). This is because the denominator of the formula for the z score for groups is the standard error of the mean, which gets smaller as the sample size gets larger. With a large enough sample size, even a slightly deviant sample mean can produce a large z score, so you should not be overly impressed by large z scores. Large z scores *do* lead to tiny p values, which allow confident rejection of the null hypothesis—but that does not imply that the deviation of the sample mean is enough to be important or interesting. I will return to this point in greater detail in Chapter 8.

Type I and Type II Errors

Even if our p value is very much under .05 and we reject the null hypothesis, we are nonetheless taking some risk. But just what is it that we are risking? I've been discussing "risk" as the risk of Dr. Null's beating our results without using our experimental manipulation and thus making us look bad. This is just a convenient story I used to make the concept of null hypothesis testing as concrete as possible. In reality no psychologist worries that someone is going to do a "null experiment" (i.e., an experiment in which the null is really true) and beat his or her results. The real risk is that Dr. Null may indeed be right; that is, the null hypothesis may really be true and we got positive-looking results (i.e., $p < .05$) by luck. In other words, we may ourselves have done a "null experiment" without knowing it and obtained our positive results by accident. If we then reject the null hypothesis in a case when it is actually true (and our experimental manipulation was entirely ineffective), we are making an error. This kind of error is called a *Type I error*. The opposite kind of error occurs when you accept the null hypothesis even though it is false (in other words, you cautiously refrain from rejecting the null hypothesis even though in reality your experiment was at least somewhat effective). This kind of error is called a *Type II error*. The distinction between the two types of errors is usually illustrated as shown in Table 5.1. Notice the probabilities indicated in parentheses. When the null hypothesis is true, there are two decisions you can make (accept or reject), and the probabilities of these two decisions must add up to 1.0. Similarly, the probabilities for these two decisions add up to 1.0 when the null hypothesis is false. (Note that it makes no sense to add all four probabilities together because they are *conditional* probabilities; see Section C.) The probability of making a Type II error is called β, but this value is not set by the researcher and can be quite difficult to determine. I will not discuss the probability of Type II errors in any detail until I deal with the topic of power in Chapter 8.

	ACTUAL SITUATION	
Researcher's Decision	**Null Hypothesis Is True**	**Null Hypothesis Is False**
Accept the Null Hypothesis	Correct Decision ($p = 1 - \alpha$)	Type II Error ($p = \beta$)
Reject the Null Hypothesis	Type I Error ($p = \alpha$)	Correct Decision ($p = 1 - \beta$)

Table 5.1

The notion of the Type II error did not come from Fisher, who did not like to think in terms of "accepting" the null hypothesis when results were not significant. Fisher felt that we should either reject the null or reserve judgment; in the latter case we are not really making an error because we are not making any firm decision. The concept embodied in Table 5.1 comes from the collaboration of two statisticians who sought to extend and systematize Fisher's method of null hypothesis testing by imagining that every null hypothesis has a complementary "alternative" hypothesis, so rejecting one provides evidence in favor of the other. One of these collaborators was the Polish mathematician Jerzy Neyman (1894–1981), and the other was Egon S. Pearson (1895–1980), son of Karl Pearson, who, among other important contributions, devised the most commonly used coefficient for the measurement of linear correlation (see Chapter 9). Although Fisher vigorously opposed the additions of the *Neyman-Pearson theory*, they became widely accepted; ironically, Fisher usually gets credit for even the aspects of modern null hypothesis testing that he fought against (Cowles, 1989).

The Trade-Off between Type I and Type II Errors

In one sense Fisher's influence has been the predominant one in that the main focus of null hypothesis testing is still the avoidance of Type I errors. The chief reason for trying not to make a Type I error is that it is a kind of false alarm. By falsely claiming statistical significance for the results of your experiment, you are sending your fellow researchers off in the wrong direction. They may waste time, energy, and money trying to replicate or extend your conclusions. These subsequent attempts may produce occasional Type I errors as well, but if the null hypothesis is actually true—if your experiment was really totally ineffective—other experiments will most likely wind up *not* rejecting the null hypothesis. However, it can take several failures to replicate the statistical significance of your study before it becomes clear that the original result was a Type I error.

The above argument should convince you to avoid Type I errors as much as possible. One way to further reduce the number of Type I errors committed by researchers in general would be to agree on an even lower alpha level, for instance, .01 or even .001. Because Type I errors are so disruptive and misleading, why not use a smaller alpha? As you will see in Chapter 8, changing the decision rule to make alpha smaller would at the same time result in more Type II errors being made. Scientists would become so cautious in trying to avoid a Type I error that they might have to ignore a lot of genuinely good experimental results because these results could very occasionally be produced by Dr. Null (i.e., produced when the null hypothesis is actually true). Also, as you will see if you read Section C, setting the Type I error *rate* at .05 does not result in as many actual Type I errors being committed as you might think. The rather arbitrary value of .05 that has been conventionally accepted as the minimal alpha level for psychology experiments represents a compromise between the potential negative consequences of Type I errors and those of Type II errors.

In some situations Type II errors can be quite serious, so in those cases we might want to use a larger alpha. Consider the case of the researcher testing a cure for a terminal disease. A Type II error in this case would mean accepting the null hypothesis and failing to say that a particular cure has some effect—when in fact the cure does have some effect, but the researcher is too cautious to say so. In such a case the researcher might consider using an alpha of .1 instead of .05. This would increase Type I errors and, therefore, false alarms, but it would reduce Type II errors and therefore lower the chances that a potential cure would be overlooked. On the other hand, if a treatment for acne is being tested, the chief concern should be Type I errors. Because acne is not life-threatening, it would be unfortunate if many people wasted their money on a totally ineffective product simply because a study accidently obtained results that made the treatment look good.

Besides changing the alpha level, another way of reducing Type II errors in certain cases is to perform a one-tailed rather than a two-tailed test. I have ignored this distinction so far to keep things simple, but now it is time to explain the difference.

One-Tailed versus Two-Tailed Tests

Suppose the group selected by the psychic had a mean math SAT score of only 400 or even 300. What could we conclude? Of course, we would not reject the null hypothesis, and we would be inclined to conclude that the psychic had no special powers for selecting people with high math aptitude. However, if the group mean were extremely low, could we conclude the psychic's powers were working in reverse, and he was picking people with low math aptitude more consistently than could be reasonably attributed to chance? In other words, could we test for statistical significance in the other direction—in the other tail of the distribution? We could if we had planned ahead of time to do a *two-tailed test* instead of a *one-tailed test*.

For the experiment I have been describing it may seem quite pointless to even think about testing significance in the other direction. What purpose could be served and what explanation could be found for the results? However, for many experiments the "other tail" cannot be so easily ignored, so you should know how to do a two-tailed as well as a one-tailed test. Fortunately, there is no difference between one- and two-tailed tests when you calculate the z score for groups. The difference lies in the *p* value. The procedure I have been describing thus far applies only to a one-tailed test. To get the *p* value for a two-tailed test, you find the area beyond z, just as for a one-tailed test, but then you double the area. For example, suppose the psychic selected a group of 25 people whose mean math SAT turned out to be 450. The z score for this group would be:

$$z = \frac{450 - 500}{\frac{100}{\sqrt{25}}} = \frac{-50}{20} = -2.5$$

If you had planned a one-tailed test, expecting the selected group to be above average, the rules of hypothesis testing would not allow you to test for statistical significance, and you would have to retain the null hypothesis. However, if you had planned a two-tailed test, you would be allowed to proceed by finding the two-tailed *p* value. First you have to look at the area beyond your calculated z score (see Figure 5.4). Because your z score is negative, take its absolute value first (remember there are no negative z scores in the normal curve table). The area beyond $z = 2.5$ is .006. To get the *p* value for a two-tailed test you simply double the area beyond z to get .012. If your alpha had been set to .05, you could then reject the null hypothesis. Of course, if you had decided to use alpha = .01, you could not then reject the null hypothesis. In our one-tailed test of the psychic's efforts, we found the *p* value to be .0668. The two-tailed *p* value would have been twice that, which is .1336. Because the one-tailed test was not significant at the .05 level, the two-tailed test would certainly not be significant either.

To understand why the *p* value is doubled when you perform a two-tailed test, think again in terms of what Dr. Null can do. When you plan a two-tailed test for one sample, you expect the mean of your sample to be *different* from the population mean, but you are not committed to saying whether it will be higher or lower. This means that Dr. Null has *two* ways to embarrass you. He can embarrass you by beating (or tying) your results *in the same direction*, or he can get results in the opposite direction that are *just as different* from the population mean as your results. Because you didn't specify which direction your results would be in (only that they

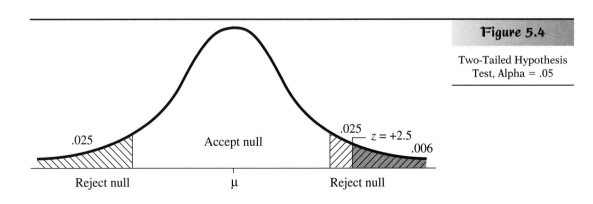

Figure 5.4

Two-Tailed Hypothesis
Test, Alpha = .05

would be different from the population mean), Dr. Null can claim to have matched your results even if his results are in the other direction, because he has matched your results in terms of deviation from the population mean. That's why when finding the p value for a two-tailed test, we have to add the same amount of area for the other tail. Doing a two-tailed test makes it a bit harder to reach statistical significance because a larger z score is required. This is the disadvantage of doing a two-tailed test. On the other hand, if you plan a one-tailed test and the results come out in the direction opposite to what you expected, it is unfair to test the significance of these unexpected results. This is the disadvantage of a one-tailed test.

Selecting the Number of Tails

So, when should you perform a one-tailed test and when a two-tailed test? Unfortunately, there is no rule that is universally agreed upon. It's a matter of judgment. However, if you plan a one-tailed test, you must state the expected direction of the results before seeing them; if the results come out in the wrong direction, no test can be performed—not even a two-tailed test. You might at first think it would be all right to plan a one-tailed test and then switch to a two-tailed test if the results come out in the wrong direction. The problem is that if this were commonly done, the percentage of Type I errors would be greater than alpha. (If alpha is .05, you are putting .05 in one tail of the null distribution and another .025 in the opposite tail, for a total alpha of .075.) Because of the potential abuse of one-tailed testing, the two-tailed test is generally considered more conservative (i.e., more likely to keep Type I errors at alpha) and is therefore less likely to be criticized. The other side of this argument is that a one-tailed test is more powerful if your prediction is correct (i.e., you are less likely to make a Type II error). However, because researchers tend to be more openly concerned about Type I errors, you are usually safer from criticism if you always do a two-tailed test.

Perhaps the most justifiable case for performing a one-tailed test is when there is a strong basis for predicting results in a particular direction (e.g., the results of previous studies) and results in the "other" tail (i.e., in the unexpected direction) are really absurd or are otherwise not worth testing. For example, suppose a medical researcher is testing an improved headache remedy, similar to others that have been effective in the past. Although it is not impossible that this new remedy could actually worsen headaches, the researcher is not interested in finding a drug that worsens headaches and could legitimately plan not to test for such a possibility. Either reason—a strong basis for prediction (based on theory or previous results) or a total lack of interest in results in the other tail—can justify performing a one-tailed test, but bear in mind that psychologists tend to be very cautious about planning one-tailed tests.

There are many cases when it may at first seem absurd to test the other tail, but not so absurd after more thought. For instance, it may seem obvious that a stimulant drug will allow subjects to produce more work in a given amount of time. But if the work requires some skill or concentration, the subjects may actually produce less because they are too stimulated to be patient or organized. Only experience in a particular field of research will tell you if a particular one-tailed test is likely to be considered acceptable by your colleagues. For psychologists in general, the two-tailed test is considered the default—the conventional thing to do—just as .05 is the conventional alpha level.

1. To perform a one-sample experiment, select a random sample of some population that interests you (or select a sample from the general population and subject it to some experimental condition), calculate the mean of the sample for the variable of interest, and compare it to the mean of the general population.
2. Even if the sample mean is different from the population mean in the direction you predicted, this result could be due to chance fluctuations in random sampling—that is, the null hypothesis could be true.
3. The null hypothesis distribution is a "map" of what results are or are not likely by chance. For the one-sample case, the null hypothesis distribution is the sampling distribution of the mean.
4. To test whether the null hypothesis can easily produce results that are just as impressive as the results you found in your experiment, find the z score of your sample mean with respect to the null hypothesis distribution. The area beyond that z score is the p value for your experiment (i.e., the probability that when the null hypothesis is true, results as good as yours will be produced).
5. When the p value for your experiment is less than the alpha level you set, you are prepared to take the risk of rejecting the null hypothesis and of declaring that your results are statistically significant.
6. The risk of rejecting the null hypothesis is that it may actually be true (even though your results look very good), in which case you are making a Type I error. The probability of making a Type I error when the null hypothesis is true is determined by the alpha level that you use (usually .05).
7. If you make alpha smaller to reduce the proportion of Type I errors, you will increase the proportion of Type II errors, which occur whenever the null hypothesis is *not* true, but you fail to reject the null hypothesis because you are being cautious. The probability of making a Type II error is not easily determined. Type II errors will be discussed thoroughly in Chapter 8.
8. One-tailed hypothesis tests make it easier to reach statistical significance in the predicted tail but rule out the possibility of testing results in the other tail. Because of the everpresent possibility of unexpected results, the two-tailed test is more generally accepted.

EXERCISES

*1. a. If the calculated z for an experiment equals 1.35, what is the corresponding one-tailed p value? The two-tailed p value?
   b. Find the one- and two-tailed p values corresponding to $z = -.7$.
   c. Find one- and two-tailed p values for $z = 2.2$.

2. a. If alpha were set to the unusual value of .08, what would be the magnitude of the critical z for a one-tailed test? What would be the values for a two-tailed test?
   b. Find the one- and two-tailed critical z values for $\alpha = .03$.
   c. Find one- and two-tailed z values for $\alpha = .007$.

*3. a. If the one-tailed p value for an experiment were .123, what would the value of z have to be?
   b. If the two-tailed p value for an experiment were .4532, what would the value of z have to be?

4. a. As alpha is made smaller (e.g., .01 instead of .05), what happens to the size of the critical z?
   b. As the calculated z gets larger, what happens to the corresponding p value?

*5. An English professor suspects that her current class of 36 students is unusually good at verbal skills. She looks up the verbal SAT score for each student and is pleased to find that the mean for the class is 540. Assuming that the general population of students has a mean verbal SAT score of 500 with a standard deviation of 100, what is the two-tailed *p* value corresponding to this class?

6. Consider a situation in which you have calculated the *z* score for a group of participants and have obtained the unusually high value of 20. Which of the following statements would be true, and which would be false? Explain your answer in each case.
   a. You must have made a calculation error because *z* scores cannot get so high.
   b. The null hypothesis cannot be true.
   c. The null hypothesis can be rejected, even if a very small alpha is used.
   d. The difference between the sample mean and the hypothesized population mean must have been quite large.

7. Suppose the *z* score mentioned in Exercise 6 involved the measurement of height for

a group of men. If $\mu = 69$ inches and $\sigma = 3$ inches, how can a group of men have a *z* score equal to 20? Give a numerical example illustrating how this can occur.

*8. Compared to a one-tailed hypothesis test, a two-tailed test requires
   a. More calculation
   b. More prior planning
   c. More integrity
   d. A larger *z* score to be significant
   e. All of the above

9. Describe a situation in which a one-tailed hypothesis test seems justified. Describe a situation in which a two-tailed test is clearly called for.

10. Describe a case in which it would probably be appropriate to use an alpha smaller than the conventional .05 (e.g., .01). Describe a case in which it might be appropriate to use an unusually large alpha (e.g., .1).

B

BASIC STATISTICAL PROCEDURES

The example in Section A of the psychic selecting people high in math ability does not depict a situation that arises often in psychological research. In fact, one-sample hypothesis tests are relatively rare in psychology, for reasons which will be mentioned later in this chapter. However, there are circumstances that can reasonably lead to the use of a one-sample test. I have tried to make the following hypothetical experiment as clear and simple as possible, but it is similar in structure to real experiments that have been published.

Dr. Sara Tonin is a psychiatrist who specializes in helping women who are depressed and have been depressed since early childhood. She refers to these patients as life-long depressives (LLD). Over the years she has gotten the impression that these women tend to be shorter than average. The general research question that interests her is whether there may be some physiological effect of childhood depression on growth function. However, to design an experiment she must translate her general research question into a more specific research hypothesis. If her research hypothesis is framed in terms of population means (probably the most common way to proceed)—such as stating that the mean height of the LLD population is less than the mean height of the general population—she will need to use the following procedures to perform a one-sample hypothesis test.

A hypothesis test can be described as a formal procedure with a series of specific steps. Although I will now describe the steps for a one-sample hypothesis test when the standard deviation for the population is known, the steps will be similar for the other hypothesis tests I will be describing in this text.

Step 1. State the Hypotheses

Based on her observations and her speculations concerning physiology, Dr. Tonin has formed the *research hypothesis* that having LLD reduces the growth rate of women as compared to women in the general population. Notice that she is not specifying how much shorter LLD women are. Her hypothesis is therefore not specific enough to be tested directly. This is common when dealing with research in the "softer" sciences, such as psychology. However, there is another, complementary hypothesis that is easy to test because it is quite specific. Dr. Tonin could hypothesize that LLD women are not shorter than average; rather, that they are exactly the same height on average as the rest of the population. This hypothesis is easy to test only if you know the average height of all women in the population—but, in this special case, that measure happens to be fairly well known.

Dr. Tonin can thus test her research hypothesis *indirectly*, by testing what she doesn't want to be true (i.e., the null hypothesis); she hopes the null hypothesis will be rejected, which would imply that the complementary hypothesis (i.e., the alternative hypothesis) is true. The alternative hypothesis is a specific, and, indirectly, testable version of the research hypothesis that motivated the study. Dr. Tonin may believe that depression produces biological changes that stunt growth, but in this example she is testing one possible, concrete manifestation of her research hypothesis: that LLD women will be shorter, on average, than the population of adult women, in general.

The null hypothesis is so named because it essentially asserts that there is "nothing going on"—in this case, with respect to the height of LLD women. Of course, this hypothesis does not imply that every LLD woman is average in height or that you cannot find many LLD women who are below average. The null hypothesis concerns the mean of the population. Stated more formally, the present null hypothesis is that the mean height for the population of LLD women is exactly the same as the mean height for the population of all women. To state this hypothesis symbolically, a capital H is used with a subscript of zero: H_0. This is usually pronounced "H sub zero" or "H nought." To symbolize the mean of the population being used as the basis of comparison we can use μ with the subscript "hyp" or "0" to indicate that this mean is associated with the null hypothesis (μ_{hyp} or μ_0). So stated symbolically, the null hypothesis (that the mean height of LLD women is the same as the mean height for all women) is $H_0: \mu = \mu_0$. However, before she can test the null hypothesis, Dr. Tonin must get even more specific and fill in the value for μ_0. Although it is not likely that anyone has ever measured the height of every woman in the country and therefore we cannot know the exact value of the population mean, height is one of those few variables for which enough data has been collected to make an excellent estimate of the true population mean (other examples include body temperature and blood pressure). For this example we will assume that $\mu_0 = 65$ inches, so Dr. Tonin can state her null hypothesis formally as $H_0: \mu = 65$ inches.

So far we have paid a lot of attention to the null hypothesis and, as you will see, that is appropriate. But what of Dr. Tonin's research hypothesis? Isn't that what we should be more interested in? As I mentioned, the research hypothesis is represented by the alternative hypothesis, which is generally defined simply as the complement, or negation, of the null hypothesis. Stated formally, the alternative hypothesis for the present example would read as follows: The mean height for the population of LLD women is *not* the same as the mean height for all women. Symbolically it would be written $H_A: \mu \neq \mu_0$, or in this case, $H_A: \mu \neq 65$ inches, where the subscript "A" stands, of course, for "alternative" (sometimes the number 1 is used as the subscript for H, in contrast to using zero for the null hypothesis).

Either the null hypothesis or the alternative hypothesis must be true (either the two means are the same or they aren't). There is, however, one complication that you may have noticed. The alternative hypothesis just described is two-tailed because there are two directions in which it could be correct. This implies that the alternative hypothesis could be true even if the LLD women were taller instead of shorter than average. This is the more conservative way to proceed (refer to Section A). However, if a one-tailed test is considered justified, the alternative hypothesis can be modified accordingly. If Dr. Tonin intends to test only the possibility that LLD women are shorter, the appropriate H_A would state that the mean height of the LLD women is *less* than the mean height of all women. Symbolically this is written $H_A: \mu < \mu_0$, or in this case, $H_A: \mu < 65$ inches. This is now a one-tailed alternative hypothesis. The other one-tailed alternative hypothesis would, of course, look the same except it would use the word *more* instead of *less* (or the *greater than* instead of the *less than* symbol).

To summarize: Before proceeding to the next step, a researcher should state the null hypothesis, decide on a one- or two-tailed test, and state the appropriate alternative hypothesis. Bear in mind that the two-tailed test is far more common in psychological research than the one-tailed test and is considered more conventional.

Step 2. Select the Statistical Test and the Significance Level

Because Dr. Tonin is comparing the mean of a single sample to a population mean, and the standard deviation of the population for the variable of interest (in this case, height) is known, the appropriate statistical test is the one-sample z test. The convention in psychological research is to set alpha, the significance level, at .05. Only in unusual circumstances (e.g., testing a preliminary condition where the null must be accepted for the remaining tests to be valid) might a larger alpha such as .1 be justified. More common than setting a larger alpha is setting a stricter alpha, such as .01 or even .001, especially if many significance tests are being performed. (This process will be discussed in detail in Chapter 13.)

Step 3. Select the Sample and Collect the Data

If Dr. Tonin wants to know if LLD women are shorter than average, there is a definitive way to answer the question that avoids the necessity of performing a hypothesis test. She simply has to measure the height of every LLD woman in the population of interest (e.g., adult women in the United States) and compare the LLD mean with the mean of the comparison population. The difficulty with this approach is a practical one and should be obvious. The solution is to select a random sample of LLD women, find the mean for that sample, and continue with the hypothesis testing procedure. How large a sample? From the standpoint of hypothesis testing, the larger the better. The larger the sample, the more accurate will be the results of the hypothesis test. Specifically, there will be fewer Type II errors with a larger sample; the proportion of Type I errors depends only on alpha and therefore does not change with sample size. However, the same practical considerations that rule out measuring the entire population tend to limit the size of the sample. The size of the sample for any particular experiment will depend on a compromise among several factors, such as the cost of measuring each subject and how large a difference between populations is expected (see Chapter 8). In the present example the sample size may be additionally limited by the number of LLD patients who can be found and their availability to the experimenter. For the purposes of illustration, let us

Table 5.2

65	59	68	63
61	62	60	66
63	63	59	67
58	66	64	64

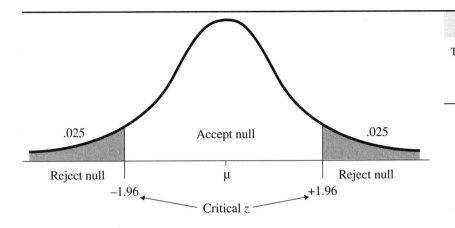

suppose that Dr. Tonin could only find 16 LLD patients, so for this example $N = 16$. The heights (in inches) of the 16 LLD women appear in Table 5.2.

Step 4. Find the Region of Rejection

The test statistic Dr. Tonin will be using to evaluate the null hypothesis is just the ordinary z score for groups, and she is assuming that these z scores follow a normal distribution. More specifically, these z scores will follow the standard normal distribution (a normal distribution with $\mu = 0$ and $\sigma = 1$) if the null hypothesis is true, so the standard normal distribution is the null hypothesis distribution. Therefore, Dr. Tonin can calculate the z score for the data in Table 5.2, look in Table A.1 to find the area beyond this z score, and then double that area to find the two-tailed p value. Finally, she can compare her p value to the alpha she set in Step 2. There is, however, an alternative procedure that is convenient to use with many types of test statistics and was the usual way of checking for statistical significance before the advent of computer data analysis.

Given that she has already set alpha to .05, Dr. Tonin can determine beforehand which z score has exactly .05 area beyond it. However, because she has chosen to perform a two-tailed test, Dr. Tonin must divide alpha in half (half of alpha will go in each tail) and then find the z score for that area. So she would look in Table A.1 for areas beyond z until she found an area that equals .025 and see what z score corresponds to it. If you look for .025 in the "Area Beyond z" column, you will see that it corresponds to a z score of 1.96. The value of the test statistic that corresponds exactly to alpha is called the *critical value* of the test statistic. In the case of z scores, the z score beyond which the area left over is equal to alpha is called a *critical z score*. So 1.96 is the critical z whenever you perform a two-tailed test at the .05 level. Actually, you have two critical values, +1.96 and −1.96, one in each tail, with .025 area beyond each one (see Figure 5.5). In a similar manner you can find the critical z scores for the .01 two-tailed test; they are +2.58 and −2.58, approximately (check the table to verify this for yourself).

Once you decide on the alpha level you will use and on your alternative hypothesis (one-tailed or two-tailed), the critical values of your z test are determined. If the z score you calculate for your experimental group is larger than the critical value, without having to look in the table you know your p value will be smaller than alpha, and your results can be declared statistically significant at the chosen alpha level. For the .05 two-tailed test,

Figure 5.6

One-Tailed Hypothesis
Test, Alpha = .05

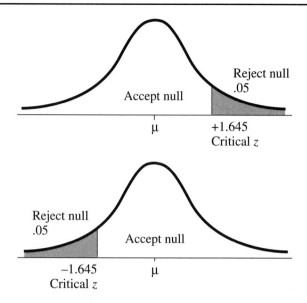

any z score larger than $+1.96$ would correspond to a two-tailed p value that is less than .05 (similarly for any z score more negative than -1.96). Thus, any z score larger than $+1.96$ would lead to the rejection of the null hypothesis at the .05, two-tailed level. That is why the area of the null hypothesis distribution above $+1.96$ is shaded and labeled "Reject null." For the same reason, the area below -1.96 is also labeled "Reject null." Each region of rejection has an area of .025, so when the two are added together, the total area of rejection equals .05, which is the alpha set for this example.

For a one-tailed test, there would be only one region of rejection on the end of the null hypothesis distribution where the experimental result is predicted to fall. If alpha is set to .05 for a one-tailed test, the region of rejection would have an area of .05 but would appear on only one side. Note that in Figure 5.6 this leads to a critical z score that is easier to beat than in the case of the two-tailed test (1.645 instead of 1.96). That is the advantage of the one-tailed test. But there is no region of rejection on the other side in case there is an unexpected finding in the opposite direction. This is the disadvantage of the one-tailed test, as discussed in Section A. Because the region of rejection for a z test depends only on alpha and whether a one- or two-tailed test has been planned, the region of rejection can and should be specified before the z test is calculated. It would certainly not be proper to change the region of rejection after you had seen the data.

Step 5. Calculate the Test Statistic

Once Dr. Tonin has collected her random sample of LLD women, she can measure their heights and calculate the sample mean ($\overline{X}$). If that sample mean happens to be exactly equal to the population mean specified by the null hypothesis, she does not need to do a hypothesis test; she has no choice but to accept (or retain, or fail to reject) the null hypothesis. If she specified a one-tailed alternative hypothesis, and it turns out that the sample mean is actually in the opposite direction from the one she predicted (e.g., H_A states $\mu < 65$, but $\overline{X}$ turns out to be more than 65), again she must accept the null hypothesis. If, however, $\overline{X}$ is in the direction predicted by H_A, or if a two-tailed alternative has been specified and $\overline{X}$ is not equal to

μ_0, a hypothesis test is needed to determine if $\overline{X}$ is so far from μ_0 (i.e., the value predicted by H_0) that the null hypothesis can be rejected.

The mean for the 16 measurements in Table 5.2 is 63 inches. (You should check this for yourself as practice.) Clearly, this mean is less than the average for all women. However, it is certainly possible to randomly select 16 non-LLD women, measure their heights, and get an average that is even lower than 63 inches. To find out the probability of randomly drawing a group of women with a mean height of 63 inches or less, we have to transform the 63 inches into a z score for groups, as described in Chapter 4. The z score of our experimental group of LLD patients with respect to the null hypothesis distribution is the test statistic.

When you have already calculated $\sigma_{\overline{X}}$, you can use Formula 5.1 to get the test statistic. However, to create a more convenient, one-step calculating formula, I will start with Formula 5.1 and replace the denominator ($\sigma_{\overline{X}}$) with Formula 4.5, to get Formula 5.2:

$$z = \frac{\overline{X} - \mu}{\dfrac{\sigma}{\sqrt{N}}}$$

Formula 5.2

Now you can plug in the values for the present example:

$$z = \frac{63 - 65}{\dfrac{3}{\sqrt{16}}} = \frac{-2}{\dfrac{3}{4}} = \frac{-2}{.75} = -2.67$$

Step 6. Make the Statistical Decision

The final step is to compare your calculated test statistic with the appropriate critical value. For a two-tailed test with alpha = .05, the critical value is ± 1.96. Because -2.67 is less than -1.96, you can reject the null hypothesis. (Most often you just ignore the sign of your test statistic and work only with positive values—keeping in mind, of course, the direction of your results.) At this point you can declare that p is less than .05. If you want to find p more exactly, Table A.1 makes this possible. From the table you can see that the area beyond the calculated z score (2.67) is .0038. So the probability of drawing a group from the normal adult female population whose mean height is as short as (or shorter than) your LLD group is .0038. This is the p value Dr. Tonin would use if she had planned a one-tailed test with H_A: $\mu < 65$ (see Figure 5.7). Because we are performing a two-tailed test, the p value would be twice .0038, or .0076, which indicates the probability of drawing a group from the non-LLD population that is at least as *extreme*

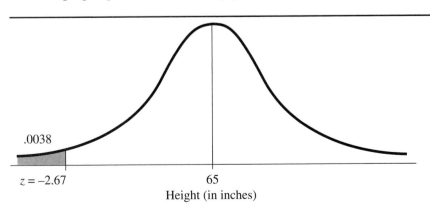

.0038

$z = -2.67$

65

Height (in inches)

Figure 5.7

Null Hypothesis Distribution Showing p Value for Experimental Group

as the LLD group. Because .0076 is less than .05, we know from this method as well that we can reject the null hypothesis at the .05 level, with a two-tailed test.

(Bear in mind that it is still possible that the two populations really have the same mean and the fact that your LLD sample is so short is just a fluke. In other words, by rejecting the null hypothesis you may be making a Type I error [i.e., rejecting a true null]. But at least this is a calculated risk. When the null hypothesis is really true for your experiment, you will make a Type I error only 5% of the time. It is also important to realize that there is no way of knowing when, or even how often the null hypothesis is actually true, so you can never have any idea of just how many Type I errors are actually being made [see Section C]. All you know is that when the null is true, you have a 95% chance of making the correct decision of accepting the null [or as some prefer to say, failing to reject the null] if you use alpha = .05.)

The six steps that I just outlined can be used to describe virtually any type of null hypothesis test, including all of the tests I will outline in the remainder of this text. That is why I took the time to describe these steps in detail. In subsequent chapters the steps will be described more briefly, and if you get confused about a step, you can return to this chapter for a more complete explanation.

Interpreting the Results

Given that there is only a small chance of finding a random sample from the general population with a mean height as low as the LLD women, Dr. Tonin can conclude that the mean for all LLD women is less than the mean for women. This fact lends some support to her theory that the stress of LLD inhibits growth. However, it doesn't prove that her theory is correct—other theories might make the same prediction for very different reasons—but it is encouraging. If there were no difference in height, or if the difference was in the other direction, Dr. Tonin would be inclined to abandon or seriously modify her theory. The amount of discouragement appropriate for results that are not significant depends on power, which will be discussed in Chapter 8. However, if the results were significant in the opposite direction, there would be a strong reason to question the original research hypothesis.

Note that the LLD "experiment" I have been describing is not a *true* experiment because the experimenter did not determine which subjects in the population would be depressed as children and which would not. The LLD women represent a *preexisting group*, that is, a group that was not created by the experimenter; the experimenter only *selected* from a group that was already in existence. It is possible for instance that being short makes the women depressed, rather than the other way around. Or that some third unknown (perhaps, hormonal) condition is responsible for both the shortness and the depression. A true experiment would be more conclusive, but it would be virtually impossible to devise (and unethical to conduct) a true experiment in this case. The LLD study is an example of correlational research, as described in Chapter 1. The distinction between "true" and "correlational" (quasi) experiments is an important concept that I will return to in subsequent chapters.

Assumptions Underlying the One-Sample z Test

To do a one-sample z test we had to know the null hypothesis distribution; that is, we had to know what group means are likely when you draw random samples of a specified size from a known population. The null hypothesis distribution will be a normal distribution with parameters that are easy to find, but only if certain conditions apply. The following are the conditions that researchers must assume are met for the one-sample z test (as I have described it) to be valid.

The Dependent Variable was Measured on an Interval or Ratio Scale

Parametric statistics depend on precise measurements of the dependent variable. If participants can be placed in order on some dimension, but truly quantitative measures cannot be obtained, a nonparametric test may be appropriate (see the last chapter of this text).

The Sample Was Drawn Randomly

For instance, referring to the example involving LLD women, a random sample must be drawn in such a way that each LLD woman in the entire population of LLD women has an equal chance of being chosen, and each of these women is selected independently of all other selections (e.g., once you have selected a particular LLD woman you cannot then decide that you will include any of her close friends who also happen to be LLDs). In reality, a researcher might try to draw a sample that is as representative as possible, but inevitably there will be limitations involving such factors as the subjects' geographical location, socioeconomic status, willingness to participate, and so forth. These factors should be kept in mind when you try to draw conclusions from a study or to generalize the results to the entire population.

The Variable Measured Has a Normal Distribution in the Population

Because of the Central Limit Theorem, this assumption is not critical when the size of the group is about 30 or more (see Chapter 4). However, if there is reason to believe that the population distribution is very far from a normal distribution, and fairly small sample sizes are being used, there is reason to be concerned. In such cases the use of nonparametric statistics should be considered (see Part VII). In all cases the researcher should look at the distribution of the sample data carefully to get an idea of just how far from normal the population distribution might be.

The Standard Deviation for the Sampled Population Is the Same as That of the Comparison Population

In the example of the psychic, I used the standard deviation of the general (i.e., comparison) population to find the standard error for the null hypothesis distribution. However, what if the population being sampled *does* have the same mean as the comparison population (as stated in the null hypothesis) but a very different standard deviation (*SD*)? Then the probability of getting a particular sample mean would *not* follow the null hypothesis distribution, and it is even possible that in some situations we could too easily conclude that the sampled population has a different mean when it has only a different *SD*. Because it would be quite unusual to find a sampled population that has exactly the same mean as the comparison population but a rather different *SD*, this possibility is generally ignored, and the *SD*s of the sampled and the comparison population are assumed to be equal. In a sense it is not a serious error if a researcher concludes that the mean of a sampled population is different from that of the general population, when it is only the *SD* that differs. In either case there is indeed something special about the population in question, which may be worthy of further exploration. I will return to this issue when I discuss the assumptions that underlie hypothesis tests involving two sample means.

Varieties of the One-Sample Test

Testing a Preexisting Group

The one-sample *z* test can be used whenever you know the mean and standard deviation of some variable of interest for an entire population and you want to know if your selected sample is likely to come from this population or from some population that has a different mean for that variable. The LLD example previously described fits this pattern. The mean height and its standard deviation for the population of women in the United States are well known (or at least a very good estimate can be made). We can then ask if based on their mean height, the sample of LLD women is likely to come from the known population. However, we cannot conclude that the feature that distinguishes our sample (e.g., depression) *causes* the difference in the dependent variable (e.g., height).

Performing a One-Sample Experiment

The example I have been discussing deals with two somewhat distinct populations: LLD and non-LLD women. The one-sample *z* test could also be used to test the effects of an experimental treatment on a sample from a population with a known mean and standard deviation for some variable. Consider the following example. You want to know if a new medication affects the heart rate of the people who take it. Heart rate is a variable for which we know the mean and standard deviation in the general population. Although it may not be obvious, testing the effect of some treatment (e.g., the effect of a drug on heart rate) can be viewed in terms of comparing populations. To understand how the heart rate experiment can be viewed as involving two different populations, you must think of the sample of people taking the new drug as representatives of a new population—not a population that actually exists, but one that would exist if everyone in the general population took the drug and their heart rates were measured. All these heart rates taken after drug ingestion would constitute a new population (not a preexisting population like the LLD women, but a population created by the experimenter's administration of the drug). Now we can ask if this new population has the same mean heart rate as the general population (those not taking this drug). If the sample of people taking the drug has a mean heart rate that is not likely to be found by choosing a random sample of people not taking the drug, you would conclude that the two populations have different means. This is the same as saying that the drug does indeed affect heart rate or that if everyone in the population took the drug, the population mean for heart rate would change.

Why the One-Sample Test Is Rarely Performed

As I mentioned at the beginning of this chapter, the one-sample test is not common in psychological research; my chief purpose in describing this test is to create a conceptual bridge to the more complex tests in later chapters. The major problem with one-sample tests is that their validity rests on the degree to which the sample obtained is random. When dealing with humans, as most psychological research does these days, it is virtually impossible for a psychologist to obtain a truly random sample. In the case of the LLD experiment, the fact that you are sampling from an existing population limits your conclusions about causation, but if your sample is not really a random sample of all possible LLDs (and imagine how hard it would be to obtain one), your conclusion is not valid at all. Regardless of your p value, you cannot even conclude that LLDs differ in height from the general population, let alone talk about causation. Any systematic bias completely invalidates your results. For instance, if only poor LLD women can be attracted to the study (perhaps because of a small reward for participating), these women may be shorter not because of depression but because of inferior nutrition that results from being poor. In general, it has been found that people who agree to be in experiments differ in various ways from those who do not (Rosenthal and Rosnow, 1975), so it is rarely possible to obtain a truly random sample.

The one-sample experiment, such as the drug/heart rate experiment described two paragraphs ago, can also be invalidated by using a sample that is not random. In addition, it is difficult to draw clear-cut causal conclusions from an experiment with only one sample. For instance, the heart rates of the people taking the drug may be affected just by being in an experiment, even if the drug has no real effect (e.g., they may fear the injection or the side effects of the drug). The best way to do the experiment is to have a second group (a *control group*) that takes a fake (i.e., totally inactive) drug and is compared to the group taking the real drug. Both groups should have the same fears, expectations, and so forth. The necessity for random samples is replaced by the need to assign subjects randomly to groups—a condition that is much easier to meet. I will discuss such two-group experiments in Chapter 7.

There are situations for which a one-group experiment is reasonable, but these are relatively rare. For instance, the SAT is usually administered in a large room with many students taking the exam at the same time. If a researcher wanted to test the hypothesis that students obtain higher scores when taking the exam in a smaller room with just a few other students, only one random group of students would have to be tested, and they would be tested in the new condition (i.e., smaller room, etc.). A control group would not be necessary because we already have a great deal of data about how students perform under the usual conditions. Moreover, it is theoretically possible to assign subjects to the new testing conditions without their knowledge or consent (of course, there are ethical considerations, but as soon as you ask the subjects to participate, you can be reasonably certain that the consenting subjects will not be a random group).

So far I have implied that the one-sample z test can only be used when the mean and standard deviation of the comparison population are known. Actually if your sample is large enough, you need only have a population mean to compare to; the unbiased standard deviation of your sample can be used in place of the population standard deviation. How large does the sample have to be? When might you have a population mean but not a corresponding population standard deviation? These questions will be answered at the beginning of the next chapter.

Publishing the Results of One-Sample Tests

Let us suppose that Dr. Tonin has written a journal article about her LLD experiment and has followed the guidelines in the *Publication Manual of the American Psychological Association* (APA), fifth edition (2001). Somewhere in her results section she will have a statement such as the following: "As expected, the LLD women sampled were on average shorter ($M = 63$ inches) than the general population ($\mu = 65$ inches); a one-sample test with alpha = .05 demonstrated that this difference was statistically significant, $z = -2.67$, $p < .05$, two-tailed." In the sentence above, M stands for the sample mean and is often written instead of $\overline{X}$ (although both have the same meaning). The designation $p < .05$ means that the p value associated with the test statistic (in this case, z) turned out to be less than .05. Note that M, z, and p are all printed in italics; according to APA style, all statistical symbols should appear in italics, except for Greek letters, such as μ. (Before personal computers made it easy to include italics in your manuscript, underlining was used to indicate which letters were supposed to be printed in italics.)

APA style requires that measurements originally made in nonmetric units (e.g., inches) also be expressed in terms of their metric equivalent. An alternative way of writing about the results of the LLD experiment would be "The mean height for the sample of LLD women was 63 inches (160 cm). A one-sample test with alpha = .05 showed that this is significantly less than the population mean of 65 inches (165 cm), $z = -2.67$, $p < .05$, two-tailed." When many tests are performed in a particular study, it is common to state the alpha and number of tails that will be used for all tests, before reporting the results.

The actual p value for Dr. Tonin's experiment was .0076, which is even less than .01. It is common to report p values in terms of the lowest alpha level at which the results were significant, regardless of the original alpha set, so Dr. Tonin would probably have written $p < .01$, instead of $p < .05$. Generally, possible alpha levels are viewed as values ending in 5 or 1 (e.g., $p < .005$ or $p < .0001$), but researchers are increasingly reporting p values more exactly as they are given by statistical software ($p = .0076$, for this example). It is especially common to report more accurate p values when describing results that are close to but not below the conventional alpha of .05 (e.g., $p < .06$ or $p = .055$). However, if the p value is much larger than alpha, so that the result is not even close to being statistically significant, it is common to state only that p is greater than alpha (e.g., $p > .05$) or that the result is not statistically significant.

B

SUMMARY

1. Null hypothesis testing can be divided into six steps:

   Step 1. State the Hypotheses
   First set up a specific null hypothesis that you wish to disprove. If your dependent variable is IQ, and the mean IQ of the general population is 100, the null hypothesis would be expressed as H_0: $\mu = 100$. The complementary hypothesis is called the alternative hypothesis, and it is the hypothesis that you would like to be true. A two-tailed alternative would be written H_A: $\mu \neq 100$; a one-tailed alternative (which is less common) would be written either H_A: $\mu < 100$ or H_A: $\mu > 100$.

   Step 2. Select the Statistical Test and the Significance Level
   If you are comparing one sample mean to a population mean, and you know the standard deviation of the population for the variable of interest, it is appropriate to perform a one-sample z test. Alpha is usu-

ally set at .05, unless some special situation requires a larger or smaller alpha.

Step 3. Select the Sample and Collect the Data

A random sample of the population of interest must be selected for your test to be valid. The larger the sample, the more accurate the results will be, but practical limitations will inevitably limit the size of the sample.

Step 4. Find the Region of Rejection

The region of rejection can be found in terms of critical z scores—the z scores that cut off an area of the normal distribution that is exactly equal to alpha. The critical z scores for a .05, two-tailed test are $+1.96$ and -1.96 (each z score cuts off .025 of the distribution in each tail).

Step 5. Calculate the Test Statistic

The first step is to calculate the mean of the sample, $\overline{X}$. Then subtract the mean of the null hypothesis distribution and divide by the standard error of the mean. (*Note*: In calculating the z score for groups, there are only a few opportunities to make errors. A common error is to forget to take the square root of N. Perhaps an even more common error is to leave N out of the formula entirely and just divide by the population standard deviation instead of the standard error of the mean. Keep the following rule in mind: If you are asking a question about a group, such as whether a particular group is extreme or unusual, you are really asking a question about the *mean* of the group and should use the z score for groups [Formula 5.2]. Only when you are concerned about one individual score, and where it falls in a distribution, should you use the simple z score for individuals, which entails dividing by the population standard deviation.)

Step 6. Make the Statistical Decision

If the z score you calculate is greater in magnitude than the critical z score, you can reject the null hypothesis. You can also find the p value by determining the amount of area beyond the calculated z score and doubling it for a two-tailed test. The p value can then be compared to alpha to make your decision; if p is less than alpha, you can reject the null hypothesis.

2. Significant results do not mean you can make causal conclusions when you are sampling from a preexisting population. Results that are not significant generally do not allow any strong conclusions at all; you cannot assert that you have proved the null hypothesis to be true.

3. The significance test introduced in this chapter depends on the null hypothesis distribution being a normal distribution centered on the mean of the general population and having a standard deviation equal to the standard error of the mean. We can be sure that this is true only if the following assumptions are met:

   a. Your dependent variable has been measured on an interval or ratio scale.

   b. Your sample mean is based on an independent, random sample of the target population (if dealing with a preexisting group, like LLD women) or an independent, random sample of the general population to which some treatment is applied.

   c. The dependent variable you are measuring is normally distributed in the general population.

   d. The standard deviation of the target population (or the general population after some treatment has been applied) is the same as it is for the general population.

4. The chief problem with one-sample tests is the difficulty involved in obtaining a truly random sample from the population of interest. A

lack of randomness threatens the validity of your statistical conclusion. Moreover, if you are applying an experimental treatment to one sample, the lack of a control group usually makes it difficult to rule out the possibility of confounding variables (e.g., the placebo effect). For these reasons, one-sample tests are rare in the psychological literature.

*1. A psychiatrist is testing a new antianxiety drug, which seems to have the potentially harmful side effect of lowering the heart rate. For a sample of 50 medical students whose pulse was measured after 6 weeks of taking the drug, the mean heart rate was 70 beats per minute (bpm). If the mean heart rate for the population is 72 bpm with a standard deviation of 12, can the psychiatrist conclude that the new drug lowers heart rate significantly? (Set alpha = .05 and perform a one-tailed test.)

*2. Can repressed anger lead to higher blood pressure? In a hypothetical study, 16 college students with very high repressed anger scores (derived from a series of questionnaires taken in an introductory psychology class) are called in to have their blood pressure measured. The mean systolic blood pressure for this sample ($\overline{X}$) is 124 mm Hg. (Millimeters of mercury are the standard units for measuring blood pressure.) If the mean systolic blood pressure for the population is 120 with a standard deviation of 10, can you conclude that repressed anger is associated with higher blood pressure? Use alpha = .05, two-tailed.

3. Suppose that the sample in Exercise 2 had been 4 times as large (i.e., 64 students with very high repressed anger scores), but the same sample mean had been obtained.
   a. Can the null hypothesis now be rejected at the .05 level?
   b. How does the calculated z for this exercise compare with that in Exercise 2?
   c. What happens to the calculated z if the size of the sample is multiplied by k but the sample mean remains the same?

*4. A psychologist has measured the IQ for a group of 30 children, now in the third grade, who had been regularly exposed to a new interactive, computerized teaching device. The mean IQ for these children is $\overline{X} = 106$.
   a. Test the null hypothesis that these children are no different from the general population of third-graders ($\mu = 100$, $\sigma = 16$) using alpha = .05.
   b. Test the same hypothesis using alpha = .01. What happens to your chances of attaining statistical significance as alpha becomes smaller (all else being equal)?

5. Referring to Exercise 4, imagine that you have read about a similar study of IQs of third-graders in which the same sample mean (106) was obtained, but the z score reported was 3.0. Unfortunately, the article neglected to report the number of participants that were measured for this study. Use the information just given to determine the sample size that must have been used.

*6. The following are verbal SAT scores of hypothetical students who were forced to take the test under adverse conditions (e.g., construction noises, room too warm, etc.): 510, 550, 410, 530, 480, 500, 390, 420, 440. Do these scores suggest that the adverse conditions really made a difference (at the .05 level)? Report your p value.

7. Suppose that an anxiety scale is expressed as T scores, so that $\mu = 50$ and $\sigma = 10$. After an earthquake hits their town, a random sample of the townspeople yields the following anxiety scores: 72, 59, 54, 56, 48, 52, 57, 51, 64, 67.
   a. Test the null hypothesis that the earthquake did not affect the level of anxiety in that town (use alpha = .05). Report your p value.
   b. Considering your decision in part a, which kind of error (Type I or Type II) could you be making?

*8. Imagine that you are testing a new drug that seems to raise the number of T cells in the blood and therefore has enormous potential for the treatment of disease. After treating 100 patients, you find that their mean ($\overline{X}$) T cell count is 29.1. Assume that μ and σ (hypothetically) are 28 and 6, respectively.

a. Test the null hypothesis at the .05 level, two-tailed.

b. Test the same hypothesis at the .1 level, two-tailed.

c. Describe in practical terms what it would mean to commit a Type I error in this example.

d. Describe in practical terms what it would mean to commit a Type II error in this example.

e. How might you justify the use of .1 for alpha in similar experiments?

9. a. Assuming everything else in the previous problem stayed the same, what would happen to your calculated z if the population standard deviation (σ) were 3 instead of 6?

b. What general statement can you make about how changes in s affect the calculated value of z?

*10. Referring to Exercise 8, suppose that $\overline{X}$ is equal to 29.1 regardless of the sample size. How large would N have to be for the calculated z to be statistically significant at the .01 level (two-tailed)?

OPTIONAL MATERIAL

The significance level, alpha, can be defined as the probability of rejecting the null hypothesis, *when the null hypothesis is true*. Alpha is called a conditional probability (as defined in Chapter 4, Section C) because it is the probability of rejecting the null *given that a particular condition is satisfied* (i.e., that the null hypothesis is true). Recall that in Table 5.1 the probabilities within each column added up to 1.0. That is because each column represented a different condition (status of H_0), and the probabilities were based on each condition separately. Specifically, the first column takes all the times that the null hypothesis is true and indicates which proportion will lead to rejection of the null (α) and which proportion will not ($1 - \alpha$). The other column deals with those instances in which the null is not true; these latter probabilities will be discussed in detail in Chapter 8.

You may find it easier to view alpha as a percentage than to view it as a probability. But if alpha is 5%, what is it 5% of? First, I will define a null experiment as one for which the null hypothesis is really true. Remember that if the null hypothesis is true, either the experimental manipulation must be totally ineffective or the sampled population must be not at all different from the general population on the given variable—and we would not want such experiments to produce statistically significant results. Alpha can thus be defined as the percentage of null (i.e., *ineffective*) experiments that nonetheless attain statistical significance. Specifically, when alpha is 5%, it means that only 5% of the null experiments will be declared statistically significant.

The whole purpose of null hypothesis testing is to keep the null experiments from being viewed as effective. However, we cannot screen out all of the null experiments without screening out all of the effective ones, so by convention we allow some small percentage of the null experiments to be called statistically significant. That small percentage is called alpha. When you call a null experiment statistically significant, you are making a Type I error, which is why we can say that alpha is the expected percentage of Type I errors.

Conditional probabilities are notoriously difficult to understand and easily lead to fallacies. For instance, it is common to forget that alpha is a conditional probability and instead to believe that 5% of all statistical tests will result in Type I errors—or that 5% of all significant results are really Type I errors (Pollard & Richardson, 1987). To combat these tendencies, I have created a concrete analogy, in which null hypothesis testing (NHT) can be viewed as a kind of spam filter.

NHT as a Spam Filter

Imagine that you are starting out with an empty inbox and an empty spam folder, and that you are about to receive one thousand e-mail messages. However, all of these messages will be scanned by a new spam filter before being sent to any of your folders. The one feature of this filter that you can control directly is the exact percentage of incoming spam messages that will be screened out (i.e., sent to the spam folder); the rest of the spam will, of course, wind up in your inbox. Why not set the screening percentage to 100% (*no* spam in your inbox)? As usual with spam filters, there is a trade-off, the more biased the filter becomes towards identifying a particular message as spam, the higher is the percentage of "real" (i.e., not spam) messages that will be misidentified as spam and sent to your spam folder, instead of the inbox, where they *should* go. Suppose that this filter is so quick to classify a message as spam that when you set the screening percentage to 95 (i.e., only 5% of the spam will be sent to the inbox), half of all the real messages get sent mistakenly to the spam folder. Setting the screening percentage to a number higher than 95 would mean that even more than half of the real messages would be sent to the spam folder, so that is not a desirable option. (You may want to complain that this imaginary spam filter is not a very good one—well, as you will see, it is a pretty fair representation of NHT as used in psychological research.) Now, with the screening percentage set to 95, and 1,000 messages on their way, I want you to guess what percentage of the messages in the inbox will actually consist of spam after all of the messages have been screened and sorted?

If you wanted to guess 5%, it is because the tendency to reverse conditional probabilities can be quite strong when dealing with abstractions. As should become clear shortly, the percentage of incoming spam that goes to the inbox will generally not be the same as the percentage of the inbox that consists of spam. The real answer to the question I posed is that I didn't give you enough information even to make a rough guess. What I didn't tell you, but you needed to know, was the proportion of the 1,000 incoming messages that really were spam messages. Let us turn this question into a specific problem by positing that 80% of the incoming messages are spam; in symbols we can write this as: $P(M_{spam}) = .8$ and therefore $P(M_{good}) = .2$. The conditional probabilities for the two types of classification errors can be written as: $P(IN|M_{spam})$ and $P(SF|M_{good})$, where the former can be read as "the probability of winding up in the Inbox *given* that the message is spam," and the latter as "the probability of going to the Spam Folder *given* that the message is good." By setting the screening percentage at 95%, we are determining that $P(IN|M_{spam})$ will equal .05. Furthermore, I simply gave you the fact that $P(SF|M_{good})$ will equal .5. However, because our focus will be on the contents of the inbox, it will be more convenient to work with the complement of $P(SF|M_{good})$, which is $P(IN|M_{good})$. The latter probability always equals $1 - P(SF|M_{good})$, so for this example, $P(IN|M_{good})$ also equals .5.

So far this has still been rather abstract. However, given an actual number of incoming messages (e.g., 1,000), we can translate the initial probabilities into concrete proportions, and those proportions into actual numbers of messages: $\#M_{spam} = 1,000 \cdot P(M_{spam}) = 1,000 \cdot .8 = 800$ and $\#M_{good} = 1,000 \cdot P(M_{good}) = 1,000 \cdot .2 = 200$. Interpreting the conditional probabilities above as proportions of these proportions, we can say that of the 800 incoming spam messages, $800 \cdot P(IN|M_{spam}) = 800 \cdot .05 = 40$ will wind up in the inbox, and therefore the other 760 will go to the spam folder. Similarly, of the 200 *good* incoming messages, $200 \cdot P(IN|M_{good}) =$

$200 \cdot .5 = 100$ will wind up in the inbox, while the other 100 will end up in the spam folder. Finally, we can determine the total number of messages in the inbox: 40 (spam) + 100 (good) = 140, and then answer the initial question that motivated this example: about 28.57% (i.e., 40/140) of the messages in the inbox will be spam.

Bayes's Theorem

Symbolically, we can write: $P(M_{spam} | IN) = .286$, which is certainly quite different in value from $P(IN | M_{spam})$, which had been set to .05. The procedure just outlined can be summarized by Bayes's Theorem for conditional probabilities, which should be a good deal less intimidating now that you have seen it dealt with in terms of concrete proportions. In terms of the spam filter example, Bayes's Theorem can be written as:

$$p(M_{spam} | IN) = \frac{P(M_{spam})P(IN | M_{spam})}{P(M_{spam})P(IN | M_{spam}) + P(M_{good})P(IN | M_{good})}$$

The formula as just written yields the proportion of the inbox messages that will fall into the spam category, by dividing the amount of spam in the inbox (as a proportion of all incoming messages) by the total amount of messages (both spam and good) in the inbox (again as a proportion of all incoming messages). The fact that nearly a third of the inbox will consist of spam can be viewed as a product of the large proportion of incoming messages that are spam, and the relatively poor job performed by this filter in directing good messages to the inbox.

Now I will reverse the proportions of spam and good messages coming in to see how these proportions affect the contents of the inbox. Also, I will illustrate the use of Bayes's theorem directly to find $P(M_{spam} | IN)$. Given that $P(M_{spam}) = .2$ and $P(M_{good}) = .8$, but that the conditional probabilities (i.e., the proportional rates of the two types of errors) have remained the same, we can find the proportion of the inbox that will consist of spam from the same formula we used previously:

$$p(M_{spam} | IN) = \frac{(.2)(.05)}{(.2)(.05) + (.8)(.5)} = \frac{.01}{.01 + .4} = \frac{.01}{.41} = .024$$

Under these conditions, less than 3% of the inbox will be spam—that is, less than a tenth of the proportion in the previous example. Now, $P(M_{spam} | IN)$ is even smaller than the filter setting—i.e., $P(IN | M_{spam})$—and it will become smaller still, as the proportion of incoming spam decreases, and the filter characteristics stay the same. This result is closer to the reality of null hypothesis testing than the previous example, as I hope to make clear shortly. I just need to draw a few parellels to show you how Bayes' Theorem can help us to understand NHT.

Applying Bayes's Theorem to NHT

Experiments for which the null hypothesis is true are the spam messages, of course, and to send a spam message to the inbox is like declaring the results of a null experiment to be statistically significant (i.e., committing a Type I error). On the other hand, sending a good message to the spam folder is like failing to obtain a significant result when the null is not true (i.e., committing a Type II error). The first example I calculated should help you to see that, ironically, NHT would be rather useless, as currently applied in psychological research, if true null hypotheses were tested often. For example, imagine an organization that tests only alternative medical remedies (e.g., crystals, acupuncture, various herbs). If the null hypothesis happens to be true for 80% of these tests, we can write that $p(H_0) = .8$ and $p(H_A) = .2$, where H_A represents an instance of a good experiment (i.e., H_0 is not true, but H_A, the alternative hypothesis is true). If obtaining a significant result is represented simply as "S," alpha (the Type I error rate) can be written as $p(S \mid H_0)$. The complement of the Type II error, which is called the power of the test, can be written as $p(S \mid H_A)$—the proportion of good experiments that are correctly declared to be significant. Again, assuming that alpha = .05, and power = .5, $p(H_0 \mid S)$, which is the same as $p(M_{spam} \mid IN)$ in the spam example, will be equal to .286. This result implies that nearly one out of every three of the significant results produced by this organization will actually be a Type I error. This is not a very acceptable state of affairs.

Because of the high proportion of nulls being tested by this fanciful organization, and the generally low power of the statistical tests (I chose a power level that is considered common in psychological research), $p(H_0 \mid S)$ is much higher than you might expect. In particular, it is much higher than alpha, which is what an unfortunately high percentage of psychologists expect it to be, because they tend to think that $p(H_0 \mid S)$ will turn out to be the same as $p(S \mid H_0)$. Expressed in words, the common fallacy is that because 5% of null tests will come out significant, people tend to think that 5% of significant results are really "nulls" (i.e., Type I errors). (This fallacy is one form of what is often referred to as the *inverse probability error*, or sometimes just the *reversal fallacy*, and it has actually been presented as true in some popular statistics texts.) As we have just seen, the percentage of siginificant results that are really Type I errors can be much higher than that, if null hypotheses are commonly tested. When applied to NHT, Bayes's Theorem looks like this:

$$p(H_0 \mid S) = \frac{p(H_0)p(S \mid H_0)}{p(H_0)p(S \mid H_0) + p(H_A)p(S \mid H_A)} \qquad \textbf{Formula 5.3}$$

The concept underlying Bayes's Theorem can be depicted concretely as in Figure 5.8.

Fortunately, there is good reason to think that the situation in psychological research is closer to the example in which $P(M_{spam})$ was equal to .2, rather than .8. When $p(H_0)$ is only .2, $p(H_0 \mid S)$ comes out to be .024, which is now less than .05. In reality, those who mistakenly think that a Type I error rate of 5% means that 5% of significant results are really Type I errors are very likely *over*estimating this percentage. Given that true nulls are

---

Figure 5.8

Total Number of
Statistically Significant
Results

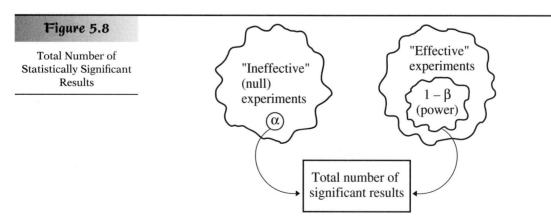

rarely tested in psychological research, it is safe to say that $p(H_0 \mid S)$ is probably much less than .05. However, some writers on this topic worry that if you consider only results that are actually published in journals, $p(H_0 \mid S)$ may be higher than you would expect, because of the strong bias in favor of publishing statistically significant results. In addition, they worry about all the good experiments that failed to reach statistical significance (remember that power is usually thought to be no more than .5 in psychology), and were therefore sent, not literally to a spam folder, but rather to a file drawer, and *not* to the inbox of some journal editor. I will discuss the so-called "file-drawer problem" after a thorough description of power analysis in Chapter 8.

On the other hand, most critics of NHT (generally the same ones who worry about the file-drawer problem) have implied that true nulls are tested so rarely in psychology that there is little point to using NHT at all. Considering that this text is largely about different ways to apply NHT (as most statistics texts in the behavioral and social sciences are), it makes sense to deal with this point next.

Is the Null Hypothesis *Ever* True in Psychological Research?

The reason I chose the psychic example to begin this chapter is that it represents one of the few arenas (e.g., parapsychology) in which there is widespread agreement that the null hypothesis could be exactly true. Imagine another math SAT experiment, in which a randomly selected group of 25 test applicants is chosen to take the SAT under conditions that differ in some specific way from the standard testing conditions. It can be argued that any condition you can imagine—even just releasing a faint, pleasant aroma into the testing room—would have an effect on SAT performance that would not be *exactly* zero in the entire population of interest, and therefore the null hypothesis would not be true.

The debates about when a null hypothesis can be true can get quite philosophical and/or technical (e.g., what if the size of the effect is less than the error involved in measuring it?), but I think that the vast majority of psychologists would concede that it seems to occur very rarely, if at all, that a significance test is applied in their field of interest in which the null hypothesis could plausibly be *exactly* true. Nonetheless, psychologists seem to believe that NHT is useful in helping them to rule out chance factors (mainly variations in sampling) as the sole explanation for their results. These essentially contradictory beliefs just add to the confusion and debate that currently surrounds the use of NHT in psychology and a number of other branches of science. The truth is that NHT has some useful side effects (e.g., experiments that are only slightly effective tend to be screened out by NHT very nearly as well as null experiments are) that are not well understood by most of the scientists who use it. My aim will be to make the (not so obvious) useful effects of NHT clearer in the chapters ahead (especially in Chapter 8).

It is important to note that deciding whether the null hypothesis is exactly true or not with respect to their results is not the only form of statistical analysis that psychologists perform. Sometimes it is important to estimate the actual size of the effect that some experimental treatment produces, rather than just declaring it to be greater than zero. Estimation is the major new topic of the next chapter.

SUMMARY

1. Alpha (α) is a *conditional* probability; it is the probability that the null hypothesis (H_0) will be rejected *given* that H_0 is true, and it is also the probability that a statistical test involving a true H_0 will result in a Type I error.

2. Alpha is a proportion that is set by convention, and does not tell us anything that is interesting, because we don't know how often the null is true when NHT is being performed in psychological research.

3. A Type I error can be thought of as a bit of spam (H_0 is true) that gets past your spam filter (null hypothesis testing), and winds up in your inbox (it attains statistical significance). Alpha, then, is the *proportion* of the spam that will land in the inbox. Using the same analogy, a Type II error occurs when a good message (H_0 is not true) is sent to the spam folder. The proportion of good messages that wind up in the spam folder is called beta (β).

4. Bayes's theorem shows us that the proportion of significant results that are really Type I errors depends on alpha, power (the complement of the Type II error rate), and the relative proportions of null and non-null hypotheses being tested. If alpha is kept at .05, and power remains around .5, the percentage of significant results that are Type I errors will rise above .05, as the proportion of true nulls becomes greater than about .3. If the proportion of true nulls remains below a more plausible maximum value of .1, the percentage of significant results that are Type I errors will be considerably below .05, even if power is as low as .3. Thus, those who succumb to the *reversal fallacy* and therefore believe that setting alpha to .05 means that the proportion of significant results that are really Type I errors will also be .05 are probably being overly suspicious of significant results (but see the discussion of the file-drawer problem in Chapter 8).

EXERCISES

*1. Alpha stands for which of the following?
   a. The proportion of experiments that will attain statistical significance
   b. The proportion of experiments for which the null hypothesis is true that will attain statistical significance
   c. The proportion of statistically significant results for which the null hypothesis is true
   d. The proportion of experiments for which the null hypothesis is true

2. In the last few years, an organization has conducted 200 clinical trials to test the effectiveness of anti-anxiety drugs. Suppose, however, that all of those drugs were obtained from the same fraudulent supplier, which was later revealed to have been sending only inert substances (e.g., distilled water, sugar pills) instead of real drugs. If alpha = .05 was used for all hypothesis tests, how many of these 200 experiments would you expect to yield significant results? How many Type I errors would you expect? How many Type II errors would you expect?

*3. Since she arrived at the university, Dr. Pine has been very productive and successful. She has already performed 20 experiments that have each attained the .05 level of statistical significance. What is your best guess for the number of Type I errors she has made so far? For the number of Type II errors?

4. Suppose that on a particular job aptitude test 20% of the unqualified applicants pass, and are therefore mistakenly identified as being qualified for the job. Also, suppose that 10% of the truly qualified applicants fail the test and are therefore considered unhireable. Assume that all two thousand of the current employees passed that test. How many of the 2,000 are probably unqualified if:
   a. half of all job applicants are unqualified?
   b. only 10% of job applicants tend to be qualified?
   c. only 10% of job applicants tend to be *un*qualified?

*5. Suppose that a new home pregnancy test gives false positive results for 10% of the nonpregnant women who take it, and misses pregnancy in 20% of the pregnant women who take the test. Of all women who get a positive result on the test, what percentage are really not pregnant, if:

    a. half of all women who take the test are pregnant?

b. only 5% of the women who take the test are pregnant?

c. 80% of the women who take the test are pregnant?

The z score for groups, used when the standard error of the mean (σ) has already been calculated:

$$z = \frac{\overline{X} - \mu}{\sigma_{\overline{X}}}$$

Formula 5.1

The z score for groups, used when the standard error of the mean (σ) has *not* already been calculated:

$$z = \frac{\overline{X} - \mu}{\dfrac{\sigma}{\sqrt{N}}}$$

Formula 5.2

Bayes's Theorem as applied to null hypothesis testing:

$$p(H_0 \mid S) = \frac{p(H_0)p(S \mid H_0)}{p(H_0)p(S \mid H_0) + p(H_A)p(S \mid H_A)}$$

Formula 5.3

KEY FORMULAS

INTERVAL ESTIMATION AND THE *t* DISTRIBUTION

You will need to use the following from previous chapters:

Symbols
μ: Mean of a population
$\overline{X}$: Mean of a sample
σ: Standard deviation of a population
s: Unbiased standard deviation of a sample
$\sigma_{\overline{x}}$: Standard error of the mean
N: Number of subjects (or observations) in a sample

Formulas
Formula 4.2: For finding X given a z score
Formula 4.5: The standard error of the mean
Formula 4.6: The z score for groups

Concepts
The null hypothesis distribution
Critical values of a test statistic

Chapter

CONCEPTUAL FOUNDATION

The previous chapter focused on variables that have been measured extensively in the general population, such as height and IQ. We have enough information about these variables to say that we know the population mean and standard deviation. However, suppose you are interested in a variable for which there is little information concerning the population; for instance, the number of hours per month each American family spends watching rented DVDs. The procedures you will learn in this chapter will make it possible to estimate the mean and standard deviation of the population from the data contained in one sample. You will also learn how to deal with a case in which you know the population mean but not its standard deviation, and you would like to know if a particular sample is likely to come from that population. This is not a common situation, but by studying it, you will develop the tools you will need to handle the more common statistical procedures in psychological research. To illustrate a one-sample hypothesis test in which we know the population mean, but not its standard deviation, I have constructed the following example.

Suppose a friend of yours is considering psychotherapy but is complaining of the high cost. (I will ignore the possibility of therapy fees based on a sliding scale to simplify the example.) Your friend says, "The cost per hour keeps going up every year. I bet therapy wasn't so expensive back in the old days." You reply that because of inflation the cost of everything keeps rising, but your friend insists that even after adjusting for inflation, it will be clear that psychotherapy is more expensive now then it was back in, for instance, 1960. The first step toward resolving this question is to find out the hourly cost of therapy in 1960. Suppose that after some library research your friend finds the results of an extensive survey of psychotherapy fees in 1960, which shows that the average hourly fee back then was $22. For the sake of the example, I will assume that no such survey has been conducted in recent years, so we will have to conduct our own survey to help settle the question.

The Mean of the Null Hypothesis Distribution

Suppose that the 1960 survey in our example was so thorough and comprehensive (it was conducted by some large national organization) that we can use the $22 average fee as the population mean for 1960. If we could conduct a current survey just as complete as the one done in 1960, we would have the answer to our question. We would only have to convert the 1960 fee to current dollars and compare the current average with the adjusted 1960 average. (For this example I will assume that 22 dollars in 1960 correspond to 63 current dollars, after adjusting for inflation.) However, considering that our resources are rather limited (as is often the case in academic research) and that there are quite a few more psychotherapists to survey today than there were in 1960, suppose that the best we can do is to survey a random sample of 100 current psychotherapists. If the mean hourly fee of this sample were $72, it would look like hourly fees had increased beyond the adjusted 1960 average (i.e., $63), but because our current figure is based on a limited sample rather than a survey of the entire population, we cannot settle the question with absolute certainty. If we were to announce our conclusion that psychotherapy is more expensive now than it was in 1960, Dr. Null would have something to say.

Dr. Null would say that the mean hourly fee is the same now as it was in 1960: $63 (in current dollars). He would say that our sample mean was just a bit of a fluke based on the chance fluctuations involved in sampling a population and that by sampling a population with a mean of $63, he could beat our $72 mean on his first try. As you learned in Chapter 5, the chance of Dr. Null beating us can be found by describing the null hypothesis distribution.

For this example, the null hypothesis distribution is what you get when you keep drawing samples of 100 psychotherapists and recording the mean hourly fee of each sample. The null hypothesis distribution will have the same mean as the population from which you are drawing the samples, in this case, $63. But the standard deviation of the null hypothesis distribution, called the standard error of the mean, will be smaller than the population standard deviation because groups of 100 do not vary from each other as much as individuals do. In fact, Formula 4.5 presented a simple formula for the standard error:

$$\sigma_{\bar{x}} = \frac{\sigma}{\sqrt{N}} \qquad \text{Formula 4.5}$$

When the Population Standard Deviation Is Not Known

If we try to apply Formula 4.5 to the present example, we immediately run into a problem. The hypothetical 1960 survey did not publish a standard deviation along with its average hourly fee. We have a population mean, but no population standard deviation, and therefore no σ to put into Formula 4.5. If there is no way to obtain the raw data from the 1960 survey, we cannot calculate σ. How can we find the null hypothesis distribution and make a decision about the null hypothesis?

The answer is that we can use the unbiased standard deviation (s) of our sample of 100 hourly fees in place of σ. (We assume that the variability has not changed from 1960 to the present; this assumption will be discussed further in Section B.) By making this substitution, we convert Formula 4.5 into Formula 6.1:

$$s_{\bar{x}} = \frac{s}{\sqrt{N}} \qquad \text{Formula 6.1}$$

where $s_{\bar{x}}$ is our estimate of $\sigma_{\bar{x}}$. By substituting Formula 6.1 into Formula 5.2 we get a modified formula for the one-sample z test, Formula 6.2:

$$z = \frac{\overline{X} - \mu}{\frac{s}{\sqrt{N}}} \qquad \textbf{Formula 6.2}$$

This formula is called the *large-sample z test*, and we can use it just as we used Formula 5.2 to do a one-sample hypothesis test. If you have already calculated an estimate of the standard error of the mean with Formula 6.1, you can use a variation of the formula above that I will label Formula 6.2A:

$$z = \frac{\overline{X} - \mu}{s_{\bar{x}}} \qquad \textbf{Formula 6.2A}$$

The large-sample z test works the same way as the one-sample z test discussed in the previous chapter and has the same assumptions, plus one added assumption: The large-sample test, which involves the use of s when σ is unknown, is only valid when the sample size is large enough. How large the sample must be is a matter of judgment, but statisticians generally agree that the lower limit for using the large-sample test is around 30 to 40 subjects. To be conservative (i.e., to avoid the Type I error rate being even slightly higher than the alpha you set), you may wish to set the lower limit higher than 40. However, most statisticians would agree that with a sample size of 100 or more, the large-sample test is quite accurate.

Calculating a Simple Example

Let us apply the large-sample z test to our example of psychotherapy fees. We can plug the following values into Formula 6.2: $\mu = 63$ (mean adjusted hourly fee in 1960), $\overline{X} = 72$ (mean hourly fee for our sample), $N = 100$ (number of therapists in the sample), and $s = 22.5$ (the unbiased standard deviation of the 100 therapists in our sample). (Normally you would have to calculate s yourself from the raw data, but to simplify matters I am giving you s for this example.)

$$z = \frac{72 - 63}{22.5/\sqrt{100}} = \frac{9}{22.5/10} = \frac{9}{2.25} = 4.0$$

As you should recall from Chapter 5, a z score this large is significant even at the .01 level with a two-tailed test. Therefore, the null hypothesis that $\mu = 63$ can be rejected. We can conclude that the mean hourly fee for the current population of therapists is greater than the mean was in 1960, even after adjusting for inflation. Dr. Null has very little chance of taking a random sample of 100 therapists from the 1960 survey and finding an adjusted mean hourly fee greater than the $72 we found for our current sample.

The *t* Distribution

So far I have done little beyond reviewing what you learned in the previous chapter. But I have set the stage for a slightly more complicated problem. What would happen if we could only find 25 therapists for our current sample? We still wouldn't have σ, and the large-sample z test would not be accurate. Fortunately, the one-sample z test can be modified for dealing with small samples. The reason the large-sample test becomes inaccurate is that it relies on the use of s as a substitute for σ. When the sample is large, s is a pretty good reflection of σ, and there is little error involved in using s as an estimate. However, the smaller the sample gets, the greater is the possibility of s being pretty far from σ. As the sample size gets below about 30, the possible error involved in using s gets too large to ignore.

This was the problem confronting William Gosset in the early 1900s when he was working as a scientist for the Guinness Brewery Company in England (Cowles, 1989). It was important to test samples of the beer and draw conclusions about the entire batch, but practical considerations limited the amount of sampling that could be done. Gosset's most significant contribution to statistics was finding a distribution that could account for the error involved in estimating σ from s with a small sample. Due to his company's restrictions about publishing (Guinness was afraid that other beer companies would make use of Gosset's work), he published his findings under the pseudonym Student. Therefore, the distribution whose usefulness Gosset discovered, for which he used the letter t, came to be known as *Student's t distribution* (or just the t distribution). Gosset's publication preceded the work of R.A. Fisher (mentioned in the previous chapter), who acknowledged the important contribution of Gosset to the study of sampling distributions (Cowles, 1989).

The t distribution resembles the standard normal distribution because it is bell-shaped, symmetrical, continues infinitely in either direction, and has a mean of zero. (The variance of the t distribution is a more complicated matter, but fortunately you will not have to deal with that matter directly.) Like the normal distribution, the t distribution is a mathematical abstraction that follows an exact mathematical formula, but can be used in an approximate way as a model for sampling situations relevant to psychology experiments.

In the previous chapter, I made use of the fact that when you are dealing with a normal distribution, the z scores also have a normal distribution, and they can therefore be found in the standard normal table. The z scores have a normal distribution because all you are doing to the original, normally distributed scores is subtracting a constant (μ) and dividing by a constant (σ or $\sigma_{\bar{x}}$). Now look at Formula 6.2 again. Instead of σ, you see s in the denominator. Unlike σ, s is *not* a constant. When the sample is large, s stays close enough to σ that it is almost a constant, and not much error is introduced. But with small samples s can fluctuate quite a bit from sample to sample, so we are no longer dividing by a constant. This means that z, as given by Formula 6.2, does not follow a normal distribution when N is small. In fact, it follows a t distribution. So when N is small, we change Formula 6.2 to Formula 6.3:

$$t = \frac{\bar{X} - \mu}{\frac{s}{\sqrt{N}}}$$ **Formula 6.3**

The only change, of course, is replacing z with t to indicate that a t distribution is now being followed. If you have already calculated an estimate of the standard error of the mean with Formula 6.1, you can use a variation of the formula above that I will label Formula 6.3A:

$$t = \frac{\bar{X} - \mu}{s_{\bar{x}}}$$ **Formula 6.3A**

Degrees of Freedom and the *t* Distribution

If you are drawing a series of samples, each of which contains only six subjects (or observations), the sample standard deviations will vary a good deal from the population standard deviation. Indeed, when *N* is only 4, there is even more variation. So it makes sense that one *t* distribution cannot correspond to all possible sample sizes. In fact, there is a different *t* distribution for each sample size, so we say that there is a family of *t* distributions. However, we do not distinguish distributions by sample size. Instead, we refer to *t* distributions according to degrees of freedom (df). In the case of one sample, df = $N - 1$, this corresponds to the fact that *s*, the unbiased sample standard deviation, is calculated using $N - 1$ (see Formula 3.6B). Therefore, when we have a sample size of 6, Formula 6.3 follows a *t* distribution with df = 5.

Before you start to think that these *t* distributions are complicated, take a look at Figure 6.1, which shows the *t* distributions for df = 3, df = 9, and df = 20. For comparison, the normal distribution is also included. Notice that as the df increase, the *t* distribution looks more like the normal distribution. By the time df equal 20 there is very little difference between the *t* and the normal distributions. So, it should be understandable that for *N* above 30 or 40 it used to be quite common (before computers became so accessible) to disregard the difference between the two distributions and just use the normal distribution. Nowadays, there is little need for the large-sample *z* test. As you look at Figure 6.1, notice that the important difference between the *t* distribution and the normal distribution is that the former has fatter tails. In terms of kurtosis, this implies that the *t* distribution is leptokurtic (see Section C in Chapter 3).

Figure 6.2 compares one of the *t* distributions (df = 3) with the normal distribution. Notice that for any *z* score in the tail area, the *p* value (the amount of area in the tail beyond that *z* score) is larger for the *t* than for the normal distribution because the *t* distribution has more area in its tails. On the other hand, consider a particular *z* score that cuts off a particu-

Figure 6.1

Comparing the *t* Distribution to the Normal Distribution

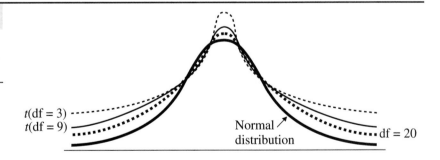

$t(df = 3)$
$t(df = 9)$
Normal distribution
$df = 20$

Figure 6.2

Areas under the *t* Distribution

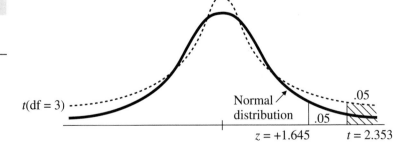

$t(df = 3)$
Normal distribution
.05
.05
$z = +1.645$ $t = 2.353$

lar amount of area in the tail. For instance, $z = 1.645$ has 5% (.05) of the area beyond it. A t value of 1.645 cuts off more than .05. To find the t value that cuts off exactly .05, we have to go further out in the tail. As shown in Figure 6.2, $t = 2.353$ has .05 area beyond it (but only if df = 3). Whereas 1.645 is the critical z for alpha = .05, one-tailed, it should be clear from Figure 6.2 that 2.353 is the critical t for alpha = .05, one-tailed, df = 3. The degrees of freedom must be specified, of course, because as the degrees of freedom increase, the tails of the t distribution become thinner, and the critical t values therefore get smaller. You can see this in Table A.2 in Appendix A, as you look down any of the columns of critical t values. Part of Table A.2 is reproduced as Table 6.1.

| df | AREA IN ONE TAIL | | | Table 6.1 |
	.05	.025	.01	
3	2.353	3.182	4.541	
4	2.132	2.776	3.747	
5	2.015	2.571	3.365	
6	1.943	2.447	3.143	
7	1.895	2.365	2.998	
8	1.860	2.306	2.896	
9	1.833	2.262	2.821	
⋮	⋮	⋮	⋮	
∞	1.645	1.960	2.326	

Critical Values of the *t* Distribution

Table A.2 lists the critical values of the t distribution for the commonly used alpha levels and for different degrees of freedom. It should be no surprise that a .025, one-tailed test corresponds to the same critical values as the .05, two-tailed test because the latter involves placing .025 area in each tail. Of course, critical t gets higher (just as critical z does) as the alpha level gets lower (look across each row in Table 6.1). And as I just mentioned, critical t values *decrease* as df increase. How small do the critical values get? Look at the bottom row of Table 6.1, where the number of degrees of freedom is indicated by the symbol ∞, which means infinity. When the df are as high as possible, the critical t values become the same as the critical values for z because the t distribution becomes indistinguishable from the normal distribution. In fact, the bottom row of the t table can be used as a convenient reference for the critical z values.

You may wonder why the t table looks so different from the standard normal distribution table (Table A.1). You could make a t table that looks like Table A.1, but a different table would have to be created for each possible number of degrees of freedom. Then you could look up the p value corresponding to any particular t value. But there would have to be at least 30 tables, each the size of Table A.1. (After about 30 df, it could be argued that the t tables are getting so similar to each other and to the normal distribution that you don't have to include any more.) By creating just one t table containing only critical values, we sacrifice the possibility of looking up the p value corresponding to any particular t, but researchers are often concerned only with rejecting the null hypothesis at a particular alpha level. However, now that computers are so readily available to perform t tests, we no longer have to sacrifice exact p values for the sake of convenience. Most computer programs that perform t tests also calculate and print the exact p value that corresponds to the calculated t.

Calculating the One-Sample *t* Test

A one-sample test that is based on the *t* distribution is called a *one-sample t test*. I will illustrate this test by returning to the example of psychotherapy fees, but this time assume that only 25 current psychotherapists can be surveyed. I will use Formula 6.3 and assume that μ, $\overline{X}$, and *s* are the same as before, but that *N* has changed from 100 to 25.

$$t = \frac{72 - 63}{22.5/\sqrt{25}} = \frac{9}{22.5/5} = \frac{9}{4.5} = 2.00$$

If the 2.00 calculated were a *z* score, it would be significant at the .05 level (critical $z = 1.96$). Because the sample size is so small, we call the calculated value a *t* value, and we must find the appropriate critical value from the *t* table. The df for this problem equal $N - 1 = 25 - 1 = 24$. Looking in Table A.2 down the column for alpha = .05, two-tailed (or .025, one-tailed), we find that the critical *t* is 2.064. The calculated *t* of 2.00 is *less* than the critical *t* of 2.064, so the calculated *t* does not fall in the rejection area—therefore the null hypothesis cannot be rejected. You can see that the *t* distribution forces a researcher to be more cautious when dealing with small sample sizes; it is more difficult to reach significance. Using the normal distribution when the *t* distribution is called for would result in too many Type I errors—more than the alpha level would indicate. Had we planned a one-tailed *t* test at the same alpha level, the critical *t* would have been 1.711 (look down the appropriate column of Table A.2), and we could have rejected the null hypothesis. However, it would be difficult to justify a one-tailed test in this case.

Sample Size and the One-Sample *t* Test

It is also important to notice that the *t* value for $N = 25$ was only half as large as the *z* score for $N = 100$ ($t = 2.00$ compared to $z = 4.00$), even though the two calculating formulas (6.2 and 6.3) are really the same. This difference has nothing to do with the change from *z* to *t*; it is due entirely to the change in *N*. To see why an increase in *N* produces an increase in the value calculated from these formulas, you must look carefully at the structure of the formulas. The denominator of each formula is actually a fraction, and the square root of *N* is in the denominator of that fraction. So *N* is in the *denominator of the denominator* of these formulas, which means that with a little algebraic manipulation we can move *N* to the numerator (keeping it under its own square root sign) without changing the value you get from the formula. Thus, making *N* larger makes the whole ratio larger. Because we are taking the square root of *N*, increasing *N* by a factor of 4 (e.g., from $N = 25$ to $N = 100$) increases the whole ratio by the square root of 4, which is 2. All other things being equal, increasing the sample size will increase the *t* value (or *z* score) and make it easier to attain statistical significance.

You have seen that increasing the sample size can help you attain statistical significance in two ways. First, a larger sample size means more df and therefore a smaller critical *t* value that you have to beat (until the sample size is large enough to use the normal distribution, after which further increases in sample size do not change the critical value). Second, increasing the sample size tends to increase the calculated *t* or *z* (no matter how large the sample is to start with). So why not always use a very large sample size? As mentioned previously, there are often practical circumstances that limit the sample size. In addition, very large sample sizes make it possible to achieve statistical significance even when there is a very small, uninteresting experimental effect taking place. This aspect of sampling will be discussed at length in Chapter 8.

Uses for the One-Sample *t* Test

The example I used to illustrate the one-sample *t* test, where we knew the population mean for the 1960 fees but not the population standard deviation, does not arise often in the real world. There are a few other types of examples, however, of variables for which there is enough information to make a good estimate of the population mean, but you have no clear idea of the standard deviation. For instance, the soap industry may be able to tell us the total amount of soap used in the United States in any given year, and from that we could calculate the mean consumption per person, without having any idea about the standard deviation. We could use the population mean to test a hypothesis about the soap usage of a sample of obsessive-compulsive patients. Or we may have a very good estimate of the average weight of 1-year-old babies, but no indication of the corresponding standard deviation. In that case, it is not hard to imagine a "special" group of babies you might want to compare to the general population. Finally, a researcher might hypothesize a particular population mean based on a theoretical prediction, and then test this value with a one-sample experiment. For instance, a task may require a musician to rehearse a particular piece of music mentally and signal when finished. The null hypothesis mean could be the amount of time normally taken to play that piece of music.

Probably the most common use of one-sample *t* tests involves participants from two groups that are matched in pairs on some basis that is relevant to the variable measured (or the pairs of scores are created by measuring each participant twice under different conditions). What would otherwise be a two-group *t* test can be transformed into a one-group (matched) *t* test by finding the difference scores for each pair. Because this is a common and important test, I will devote an entire chapter to it. However, to understand matched *t* tests, it is important to understand the concept of correlation, so the topic of matched *t* tests will be postponed until Chapter 11.

Cautions Concerning the One-Sample *t* Test

Remember that even if statistical significance is attained, your statistical conclusion is valid only to the extent that the assumptions of your hypothesis test have been met. Probably the biggest problem in conducting any one-sample test, as mentioned in the previous chapter, is ensuring the randomness of the sample that is drawn. In the example of psychotherapy fees, all psychotherapists in the population being investigated must have an equal chance of being selected and reporting their fee, or the sample mean may not accurately represent the population. However, it is easy for sampling biases to creep into such a study. Therapists working at a clinic may be easier to find, therapists practicing a certain style of therapy may be more cooperative, and therapists charging unusually low or high fees may be less willing to disclose their fees. Any such bias in the sampling could easily lead to a sample mean that misrepresents the true population mean and therefore results in an invalid conclusion. This situation certainly does not improve when small sample sizes are used, so this is one reason why one-sample *t* tests are usually undesirable. As I also mentioned in the previous chapter, when evaluating the results of a one-sample experiment, even if a truly random sample has been obtained, the lack of a comparison group can prevent you from ruling out various alternative explanations (i.e., confounding variables). You usually need a second group that is treated and measured the same way, except for the critical ingredient that you are testing. Two-group experiments will be discussed in the next chapter.

Estimating the Population Mean

Situations in which we know the population mean, or have a basis for hypothesizing about it, are relatively rare. A much more common situation is one in which you would like to know the population mean for some variable and there is no previous information at all. In terms of our previous example, even if there were no prior survey of psychotherapy fees, we might have a good reason to want to know the average fee right now. Some variables pertain to phenomena that are so new that there is little information to begin with—for example, the number of hours per month spent text messaging from a cell phone or listening to music on an MP3 player. This kind of information can be of great interest to market researchers.

A psychologist might want to know how many hours married couples converse about marital problems each week or how many close friends the average person has. Or, a psychologist might be interested in ordinary variables in a particular subpopulation (e.g., the average blood pressure of African-Americans or the average IQ of left-handed people). The procedure to find the population mean for any of these variables is not complicated or difficult to understand. All you need to do is to measure every individual in the population of interest and then take the mean of all these measures. The problem, of course, is a practical one. The most practical solution involves taking a random sample of the population and measuring all the individuals in the *sample*.

We can use a random sample to estimate the mean of the population (μ). Not surprisingly, the best estimate of the population mean that you can get from a sample is the mean of that sample ($\overline{X}$). As you know from what happens to the sampling distribution of the mean, the larger the sample, the closer the sample mean is likely to be to the population mean. This implies that larger samples give better (i.e., more accurate) estimates of the population mean.

Interval Estimation and the Confidence Interval

When using the sample mean as an estimate of the population mean we are making a *point estimate*, suggesting a single value or number—a point— where the population mean is expected to be. In the example concerning psychotherapy fees, the point estimate of the current hourly fee was the mean of our sample—that is, $72. Because larger samples give more accurate point estimates, the larger the sample, the greater our confidence that the estimate will be near the actual population mean. However, the point estimate alone cannot tell us how much confidence to invest in it. A more informative way to estimate the population mean is through *interval estimation*. By using an interval (a range of values instead of just one point) to estimate the population mean, it becomes easy to express how much confidence we have in the accuracy of that interval. Such an interval is therefore called a *confidence interval*.

The common way of constructing a confidence interval (and the only one I will discuss) is to place the point estimate in the center and then mark off the same distance below and above the point estimate. How much distance is involved depends on the amount of confidence we want to have in our interval estimate. Selecting a confidence level is similar to selecting an alpha (i.e., significance) level. For example, one of the most common confidence levels is 95%. After constructing a 95% confidence interval, you can feel 95% certain that the population mean lies within the interval specified. To be more precise, suppose that Clare Inez constructs 95% confidence intervals for a living. She specifies hundreds of such intervals each year. If all the necessary assumptions are met (see Section B), the laws of statistics tell us that about 95% of her intervals will be hits; that is, in 95% of the cases, the population mean will really be in the interval Clare specified. On the other hand, 5% of her intervals will be misses; the population mean will not be in those intervals at all.

The confidence intervals that are misses are something like the Type I errors in null hypothesis testing; in ordinary practice one never knows which interval is a miss, but the overall percentage of misses can be controlled by selecting the degree of confidence. The 95% confidence interval is popular because in many circumstances a 5% miss rate is considered tolerable, just as a 5% Type I error rate is considered tolerable in null hypothesis testing. If a 5% miss rate is considered too large, you can construct a 99% confidence interval to lower the miss rate to 1%. The drawback of the 99% confidence interval, however, is that it is larger and thus identifies the probable location of the population mean less precisely.

Confidence intervals based on large samples tend to be smaller than those based on small samples. If we wanted to estimate the current hourly fee for psychotherapy by constructing a confidence interval, the interval would be centered on our point estimate of \$72. In the case where $N = 100$, the interval would be smaller (the limits of the interval would be closer to the point estimate) than in the case where $N = 25$. In fact, any confidence interval can be made as small as desired by using a large enough sample. This concept should be more understandable when I demonstrate the procedure for calculating the confidence interval in Section B.

SUMMARY

1. If you have a sample (e.g., hourly fees of current psychotherapists) and you want to know if it could reasonably have come from a population with a particular mean (e.g., 1960 population of psychotherapists), but you don't know the population standard deviation (σ), you can still conduct a one-sample z test. To conduct the one-sample z test when you don't know σ, you must have a large enough sample (preferably 100 or more). In that case you can use the unbiased sample standard deviation (s) in place of σ.

2. If the sample size is fairly small (and you still don't know the population standard deviation), you can nonetheless conduct a one-sample test, but you must use the t distribution to find your critical values. The t distribution is actually a family of distributions that differ depending on the number of degrees of freedom (df equals the sample size $- 1$).

3. The t distribution has fatter tails than the normal distribution, which means that the critical value for t will be larger than the critical value for z for the same alpha level (and, of course, the same number of tails). As the df increase, the t distribution more closely resembles the normal distribution. By the time the df reach about 100, the difference between the two types of distributions is usually considered negligible.

4. Larger sample sizes tend to improve the likelihood of significant results in two ways: by increasing the df and therefore reducing the critical value for t, and by causing the formula for t to be multiplied by a larger number. The latter effect is a consequence of the fact that the one-sample t formula can be manipulated so that the square root of the

sample size appears in the numerator (e.g., if all else stays the same, the calculated *t* is doubled when the sample size is multiplied by 4).

5. The sample mean can be used as a *point estimate* of the population mean. However, *interval estimation* provides more information than point estimation. A *confidence interval* (CI) for the population mean can be constructed with the point estimate in the center.

6. The greater the degree of confidence, the larger the confidence interval must be. The 95% CI is the most common. For a given level of confidence, increasing the sample size will tend to make the CI smaller.

EXERCISES

*1. The unbiased variance (s^2) for a sample of 200 participants is 55.
   a. What is the value of the estimated standard error of the mean ($s_{\bar{x}}$)?
   b. If the variance were the same but the sample were increased to 1800 participants, what would be the new value of $s_{\bar{x}}$?

2. A survey of 144 college students reveals a mean beer consumption rate of 8.4 beers per week, with a standard deviation of 5.6.
   a. If the national average is seven beers per week, what is the *z* score for the college students? What *p* value does this correspond to?
   b. If the national average were four beers per week, what would the *z* score be? What can you say about the *p* value in this case?

*3. a. In a one-group *t* test based on a sample of ten participants, what is the value for df?
   b. What are the two-tailed critical *t* values for alpha = .05? For alpha = .01?
   c. What is the one-tailed critical *t* for alpha = .05? For alpha = .01?

4. a. In a one-group *t* test based on a sample of 20 participants, what is the value for df?
   b. What are the two-tailed critical *t* values for alpha = .05? For alpha = .01?
   c. What is the one-tailed critical *t* values for alpha = .05? For alpha = .01?

*5. a. Twenty-two stroke patients performed a maze task. The mean number of trials $\bar{X}$ was 14.7 with s = 6.2. If the population mean (μ) for this task is 6.5, what is the calculated value for *t*? What is the critical *t* for a .05, two-tailed test?
   b. If only 11 patients had been run but the data were the same as in part a, what would be the calculated value for *t*? How does this value compare with the *t* value calculated in part a?

6. a. Referring to part a of Exercise 5, what would the calculated *t* value be if s = 3.1 (all else remaining the same)?
   b. Comparing the *t* values you calculated for Exercises 5a and 6a, what can you say about the relation between *t* and the sample standard deviation?

*7. The data from exercise 3A10a are reprinted here: 3, 8, 13, 23, 26, 26, 26, 27, 28, 28, 29, 30, 32, 41, 49, 56. Perform a one-group *t* test to determine whether you can reject the null hypothesis that the population mean for these data is 34 (use alpha = .05, two-tailed).

8. The data from exercise 2B4 were based on the speeds (in miles per hour) of 25 drivers, and are reprinted here: 65, 57, 49, 75, 82, 60, 52, 63, 49, 75, 58, 66, 54, 59, 72, 63, 85, 69, 74, 48, 79, 55, 45, 58, 51. Test the null hypothesis, at the .01 level (two-tailed), that the mean of the population from which these data were drawn is 55 mph.

*9. Imagine that the *t* value has been calculated for a one-sample experiment. If a second experiment used a larger sample but resulted in the same calculated *t*, how would the *p* value for the second experiment compare to the *p* value for the first experiment? Explain.

10. The calculated *t* for a one-sample experiment was 1.1. Which of the following can you conclude?
   a. The sample mean must have been close to the mean of the null hypothesis distribution.
   b. The sample variance must have been quite large.
   c. The sample size (*N*) could *not* have been large.
   d. The null hypothesis *cannot* be rejected at the .05 level.
   e. None of the above can be concluded without further information.

The formal step-by-step procedure for conducting a large-sample z test is identical to the procedure for conducting a one-sample z test described in Section B of Chapter 5. The only difference is that z is calculated using s instead of σ. The procedure for the one-sample t test is identical to the procedure for conducting a large-sample z test, except for the use of the t table to find the critical values. Therefore, I will not repeat the steps of null hypothesis testing as applied to the one-sample t test in this section. The statistical procedure on which I will concentrate in this section is the calculation of a confidence interval (CI) for the population mean. (This procedure is common, but probably more so in fields other than psychology, such as business and politics.) For the sake of simplicity I will continue with the example of psychotherapy fees. Rather than comparing current fees with those of preceding years, however, I will modify the psychotherapy example and assume that a national organization of therapists is trying to determine the current mean hourly fee for psychotherapy in the United States. The following four-step procedure can be used to find the CI for the population mean.

Step 1. Select the Sample Size

The first step is to decide on the size of the random sample to be drawn. For any given level of confidence, the larger the sample, the smaller (and more precise) will be the confidence interval. (Of course, if your sample was as large as the entire population, the sample mean would equal the population mean, and the confidence interval would shrink to a single point, giving an exact answer.) However, at some point, the goal of increasing the accuracy of the confidence interval will not justify the added expense in time or money of increasing the sample size. If a particular size confidence interval is desired and the population standard deviation is known, the required sample size can be determined (as I will show later in this section). For now, in keeping with the previous example, I will assume that our sample size has been set to 100.

Step 2. Select the Level of Confidence

Next, the level of confidence must be determined for the interval estimate. This is analogous to choosing an alpha level for hypothesis testing. As I stated before, the most common choice is 95%, so I will begin with this level.

Step 3. Select the Random Sample and Collect the Data

Once you have some idea of the desired sample size, a random sample must be collected. The accuracy of the confidence interval depends upon the degree to which this sample is truly random. With a vast, diversified population some form of *stratified* sampling may be used to help ensure that each segment of the population is proportionally represented. This form of sampling, which is often used for public opinion polls and marketing research, requires that the population be divided and sampled according to mutually exclusive strata (e.g., men versus women, levels of socioeconomic status). Stratified sampling and related methods, which are discussed more extensively in texts devoted to research design, often require the resources of a large organization. Individual psychological researchers are less likely to have the resources for this kind of sampling, but they are also less likely to be interested in estimating population means.

Step 4. Calculate the Limits of the Interval

Formula for Large Samples

The formula that is used to find the limits of a confidence interval should be quite familiar because it is just a variation of the one used in null hypothesis testing. In fact, the two procedures are closely related. Recall that in null hypothesis testing, we start with a particular population mean (e.g., the 1960 adjusted mean of $63) and want to know what kind of sample means are likely, so we find the appropriate sampling distribution and center it on the population mean (see Figure 6.3). To find a confidence interval the problem must be reversed. Starting with a particular *sample* mean, the task is to find which *population* means are likely to produce that sample mean and which are not. Therefore, we center the null hypothesis distribution on the *sample mean*, which is our point estimate of the population mean.

In our example, the mean for the random sample of 100 psychotherapists was $72. A normal distribution centered on the sample mean can then indicate which population means are likely and which are not (as in Figure 6.4). For instance, if we want to find the most likely 95% of the possible population means, we just have to find the z scores that mark off the middle 95% of the normal distribution. You will recall that $z = -1.96$ and $z = +1.96$ enclose the middle 95% of a normal distribution. These are the limits of the 95% CI in terms of z scores, as Figure 6.4 shows.

Although Figure 6.4 can be a useful teaching device to clarify the procedure for constructing a CI, it must not be taken literally. Population means do *not* vary around a sample mean—in fact, they do not vary at all. For any given problem, there is just one population mean, and it stays the

Figure 6.3

Null Hypothesis
Distribution

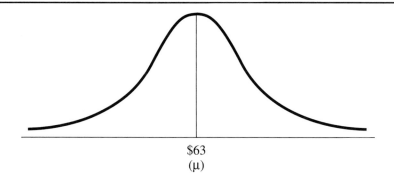

$63
(μ)

Figure 6.4

Ninety-five percent
Confidence Interval for
the Population Mean

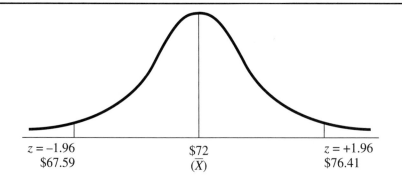

$z = -1.96$ $72 $z = +1.96$
$67.59   ($\overline{X}$) $76.41

same no matter what we do to look for it. However, we do not know the value of the population mean for most problems and never will. Therefore, it can be useful to think in terms of possible population means as distributed around the sample mean; the closer the possible population mean is to the sample mean, the more likely a candidate it is to be the actual population mean.

Knowing that the limits of the CI are $z = \pm 1.96$, we know that we have to go about two standard deviations above and below the point estimate of 72. If we knew the size of the standard deviation of this distribution, we could easily find the limits of the CI. Fortunately, the standard deviation of this imaginary distribution around the sample mean is the same as for the null hypothesis distribution; it is the standard error of the mean given by Formula 6.1, which we computed as part of our calculations for the null hypothesis test in Section A:

$$s_{\bar{x}} = \frac{s}{\sqrt{N}} = \frac{22.5}{\sqrt{100}} = \frac{22.5}{10} = 2.25$$

Now that we know the standard error, we can go about two standard errors (1.96 to be more exact) above and below our point estimate of 72 to find the limits of our confidence interval. Two standard errors = $2 \times 2.25 =$ 4.5, so the limits are approximately $72 - 4.5 = 67.5$ and $72 + 4.5 = 76.5$. We have just found, in an informal way, approximate limits for the 95% confidence interval for the population mean. There is a formula for this procedure, but I wanted you to get the general idea first.

The purpose of Formula 6.2 is to find the z score corresponding to a sample mean, but it can be used in a backward way. When constructing a CI, you start by knowing the z scores (once the confidence percentage is picked, the z score limits are determined). Then those z scores must be converted back to raw scores, which in this case are limits on the population mean (see Figure 6.4). Formula 6.2 can be used for this purpose by filling in the z score and solving for μ. (You must perform this calculation twice, once with the positive z score and again with the corresponding negative z score.) Or you can solve Formula 6.2 for μ *before* inserting any particular values. The formula I'm trying to create will look neater, however, if I start with Formula 6.2A. Multiplying both sides of Formula 6.2A by the standard error, then adding the sample mean to both sides and adjusting the signs to indicate both positive and negative z scores will result in the following formula for the CI:

$$\mu = \bar{X} \pm z_{\text{crit}} s_{\bar{x}}$$

If this formula has a familiar look, that's because it resembles Formula 4.2, which is used when you know the z score for an individual and want to find the corresponding raw score. The above formula is also used when you know the z score, except in this case it is the z score for groups, and you are looking for the population mean instead of for a raw score. This formula gives two values for μ: one above $\bar{X}$ and the other the same distance below $\bar{X}$. These two values are the upper and lower limits of μ, respectively, and together they define the confidence interval. So the formula above can be rewritten in two parts, which together will be designated Formula 6.4:

$$\mu_{\text{lower}} = \bar{X} - z_{\text{crit}} s_{\bar{x}} \qquad \textbf{Formula 6.4}$$

$$\mu_{\text{upper}} = \bar{X} + z_{\text{crit}} s_{\bar{x}}$$

The term μ_{lower} in this formula is just a shorthand way of indicating the lower limit of the confidence interval for the population mean. Any value for μ that is even lower than μ_{lower} would not be considered as being in the confidence interval; an analogous interpretation applies to μ_{upper}. Formula 6.4 can be used to find the exact 95% CI for the example of psychotherapy fees, as follows:

$$\mu_{lower} = 72 - (1.96)(2.25) = 72 - 4.41 = 67.59$$
$$\mu_{upper} = 72 + (1.96)(2.25) = 72 + 4.41 = 76.41$$

Based on the preceding calculations, we can state with a confidence level of 95% that the mean hourly fee for the entire population of current psychotherapists is somewhere between $67.59 and $76.41.

The 99% Confidence Interval

For confidence intervals to be correct more often, say, 99% of the time, the appropriate z_{crit} must be used in Formula 6.4. For the 99% CI, z_{crit} is the z that corresponds to alpha = .01, two-tailed, so $z_{crit} = \pm 2.58$. Plugging the appropriate z_{crit} into the formula, we get:

$$\mu_{lower} = 72 - (2.58)(2.25) = 72 - 5.80 = 66.20$$
$$\mu_{upper} = 72 + (2.58)(2.25) = 72 + 5.80 = 77.80$$

Based on these calculations, we can say with 99% confidence that the mean hourly fee for psychotherapy is currently between $66.20 and $77.80. Note that this interval is larger than the 95% CI, but there is also more confidence associated with this interval estimate. It should be clear that greater confidence requires a larger z score, which in turn results in a larger interval. Whereas 95% confidence is the most conventional level, practical considerations can dictate either a larger or smaller level of confidence. If you are designing a piece of equipment (e.g., a protective helmet) that must be able to accommodate nearly everyone in the population, you may want a 99%, or even a 99.9%, confidence interval for the population mean (e.g., head circumference). On the other hand, if a marketing researcher needs a rough estimate of the average yearly per capita beer consumption in the United States, a 90% CI may suffice.

Interval Width as a Function of Sample Size

There is a way to make the interval estimate smaller and more precise without sacrificing confidence, and that is to increase the sample size. Enlarging N has the effect of reducing $s_{\bar{x}}$, and according to Formula 6.4, the reduction of $s_{\bar{x}}$ will reduce the size of the interval proportionally (e.g., cutting $s_{\bar{x}}$ in half will halve the size of the interval). A glance at Formula 6.1 confirms that the multiplication of N by some factor results in s being divided by the square root of that factor (e.g., multiplying N by 4 divides $s_{\bar{x}}$ by 2). Therefore, the confidence interval is reduced by the square root of whatever factor N is multiplied by (assuming that the sample standard deviation stays the same, which is a reasonable approximation when dealing with large samples).

Because the size of the CI depends on $s_{\bar{x}}$, which depends on N, you can specify any desired width for the interval and figure out the approximate

N required to obtain that width, assuming that the population standard deviation can be reliably estimated. When dealing with large sample sizes, the width of the CI will be about four times the standard error of the mean (1.96 in each direction). If the width of the CI is represented by W (e.g., if you are dealing with height in inches, W is a number of inches), $W = 4s_{\bar{x}} = 4s/\sqrt{N}$. Therefore, $\sqrt{N} = 4s/W$. Finally, by squaring both sides, I can create a formula for the desired sample size:

$$N = \left(\frac{4s}{W}\right)^2 \qquad\qquad \textbf{Formula 6.5}$$

If you want to estimate the mean IQ for left-handed students (assuming $\sigma = 16$ for this IQ test, we can expect s to be about the same), and you want the width of the interval to be five IQ points, the required $N = (4 \times 16/5)^2 = 12.8^2 = 164$ (approx.). Of course, this formula is not very accurate when the sample size is small and you must use s as an estimate of σ.

Confidence Intervals Based on Small Samples

Formula 6.4 is only valid when using large samples or when you know the standard deviation of the population (σ). (In the latter case, $\sigma_{\bar{x}}$ is used in place of $s_{\bar{x}}$.) I did not consider the situation in which you know σ but you are trying to estimate μ because that situation is uncommon in psychological research. Generally, if you know enough about a population to know its standard deviation, you also know its mean. (In the case of physical measurement, the standard deviation may be based entirely on errors of measurement that are well known, but this is not likely in psychology, where individual differences among subjects can be quite unpredictable.)

I discussed the problem of using s to estimate σ when sample size is not large in connection with one-sample hypothesis tests; the same discussion applies to interval estimation as well. For smaller sample sizes (especially those below 30), we must substitute t_{crit} for z_{crit} in Formula 6.4 to create Formula 6.6 (or we can twist around Formula 6.3A, like we did with Formula 6.2A):

$$\mu_{lower} = \bar{X} - t_{crit}s_{\bar{x}} \qquad\qquad \textbf{Formula 6.6}$$
$$\mu_{upper} = \bar{X} + t_{crit}s_{\bar{x}}$$

The appropriate t_{crit} corresponds to the level of confidence in the same way that z_{crit} does: You subtract the confidence percentage from 1 to get the corresponding alpha (e.g., $1 - 95\% = 1 - .95 = .05$) and find the two-tailed critical value. In the case of t_{crit} you also must know the df, which equals $N - 1$, just as for one-sample hypothesis testing. For the small-sample example from Section A, in which only 25 therapists could be polled, df $= N - 1 = 25 - 1 = 24$. Therefore, t_{crit} for a 95% CI would be 2.064—the same as the critical value used for the .05, two-tailed hypothesis test. For the small sample, $s_{\bar{x}}$ is $22.5/\sqrt{25} = 22.5/5 = 4.5$. To find the 95% CI for the population mean for this example we use Formula 6.6:

$$\mu_{lower} = 72 - (2.064)(4.5) = 72 - 9.29 = 62.71$$
$$\mu_{upper} = 72 + (2.064)(4.5) = 72 + 9.29 = 81.29$$

Notice that the 95% CI in this case—62.71 to 81.29—is considerably larger than the 95% CI found earlier for $N = 100$. This occurs for two reasons. First, t_{crit} is larger than the corresponding z_{crit}, reflecting the fact that there is more error in estimating σ from s when the sample is small, and therefore the CI should be larger and less precise. Second, the standard error ($s_{\bar{x}}$) is larger with a small sample size, because s is being divided by a smaller (square root of) N.

Relationship between Interval Estimation and Null Hypothesis Testing

By now I hope the similarities between the procedures for interval estimation and for null hypothesis testing have become obvious. Both procedures involve the same critical values and the same standard error. The major difference is that $\bar{X}$ and μ exchange roles, in terms of which is the center of attention. Interestingly, it turns out that once you have constructed a confidence interval you do not have to perform a separate procedure to conduct null hypothesis tests. In a sense, you get null hypothesis testing for free.

To see how this works, let's look again at the 95% CI we found for the $N = 100$ psychotherapists example. The interval ranged from $67.59 to $76.41. Now recall the null hypothesis test in Section A. The adjusted 1960 population mean we were testing was $63. Notice that $63 does not fall within the 95% CI. Therefore, $63 would not be considered a likely possibility for the population mean, given that the sample mean was $72 (and given the values of s and N in the problem). In fact, the 1960 population mean of $63 was rejected in the null hypothesis test in Section A (for $N = 100$). The 95% CI tells us that $67 or $77 would also have been rejected, but $68 or $76 would have been accepted as null hypotheses. So interval estimation provides a shortcut to null hypothesis testing by allowing you to see at a glance which population means would be accepted as null hypotheses and which would not.

Because the 95% CI gives the range of possible population means that has a 95% chance of containing the population mean, any population mean outside of that range can be rejected as *not* likely at the $1 - 95\% = 5\%$ significance level. In general, if you want to perform a two-tailed hypothesis test with alpha (α) as your criterion, you can construct a CI with confidence equal to $(1 - \alpha) \times 100\%$ and check to see whether the hypothesized μ is within the CI (accept H_0) or not (reject H_0). Thus a 99% CI shows which population means would be rejected at the .01 level. Looking at the 99% CI for $N = 100$ psychotherapists, which ranged from $66.20 to $77.80, we see that $63 falls outside the 99% CI. Therefore, that population mean would be rejected not only at the .05 level but at the .01 level as well. Finally, we turn to the 95% CI for the $N = 25$ example, which ranged from $62.71 to $81.29. In this case, $63 falls *within* the 95% CI, implying that a population mean of $63 would have to be *accepted* at the .05 level when dealing with the smaller sample. This decision agrees with the results of the corresponding null hypothesis test performed in Section A.

The Capture Percentage of a Confidence Interval

One very practical way to think of a CI is in terms of the probability that an exact replication of your study will produce a second sample mean that falls within the CI that you constructed around the original sample mean. Stated another way, you may want to know, on average, what percentage of the sample means produced by repeated replications of your study can be expected to fall within the limits of the CI for your original study. Cumming and Maillardet (2006) refer to this percentage as the *capture percentage* (CP), and they calculated its value over a wide range of circumstances. Note that the CP for a 95% CI is *not* 95%. The 95% refers to the capture percentage for the population mean and not for the next sample mean in a replication study. Because the replication sample mean can be way off relative to the true population mean, the CP for a 95% CI will be considerably less than 95%—Cumming and Maillardet (2006) found it to be closer to 83%. It takes a 99% CI to have a CP that begins to approach 95%. Cumming and Maillardet also found that they could obtain comparable CPs when dealing with rather strange distributions, if they calculated robust CIs based on 20% trimming of their samples. I will discuss robust CIs further in Section C.

Assumptions Underlying the One-Sample *t* Test and the Confidence Interval for the Population Mean

The assumptions that justify the use of the *t* test and the confidence interval are basically the same as those for the one-sample *z* test described in the previous chapter.

Independent Random Sampling

Allow me to include a reminder here about the independence of the sampling. The results would be biased if groups of psychotherapists who work together were deliberately included in the sampling (this could cause you to underestimate the variability among therapists' fees). Some grouping could occur by accident, but any attempt to sample the associates of therapists already included would violate the principle of independence.

Normal Distribution

When a large sample is being used, there is little need to worry about whether the variable of interest (e.g., therapy fees) is normally distributed in the population. The Central Limit Theorem implies that unless the population distribution has a very extreme skew or other bizarre form, *and* the sample is small, the null hypothesis distribution will be closely approximated by a *t* distribution with the appropriate df. (The *t* test is considered "robust" with respect to the normal distribution assumption.) Whereas some statisticians consider 30 a large enough sample to justify obtaining your critical values from the standard normal distribution, the more cautious approach is to use the *t* distribution until the sample size is over about 100. When in doubt, use the *t* distribution; no matter how large (or small) the sample is, the *t* distribution will always yield a valid critical value.

Standard Deviation of the Sampled Population Equals That of the Comparison Population

In the formula for the large-sample test and the formula for the large-sample CI, the standard deviation of the sample is used as an estimate of the standard deviation for the comparison population. This only makes sense if we assume that the two populations have the same standard deviation, even though they may or may not have different means. In terms of the psychotherapy example, we are assuming that the standard deviation in 1960 was the same as it is now, and we are only testing to see if the mean is different. As I mentioned in the previous chapter, a problem arises only if the standard deviation has changed but the mean has not. This is a rather unlikely possibility that is usually ignored.

Use of the Confidence Interval for the Population Mean

One-sample experiments are not common mainly because comparison of the experimental group to some second (control) group is nearly always helpful, as discussed in Section A. On the other hand, constructing confidence intervals for the population mean is quite common. For example, various ratings services, such as the one provided by the A.C. Nielsen Company, periodically estimate the number of televisions in use at a given moment or the mean number of hours that each family has the television turned on during a particular week. These estimates come with "error bars" that create a confidence interval. Also common are opinion polls and election polls. Here the results are usually given as proportions or percentages, again with error bars that establish a confidence interval. Stratified sampling is usually used to ensure that the proportions of various subgroups in the sample (e.g., teen-aged boys, women over 65) reflect the corresponding proportions in the population of interest.

When finding a confidence interval for the mean, it is preferable to use a sample size of at least 100, but there are circumstances that limit sample size so that the *t* distribution must be used. An example would be the estimation of jaw size for a particular prehistoric ape. An anthropologist wishing to make the estimate may have only a dozen jawbones of that ape available for measurement. Or, a medical researcher may have access to only 20 patients with a particular genetic abnormality but would like to estimate that group's population mean for some physiological variable, such as the level of a particular enzyme found in their blood.

Publishing the Results of One-Sample *t* Tests

In Chapter 5, I described the reporting of one-sample tests when the population standard deviation is known, and that description applies here as well, with a couple of simple additions. In the case of the large-sample *z* test when the population standard deviation is *not* known, it is a good idea to include the sample size in parentheses to help the reader judge to what extent the large-sample test is appropriate. For instance: "The hourly fee ($M = \$72$) for our sample of current psychotherapists is significantly greater, z (100) = 4.0, $p < .001$, than the 1960 hourly rate ($\mu = \$63$, in current dollars)."

In reporting the results of a *t* test, the number of degrees of freedom should always be put in parentheses after the *t* because that information is required to determine the critical value for *t*. Also, the sample standard deviation should be included along with the mean, in accord with the guidelines listed in the fifth edition of the APA *Publication Manual*. For instance: "Although the mean hourly fee for our sample of current psychotherapists was considerably higher ($M = \$72$, SD $= 22.5$) than the 1960 population mean ($\mu = \$63$, in current dollars), this difference only approached statistical significance, $t(24) = 2.00, p < .06$."

1. The construction of a confidence interval for the mean of a population can be expressed as a series of four steps:

   Step 1. Select the Sample Size
   Larger samples lead to smaller (i.e., more precise) confidence intervals, but generally cost more to obtain.

   Step 2. Select the Level of Confidence
   The most common level of confidence is 95%, but if there is little tolerance for error, a 99% CI may be preferred. If less confidence is needed, a 90% CI has the advantage of being smaller without increasing the sample size.

   Step 3. Select the Random Sample and Collect the Data
   The randomness of the sample is critical to its accuracy. Stratified sampling can help ensure that a sample is a good representation of the population.

   Step 4. Calculate the Limits of the Interval
   First find the z score limits for your CI (e.g., ± 1.96 for 95%), and then use the formula to find the means corresponding to those z scores. In general, the sample mean is in the middle of the CI, and the limits are found by adding and subtracting the standard error multiplied by the magnitude of the z score limits. For a 99% CI, the z score limits correspond to the .01, two-tailed critical values, ± 2.58.

2. If you want the width of your CI to be W units wide, the sample size required is 4 times the standard deviation divided by W and then squared. This estimate is not very accurate if the sample size is less than 40.

3. If your sample size is less than 40, the z score limits of your CI should be replaced by the appropriate critical values from the t distribution, with df $= N - 1$. Of course, you can use the t instead of the normal distribution for your limits regardless of your sample size; the t value will always be more accurate, although the difference from z becomes negligible for large sample sizes.

4. Once the limits of a CI have been found, it is easy to use the interval for hypothesis testing. Any value for a population mean that is not contained in a 95% CI would be rejected as a null hypothesis at the .05 level, two-tailed. Similarly, a 99% CI allows hypothesis testing at the .01 level, two-tailed.

5. The assumptions for the one-sample t test are the same as those for the one-sample z test: independent random sampling, a normally distributed variable, and the same standard deviation for both the general (null hypothesis) population and the treated population (or the special population being sampled). The assumptions involved in confidence intervals are the same as those for the corresponding hypothesis tests.

6. Confidence intervals for the population mean are frequently constructed for marketing and political purposes and usually involve stratified random sampling to ensure a proportional representation of the population. Occasionally, CIs must be constructed from small sample sizes, in which case the t distribution should be used to find the limits.

*1. A high school is proud of its advanced chemistry class, in which its 16 students scored an average of 89.3 on the statewide exam, with $s = 9$.
  a. Test the null hypothesis that the advanced class is just a random selection from the state population ($\mu = 84.7$), using alpha = .05 (two-tailed).
  b. Test the same hypothesis at the .01 level (two-tailed). Considering your decision with respect to the null hypothesis, what type of error (Type I or Type II) could you be making?

2. Are serial killers more introverted than the general population? A sample of 14 serial killers serving life sentences was tested and found to have a mean introversion score ($\overline{X}$) of 42 with $s = 6.8$. If the population mean (μ) is 36, are the serial killers significantly more introverted at the .05 level? (Perform the appropriate one-tailed test, although normally it would not be justified.)

*3. A researcher is trying to estimate the average number of children born to couples who do not practice birth control. The mean for a random sample of 100 such couples is 5.2 with $s = 2.1$.
  a. Find the 95% confidence interval for the mean (μ) of all couples not practicing birth control.
  b. If the researcher had sampled 400 couples and found the same sample mean and standard deviation as in part a, what would be the limits of the 95% CI for the population mean?
  c. Compare the width of the CI in part a with the width of the CI in part b. What is the general principle relating changes in sample size to changes in the width of the CI?

*4. A psychologist studying the dynamics of marriage wanted to know how many hours per week the average American couple spends discussing marital problems. The sample mean ($\overline{X}$) of 155 randomly selected couples turned out to be 2.6 hours, with $s = 1.8$.
  a. Find the 95% confidence interval for the mean (μ) of the population.
  b. A European study had already estimated the population mean to be 3 hours per week for European couples. Are the American couples significantly different from the European couples at the .05 level? Show how your answer to part a makes it easy to answer part b.

5. If the psychologist in exercise 4 wanted the width of the confidence interval to be only half an hour, how many couples would have to be sampled?

6. A market researcher needed to know how many blank videotapes are purchased by the average American family each year. To find out, 22 families were sampled at random. The researcher found that $\overline{X} = 5.7$, with $s = 4.5$.
  a. Find the 95% confidence interval for the population mean.
  b. Find the 99% CI.
  c. How does increasing the amount of confidence affect the width of the confidence interval?

*7. A study is being conducted to investigate the possibility that autistic children differ from other children in the number of digits from a list that they can correctly repeat back to the experimenter (i.e., digit retention). A sample of ten 12-year-old autistic children yielded the following number of digits for each child: 10, 15, 14, 8, 9, 14, 6, 7, 11, 13.
  a. If a great deal of previous data suggest that the population mean for 12-year-old children is 7.8 digits, can the researcher conclude that autistic children are significantly different on this measure at the .05 level (two-tailed)? At the .01 level (two-tailed)?
  b. Find the 95% confidence interval for the mean digit retention of all 12-year-old autistic children.

8. A psychologist would like to know how many casual friends are in the average person's social network. She interviews a random sample of people and determines for each the number of friends or social acquaintances they see or talk to at least once a year. The data are as follows: 5, 11, 15, 9, 7, 13, 23, 8, 12, 7, 10, 11, 21, 20, 13.
  a. Find the 90% confidence interval for the mean number of friends for the entire population.
  b. Find the 95% CI.

c. If a previous researcher had predicted a population mean of 10 casual friends per person, could that prediction be rejected as an hypothesis at the .05 level, two-tailed?

*9. Find the 95% CI for the data in exercise 6A7. Is this CI consistent with the conclusion you drew with respect to the null hypothesis in that exercise? Explain.

10. A study of 100 pet owners revealed that this group has an average heart rate ($\overline{X}$) of 70 beats per minute. By constructing confidence intervals (CI) around this sample mean a researcher discovered that the population mean of 72 was contained in the 99% CI but not in the 95% CI. This implies that

a. Pet owners differ from the population at the .05 level, but not at the .01 level.
b. Pet owners differ from the population at the .01 level, but not at the .05 level.
c. Pet owners differ from the population at both the .05 and the .01 level.
d. Pet owners differ from the population at neither the .05 nor the .01 level.

At the end of Section A I asserted that the mean of a sample ($\overline{X}$) is the best point estimate of the population mean (μ). Actually, there are a number of ways to evaluate how good a particular method is for estimating a population parameter, so it can be difficult to say that any one approach is the best. Fortunately, the sample mean turns out to be a good estimator by several important criteria. There is a whole branch of mathematical statistics that deals with estimation and estimators, but I will only touch on the subject briefly in this section. You can read more about this topic in an advanced statistics text (e.g., Hays, 1994).

OPTIONAL MATERIAL

An *estimator* is defined as a formula, that when applied to sample data produces an estimate of a population parameter. The formula for the sample mean is an estimator, as are Formulas 3.4A and 3.6A. One property of an estimator is whether or not it is biased. As I explained in Chapter 3, both the sample mean ($\Sigma X/N$) and Formula 3.6A are unbiased, whereas Formula 3.4A is a biased estimator of the population variance. In general, estimators that are unbiased are more desirable and are considered better estimators than those that are biased. However, bias is not the only important property of estimators. In fact, a biased estimator can be better than an unbiased one if it is superior in certain other properties. I will briefly mention a few other properties of estimators next.

Some Properties of Estimators

In addition to being unbiased, another desirable property of an estimator is *consistency*. An estimator is consistent if it has a greater chance of coming close to the population parameter as the sample size (N) increases. We know that the sample mean is consistent beacuse the sampling distribution of the mean gets narrower as N gets larger. Both the biased and the unbiased sample variances are consistent estimators, as well.

One reasonable way of determining which of two estimators of a particular population parameter is better is based on *relative efficiency*. If there are two different unbiased estimators of the same population parameter, the estimator that has the smaller sampling variance is the more efficient (assuming the same sample size for the two estimators).

Maximum Likelihood Principle

In Chapter 3, I emphasized that the mean is the sample statistic that minimizes squared errors; if you subtract any other particular value from each of the scores in the sample, and add all of the squared differences, you will get a larger *sum of squares* than you would get when using the mean as the value subtracted. Therefore, the sample mean is sometimes referred to as a *least-squares* estimator. Another important property that an estimator can have is that a particular estimator may yield the value of a population parameter that makes your sample data appear more likely than does any other value for that population parameter. Explaining the *maximum-likelihood* principle of estimation would require going well beyond the level of this text, but I would like to at least introduce you to this important statistical concept.

Given a particular mean for a normal distribution, we can say that some sample values are more likely to occur than others (i.e., values closer to the mean). Now imagine that there is a mathematical function (and I'm not pretending that this is easy to do) that gives you a probability value for your entire sample as a whole, as a function of some unknown population parameter. That is, for any given value of the population parameter of interest (e.g., the population mean), this function tells you how likely it would be to obtain the set of values in your particular sample. This function is therefore called a *likelihood function*. The value of the unknown parameter that yields the largest likelihood for your sample data can be considered the best point estimate of that parameter. Certainly, a value for the population mean that makes your sample values as likely as possible, compared to any other value that the population mean could have, is a good estimate of the population mean. It turns out, rather conveniently, that the sample mean always gives you the value of the population mean that maximizes the likelihood of your sample data. Therefore, among its other useful properties, the sample mean is the *maximum-likelihood estimator* of the population mean.

Robustness

Despite all of its useful properties, the sample mean has its weaknesses. As we have seen in Chapter 3, the mean can be affected a great deal by a few extreme outliners on one side of a distribution. Whereas the mean is a particularly good measure of central tendecy for any roughly symmetrical, bell-shaped curve, it is not necessarily the best measure to use when your population distribution is strongly skewed. In Section B of this chapter, I mentioned that the *t* test is considered *robust* with respect to the normality assumption—that is, the *p* values it yields are reasonably accurate even when the dependent variable under study does not follow a normal distribution in the population. However, the *t* test can be made more robust by basing it on a measure of central tendency that is more robust that the ordinary sample mean. I will consider such a possibility next.

A More Robust *t* Test

In Section C of Chapter 3, I described "trimming" as one way to deal with anticipated outliers. The mean of the values that remain after trimming, called the *trimmed mean,* is a more robust estimator of the population mean than is the mean of all of the values in the sample. Of course, even a trimmed mean can be influenced by outliers, if the distribution being sampled is such that a large proportion of the scores can be considered outliers. The greater the percentage of trimming, the greater is the protection against outliers. However, this protection comes at a cost—as more and more data are thrown away, power is reduced. The percentage of trimming that should be chosen is based on a balancing of the need for robustness and the need for power. Wilcox (2001) makes a strong case for 20% trimming as a compromise that can be used effectively in a wide range of data situations.

The Effects of an Outlier on a One-Sample *t* Test

To motivate the need for robust statistics in general, I will demonstrate the devastating effect that a single not-so-distant outlier can have on the one-sample *t* test. For a convenient data set, I will use the heights of 16 LLD women as given in Table 5.2, and reprinted here in numerical order: 58, 59, 59, 60, 61, 62, 63, 63, 63, 64, 64, 65, 66, 66, 67, 68. The ordinary *t* test against a hypothesized population mean of 65 yields the following results:

$$t = \frac{63.0 - 65.0}{\frac{3.011}{\sqrt{16}}} = \frac{-2.0}{.7528} = -2.66$$

(This result differs slightly from the *z* value found in Chapter 5, Section B, not because of rounding error, but because the *t* test uses the *s* from the sample, which is 3.011, rather than σ—which was assumed to be 3.0, but is usually not known.)

The calculated *t* value easily exceeds the critical value: $2.66 > t_{.05}(15) = 2.131$ (all of the tests in this section are two-tailed, so I will be ignoring the minus sign when testing for significance). Now, I will replace the lowest score in this data set, 58 inches, with an outlying value of 45 inches. You might think that this change would increase the magnitude of your *t* value. After all, reducing the lowest score by 13 inches pulls down the mean from 63.0 to about 62.2 inches. However, the standard deviation increases from 3.011 to 5.32. The increase in *s* outweighs the reduction in $\overline{X}$, as you can see from the following calculation:

$$t = \frac{62.1875 - 65.0}{\frac{5.32}{\sqrt{16}}} = \frac{-2.8125}{1.33} = -2.11$$

The *t* value is no longer significant. The problem is that due to the squaring involved in its calculation, the standard deviation is even more sensitive to the effects of outliers than the mean. In reality, you would probably be able to exclude the score of 45 inches on some independent basis, but in general you cannot go around discarding any outliers that hurt your test statistic and retain any that help. If you suspect that the distribution of your dependent variable is markedly skewed or "heavy-tailed" (i.e., leptokurtic—having an unusually large proportion of extreme scores on both sides of the distribution), you can increase the accuracy (and hence the validity) of your *t* test by planning a robust test based on trimming your sample.

One-Sample *t* Test for a 20% Trimmed Mean

Let us look at how the one-sample *t* test changes when applied to a trimmed sample. I will begin by applying 20% trimming to the LLD data reprinted in the previous example. This means that both the lowest *and* highest 20% of the scores in the sample should be deleted—though in practice somewhat less than 40% of the scores are usually trimmed, due to rounding off the percentage. Here's how it works. To determine how many scores to delete, first multiply N by .2; for this example, .2N equals .2 · 16 = 3.2. Then, drop any fraction. In this case, the number of scores to be deleted from each end of the array is 3 (even if .2N comes out to 3.9, the deletion number is just 3). The trimmed sample looks like this: 60, 61, 63, 63, 63, 64, 64, 65, 66. The trimmed mean, $\overline{X}_t$, is the mean of these 10 values, so $\overline{X}_t = 63.1$.

To complete the t test we need to estimate the standard error of the mean. You might be tempted to calculate the unbiased standard deviation of the remaining ten scores and divide it by the square root of 10, but that would not be fair. Having tossed out the highest and lowest scores from your sample, calculating s on the remaining scores would severely underestimate σ. On the other hand, extreme scores have an unduly strong influence on s, so there would be little point to trimming if we were going to include all of the original scores in our calculation of s. A reasonable compromise is to "Winsorize" the original sample, as described in Section C of Chapter 3, and then calculate s for the Winsorized sample (Wilcox, 2003). To create the Winsorized data set, just start with the trimmed data, and repeat the lowest and highest values until the number of scores that were deleted on each end of the array have been replaced. For our example, the lowest score, 60, is added back to the sample three times, as is the highest score, 66, to yield: 60, 60, 60, 60, 61, 62, 63, 63, 63, 64, 64, 65, 66, 66, 66, 66. Note that we are back to having an N of 16, but with somewhat less dispersion. The (unbiased) SD of these numbers—I'll call it s'—is 2.35, which is distinctly less than the SD of the original sample (3.01), but considerably larger than the SD of the trimmed data ($s_t = 1.79$).

In order to test the difference of the trimmed mean (63.1 inches) from the population mean of all women (65 inches), we have to use s' to estimate the standard error of the trimmed mean, $s'_{\overline{X}}$. The formula for 20% trimming is as follows: $s'_{\overline{X}} = s'/(.6\sqrt{N})$, where N is the number of scores in the original (also the Winsorized) sample, and .6 is the (rounded) proportion of the original scores that remain after trimming. Therefore, $s'_{\overline{X}} = 2.35/(.6 \cdot 4) = 2.35/2.4 = .98$, and the t for testing the trimmed mean is: $t' = (63.1 - 65)/.98 = -1.9/.98 = 1.94$. The full formula for testing one 20% trimmed mean is:

$$t = \frac{\overline{X}_t - \mu_0}{\dfrac{s'}{.6\sqrt{N}}}$$

Formula 6.7

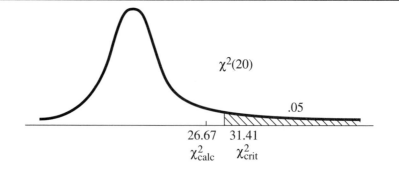

$\chi^2(20)$

.05

26.67   31.41

χ^2_{calc} χ^2_{crit}

standard deviation of the Winsorized sample, P is the trimming proportion (e.g., for 20% trimming, $P = .2$), and N is the size of the original sample. For the 20% trimmed LLD data, all of these quantities were calculated for the trimmed t test, so I can just plug those values into Formula 6.8 to find the 95% CI for the 20% trimmed population mean:

$$\mu_{lower} = \overline{X}_t - t_{crit} \frac{s'}{(1 - 2P)\sqrt{N}} = 63.1 - 2.262 \frac{2.35}{(1 - 2 \cdot .2) \cdot 4}$$

$$= 63.1 - 2.262 \cdot .98 = 63.1 - 2.22$$

Similarly, $\mu_{upper} = 63.1 + 2.22$, so the limits of the interval are 60.88 and 65.32. Note that the hypothesized population mean of 65.0 is contained within the 95% CI, confirming that the trimmed mean of 63.1 does *not* differ significantly from 65.0 at the .05 level (two-tailed).

A Bootstrapped Confidence Interval

The confidence interval calculated above makes use of the critical values of the t distribution; therefore, the accuracy of that CI depends to some extent on the normality assumption, as well as on the appropriateness of our trimming percentage. The method I will describe next can free you from the need to make such assumptions. In Chapter 5, I mentioned that, *in theory*, you could repeat Dr. Null's experiment (i.e., just random sampling) very many times, until the means of all those samples pile up into a distribution that you could then use as the null hypothesis distribution (NHD). It is, of course, infinitely more convenient to make a few simplifying assumptions, and then use a ready-made distribution, such as one of the t distributions, as your NHD. However, whereas it is not feasible to perform many null experiments in the real world, modern computers now make it easy to do what may be the next best thing: *resampling*.

The basic principle of resampling is to use the distribution of the data you have already collected as your best guess for the distribution of the population from which you are sampling. Then you can use a statistical program to select many independent random samples from your own data to create an empirical NHD. It should not be hard to see why resampling procedures are often referred to as *computer-intensive methods* (Kline, 2004). There is more than one resampling method, but probably the simplest to understand is the one that is known as *bootstrapping*. This name comes from the expression "to pull yourself up by your bootstraps." This, of course, is physically impossible to do, but it has come to symbolize the goal of being independent. In statistical terms, bootstrapping allows you to be independent from making assumptions about the distribution of your dependent variable. To show how bootstrapping can be used to create a CI for a population mean, I will describe the procedure that Wilcox (2001) refers to as *the percentile t bootstrap*.

The bootstrap procedure could hardly be more straightforward. To apply this procedure to the LLD data, the computer program would begin by drawing a sample of $N = 16$ (the same size as your dataset) *with replacement*. That last stipulation, "with replacement," is crucial; it is the only way you could get a collection of 16 numbers that differs from your original dataset. For instance, because each score selected for the new sample of 16 scores is replaced in the dataset before the next selection, you could wind up with a strange sample consisting of sixteen 58s (or sixteen 68s). This would not happen very often, but it would not be impossible (the probability of drawing such a sample is one sixteenth raised to the 16th power, a very tiny fraction indeed). As a somewhat more realistic example, let's suppose the computer selects eight 63s and eight 64s for its first sample. The next step is to calculate a *t* value for this sample, based on the sample's mean and standard deviation, and using the mean of the original data set as the population mean. For the sample just described the *t* value is:

$$t' = \frac{\overline{X}_b - \overline{X}}{\frac{s_b}{\sqrt{N}}} = \frac{63.5 - 63}{\frac{.5164}{4}} = \frac{.5}{.1291} = 3.873$$

I labeled this value as t' instead of t, because there is no guarantee that these values will follow a distribution that even resembles the *t* distribution (the subscript b indicates that a statistic is from the bootstrap sample). Then, another sample of 16 values is drawn, and another value for t' is calculated. Wilcox (2001) recommends that 999 values for t' should be calculated in this way, and a frequency distribution of these t' values created. This frequency distribution serves as your NHD; it is from this distribution that you obtain the critical t' values for your CI. For the 95% CI, find the 2.5 and 97.5 percentiles of the t' distribution you generated (if you created about 1,000 t' values, you would be looking for the *t*s that are 25 scores above the lowest t', and 25 scores below the highest t'. Finally, convert these t's into raw scores by using the ordinary CI formula for the population mean, in which the mean and standard error are based on the original sample. (The mean and standard error of the LLD data are the numerator and denominator, respectively, of the first *t* test calculated in this section: 63.0 and .7528.) Note that rather than calculating ordinary *t* values as the basis for your percentile bootstrap, you can use Formula 6.7, and thus base your bootstrap procedure on robust *t* values, instead.

When Should You Use Robust Methods and Which Ones?

Some quantitative psychologists (e.g., Rand Wilcox, on whose work much of this section is based) suggest that you should always use robust statistical methods, unless you are quite sure that the distribution you are dealing with does not vary substantially from the normal distribution. For instance, it is easy to see when the scores in your sample are strongly skewed in one direction or the other, and this is a good indication that you are dealing with a skewed population distribution, rather than a normal distribution. For a distribution with an extreme positive skew, it is fairly common to apply a log transformation to all of the values in the sample, and then to perform inferential statistics on the log-transformed values. The difficulty with this approach occurs when you have to extrapolate from inferences about your log-transformed values to inferences about your original data. The trimmed mean or one of the *M*-estimators may yield results that are easier to interpret.

A departure from normality that can be hard to detect in your sample data involves kurtosis, as defined in Chapter 3, Section C. In particular, dealing with a leptokurtic (i.e., heavy-tailed) distribution means that you may have an unusual number of extreme scores on both the low and high ends of your sample data, which can greatly inflate your standard deviation, and thus reduce your chances of attaining significant results. Trimming is a particularly appropriate and straightfoward way to deal with this potential problem. Given the difficulty in detecting leptokurtosis, it can be argued that you should always trim your data to be on the safe side.

However, as I will remind you again and again in the upcoming chapters, when there are such choices to make in handling your data, your best bet will usually be to look for articles—published in a journal in which you would like to be published—that deal with data resembling your own. Researchers in a specific branch of psychology often evolve their own consensus with respect to the handling of particular data problems, which may disagree with the customs of other research areas, as well as the advice of many statistical experts. Given that publication is usually an important goal of collecting data, it makes sense to familiarize yourself with the statistical practices in your specific area of research. Although the exercises in this text will always tell you which statistical methods to apply, it would be wise to heed any advice that your course instructor may give whenever this text presents more than one alternative approach to the same statistical analysis.

C

SUMMARY

1. *Estimators* are formulas, which when applied to sample data, can be used to estimate population parameters. Good estimators are those that tend to be *unbiased* (their average values are equal to the population parameter, rather than slightly above or below it), *consistent* (they are more likely to be close to the parameter as the sample size increases), and *relatively efficent* (they exhibit less variability than other estimators for a given sample size).

2. The arithmetic mean of a sample happens to be not only a *least-squares* estimator of the population mean, it is also the *maximum-likelihood* estimator. However, it is not very robust; it is affected rather easily by extreme scores.

3. A trimmed mean is more robust than an ordinary mean. The larger the percentage of trimming, the more robust the trimmed mean. When trimming is maximized, the trimmed mean is equal to the median.

4. A *t* test based on the trimmed mean uses the variance of the Winsorized sample to estimate the standard error of the mean. As the trimming percentage is increased, the variance tends to decrease, but the reduction in the effective sample size can lead to a net loss of power. For large samples, 20% trimming has been found to be optimal. A robust CI can be found by rearranging the terms of the robust *t* test.

5. *M*-estimators are similar to trimmed means; however, a complex formula is used to determine which scores appear to be outliers and should therefore be deleted.

6. *Resampling* is a computer-intensive alternative to trimming one's samples, which does not require assumptions about the population distribution. A straightforward form of resampling, especially when dealing with a single sample, is *bootstrapping*. The computer draws samples from your original data, with replacement, and calculates a *t* value for each sample. After about a thousand such *t* values have been calculated, the obtained (empirical) frequency distribution can be used to find appropriate "critical *t*" values on which to base a confidence interval.

*1. Recalculate the one-group *t* test and 95% CI for the data in Exercise 6B7, using 10% trimming. Does the trimmed sample mean differ significantly from the population mean of 7.8 at the .05 level (two-tailed)?

2. Recalculate the one-group *t* test and 95% CI for the data in Exercise 6B8, using 20% trimming. Does the trimmed sample mean differ significantly from the population mean of 10 at the .05 level (two-tailed)?

*3. Redo the *t* test you performed in Exercise 6A7, and the CI you constructed for the same data in exercise 6B9, after 20% trimming. Explain the difference in the *t* value due to trimming in terms of the tails of the sample distribution before and after trimming.

4. Redo the *t* test you performed in Exercise 6A8, and construct the 99% CI after 20% trimming. Is your CI consistent with the conclusion from your trimmed *t* test?

KEY FORMULAS

Estimate of the standard error of the mean (when the population standard deviation is not known):

$$s_{\bar{x}} = \frac{s}{\sqrt{N}}$$

Formula 6.1

The large-sample *z* test. This formula can be used to conduct a one-sample *z* test when the population standard deviation is not known but the sample size is sufficiently large:

$$z = \frac{\bar{X} - \mu}{\frac{s}{\sqrt{N}}}$$

Formula 6.2

Large-sample *z* test (estimate of standard error has already been calculated):

$$z = \frac{\bar{X} - \mu}{s_{\bar{x}}}$$

Formula 6.2A

One-sample *t* test (population standard deviation is not known):

$$t = \frac{\bar{X} - \mu}{\frac{s}{\sqrt{N}}}$$

Formula 6.3

One-sample *t* test (estimate of standard error has already been calculated):

$$t = \frac{\bar{X} - \mu}{s_{\bar{x}}}$$

Formula 6.3A

Confidence interval for the population mean (sample size is sufficiently large):

$$\mu_{lower} = \bar{X} - z_{crit}s_{\bar{x}}$$

Formula 6.4

$$\mu_{upper} = \bar{X} + z_{crit}s_{\bar{x}}$$

Formula for finding the sample size required to attain a particular width (*W*) for your confidence interval, given the standard deviation for the variable measured (assumes that you are dealing with large enough sample sizes that *s* is a good estimate of the population standard deviation):

$$N = \left(\frac{4s}{W}\right)^2$$

Formula 6.5

Confidence interval for the population mean (using t distribution):

$$\mu_{\text{lower}} = \overline{X} - t_{\text{crit}} s_{\overline{x}}$$

Formula 6.6

$$\mu_{\text{upper}} = \overline{X} + t_{\text{crit}} s_{\overline{x}}$$

The one-sample t test for a 20% trimmed mean:

$$t = \frac{\overline{X}_t - \mu_0}{\dfrac{s'}{.6\sqrt{N}}}$$

Formula 6.7

The confidence interval for one trimmed population mean (where P is the proportion trimmed from each end of the distribution):

$$\mu_{\text{lower}} = \overline{X}_t - t_{\text{crit}} \frac{s'}{(1 - 2P)\sqrt{N}} \, ;$$

$$\mu_{\text{upper}} = \overline{X}_t + t_{\text{crit}} \frac{s'}{(1 - 2P)\sqrt{N}}$$

Formula 6.8

THE t TEST FOR TWO INDEPENDENT SAMPLE MEANS

You will need to use the following from previous chapters:

Symbols
SS: Sum of squared deviations from the mean
σ: Standard deviation of a population
$\sigma_{\bar{x}}$: Standard error of the mean
s: Unbiased standard deviation of a sample
$s_{\bar{x}}$: Sample estimate of the standard error of the mean

Formulas
Formula 4.5: The standard error of the mean
Formula 6.3: The one-sample t test
Formula 6.6: The confidence interval for the population mean

Concepts
The t distribution
Degrees of freedom (df)

Chapter

CONCEPTUAL FOUNDATION

The purpose of this chapter is to explain how to apply null hypothesis testing when you have two samples (treated differently) but do not know any of the population means or standard deviations involved. For the procedures of this chapter to be appropriate, the two samples must be completely separate, independent, random samples, such that the individuals in one sample are not in any way connected or related to individuals in the other sample. It is sometimes desirable to use two samples that are somehow related, but I will not deal with this case until Chapter 11. As usual, I begin with an imaginary example to explain the concepts involved.

Suppose you have a friend who loves to exercise and is convinced that people who exercise regularly are not sick as often, as long, or as severely as those who do not exercise at all. She asks for your advice in designing a study to test her hypothesis, and the two of you agree that a relatively easy way to test the hypothesis would be to select at random a group of regular exercisers and an equal-sized group of nonexercisers and follow the individuals in both groups for a year, adding up the days each person would be considered sick. (You both agree that it would be too difficult to measure the severity of sickness in this informal study.) Now suppose that the year is over, and your friend is delighted to find that the mean number of sick days for the exercisers is somewhat lower than the mean for the nonexercisers. She starts to talk about the possibility of publishing the findings. It's time to tell your friend about Dr. Null.

If your friend tries to publish her results, Dr. Null is sure to know about it. He'll say that if the yearly sick days of the entire population of exercisers were measured, the mean number of sick days would be exactly the same as the mean for the entire population of nonexercisers. This, of course, is the null hypothesis applied to the two-group case. Dr. Null is also prepared to humiliate your friend in the following manner: He is ready to conduct his own experiment using two random samples (the same size as your friend's samples), but he will take both random samples from the nonexerciser population. He arbitrarily labels one of the samples exercisers and the other nonexercisers. (The subjects in Dr. Null's experiments wouldn't know this, of course.) Dr. Null claims that in his bogus experiment he will find that his phony exercise group has fewer sick days than the nonexercise group and that the difference will be just as large as, if not larger than, the difference your friend found.

Your friend admits that it would be quite humiliating if Dr. Null could really do that. "But," she insists, "this Dr. Null is relying solely on chance, on sampling fluctuations. It's very unlikely that he will find as large a difference as I found, and half the time his results will be in the wrong direction." At this point you must explain that researchers cannot rely on their subjective estimates of probability; the only way to have any confidence about what Dr. Null can and cannot do is to draw a map of what he can do—a probability map—called the null hypothesis distribution. Once you know the parameters of this distribution, you truly know your enemy; you have Dr. Null's "number." You know when you can reject him with confidence and when you are better off playing it safe. Finding the null hypothesis distribution in the two-group case is a bit trickier than in the one-group case. But considering that hypothesis tests involving two sample means are extremely common, it is well worth the effort to understand how these tests work.

Null Hypothesis Distribution for the Differences of Two Sample Means

To find the null hypothesis distribution in the two-group case, we have to consider what Dr. Null is doing when he performs his bogus experiment. All he is doing is selecting two random samples from the same population (actually, from two populations that have the same mean, but the distinction is not important at this point) and measuring the difference in their means, which can be negative or positive, depending on whether the first or second sample has the higher mean. If he conducts this experiment very many times, these difference scores will begin to pile up and form a distribution. If the variable being measured happens to have a normal distribution, our task is fairly easy. A statistical theorem states that the differences (from subtracting the two sample means) will also form a normal distribution.

Finding the mean of this distribution is particularly easy in the case we've been describing. If Dr. Null's hypothesis is that the exercisers and the nonexercisers have the same population mean, sometimes the exerciser sample will have a higher mean and sometimes the nonexerciser sample mean will be higher, so some of the differences will be positive and some negative. However, there's no reason for the positives to outweigh the negatives (or vice versa) if the two population means are indeed equal, so Dr. Null's distribution will be a normal distribution centered at zero (see Figure 7.1).

Small differences will be the most common, with larger differences becoming increasingly less likely. (Differences that are *exactly* zero will not

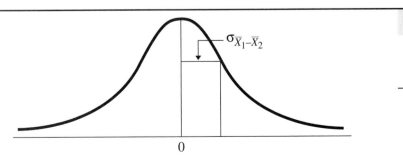

Figure 7.1

Normal Distribution of Differences of Sample Means

be common; see Section C.) The only thing we still need to know about the null hypothesis distribution is its standard deviation. Given the standard deviation, we will know everything we need to know to make judgments about the null hypothesis in the case of two independent groups.

Standard Error of the Difference

The standard deviation of Dr. Null's difference scores is called the *standard error of the difference* and is symbolized as $\sigma_{\bar{x}_1 - \bar{x}_2}$; the subscript shows that each difference score is the subtraction of two sample means. You may recall that finding the standard error of the mean ($\sigma_{\bar{x}}$), as we did in Chapter 4, was rather easy—it is just the standard deviation of the individual scores divided by the square root of N. The formula for finding the standard error of the differences is a little more complicated but not much. To show the similarity between $\sigma_{\bar{x}}$ and $\sigma_{\bar{x}_1 - \bar{x}_2}$, I will rewrite Formula 4.5 in a different but equivalent form. Formula 4.5 looks like this:

$$\sigma_{\bar{x}} = \frac{\sigma}{\sqrt{N}}$$

Now I square both sides to get:

$$\sigma_{\bar{x}}^2 = \frac{\sigma^2}{N}$$

Finally, I take the square root of both sides to get an alternate Formula 4.5 that looks like this:

$$\sigma_{\bar{x}} = \sqrt{\frac{\sigma^2}{N}}$$

Note that this modified version of Formula 4.5 gives you the same value as the original version if you plug in numbers, but it is expressed in terms of the population variance instead of the standard deviation. The reason I have bothered to change the appearance of Formula 4.5 is that this alternate version looks a lot like the formula for $\sigma_{\bar{x}_1 - \bar{x}_2}$, which I will refer to as Formula 7.1:

$$\sigma_{\bar{x}_1 - \bar{x}_2} = \sqrt{\frac{\sigma_1^2}{n_1} + \frac{\sigma_2^2}{n_2}} \qquad \textbf{Formula 7.1}$$

Formula 7.1 gives us just what we want. This formula tells you how large a difference is "typical" when you draw two samples from populations with the same mean and subtract the two sample means. (One sample has n_1 subjects and comes from a population with variance equal to σ_1^2; the other has n_2 subjects and is from a population with variance equal to σ_2^2.) Of course, *on the average* you would expect *no* difference because the two samples come from populations with the same mean. However, you know that because of chance fluctuations the sample means will virtually never be the same. When you calculate the preceding formula for $\sigma_{\bar{x}_1 - \bar{x}_2}$ you know that about two thirds of the differences of sample means will be less (in magnitude) than the value you get from the formula. (Recall that about two thirds of any normal distribution falls within one standard deviation on either side of the mean.)

Because Formula 7.1 refers to what happens when Dr. Null is right, and the two populations (e.g., exercisers and nonexercisers) have the same mean, you may wonder why the formula allows for the possibility that the two populations have different variances. This is a somewhat strange possibility, so I will be returning to it later in this section and again in Section C. You may also wonder why it is that to get the standard error of the differences we *add* the two variances in the formula (after dividing each by its corresponding n) instead of subtracting them. I am hoping that my explanation in Section C will give you some intuitive feeling for why the variances are added.

Finally, you may wonder why I'm using lowercase ns for the sample sizes in Formula 7.1. This is to avoid possible confusion later on between the total number of subjects in the two groups added together and the number of subjects in each group. I will reserve the use of capital N for the total number of subjects, and use a subscript T for total to make the distinction even more obvious. For a two-sample experiment, $N_T = n_1 + n_2$. If the two samples are the same size ($n_1 = n_2$), I will use n without a subscript to represent the number of subjects *in each* sample. Of course, in the one-sample case, there is no possibility of confusion, so it is reasonable to use N for the sample size.

Formula for Comparing the Means of Two Samples

Now that we know the mean and standard deviation of Dr. Null's distribution, and we know that it is a normal distribution, we have a map of what Dr. Null can do, and we can use it to get a p value and make a decision about whether we can reject the null hypothesis and be reasonably confident about doing it (given the usual assumptions, to be reviewed later). As in Chapter 5, we must take our experimental results and see where they fall on Dr. Null's distribution. We need to find the z score of our results with respect to the null hypothesis distribution because that will allow us to find the p value, which in turn tells us the chance that Dr. Null will beat our results without actually doing our experiment—that is, just by performing random sampling. The z score for the case of two independent groups is very similar to the z score for one group. Examine the one-group formula (Formula 5.1), which follows for comparison:

$$z = \frac{\overline{X} - \mu_0}{\sigma_{\overline{x}}}$$ **Formula 5.1**

Now consider the formula for the two-group case, Formula 7.2:

$$z = \frac{(\overline{X}_1 - \overline{X}_2) - (\mu_1 - \mu_2)}{\sigma_{\overline{x}_1 - \overline{x}_2}}$$ **Formula 7.2**

Note that the denominator in Formula 7.2 is given by Formula 7.1.

Notice that in both cases the z score formula follows the same basic structure: The numerator is the difference between your actual experimental results and what Dr. Null expects your experimental results to be. The denominator is the amount of difference that is typical when the null hypothesis is true, the kind of difference that can arise from chance fairly easily. If the difference you found for your experiment (i.e., the numerator) is less than or equal to the standard error (i.e., the denominator), the z score will be less than or equal to 1, and your results will fall near the middle of the null hypothesis distribution, indicating that Dr. Null can beat your results rather easily. If the experimental difference you found is at least twice the standard error, you can conclude (using $\alpha = .05$) that results as extreme as yours do not come up often enough by chance to worry.

Null Hypothesis for the Two-Sample Case

I will soon present numerical examples to make the use of these formulas more concrete, but for the moment take a closer look at the numerator of Formula 7.2. The first part, $\overline{X}_1 - \overline{X}_2$, represents your experimental results—for instance, the difference in mean sick days you found between exercisers and nonexercisers. The second part, $\mu_1 - \mu_2$, represents the difference Dr. Null expects, that is, the null hypothesis. You might expect this term always to be zero and that you needn't bother to include it in the formula. After all, doesn't Dr. Null always expect there to be no difference between the two populations? Actually, the second part of the numerator often *is* left out—it is almost always zero. I have left the term $\mu_1 - \mu_2$ in Formula 7.2 for the sake of completeness because it is possible (even though extremely rare) to consider a null hypothesis in which $\mu_1 - \mu_2$ does not equal zero. The following is an example of a scientifically plausible (albeit unethical) case in which the null hypothesis would not involve a zero difference.

Suppose a group of boys and a group of girls are selected and each child is given a growth hormone daily from the age of 5 until the age of 7. Then the mean height for each group is measured at age 18. The purpose of the experiment may be to show that the growth hormone is more effective on boys than on girls, but even if the hormone has the same effect on boys and girls, the boys will be taller on average than the girls at age 18. Knowing this, Dr. Null doesn't expect the mean heights for boys and girls to be equal. Rather, because he expects the growth hormone to have no effect (or the same effect on both groups), he expects the difference in the heights of boys and girls to be the difference normally found at 18 when no growth hormone is administered. In this unusual case, the null hypothesis is not $\mu_1 - \mu_2 = 0$ but rather $\mu_1 - \mu_2 = d$, where d equals the normal height difference at 18. So in this special case, the experimenter is hoping that the $\overline{X}_1 - \overline{X}_2$ difference found in the experiment is not only greater than zero but greater than d (or less than d if the experimenter expects the hormone to be more effective with girls).

Now let's return to the use of Formula 7.2, using our example about exercise and sick days. First I will substitute Formula 7.1 into the denominator of Formula 7.2 to obtain Formula 7.3:

$$z = \frac{(\overline{X}_1 - \overline{X}_2) - (\mu_1 - \mu_2)}{\sqrt{\dfrac{\sigma_1^2}{n_1} + \dfrac{\sigma_2^2}{n_2}}}$$

Formula 7.3

For the example involving exercise and sick days, the null hypothesis would be that $\mu_1 - \mu_2 = 0$—that the population mean for exercisers is the same as the population mean for nonexercisers. So when we use Formula 7.3, the second part of the numerator will drop out. Suppose the mean annual sick days for the exerciser sample is 5, and for the nonexercisers the mean is 7. Thus, the numerator of Formula 7.3, $\overline{X}_1 - \overline{X}_2$, is $5 - 7 = -2$. This result is in the predicted direction—so far so good. Next assume that the two samples are the same size, $n_1 = n_2 = 15$. Now all we need to know is the variance for each of the two populations.

Actually, that's quite a bit to know. If we knew the variance of each population, wouldn't we also know the mean of each population? In general, yes; if you have enough information to find the variance for a population, you have enough information to find the mean. (It is possible to devise an example in which the amount of variance is known from past experience but the population mean is not, but such situations are very rare in psychological research.) However, if we knew the mean of each population, there would be no reason to select samples or do a statistical test. We would have our answer. If the population mean for exercisers turned out to be less than that for nonexercisers, that could be reported as a fact and not as a decision with some chance of being in error. (For instance, we do not need to perform a statistical test to state that women in the United States are shorter than men because we essentially know the means for both populations.)

The z Test for Two Large Samples

In Chapter 5 we considered a case in which we knew the mean and variance for one population (i.e., the height of adult women in the United States) and wanted to compare the known population to a second, unknown population (i.e., LLD women). In that situation a one-sample test was appropriate. However, when you want to compare two populations on a particular variable and you don't know the mean of either population, it is appropriate to conduct a two-sample test. That is the case with the exercise and sick days example. But for any case in which you don't know either population mean, it is safe to say that you won't know the population variances either. Therefore you won't be able to use Formula 7.3. In fact, it looks like any time you have enough information to use Formula 7.3, you won't need to, and if you don't have the information, you won't be able to. At this point you must be wondering why I introduced Formula 7.3 at all. Fortunately, there is a fairly common situation in which it is appropriate to use Formula 7.3: when you are dealing with large samples.

If your samples are very large, the variances of the samples can be considered excellent estimates of the corresponding population variances—such good estimates, in fact, that the sample variances can be used in Formula 7.3 where population variances are indicated. Then the z score can be calculated and tested against the appropriate critical z, as described in Chapter 5. Therefore, Formula 7.3 can be called the *large-sample test for two independent means*. How large do the samples have to be? There is no exact answer. The error involved in estimating the population variances from the sample variances continues to decrease as sample size increases. However, once each sample contains more than 100 subjects, the decrease in error due to a further increase in sample size becomes negligible. Before computers made it so easy to deal with the t distribution, statisticians would often suggest using Formula 7.3 even when there were only about 30 to 40 subjects per group. However, the more conservative approach (i.e., an approach that emphasizes minimizing Type I errors) has always been to consider only samples that are at least in the hundreds as large. What can you do if your sample sizes cannot be considered large?

Separate-Variances t Test

When sample sizes are not large, there can be considerable error in estimating population variances from sample variances. The smaller the samples, the worse the error is likely to get. In Chapter 6 you learned how to compensate for this kind of error. For the current example we must again use the t distribution. If we take Formula 7.3 and substitute each sample variance (s^2) for its corresponding population variance (σ^2), we get a formula that we would expect to follow the t distribution:

$$t = \frac{(X_1 - X_2) - (\mu_1 - \mu_2)}{\sqrt{\dfrac{s_1^2}{n_1} + \dfrac{s_2^2}{n_2}}}$$

Formula 7.4

This formula is called the separate-variances *t* test, because each sample variance is separately divided by its own sample size. Unfortunately, this formula does not follow the *t* distribution in a simple way, and its use is somewhat controversial. Therefore, I will postpone a more detailed discussion of this formula until Section C. In the meantime, by assuming *homogeneity of variance* (i.e., assuming that the two population variances are equal), we can modify Formula 7.4 and make it easy to use the *t* distribution for our critical values. This modification concerns only the denominator of the formula, and requires a different way of estimating the standard error of the difference, $\sigma_{\bar{x}_1 - \bar{x}_2}$. Because this modified way of estimating $\sigma_{\bar{x}_1 - \bar{x}_2}$ involves pooling together the two sample variances, the resulting formula is referred to as the *pooled-variances t test*. Without a computer, this test is easier to perform than the separate-variances *t* test, which probably explains, at least in part, why the latter test never became very popular.

The Pooled-Variances Estimate

If we are willing to assume that $\sigma_1^2 = \sigma_2^2$ the two sample variances can be pooled together to form a single estimate of the population variance (σ^2). The result of this pooling is called the *pooled variance*, s_p^2, and it is inserted into Formula 7.4 as a substitute for both sample variances, which produces Formula 7.5A. The pooled variance can be factored out of each fraction in the denominator of Formula 7.5A to produce the algebraically equivalent Formula 7.5B.

$$t = \frac{(\bar{X}_1 - \bar{X}_2) - (\mu_1 - \mu_2)}{\sqrt{\dfrac{s_p^2}{n_1} + \dfrac{s_p^2}{n_2}}}$$

Formula 7.5A

$$t = \frac{(\bar{X}_1 - \bar{X}_2) - (\mu_1 - \mu_2)}{\sqrt{s_p^2\left(\dfrac{1}{n_1} + \dfrac{1}{n_2}\right)}}$$

Formula 7.5B

The rationale for pooling the two sample variances is based on the assumption that the two populations have the same variance, so both sample variances are estimates of the same value. (I will discuss the consequences of making this homogeneity assumption in Section B.) However, when the sample sizes are unequal, it is the larger sample that is the better estimate of the population variance, and it should have more influence on the pooled-variance estimate. We ensure that this is the case by taking a weighted average (as defined in Chapter 3) of the two sample variances.

When dealing with sample variances, the weight of a particular variance is just one less than the number of observations in the sample (i.e., $N - 1$)—in other words, the weight is the degrees of freedom (df) associated with that variance. Recall Formula 3.7A: $s^2 = SS/(N - 1)$. Notice that the weight associated with each variance is actually the denominator that was used to calculate the variance in the first place. To take the weighted average of the two sample variances, multiply each variance by its weight (i.e., its df), add the two results together, and then divide by the total weight (the df of the two variances added together). The resulting weighted average is the pooled variance, as shown in Formula 7.6A:

$$s_p^2 = \frac{(n_1 - 1)s_1^2 + (n_2 - 1)s_2^2}{n_1 + n_2 - 2}$$

Formula 7.6A

Formula 3.7A provides a way of rewriting the numerator of Formula 7.6A. If we multiply both sides of Formula 3.7A by $N - 1$, we get $SS = (N - 1)s^2$. This relationship allows us to rewrite Formula 7.6A as Formula 7.6B:

$$s_p^2 = \frac{SS_1 + SS_2}{N_1 + N_2 - 2} \qquad \textbf{Formula 7.6B}$$

Formula 7.6B gives us another way to think of the pooled variance. You simply add the SS (the top part of the variance formula) for the two groups together and then divide by the total degrees of freedom (the df for both groups added together).

The Pooled-Variances *t* Test

We are now ready to put the pieces together to create one complete formula for the pooled-variances *t* test. By inserting Formula 7.6A into Formula 7.5B, we get a formula that looks rather complicated but is easy to understand if you recognize each of its parts:

$$t = \frac{(\overline{X}_1 - \overline{X}_2) - (\mu_1 - \mu_2)}{\sqrt{\dfrac{(n_1 - 1)s_1^2 + (n_2 - 1)s_2^2}{n_1 + n_2 - 2}\left(\dfrac{1}{n_1} + \dfrac{1}{n_2}\right)}} \qquad \textbf{Formula 7.7A}$$

As an alternative you can instead insert Formula 7.6B into Formula 7.5B to get Formula 7.7B:

$$t = \frac{(\overline{X}_1 - \overline{X}_2) - (\mu_1 - \mu_2)}{\sqrt{\dfrac{SS_1 + SS_2}{n_1 + n_2 - 2}\left(\dfrac{1}{n_1} + \dfrac{1}{n_2}\right)}} \qquad \textbf{Formula 7.7B}$$

Of course, both Formula 7.7A and Formula 7.7B will give you exactly the same *t* value. Formula 7.7A is more convenient to use when you are given, or have already calculated, either the standard deviation or the variance for each sample. Formula 7.7B may be more convenient when you want to go straight from the raw data to the *t* test. Because I recommend always calculating the means and standard deviations of both groups and looking at them carefully before proceeding, I favor Formula 7.7A when performing a pooled-variances *t* test.

Formula for Equal Sample Sizes

If both samples are the same size, that is, $n_1 = n_2$, the weighted average becomes a simple average, so:

$$s_p^2 = \frac{s_1^2 + s_2^2}{2}$$

Instead of writing n_1 and n_2, you can use n without a subscript to represent the number of participants (or more generally speaking, the number of observations) in each sample. In this case, Formula 7.5B becomes:

$$t = \frac{(\overline{X}_1 - \overline{X}_2) - (\mu_1 - \mu_2)}{\sqrt{s_p^2\left(\dfrac{2}{n}\right)}}$$

After substituting the simple average for s_p^2 in the above formula, we can cancel out the factor of 2 and get the formula for the pooled-variance *t* test for two groups of equal size, Formula 7.8:

$$t = \frac{(\overline{X}_1 - \overline{X}_2) - (\mu_1 - \mu_2)}{\sqrt{\dfrac{s_1^2 + s_2^2}{n}}}$$

                                                                                     Formula 7.8

Formula 7.8 is also the separate-variances *t* test for two groups of equal size. If you replace both n_1 and n_2 with n in Formula 7.4, you have a common denominator in the bottom part of the formula, as in Formula 7.8. This means that when the two groups are the same size, it doesn't matter whether a pooled or a separate-variances *t* test is more appropriate; in either case you can use Formula 7.8 to calculate your *t* value. The critical *t*, however, will depend on whether you choose to perform a separate- or a pooled-variances *t* test; the latter case is the easier and will therefore be described first.

Calculating the Two-Sample *t* Test

Let us return to our example involving exercisers and nonexercisers. I deliberately mentioned that both samples contained 15 subjects because I want to deal numerically with the simpler case first, in which the two samples are the same size. (In Section B, I will tackle the general case.) Because the two samples are the same size, we can use Formula 7.8. As previously mentioned, the difference between the two sample means for this example is -2, and the null hypothesis is that the two population means are equal, so the value of the numerator of Formula 7.8 is -2. To calculate the denominator we need to know the variances of both samples. This would be easy to calculate if you had the raw data (i.e., 15 numbers for each group). For the sake of simplicity, assume that the standard deviation for the 15 exercisers is 4 and the standard deviation for the 15 nonexercisers is 5. Now to get the sample variances all you have to do is square the standard deviations: $s_1^2 = 16$ and $s_2^2 = 25$. Plugging these values into Formula 7.8 gives a *t* ratio, as follows:

$$t = \frac{-2}{\sqrt{\dfrac{16 + 25}{15}}} = \frac{-2}{\sqrt{2.733}} = \frac{-2}{1.65} = -1.21$$

Recall that the n in Formula 7.8 refers to the size of each group, not the total of both groups (which would have been represented by a capital letter). Also, the preceding *t* value is a negative number, but normally you ignore the sign of the calculated *t* when testing for significance. In fact, it is common to subtract the means in the numerator in whatever order makes the *t* value come out positive. Of course, the very first step would be to compare the two sample means to determine the direction of the results. (In the case of a one-tailed test you are at the same time determining whether a statistical test will be performed at all.) Then when you calculate the *t* value, you need only be concerned with its magnitude to see if it is extreme enough that you can reject the null hypothesis. The magnitude of the *t* value in our example is $t = 1.21$. To determine whether this value is statistically significant, we have to find the appropriate critical value for *t*.

Interpreting the Calculated *t*

When we attempt to use the *t* table (Table A.2 in Appendix A), we are reminded that we must first know the number of degrees of freedom that apply. For the pooled-variances t test of two independent groups, the degrees of freedom are equal to the denominator of the formula that was used to calculate the pooled variance (Formula 7.6A or Formula 7.6B)—that is, $n_1 + n_2 - 2$. If the two groups are the same size, this formula simplifies to $n + n - 2$, which equals $2n - 2$. For the present example, df $= 2n - 2 = 30 - 2 = 28$. If we use the conventional alpha $= .05$ and perform a two-tailed test, under df $= 28$ in the t table we find that the critical t equals 2.048. Thus our calculated t of 1.21 is smaller than the critical t of 2.048. What can we conclude?

Because our calculated t is smaller than the critical t, we cannot reject the null hypothesis; we cannot say that our results are statistically significant at the .05 level. Our calculated t has landed too near the middle of the null hypothesis distribution. This implies that the difference we found between exercisers and nonexercisers is the kind of difference that can occur fairly often by chance (certainly more than .05, or 1 time out of 20, which is the largest risk we are willing to take). So we must be cautious and concede that Dr. Null has a reasonable case when he argues that there is really no difference at all between the two populations being sampled. We may not be completely convinced that Dr. Null is right about our experiment, but we have to admit that the null hypothesis is not far-fetched enough to be dismissed. If the null hypothesis is *not* really true for our experiment, by failing to reject the null hypothesis we are committing a Type II error. But we have no choice—we are following a system (i.e., null hypothesis testing) that is focused on minimizing Type I errors.

Limitations of Statistical Conclusions

What if our calculated t were larger than the critical t, and we could therefore reject the null hypothesis? What could we conclude in that case? First, we should keep in the back of our minds that in rejecting the null hypothesis we could be making a Type I error. Perhaps the null hypothesis is true, but we just got lucky. Fortunately, when the calculated t is larger than the critical t, we are permitted to ignore this possibility and conclude that the two population means are not the same—that in fact the mean number of sick days for all exercisers in the population is lower than the mean for all the nonexercisers. However, rejecting the null hypothesis does not imply that there is a large or even a meaningful difference in the population means. Rejecting the null hypothesis, regardless of the significance level attained, says nothing about the size of the difference in the population means. That is why it can be very informative to construct a confidence interval for the difference of the means. This topic will be covered in detail in Section B.

An important limitation of the experiment described above should be made clear. Had the null hypothesis been rejected, the temptation would have been to conclude that regular exercise somehow made the exercisers less susceptible to getting sick. Isn't that why the experiment was performed? However, what we have been calling an experiment—comparing exercisers to nonexercisers—is not a true experiment, and therefore we cannot conclude that exercise *causes* the reduction in sick days. The only way to be sure that exercise is the factor that is causing the difference in health is to assign subjects randomly to either a group that exercises regularly or one that does not. The example we have been discussing is a *quasi-experiment* (or correlational study, as discussed in Chapter 1) because it was the subjects themselves who decided whether to exercise regularly or not. The problem with this design is that we cannot be sure that exercising is the only difference between the two groups. It is not unreasonable to suppose that healthy people (perhaps because of genetic factors or diet) are more likely to become exercisers and that they would be healthier even if they didn't exercise. Or there could be personality differences between those who choose to exercise and those who do not, and it could be those personality factors, rather than the exercise itself, that are responsible for their having fewer sick days.

It is important to know whether it is the exercise or other factors that is reducing illness. If it is the exercise that is responsible, we can encourage nonexercisers to begin exercising because we have reason to believe that the exercise improves health. But if exercisers are healthier because of genetic or personality factors, there is no reason for nonexercisers to begin exercising; there may be no evidence that exercise will change their genetic constitution or their personality. To prove that it is the exercise that reduces the number of sick days, you would have to conduct a true experiment, randomly assigning subjects to either a group that must exercise regularly or a group that does not.

SUMMARY

1. In a two-sample experiment, the two sample means may differ in the predicted direction, but this difference may be due to chance factors involved in random sampling. To find the null hypothesis distribution that corresponds to the two-sample case, we could draw two random samples (of appropriate size) from populations with the same mean and find the difference between the two sample means. If we did this many times, the differences would tend to pile up into a normal distribution.

2. If the null hypothesis is that the two populations have the same mean, the null hypothesis distribution will have a mean of zero. The standard deviation for this distribution is called the standard error of the difference.

3. If both samples are large, a *z* score can be used to make a decision about the null hypothesis. The denominator of the formula for the *z* score is the separate-variances formula for the standard error of the difference.

4. If the samples are not large and the population variances can be assumed equal, a pooled-variance *t* test is recommended. The standard error of the difference is based on a weighted average of the two sample variances. The critical values come from the *t* distribution with df $= n_1 + n_2 - 2$. If the population variances cannot be assumed equal, a separate-variances *t* test should be considered (see Section B). If the two sample sizes are equal, you need not worry about homogeneity of variance—a simplified formula for *t* can be applied.

5. If the two populations being sampled were not created by the experimenter, the possible conclusions are limited. Also, concluding that the population means differ significantly does not imply that the difference is large enough to be interesting or important. In some cases, a confidence interval for the difference in population means can provide helpful information.

EXERCISES

*1. Hypothetical College is worried about its attrition (i.e., drop-out) rate, so it measured the entire freshman and sophomore classes on a social adjustment scale to test the hypothesis that the better-adjusted freshmen are more likely to continue on as sophomores. The mean for the 150 freshmen tested was 45.8, with *s* = 5.5; the mean for the 100 sophomores was 47.5, with $s = 4$.

a. Use the appropriate formula to calculate a test statistic to compare the means of the freshman and sophomore classes.

b. What is the p value associated with your answer to part a?

2. a. For a two-group experiment, $n_1 = 14$ and $n_2 = 9$. How many degrees of freedom would be associated with the pooled-variances t test?

b. What are the two-tailed critical t values, if alpha = .05? If alpha = .01?

*3. A group of 101 participants has a variance (s_1^2) of 12, and a second group of 51 participants has a variance (s_2^2) of 8. What is the pooled variance, s_p^2, of these two groups?

4. The weights of 20 men have a standard deviation $s_M = 30$ lb., and the weights of 40 women have $s_W = 20$ lb. What is the pooled variance, s_p^2, of these two groups?

*5. A particular psychology experiment involves two equal-sized groups, one of which has a variance of 120 and the other of which has a variance of 180. What is the pooled variance? (*Note*: It doesn't matter how large the groups are, as long as they are both the same size.)

6. The two groups in a psychology experiment both have the same variance, $s^2 = 135$. What is the pooled variance? (*Note*: It doesn't matter how large the groups are or even whether they are the same size.)

*7. In a study of a new treatment for phobia, the data for the experimental group were $\overline{X}_1 = 27.2$, $s_1 = 4$, and $n_1 = 15$. The data for the control group were $\overline{X}_2 = 34.4$, $s_2 = 14$, and $n_2 = 15$.

a. Calculate the separate-variances t value.

b. Calculate the pooled-variances t value.

8. a. Design a true experiment involving two groups (i.e., the experimenter decides, at random, in which group each participant will be placed).

b. Design a quasi-experiment (i.e., an observational or correlational experiment) involving groups not created, but only selected, by the experimenter. How are your conclusions from this experiment limited, even if the results are statistically significant?

B

BASIC STATISTICAL PROCEDURES

The previous section described an experiment that was observational or correlational in nature; the subjects were sampled from groups that were already in existence. In this section I will describe a true experiment, employing the same dependent variable as in the previous section but a different independent variable. Another difference in the example in this section is that in this true experiment, the two groups are *not* the same size. This gives an opportunity to demonstrate the calculation of the two-group t test when $n_1 \neq n_2$. There is absolutely no difference between the statistical analysis of a true experiment and the corresponding analysis of an observational experiment. I illustrate both types of experimental designs so that I can comment on the different conclusions that can be drawn *after* the statistical analysis.

Another factor that may affect the number of days an individual is sick, and one that received much publicity several years ago, is taking very large doses (i.e., megadoses) of vitamin C daily. Although quite a few studies have already been conducted in this area, the conclusions remain controversial. However, this topic serves our purposes well at this point because it is relatively easy to design a true experiment to test the effects of vitamin C. All we need to do is select a group of volunteers who are not already taking large amounts of vitamin C and form two groups by randomly assigning half the subjects to the vitamin C group and the rest to the placebo group. (A placebo is some harmless substance that we can be sure won't have any direct physiological effect on the participants at all.)

To prevent the biases of the participants or the experimenters from affecting the results, the study should follow a "double-blind" design—that is, the vitamin C and placebo capsules should be indistinguishable by the participants and coded in such a way that the experimenters who interact with the participants also have no idea which participants are taking vitamin C and which the placebo. (The code is not broken until the end of the experiment.) Because the participants have been randomly assigned to groups, and the effects of expectations are the same for both, any statistically significant difference between the two groups can be attributed specifically to the action of vitamin C. If we tried to perform this experiment with only one sample, we would not have a placebo group for comparison. In that case, any reduction in sick days for the vitamin C group could be either from vitamin C or from the expectations of the participants (i.e., the "placebo effect").

To set up and analyze a two-sample experiment we can use the same six steps that we used in Chapter 5 to test hypotheses involving one sample. In fact, having learned and practiced the use of these six steps for one-sample problems, you should find the steps below quite familiar; it is only the calculation formula that looks different. And I hope that Section A helped you to see that whereas the two-group *t* formula can look rather complicated, it is conceptually very similar to the one-sample *t* and one-sample *z* formulas.

Step 1. State the Hypotheses

The research hypothesis is that subjects who take the vitamin C will be sick fewer days than the subjects who take the placebo. However, this hypothesis is not specific enough to be tested directly. The common solution to this problem, as you saw in the previous two chapters, is to set up a specific null hypothesis, which the researcher hopes to disprove. The appropriate null hypothesis in this case is that it makes no difference whether the subjects take vitamin C or the placebo—the average number of sick days will be the same. Next we need to translate this idea into a statement about populations. If everyone in the population of interest had been in the vitamin C group, the mean of that distribution could be designated μ_C. Similarly, if the entire population had been in the placebo group, the mean of the resulting distribution could be designated μ_P. The null hypothesis (H_0) for this experiment is that these two hypothetical population means are equal: $\mu_C = \mu_P$, or $\mu_C - \mu_P = 0$.

The alternative hypothesis is the complement of the null hypothesis: H_A: $\mu_C \neq \mu_P$, or $\mu_C - \mu_P \neq 0$. If a one-tailed alternative hypothesis is considered justified, there are, of course, two possibilities: either H_A: $\mu_C < \mu_P$ (or $\mu_C - \mu_P < 0$) or H_A: $\mu_C > \mu_P$ (or $\mu_C - \mu_P > 0$). According to the research hypothesis that inspired the present example, if a one-tailed hypothesis were to be used, it would be H_A: $\mu_C < \mu_P$ because with vitamin C we expect *fewer* sick days. Whether it is appropriate to conduct a one-tailed test in a two-group experiment depends on the same issues that were already discussed for one-sample experiments (see Chapter 5, Section A). For the present example I will follow the more conservative approach and use the two-tailed alternative hypothesis.

Step 2. Select the Statistical Test and the Significance Level

We want to draw an inference about two population means from the data in two fairly small samples. If we are willing to assume that the populations have the same variance (and we are for this example), we can perform the pooled-variance *t* test for two independent samples. There is nothing different about setting alpha for a two-group test as compared to the one-group test. The most common alpha level is .05, and only unusual circumstances would justify a different alpha.

Step 3. Select the Samples and Collect the Data

If the populations already exist, as in the study of exercisers and nonexercisers in Section A, a random sample of each should be taken. For an experimental study, the ideal is to select two truly random samples—one for each condition. In reality, it is very likely that only one sample would be drawn; the researcher might try to make the sample as representative of the population as feasible, but most likely he or she would use a sample of convenience. Then the participants would be *randomly assigned* to one condition or the other. The random assignment of participants would ensure that there were no systematic differences between the two groups before the experimental conditions could be imposed; the only differences between the groups would be those due to chance. The implications of random assignment will be discussed further under the section "Assumptions of the *t* Test for Two Independent Samples."

The larger the samples, the more accurate will be the conclusions. Samples large enough to allow the use of the large-sample *z* test of two independent means (Formula 7.3) are preferable because in that case there is little reason to worry about the shapes of the population distributions. Such large samples may not be available or affordable, and if you are looking for a rather large experimental effect, they may not be necessary. (This latter point will be explained in the next chapter.) When using fairly small samples (each less than about 30 or 40), an effort should be made to ensure that the two groups are the same size so that the more complicated separate-variances *t* test need not be considered (we will deal with the more complicated case at the end of this section). For the present example, suppose that 12 participants were assigned to each group, but at the end of the experiment two participants confessed that they frequently forgot to take their pills, and therefore their data had to be eliminated. Imagine that, coincidentally, both of these participants were from the placebo group, so $n_C = 12$ and $n_P = 10$. The data for the vitamin C experiment would thus consist of 22 numbers, each of which would represent the number of sick days for a particular participant during the course of the experiment. To streamline the calculations I will give only the means and standard deviations for each group.

Step 4. Find the Region of Rejection

Because the sample sizes in our example are not large enough to justify a large-sample *z* test, we will have to use the *t* distribution. Having selected alpha = .05 and a two-tailed test, we only need to know the degrees of freedom to look up the critical *t* value. The df for the two-group (pooled-

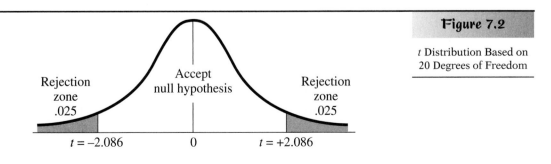

Figure 7.2

t Distribution Based on 20 Degrees of Freedom

variance) *t* test is $n_1 + n_2 - 2$. Therefore, the df for the present example is $12 + 10 - 2 = 22 - 2 = 20$. Table A.2 shows that the critical $t = 2.086$. Because we are planning for a two-tailed test, there is a rejection region in each tail of the *t* distribution, at -2.086 and at $+2.086$, as shown in Figure 7.2.

Step 5. Calculate the Test Statistic

Suppose that $\overline{X}_C = 4.25$ sick days, $s_C = 3$, $\overline{X}_P = 7.75$ sick days, and $s_{\text{plac}} = 4$. (I'm using "plac" to represent placebo for this example to avoid confusion with s_p^2, which represents the pooled variance.) First you would check to see which mean is higher. Even though a two-tailed test is planned, it is clear that the designer of this experiment is hoping that the vitamin C mean is lower. In fact, had the means come out in the direction opposite the expected one, the researcher might well have decided not to even bother testing the results for significance. (If a one-tailed test had been planned and the results had come out in the direction opposite what he or she expected, the researcher would have been ethically bound to refrain from testing the results.) Because the samples are not large and not equal in size, the appropriate formula to use is either Formula 7.5A or Formula 7.5B. In either case, we would calculate the numerator like this:

$$(\overline{X}_1 - \overline{X}_2) - (\mu_1 - \mu_2) = (7.75 - 4.25) - 0 = 3.5 - 0 = 3.5$$

Note that because our null hypothesis is that $\mu_1 - \mu_2 = 0$, that term becomes zero. Note also that I deliberately arranged the sample means in the formula so that the smaller would be subtracted from the larger, giving a positive number. This is often done to avoid the added complexity of negative numbers. Because we have already taken note of which mean was larger, we do not have to rely on the sign of the *t* value to tell us the direction of our results. It is only when you are performing a series of *t* tests on the same two groups—for instance, comparing men and women on a battery of cognitive tests—that you would probably want to subtract the means of the two groups in the same order every time, regardless of which mean is larger in each case. Then the sign of the *t* value would tell you which group scored more highly on each variable, making it less confusing to compare the various *t* tests to each other.

The next step is to pool the two sample variances together to get s_p^2. Because we already have the standard deviations and can therefore easily get the variances, we will use Formula 7.6A. First we square each of the standard deviations to get $s_C^2 = 3^2 = 9$ and $s_{\text{plac}}^2 = 4^2 = 16$, and then we plug these values into the formula:

$$s_p^2 = \frac{(n_1 - 1)s_1^2 + (n_2 - 1)s_2^2}{n_1 + n_2 - 2} = \frac{11(9) + 9(16)}{12 + 10 - 2}$$

$$s_p^2 = \frac{99 + 144}{20} = \frac{243}{20} = 12.15$$

Note that the pooled variance falls between the two sample variances, but not exactly midway between. Had the two samples been the same size, the pooled variance would have been 12.5, which is halfway between 9 and 16. When the two samples are not the same size, the pooled variance will be closer to the variance of the larger sample; in this case the larger sample has variance $= 9$, so the pooled variance is closer to 9 than it is to 16. This is a consequence of taking a weighted average. Now we can insert our value for s_p^2 into Formula 7.5B (which is just a tiny bit simpler mathematically that Formula 7.5A):

$$t = \frac{(\overline{X}_1 - \overline{X}_2) - (\mu_1 - \mu_2)}{\sqrt{s_p^2\left(\dfrac{1}{n_1} + \dfrac{1}{n_2}\right)}} = \frac{3.5}{\sqrt{12.15\left(\dfrac{1}{12} + \dfrac{1}{10}\right)}}$$

$$t = \frac{3.5}{\sqrt{12.15(.1833)}} = \frac{3.5}{\sqrt{2.23}} = \frac{3.5}{1.49} = 2.345$$

Step 6. Make the Statistical Decision

The calculated t equals 2.345, which is larger than the critical t (2.086), so the null hypothesis can be rejected. By looking at Figure 7.2 you can see that $t = 2.345$ falls in the region of rejection. We can say that the difference between our two samples is statistically significant at the .05 level, allowing us to conclude that the means of the two populations (i.e., the vitamin C population and the placebo population) are not exactly the same.

Interpreting the Results

Because our vitamin C experiment was a true experiment, involving the random assignment of subjects to the two groups, we can conclude that it is the vitamin C that is responsible for the difference in mean number of sick days. We can rule out various alternative explanations, such as placebo effects (subjects in both groups thought they were taking vitamin C), personality differences, and so forth. However, bear in mind that no experiment is perfect; there can be factors at play that the experimenters are unaware of. For example, researchers conducting one vitamin C experiment found that some subjects had opened their capsules and tasted the contents to try to find out if they had the vitamin or the placebo (these subjects knew that vitamin C tastes sour). But even if the experiment *were* perfect, declaring statistical significance is not in itself very informative. By rejecting the null hypothesis, all we are saying is that the effect of vitamin C is not identical to the effect of a placebo, a totally inactive substance. We *are* saying that vitamin C produces a reduction in sick days (relative to a placebo), but we are *not* saying by how much. With sufficiently large groups of subjects, statistical significance can be obtained with mean differences too small to be of any practical concern.

If you are trying to decide whether to take large amounts of vitamin C to prevent or shorten the common cold or other illnesses, knowing that the null hypothesis was rejected in the study described above will not help you much. What you need to know is the *size* of the reduction due to vitamin C; in other words, how many sick days can you expect to avoid if you go to the trouble and expense of taking large daily doses of vitamin C? For an average reduction of half a day per year, you may decide not to bother. However, if the expected reduction is several days, it may well be worth considering. There are many cases when rejecting the null hypothesis in a two-group experiment is not in itself very informative, and a researcher needs additional information concerning the difference between the means. In such cases, a confidence interval can be very helpful.

Confidence Intervals for the Difference between Two Population Means

If an experiment is conducted to decide between two competing theories that make opposite predictions, simply determining which population mean is larger may be more important than knowing the magnitude of the difference. On the other hand, in many practical cases, it would be useful to know just how large a difference there is between the two population means. In Chapter 6, when we wanted to estimate the mean of a population, we used the sample mean, $\overline{X}$, as a *point estimate* of μ. A similar strategy is used to estimate the difference of two population means, $\mu_1 - \mu_2$; the best point estimate from our data would be $\overline{X}_1 - \overline{X}_2$. For the vitamin C experiment, our point estimate of $\mu_C - \mu_P$ is $\overline{X}_C - \overline{X}_P = 7.75 - 4.25 = 3.5$ sick days. If this experiment were real, a reduction of 3.5 sick days per year would be worthy of serious consideration. However, even if the experiment were real, this would be just an estimate. As you know, there is a certain amount of error associated with this estimate, and it would be helpful to know just how much error is involved before we take any such estimate too seriously.

As you learned in Chapter 6, a point estimate can be supplemented by an interval estimate, based on a *confidence interval*. We can use the same procedure described in that chapter, modified slightly for the two-group case, to construct a confidence interval for the difference between two population means. First, recall that in testing the null hypothesis in a two-group experiment, we center the null hypothesis distribution on the value of $\mu_1 - \mu_2$ that is predicted by the null hypothesis, usually zero. Then, in a two-tailed test, we mark off the critical values (of *t* or *z*, whichever is appropriate) symmetrically on either side of the expected value (as in Figure 7.2). To find the confidence interval, the process is similar, except that we center the distribution on the point estimate for $\mu_1 - \mu_2$ that is, $\overline{X}_1 - \overline{X}_2$. Critical values are then marked off symmetrically just as in a null hypothesis test. The final step is to translate those critical values back into upper and lower estimates for $\mu_1 - \mu_2$. I will illustrate this procedure by constructing a confidence interval around the difference in sample means that we found in the vitamin C experiment.

Constructing a 95% Confidence Interval in the Two-Sample Case

We start by using the same distribution we used to test the null hypothesis: a *t* distribution with 20 degrees of freedom. However, the value at the center of the distribution is not zero, but $\overline{X}_C - \overline{X}_P = 3.5$. Then we must decide what level of confidence is desired. If we choose to construct a 95% confidence interval (the most common), the critical values are the same as in the .05, two-tailed test: -2.086 and $+2.086$ (see Figure 7.3). To convert these critical values into upper and lower estimates of $\mu_1 - \mu_2$ we need to turn the *t* formula around as we did in the one-group case. This time we start with a generic version of the two-group *t* formula, as follows:

$$t = \frac{(\overline{X}_1 - \overline{X}_2) - (\mu_1 - \mu_2)}{s_{\overline{x}_1 - \overline{x}_2}}$$

We already know the value of *t*—it is equal to the critical value. What we want to do is solve the formula for $\mu_1 - \mu_2$:

$$(\overline{X}_1 - \overline{X}_2) - (\mu_1 - \mu_2) = t_{\text{crit}} s_{\overline{x}_1 - \overline{x}_2}$$

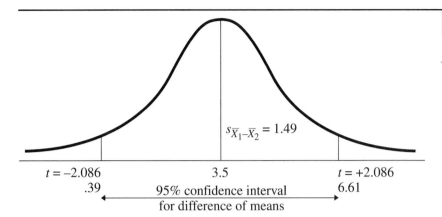

Figure 7.3

t Distribution (df = 20)

Note that t_{crit} can be positive or negative. A final rearrangement to isolate $\mu_1 - \mu_2$ yields Formula 7.9:

$$\mu_1 - \mu_2 = (\overline{X}_1 - \overline{X}_2) \pm t_{\text{crit}} s_{\overline{x}_1 - \overline{x}_2} \qquad \textbf{Formula 7.9}$$

Note the resemblance between Formulas 7.9 and 6.6. Recall that $\overline{X}_1 - \overline{X}_2 = 3.5$, and $t_{\text{crit}} = 2.086$; $s_{\overline{x}_1 - \overline{x}_2}$ is the denominator of the t value we calculated in step 5, which was equal to 1.49. Plugging these values into Formula 7.9, we get:

$$\mu_1 - \mu_2 = 3.5 \pm 2.086(1.49) = 3.5 \pm 3.11$$

So the upper estimate for the population mean difference is 3.5 + 3.11 = 6.61, and the lower estimate is 3.5 − 3.11 = 0.39. What does this tell us? First, notice that because both boundaries are positive numbers, zero is not included in the confidence interval. This means that based on our sample data, we can say, with 95% confidence, that $\mu_C - \mu_P$ is not zero (because zero is not in the interval that represents the 95% CI). You should notice that this conclusion is equivalent to rejecting the null hypothesis that $\mu_C - \mu_P = 0$, at the .05 level in a two-tailed test. As I pointed out in Chapter 6, whatever the value predicted by the null hypothesis, if it is included in the 95% confidence interval, the null hypothesis cannot be rejected at the .05 level (two-tailed).

Something else you may have noticed about the confidence interval we just calculated is that it is so wide that it is not very helpful. To be 95% certain, we must concede that the true reduction in sick days due to vitamin C may be as low as .39 or as high as 6.61. Of course, if we are willing to be less certain of being right, we can provide a narrower confidence interval (e.g., 90% or even 80%). However, to make the confidence interval narrower without reducing confidence we need to increase the number of participants. Increasing the number of participants reduces the standard error of the difference and in general provides more accurate information concerning population parameters.

Constructing a 99% Confidence Interval in the Two-Sample Case

A 99% confidence interval therefore will be even larger than the 95% CI, but I will construct one to further illustrate the use of Formula 7.9. The critical *t* is the one that corresponds to an alpha of .01 (two-tailed); the df are still 20. Table A.2 shows that $t_{crit} = 2.845$. The rest of the values in Formula 7.9 remain the same:

$$\mu_1 - \mu_2 = 3.5 \pm 2.845(1.49) = 3.5 \pm 4.24$$

The 99% confidence interval therefore ranges from $-.74$ to 7.74. Notice that zero *is* included in this interval, which tells us that the null hypothesis of zero difference between population means cannot be rejected at the .01 level, two-tailed. To be 99% certain of our statement, we would have to concede that whereas vitamin C may reduce sick days by more than a week, it also may actually *increase* sick days slightly. The latter possibility cannot be ruled out if we are to have 99% confidence.

Assumptions of the *t* Test for Two Independent Samples

Independent Random Sampling

Ideally, both groups should be simple random samples that are completely independent of each other. For example, in an observational study comparing men and women, independence would be violated if a group of married men was selected and then all of their wives were selected to be the group of women. The *t* test formulas in this chapter apply only when the two samples are independent of each other. (Sometimes it is advantageous to create a systematic relationship or dependence between the two samples; the analysis of that type of experiment is the topic of Chapter 11.)

Even for experimental studies, true random sampling is virtually impossible, so for the experimental conclusions to have any validity, the participants must at least be *randomly assigned* to the two experimental conditions. For example, the results of the vitamin C experiment would have no validity if the experimenter had information about the general health of each participant which he or she could use (even if unconsciously) when assigning participants to groups. However, even if the assignment of the participants is random (actually, quasi-random, given the usual constraint that we begin with an even number of participants and assign half to each condition; this precludes flipping a coin for each person because the coin doesn't have to come up heads for exactly half of the flips), there is no guarantee that you will be able to generalize your results to the larger population in which you are interested. For instance, if you begin with a sample of convenience such as a group of college sophomores and then randomly assign them to conditions, you may find that labeling a particular task as an indication of IQ is highly motivating, but this effect may not replicate when applied to a sample of older, successful adults.

Whether your result can be generalized beyond the demographics of your sample is less a statistical question than a scientific one. For instance, I would expect experimental manipulations involving basic sensory processes to generalize more easily to the overall human population than those involving social attitudes because the former are more likely to involve relatively universal principles of biology, whereas the latter are more subject to cultural influences. This is a question each subarea of psychology must deal with for itself.

Normal Distributions

Ideally, the variable being measured (e.g., number of sick days) should be normally distributed in both populations (and, of course, it should be measured on an interval or ratio scale). This assures that the differences of sample means will also follow a normal distribution so that we can perform a valid two-group hypothesis test using z scores (for large samples) or the t distribution (when the population variances must be estimated from small samples). However, as I have mentioned, although many variables follow the normal distribution pretty closely (though none perfectly), many of the variables in psychology experiments do not. In fact, in the example I have been using, the variable (sick days) is not likely to have a normal distribution; its distribution is likely to be positively skewed, with most people not very far from zero and a few people far above the others.

Fortunately, the Central Limit Theorem can be generalized to imply that even when two populations are not normally distributed, the distribution of sample mean differences will approach the normal distribution as the sample sizes increase. In fact, for most population distributions, the sample sizes do not have to be very large before the methods of this chapter can be used with acceptable accuracy. As with the one-sample t test, we can say that the two-group t test is *robust* with respect to the assumption of normal distributions. Nonetheless, you should observe how the data are distributed in each sample (using the techniques described in Chapter 2). If the data are extremely skewed or otherwise strangely distributed, and sample sizes are not large, you should consider using a data transformation (see Chapter 3), a nonparametric procedure (such as the Mann-Whitney rank-sum test, described in Chapter 21), or trimming (see Section C).

Homogeneity of Variance

The pooled-variances t test described in this chapter is strictly valid only if you assume that the two populations (e.g., exercisers and nonexercisers) have the same variance—a property called *homogeneity of variance* (HOV)—whether or not they have the same mean. This assumption is reasonable when we expect that exposing a population to a treatment (e.g., regular exercise, regular doses of vitamin C) is likely to change the mean of the population (shifting everyone in the population by about the same amount) without appreciably changing its variance. However, there are situations in which a treatment could increase the variance of a population with or without affecting the mean, so the possibility of *heterogeneity of variance* should always be considered.

Fortunately, there are three situations in which you can perform a two-group hypothesis test without having to worry about homogeneity of variance. The first situation is when both samples are quite large (at least about 100 subjects in each). Then you can use Formula 7.3 and the critical values for z. The second situation is when the two samples are the same size; then you can use Formula 7.8 and the df for the pooled-variances test. Statisticians have demonstrated that even large discrepancies in population variances have little influence on the distribution of calculated t when the two samples are the same size. The third situation is when your two sample variances are very similar; in that case, it is reasonable to assume HOV without testing for it and to proceed with the pooled-variances test. One "rule of thumb" is that if one sample variance is no more than twice as large as the other, you can safely assume HOV in the population.

If your samples are not very large, not equal in size, and one sample variance is more than twice the other, you should consider the possibility of performing a separate-variances *t* test. One procedure that can help you decide between performing the pooled- and separate-variances *t* tests is a test for homogeneity of variance.

HOV Tests and the Separate-Variances *t* Test

Several procedures have been suggested for testing homogeneity of variance. The simplest is the *F* text, in which two sample variances are divided to obtain an *F* ratio; I will return to the *F* test in greater detail after I have introduced the *F* distribution in Chapter 12. Although it is the simplest, the *F* test as just described is too easily thrown off when the dependent variable is not normally distributed. Several more robust alternatives have been proposed—some of which are calculated automatically when a statistical computer package is used to calculate a *t* test (e.g., Levene's test is given by SPSS). Unfortunately, homogeneity of variance tests tend to lose their power and validity with small sample sizes, which is when they are most needed. Indeed, the whole matter of testing HOV in order to decide whether to apply the separate-variances *t* test remains controversial. Fortunately, there are situations in which this decision can be avoided entirely, as described next.

The calculation of the separate-variances *t* value is not difficult; the problem is that the *t* calculated from Formula 7.4 does not follow the *t* distribution with $n_1 + n_2 - 2$ degrees of freedom. Finding the null hypothesis distribution that corresponds to the separate-variances *t* has come to be known as the *Behrens-Fisher problem*, after the two statisticians who pioneered the study of this topic. Fortunately the separate-variances *t* follows *some t* distribution fairly well, and one way to view the problem boils down to finding the best estimate for the df of that *t* distribution. Estimating that df requires a rather complicated formula that I will save for Section C. Next, I will describe a situation in which you would be able to avoid using that formula entirely.

The usual formula for adjusting the df for a separate-variances *t* test yields a df (df_{s-v}) that will be lower than the df for the corresponding pooled-variances *t* test; df_{s-v} can be equal to, but never higher than, $n_1 + n_2 - 2$. It is also true, and useful to note, that df_{s-v} can never come out smaller than $n_s - 1$, where n_s is the smaller of n_1 and n_2. For example, if $n_1 = 10$ and $n_2 = 12$, n_s is equal to 10, and the lower limit for df_{s-v} is $10 - 1 = 9$. Therefore, you can always check your separate-variances *t* value (from Formula 7.4) against the "worst-case scenario" critical *t*, based on the lowest possible value for df_{s-v} (recall that as df gets smaller, t_{crit} gets larger). If your separate-variances *t* exceeds the worse-case critical *t*, you do not need a special formula to adjust your df; this is a situation in which you know your *t* test will be significant, for any possible df adjustment. It is also the case that, once you have decided to perform the separate-variances *t* test, if your s-v *t* value does not exceed the critical *t* for the pooled-variances *t* test, you can give up on attaining significance with your *t* test. Any adjustment to your df can only make it smaller, and therefore move you further from being able to reject the null hypothesis.

The one situation in which you should seriously consider using the df-adjustment formula presented in Section C occurs when your samples' sizes are unequal and not large, an HOV test attains significance (implying that the population variances are different, and therefore the sample variances should not be pooled), and your separate-variances *t* value is significant with

$n_1 + n_2 - 2$ degrees of freedom, but not with $n_s - 1$ df. Conveniently, many statistical software packages (e.g., SPSS) provide exact p values for both separate- and pooled-variances t tests, as well as an HOV test, whenever you request an independent-samples t test. I present the df-adjustment formula in Section C not because I expect you to ever need to calculate it by hand, but rather to help you understand just what the adjustment is based on.

When to Use the Two-Sample t Test

The two-sample t test can be used to analyze a study of samples from two preexisting populations—a quasi-experiment—or to analyze the results of subjecting two randomly assigned samples to two different experimental conditions—a true experiment. However, a t test is appropriate only when the *dependent variable* in a two-group experiment has been measured on an interval or ratio scale (e.g., number of sick days per year, weight in pounds). If the dependent variable has been measured on an ordinal or categorical scale, the nonparametric methods described in Part VII of this text will be needed. On the other hand, the *independent variable*, which has only two values, is usually measured on a categorical scale (e.g., exercisers and nonexercisers; psychotics and normals). In some cases, the independent variable may have been measured originally using an interval or ratio scale, which was later converted into a dichotomy (the simplest form of categorical scale) by the researcher.

For example, a teacher may give out a social anxiety questionnaire at the beginning of a semester and then keep track of how much time each student speaks during class. To analyze the data, the teacher could use the anxiety measure to divide the class into two equal groups: those "high" in anxiety and those "low" in anxiety. (This arrangement is called a median split; the quote marks indicate that these are just relative distinctions within this one class.) A t test can be performed to see if the two groups differ significantly in class speaking time. However, if the social anxiety scores form a nearly normal distribution, it is probably better to use a correlational analysis (described in Chapter 9) than to throw away information by merely classifying students as either high or low in anxiety. In fact, the exercise example in Section A might be better designed as a correlation between amount of exercise and the number of sick days in a sample of subjects varying widely in their degree of daily or weekly exercise. It is when the two groups are distinctly different, perhaps clustered at the two opposite extremes of a continuum or representing two qualitatively different categories, that the t test is particularly appropriate.

When to Construct Confidence Intervals

If I have done a good job of selling the importance of constructing confidence intervals for a two-group experiment, you may feel that they should always be constructed. However, for many two-group experiments, confidence intervals for the difference in means may not be very meaningful. Consider, for example, an experiment in which one group of subjects watches a sad movie, and the other group watches a happy movie. The dependent variable is the number of "sad" words (e.g., funeral) subjects can recall from a long list of words studied right after the movie. A difference of two more words recalled by those who watched a sad movie could turn out to be statistically significant, but a confidence interval would not be easy to interpret. The problem is that the number of words recalled is not a universal measure that can easily be compared from one experiment to another; it depends on the specifics of the experiment: the number of words in each list, the time spent studying the list, and so forth. On the other hand, days lost from work each year because of illness is meaningful in itself. When the units of the dependent variable are not universally meaningful, the confidence interval for the difference of means may not be helpful; a standardized measure of effect size or strength of association may be preferred. One such measure will be described in the next chapter.

Heterogeneity of Variance as an Experimental Result

Besides testing an assumption of the pooled-variances *t* test, the homogeneity of variance test can sometimes reveal an important experimental result. For instance, one study found that just filling out a depression questionnaire changed the mood of most of the subjects. Subjects who were fairly depressed to begin with were made even more depressed by reading the depressing statements in the questionnaire. On the other hand, nondepressed subjects were actually happier after completing the questionnaire because of a contrast effect: They were happy to realize that the depressing statements did not apply to them. The subjects made happier balanced out the subjects made sadder, so the mean mood of the experimental group was virtually the same as that of a control group that filled out a neutral questionnaire. Because these two sample means were so similar, the numerator of the *t* test comparing the experimental and control groups was near zero, and therefore the null hypothesis concerning the population means was *not* rejected. However, the *variance* of the experimental group increased compared to that of the control group as a result of subjects becoming either more sad or more happy than usual. In this case, accepting the null hypothesis about the population means is appropriate, but it does not indicate that the experimental treatment *did* do *something*. However, the homogeneity of variance test can reveal that the experimental treatment had some effect worthy of further exploration. Although researchers usually hope that the homogeneity of variance test will fail to reach significance so that they can proceed with the pooled-variance *t* test, the above example shows that sometimes a significant difference in variance can be an interesting and meaningful result in itself.

Publishing the Results of the Two-Sample *t* Test

If we were to try to publish the results of our vitamin C experiment, we would need to include a sentence like the following in the results section: "Consistent with our predictions, the vitamin C group averaged fewer days sick ($M = 4.25$, $SD = 3$) than did the placebo group ($M = 7.75$, $SD = 4$), $t(20) = 2.34$, $p < .05$, two-tailed." The number in parentheses following *t* is the number of degrees of freedom associated with the two-group test ($n_1 + n_2 - 2$).

The following is an excerpt from the results section of a published journal article titled "Group Decision Making under Stress" (Driskell and Salas, 1991), which adheres to APA style rules. For this experiment, a two-group *t* test was the most appropriate way to test two preliminary hypotheses:

> Results also indicated that subjects in the stress conditions were more likely than subjects in the no-stress conditions to report that they were excited ($Ms = 3.86$ vs. 4.78), $t(72) = 2.85$, $p < .01$), and that they felt panicky ($Ms = 4.27$ vs. 5.08), $t(72) = 2.64$, $p = .01$ (p. 475).

Note that the second p value stated was exactly .01 and was therefore expressed that way. Also, note that the hypotheses being tested were of a preliminary nature. To study the effects of stress on decision making, the authors had to induce stress in one random group but not the other. Before comparing the two groups on decision-making variables, the authors checked on the effectiveness of the stress manipulation by comparing the responses of subjects in the two groups on a self-report questionnaire (hence, this is called a manipulation check). The significant differences on the questionnaire (as noted in the quote above) suggest that the stress manipulation was at least somewhat effective and that stress could be the cause of the group differences in the main (dependent) variables being measured.

B

SUMMARY

1. I recommend calculating the mean and (unbiased) standard deviation for each group first for descriptive purposes and then using Formula 7.6A to obtain the pooled variance estimate and Formula 7.5B to complete the t test (assuming the pooled-variance test is justified). However, when using this approach, don't forget to square s to get the variance, and don't forget to take the square root of the denominator of Formula 7.5B.

2. After you reject the null hypothesis for a two-group study, you usually want more specific information about the difference of the population means. If your dependent variable was measured in units that are meaningful outside your own experiment (e.g., heart rate in beats per minute rather than number of words remembered from a list created for your experiment), a confidence interval (CI) for the difference of the population means can give a sense of how effective your experimental manipulation was and can tell you whether your results are likely to have practical applications.

3. The CI is centered on the difference between your two sample means. The distance of the limits above and below the center is the standard error of the difference times the appropriate critical t value (e.g., .05, two-tailed critical values are used for a 95% CI). Given the usual null hypothesis, if zero is contained in the CI (one limit is positive and the other is negative), the null hypothesis cannot be rejected at the alpha level corresponding to the CI (e.g., the 99% CI corresponds to the .01 alpha level).

4. The following assumptions are required for the t tests in this chapter:
   a. *Independent random sampling.* Technically, the samples should be drawn randomly from the population of interest, with each selection independent of all the others. Moreover, the selections in one group should be entirely independent of the selections in the other group. The condition that is almost always substituted for this assumption is the random assignment to two conditions of participants from a sample of convenience.
   b. *Normality.* As in the case of one sample, the two-group t test is robust with respect to violations of this assumption. With samples over 30, there is little concern about the shape of the distribution, except in extreme cases.
   c. *Homogeneity of variance.* This assumption is only required if you wish to use the pooled, rather than the separate, variance t test. If your samples are very large, use the large-sample test (Formula 7.3; you can use the normal distribution for your critical values); if your samples are the same size, or your sample variances are very similar, you can use the pooled-variance test without worrying about

testing this assumption. If your samples are fairly small and unequal in size, and their variances are quite different, you should consider the separate-variance test.

5. A test for homogeneity of variance can be performed to determine whether it is valid to pool the variances. If the test is significant (i.e., $p < .05$), the equality of the population variances cannot be assumed, and a separate-variances *t* test should be performed. However, note that there is more than one legitimate test for homogeneity of variance, and there is some debate about the validity of these tests with small samples (which is when they are needed most).

6. Calculating the separate-variances *t* value is easy, but finding the appropriate critical *t* value to compare it to is not. A simplified solution for determining the significance of the separate-variances *t* test will work in many cases. It is based on determining the critical *t* for the smallest (one less than the smaller sample size) and largest $(n_1 + n_2 - 2)$ df that may be associated with this test. If the separate-variances *t* is greater than the critical *t* corresponding to the smallest possible df, the result is significant. If the separate-variances *t* is smaller than the critical *t* corresponding to the largest possible df, the result is not significant. If neither of these situations applies, a more exact determination of the df is required (see Section C).

7. In certain experimental situations, differences in variance may be more dramatic or interesting than the difference in the means. For such situations, the homogeneity of variance test may have more important practical or theoretical implications.

EXERCISES

*1. Seven acute schizophrenics and ten chronic schizophrenics have been measured on a clarity of thinking scale. The mean for the acute sample was 52, with $s = 12$, and the mean for the chronic sample was 44, with $s = 11$. Perform a pooled-variance *t* test with alpha $= .05$ (two-tailed), and state your statistical conclusion.

2. A group of 30 participants is divided in half based on their self-rating of the vividness of their visual imagery. Each participant is tested on how many colors of objects they can correctly recall from a briefly seen display. The more vivid visual imagers recall an average $(\overline{X}_1)$ of 12 colors with $s_1 = 4$; the less vivid visual imagers recall an average $(\overline{X}_2)$ of 8 colors with $s_2 = 5$.

a. Perform a two-group *t* test of the null hypothesis that vividness of visual imagery does not affect the recall of colors; use alpha $= .01$, two-tailed. What is your statistical conclusion?

b. What are some of the limitations that prevent you from concluding that vivid

visual imagery *causes* improved color recall in this type of experiment?

*3. On the first day of class, a third-grade teacher is told that 12 of his students are "gifted," as determined by IQ tests, and the remaining 12 are not. In reality, the two groups have been carefully matched on IQ and previous school performance. At the end of the school year, the gifted students have a grade average of 87.2 with $s = 5.3$, whereas the other students have an average of 82.9, with $s = 4.4$. Perform a *t* test to decide whether you can conclude from these data that false expectations can affect student performance; use alpha $= .05$, two-tailed.

*4. A researcher tested the diastolic blood pressure of 60 marathon runners and 60 non-runners. The mean for the runners was 75.9 mm Hg with $s = 10$, and the mean for the nonrunners was 80.3 mm Hg with $s = 8$.

a. Find the 95% confidence interval for the difference of the population means.

b. Find the 99% confidence interval for the difference of the population means.

c. Use the confidence intervals you found in parts a and b to test the null hypothesis that running has no effect on blood pressure at the .05 and .01 levels, two-tailed.

5. Imagine that the study described in Exercise 4 was conducted with 4 times as many participants.
   a. Find both the 95% and 99% confidence intervals.
   b. Compare the widths of these intervals to their counterparts in Exercise 4, and state the general principle illustrated by the comparison. (Your comparison should be an approximate one because in addition to rounding error, a slight change in the critical t from Exercise 4 to Exercise 5 will influence the comparison.)

6. A psychologist is studying the concentration of a certain enzyme in saliva as a possible indicator of chronic anxiety level. A sample of 12 anxiety neurotics yields a mean enzyme concentration of 3.2 with $s = .7$. For comparison purposes, a sample of 20 subjects reporting low levels of anxiety is measured and yields a mean enzyme concentration of 2.3, with $s = .4$.
   a. Perform a t test (alpha = .05, two-tailed) to determine whether the two populations sampled differ with respect to their mean saliva concentration of this enzyme.
   b. Based on your answer to part a, what type of error (Type I or Type II) might you be making?

*7. Will students wait longer for the arrival of an instructor who is a full professor than for one who is a graduate student? This question was investigated by counting how many minutes undergraduate students waited in two small seminar classes, one taught by a full professor and one taught by a graduate student. The data (in minutes) are as follows:

   Graduate Student Instructor: 9, 11, 14, 14, 16, 19, 37
   Full Professor: 13, 15, 15, 16, 18, 23, 28, 31, 31

   a. Use the pooled-variances t test to test the null hypothesis at the .05 level, two-tailed.
   b. Find the limits of the 95% confidence interval for the difference of the two population means.

*8. Suppose that the undergraduate in Exercise 7 who waited 37 minutes for the graduate student wasn't really waiting but had simply fallen asleep. Eliminate the measurement for that particular participant.
   a. Retest the null hypothesis. What can you say about the susceptibility of the t test to outliers?
   b. On the average, how much more time did students wait for the professor than for the graduate student? Construct the 95% confidence interval for the difference in mean waiting times.
   c. Would a separate-variances t test have been appropriate to answer part a? Explain.

*9. An industrial psychologist is investigating the effects of different types of motivation on the performance of simulated clerical tasks. The 10 participants in the "individual motivation" sample are told that they will be rewarded according to how many tasks they successfully complete. The 10 participants in the "group motivation" sample are told that they will be rewarded according to the average number of tasks completed by all the participants in their sample. The number of tasks completed by each participant are as follows:

   Individual Motivation: 11, 17, 14, 10, 11, 15, 10, 8, 12, 15
   Group Motivation: 10, 15, 14, 8, 9, 14, 6, 7, 11, 13

   a. Perform a pooled-variances t test. Can you reject the null hypothesis at the .05 level?
   b. Would the separate-variances t value differ from the t you found in part a? Explain why or why not.
   c. What is the largest critical t possible for a separate-variances t test on these data? Would using this worst-case critical t change your conclusion in part a? Explain.

10. Suppose that a second industrial psychologist performed the same experiment described in Exercise 9 but used considerably larger samples. If the second experimenter obtained the same sample variances and the same calculated t value, which experimenter obtained the larger difference in sample means? Explain how you arrived at your answer.

Zero Differences between Sample Means

The virtual impossibility of obtaining a zero difference between two sample means involves a mathematical paradox. If you look at the null hypothesis distribution for the study of exercisers versus nonexercisers (see Figure 7.1), it is easy to see that the mode (the highest vertical point of the distribution) is zero—implying that zero is the most common difference that Dr. Null will find. But it is extremely unlikely to get a zero difference between two sample means, even when doing the bogus experiments that Dr. Null does. (Similarly, recall that in describing one-sample tests I mentioned that it is very unlikely that the sample mean selected by Dr. Null will be exactly equal to the population mean.) The problem is that when dealing with a smooth mathematical distribution, picking any one number from the distribution *exactly*, including exactly zero, has an infinitely small probability. This is why probability was defined in terms of intervals or ranges of a smooth distribution in Chapter 4. So, although numbers around zero are the most common means for Dr. Null's experiments, getting *exactly* zero (i.e., an infinite number of zeros past the decimal point) can be thought of as virtually impossible.

Adding Variances to Find the Variance of the Difference

To get some feeling for why you add the variances of two samples when calculating the standard error of the differences, first imagine randomly selecting one score at a time from a normal distribution. Selecting a score within one standard deviation from the mean would be fairly likely. On the other hand, selecting a score about two standard deviations from the mean would be quite unusual. Next, imagine randomly selecting two scores at a time. It would not be unusual if each score was about one standard deviation from the mean; however, it would be just as likely for both scores to be on the same side of the mean as it would be for the two scores to be on opposite sides. If both are on the same side of the mean, the difference between them will be about zero. But if the scores are on opposite sides, the difference between them would be about two standard deviations. The point of this demonstration is to suggest that it is easier to get a *difference of two scores* that is equal to two standard deviations than it is to get a single score that is two standard deviations from the mean. The reason is that when you have two scores, even though each could be a fairly typical score, they could easily be from opposite sides of the distribution, and that increases the difference between them. Although this explanation is not very precise, it is meant merely to give you some feeling for why you could expect difference scores to have more, rather than less, variability from the mean than single scores. The statistical law in this case states that when you subtract (or add) two random scores from two normal distributions, the variance of the difference (or sum) is the sum of the two variances. (Note that you can add variances but not standard deviations.)

The Critical Value for the Separate-Variances *t* Test

As I mentioned in Section B, the difficult part of performing the separate-variances *t* test is finding the most appropriate value of df with which to look up a critical value from the t distribution. By far, the most common formula for estimating the df for the s-v test is the one first presented by Welch (1947). In Chapter 12, I will show how Welch's df formula can be applied to the multigroup case, but it will be considerably easier to understand this formula in the simplest (i.e., two-group) case. A convenient way to look at the Welch formula is in terms of the *variance of means* corresponding to each sample. The variance of means is just the unbiased variance of a sample divided by its size: $s_{\overline{X}}^2 = s^2/N$ (note that this is what you get if you take the standard error of the mean and square it). To simplify the notation I will use w to stand for the variance of means , so that $w_1 = s_1^2/n_1$ and $w_2 = s_2^2/n_2$. In terms of these ws, the Welch-adjusted df is:

$$\text{df}_{\text{Welch}} = \frac{(w_1 + w_2)^2}{\dfrac{w_1^2}{n_1-1} + \dfrac{w_2^2}{n_2-1}}$$

Formula 7.10

The separate-variance t test can also be written in terms of the ws, as follows:

$$t = \frac{\overline{X}_1 - \overline{X}_2}{\sqrt{w_1 + w_2}}$$

Formula 7.11

Even though this adjustment is virtually never used when $n_1 = n_2$, it could be. It is instructive to look at what happens to Formula 7.10 when the ns are equal; df_{Welch} reduces to:

$$(n - 1)\frac{(s_1^2 + s_2^2)^2}{s_1^4 + s_2^4}$$

where s^4 is the square of the varience—that is , $(s^2)^2$. It is not obvious from looking at this formula, but the ratio being multiplied by n-1 reaches a maximum of 2 when the variances are equal, and approaches a minimum of 1 as the variances become extremely different.

Similarly, when the ns are not equal the maximum value for df_{Welch}, as given by Formula 7.10, is always $n_1 + n_2 - 2$ (which reduces to $2(n - 1)$, when the ns are equal); df_{Welch} will never be larger than the df you would use for the pooled-variance t test on the same data. When the ns are *not* equal, the minimum value for df_{Welch} is $n_s - 1$, where n_s is the smaller of the two ns. Knowing the minimum and maximum possible values for df_{Welch} is what led to the simple strategy, described in Section B, that can enable you to avoid using Formula 7.10 in many cases.

When the two sample variances are the same, df_{Welch} becomes a function of the two sample sizes; the greater the discrepancy between the two ns, the more closely df_{Welch} approaches its minimum value of $n_s - 1$. However, as in the case of equal ns, the separate- and pooled-variance t tests yield the *same* t value when the two variances are the same (this follows from the fact that if both variances equal, say 20, then s_{pooled}^2 will equal 20, as well), and there would be no reason to suspect a lack of homogeneity of variance in that case. When both the variances and the ns differ between the two samples it makes sense to think of the separate-variance t test in terms of the two values for w.

It is when the larger sample has the smaller variance that the ws are more discrepant, and this pattern has the effect of making the denominator of the s-v test larger. Thus, this pattern produces an s-v t value that is *smaller* than the corresponding p-v t value. As an extreme example, imagine that the control group has $n_1 = 10$ and $s_1^2 = 100$, while the experimental group has $n_2 = 100$ and $s_2^2 = 10$. In this case, $w_1 = 10$ and $w_2 = .1$, so $\sqrt{(w_1 + w_2)} = \sqrt{10.1} = 3.18$. However, $s_{\text{pooled}}^2 = (900 + 990) / 108 = 17.5$, so the denominator of the p-v t test equals $\sqrt{[17.5 (.1 + .01)]} = \sqrt{1.925} = 1.39$. Because the denominator of the p-v test (1.39) is much smaller than the denominator of the corresponding s-v test (3.18), and the numerators of the two tests are always the same, the p-v *t* value will be considerably *larger* than the s-v *t* value. Moreover, there is considerable adjustment of the df for the s-v test in this case:

$$df_{Welch} = \frac{(10 + .1)^2}{\frac{100}{9} + \frac{.01}{99}} = \frac{102.01}{11.1112} = 9.18$$

Note that df_{Welch} is close to its minimum value of 9, whereas the df for the p-v test would be 108!

It is when the larger sample has the *larger* variance that the s-v *t* value comes out larger than the corresponding p-v *t* value, and there is less adjustment of the df. Reversing sample sizes in the previous example yields equal *w*s: $w_1 = w_2 = 1.0$. The s-v denominator ($\sqrt{2} = 1.41$), is now less than half the size of the p-v denominator, which is 3.19 (the new s^2_{pooled} is 92.5). Furthermore, the adjustment of the df for the s-v test is less severe:

$$df_{Welch} = \frac{(1 + 1)^2}{\frac{1}{99} + \frac{1}{9}} = \frac{4}{.1212} = 33$$

The df for the s-v test (33) is still considerably less than the df for the p-v test (108) due to the discrepancy of the sample sizes, but the adjustment is mitigated by the lack of discrepancy between the two *w*s. This comparison suggests that you will have a better chance of attaining significance if you use the pooled-variance *t* test when the larger sample has the smaller variance, and the separate-variance *t* test when the reverse is true. However, given the greater concern with controlling Type I rather than Type II errors, the recommendation of statisticians is just the opposite: To be conservative, you should perform the pooled-variance *t* test only when the larger sample has the larger variance.

When the variances of two samples are very different, there are usually other problems with the data, such as skewed distributions, or the presence of outliers. In such cases, researchers often apply data transformations to make their distributions more normal, or they give up on the normal distribution entirely and convert their data to ranks, in order to apply a nonparametric analog of the *t* test, the most popular of which is described in Chapter 21. However, a promising alternative is to trim the samples, as described in the previous chapter, and then conduct a two-group *t* test on the trimmed samples.

Random Assignment and the Separate-Variances *t* Test

As I mentioned in Section B, the *t* test formulas of this chapter are based on the assumption that two random samples are being drawn independently from two normal distributions that share the same mean. However, nearly all psychology experiments are conducted by first dividing a single sample of convenience into two different groups as randomly as possible. The random assignment of participants to the two groups prevents bias and possible confounding variables from ruining the validity of the experimental results, but this procedure tends to produce different standard errors from what you would expect from two independent random samples, both drawn directly from the populations of interest. Some statisticians had been concerned that random assignment could lead to a higher rate of Type I errors than otherwise expected, but Reichardt and Gollob (1999) showed that this is not generally the case, particularly when the samples are the same size. However, when a discrepancy in sample sizes occurs, an inflation of the Type I error rate can result—but only when the smaller sample has the larger variance and the pooled-variance test is used. In such cases, Reichardt and Gollob (1999) found that the separate-variances t test was sufficiently conservative to correct the potential problem. Thus, the recommendations of these authors with respect to the use of the separate-variances t test are in line with those of other statisticians: To maximize power, always use the pooled-variance test in the equal-n case. When the ns are not equal, protect your Type I error rate by using the pooled-variance test when the larger sample has the larger variance, and using the separate-variances test when the larger sample has the *smaller* variance.

Whereas some statisticians argue that a good conservative approach to the t test involves always using the separate-variances version, an argument can be made in favor of always using the pooled-variances version for its extra power and ignoring the possibility of making some extra Type I errors due to discrepancies in the variances of the two populations, when their means do not differ. So, what if a difference in population variances leads to an erroneous rejection of the null hypothesis that the population means are the same? If the population variances are truly different, that argues against the notion that the treatment being tested is truly ineffective—why should the variances differ if participants are randomly assigned and the treatment is ineffective? Although it is *some* kind of error to state that the population means differ when it is really the variances that differ, it is not a total false alarm—the treatment in question may be worthy of further exploration. If the hypothesis you are really concerned with is whether or not the treatment has any effect whatsoever, and not just whether the population means differ, I can imagine relaxing concerns about the homogeneity assumption underlying the pooled-variance test.

The t Test for Two Trimmed Means

A robust form of the two-group t test, recommended by Wilcox (2003), begins with 20% trimming of both samples (see Chapter 6, Section C). The numerator of this t test would be the difference of the two trimmed means $(\overline{X}_{t1} - \overline{X}_{t2})$. Because we want our t test to be as robust as possible we will not assume homogeneity of variance, and will therefore conduct a separate-variances t test on the trimmed samples. To make our ws robust we will base them on the Winsorized variances, symbolized as $s^{2'}$ (the Winsorized s^2 is just the square of the Winsorized s, as described in the previous chapter). For 20% trimming, a robust w could be defined as $w' = s^{2'}/.36N$, but I will present the somewhat more refined procedure proposed by Yuen and Dixon (1973). Let us define n_t as the number of scores that remain in a sample after trimming. For example, if $N = 12$ and you want to apply 20% trimming, you will remove two scores from both the low and high ends of the scores, so $n_t = 8$. (The number of scores trimmed away on each end is PN, where P is the trimming proportion, and any fraction is dropped. For this example, $PN = .2 \times 12 = 2.4$, which rounds down to 2.) Now we can define w' as $s^{2'}/n'_t$, where n'_t is given by the following formula:

$$n'_t = \left(\frac{n_t - 1}{n - 1} \right) n_t \qquad \text{**Formula 7.12**}$$

As compared to $.36n$, n'_t will often be larger (for this example, $n'_t = 5.091$, whereas $.36n = 4.32$), but the difference becomes quite small for large ns. The robust version of Formula 7.11 can be written quite simply as:

$$t'_{s-v} = \frac{\overline{X}_{t1} - \overline{X}_{t2}}{\sqrt{w'_1 + w'_2}} \qquad \text{**Formula 7.13**}$$

The df for this test is found from Formula 7.10, substituting w' for each w, and n_t (not n_t') for each n.

For an example of a two-sample robust t test, imagine that the raw data for the vitamin C experiment described in Section B were as follows: vitamin C: 10, 7, 7, 6, 5, 5, 3, 3, 3, 1, 1, 0; placebo: 14, 12, 11, 10, 9, 6, 4, 4, 4, 3. The 20% trimmed data would look like this: vitamin C: 7, 6, 5, 5, 3, 3, 3, 1; placebo: 11, 10, 9, 6, 4, 4. The Winsorized samples would look like this: vitamin C: 7, 7, 7, 6, 5, 5, 3, 3, 3, 1, 1, 1; placebo: 11, 11, 11, 10, 9, 6, 4, 4, 4, 4. From the trimmed samples we obtain $\overline{X}_{t1} = 4.125$, $\overline{X}_{t2} = 7.333$, $n_{t1} = 8$, $n_{t2} = 6$. From the Winsorized samples we find that $s_1^{2'} = 5.72$ and $s_2^{2'} = 10.71$. Applying Formula 7.12 to the n_ts, we get $n_{t1}' = 5.091$ and $n_{t2}' = 3.333$. Finally, the robust ws are $w_1' = 5.72 / 5.091 = 1.124$, and $w_2' = 10.71 / 3.333 = 3.213$. Inserting these values into Formula 7.13 we obtain the following robust t value:

$$t_{s-v}' = \frac{4.125 - 7.333}{\sqrt{1.124 + 3.213}} = \frac{-3.208}{\sqrt{4.337}} = \frac{-3.208}{2.0825} = -1.54$$

You don't need to look up a critical value to know that this t will not be significant at the .05 level. As you can see, the preceding example does not seem to be one in which trimming provides an advantage. When your data do not appear markedly inconsistent with the assumptions of the ordinary pooled-variance t test, that test is likely the most powerful one you can use to detect a difference in population means, especially when your sample sizes are not very large. However, to complete the example, and illustrate the use of Formula 7.10, I will calculate df$_{Welch}$ for this case.

$$df_{Welch} = \frac{(1.124 + 3.213)^2}{\dfrac{1.124^2}{7} + \dfrac{3.213^2}{5}} = \frac{4.337^2}{.1805 + 2.065} = \frac{18.81}{2.2455} = 8.38$$

Given the appropriate statistical function (e.g., SPSS calls it CDF.T), you could obtain the critical t corresponding to 8.38 degrees of freedom, and use it to find a confidence interval for the vitamin C/placebo population difference ($\mu_1 - \mu_2 = 3.208 \pm t_{crit} \times 2.0825$). (Without the availability of software, you would have to round off 8.38 to 8, and look in Table A.2.) Bear in mind that you can avoid using the t distribution entirely by applying a percentile bootstrap procedure to the differences in pairs of trimmed means, found by resampling from your own two samples, as described briefly in the previous chapter, and more extensively in Wilcox (2003).

Resampling Methods

In the previous chapter, I described how a bootstrapping procedure could generate an empirical approximation of the null hypothesis distribution based on your actual data. In the two-group case, you could resample *with replacement* separately from each of your two samples, and calculate the two-group t values until they pile up into a reasonably smooth distribution. However, the type of bootstrapping just decribed is most appropriate when each of the two samples has been obtained randomly and independently from the population of interest. As I mentioned earlier, the quasi-random assignment of participants from one sample of convenience to two experimental conditions is the more usual procedure. A resampling method that mirrors this reality is called a *rerandomization* or *permutation* procedure. This is the final alternative to the ordinary two-group t test that I will describe.

To perform a rerandomization test on the data from the vitamin C example, you would take your original 22 values and randomly resort them into two new sets of 12 and 10 scores each. (This is equivalent to drawing a random sample of 12 scores *without replacement* from the original 22; the remaining 10 scores are automatically the other sample.) Calculate the difference in the means of these two samples; then, randomly divide your 22 values into a set of 12 values and a set of 10 values again, and calculate another difference of sample means. The reason why this resampling procedure is often called a permutation test is that you can go about it by placing your 22 scores in every possible different order (i.e., permutation), each time taking the mean of the first 12 and the mean of the remaining 10. However, even with a total number of only 22 scores, the number of possible permutations is amazingly high—and too time consuming to deal with, even for modern personal computers. Fortunately, after just a few thousand repetitions of this resampling process, you can look at the resulting frequency distribution of the means differences to make a decision about the null hypothesis. If the difference of means from your original experiment is not included in the middle 95% of this distribution, you can reject the null hypothesis of equal population means at the .05 level; otherwise, you cannot (Wilcox, 2001).

The big advantage of the permutation test is that it makes no simplifying assumptions about the shape of the population distributions, such as assuming normal distributions. On the other hand, its disadvantage is that it assumes that both populations follow the same distribution—as represented by the data of your two samples combined. Although it seems that you are testing only whether the two populations have the same mean, you are really testing whether the two population distributions are identical in every respect. A diiference in variance or kurtosis between the two distributions could lead to the rejection of the null hypothesis, even when the populations have the same mean. This is not a serious problem if your primary concern is whether a particluar treatment makes any difference at all, relative to a control group, but if your focus is on discerning the actual difference of the experimental and control population means, this is not the test you want.

The simplest form of the permutation test, applied to the vitamin C example, would require putting all 22 scores in size order, and then replacing the original scores with their ranks (1 to 22). Then, those ranks are summed separately for each of the original samples. If one sample has mostly high ranks, and the other mostly low ranks, you may be able to reject the null hypothesis that the two underlying populations are identical. This *rank-sum* test will be discussed in greater detail in the last chapter of this text.

SUMMARY

1. To find the critical value of t for a separate-variances t test, the df should first be adjusted by Welch's formula. This formula can be conveniently expressed in terms of two ratios, such that each ratio is the variance of one of the samples divided by its size (i.e., $w = s^2/n$). As the degree to which the larger sample has the smaller variance increases, the ws increasingly diverge, the denominator of the s-v test grows larger (making the s-v t value smaller), and the adjusted df (i.e., df_{Welch}) decreases towards a minimum of $n_s - 1$, where n_s is the smaller of the two ns (or n, if the ns are equal).

2. To be conservative (i.e., more concerned with Type I than Type II errors), you should never perform the pooled-variance t test when the larger sample has the *smaller* variance.

3. A robust form of the two-group t test compares the two trimmed means in terms of the two Winsorized variances (each separately divided by the appropriate trimmed n), and uses the Welch formula to find the df for the critical t value. Another robust alternative to the two-

group *t* test that avoids the normal-distribution assumption is the permutation (or rerandomization) test. The scores of both samples are combined into one group and randomly reshuffled into two new groups thousands of times. The mean differences produced in this way form a distribution that can be used to decide if the mean difference from the two original samples is sufficiently large to reject the null hypothesis.

Note: For purposes of comparison, I sometimes refer to exercises from previous sections of a chapter or from earlier chapters. A shorthand notation, consisting of the chapter number and section letter followed by the problem number, will be used to refer to exercises. For example, below, Exercise 7B1 refers to Exercise 1 at the end of Section B of Chapter 7.

*1. Calculate the separate-variances *t* test for Exercise 7B1. Round off df_{Welch} to the nearest integer to look up the critical *t*, and state your decision with respect to the null hypothesis.

2. a. Repeat the steps of Exercise 1 (i.e., the previous exercise) with the data from Exercise 7B6.
   b. Repeat the steps of Exercise 1 with the data from Exercise 7B8.

*3. Suppose the results of a study were as follows: $\overline{X}_1 = 24$, $s_1 = 4.5$, $N_1 = 5$; $\overline{X}_2 = 30$, $s_2 = 9.4$, $N_2 = 15$.
   a. Calculate the pooled-variances *t* test.

b. Calculate the separate-variances *t* test, and compare the result to your answer for part a.

4. Imagine that the standard deviations from Exercise 3 are reversed, so the data are $\overline{X}_1 = 24$, $s_1 = 9.4$, $N_1 = 5$; $\overline{X}_2 = 30$, $s_2 = 4.5$, $N_2 = 15$.
   a. Calculate the pooled-variances *t* test.
   b. Calculate the separate-variances *t* test, and compare the result to your answer for part a.
   c. When the larger group has the larger standard deviation, which *t* test yields the larger value?

*5. Perform a 20% trimmed *t* test on the data from Exercise 7B9, approximating *w'* as $s^{2'}/.36N$ (instead of using Formula 7.12). Round off df_{Welch} to the nearest integer to find your critical *t*.

6. Redo Exercise 7B7 with 20% trimming, according to Yuen's procedure (i.e., adjust n_t with Formula 7.12). Round off df_{Welch} to the nearest integer to find your critical *t*.

The *z* test for two samples (use only when you know both of the population variances or when the samples are sufficiently large):

$$z = \frac{(\overline{X}_1 - \overline{X}_2) - (\mu_1 - \mu_2)}{\sqrt{\frac{\sigma_1^2}{n_1} + \frac{\sigma_2^2}{n_2}}}$$ **Formula 7.3**

The separate-variances *t* test (may *not* follow a *t* distribution with df = $n_1 + n_2 - 2$; see Section C):

$$t = \frac{(\overline{X}_1 - \overline{X}_2) - (\mu_1 - \mu_2)}{\sqrt{\frac{s_1^2}{n_1} + \frac{s_2^2}{n_2}}}$$ **Formula 7.4**

Pooled-variances *t* test (pooled-variance estimate has already been calculated):

$$t = \frac{(\overline{X}_1 - \overline{X}_2) - (\mu_1 - \mu_2)}{\sqrt{s_p^2\left(\frac{1}{n_1} + \frac{1}{n_2}\right)}}$$ **Formula 7.5B**

Pooled-variance estimate of the population variance:

$$s_p^2 = \frac{(n_1 - 1)s_1^2 + (n_2 - 1)s_2^2}{n_1 + n_2 - 2}$$

Formula 7.6A

Pooled-variances t test (when the variances of both samples have already been calculated):

$$t = \frac{(\overline{X}_1 - \overline{X}_2) - (\mu_1 - \mu_2)}{\sqrt{\dfrac{(n_1 - 1)s_1^2 + (n_2 - 1)s_2^2}{n_1 + n_2 - 2}\left(\dfrac{1}{n_1} + \dfrac{1}{n_2}\right)}}$$

Formula 7.7A

Pooled-variances t test (raw-score version; SS has already been calculated for each sample):

$$t = \frac{(\overline{X}_1 - \overline{X}_2) - (\mu_1 - \mu_2)}{\sqrt{\dfrac{SS_1 + SS_2}{n_1 + n_2 - 2}\left(\dfrac{1}{n_1} + \dfrac{1}{n_2}\right)}}$$

Formula 7.7B

The t test for equal-sized samples (note that n in the formula is the number of subjects in *each* group and not the total in both groups combined):

$$t = \frac{(\overline{X}_1 - \overline{X}_2) - (\mu_1 - \mu_2)}{\sqrt{\dfrac{s_1^2 + s_2^2}{n}}}$$

Formula 7.8

The confidence interval for the difference of two population means ($s_{\bar{x}_1 - \bar{x}_2}$ has already been calculated using the denominator of Formula 7.4, 7.7A, 7.7B, or 7.8, as appropriate):

$$\mu_1 - \mu_2 = (\overline{X}_1 - \overline{X}_2) \pm t_{\text{crit}}\, s_{\bar{x}_1 - \bar{x}_2}$$

Formula 7.9

The Welch-adjusted df for the separate-variances t test (where $w_i = s_i^2/n_i$):

$$\mathrm{df}_{\text{Welch}} = \frac{(w_1 + w_2)^2}{\dfrac{w_1^2}{n_1 - 1} + \dfrac{w_2^2}{n_2 - 1}}$$

Formula 7.10

The separate-variances t test expressed in terms of w (where $w_i = s_i^2/n_i$):

$$t = \frac{\overline{X}_1 - \overline{X}_2}{\sqrt{w_1 + w_2}}$$

Formula 7.11

The trimmed n, as adjusted for use in Yuen and Dixon's (1973) trimmed t test method (where $w_i = s_i^2/n_i$):

$$n_t' = \left(\frac{n_t - 1}{n - 1}\right)n_t$$

Formula 7.12

Yuen and Dixon's separate-variances trimmed t test ($w' = s^{2'}/n_t'$, where $s^{2'}$ is the Winsorized variance and n_t' is the trimmed n after being adjusted by Formula 7.12):

$$t_{s-v}' = \frac{\overline{X}_{t1} - \overline{X}_{t2}}{\sqrt{w_1' + w_2'}}$$

Formula 7.13

Note: The $\mathrm{df}_{\text{Welch}}$ for this test is found from the w' and n_t (*not* n_t') for each group.

STATISTICAL POWER AND EFFECT SIZE

8

Chapter

CONCEPTUAL FOUNDATION

You will need to use the following from previous chapters:

Symbols
μ: Mean of a population
$\overline{X}$: Mean of a sample
σ: Standard deviation of a population
s_p: Square root of the pooled variance estimate

Formulas
Formula 7.8: The t test for two equal-sized groups

Concepts
The null hypothesis distribution
Type I and type II errors

In this chapter I will discuss how to estimate the probability of making a Type II error and thus the probability of *not* making a Type II error, which is called the *power* of a statistical test. I will also introduce the concept of effect size and show how effect size, sample size, and alpha combine to affect power. Although the usual purpose of power analysis is to plan an experiment, a thorough understanding of this topic will enable you to be much more astute as an interpreter of experimental results already obtained. Most importantly, the discussion of power and effect size makes the difference between statistical significance and practical importance clear. It will be easiest to introduce power in the context of the two-sample t test described in the previous chapter; the topic of power will be revisited in subsequent chapters.

The Alternative Hypothesis Distribution

In the previous three chapters we looked carefully at the null hypothesis distribution (NHD), which helped clarify how Type I errors are controlled. However, Type II errors cannot occur when the null hypothesis is true; a Type II error occurs only when the null hypothesis is *not* true, but it is accepted anyway. (Take another look at Table 5.1.) Because R.A. Fisher was not interested in the concept of Type II errors, we owe the topic of Type II error rates and power to the work of Neyman and Pearson (1928).

To understand the factors that determine the rate of Type II errors, you need a picture of what can happen when the null hypothesis (H_0) is false and the alternative hypothesis (H_A) is true. Unfortunately, this approach is a bit more complicated than it sounds. The null hypothesis distribution was fairly easy to find because the null hypothesis is stated specifically (e.g., $\mu_0 = 65$ inches, or $\mu_1 = \mu_2$). If we try to find the *alternative hypothesis distribution* (AHD), the problem we encounter is that the alternative hypothesis is usually stated as the complement of the null hypothesis and is therefore *not* specific (e.g., $\mu_0 \neq 65$ inches, or $\mu_1 \neq \mu_2$). If the H_A states that the population mean is *not* 65 inches, there are many possibilities left to choose from (e.g., $\mu_0 = 66$ inches, $\mu_0 = 60$ inches, etc.). To introduce the study of Type II errors, and therefore the topic of power analysis, I will need to begin with an alternative hypothesis that is stated specifically.

Suppose that a Martian scientist has been studying the earth for some time and has noticed that adult humans come in two varieties: men and women. The Martian has also noticed that the women seem to be shorter than the men, but being a scientist, the Martian does not trust its own judgment. So the Martian decides to perform an experiment. The Martian's null hypothesis is that the mean height for all men is the same as the mean height for all women. But we on earth know that the null hypothesis is not true, and moreover, we know which of the possible alternatives *is* true. For the sake of simplicity I will assume that the mean height for men is exactly 69 inches, and for women it is 65 inches, with a standard deviation of 3 inches for both genders.

Because in this case the H_0 is not true, the Martian cannot make a Type I error; it will either reject H_0 correctly or retain H_0 and make a Type II error. To find the probability of committing a Type II error, we need to draw the alternative hypothesis distribution. Suppose that the Martian plans to select at random four men and four women and then perform a *t* test for two sample means. It would be logical to assume that the AHD will be a *t* distribution with $4 + 4 - 2 = 6$ degrees of freedom, just as the NHD would be. However, the difference is that the NHD would be centered on zero, which is the *t* value that we would expect to get on the average if men and women were equal in height, but the AHD is centered on some value other than zero. A *t* distribution that is not centered on zero is called a *noncentral t distribution*. The noncentral *t* distribution is awkward to deal with, so to simplify matters the AHD will be assumed to be a normal distribution. This assumption involves some error, especially for small sample sizes, but the concept is basically the same whether we use a normal distribution or a *t* distribution—and as you will see, power analysis is a matter of approximation anyway.

To find the value upon which the AHD is centered, we need to know the *t* value that the Martian would get, on the average, for its experiment. (I will continue to refer to *t* instead of *z*, even though I will use the normal distribution to simplify the calculations.) It will be helpful to look at the formula for a two-sample test when the sample sizes are equal (Formula 7.8):

$$t = \frac{(\overline{X}_1 - \overline{X}_2)}{\sqrt{\dfrac{s_1^2 + s_2^2}{n}}}$$

Formula 7.8

To get the average *t* value, we replace each variable in Formula 7.8 with its average value; for the present example, the average for the group of men will be 69 inches and for the women, 65 inches, and the two variances would each average $3^2 = 9$. Recall that *n* is the number of subjects in *each* group:

$$\text{Average } t = \frac{69 - 65}{\sqrt{\dfrac{9 + 9}{4}}} = \frac{4}{\sqrt{4.5}} = \frac{4}{2.12} = 1.89$$

From this calculation you can see that, on average, the Martian will get a *t* value of 1.89. However, if we assume that the AHD is normal, the critical value needed for significance ($\alpha = .05$, two-tailed) is 1.96, so most of the time the Martian will not be able to reject the null hypothesis, and therefore it will commit a Type II error. From Figure 8.1 it appears that the Martian will make a Type II error (falsely concluding no height difference between men and women) about 53% of the time and will correctly reject the null hypothesis about 47% of the time. The proportion of the AHD that results in Type II errors (i.e., the Type II error rate) is symbolized by the Greek let-

Figure 8.1

Alternative Hypothesis
Distribution
(alpha = .05)

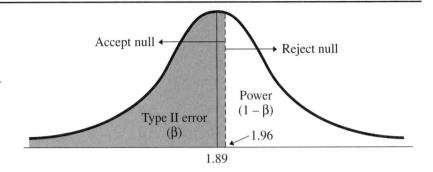

Accept null ← → Reject null

Power
(1 − β)

Type II error
(β)

1.96

1.89

Figure 8.2

Alternative Hypothesis
Distribution
(alpha = .10)

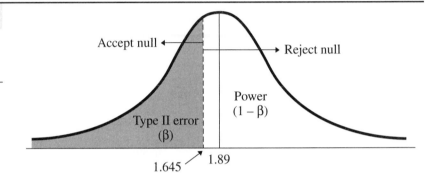

Accept null ← → Reject null

Power
(1 − β)

Type II error
(β)

1.645   1.89

ter *beta* (β); in this case β equals about .53. The proportion of the AHD that results in rejecting the null hypothesis is $1 - β$ (as shown in Table 5.1), and it is called the *power* of the test; in this case, power is about $1 - .53 = .47$.

It is important to realize that even though there is a considerable height difference (on average) between men and women, the Martian's experiment most often will not provide sufficient evidence for it. This is the price of null hypothesis testing. The critical *t* in this case is determined by the need to prevent 95% of "null experiments" from being significant. But that also means that many experiments for which the null is *not* true (like the Martian's experiment, in this case) will nonetheless fail to beat the critical *t* and be judged not significant. In fact, the power of the Martian's experiment is so low (because of the small sample sizes) that it hardly pays to perform such an experiment. One simple way to increase the power would be to increase the sample sizes, as will be explained. Another way to increase power is to raise alpha from .05 to some larger value such as .1. This would change the critical value to 1.645 (so would performing a one-tailed test at the .05 level). As you can see from Figure 8.2, power is now close to 60%, and the proportion of Type II errors has been reduced accordingly. However, this increase in power comes at the price of increasing Type I errors from 5% to 10%, which is usually considered too high a price.

The Expected *t* Value (Delta)

To find the percentages of Figure 8.1 and Figure 8.2 that were below the critical values, I made a rough approximation based on visual inspection of the areas marked off in the distributions. Section B will present a method involving the use of tables to determine the percentages more accurately.

For now, the concept of power needs further explanation. Thus far, power has been described as dependent on the average t value for some specific AHD. This average t is often symbolized by the Greek letter *delta* (δ). The delta corresponding to the null hypothesis equals zero (when H_0 is $\mu_1 = \mu_2$). If delta happens to just equal the critical value needed to attain statistical significance, power (as well as β) will equal 50%; half the time the t value will be above its average and therefore significant, and half the time, below.

Delta can also be thought of as the *expected t* value that corresponds to a particular AHD. The expected t can be calculated for any specific alternative hypothesis (provided that the population standard deviation is also specified); if that alternative hypothesis is true, the actual t values from experiments would fluctuate around, but would average out to, the expected t. Based on the formula for a two-sample t test with equal-sized groups (Formula 7.8), a general formula can be derived for calculating δ. First, we replace each term in Formula 7.8 by its expected or average value, which is the value for that statistic in the whole population. If we make the common assumption that $\sigma_1^2 = \sigma_2^2$ (homogeneity of variance), the result is as follows:

$$\delta = \frac{\mu_1 - \mu_2}{\sqrt{\dfrac{\sigma_1^2 + \sigma_2^2}{n}}} = \frac{\mu_1 - \mu_2}{\sqrt{\dfrac{2\sigma^2}{n}}}$$

Next, it will serve my purpose to use the laws of algebra to rearrange this formula a bit. The denominator can be separated into two square roots, and then the square root involving n can be moved to the numerator, which causes it to be turned upside down, as follows:

$$\delta = \frac{\mu_1 - \mu_2}{\sqrt{\dfrac{2}{n}}\sqrt{\sigma^2}} = \frac{\mu_1 - \mu_2}{\sigma\sqrt{\dfrac{2}{n}}} = \frac{\mu_1 - \mu_2\sqrt{\dfrac{n}{2}}}{\sigma}$$

Finally, the square root part can be separated from the fraction to yield Formula 8.1:

$$\delta = \frac{(\mu_1 - \mu_2)}{\sigma}\sqrt{\frac{n}{2}} \qquad\qquad \textbf{Formula 8.1}$$

Formula 8.1 shows that δ can be conceptualized as the product of two easily understood terms. The first term depends only on the size of the samples. The second term is the separation of the two population means in terms of standard deviations; it is like the z score of one population mean with respect to the other. This term is called the *effect size*, and, due to the pioneering work of Jacob Cohen (1988), it is often symbolized by the letter **d**. (Many statisticians use δ for the population effect size, in keeping with the use of a Greek letter for a characteristic of the population, but then a different letter must be used for the expected t [e.g., Δ, the upper-case delta]. I will stick with the use of **d** for effect size, which has become quite popular in psychology, but I will use boldface to remind you that it is a population parameter). The formula for **d** will be designated Formula 8.2:

$$\mathbf{d} = \frac{\mu_1 - \mu_2}{\sigma} \qquad\qquad \textbf{Formula 8.2}$$

Expressing Formula 8.1 in terms of **d** gives us Formula 8.3:

$$\delta = \mathbf{d}\sqrt{\frac{n}{2}} \qquad\qquad \textbf{Formula 8.3}$$

The separation of the expected t value into the effect size and a term that depends only on sample size was one of the major contributions of Jacob Cohen to psychological statistics, and it is both useful and instructive. However, the concept of effect size requires some further explanation.

The Effect Size

The term *effect size* suggests that the difference in two populations is the effect of something; for instance, the height difference between males and females can be thought of as just one of the effects of gender. The effect size for the male–female height difference can be found by plugging the appropriate values into Formula 8.2:

$$\mathbf{d} = \frac{\mu_1 - \mu_2}{\sigma} = \frac{69 - 65}{3} = \frac{4}{3} = 1.33$$

An effect size of 1.33 is considered quite large, for reasons soon to be made clear. One way to get a feeling for the concept of effect size is to think of **d** in terms of the amount of overlap between two population distributions. Figure 8.3 depicts four pairs of overlapping population distri-

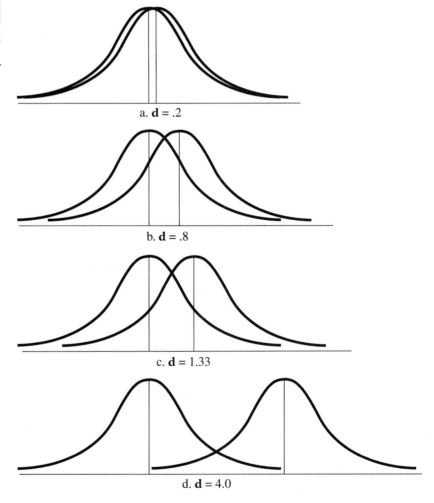

Figure 8.3

Overlap of Populations as a Function of Effect Size

a. **d** = .2

b. **d** = .8

c. **d** = 1.33

d. **d** = 4.0

butions, each pair corresponding to a different effect size. Notice that when **d** = .2, there is a great deal of overlap. If, for instance, the two populations consist of participants who were instructed to use imagery for memorization and participants who received no such instruction, and the variable being measured is amount of recall, when **d** = .2, many participants who received no instructions will recall more than participants who *were* instructed. On the other hand, when **d** = 4.0, there is very little overlap. In this case, very few of the participants who received no instructions will perform better than even the worst of the participants who did receive instructions. It is rare to find an experiment in which **d** = 4.0 because the effect being measured would be so obvious it would be noticed without the need for an experiment. (For example, in comparing the reading skills of first- and fourth-graders the effect of age on reading ability is quite obvious.) In fact, even **d** = 1.33 is generally too large to require an experiment. Although there is a fair amount of overlap between the heights of men and women, the difference is quite noticeable. An experiment in which **d** = 1.33 would most likely be performed only in a situation in which there had been little opportunity to observe the effect (such as a test of a new and effective form of therapy).

According to guidelines suggested by Cohen (1988), **d** = .8 is considered a large effect size: not so large as to be completely obvious from casual observation, but large enough to have a good chance of being found statistically significant with a modest number of participants. By contrast, **d** = .2 is considered a small effect; effect sizes much smaller would usually not be worth the trouble of investigating. The difference in overlap between **d** = .2 and **d** = .8 can be seen in Figure 8.3. Finally, **d** = .5 is considered to be a medium effect size.

Power Analysis

Power analysis is the technique of predicting the power of a statistical test before an experiment is run. If the predicted power is too low, it would not be worthwhile to conduct the experiment unless some change could be made to increase the power. How low is too low? Most researchers would agree that .5 is too low; before investing the time, money, and effort required to perform an experiment, one would prefer to have better than a 50% chance of attaining significant results. A power of .7, which corresponds to a Type II error rate of .3, is often considered minimal. On the other hand, it is not generally considered important to keep the Type II error rate as low as the Type I error rate. Keeping the Type II error rate as low as the usual .05 Type I error rate would result in a power of .95, which is generally considered higher than necessary. A power of about .8 is probably the most reasonable compromise in that it yields a high enough chance of success to be worth the effort without the drawbacks involved in raising power any higher. (The problems associated with increasing power will be discussed shortly.)

Because the experimenter can (at least theoretically) choose the alpha and sample size to be used, the tricky part of power analysis is predicting the effect size of the intended experiment. You cannot always predict effect size but with power analysis you can find the power associated with each possible effect size for a given combination of alpha and sample size. This kind of analysis can help the experimenter decide whether to run the experiment as planned, try to increase the power, or abandon the whole enterprise. Before we consider the difficult task of predicting effect sizes, let us look further at how sample size combines with effect size to determine power.

Once α and the number of tails are set, power depends only on δ, the expected t value. Looking at Formula 8.3, you see that δ is determined by multiplying the effect size by a term that depends on sample size. This means that power can be made larger either by making the effect size larger or by making the sample size larger. Because it is often not possible to influence effect size (e.g., the gender difference in height), power is usually manipulated by changing the sample size. Theoretically, power can always be made as high as desired, *regardless of the effect size*, by sufficiently increasing the sample size (unless the effect size is exactly zero). This fact has important implications for the interpretation of statistical results, as do practical limitations on sample size.

The Interpretation of t Values

Suppose that a particular two-sample experiment produces a very large t value, such as 17. The p value associated with that t would be extremely low; the results would be significant even with alpha as low as .0001. What does this imply about effect size? Actually, it tells us nothing about effect size. The large t does appear to tell us something about δ, however. It seems quite unlikely that δ could be zero, and yet yield such a large t value; that is why we can reject the null hypothesis with such a high degree of confidence. In fact, δ is most likely to be somewhere near 17; if the expected t were as low as, say, 5 or as high as about 30, there would be little chance of obtaining a t of 17. So, although there is always some chance that the obtained t of 17 is a fluke that is either much higher or lower than δ, the chances are that δ is indeed in the vicinity of 17. The reason this tells us nothing about the probable effect size is that we have not yet considered the sample size involved. It is possible for δ to be 17 when the effect size is only .1 (in which case there would have to be about 58,000 participants in each group) or when the effect size is about 10 (in which case there would need to be only about 6 in each group).

The important message here is that no matter how certain we may be that the effect size is not zero, a large t value does not imply that the effect size must be fairly large, because even a very tiny effect size can lead to a large expected t if very large samples are used. It is important to remember that statistical significance does not imply by itself that the effect size is large enough to be interesting or of any practical importance.

On the other hand, a small obtained t value does not allow the rejection of the null hypothesis; an effect size of zero cannot be ruled out and must be considered a reasonable possibility. Again, nothing definitive can be said about the effect size simply because the t value is small, but a small obtained t implies that δ is probably small. If a small t is obtained using large sample sizes, a smaller effect size is implied than if the same t value were obtained with small samples. This principle has implications for the interpretation of negative (i.e., not significant) results.

When testing a new drug, therapy, or any experimental manipulation, negative results (i.e., acceptance of the null hypothesis) seem to suggest that the treatment is not at all effective. However, a nonsignificant result could actually be a Type II error. The smaller the samples being used, the lower the power for a given effect size, and the more likely is a Type II error. Negative results based on small samples are less conclusive and less trustworthy than negative results based on large samples. Although we cannot prove that the null hypothesis is true by obtaining a small and nonsignificant t value, even with very large samples, a very small t obtained with very large samples does provide strong evidence that the effect size is likely to be small. Negative results with small samples, on the other hand, tell us very little. Occasionally, a researcher may test a well-known theory using small sample sizes, obtain negative results, and conclude that the theory has been disproved. One advantage of a good understanding of power analysis is that you would be unlikely to make that mistake or be taken in by the researcher who does.

Comparing *t* Values from Different Studies

Another lesson to be learned through power analysis is that two experimental results, based on the same sample sizes, but with one leading just barely to a significant *t* test (e.g., $p = .049$) and the other just missing (e.g., $p = .051$), are not very different. If you think only in terms of null hypothesis testing, you are likely to emphasize the difference between the results, as though the two results are telling us very different things. If you tend to think in terms of effect sizes, however, you will notice that the two results lead to very similar estimates of effect size. Also, it is important to realize that the fact that one result is significant and the other is not does not imply that these two results differ significantly from each other. If one result were for two groups of women (i.e., an experimental group and a control group) and the other for two corresponding groups of men, you could not conclude that men and women differ with respect to their sensitivity to whatever treatment you are using (to compare the *t* tests from the men and women you could use the methods of Chapter 14). As long as the male groups are reasonably close in size to the female groups, it actually looks like the results were very similar for the two genders. Much confusion in the psychological literature has occurred because of a lack of attention to effect sizes. This situation has definitely improved in the past few decades, especially due to the increasing popularity of a method called meta-analysis (see Section C).

Although *p* values, like *t* values, do not by themselves indicate anything about effect size, if you are looking at a series of similar tests all based on the same sample sizes, the smaller *p* values will indeed tend to be associated with the larger estimates of effect size. Moreover, whereas the *t* value does not tell you directly about the effect size in the population, when combined with the sample size, *t* can tell you the effect size in a sample, which can serve as an estimate of the effect size in the population, as I will describe next.

Estimating Effect Size

I began this chapter with an example for which **d** is well known (the male–female height difference), so power could be estimated easily. Of course, if you actually knew the value of **d** for a particular experimental situation, there would be no reason to conduct an experiment. In fact, most experiments are conducted to decide whether **d** may be zero (i.e., whether the null hypothesis is true). Usually the experimenter believes **d** is not zero and has a prediction regarding the direction of the effect but no exact prediction about the size of **d**. However, to estimate power and determine whether the experiment has a good chance of producing positive (i.e., significant) results, the researcher must make some guess about **d**. One way to estimate **d** is to use previous research or theoretical models to arrive at estimates of the two population means and the population standard deviation that you are dealing with and then calculate an estimated **d** according to Formula 8.2. The estimate of power so derived would only be true to the extent that the estimates of the population parameters were reasonable.

It can be especially difficult to estimate population parameters if you are dealing with newly created measures. For instance, you might want to compare memory during a happy mood with memory during a sad mood, measuring memory in terms of the number of words recalled from a list devised for the experiment. Because this word list may never have been used before, there would be no way to predict the number of words that would be recalled in each condition. However, even when a previous study cannot help you estimate particular population parameters, the sample statistics from a previous study can be used to provide an overall estimate of **d**. A simple estimate of **d** that could be derived from a previous two-group study is based on the difference of the two sample means divided by the square root of the pooled variance (s_p), as shown in Formula 8.4:

$$g = \frac{\overline{X}_1 - \overline{X}_2}{s_p}$$

Formula 8.4

Note the similarity to Formula 8.2; each population parameter from Formula 8.2 has been replaced by a corresponding sample statistic. My use of the letter g to represent a point estimate of **d** is consistent with Hedges (1981) and is in keeping with the convention of using English letters for sample statistics (and their combinations). The value of g from a similar study performed previously can then be used to estimate the likely value of **d** for your proposed study. (As calculated above, g is a somewhat biased estimator of **d**, but because the correction factor is complicated, rarely used, and the bias becomes very slight for large samples [Hedges, 1981] I will save the less biased formula for Section C. The value obtained from Formula 8.4 will suffice as an estimate of **d** for most purposes.)

The ordinary t test with equal sample sizes can be expressed in terms of g, as follows:

$$t = g \sqrt{\frac{n}{2}}$$

Note the similarity to Formula 8.3. If you need to estimate g from a published study, you are likely to find the means of the two groups but are not likely to find s_p. But as long as you have a t value and the sample size, you can turn around the formula above to find g:

$$g = t \sqrt{\frac{2}{n}}$$

Formula 8.5

If the sample sizes are not equal, you will need to find the harmonic mean of the two ns, as shown in Section B.

Of course, there may not be any previous studies similar enough to provide an estimate of **d** for your proposed study. If you must take a guess at the value of **d**, you can use the guidelines established by Cohen (1988), in which .2, .5, and .8 represent small, medium, and large effect sizes, respectively. To make such an estimate, you would have to know the subject area well in terms of the variables being measured, experimental conditions involved, and so forth.

Manipulating Power

The most common way of manipulating power is the regulation of sample size. Although power can be altered by changing the alpha level, this approach is not common because of widespread concern about keeping Type I errors to a fixed, low level. A third way to manipulate power is to change the effect size. This last possibility may seem absurd at first. If we consider the male–female height example, it *is* absurd; we cannot change the difference between these populations. However, if the two populations represent the results of some treatment effect and a control condition, respectively, there is the possibility of manipulation. For example, if a drug raises the heart rate, on average, by 5 beats per minute (bpm) relative to a placebo, and the common standard deviation is 10 bpm, the effect size will be 5/10 = .5. It is possible that a larger dose of the drug could raise the heart rate even more, without much change in the standard deviation. In that case, the larger dose would be associated with a larger effect size. Other treatments or therapies could be intensified in one way or another to increase the relative separation of the population means and thus increase the effect size. Of course, it is not always possible to intensify an experimental manipulation, and sometimes doing so could be unpleasant, if not actually dangerous, to the participants. From a practical standpoint it is often desirable to keep the effect in the experiment at a level that is normally encountered in the real world (or would be if the new treatment were to be adopted).

There is yet another potential way to increase the effect size without changing the difference between the population means: the standard deviation (σ) could be decreased. (Because σ is in the denominator of the formula for **d**, lowering σ will increase **d**.) Normally, researchers try to keep σ as low as possible by maintaining the same experimental conditions for each participant. But even if very little random error is introduced by the experiment, there will always be the individual differences of the participants contributing to σ. For some experiments, the person-to-person variability can be quite high, making it difficult to have sufficient power without using a prohibitively large number of participants. A very useful way to avoid much of the person-to-person variability in any experiment is to measure each participant in more than one condition, or to match similar participants and then place them in different experimental groups. This method can greatly increase the power without intensifying the experimental manipulation or increasing the sample size. However, techniques for matching participants will be better understood after correlation is explained, so this topic will be delayed until Chapter 11.

SUMMARY

1. To estimate the Type II error rate (β) for a particular experiment, it is helpful to choose a specific alternative hypothesis and then to draw the alternative hypothesis distribution (AHD). For the two-group case, the null hypothesis distribution is usually centered on zero, but the AHD is centered on the expected *t* value, which depends on the specific alternative hypothesis.

2. Once the AHD is determined, the critical value is drawn on the same distribution. The proportion of the AHD below (to the left of) the critical value is β, and the proportion above (to the right) is 1 − β, which is called *power*.

3. Changing α changes the critical value, which, in turn, changes the power. A smaller α is associated with fewer Type I errors but more Type II errors, and therefore lower power.

4. The expected *t* value is called *delta* (δ), and it can be calculated as the product of two terms: one that depends on the sample size and one that is called the population effect size. The measure of effect size, symbolized as **d**, is related to the separation (or conversely, the degree of overlap) of the two population distributions; for many purposes **d** = .2 is considered a small effect size, .5 is medium, and .8 is large.

5. If **d** is known or can be accurately estimated (and both α and the sample size have been chosen), it is easy to determine power. However, **d** must often be estimated roughly from previous results or theoretical considerations. Once **d** has been estimated, you can find the sample sizes necessary to achieve a reasonable amount of power; .8 is usually considered a reasonable level for power.

6. A large obtained t value implies that the expected t is probably large, but does not say anything about effect size until sample size is taken into account. Nonsignificant results associated with small sample sizes are less informative than nonsignificant results based on large sample sizes.

7. The usual way of increasing power is to increase the sample sizes. Sometimes it is possible to increase power by increasing **d**, but this requires either increasing the effectiveness or strength of the treatment (or whatever distinguishes the two populations) or reducing the person-to-person variability (this can be done by matching, which will be explained in later chapters). Finally, power can be increased by making α larger, but this can rarely be justified.

EXERCISES

1. What is the Type II error rate (β) when power is (a) .65? (b) .96? What is the power when β is (c) .12? (d) .45?

*2. Suppose that the mean heart rate of all pregnant women (μ_P) is 75 bpm, and the mean heart rate for nonpregnant women (μ_N) is 72. If the standard deviation for both groups is 10 bpm, what is the effect size (**d**)?

3. If the mean verbal SAT score is 510 for women and 490 for men, what is the **d**?

*4. In Exercise 2, if a t test were performed to compare 28 pregnant women with an equal number of nonpregnant women, what would be the expected t value (δ)? If a two-tailed test were planned with α = .05, would the results come out statistically significant more than half the time or not?

5. Suppose the experiment in Exercise 4 were performed with 4 times as many women in each group. What would be the new expected t? How does this compare with the answer you found in Exercise 2? Can you state the general rule that applies?

*6. If two population means differ by one and a half standard deviations, what is the value of **d**? If a t test is performed using 20 participants from each population, what will δ be?

7. If two population means differ by three standard deviations, and a t test is per-

formed with 20 participants in each group, what will δ be? Compare this value to your answer for the previous exercise. What general rule is being demonstrated?

*8. Calculate g, the sample estimate of effect size for
   a. Exercise 7B1
   b. Exercise 7B6.

9. The t value calculated for a particular two-group experiment was −23. Which of the following can you conclude? Explain your choice.
   a. A calculation error must have been made.
   b. The number of participants must have been large.
   c. The effect size must have been large.
   d. The expected t was probably large.
   e. The alpha level was probably large.

*10. Suppose you are in a situation in which it is more important to reduce Type II errors than to worry about Type I errors. Which of the following could be helpful in reducing Type II errors? Explain your choice.
   a. Make alpha unusually large (e.g., .1).
   b. Use a larger number of participants.
   c. Try to increase the effect size.
   d. All of the above.
   e. None of the above.

In Section A, I mentioned that the alternative hypothesis distribution (AHD) for a *t* test involving two small groups is actually a *noncentral t distribution*. The value that it is centered on, δ, is therefore called the *noncentrality parameter*. To find power accurately, we need to find the proportion of the appropriate noncentral *t* distribution that is in the rejection zone of the test. Finding these exact proportions requires sophisticated software, but fortunately such software is becoming increasingly available on the web.

B

**BASIC
STATISTICAL
PROCEDURES**

Using Power Tables

Before there were personal computers, two-group power analysis required the use of a whole series of noncentral *t* tables, as presented by Cohen (1988). As an alternative, you might have consulted a graph containing a series of "power curves" that depict power as a function of effect size for different possible sample sizes. A simpler (albeit less accurate) alternative, and the method used in Section A, is to use the normal distribution as an approximation, regardless of sample size. (More accurate power tables that can be used for the *t* test will be described in the context of analysis of variance in Chapter 12, Section C, and the use of the appropriate function in SPSS to obtain *p* values from noncentral *t* distributions will be explained in Section D—the web supplement for this chapter.) This approach teaches the concepts of power and requires only one simple table, which has been included as Table A.3. (Also, the reverse of Table A.3 has been included as Table A.4.) The following examples will demonstrate how the tables work.

Assume that δ for a particular experiment is 3.0. The possible experimental outcomes can be approximated as a normal distribution centered on the value 3.0, as shown in Figure 8.4. The critical value for α = .05, two-tailed is 1.96, so you'll notice a vertical line at this value. The area to the left of the critical value contains those experimental results for which the null hypothesis must be accepted (i.e., Type II errors). Because the critical value is about one standard deviation below the mean (3.0 − 1.96 = 1.04), the area to the left of the critical value is about 16% (i.e., the area beyond $z = 1.0$). Therefore, for this problem, the Type II error rate (β) = .16, and power = 1 − .16 = .84 (see Figure 8.4). Table A.3 can be used to obtain the same result. Look down the column for the .05, two-tailed test until you get to δ = 3.0. The entry for power is .85, which is close to the value of .84 approximated above (1.96 is a little *more* than one standard deviation below the mean of 3, leaving a bit more area to the right). Also notice that the power associated with δ = 2.0 is just slightly more than .50 (because 2.0

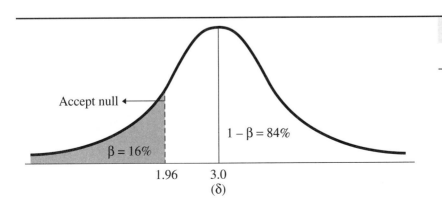

Figure 8.4

Power for Delta = 3.0

is slightly greater than 1.96). The table makes the determination of power easier because the subtraction of the critical value from δ and the calculation of the area have been done for you.

Table A.3 can also be used to find the δ that is required to yield a particular level of power. If, for instance, you'd like a power of .80 at $\alpha = .05$, two-tailed, you could look down the appropriate column within the table to find that level of power (or the closest value in the table) and then look across for the corresponding δ; in this case, the exact value of .80 happens to appear in the table, corresponding to $\delta = 2.8$. Table A.4 makes this task easier by displaying the values of δ that correspond to the most convenient levels of power. Notice that next to power $= .8$ under the appropriate column heading is, of course, $\delta = 2.8$. Also notice that if you read across from power $= .50$, the δ in each column is the corresponding critical z. This again shows that when the critical value is the expected value, half the time you will obtain significant results (power $= .5$) and half the time you will not ($\beta = .5$).

Using the Formula and the Table Together

Let us return to the example of the Martian scientist trying to prove that men and women differ in height. If the Martian were to compare eight randomly selected men with eight randomly selected women, and it knew that **d** $= 1.33$, it could calculate δ using Formula 8.3:

$$\delta = d\sqrt{\frac{n}{2}} = 1.33\sqrt{\frac{8}{2}} = (1.33)\sqrt{4} = (1.33)(2) = 2.66$$

The Martian needs to round off the calculated δ, 2.66, to 2.7 to use Table A.3. For $\alpha = .05$, two-tailed, the power is a reasonable .77.

The Relationship between Alpha and Power

Looking again at Table A.3, you can see the effect of changing the alpha level. Pick any value for δ, and the row of power values corresponding to that δ will show how power changes as a function of α. (For now we will consider only the two-tailed values.) If $\delta = 2.6$, for example, power is a rather high .83 for $\alpha = .1$. However, if α is lowered to the more conventional .05 level, power is lowered as well, to .74. And if α is lowered further, to .01, power is reduced to only .5. Figure 8.5 shows graphically that as α is decreased, the critical value moves further to the right on the alternative hypothesis distribution, so there is less area above (i.e., to the right of) the critical value, representing fewer statistically significant results (i.e., lower power).

Figure 8.5

Power as a Function of Alpha for a Fixed Value of Delta

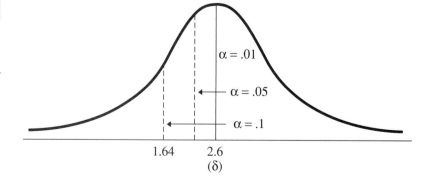

Considering the greater power associated with α = .05, one-tailed, as compared to α = .05, two-tailed, the advantage of the one-tailed test should be obvious. However, the one-tailed test, as discussed earlier, involves a promise on the part of the researcher not to test results in the direction opposite to what is predicted. Whether such a promise is considered appropriate or acceptable depends on circumstances and conventions (see Chapter 5). Note also that if α is reduced to be more conservative (i.e., more cautious about Type I errors), power will be reduced as well. On the other hand, in situations when a Type II error may have serious consequences, and there is less than the usual concern about Type I errors, α can be increased to raise power. Such situations, however, are quite rare. Typically, researchers try to manipulate power by manipulating factors that affect δ, most commonly the sample sizes. However, in some situations, the researcher may have little control over the sizes of his or her samples.

Power Analysis with Fixed Sample Sizes

For a situation in which the sample sizes are to some extent fixed by circumstances, the researcher may find it useful to estimate the power for a range of effect sizes, to determine the smallest effect size for which the power is reasonable (i.e., worth performing the experiment). If the smallest effect size that yields adequate power is considerably larger than could be optimistically expected, the researcher would probably decide not to conduct the experiment as planned.

Instead of finding power for a range of effect sizes, we can find the smallest effect size that yields adequate power by solving Formula 8.3 for **d** to create Formula 8.6:

$$\delta = \mathbf{d}\sqrt{\frac{n}{2}}$$

$$\mathbf{d} = \frac{\delta}{\sqrt{\frac{n}{2}}}$$

$$\mathbf{d} = \delta\sqrt{\frac{2}{n}} \qquad\qquad \textbf{Formula 8.6}$$

Note the similarity to Formula 8.5 (Just replace the sample statistics in Formula 8.5 with the population parameters to which they correspond).

Find δ by looking in Table A.4 under the lowest value of power considered worthwhile (this is a matter of judgment and depends on the cost of doing the experiment, etc.). Then use the value of δ found in the table, along with whatever sample sizes you are stuck with, in Formula 8.6. For instance, suppose .6 is the lowest power you are willing to consider. This corresponds to δ = 2.21 for α = .05, two-tailed. If you have only 10 subjects in each group, n = 10. Plugging these values into Formula 8.6, we get:

$$\mathbf{d} = \sqrt{\frac{2}{10}}(2.21) = (.447)(2.21) = .99$$

An effect size of .99 is needed to attain power = .6. Any smaller effect size would yield less than adequate power in this example. An effect size of .99 is fairly large, and if such a large effect size cannot be expected for this hypothetical experiment, the experiment is not worth doing.

Sample Size Determination

Usually a researcher has a fair amount of flexibility in determining the size of the groups to be used in an experiment. However, there is invariably some cost per participant that must be considered. The cost can be quite high if the participant is being tested with a sophisticated medical procedure (e.g., PET scan of blood flow in the brain) or quite low if the participant is filling out a brief questionnaire in a psychology class. But there is always some cost in time and effort to run the participant, analyze the data, and so forth. Even if the participants are volunteers, remember that the supply is not endless, and often the larger the number of participants used for one experiment, the fewer that are immediately available for future experiments. With this in mind, most researchers try to use the smallest number of participants necessary to have a reasonable chance of obtaining significant results with a meaningful effect size. Power analysis can aid this endeavor.

First, you must decide on the lowest acceptable level for power. As mentioned before, this is based on each researcher's own situation and judgment. However, power = .8 is often considered desirable, so this value will be used for our example. Deciding on power, as well as α, allows us to find δ from Table A.4. Next, an estimate for **d** must be provided. For our first example, we will use the male–female height difference, for which **d** = 1.33. The values for δ and **d** can be inserted into Formula 8.3 to find the required sample size, but it is a bit more convenient to solve Formula 8.3 for *n* to create Formula 8.7:

$$\delta = \mathbf{d}\sqrt{\frac{n}{2}}$$

$$\sqrt{\frac{n}{2}} = \frac{\delta}{\mathbf{d}}$$

$$\frac{n}{2} = \left(\frac{\delta}{\mathbf{d}}\right)^2$$

$$n = 2\left(\frac{\delta}{\mathbf{d}}\right)^2 \qquad\qquad \textbf{Formula 8.7}$$

Inserting the values for **d** and δ into Formula 8.7, we get:

$$n = 2\left(\frac{2.8}{1.33}\right)^2 = 2(2.1)^2 = (2)(4.41) = 8.82$$

This above calculation indicates that nine subjects are needed in each of the two groups to have acceptable power. The required sample size in this example is unusually low because **d** is unusually high.

Setting Limits on the Sample Size

Determining the sample size is very straightforward if **d** can be estimated. However, power analysis can still be useful in terms of setting limits on the sample size, even if no estimate for **d** is at hand. Sometimes, participants are fairly easy to come by (e.g., handing out a brief questionnaire to a large psychology class), but even in such a case, it can be useful to determine the largest sample size worth testing. A reasonable way to begin this determination is by deciding on the smallest effect size worth dealing with. There is no point in running so many participants that you can easily obtain statistical significance with an effect size too small to be important. How small is too small when it comes to **d**? One way to decide this question involves focusing on the numerator of the formula for **d**—that is, $\mu_1 - \mu_2$. In many situations it is not difficult to determine when the difference between the population means is too small to be of any consequence. For instance, if a new reading technique increases reading speed on average by only one word per minute, or a new therapy for overeating results in the loss of only 3 pounds a year, these differences may be too small to be worth finding statistically significant. Having decided on the smallest $\mu_1 - \mu_2$ that is worth testing, you must then divide this number by an estimate of σ to derive a minimal value for **d**. For most variables, there is usually enough prior information to make at least a crude guess about σ. However, if there is insufficient basis to speculate about $\mu_1 - \mu_2$ or σ individually, it is still possible to decide when **d** is too low to be important. Although the minimal effect size depends on the specific situation, **d** = .2 is generally considered a small effect size. For the following example, I will suppose that an effect size less than .2 is not worth the trouble of testing. To determine the upper limit for sample size, insert the minimal value of **d** in Formula 8.7, along with the value of δ that corresponds to the highest level of power that you feel is necessary (I'll use power = .8, again). Using **d** = .2 and δ = 2.8, we find the following:

$$n = 2\left(\frac{\delta}{\mathbf{d}}\right)^2 = 2\left(\frac{2.8}{.2}\right)^2 = (2)(14)^2 = (2)(196) = 392$$

This calculation demonstrates that to have a good chance (.8) of obtaining significant results when the effect size is only .2, nearly 400 participants are needed in each group. It should also be clear that there would be little point to ever using more than 400 participants per group in this situation because that would result in a good chance of obtaining significant results when **d** is so tiny that you are not interested in it.

On the other hand, there are times when participants are very expensive to run or hard to find because they must meet strict criteria (e.g., left-handed people with no left-handed relatives). In such a case, you may be interested to know the smallest sample size that should be considered. This time you begin by deciding on the largest that **d** might possibly be. You might feel that a **d** larger than .8 would have already been noticed for your experiment and that there is little chance that **d** is actually larger than this. You must also decide on the lowest level of power at which you would still be willing to conduct the experiment. Let us say that this power level is .7, so δ = 2.48 for α = .05 (two-tailed). To determine the lower limit for sample size, I'll insert the above values for **d** and δ into Formula 8.7:

$$n = 2\left(\frac{2.48}{.8}\right)^2 = (2)(3.1)^2 = (2)(9.6) = 19.2$$

This calculation indicates that, regardless of the expense to run or find each participant, you must run at least (about) 20 participants in each group. Using fewer participants will lead to inadequate power, even when the effect size is as large as you think can be reasonably expected.

The Case of Unequal Sample Sizes

All of the power formulas in this section that deal with two groups assume that the two groups are the same size: The n that appears in each formula is the n of each group. When the two samples are not the same size, these formulas can still be used by averaging the two different ns together. However, the ordinary average, or mean, does not provide an accurate answer. The power formulas are most accurate when n is found by calculating the *harmonic mean* of n_1 and n_2, which I will symbolize as n_h. Formula 8.8 for the harmonic mean of two numbers is as follows:

$$\text{Harmonic mean} = n_h = \frac{2n_1n_2}{n_1 + n_2}$$ **Formula 8.8**

Suppose a neuropsychologist has available ten patients with Alzheimer's disease and wants to compare them on a new test with patients suffering from Korsakoff's syndrome. Unfortunately, the neuropsychologist can find only five Korsakoff patients to take the new test. Eliminating half of the Alzheimer's patients would simplify the statistical analysis, but would produce a considerable loss of power. Fortunately, the usual power formulas can be used by first calculating the harmonic mean of the two sample sizes as follows:

$$n_h = \frac{(2)(10)(5)}{10 + 5} = \frac{100}{15} = 6.67$$

Again assuming that **d** = 1.0, we find the value of δ from (a slightly modified version of) Formula 8.3:

$$\delta = \mathbf{d}\sqrt{\frac{n_h}{2}} = 1.0\sqrt{\frac{6.67}{2}} = \sqrt{3.33} = 1.83$$

If we round off δ to 1.8 and look in Table A.3 (α = .05, two-tailed), we find that power is only .44. However, this is nonetheless an improvement over the power that would remain (i.e., .36) if we threw away half the Alzheimer's patients to attain equal-sized groups.

A situation that may arise occasionally is the need to find g from published results in which the two samples are not the same size. Consider the vitamin C/placebo experiment from the last chapter. If we did not have access to the original data, but read that t was equal to 2.345, and knew that the two sample sizes were 12 and 10, respectively, we could nonetheless calculate g accurately by using an appropriately modified version of Formula 8.5, as follows: $g = t\sqrt{2/n_h}$. In fact, we can insert Formula 8.8 into the preceding formula and rearrange the terms algebraically (e.g., the factor of 2 cancels out) to create a convenient, all-purpose formula for obtaining g from a published t value.

$$g = t\sqrt{\frac{n_1 + n_2}{n_1 n_2}}$$ **Formula 8.9**

For the vitamin C experiment,

$$g = 2.345\sqrt{\frac{12 + 10}{(12)(10)}} = 2.345\sqrt{.1833} = 2.345(.428) = 1.004$$

The Power of a One-Sample Test

Because one-sample experiments are so much less common than two-sample experiments, power analysis thus far has been described in terms of the latter rather than the former. Fortunately, there is very little difference between the two cases. The chief difference is in the formula for δ and its variations. For the one-sample case:

$$\delta = \frac{\mu_1 - \mu_2}{\dfrac{\sigma}{\sqrt{n}}}$$

which leads to Formula 8.10:

$$\delta = \mathbf{d}\sqrt{n} \hspace{4cm} \textbf{Formula 8.10}$$

Formula 8.10 can be solved for n to create Formula 8.11 for finding the sample size in the one-sample case:

$$n = \left(\frac{\delta}{\mathbf{d}}\right)^2 \hspace{4cm} \textbf{Formula 8.11}$$

Notice that the one-sample formulas differ from the two-sample formulas only by a factor of 2; all of the procedures and concepts are otherwise the same. The absence of the factor of 2 increases the power of the one-sample test, relative to the two-sample test, when all else is equal. This is because greater variability is possible (which hurts power) when two samples are compared to each other than when one sample is compared to a population (which does not vary). In later chapters, I will show how to calculate power for other statistical procedures.

SUMMARY

1. Table A.3 uses the normal distribution to approximate power for given values of expected $t(\delta)$ and α (one- or two-tailed). Table A.4 is similar, except that you can begin with one of the usual levels of power desired and find the corresponding δ. You can see from Table A.3 that for a given value of δ, power decreases as α is made smaller.
2. Power analysis for a two-sample experiment can be separated into two categories, depending on whether your sample sizes are fixed or flexible.
   a. When the sample sizes are fixed by circumstance (e.g., a certain number of patients with a particular condition are available), you can estimate your power if you can estimate the effect size that you are dealing with (either by estimating the population parameters that make up $\mathbf{d}$ or by estimating $\mathbf{d}$ directly). If the power thus estimated is too low, you may choose on that basis not to perform the experiment as designed. On the other hand, you can work backward by deciding on the lowest acceptable level for power, find the corresponding δ, and then put the fixed sample size into the equation to solve for $\mathbf{d}$. Because the $\mathbf{d}$ thus found is the lowest $\mathbf{d}$ that yields an acceptable level of power with the sample sizes available, you may decide not to perform the experiment if you expect the true value of $\mathbf{d}$ to be less than the $\mathbf{d}$ you calculated.
   b. When there is a fair amount of flexibility in determining sample size, you can begin by estimating $\mathbf{d}$ and then looking up the δ that corresponds to your desired level of power. The appropriate equation can then be used to find the sample sizes needed to attain the desired power level with the effect size as estimated. Power analysis can also be used to set limits on the sample size.
   c. With flexible sample sizes, you can begin by deciding on the smallest $\mathbf{d}$ that is worth testing and then look up the δ that corresponds to the highest level of power that you feel you might need. The sample size you would then calculate is the *largest* you would consider using. There would be no point to attaining even greater power with

effect sizes that you consider to be trivial. On the other hand, you could decide on the largest **d** that can reasonably be expected for the proposed experiment and look up δ for the lowest level of power that you would consider acceptable. The sample size that you would calculate in this case is the bare minimum that is worth employing. Using any fewer participants would mean that the power would be less than acceptable even in the most optimistic case (i.e., **d** as large as can be expected).

3. The harmonic mean of two numbers (n_h) is equal to twice the product of the two numbers divided by their sum. The harmonic mean of two differing sample sizes can be used in place of n in power formulas for two groups.

4. The principles concerning power that were described for the two-sample experiment apply, as well, to the one-sample experiment. The effect size measure, **d**, is still defined in terms of the difference between two populations, although only one of the populations is being sampled. The formulas are basically the same except for a factor of 2.

EXERCISES

*1. a. What is the Type II error rate (β) and the power associated with δ = 1.5 (α = .05, two-tailed)? With δ = 2.5?
   b. Repeat part a for α = .01, two-tailed. What is the effect on the Type II error rate of reducing alpha?
   c. What δ is required to have power = .4? To have power = .9? (Assume α = .05, two-tailed for this exercise.)

2. a. A researcher has two sections of a psychology class, each with 30 students, to use as participants. The same puzzle is presented to both classes, but the students in one of the classes are given a helpful strategy for solving the puzzle. If the **d** is .9 in this situation, what will be the power at α = .05, two-tailed?
   b. If you had no estimate of **d**, but considered power less than .75 to be unacceptable, how high would **d** have to be for you to run the experiment?

*3. a. To attain power = .7 with an effect size that also equals .7, how many participants are required in each group of a two-group experiment (use α = .01, two-tailed)?
   b. How many participants are required per group to attain power = .85 (all else the same as in part a)?

4. In Exercise 3, how many participants would be required in a one-group experiment? How does this compare to your previous answer?

*5. If the smallest **d** for which it is worth showing significance in a particular experiment is .3, and power = .8 is desired,
   a. What is the largest number of participants per group that should be used, when α = .05, one-tailed?
   b. When α = .01, one-tailed?

6. A drug for treating headaches has a side effect of lowering diastolic blood pressure by 8 mm Hg compared to a placebo. If the population standard deviation is known to be 6 mm Hg,
   a. What would be the power of an experiment (α = .01, two-tailed) comparing the drug to a placebo using 15 participants per group?
   b. How many participants would you need per group to attain power = .95, with α = .01, two-tailed?

*7. a. What are the ordinary (i.e., arithmetic) and harmonic means of 20 and 10?
   b. If you are comparing 20 schizophrenics to 10 manic depressives on some physiological measure for which **d** is expected to be about .8, what would be the power of a *t* test between these two groups, using alpha = .05, two-tailed? Would this experiment be worth doing?

8. Assume that for the experiment described in Exercise 7 power = .75 is considered the lowest acceptable level. How large would **d** have to be to reach this level of power?

*9. In Exercise 7B3, if the effect size in the experiment were equal to the value of g obtained,

    a. How many participants would have been needed to attain power = .7 with α = .05, two-tailed?

    b. Given the number of participants in Exercise 7B3, how large an effect size would be needed to have a power of .9, with α = .05, two-tailed?

10. a. If d less than .2 was always considered too small to be of interest, power = .7 was always considered acceptable, and you always tested with α = .05, two-tailed, what is the largest number of participants in a one-sample experiment that you would ever need?

    b. If you never expect d to be more than .8, you never want power to be less than .8, and you always test at α = .05, two-tailed, what is the smallest number of participants in a one-sample experiment that you would ever use?

*11. a. In Exercise 7B1, if d were 1.1, what would be the power of this experiment when tested with alpha = .05, two-tailed?

    b. In Exercise 7B6, how large would d have to be to attain power = .7, with alpha = .05, two-tailed?

12. Given that a t value is reported to be 2.4, how large is g if:

    a. n_1 = 5 and n_2 = 10.

    b. n_1 = 40 and n_2 = 20.

Calculating Power Retrospectively

C

OPTIONAL MATERIAL

Regardless of whether a researcher conducted a power analysis before running her study, it would not be unusual for that researcher to look at her results and wonder just how much power her test actually had. To calculate the power of a t test that has already been run (sometimes referred to as a post hoc or *retrospective power*) you can use the value of g from your data as your estimate of d, and combine that with your sample sizes to calculate δ (delta). This amounts to the same thing as using your calculated value of t as your estimate of δ. However, as I mentioned earlier, g is a biased estimator of d. In fact, it slightly *over*estimates d, but because the bias is quite small for reasonable sample sizes, you will rarely see the corrected formula for g. I will show it here for completeness, and because I want to use an unbiased estimate of d for the retrospective power calculation. The formula that follows does not eliminate the bias of g entirely, but it is considered sufficiently accurate for practical purposes (Hedges, 1982):

$$\text{est. } \mathbf{d} = g\left(1 - \frac{3}{4df - 1}\right) \qquad \textbf{Formula 8.12}$$

where df equals $n_1 + n_2 - 2$.

As an example, I will use Formula 8.12 to adjust the value for g that I calculated for the vitamin C experiment at the end of Section B. The unbiased estimate of d comes out to:

$$\text{est. } \mathbf{d} = 1.004\left[1 - \frac{3}{(4 * 20) - 1}\right] = 1.004\left(1 - \frac{3}{79}\right) = 1.004(.962) = .966$$

Now I can use Formula 8.3 (as modified for unequal ns) to estimate δ (Note that by using Formula 8.9 I avoided a separate calculation for the harmonic mean in finding g at the end of Section B, but you can use Formula 8.8 to verify for yourself that n_h for 10 and 12 is 10.91.)

$$\delta = \mathbf{d}\sqrt{\frac{n_h}{2}} = .966\sqrt{\frac{10.91}{2}} = .966 \cdot 2.335 = 2.26$$

Interpolating in Table A.3, power is approximately .615. Note that the delta we just calculated (2.26) is slightly less than the t value we calculated in the previous chapter (2.345) due to the slight adjustment in g. Without the adjustment, our power estimate would have been a little higher (the interpolated power value for 2.345 is about .65).

Constructing Confidence Intervals for Effect Sizes

The calculations in the preceding section show that our best point estimate for **d** for the vitamin C experiment is .966. However, as you have seen by now, confidence intervals (CI) can be much more informative than point estimates, and such is the case with effect sizes. You may recall that I have already calculated a CI for the difference of the means in the vitamin C experiment; that CI can be used as the basis for constructing an approximate CI for **d**.

The limits of the 95% CI calculated for the vitamin C experiment were .39 fewer days spent sick, and 6.61 fewer days of sickness each year. These are the limits for $\mu_1 - \mu_2$. If we want approximate limits for **d**, we need to divide each limit for $\mu_1 - \mu_2$ by an estimate of σ (because $\mathbf{d} = (\mu_1 - \mu_2) / \sigma$). Our best estimate for σ is s_p, which you may recall from the power example comes out to 3.486, so the rough CI for **d** goes from .112 (i.e., .39 / 3.486) to 1.90 (i.e., 6.61 / 3.486). This is a very wide range of possible effects sizes; to have 95% confidence we have to concede that the true effect size can be anywhere from quite small to very large. At least the CI for **d** does not include zero, but we knew that already, because the t value was statistically significant (or because the CI for $\mu_1 - \mu_2$ did not include zero).

The problem with the CI that I just calculated—the reason it is so rough and approximate—is that I used s_p as an estimate for σ. The smaller the sample sizes, the more inaccurate the CI based on this estimate becomes (the same problem exists when using Tables A.3 and A.4 for power calculations). However, the more accurate method is surprisingly inconvenient without specialized software. Fortunately, SPSS has a function that gives you values for any noncentral t distribution, and although it is a bit tedious, you can obtain an accurate CI with the assistance of this function. The details of the procedure are described in Section D of this chapter on my Web site.

When will you want to construct CIs for **d**? As I mentioned in Section B of the previous chapter, CIs for the difference of the population means are informative when the units of your dependent variable are familiar and easy to interpret (e.g., pounds lost, days sick), but CIs for effect size can be more meaningful when the DV is specific to the experiment (e.g., number of words correctly recalled from a list prepared for a particular experiment). Moreover, effect size estimates from experiments using different DVs to measure the same construct (e.g., memory measured by number of words recalled, or number of faces correctly recognized) can be combined to create a more accurate CI for effect size than is available from any one study. I will return to this notion, when I discuss meta-analysis at the end of this section.

Robust Estimates of Effect Size

In Section C of the previous chapter, I described in some detail how to calculate a robust t test based on trimmed means. If you calculate the trimmed t test, you will of course want to present a correspondingly robust measure of the effect size in your trimmed samples. One obvious soluton is to calculate robust g (g') in terms of the trimmed means and Winsorized variances. The correction factor for 20% trimming is included in the following formula, taken from Algina, Keselman, and Penfield (2005):

$$g' = .642\left(\frac{\overline{X}_{t1} - \overline{X}_{t2}}{s'_p}\right) \qquad \text{Formula 8.13}$$

where s'_p is the square root of the pooled Winsorized variances. Algina, Keselman, and Penfield (2005) also explain how to find a robust CI for the effect-size measure given by Formula 8.13.

Refining the Spam-Filter Analogy

My main purpose in this section is to help you use what you have learned so far about power to increase your understanding of NHT. Let us revisit the spam-filter analogy of Chapter 5, this time defining spam in terms of effect size. Thus, *spam* can be thought of as corresponding to the results of a study for which the corresponding population effect size is truly zero (i.e., **d** = 0). With alpha set to .05, 95% of this spam is screened out in the sense that the results will not be labeled as statistically significant, and are therefore not so likely to mislead other researchers into thinking that the population effects they represent are other than zero. In order to extend the spam-filter analogy to more realistic circumstances, I need to define some messages as *near-spam*; you can think of these as mass-mailings in which you might have some slight interest (or at least they are not totally irrelevant or offensive). In terms of NHT, I will arbitrarily define near-spam as the results of studies for which the corresponding population effect size is very tiny—let's say **d** = .01 (i.e., the mean of one population is one-hundredth of a standard deviation higher or lower than the mean of the other population). As I will demonstrate, NHT is very nearly as good at screening out near-spam as it is with genuine spam.

For example, suppose that you are conducting a fairly large near-spam study with 200 participants in each group. What is your power?

$$\delta = \mathbf{d}\sqrt{\frac{n}{2}} = .01\sqrt{100} = .1$$

You may have noticed that I did not include values for delta this low in Table A.3. Power will not be zero, because delta is not zero, but the proportion of results screened out will be very close to 1 – alpha. A major complication I will have to deal with, however, is that when **d** = 0, the results being screened out involve correct decisions, but when **d** = .01, all of the results being screened out lead to Type II errors (i.e., by not rejecting the null we are acting as though **d** equals 0, when it really equals .01) Whereas most effect sizes around .01 can be fairly characterized as near-spam, some of them may be properly considered as tiny, but potentially important effects, so some of these Type II errors may be problematic. I will return to this issue later in this section.

Note that when **d** is as small as .01, power changes very little—until the sample sizes get extremely large. For instance, even with samples of 20,000 participants each, the power to detect near-spam is only about .17. For better or worse, NHT can be a very effective filter against near-spam. Conversely, when dealing with modest sample sizes, changes in effect sizes on the small end have surprisingly little impact on power. For instance, with 200 participants per group, power is no more than .06 for near-spam. However, for effect sizes ten times as large (i.e., **d** = .1), power increases only to .17 (as in the preceding example, δ comes out to a value of 1.0).

The Type III Error

Before plunging further into the controversy surrounding NHT, I would like to point out an interesting complication that occurs in defining and measuring power at very low levels. This complication involves the rarely discussed *Type III error* (or "error of the third kind" as it was first called by Mosteller, 1948). Although other definitions have occasionally been used, I will follow the traditional convention of defining a Type III error as a *directional* error that attains statistical significance (Leventhal & Huynh, 1996). For example, suppose that, in reality, women have slightly better hearing ability than men, at a certain frequency, but that an experiment yields statistically significant results in the opposite direction. This cannot be a Type I error, because the null hypothesis is not true, but neither is it a Type II error, because statistical significance is being declared. Yet, as I have described the situation, an error is certainly being committed.

The reason the Type III error is so rarely mentioned is that the Type III error *rate* is thought to be always very small. However, when power is very low, as occurs when dealing with near-spam and ordinary-sized samples, nearly half of all *significant* results will actually be Type III errors. I have illustrated the Type III error rate for two values of delta in Figure 8.6.

Along the horizontal axis in Figure 8.6 are represented all the *t* values that could possibly arise from the male/female hearing experiment (note that positive values indicate experimental results in the correct direction [i.e., the same direction as the population means]). You can see that when delta is equal to 1.0 (Figure 8.6a) about 17% of the results are significant in the correct direction, and only about one-tenth of one percent (.1%) are significant in the wrong direction—these are the Type III errors. Notice that it makes little numerical difference whether you define power in terms of *any* rejections of the null hypothesis, or just in terms of rejections in the correct direction. The situation is different, however, in Figure 8.6b.

When δ equals .1 (Figure 8.6b), the percentage of results that attain significance in the correct direction is only about 3%. Moreover, the Type III

Figure 8.6

The Type III Error Rates (shaded areas) for Two Values of Delta

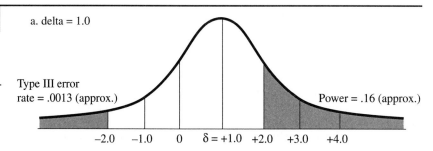

a. delta = 1.0

Type III error rate = .0013 (approx.)

Power = .16 (approx.)

−2.0   −1.0   0   δ = +1.0   +2.0   +3.0   +4.0

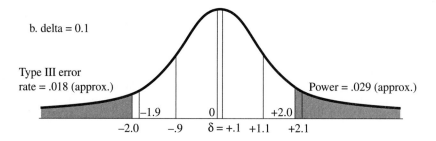

b. delta = 0.1

Type III error rate = .018 (approx.)

Power = .029 (approx.)

−1.9   0   +2.0

−2.0   −.9   δ = +.1   +1.1   +2.1

error rate is almost as large—it is just about 2%. (Note that all of these power values are based on .05, two-tailed tests.) When I stated at the end of the previous subsection that power is "no more than .06" for near-spam with 200 participants per group, I was referring to this situation. If power is defined in terms of *any* rejections of the null hypothesis, it is a bit more than .05 for $\delta = .1$, but if power is defined more carefully in terms of *directionally correct* rejections of the null, it is equal to no more than .03. Without using NHT at all, however, note that nearly half of the results in Figure 8.6b are in the wrong direction (look at the proportion of the distribution's area that is to the left of zero). I will return to this potential benefit of NHT shortly.

The non-directional two-tailed test. The concept of the Type III error rests on the fact that what most statistics texts and researchers refer to as a *two-tailed test* can be more realistically portrayed in common practice as "two simultaneous one-tailed tests predicting opposite directions . . . [such that] each one-tailed test would normally use .5α." (Leventhal and Huynh, 1996, p. 279). That is, the H_0 for a two-group test is almost never just rejected—it is rejected in one direction or the other. By contrast, a truly nondirectional two-tailed test would lead to one of only two possible conclusions: There is no difference between population means (or, at least, we do not have sufficient evidence to say that there is a difference), or the two population means are different (with no direction specified). The latter conclusion would only be useful when it is surprising or interesting that the two conditions differ at all—e.g., the overall performance of participants in the same experiment as run by two different research assistants. If you have no interest in drawing a conclusion about which assistant elicited the better performance, you cannot make a Type III error. As you might imagine, nondirectional two-tailed tests are rarely performed (I have never seen one reported). What psychological researchers usually perform are two one-tailed tests with alphas of .025 each (this is the procedure commonly called the two-tailed test), and occasionally single one-tailed tests with an alpha of .05.

Are Near-Spam Studies *Ever* Performed in Psychological Research?

The title of this section is meant to echo a similar question in Chapter 5 about the null hypothesis. The answer to the question of this section is quite different, however. Whereas it can be argued that the null hypothesis is specified so exactly that it has an infinitesimal chance of being true, or that psychologists almost never test effects that are physically impossible, it is easy to imagine that near-spam studies are being performed regularly. In fact, a major criticism of NHT is that the null hypothesis is virtually never true, because changing any one variable that affects people is bound to have some nonzero effect on just about any other variable you might measure on the same people—and therefore any study you can imagine will produce statistically significant results, if the samples just happen to be large enough. The indirect connection between every human variable and every other leads to a phenomenon that has been called the *crud* (or correlation background noise) factor, the naming of which was attributed to David Lykken by Meehl (1990).

The critics of NHT warn that the use of large samples (especially in conjunction with other power-boosting methods that I will be discussing later in this text, such as taking repeated measures, matching participants across conditions, or using analysis of covariance) would create a great deal of chaos, because so may of the crud studies (what I have been calling near-spam) would start producing statistically significant results. In a way, then, it may be fortunate that economic and other practical constraints conspire to keep power low for tiny population effects in psychological research. Admittedly, it is more by happenstance than design that the use of NHT serves as an effective filter against crud (i.e., near-spam) studies. Moreover, NHT would not be very helpful if the *majority* of psychological studies involved tiny effects. Recall from Chapter 5 that even if only, say, 7% of the tests of tiny effects were to come out significantly, a large proportion of all significant results would be coming from tiny effects, if tiny effects were very commonly tested. Fortunately, it seems that although psychologists may sometimes test tiny effects, they more often test effects that are small to moderate in size.

The Main Practical Advantage of NHT: Filtering Out Tiny Effects

If statistical significance tells you little about whether the null hypothesis is really true (recall from Chapter 5 that $p(S \mid H_0)$ does not equal $p(H_0 \mid S)$), and "significant" does not necessarily mean that the effect size seems to be fairly large, why bother labeling some results as significant at all? Indeed, since the use of NHT first became popular, it has been vigorously argued that NHT is so illogical and counterproductive that it should not be used by psychologists at all (e.g., Berkson, 1938; Rozeboom, 1960; Carver, 1978; Schmidt, 1996). To understand the chief practical advantage of NHT, which makes its use so resistant to the logical criticisms of statisticians, we need to return to the spam-filter analogy and take it even more literally.

Imagine that inboxes are not for e-mail, but are the actual inboxes of journal editors, and the spam folders are file drawers that belong to individual researchers in the field. The researchers operate spam filters themselves by performing null hypothesis tests on the results of their own studies. Of course, very few of the studies would actually be spam, but there could be a good deal of near-spam (especially from exploratory studies) mixed in with the good studies. For simplicity, let us say that when the main result of a study attains statistical significance it is sent to the inbox of a journal, and when it does not, the study's results are placed in one of the researcher's file drawers. Note that the vast majority of the near-spam studies (as well as any spam studies) will not produce significant results, and therefore their reports will not end up cluttering the inboxes of overworked journal editors. That these results wind up in file drawers, often never to be seen again, produces what Rosenthal (1979) famously called the "file drawer problem"; I will return to this problem near the end of this section.

If the main advantage of NHT is that, as Abelson (1997) put it, it "acts as a (mediocre) filter for separating effects of different magnitude" (p. 14), why not just cut out the "middle person" and declare statistical significance based directly on accurate estimates of effect size? (This is essentially what many statistical reformers are calling for.) This is the hardest part of NHT to explain, because it has little to do with science, and much to do with the practicalities of a community of human beings competing for limited resources. The truth is that it would be very hard for researchers to agree on some arbitrary effect size (e.g., **d** = .1) as a universal cutoff for significance, and then adhere strictly to that criterion with respect to sending in manuscripts for publication in prominent journals. However, without such a firm agreement, the spam-filter that is NHT would move from the hands of individual researchers to the control of journal editors, who could be easily swamped by the task of rendering judgments on an enormous number of articles reporting a wide variety of effect sizes, all vying for the attention of the peer reviewers (i.e., fellow researchers) of top-tier journals. As it is now—even with the widespread use of NHT—there is so much research being submitted for publication in psychology that rejection rates are very high at the best journals, and the lag between submitting a manuscript and seeing it published is often more than a year.

The curious advantage of NHT is that its arbitrary use of .05 for alpha is so deeply ingrained in the culture of psychological research, having been popularized by Fisher (1955) at a time when psychology was still a rather small field of inquiry, that only under rare or special circumstances is an alpha larger than .05 considered acceptable (e.g., testing the homogeneity-of-variance assumption before conducting a pooled-variance t test). It is common practice, of course, to report results with p less than .1 as "approaching significance," but this is itself an indication of the high respect with which the .05 level is regarded.

As you will see in later chapters, alpha for a particular statistical test may be reduced considerably below .05, but almost always with the larger goal of keeping the alpha for the entire experiment from creeping above the sacrosanct .05 level. In sum, the combination of NHT at this alpha level with the usual sample sizes (and power-boosting analyses) available has led, somewhat accidentally, to a fairly acceptable compromise among Type I, Type II, and Type III errors. It is certainly not an ideal compromise, and in principle NHT could be greatly improved, but the field of psychology these days seems just too vast to allow for some new compromise agreement on such a fundamental issue.

It is important to note that not all of the tiny effects that psychologists study can be viewed as crud or near-spam. In fact, some of these tiny effects can be of great theoretical interest, and it can be considered a drawback of NHT that such effects are almost always labeled as not being significant. However, there is no simple system I know of that can identify which of the tiny effects are real, and which are part of the background noise. The challenge of those who want to end the practice of NHT is to devise a system that can sort apparently small effects into those that are due almost entirely to sampling error from those that signal something important. In the meantime, it would help if researchers stuck more closely to Fisher's approach to NHT, and not Neyman and Pearson's, in their attitude towards results that fail to attain significance—that is, researchers should reserve judgment in such cases, and not act as though they have shown the null hypothesis to be true.

When the Use of NHT Can Be Most Harmful

If you take a look at Figure 8.7, you can see graphically why NHT is such an effective filter against very small effect sizes. Even when **d** is as large as .1, notice that beta (and therefore power) changes rather little over a wide range of sample sizes (and hardly at all for **d** = .01). However, when **d** is in the neighborhood of .5 (what J. Cohen called a medium-sized effect), power changes dramatically over the range of sample sizes that are commonly used in psychological research. A t test involving 30 participants in each group and a medium population effect size has a bit less than a 50% chance of yielding significant results. Out of ten researchers performing such a test on similar experiments, it would be common for, say, four to attain significant results, while the other six do not. That alone can lead to much confusion; the differences with respect to significance could easily be attributed to minor differences in the studies being conducted, when in fact the population effect size may be nearly identical for all ten studies. Moreover, if an eleventh researcher uses 60 participants per group (not an unreasonable number) for essentially the same study, his or her test will have power of about .75, and therefore a much greater chance of attaining significance, even though the effect in the population has not changed. Let us take a look at some of the problems that uncritical reliance on NHT can cause in these situations.

Perhaps the most serious misuse of NHT is the all-too-common tendency of researchers to interpret a failure to reject H_0 as a license to act as though H_0 were literally true, or even approximately true. For example, many potentially confounding variables (e.g., an age difference between two patient groups) have too readily been dismissed as inconsequential, because an initial t test on that variable yielded a p value not very close to the traditional value for alpha (i.e., somewhat larger than .1). With fairly small sample sizes, a confounding variable could account for a substantial amount of the effect of interest without itself yielding a significant difference. Part of the problem is that whereas researchers have NHT to help them decide when to reject a null hypothesis, there are no agreed-upon guidelines for deciding when your apparent effect size is small enough to ignore completely (i.e., to functionally accept H_0). Although it is sometimes innocuous to use a lack of statistical significance as evidence that the difference between two conditions can be ignored for some *practical* purposes, it is quite a different matter, and seriously problematic, to use a lack of con-

Figure 8.7

Power $(1 - \beta)$ as a Function of Sample Size (n) for Different Effect Sizes (**d**)

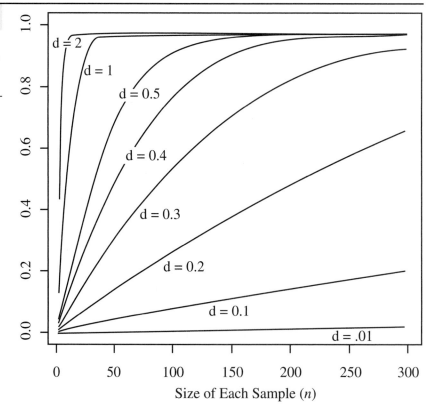

ventional statistical significance as justification to assume that the population effect size is essentially zero, with little attention paid to the size of the samples involved in the research.

A related form of NHT abuse occurs when several dependent variables (DV) are tested, but the researcher's theory suggests that only some of them should be affected by the experimental condition. It is quite misleading to imply that the independent variable is having very different effects on two particular DVs, when testing the first one yields a p of .14, and its effect is therefore dismissed as not even approaching significance (suggesting no effect at all), whereas the second DV is touted as significant because p equals about .04. This misleading categorical distinction has been referred to as the "p-value fallacy" by Dixon (2003); a similar point had been made earlier by Loftus (1996).

The opposite form of NHT abuse occurs when statistical significance is used as automatic evidence of an adequately large effect size. This form of abuse is particularly insidious when applied to variables that serve as manipulation checks. For instance, in order to establish that a particular mood-changing manipulation produced enough of a change in affect to account for some aspect of the behavioral results, many researchers apparently believe that it is sufficient to report that the change in affect was statistically significant. Of course, in this application, it is the size of the effect that is truly relevant, not whether we can have some confidence in stating that it is not likely to be zero. To combat such misuses of NHT, the fifth edition of the Publication Manual of the American Psychological Association (APA, 2001) has added this statement: "For the reader to fully understand the importance of your findings, it is almost always necessary to include some index of effect size or strength of relationship in your Results section." (p. 25). (I will specifically discuss the concept of *strength of relationship* in later chapters.) I completely agree, and want to emphasize that my arguments that the conventional use of NHT can have some practical benefits should in no way be taken as a suggestion that p values should not *always* be accompanied by estimates of effect size, confidence intervals, or confidence intervals for effect size, as I have described earlier.

Another Advantage of NHT: Indicating the Probability of Replication

The use of NHT reduces the chances of making directional errors. Recall that when a tiny population effect size is tested with small samples, power is low, and you are almost as likely to obtain results in the opposite direction as the correct one (e.g., obtaining a slightly higher sample mean for men, when the population mean is slightly larger for women). However, if you tend to ignore the direction of results that do not attain significance, you will rarely make a directional error (i.e., NHT keeps the Type III error rate very low). Moreover, the p value associated with NHT can give you some idea of the replicability of your results, if you know how to interpret it. We have already seen from our discussion of power that an exact replication of your study is just about as likely to give you a larger p as it is to give you a smaller p than the one you got the first time, so a p of .049 indicates no more than a 50/50 chance of attaining significance at the .05 level with an exact replication (and a 50/50 chance that the next p will be somewhat larger than .05, and therefore not significant). However, whereas a p of .01 indicates only about a .5 chance that an exact replication will also yield a p as small as .01, a p of .01 tells us that the probability is considerably larger than .5 that a replication will produce results that are significant at the .05

level in the same direction (as long as the samples are not very small, this probability will be a bit more than .7). The further our p value drops below .05, the more confident we can be that future replications of our study will yield results that are significant, at least at the .05 level.

The p value can also tell you the probability that a replication of your study will produce *results in the same direction*, though not necessarily significantly so. Recently, Killeen (2005) devised an alternative to the p value called p_{rep}, so named because it estimates the probability that an exact replication of a two-group experiment will yield results in the same direction. For example, if your study yields a p_{rep} of .9, you can expect that 90% of exact replications of your study would produce results in the same direction (in terms of which sample mean is the larger). Killeen (2005) provides a formula that you can use to convert your p to a p_{rep}. For instance, a p of exactly .05 corresponds to a p_{rep} of about .9; so, whereas a p of .05 indicates only about a 50% chance of getting significance in the same direction upon replication, it indicates a 90% chance that the results of a replication will at least be in the same direction.

Meta-Analysis

For most two-group studies in psychology, it is **d**, the population effect size, which the researcher would most like to know. Assuming that both population distributions involved are roughly bell-shaped, **d** tells you approximately how much the population distributions overlap (as depicted in Figure 8.3). To grasp the magnitude of a gender difference, say, or the effectiveness of a new teaching method, **d** just might tell you all you need to know. Null hypothesis testing, on the other hand, is only about trying to decide whether **d** is zero or not, and about determining which population mean is the larger. That is why it is so often helpful to supplement NHT by reporting g (or, even better, a CI for **d** that is based on g). It should also be clear to you by now that two studies with similar gs—for example, .4 and .6—provide similar information, even if it turns out that the study with the higher g is statistically significant (e.g., $p = .03$), and the lower-g study fails to reach significance (e.g., $p = .08$). On the other hand, two studies that are both significant at the .05 level could have gs that differ very widely. Moreover, the larger the sample sizes of the two studies, the more confidence you can have that widely different gs represent a real difference in effectiveness between the two studies.

When a series of similar studies all result in similar values for g, it is very reasonable to have a good deal more confidence in the true value for **d** than you would from knowing the g for just one or two studies. There ought to be a way to combine the information from any number of similar studies in order to make a stronger assertion about the underlying value for **d**. That was what Gene Glass (1976) must have been thinking when he first proposed the term *meta-analysis* for a method of combining effect sizes across studies that all manipulate essentially the same independent variable, and collect similar measures of the outcome. Over the last 3 decades, several systems have been proposed for meta-analysis, but the one that is probably the most straightforward and easiest to explain is the one that was presented in great detail by Hedges and Olkin (1985). Furthermore, Hedges, Cooper, and Bushman (1992) have presented a strong case in favor of that system, which I will describe briefly to conclude this section.

Any meta-analysis must begin with a careful selection of studies, based on asking such questions as: Are the participants comparable across the studies? Are the manipulations really similar? Are the dependent variables of different studies measuring essentially the same ability, trait, and so forth? Then, before proceeding, a preliminary test can be conducted to determine whether the *g*s of all the studies are similar enough to justify being pooled. If the *g*s vary significantly from each other, it is time to rethink the meta-analysis, and consider more carefully which studies should be included. If the *g*s do not differ too much, Hedges and Olkin (1985) proposed that a weighted average of the *g*s be taken. The weight given to each *g* depends (inversely) on its estimated variance; more weight is given, of course, to *g*s that come from larger samples, as well as to *g*s that are smaller in magnitude (and therefore tend to vary less). The resulting *weighted mean effect size* can then form the basis of a much narrower (i.e., more precise) CI for **d** than could be derived from any one ordinary study, because the standard error of this meta CI is based on the combined sample sizes of all of the studies whose *g*s were averaged together. If preferred, the mean effect size can be tested for statistical significance; given that this test would be based on the total *N* of all the studies, there would be much less chance of committing a Type II error than when working with a single study. Meta-analysis can provide an enormous boost in power relative to individual studies.

The other two major meta-analytic approaches (Hunter & Schmidt, 1990; Rosenthal, 1991) are based on the use of correlation coefficients as measures of effect size. The most commonly used correlation coefficient, Pearson's *r*, is the topic of the next chapter.

SUMMARY

1. Retrospective power can be calculated for a study already conducted. In that case, **d** can be estimated by *g* from the study conducted, but accuracy can be improved by using a formula that corrects the bias of *g* as an estimate of **d**.

2. Especially when the dependent variable is measured in units that are not universally meaningful, it can be useful to construct a CI for the effect size, rather than the difference of the means. An approximate CI can be created by dividing the upper and lower limits of the CI for the difference of means by s_p, the square root of the pooled variance. A more accurate CI requires values from the appropriate noncentral *t* distribution (see Section D for this chapter on the Web).

3. A robust measure of effect size can be derived by subtracting one trimmed mean from the other, and dividing this difference by the square root of the pooled Winsorized variances of the two samples.

4. The use of NHT creates a very effective filter that prevents tiny effect sizes from yielding statistically significant results. Any effect size larger than zero that is thus filtered out is a Type II error, but fortunately, Type II errors are much more likely to occur when dealing with very small effects, as compared to effects that are large enough to be practically useful or worth investigating further. It might be better if we adopted Fisher's attitude toward results that fail to reach significance, in which case we would reserve judgment, and could therefore not be accused of committing Type II errors (a concept invented by Neyman and Pearson as a modification of Fisher's system).

5. The use of NHT greatly reduces the number of directional errors, which could otherwise easily occur whenever very small effect sizes yielded sample means in the direction opposite to the population means. The rate of Type III errors is very low with NHT, because such

directional errors can be said to occur only when a result attains statistical significance.
6. A major practical advantage of the widespread use of NHT is that researchers screen out their own nonsignificant results by not submitting them for publication, thus greatly relieving the burden of journal editors and reviewers.
7. A major disadvantage of NHT is that researchers too often appear to confuse a lack of statistical significance with positive evidence that the population effect size is zero, or very close to zero, and not infrequently seem to believe that statistical significance guarantees an effect size large enough to be useful.
8. Another advantage of NHT is that the p value it yields provides useful (albeit indirect) information about the probability that a replication of a two-group experiment will produce results in the same direction, or even significant results (at, say, the .05 level) in the same direction.
9. Another disadvantage of NHT is related to the so-called "file-drawer problem"; studies in which the effect sizes come out accidentally larger than they should are more likely to get published, leading to an overestimation of true population effect sizes.
10. Meta-analysis consists of statistical procedures for averaging effect-size estimates across separate, but similar, studies, and for deciding whether the effect sizes are too disparate to be averaged. Meta-analysis has the potential to greatly improve our estimation of effect sizes and the accuracy of our statistical decisions, but a good deal of scientific judgment is required to decide which studies can be reasonably combined. To avoid bias in these estimates, it is particularly important to gain access to the results of relevant individual studies that did not attain statistical significance, and would therefore not be very likely to be published. The need for a web repository for nonsignificant results remains a problem for psychological researchers.

EXERCISES

*1. Redo both parts of Exercise 8A8, using Formula 8.12 to obtain a less-biased version of g.
2. Redo both parts of Exercise 8B12, using Formula 8.12 to obtain a less-biased version of g.
*3. Estimate retrospective power for Exercise 7A7 for a .05 test that is:
a. one tailed   b. two tailed
4. Estimate retrospective power for the following previous exercises:
a. Exercise 7B4   b. Exercise 7B6

*5. Construct approximate 99% CIs for the population effect sizes in the following previous exercises:
a. Exercise 7B4   b. Exercise 7B6
6. Find the robust value for g associated with the 20% trimmed t you found in each of the following previous exercises:
a. Exercise 7C1   b. Exercise 7C2

KEY FORMULAS

Note: Formulas 8.1, 8.3, 8.5, 8.6, and 8.7 assume equal sample sizes in a two-group study.

Delta (expected z or t), in terms of population parameters, and the proposed sample size:

$$\delta = \frac{\mu_1 - \mu_2}{\sigma}\sqrt{\frac{n}{2}}$$

Formula 8.1

Effect size, **d**, in terms of population parameters:

$$\mathbf{d} = \frac{\mu_1 - \mu_2}{\sigma}$$ **Formula 8.2**

Delta, in terms of the population effect size (useful when **d** can be estimated):

$$\delta = \mathbf{d}\sqrt{\frac{n}{2}}$$ **Formula 8.3**

Estimate of effect size based on sample statistics from a two-group experiment:

$$g = \frac{\overline{X}_1 - \overline{X}_2}{s_p}$$ **Formula 8.4**

Estimate of effect size when t has been calculated and sample size is known:

$$g = t\sqrt{\frac{2}{n}}$$ **Formula 8.5**

Effect size in terms of δ and the sample size (useful for finding the effect size required to obtain adequate power with fixed sample sizes):

$$\mathbf{d} = \delta\sqrt{\frac{2}{n}}$$ **Formula 8.6**

The required sample size (of each group in a two-group experiment) to attain a given level of power for a particular value for the effect size:

$$n = 2\left(\frac{\delta}{\mathbf{d}}\right)^2$$ **Formula 8.7**

Harmonic mean of two numbers (gives the value of n to be used in two-group power formulas when the two sample sizes are unequal):

$$n_h = \frac{2n_1 n_2}{n_1 + n_2}$$ **Formula 8.8**

The sample effect size (g) in terms of the pooled-variances t value, when the two samples differ in size:

$$g = t\sqrt{\frac{n_1 + n_2}{n_1 n_2}}$$ **Formula 8.9**

Delta in terms of effect size for the one-group case (corresponds to Formula 8.3):

$$\delta = \mathbf{d}\sqrt{n}$$ **Formula 8.10**

Required sample size for the one-group case (corresponds to Formula 8.7).

$$n = \left(\frac{\delta}{\mathbf{d}}\right)^2$$ **Formula 8.11**

An (almost) unbiased estimate of the population effect size, based on g (df $= n_1 + n_2 - 2$):

$$\text{est. } \mathbf{d} = g\left(1 - \frac{3}{4\text{df} - 1}\right)$$

Formula 8.12

A robust measure of effect size, based on two 20% trimmed samples:

$$g' = .642\left(\frac{\overline{X}_{t1} - \overline{X}_{t2}}{s_p'}\right)$$

Formula 8.13

LINEAR CORRELATION

You will need to use the following from previous chapters:

Symbols
μ: Mean of a population
$\bar{X}$: Mean of a sample
σ: Standard deviation of a population
s: Unbiased standard deviation of a sample
SS: Sum of squared deviations from the mean

Formulas
Formula 3.4A: Variance of a population
Formula 4.1: The z score
Formula 6.3: The t test for one sample

Concepts
Properties of the mean and standard deviation
The normal distribution

9

Chapter

In this chapter, I will take a concept that is used in everyday life, namely, correlation, and show how to quantify it. You will also learn how to test hypotheses concerning correlation and how to draw conclusions about correlation, knowing the limitations of these methods.

A

**CONCEPTUAL
FOUNDATION**

People without formal knowledge of statistics make use of correlations in many aspects of life. For instance, if you are driving to a supermarket and see many cars in the parking lot, you automatically expect the store to be crowded, whereas a nearly empty parking lot would lead you to expect few people in the store. Although the correlation between the number of cars and the number of people will not be perfect (sometimes there will be more families and at other times more single people driving to the store, and the number of people walking to the store may vary, as well), you know intuitively that the correlation will be very high.

We also use the concept of correlation in dealing with individual people. If a student scores better than most others in the class on a midterm exam, we expect similar relative performance from that student on the final. Usually midterm scores correlate highly with final exam scores (unless the exams are very different—for example, multiple choice vs. essays). However, we do not expect the correlation to be perfect; many factors will affect each student's performance on each exam, and some of these factors can change unexpectedly. To explain how to quantify the degree of correlation, I will begin by describing perfect correlation.

Perfect Correlation

Consider the correlation between midterm and final exam scores for a hypothetical course. The simplest way to attain *perfect correlation* would be if each student's score on the final was identical to that student's score on the midterm (student A gets 80 on both, student B gets 87 on both, etc.). If you know a student's midterm score, you know his or her final exam score, as well. But this is not the only way to attain perfect correlation. Suppose that each student's final score is exactly 5 points less than his or her midterm score. The correlation is still perfect; knowing a student's midterm score means that you also know that student's final score.

It is less obvious that two variables can be perfectly correlated when they are measured in different units and the two numbers are not so obviously related. For instance, theoretically, height measured in inches can be perfectly correlated with weight measured in pounds, although the numbers for each will be quite different. Of course, height and weight are not perfectly correlated in a typical group of people, but we could find a group (or make one up for an example) in which these two variables *were* perfectly correlated. However, height and weight would not be the same number for anyone in the group, nor would they differ by some constant. Correlation is not about an individual having the same *number* on both variables; it is about an individual being in the same position on both variables relative to the rest of the group. In other words, to have perfect correlation someone slightly above average in height should also be slightly above average in weight, for instance, and someone far below average in weight should also be far below average in height.

To quantify correlation, we need to transform the original score on some variable to a number that represents that score's position with respect to the group. Does this sound familiar? It should, because the z score is just right for this job. Using z scores, perfect correlation can be defined in a very simple way: If both variables are normally distributed, perfect *positive correlation* can be defined as each person in the group having the same z score on both variables (e.g., subject A has $z = +1.2$ for both height and weight; subject B has $z = -.3$ for both height and weight, etc.).

Negative Correlation

I have not yet mentioned *negative correlation*, but perfect negative correlation can be defined just as simply: Each person in the group has the same z score in magnitude for both variables, but the z scores are opposite in sign (e.g., someone with $z = +2.5$ on one variable must have $z = -2.5$ on the other variable, etc.). An example of perfect negative correlation is the correlation between the score on an exam and the number of points taken off: If there are 100 points on the exam, a student with 85 has 15 points taken off, whereas a student with 97 has only 3 points taken off. Note that a correlation does not have to be based on individual people, each measured twice. The individuals being measured can be schools, cities, or even entire countries. As an example of less-than-perfect negative correlation, consider the following. If each country in the world is measured twice, once in terms of average yearly income and then in terms of the rate of infant deaths, we have an example of a negative correlation (more income, fewer infant deaths, and vice versa) that is *not* perfect. For example, the United States has one of the highest average incomes, but not as low an infant death rate as you would expect.

The Correlation Coefficient

In reality, the correlation between height and weight, or between midterm and final scores, will be far from perfect, but certainly not zero. To measure the amount of correlation, a *correlation coefficient* is generally used. A coefficient of $+1$ represents perfect positive correlation, -1 represents perfect negative correlation, and 0 represents a total lack of correlation. Numbers between 0 and 1 represent the relative amount of correlation, with the sign (i.e., $+$ or $-$) representing the direction of the correlation. The correlation coefficient that is universally used for the kinds of variables dealt with in this chapter is the one first formally presented in an 1896 paper by Karl

Pearson on the statistics of heredity (Cowles, 1989); *Pearson's correlation coefficient* is symbolized by the letter *r* and is often referred to as "Pearson's *r*." (It is also sometimes referred to as Pearson's "product-moment" correlation coefficient, for reasons too obscure to clarify here.) In terms of *z* scores, the formula for Pearson's *r* is remarkably simple (assuming that the *z* scores are calculated according to Formula 4.1). It is given below as Formula 9.1:

$$r = \frac{\sum z_x z_y}{N}$$ **Formula 9.1**

This is not a convenient formula for calculating *r*. You would first have to convert all of the scores on each of the two variables to *z* scores, then find the cross product for each individual (i.e., the *z* score for the *X* variable multiplied by the *z* score for the *Y* variable), and finally find the mean of all of these cross products. There are a variety of alternative formulas that give the same value for *r* but are easier to calculate. These formulas will be presented in Section B. For now I will use Formula 9.1 to help explain how correlation is quantified.

First, notice what happens to Formula 9.1 when correlation is perfect. In that case, z_x always equals z_Y, so the formula can be rewritten as:

$$r = \frac{\sum z_x z_x}{N} = \frac{\sum z_x^2}{N}$$

The latter expression is the variance for a set of *z* scores, and therefore it always equals +1, which is just what *r* should equal when the correlation is perfect. If you didn't recognize that $\sum z_x^2/N$ is the variance of *z* scores, consider the definitional formula for variance (Formula 3.4A):

$$\sigma^2 = \frac{\sum (X - \mu)^2}{N}$$ **Formula 3.4A**

Expressed in terms of *z* scores, the formula is:

$$\frac{\sum (z - \mu)^2}{N}$$

Because the mean of the *z* scores is always zero, the μ_z term drops out of the second expression, leaving $\sum z_x^2/N$. For perfect negative correlation, the two *z* scores are always equal but opposite in sign, leading to the expression:

$$r = \frac{\sum - (z^2)}{N} = \frac{-\sum z^2}{N} = -1$$

When the two *z* scores are always equal, the largest *z* scores get multiplied by the largest *z* scores, which more than makes up for the fact that the smallest *z* scores are being multiplied together. Any other pairing of *z* scores will lead to a smaller sum of cross products and therefore a coefficient less than 1.0 in magnitude. Finally, consider what happens if the *z* scores are randomly paired (as in the case when the two variables are not correlated at all). For some of the cross products the two *z* scores will have the same sign, so the cross product will be positive. For just about as many cross products, the two *z* scores would have opposite signs, producing negative cross products. The positive and negative cross products would cancel each other out, leading to a coefficient near zero.

Linear Transformations

Going back to our first example of perfect correlation—having the same score on both the midterm and the final—it is easy to see that the *z* score would also be the same for both exams, thus leading to perfect correlation. What if each final exam score is 5 points less than the corresponding midterm score? You may recall that subtracting a constant from every score in a group changes the mean but not the z scores, so in this case the z scores for the two exams are still the same, which is why the correlation is still perfect. The midterm scores were converted into the final scores by subtracting 5 points, a conversion that does not change the z scores. In fact, there are other ways that the midterm scores could be changed into final scores without changing the z scores—in particular, multiplying and dividing. It turns out that you can do any combination of adding, subtracting, multiplying, and dividing by constants, and the original scores will be perfectly correlated with the transformed scores. This kind of transformation, which does not change the z scores, is called a *linear transformation* for reasons that will soon be made graphically clear.

A good example of a linear transformation is the rule that is used to convert Celsius (also called centigrade) temperatures into Fahrenheit temperatures. The formula is $°F = \frac{9}{5} °C + 32$ (notice that this resembles the general formula for a straight line: $Y = mX + b$). If you measure the high temperature of the day in both Fahrenheit and Celsius degrees for a series of days, the two temperatures will, of course, be perfectly correlated. Similarly, if you calculate the correlation between height in inches and weight in pounds and then recalculate the correlation for the same people but measure height in centimeters and weight in kilograms, the correlation will come out exactly the same both times. It is the relative positions of the measures of the two variables, and not the absolute numbers, that are important. The relative positions are reflected in the z scores, which do not change with simple (i.e., linear) changes in the scale of measurement. Any time one variableis a linear transformation of the other, the two variables will be perfectly correlated.

Graphing the Correlation

The correlation coefficient is a very useful number for characterizing the relationship between two variables, but using a single number to describe a potentially complex relationship can be misleading. Recall that describing a group of numbers in terms of the mean for the group can be misleading if the numbers are very spread out. Even adding the standard deviation still does not tell us if the distribution is strongly skewed or not. Only by drawing the distribution could we be sure if the mean and standard deviations were good ways to describe the distribution. Analogously, if we want to inspect the relationship between two variables we need to look at some kind of bivariate distribution. The simplest way to picture this relationship is to draw what is called a *scatterplot*, or scatter diagram (*scattergram* for short).

A scatterplot is a graph in which one of the variables is plotted on the X axis and the other variable is plotted on the Y axis. Each individual is represented as a single dot on the graph. For instance, Figure 9.1 depicts a scatterplot for height versus weight in a group of 10 people. Because there are 10 people, there are 10 dots on the graph. If an eleventh person were 65 inches tall and 130 pounds, you would go along the X axis to 65 inches and then up from that point to 130 pounds on the Y axis, putting an eleventh dot at that spot. Now look at Figure 9.2. This scatterplot represents 10 people for whom height and weight are perfectly correlated. Notice that all of the dots fall on a single straight line. This is the way a scatterplot always looks when there is perfect (linear) correlation (for perfect negative correlation, the line slants the other way). As mentioned in the preceding, if the scores on one variable are a linear transformation of the scores on the other

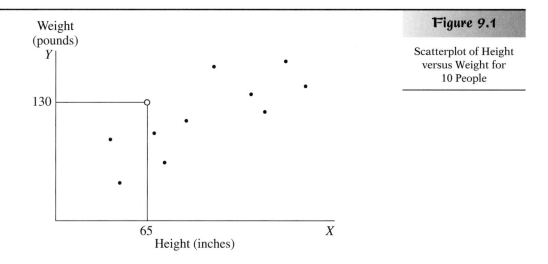

Figure 9.1

Scatterplot of Height
versus Weight for
10 People

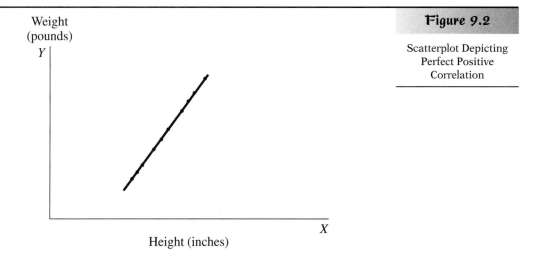

Figure 9.2

Scatterplot Depicting
Perfect Positive
Correlation

variable, the two variables will be perfectly correlated. You can see from Figure 9.2 why this is called a *linear* transformation.

The straight line formed by the dots in Figure 9.2 tells us that a linear transformation will convert the heights into the weights (and vice versa). That means the same simple formula can give you the weight of any subject once you put in his or her height (or the formula could be used in reverse to get height from weight). In this case the formula is $W = 4H - 120$, where W represents weight in pounds and H is height in inches. Of course, for any 10 people it is very unlikely that the height–weight correlation will be perfect. In the next chapter you will learn how to find the best linear formula (i.e., the closest straight line) even when you don't have perfect correlation. At this point, however, we can use scatterplots to illustrate the characteristics and the limitations of the correlation coefficient.

Dealing with Curvilinear Relationships

An important property of Pearson's r that must be stressed is that this coefficient measures only the degree of *linear* correlation. Only scatterplots in which all of the points fall on the same straight line will yield perfect correlation, and for less-than-perfect correlation Pearson's r only measures the tendency for all of the points to fall on the same straight line. This characteristic of Pearson's r can be viewed as a limitation in that the relation between two variables can be a simple one, with all of the points of the scatterplot falling on a smooth curve, and yet r can be considerably less than 1.0. An example of such a case is depicted in Figure 9.3. The curve depicted in Figure 9.3 is the scatterplot that would be obtained for the relationship between the length (in feet) of one side of a square room and the amount of carpeting (in square yards) that it would take to cover the floor. These two variables are perfectly correlated in a way, but the formula that relates Y to X is not a linear one (in fact, $Y = \frac{X^2}{9}$). To demonstrate the perfection of the relationship between X and Y in Figure 9.3 would require a measure of *curvilinear correlation*, which is a topic beyond the scope of this text. Fortunately, linear relationships are common in nature (including in psychology), and for many other cases a simple transformation of one or both of the variables can make the relationship linear. (In Figure 9.3, taking the square root of Y would make the relation linear.)

A more extreme example of a curvilinear relation between two variables is pictured in Figure 9.4. In this figure, age is plotted against running speed for a group of people who range in age from 5 to 75 years old (assuming none of these people are regular runners). If Pearson's r were calculated for this group, it would be near zero. Notice that if r were calculated only for the range from 5 years old to the optimum running age (probably about 20 years old), it would be positive and quite high. However, r for the range past the optimum point would be of the same magnitude, but negative. Over the whole range, the positive part cancels out the negative part to produce a near-zero linear correlation. Unfortunately, the near-zero r would not do justice to the degree of relationship evident in Figure 9.4. Again, we would need a measure of curvilinear correlation to show the high degree of predictability depicted in Figure 9.4. If you preferred to work with measures of linear relationship, it might be appropriate to restrict the range of one of the variables so that you were dealing with a region of the scatterplot in which the relationship of the two variables was fairly linear. Bear in mind that when psychologists use the term *correlation* without specifying

---

Figure 9.3

Scatterplot Depicting
Perfect Curvilinear
Correlation

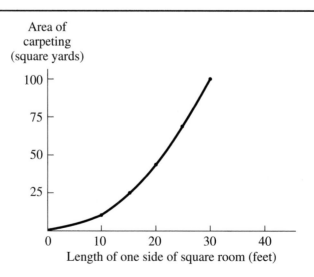

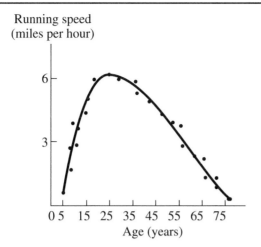

Figure 9.4

Scatterplot Depicting
Curvilinear Correlation
(Linear Correlation
Near Zero)

linear or curvilinear, you can assume that linear correlation is being discussed; I will adhere to this convention, as well.

Problems in Generalizing from Sample Correlations

When psychologists wonder whether IQ is correlated with income, or
whether income is correlated with self-esteem, they are speculating about
the correlation coefficient you would get by measuring everyone in the population of interest. The Pearson's r that would be calculated if an entire population had been measured is called a *population correlation coefficient* and
is symbolized by ρ, the lower case Greek letter rho. Of course, it is almost
never practical to calculate this value directly. Instead, psychologists must
use the r that is calculated for a sample to draw some inference about ρ. A
truly random sample should yield a sample r that reflects the ρ for the population. Any kind of biased sample could lead to a sample r that is misleading.
One of the most common types of biased samples involves a narrow (i.e.,
truncated) range on one (and often both) of the variables, as described next.

Restricted, or Truncated, Ranges

Data from a sample that is certainly not random are pictured in Figure 9.5.
This is a scatterplot of math anxiety versus number of careless errors made
on a math test by a group of "math phobics." Most students, when asked to
guess at the value of r that would be calculated for the data in Figure 9.5, say
that the correlation will be highly positive; actually, r will be near zero. The
reason r will be near zero is that the points in this figure do not fall on a
straight line or even nearly so. The reason students often expect a high positive correlation is that they see that for each individual a high number
on one variable is paired with a high number on the other variable. The problem is that they are thinking in terms of absolute numbers when they say that
all individuals are high on both variables. However, correlation is based not
on absolute numbers, but on relative numbers (i.e., z scores) within a particular group. If you calculated Pearson's r, you would find that all of the points
in Figure 9.5 do *not* have high positive z scores on both variables. The math
phobic with the lowest anxiety in the group (the point furthest left on the
graph) will have a negative z score because he or she is relatively low in anxiety in *that* group, despite being highly anxious in a more absolute sense.

Figure 9.5

Scatterplot Depicting
Restricted Range

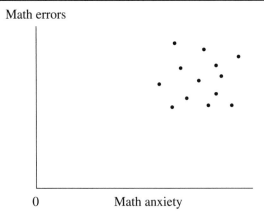

For the group of math phobics to produce a high value for r, the least anxious one should also have the fewest errors (and vice versa). In fact, all of the math phobics should form a straight, or nearly straight, line. Intuitively, you would be correct in thinking that small differences in anxiety are not going to be reliably associated with small changes in the number of errors. But you would expect large differences in one variable to be associated with fairly large differences in the other variable. Unfortunately, there are no large differences within the group depicted in Figure 9.5. This situation is often referred to as a *truncated (or restricted) range*. If some people low in anxiety were added to Figure 9.5, and these people were also low in errors committed, all of the math phobics *would* have high positive z scores on both variables (the newcomers would have negative z scores on both variables), and the calculated r would become highly positive.

A truncated range usually leads to a sample r that is lower than the population ρ. However, in some cases, such as the curvilinear relation depicted in Figure 9.4, a truncated range can cause r to be considerably higher than ρ. Another sampling problem that can cause the sample r to be much higher or lower than the correlation for the entire population is the one created by outliers, as described next.

Bivariate Outliers

A potential problem with the use of Pearson's r is its sensitivity to outliers (sometimes called outriders). You may recall that both the mean and standard deviation are quite sensitive to outliers; methods for dealing with this problem were discussed in Chapter 3, Section C. Because the measurement of correlation depends on *pairs* of numbers, correlation is especially sensitive to *bivariate outliers*. A bivariate outlier need not have an extreme value on either variable, but the combination it represents must be extreme (e.g., a 74-inch tall man who weighs only 140 pounds or a 65-inch tall man who weighs as much as 280 pounds). A graphical example of a bivariate outlier is shown in Figure 9.6. Notice that without the outlier there is a strong negative correlation; with the outlier the correlation actually reverses direction and becomes slightly positive.

Where do outliers come from? Sometimes their origins can be pretty obvious. If you are measuring performance on a task, an outlier may arise because a participant may have forgotten the instructions or applied them incorrectly, or may even have fallen asleep. If a self-report questionnaire is involved, it is possible that a participant was not being honest or didn't

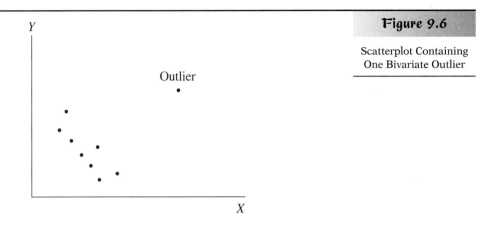

Figure 9.6

Scatterplot Containing
One Bivariate Outlier

quite understand what was being asked. If you can find an independent basis for eliminating an outlier from the data (other than the fact that it is an outlier), you should remove the outlier before calculating Pearson's r. On the other hand, sometimes you simply have to live with an outlier. Once in a while the outlier represents a very unlucky event that is not likely to appear in the next sample. (This possibility highlights the importance of replication.) Or the outlier may represent the influence of as yet unknown factors that are worthy of further study.

In any case, the scatterplot should always be inspected before any attempt is made to interpret the correlation. It is always possible that your Pearson's r has been raised or lowered in magnitude, or even reversed in direction, by a truncated range or a single outlier. An unexpected curvilinear trend can be discovered. Finally, the spread of the points may not be even as you go across the scatterplot. (The importance of this last problem—heteroscedasticity—will be discussed in Chapter 10.)

Correlation Does Not Imply Causation

A large sample r suggests that ρ for the population may be large as well, or at least larger than zero. In Section B you will see how the sample r can be used to draw an inference about ρ. Once you have concluded that two variables are indeed closely related, it can be tempting to believe that there must be a causal relationship between the two variables. But it is much more tempting in some situations than others. If you take a sample of pupils from an elementary school, you will find a high correlation between shoe size and vocabulary size, but it is safe to say that you would *not* be tempted to think that having larger feet makes it easier to learn new words (or vice versa). And, of course, you would be right in this case. The high correlation could result from having a range of ages in your sample. The older the pupil, the larger his or her feet tend to be, and the larger his or her vocabulary tends to be, as well. Shoe size and vocabulary size are not affecting each other directly; both are affected by a third variable: age.

On the other hand, if you observed a high negative correlation between number of hours spent per week on exercise and number of days spent sick each year, you probably would be tempted to believe that exercise was *causing* good health. However, it is possible that the flow of causation is in the opposite direction; healthier people may have a greater inclination to exercise. It is also possible that, as in the preceding example, some unnoticed third variable is responsible for both the increased exercise and the decreased illness. For instance, people who are more optimistic may have a greater inclination to exercise as well as better psychological defenses against stress that protect them from illness. In other words, the optimism may be causing both the inclination to exercise and the improved health. If this latter explanation were true, encouraging people to exercise is not the way to improve their health; you should be trying to find a way to increase their optimism. Although sometimes the causal link underlying a strong correlation seems too obvious to require experimental evidence (e.g., the correlation between number of cigarettes smoked and likelihood of contracting lung cancer), the true scientist must resist the temptation of relying solely on theory and logic. He or she should make every effort to obtain confirmatory evidence by means of true experiments.

If the preceding discussion sounds familiar, it is because the same point was made with respect to the two-group experiment. If a group of exercisers is compared to a group of nonexercisers with respect to sick days, this is a quasi-experimental design (sometimes, confusingly called a correlational design) that is open to the third variable (e.g., optimism) explanation just mentioned. It is by randomly assigning participants to an exercise or no-exercise group that the experimenter can definitively rule out third variable explanations and conclude that exercise directly affects health.

True Experiments Involving Correlation

If instead of forming two distinct groups, an experimenter randomly assigned each participant to a different number of exercise hours per week, it would then not be appropriate to calculate a t test; it would make more sense to calculate the correlation coefficient between amount of exercise and number of sick days. However, in this special case, a significant correlation could be attributed to a causal link between exercise and health. Third variables, such as optimism, could be ruled out because the *experimenter* determined the amount of exercise; personality and constitutional variables were not allowed to determine the participants' amount of exercise. Such experiments, which lead directly to a correlation coefficient instead of a t value, are rare, but they will be discussed in more detail in the context of linear regression in Chapter 10.

SUMMARY

1. If each individual has the same score on two different variables (e.g., midterm and final exams), the two variables will be perfectly correlated. However, this condition is not necessary. Perfect correlation can be defined as all individuals having the same z score (i.e., the same relative position in the distribution) on both variables.

2. Negative correlation is the tendency for high scores on one variable to be associated with low scores on a second variable (and vice versa). Perfect negative correlation occurs when each individual has the same magnitude z score on the two variables, but the z scores are *opposite in sign*.

3. Pearson's correlation coefficient, r, can be defined as the average of the cross products of the z scores on two variables. Pearson's r ranges from -1.0 for perfect negative correlation to 0 when there is no linear relationship between the variables to $+1.0$ when the correlation is perfectly positive.

4. A *linear transformation* is the conversion of one variable into another by only arithmetic operations (i.e., adding, subtracting, multiplying, or dividing) involving constants. If one variable is a linear transformation of another, each individual will have the same z score on both variables, and the two variables will be perfectly correlated. Changing the units of

measurement on either or both of the variables will not change the correlation as long as the change is a linear one (as is usually the case).
5. A *scatterplot*, or *scattergram*, is a graph of one variable plotted on the X axis versus a second variable plotted on the Y axis. A scatterplot of perfect correlation will be a straight line that slopes up to the right for positive correlation and down to the right for negative correlation.
6. One important property of Pearson's r is that it assesses only the degree of *linear* relationship between two variables. Two variables can be closely related by a very simple curve and yet produce a Pearson's r near zero.
7. Problems can occur when Pearson's r is measured on a subset of the population but you wish to extrapolate these results to estimate the correlation for the entire population (ρ). The most common problem is having a *truncated*, or *restricted*, range on one or both of the variables. This problem usually causes the r for the sample to be considerably less than ρ, although in rare cases the opposite can occur (e.g., when you are measuring one portion of a curvilinear relationship).
8. Another potential problem with Pearson's r is that a few *bivariate outliers* in a sample can drastically change the magnitude (and, in rare instances, even the sign) of the correlation coefficient. It is very important to inspect a scatterplot for curvilinearity, outliers, and other aberrations before interpreting the meaning of a correlation coefficient.
9. Correlation, even if high in magnitude and statistically significant, does not prove that there is any causal link between two variables. There is always the possibility that some third variable is separately affecting each of the two variables being studied. An experimental design, with random assignment of participants to conditions (or different quantitative levels of some variable), is required to determine whether one particular variable is *causing* changes in a second variable.

EXERCISES

1. Describe a realistic situation in which two variables would have a high positive correlation. Describe another situation for which the correlation would be highly negative.
2. A recent medical study found that the moderate consumption of alcoholic beverages is associated with the fewest heart attacks (as compared to heavy drinking or no drinking). It was suggested that the alcohol caused the beneficial effects. Devise an explanation for this relationship that assumes there is no direct causal link between drinking alcohol and having a heart attack. (*Hint*: Consider personality.)
*3. A study compared the number of years a person has worked for the same company (X) with the person's salary in thousands of dollars per year (Y). The data for nine employees appear in the following table. Draw the scatterplot for these data and use it to answer the following questions.

Years (X)	Annual Salary (Y)
5	24
8	40
3	20
6	30
4	50
9	40
7	35
10	50
2	22

a. Considering the general trend of the data points, what direction do you expect the correlation to be in (positive or negative)?
b. Do you think the correlation coefficient for these data would be meaningful or misleading? Explain.
*4. A psychologist is studying the relation between anxiety and memory for unpleasant events. Participants are measured with an anxiety questionnaire (X) and with a test of

their recall of details (*Y*) from a horror movie shown during the experimental session. Draw the scatterplot for the data in the following table and use it to answer the following questions.

Anxiety (*X*)	No. Details Recalled (*Y*)
39	3
25	7
15	5
12	2
34	4
39	3
22	6
27	7

a. Would you expect the magnitude of the correlation coefficient to be high or low? Explain.
b. Do you think the correlation coefficient for these data would be meaningful or misleading? How might you explain the results of this study?

*5. A clinical psychologist believes that depressed people speak more slowly than others. He measures the speaking rate (words per minute) of six depressed patients who had already been measured on a depression inventory. Draw the scatterplot for the data in the following table and use it to answer the following questions.

Depression (*X*)	Speaking Rate (*Y*)
52	50
54	30
51	39
55	42
53	40
56	31

a. What direction do you expect the correlation to be in (i.e., what sign do you expect the coefficient to have)?
b. What expectations do you have about the magnitude of the correlation coefficient (very low, very high, or moderate)?

6. During the first recitation class of the semester, a teaching assistant asked each of the 12 students in his statistics section to rate his or her math phobia on a scale from 0 to 10.

Later, the TA paired those phobia ratings with the students' scores on their first exam. The data appear in the following table; graph the scatterplot in order to answer the questions that follow the table.

Phobia Rating (*X*)	Exam 1 (*Y*)
5	94
1	96
5	87
7	75
0	91
1	93
3	90
8	77
5	97
10	78
6	89
2	88

a. Do you expect a positive or negative correlation? Do you expect the magnitude of the correlation to be fairly low, rather high, or moderate?
b. What characteristic of the pattern that you can see in the scatterplot prevents the correlation from being very close to perfect?

*7. A psychologist is studying the relationship between the reported vividness of visual imagery and the ability to rotate objects mentally. A sample of graduate students at a leading school for architecture is tested on both variables, but the Pearson's *r* turns out to be disappointingly low. Which of the following is the most likely explanation for why Pearson's *r* was not higher?
a. One or both of the variables has a restricted range.
b. The relationship between the two variables is curvilinear.
c. The number of degrees of freedom is too small.
d. One variable was just a linear transformation of the other.

8. A college admissions officer is tracking the relationship between students' verbal SAT scores, and their first-year grade point averages (GPA). The data for the first 10 students she looked at appear in the following table. Graph the scatterplot in order to answer the questions that follow the table.

Verbal SAT (*X*)	GPA (*Y*)
510	2.1
620	3.8
400	2.2
480	3.1
580	3.9
430	2.4
530	3.6
680	3.5
420	3.3
570	3.4

a. Do you expect a positive or negative correlation? Do you expect the magnitude of the correlation to be fairly low, rather high, or moderate?
b. What characteristic of the pattern that you can see in the scatterplot prevents the correlation from being very close to perfect?

*9. The correlation between scores on the midterm and scores on the final exam for students in a hypothetical psychology class is .45.
a. What is the correlation between the midterm scores and the number of points taken off for each midterm?
b. What is the correlation between the number of points taken off for each midterm and the final exam score?

10. Suppose there is a perfect negative correlation between the amount of time spent (on average) answering each question on a test and the total score on the test. Assume that $\overline{X} = 30$ seconds and $s_x = 10$ for the time per item, and that $\overline{Y} = 70$, with $s_y = 14$ for the total score.
a. If someone spent 10 seconds per item, what total score would he or she receive?
b. If someone's score on the test was 49, how much time did he or she spend on each item?

We now turn to the practical problem of calculating Pearson's correlation coefficient. Although Formula 9.1 is easy to follow, having to convert all scores to *z* scores before proceeding makes it unnecessarily tedious. A formula that is easier to use can be created by plugging Formula 4.1 for the *z* score into Formula 9.1 and rearranging the terms algebraically until the formula assumes a convenient form. The steps below lead to Formula 9.2:

BASIC STATISTICAL PROCEDURES

$$r = \frac{\sum z_x z_y}{N} = \frac{\sum \left(\dfrac{X - \mu_x}{\sigma_x} \right) \left(\dfrac{Y - \mu_y}{\sigma_y} \right)}{N} = \frac{\sum (X - \mu_x)(Y - \mu_y)}{N \sigma_x \sigma_y}$$

$$= \frac{\sum XY - N\mu_x\mu_y}{N\sigma_x\sigma_y} \qquad \text{therefore} \qquad r = \frac{\dfrac{\sum XY}{N} - \mu_x\mu_y}{\sigma_x\sigma_y} \qquad \textbf{Formula 9.2}$$

Notice that the first term in Formula 9.2, $\Sigma\,XY/N$, is in itself similar to Formula 9.1, except that it is the average cross product for the original scores rather than for the cross products of *z* scores. Instead of subtracting the mean and dividing by the standard deviation for each score to obtain the corresponding *z* score, with Formula 9.2 we need only subtract once ($\mu_x\mu_y$) and divide once ($\sigma_x\sigma_y$). Note also that we are using population values for the mean and standard deviation because we are assuming for the moment that Pearson's *r* is being used only as a means of describing a set of scores (actually pairs of scores) at hand. So, for our purposes that set of scores is a population. Of course, for the mean it does not matter whether we refer to the population mean (μ) or the sample mean ($\overline{X}$); the calculation is the same. There is a difference, however, between σ and *s*—the former is calculated with *N* in the denominator, whereas the latter is calculated with $N - 1$ (see Chapter 3). For Formula 9.2 to yield the correct answer, σ must be calculated rather than *s*.

The Covariance

The numerator of Formula 9.2 has a name of its own; it is called the *covariance*. This part gets larger in magnitude as the two variables show a greater tendency to "covary"—that is, to vary together, either positively or negatively. Dividing by the product of the two standard deviations ensures that the correlation coefficient will never get larger than $+1.0$ or smaller than -1.0, no matter how the two variables covary. Formula 9.2 assumes that you have calculated the biased standard deviations, but if you are dealing with a sample and planning to draw inferences about a population, it is more likely that you have calculated the unbiased standard deviations. It is just as easy to calculate the correlation coefficient in terms of s rather than σ, but then the numerator of Formula 9.2 must be adjusted accordingly. Just as the denominator of Formula 9.2 can be considered biased (when extrapolating to the larger population), the numerator in this formula has a corresponding bias, as well. These two biases cancel each other out so that the value of r is not affected, but if you wish to use unbiased standard deviations in the denominator you need to calculate an unbiased covariance in the numerator.

The Unbiased Covariance

The bias in the numerator of Formula 9.2 is removed in the same way we removed bias from the formula for the standard deviation: We divide by $N - 1$ instead of N. Adjusting the covariance accordingly and using the unbiased standard deviations in the denominator gives Formula 9.3:

$$r = \frac{\frac{1}{N-1}\left(\sum XY = N\overline{X}\,\overline{Y}\right)}{s_x s_y} \qquad \textbf{Formula 9.3}$$

I used the symbol for the sample mean in Formula 9.3 to be consistent with using the sample standard deviation. It is very important to realize that Formula 9.3 always produces the same value for r as Formula 9.2, so it does not matter which formula you use. If you have already calculated σ for each variable, or you plan to anyway, you should use Formula 9.2. If you have instead calculated s for each variable, or plan to, it makes sense to use Formula 9.3.

An Example of Calculating Pearson's *r*

To make the computation of these formulas as clear as possible, I will illustrate their use with the following example. Suppose that a researcher has noticed a trend for women with more years of higher education (i.e., beyond high school) to have fewer children. To investigate this trend, she selects six women at random and records the years of higher education and number of children for each. The data appear in Table 9.1, along with the means and standard deviations for each variable and the sum of the cross products.

We can apply Formula 9.2 to the data in Table 9.1 as follows (note the use of σ instead of s):

$$r = \frac{\frac{\sum XY}{N} - \mu_x \mu_y}{\sigma_x \sigma_y} = \frac{\frac{30}{6} - (3.5)(2.5)}{(2.99)(1.71)} = \frac{5 - 8.75}{5.1} = \frac{-3.75}{5.1} = -.735$$

Notice that r is highly negative, indicating that there is a strong tendency for more years of education to be associated with fewer children

X (Years of Higher Ed)	Y (No. of Children)	XY	
0	4	0	Table 9.1
9	1	9	
5	0	0	
2	2	4	
4	3	12	
1	5	5	
$\overline{X} = 3.5$	$\overline{Y} = 2.5$	$\Sigma\, XY = 30$	
$\sigma_x = 2.99$	$\sigma_y = 1.71$		
$s_x = 3.27$	$s_y = 1.87$	$\Sigma\, XY/N = 30/6 = 5$	

in this particular sample. Notice that we get exactly the same answer for r if we use Formula 9.3 instead of Formula 9.2 (note the use of s rather than σ):

$$r = \frac{\dfrac{1}{N-1}\left(\sum XY - N\overline{X}\,\overline{Y}\right)}{s_x s_y} = \frac{\dfrac{1}{5}[30 - (6)(3.5)(2.5)]}{(3.27)(1.87)} = \frac{-4.5}{6.115} = -.735$$

Alternative Formulas

There are other formulas for Pearson's r that are algebraically equivalent to the preceding formulas and therefore give exactly the same answer. I will show some of these in the following because you may run into them elsewhere, but I do not recommend their use. My dislike of these formulas may be more understandable after you have seen them.

A traditional variation of Formula 9.2 is shown as Formula 9.4:

$$r = \frac{\sum (X - \overline{X})(Y - \overline{Y})}{\sqrt{SS_x SS_y}} = \frac{SP}{\sqrt{SS_x SS_y}} \qquad \textbf{Formula 9.4}$$

The numerator of Formula 9.4 is sometimes referred to as the sum of products of deviations, shortened to the sum of products, and symbolized by SP. The SS terms simply represent the numerators of the appropriate variance terms. (You'll recall that Formula 3.7A states that $s^2 = SS/(N-1) = SS/df$.) The chief drawback of Formula 9.4 is that it requires calculating a deviation score for each and every raw score—which is almost as tedious as calculating all of the z scores. To simplify the calculations, Formula 9.4 can be algebraically manipulated into the following variation:

$$\frac{\sum XY - \dfrac{(\sum X)(\sum Y)}{N}}{\sqrt{\left(\sum X^2 - \dfrac{(\sum X)^2}{N}\right)\left(\sum Y^2 - \dfrac{(\sum Y)^2}{N}\right)}}$$

This formula can be made even more convenient for calculation by multiplying both the numerator and the denominator by N to yield Formula 9.5:

$$r = \frac{N \sum XY - (\sum X)(\sum Y)}{\sqrt{\left[N \sum X^2 - (\sum X)^2 \right]\left[N \sum Y^2 - (\sum Y)^2 \right]}} \qquad \textbf{Formula 9.5}$$

This formula is often referred to as a "raw-score" computing formula because it calculates r directly from raw scores without first producing intermediate statistics, such as the mean or *SS*.

Which Formula to Use

At this point I want to emphasize that all of the formulas for r in this chapter will give exactly the same answer (except for slight deviations involved in rounding off numbers before obtaining the final answer), so which formula you use is merely a matter of convenience. In the days before electronic calculators or computers, it was of greater importance to use a formula that minimized the steps of calculation, so Formula 9.5 was generally favored. I dislike Formula 9.5, however, because it is easy to make a mistake on one of the sums or sums of squares without noticing it, and it can be hard to track down the mistake once you realize that the value for r is incorrect (for example, when you obtain an r greater than 1.0).

In contrast, the means and standard deviations that go into Formula 9.2 or Formula 9.3 can each be checked to see that they are reasonable before proceeding. Just be sure to keep at least four digits past the decimal point for each of these statistics when inserting them in Formula 9.2 or Formula 9.3; otherwise the error introduced by rounding off these statistics can seriously affect the final value of r you obtain. The only term whose value can be way off without looking obviously wrong is $\Sigma XY/N$, so this term must be checked very carefully. Of course, in actual research, the correlation will be most often calculated by a computer. However, for teaching the concepts of correlation, I prefer Formula 9.2 or Formula 9.3 because their structures reveal something about the quantity being calculated.

Testing Pearson's r for Significance

Using the t Distribution

The r that we calculated for the example about women's education and number of children was quite high ($r = -.735$), and it would appear that the hypothesis $\rho = 0$ must be quite far-fetched. On the other hand, recall that our correlation was based on only six randomly selected women, and you should be aware that fairly large correlations can occur easily by chance in such a tiny sample. We need to know the null hypothesis distribution that corresponds to our experiment. That is, we need to know, in a population where $\rho = 0$ for these two variables, how the sample rs will be distributed when randomly drawing six pairs of numbers at a time.

Given some assumptions that will be detailed shortly, the laws of statistics tell us that when $\rho = 0$ and the sample size is quite large, the sample rs will be normally distributed with a mean of 0 and a standard error of about $1/\sqrt{N}$. For sample sizes that are not large, the standard error can be estimated by using the following expression:

$$\sqrt{\frac{1 - r^2}{N - 2}}$$

The significance of r can then be tested with a formula that resembles the one for the one-group t test:

$$t = \frac{r - \rho_0}{\sqrt{\dfrac{1 - r^2}{N - 2}}}$$

where ρ_0 represents the value of the population correlation coefficient according to the null hypothesis. In the most common case, the null hypothesis specifies that $\rho_0 = 0$, and this simplifies the formula. Also, it is common to rearrange the formula algebraically into the following form:

$$t = \frac{r\sqrt{N - 2}}{\sqrt{1 - r^2}} \qquad\qquad \textbf{Formula 9.6}$$

Testing a null hypothesis other than $\rho = 0$, or constructing a confidence interval around a sample r, requires a transformation of r, which is discussed in Section C. As an example, let's test the correlation from Table 9.1 against $\rho = 0$:

$$t = \frac{-.735\sqrt{6 - 2}}{\sqrt{1 - (-.735)^2}} = \frac{-.735\sqrt{4}}{\sqrt{1 - .54}} = \frac{-.735(2)}{\sqrt{.46}} = \frac{-1.47}{.678} = -2.17$$

The df for correlations is $N - 2$, where N is the number of different participants (i.e., the number of *pairs* of scores), so we need to find the critical t for $6 - 2 = 4$ df. Because -2.17 is not less than -2.776, H_0 cannot be rejected at the .05 level, two-tailed.

Using the Table of Critical Values for Pearson's r

There is a simpler way to test r for statistical significance. The t tests have already been performed to create a convenient table (see Table A.5 in Appendix A) that allows you to look up the critical value for r as a function of alpha and df. For df $= 4$ and alpha $= .05$ (two-tailed), we see from Table A.5 that the critical $r = .811$. Because the magnitude of our calculated r is .735 (the sign of the correlation is ignored when using Table A.5), which is less than .811, we must retain (i.e., we cannot reject) the null hypothesis that $\rho = 0$.

The Critical Value for r as a Function of Sample Size

The statistical conclusion for $r = -.735$ may be surprising, considering how large this sample r is, but Table A.5 reveals that rather high rs can be commonly found with samples of only six subjects, even when there is no correlation at all in the population. By looking down the column for the .05, two-tailed alpha in Table A.5, you can see how the critical r changes with df (and therefore with the sample size). With a sample of only four subjects (df $= 2$), the sample r must be over .95 in magnitude to be statistically significant. With 10 subjects (df $= 8$) the critical value reduces to .632, whereas with 102 subjects (df $= 100$), a sample r can be significant just by being larger than .195. Also, it should not be surprising to see that the critical values for r become larger as alpha gets smaller (looking across each row toward the right).

An important point to remember is that, unlike z scores or t values, any sample r, no matter how small (unless it is exactly zero), can be statistically significant if the sample size is large enough (I arbitrarily ended Table A.5 at df $= 1000$, but I could have continued indefinitely—at least until I ran out of paper). Conversely, even a correlation coefficient close to 1.0 can fail to attain significance if the sample is too small. One way to understand the dependence of the critical r on sample size is to realize that the sample rs are clustered more tightly around the population ρ as sample size increases—it becomes more and more unlikely to obtain a sample r far from ρ as N gets larger. Another important point is that the r you calculate for a sample is just as likely to get smaller as larger when you increase the sample size. I point this out because some students get the notion that a larger r is likely with a larger sample; this misconception may arise either because the t value tends to increase with a larger sample or because a given value for r is more likely to be significant with a larger sample. What does happen as the sample gets larger is that the sample r tends to be closer to ρ.

Table A.5 assumes that $\rho = 0$, which is why we can ignore the sign of the sample when testing for significance; the distribution of sample rs around ρ will be symmetric when $\rho = 0$. This will not be the case, however, when the null hypothesis specifies any ρ other than zero. This latter case is more complex, and therefore I will postpone that discussion until Section C.

Understanding the Degrees of Freedom

To give you an idea why the df should be $N - 2$, consider what happens when you try to calculate r for only two pairs of numbers. The scatterplot of the two pairs tells you that the correlation must be either $+1$ or -1 because the scatterplot consists of only two points, which therefore can always be made to fall on the same straight line. (The line will either slant up [positive] or down [negative] as you move to the right. If the two points form a horizontal or vertical line, one of the standard deviations is zero, and r cannot be calculated.) If you are sampling only two cases from a population, the correlation you calculate will be perfect regardless of the two variables being measured and regardless of the magnitude of the population correlation for those variables (if you gather two friends, you can pick any ridiculous combination of two variables and show that they are perfectly correlated in your sample of two). Consequently, an N of 2 gives us no information about the magnitude of the population correlation. If you think of the degrees of freedom as the number of pieces of information you have about the population correlation, it makes sense that df $= N - 2 = 2 - 2 = 0$ when you have only two cases. You need a sample of three to have a single piece of information about the size of the population correlation.

Imagine that you are sampling from a population in which your two variables have a zero correlation. You know from the preceding discussion that your sample r will be ridiculously inflated (in fact, it will be 1.0 in magnitude) if your sample size is two. That inflation does not completely disappear if your sample size is three. In fact, the sample r you can expect when $\rho = 0$ is not 0 but rather $\sqrt{1/(N - 1)}$. So if your sample size is three, you can expect (on the average) a sample r of $\sqrt{1/2} = .707$. Even with a sample size of 11, your expected sample r (when $\rho = 0$) is considerably more than zero; it is $\sqrt{.1} = .316$. Only as your sample size gets large does your sample r begin to accurately reflect the correlation in the population. The inflation of r when the degrees of freedom are few becomes a more insidious problem when you are calculating a multiple regression. Appropriate adjustments for this overestimation will be presented in Chapter 17.

Assumptions Associated with Pearson's r

Pearson's r is sometimes used purely for descriptive purposes. However, more often a sample r is used to draw some inference concerning the population correlation, such as deciding whether or not ρ = 0. Such inferences are based on the following assumptions.

Independent Random Sampling

This assumption applies to all of the hypothesis tests in this text. In the case of correlation, it means that even though a relation may exist between the two numbers of a pair (e.g., a large number for height may be consistently associated with a large number for weight), each pair should be independent of the other pairs, and all of the pairs in the population should have an equal chance of being selected.

Normal Distribution

Each of the two variables should be measured on an interval or ratio scale and be normally distributed in the population.

Bivariate Normal Distribution

If the assumption of a bivariate normal distribution is satisfied, you can be certain that the preceding assumption will also be satisfied, but it is possible for each variable to be normally distributed separately without the two variables jointly following a bivariate normal distribution (which is a stricter assumption than the one discussed in the preceding).

To picture a *bivariate distribution*, begin by imagining a univariate distribution. In the univariate distribution (the kind that was introduced in Chapter 2), one variable is placed along the horizontal axis, and the (vertical) height of the distribution represents the likelihood of each value. In the bivariate distribution, two axes are needed to represent the two variables, and a third axis is needed to represent the relative frequency of each possible pair of values. How can all three dimensions be represented? One way would be to use an axis "going into the page" to represent the second variable (the first would be on the horizontal axis), and use the vertical axis to represent the frequency of each bivariate score. A common bivariate distribution would come out looking like a wide-brimmed hat, as shown in Figure 9.7. Values near the middle are more common, whereas values that are far away in any direction, including diagonally, are less common.

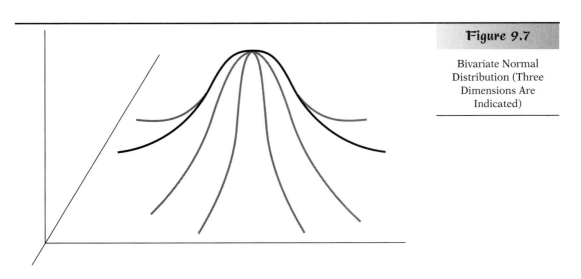

Figure 9.7

Bivariate Normal
Distribution (Three
Dimensions Are
Indicated)

A bivariate normal distribution can involve any degree of linear relationship between the two variables from $r = 0$ to $r = +1.0$ or -1.0, but a curvilinear relationship would not be consistent with a bivariate normal distribution. If there is good reason to suspect a nonlinear relationship, Pearson's r should not be used; instead, you should consult advanced texts for alternative techniques. If the bivariate distribution in your sample is very strange, or if one of the variables has been measured on an ordinal scale, the data should be converted to ranks before applying the Pearson correlation formula. When the correlation is calculated for ranked data, the *Spearman rank-order* correlation formula is commonly used, and the resulting coefficient is often called the Spearman rho (r_S). This correlation coefficient is interpreted in the same way as any other Pearson's r, but the critical values for testing significance are different (see Chapter 21).

As with the assumption of a normal distribution, the assumption of a bivariate normal distribution becomes less important as sample size increases. For very large sample sizes, the assumption can be grossly violated with very little error. Generally, if the sample size is above about 30 or 40, and the bivariate distribution does not deviate a great deal from bivariate normality, the assumption is usually ignored.

Uses of the Pearson Correlation Coefficient

Reliability and Validity

A very common use of Pearson's r is in the measurement of reliability, which occurs in a variety of contexts. For instance, to determine whether a questionnaire is assessing a personality trait that is stable over time, each participant is measured twice (with a specified time interval in between), and the correlation of the two scores is calculated to determine the *test-retest reliability*. There should be a strong tendency for participants to have the same score both times; otherwise, we may not be measuring any particular trait in a stable way. The internal consistency of the questionnaire can also be checked; separate subscores for the odd- and even-numbered items can be correlated to quantify the tendency for all items in the questionnaire to measure the same trait. This is called *split-half reliability*. (Usually more sophisticated statistics are preferred for measuring internal reliability.)

Sometimes a variable is measured by having someone act as a judge to rate the behavior of a person (e.g., how aggressive or cooperative a particular child is in a playground session). To feel confident that these ratings are not peculiar to the judge used in the study, a researcher may have two judges rate the same behavior so that the correlation of these ratings can be assessed. It is important to have high *inter-rater reliability* to trust these ratings. In general, correlation coefficients for reliability that are below .7 lead to a good deal of caution and rethinking.

Another frequent use of correlation is to establish the *criterion validity* of a self-report measure. For instance, subjects might fill out a questionnaire containing a generosity scale and then later, in a seemingly unrelated experiment for which they are paid, be told that the experimenter is running low on funds and wants them to give back as much of their payment as they are comfortable giving. A high correlation between the self-reported generosity score and the actual amount of money subsequently donated would help to validate the self-report measure. It is also common to measure the degree of correlation between two questionnaires that are supposed to be measuring the same variable, such as two measures of anxiety or two measures of depression.

Relationships between Variables

The most interesting use for correlation is to measure the degree of association between two variables that are not obviously related but are predicted by some theory or past research to have an important connection. For instance, one aspect of Freudian theory might give rise to the prediction that stinginess will be positively correlated with stubbornness because both traits are associated with the anal retentive personality. As another example, the correlation that has been found between mathematical ability and the ability to imagine how objects would look if rotated in three-dimensional space supports some notions about the cognitive basis for mathematical operations. On the other hand, some observed correlations were not predicted but can provide the basis for future theories, such as the correlations that are sometimes found between various eating and drinking habits and particular health problems.

Finally, correlation can be used to evaluate the results of an experiment when the levels of the manipulated variable come from an interval or ratio scale. For example, the experimenter may vary the number of times particular words are repeated in a list to be memorized and then find the correlation between number of repetitons and the probability of recall. A social psychologist may vary the number of bystanders (actually, confederates of the experimenter) at an accident scene to see if an individual's response time for helping someone who seems to have fainted (another confederate, of course) is correlated with the number of other onlookers.

Publishing the Results of Correlational Studies

If the results of the study of years of education versus number of children were published, the results section of the article might contain a sentence such as the following: "Although a linear trend was observed between the number of years of education and the number of children that a woman has had, the correlation coefficient failed to reach significance, $r(6) = -.735, p > .05$." The number in parentheses following r is the sample size. Of course, a considerably larger sample would usually be used, and such a high correlation would normally be statistically significant.

An Excerpt from the Psychological Literature

An example of the use of Pearson's r in the psychological literature is found in the following excerpt from a study of the subjective responses of undergraduate psychology students who reported marijuana use (Davidson & Schenk, 1994). Subjects completed two scales concerning their first experience with marijuana, indicating amount of agreement with each of several statements like the following: "Marijuana made small things seem intensely interesting" (Global Positive 1) and "Marijuana caused me to lose control and become careless" (Global Negative 1). Among other results, the authors found that "Global Positive 1 and Global Negative 1 scores were related to each other $[r(197) = .31, p < .01]$." Notice the large sample size, which makes it relatively easy to attain statistical significance without a large correlation coefficient.

This is a good example of a situation in which a one-tailed significance test would not be justified. You might have expected a negative correlation (more positive features of marijuana use associated with fewer negative features, and vice versa), but it is also understandable that some students would be generally affected more by their first use of marijuana and would experience more of both the positive and negative features, whereas other students would be relatively unaffected.

B

SUMMARY

1. To calculate Pearson's r without transforming to z scores, you can begin by calculating the means and biased standard deviations of both variables, as well as the cross products (i.e., X times Y) for all of the individuals. Pearson's r is equal to the mean of the cross products minus the cross product of the two means, divided by the product of the two biased SDs. The numerator of this ratio is the biased covariance. If you wish to use the unbiased standard deviations in the denominator, the numerator must be adjusted to correct the bias.

2. The preceding calculation method is not very accurate unless you retain at least four digits past the decimal point for the means, SDs, and the mean of cross products. If a computer is not available (or a calculator with Pearson's r as a built-in function), there is a raw-score formula you can use that is less susceptible to rounding error.

3. A correlation coefficient can be tested for significance with a t test or by looking up a critical value for r in Table A.5 (the df equals 2 less than the sample size or number of paired scores). As the sample size increases, smaller rs become significant. Any r other than zero can become significant with a large enough sample size. However, sample r does not tend to get larger just because the sample size is increased—rather, it tends to get closer to ρ.

4. The following assumptions are required for testing the significance of a Pearson correlation coefficient:
   a. *Both variables are measured on interval or ratio scales.*
   b. *The pairs of scores have been sampled randomly and independently* (i.e., within a pair, the scores may be related, but one pair should not be related to any of the others).
   c. *The two variables jointly follow a bivariate normal distribution* (this implies that each variable separately will be normally distributed).

5. Possible uses for correlation coefficients include:
   a. Reliability (e.g., test-retest, split-half, inter-rater).
   b. Validity (e.g., construct).
   c. Observing relations between variables as they naturally occur in the population (as reflected in your sample).
   d. Measuring the causal relation between two variables after assigning participants randomly to different quantitative levels of one of the variables.

EXERCISES

*1. A professor has noticed that in his class a student's score on the midterm is a good indicator of the student's performance on the final exam. He has already calculated the means and (biased) standard deviations for each exam for a class of 20 students: $\mu_M = 75$, $\sigma_M = 10$; $\mu_F = 80$, $\sigma_F = 12$; therefore, he only needs to calculate ΣXY to find Pearson's r.
   a. Find the correlation coefficient using Formula 9.2, given that $\Sigma XY = 122{,}000$.
   b. Can you reject the null hypothesis (i.e., $\rho = 0$) at the .01 level (two-tailed)?
   c. What can you say about the degree to which the midterm and final are linearly related?

2. Calculate Pearson's r for the data in Exercise 9A4, using Formula 9.2. Test the significance of this correlation coefficient by using Formula 9.6.

3. a. Calculate Pearson's r for the data in Exercise 9A5, using Formula 9.3.
   b. Recalculate the correlation you found in part a, using Formula 9.5 (*Hint*: The calculation can be made easier by subtracting a constant from all of the scores on the first variable.)
   c. Are the two answers exactly the same? If not, explain the discrepancy.

*4. a. Calculate Pearson's r for the data in Exercise 9A3 and test for significance with alpha = .05 (two-tailed).
   b. Delete the employee with 4 years at the company and recalculate Pearson's r. Test for significance again.
   c. Describe a situation in which it would be legitimate to make the deletion indicated in part b.

*5. A psychiatrist has noticed that the schizophrenics who have been in the hospital the longest score the lowest on a mental orientation test. The data for 10 schizophrenics are listed in the following table:

Years of Hospitalization (X)	Orientation Test (Y)
5	22
7	26
12	16
5	20
11	18
3	30
7	14
2	24
9	15
6	19

   a. Calculate Pearson's r for the data.
   b. Test for statistical significance at the .05 level (two-tailed).

*6. If a test is reliable, each participant will tend to get the same score each time he or she takes the test. Therefore, the correlation between two administrations of the test (test-retest reliability) should be high. The reliability of the verbal GRE score was tested using five participants, as shown in the following table:

Verbal GRE (1)	Verbal GRE (2)
540	570
510	520
580	600
550	530
520	520

   a. Calculate Pearson's r for the test-retest reliability of the verbal GRE score.
   b. Test the significance of this correlation with alpha = .05 (one-tailed). Would this correlation be significant with a two-tailed test?

7. A psychologist wants to know if a new self-esteem questionnaire is internally consistent. For each of the nine participants who filled out the questionnaire, two separate scores were created: one for the odd-numbered items and one for the even-numbered items. The data appear in the following table. Calculate the split-half reliability for the self-esteem questionnaire using Pearson's r.

Subject No.	Odd Items	Even Items
1	10	11
2	9	15
3	4	5
4	10	6
5	9	11
6	8	12
7	5	7
8	6	11
9	7	7

*8. A psychologist is preparing stimuli for an experiment on the effects of watching violent cartoons on the play behavior of children. Each of six cartoon segments is rated on a scale from 0 (peaceful) to 10 (extremely violent) by two different judges, one male and one female. The ratings follow.

Segment No.	Male Rater	Female Rater
1	2	4
2	1	3
3	8	7
4	0	1
5	2	5
6	7	9

   a. Calculate the interrater reliability using Pearson's r.
   b. Test the significance of the correlation coefficient at the .01 level (one-tailed).

*9. Does aerobic exercise reduce blood serum cholesterol levels? To find out, a medical researcher assigned volunteers who were not already exercising regularly to do a randomly selected number of exercise hours per week. After 6 months of exercising the prescribed number of hours, each participant's cholesterol level was measured, yielding the data in the following table:

Subject No.	Hours of Exercise per Week	Serum Cholesterol Level
1	4	220
2	7	180
3	2	210
4	11	170
5	5	190
6	1	230
7	10	200
8	8	210

a. Calculate Pearson's correlation coefficient for the data in the table.
b. Test the significance of the correlation coefficient at the .05 level (two-tailed).
c. What conclusions can you draw from your answer to part b?

10. One of the most common tools of the cognitive psychologist is the lexical decision task, in which a string of letters is flashed on a screen and an experimental participant must decide as quickly as possible whether those letters form a word. This task is often used as part of a more complex experiment, but this exercise considers reaction time as a function of the number of letters in a string for a single participant. The following data represent the participant's reaction times in response to three strings of each of four lengths: 3, 4, 5, or 6 letters. Calculate Pearson's correlation coefficient for the data and test for statistical significance.

Trial No.	Numbers of Letters in String	Reaction Time in Milliseconds
1	6	930
2	5	900
3	3	740
4	5	820
5	4	850
6	4	720
7	3	690
8	6	990
9	4	810
10	3	830
11	6	880
12	5	950

*11. Calculate Pearson's r for the data in Exercise 9A6. Test for significance at both the .05 and .01 levels (two-tailed).

12. Calculate Pearson's r for the data in Exercise 9A8. Test for significance at both the .05 and .01 levels (two-tailed).

The Power Associated with Correlational Tests

OPTIONAL
MATERIAL

It is not difficult to apply the procedures of power analysis that were described in the previous chapter to statistical tests involving Pearson's r. Just as the null hypothesis for the two-group t test is almost always $\mu_1 - \mu_2 = 0$, the null hypothesis for a correlational study is almost always $\rho_0 = 0$. The alternative hypothesis is usually stated as $\rho_A \neq 0$ (for a one-tailed test, $\rho_A < 0$ or $\rho_A > 0$), but to study power it is necessary to hypothesize a particular value for the population correlation coefficient. Given that ρ_A does not equal 0, the sample rs will be distributed around whatever value ρ_A does equal. The value for ρ_A specified by the alternative hypothesis can be thought of as an "expected" r. How narrowly the sample rs are distributed around the expected r depends on the sample size.

To understand power analysis for correlation, it is important to appreciate a fundamental difference between Pearson's r and the t value for a two-group test. The expected $r(\rho_A)$ is a measure similar to **d** (the effect size associated with a t test); it does not depend on sample size. Rather, the expected r describes the size of an effect in the population, and, by itself, its size does not tell you whether a particular test of the null hypothesis is likely to come out statistically significant. By contrast, the expected t value is a reflection of both **d** *and* sample size, and it does give you a way to determine the likelihood of attaining statistical significance. The point is that when performing power analysis for correlation, the expected r plays the same role as **d** in the power analysis of a t test, and not the role of expected t. We still need to transform ρ_A (i.e., the expected r) into a delta value that can be looked up in the power table. This transformation is done with Formula 9.7:

$$\delta = \rho_A \sqrt{N - 1} \qquad \qquad \textbf{Formula 9.7}$$

Notice the similarity to the formula for δ in the one-group t test, and keep in mind that ρ_A plays the same role as **d**. For instance, if we have reason to expect a correlation of .35, and we have 50 participants available:

$$\delta = .35\sqrt{50 - 1} = .35\sqrt{49} = .35 \cdot 7 = 2.45$$

Assuming alpha = .05, two-tailed, Table A.3 tells us that for $\delta = 2.45$, power is between .67 and .71. Chances are considerably better than 50% that the sample of 50 participants will produce a statistically significant Pearson's r (if we are right that the true $\rho = .35$), but the chances may not be high enough to justify the expense and effort of conducting the study.

If we still expect that $\rho = .35$, but we desire power to be .85, we can calculate the required N by solving Formula 9.7 for N to create Formula 9.8:

$$N = \left(\frac{\delta}{\rho_A}\right)^2 + 1 \qquad\qquad \textbf{Formula 9.8}$$

From Table A.4 we see that to obtain power = .85, we need $\delta = 3.0$ ($\alpha = .05$, two-tailed). Plugging this value into the formula above, we find that:

$$N = \left(\frac{3}{.35}\right)^2 + 1 = (8.57)^2 + 1 = 73.5 + 1 = 74.5$$

Therefore, a sample of 75 participants is required to have power = .85, if $\rho = .35$.

As demonstrated in Chapter 8, power analysis can be used to determine the maximum number of participants that should be used. First determine δ based on the desired levels for power and α. Then plug the smallest correlation of interest for the variables you are dealing with into Formula 9.8 in place of ρ_A. The N given by the calculation is the largest sample size you should use. Any larger N will have an unnecessarily high chance of giving you statistical significance when the true correlation is so low that you shouldn't care about it. Similarly, in place of ρ_A you can plug in the largest correlation that can reasonably be expected. The N that is thus calculated is the minimum sample size you should employ; any smaller sample size will give you a less than desirable chance of obtaining statistical significance.

Sometimes the magnitude of the correlation expected in a new study can be predicted based on previous studies (as with **d**). At other times the expected correlation is characterized roughly as small, medium, or large. The conventional guideline for Pearson's r (Cohen, 1988) is that .1 = small, .3 = medium, and .5 = large. Correlations much larger than .5 usually involve two variables that are measuring the same thing, such as the various types of reliability described earlier or, for instance, two different questionnaires that are both designed to assess a patient's current level of depression.

You may have noticed that whereas $r = .5$ is considered a large correlation, it is only a medium value for **d**. This difference arises from the fact that even though ρ and **d** are (conceptually) similar in terms of what they are assessing in the population, they are measured on very different scales. Correlation can only range between 0 and 1 in magnitude, whereas **d** is like a z score, with no limit to how large it can get. Although it can be used as an alternative measure of effect size, correlation is more often referred to as a measure of the strength of association between two variables.

Once you have determined that a particular sample r is statistically significant, you are permitted to rule out that $\rho = 0$—that the two variables have no linear relationship in the population. More than that, your sample r provides a point estimate of ρ; if $r = .4$, .4 is a good guess for ρ. However, as I first discussed in Chapter 6, the accuracy of the point estimate depends on the sample size. You can be more confident that ρ is close to .4 if the r of .4 comes from a sample of 100 participants than if the .4 was calculated for only 10 participants. An interval estimate would be more informative than the point estimate, providing a clearer idea of what values are likely for the population correlation coefficient. Interval estimation for ρ is not performed as often as it is for μ, probably because it has fewer practical implications—but it should be performed more often than it is. Unfortunately, constructing a confidence interval for ρ involves a complication that does not arise when dealing with μ, as you will see next.

Fisher Z Transformation

The complication in constructing a confidence interval for ρ concerns the distribution of sample rs around ρ. When ρ = 0, the sample rs form a symmetrical distribution around 0, which can be approximated by a normal distribution (especially as the sample size gets fairly large). However, when ρ equals, for example, +.8, the sample rs are not going to be distributed symmetrically around +.8. There is a ceiling effect; *a sample r cannot get higher than +1.0*, so there is much more room for r to be lower than +.8 (as low as −1.0) than to be higher than +.8. The distribution will be negatively skewed (it would be positively skewed for ρ = −.8); the closer ρ gets to +1.0 or −1.0, the more skewed the distribution becomes. Fortunately, Fisher (1970) found a way to transform the sample rs so that they will follow an approximate normal distribution, regardless of ρ. This method allows us once again to use the simplicity of z scores in conjunction with the standard normal table (Table A.1). But first we have to transform our sample r. Although the transformation is based on a formula that requires finding logarithms, I have already performed the transformations for regularly spaced values for r and placed them in a convenient table (see Table A.6). The transformed rs are called Zs (the capital Z is usually used to avoid confusion with z scores), and the process is referred to as the Fisher Z transformation. I will make use of this transformation below.

The Confidence Interval for ρ

To construct a confidence interval for ρ, we would expect to begin by putting a point estimate for ρ (e.g., the sample r) in the middle of the interval. But to ensure a normal distribution, we first transform r by finding the corresponding Z_r in Table A.6. We must also select a level of confidence. We'll begin with the usual 95% CI. Assuming a normal distribution around the transformed point estimate, we know which z scores will mark off the middle 95%: +1.96 and −1.96. This means that we need to go about two standard errors above and below Z_r. Finally, we need to know the standard error for Z_r. Fortunately, this standard error is expressed in an easy formula:

$$\text{Standard Error } (Z_r) = \sqrt{\frac{1}{N-3}} = \frac{1}{\sqrt{N-3}}$$

We can now work out a numerical example. The Pearson's r for the study of years of education and number of children was −.735. Looking in Table A.6 we find that the corresponding Z_r for +.735 is .94, so we know that for −.735, Z_r equals −.94. Next we find the standard error. Because $N = 6$ in our example, the standard error $= 1/\sqrt{(N-3)} = 1/\sqrt{(6-3)} = 1/\sqrt{3} = 1/1.732 = .577$. Having chosen to construct a 95% CI, we add and subtract 1.96 × .577 from −.94. So the upper boundary is −.94 + 1.13 = +.19 and the lower boundary is −.94 − 1.13 = −2.07.

Perhaps you noticed something peculiar about one of these boundaries. Correlation cannot be less than −1, so how can the lower boundary be −2.07? Remember that these boundaries are for the *transformed* correlations, not for ρ. To get the upper and lower limits for ρ, we have to use Table A.6 in reverse. We look up the values of the boundaries calculated above in the Z_r column to find the corresponding rs. Although we cannot find .19 as an entry for Z_r, there are entries for .187 and .192, corresponding to rs of .185 and .190, respectively. Therefore, we can estimate that

$Z_r = .190$ would correspond approximately to $r = .188$. That is the upper limit for r. To find the lower limit we look for 2.07 under the Z_r column. (As in the normal distribution table, there is no need for negative entries—the distribution is symmetrical around zero.) There are entries for 2.014 and 2.092, corresponding to rs of .965 and .970, respectively. Therefore, we estimate that the lower limit for r is approximately $-.968$. For this example, we can state with 95% confidence that the population correlation coefficient for these two variables is somewhere between $-.968$ and $+.188$. Note that $\rho = 0$ is included in the 95% CI. That tells us that the null hypothesis (i.e., $\rho = 0$) cannot be rejected at the .05, two-tailed level, confirming the hypothesis test that we conducted in Section B.

I hope you noticed that the CI we just found is so large as to be virtually useless. This result is due to the ridiculously small sample size. Normally, you would not bother to calculate a CI when dealing with such a small N. In fact, the use of $1/\sqrt{(N - 3)}$ to represent the standard error of Z is not accurate for small sample sizes, and the inaccuracy gets worse as the sample r deviates more from zero. It is also assumed that the variable in question has a bivariate normal distribution in the population. If your distribution seems strange and/or the sample r is quite high in magnitude, it is important to use a fairly large sample size. As usual, 30 to 40 would be considered minimal in most situations, but an N of more than 100 may be required for accuracy in more extreme situations.

The process outlined in the preceding for constructing a confidence interval for r can be formalized in the following formula:

$$Z_\rho = Z_r \pm z_{\text{crit}} s_r$$

Note how similar this formula is to Formula 6.4 for the population mean. If we insert the equation for the standard error into the preceding formula and separate the formula into upper and lower limits, we obtain Formula 9.9:

$$\text{Upper } Z_\rho = Z_r + z_{\text{crit}} \frac{1}{\sqrt{N - 3}} \qquad \textbf{Formula 9.9}$$

$$\text{Lower } Z_\rho = Z_r - z_{\text{crit}} \frac{1}{\sqrt{N - 3}}$$

Don't forget that Formula 9.9 gives the confidence interval for the transformed correlation. The limits found with Formula 9.9 must be converted back to ordinary correlation coefficients by using Table A.6 in reverse.

Testing a Null Hypothesis Other Than $\rho = 0$

I mentioned in the previous section that the null hypothesis for correlation problems is usually $\rho = 0$ and that matters get tricky if $\rho \neq 0$. The problem is that when $\rho \neq 0$, the sample rs are not distributed symmetrically around r, making it difficult to describe the null hypothesis distribution. This is really the same problem as finding the confidence interval around a sample r that is not zero, and it is also solved by the Fisher Z transformation. Let us once more consider the correlation between education and number of children. Suppose that a national survey 50 years ago showed the correlation (ρ) to be $-.335$, and we want to show that the present correlation, $r = -.735$, is significantly more negative. Also suppose for this problem that the present correlation is based on 19 participants instead of only 6.

Now the null hypothesis is that $\rho_0 = -.335$. To test the difference between the present sample r and the null hypothesis, we use the formula for a one-group z test, modified for the transformed correlation coefficients:

$$z = \frac{Z_r - Z_\rho}{\sigma_r}$$

If we insert the equation for the standard error (the same equation used for the standard error in the confidence interval) into the above equation, we obtain Formula 9.10:

$$z = \frac{Z_r - Z_\rho}{\sqrt{\dfrac{1}{N - 3}}}$$ **Formula 9.10**

Before we can apply this formula to our example, both the sample r and ρ_0 must be transformed by Table A.6. As we found before, $-.735$ corresponds to $-.94$. For $-.335$ (ρ_0) the transformed value is $-.348$. Inserting these values into Formula 9.10, we get

$$z = \frac{-.94 - (-.348)}{\sqrt{\dfrac{1}{16}}} = \frac{-.592}{.25} = -2.37$$

Because $z = -2.37$ is less than -1.96, the null hypothesis that $\rho_0 = .335$ can be rejected at the .05 level. We can conclude that the education/children correlation is more negative in today's population than it was 50 years ago.

Testing the Difference of Two Independent Sample rs

The procedures described in the preceding subsection can be modified slightly to handle one more interesting test involving correlations. If a sample r can be tested in comparison to a population ρ that is not zero, it is just a small step to creating a formula that can test the difference between two sample rs. For example, the education/children correlation we have been dealing with was calculated on a hypothetical random group of six women. Suppose that for the purpose of comparison, a group of nine men is randomly selected and measured on the same variables. If the correlation for men were in the reverse direction and equal to $+.4$, it would seem likely that the correlation for men would be significantly different from the correlation for women. However, a hypothesis test is required to demonstrate this.

The formula for comparing two sample rs is similar to the formula for a two-group z test:

$$z = \frac{Z_{r_1} - Z_{r_2}}{\sigma_{r_1 - r_2}}$$

The standard error of the difference of two sample rs is a natural extension of the standard error for one correlation coefficient, as the following formula shows:

$$\sigma_{r_1 - r_2} = \sqrt{\frac{1}{N_1 - 3} + \frac{1}{N_2 - 3}}$$

Combining this formula with the previous one, we obtain a formula for testing the difference of two sample rs:

$$z = \frac{Z_{r_1} - Z_{r_2}}{\sqrt{\dfrac{1}{N_1 - 3} + \dfrac{1}{N_2 - 3}}}$$ **Formula 9.11**

To apply Formula 9.11 to our present example, we must first transform each of the sample rs and then insert the values, as follows:

$$z = \frac{-.94 - (+.424)}{\sqrt{\frac{1}{6-3} + \frac{1}{9-3}}} = \frac{-1.364}{\sqrt{\frac{1}{2}}} = \frac{-1.364}{.707} = -1.93$$

Surprisingly, the null hypothesis cannot be rejected at the .05 level (two-tailed); -1.93 is not less than -1.96. The reason that two sample rs that are so discrepant are not significantly different is that the sample sizes are so small, making the power quite low.

It is sometimes interesting to compare two sample rs calculated on the same group of people but with different variables. For instance, you may wish to test whether the correlation between annual income and number of children is larger (in magnitude) than the correlation between years of education and number of children for a particular group of women. Because the two correlation coefficients involve the same people, they are not independent, and you cannot use Formula 9.11. A statistical solution has been worked out to test the difference of two nonindependent rs, and the interested reader can find this solution in some other texts, such as the one by Howell (2007).

SUMMARY

1. The calculation of power for testing the significance of a correlation coefficient is similar to the calculation that corresponds to a one-sample t test. The hypothesized population correlation coefficient (ρ_A) plays the same role played by d; ρ_A must be combined with the proposed sample size to find a value for δ, which can then be used to look up power.
2. A confidence interval, centered on the sample r, can be constructed to estimate ρ. However, this distribution tends to be more skewed as r deviates more from zero. Fisher's Z transformation must be applied to the sample r to create an approximate normal distribution. The CI formula is otherwise similar to the one for the population mean, except that the limits found must be converted back to limits for ρ by using the Fisher transformation in reverse.
3. The Fisher Z transformation is also needed for significance testing when the null hypothesis specifies that ρ is some particular value other than zero and for testing the significance of a difference between two sample rs.

EXERCISES

*1. Given the size of the sample in Exercise 9B11, how high would the population correlation (ρ) have to be between phobia ratings and stats exam scores in order to attain a power level equal to .85 for a .05, two-tailed test?

2. If the population correlation (ρ) for verbal SAT scores and GPAs were equal to the Pearson r you calculated in Exercise 9B12, how many students would you need in your sample to have power equal to .95 for a .01, one-tailed test?

*3. What is the value of Fisher's Z transformation for the following Pearson correlation coefficients?
  a. .05
  b. .1

  c. .3
  d. .5
  e. .7
  f. .9
  g. .95
  h. .99

  What is the value of Pearson's r that corresponds most closely with each of the following values for Fisher's Z transformation?
  i. .25
  j. .50
  k. .95
  l. 1.20
  m. 1.60
  n. 2.00
  o. 2.50

4. a. As Pearson's r approaches 1.0, what happens to the discrepancy between r and Fisher's Z?
   b. For which values of Pearson's r does Fisher's Z transformation seem unnecessary?

*5. In Exercise 9B5,
   a. What would the power of the test have been if the correlation for the population (ρ) were .5?
   b. What would the power of the test have been if ρ were equal to the sample r found for that problem?
   c. How many schizophrenics would have to be tested if ρ were equal to the sample r and you wanted power to equal .90?

6. For Exercise 9B5,
   a. Find the 95% confidence interval (CI) for the population correlation coefficient relating years of hospitalization to orientation score.
   b. Find the 99% CI for the same problem. Is the sample r significantly different from zero if alpha = .01 (two-tailed)? Explain how you can use the 99% CI to answer this question.

*7. In Exercise 9B5, suppose the same two variables were correlated for a sample of 15 prison inmates and the sample r were equal to .2. Is the sample r for the schizophrenics significantly different (alpha = .05, two-tailed) from the sample r for the inmates?

8. a. If a correlation less than .1 is considered too small to be worth finding statistically significant and power = .8 is considered adequate, what is the largest sample size you should use if a .05 two-tailed test is planned?
   b. If you have available a sample of 32 participants, how highly would two variables have to be correlated to have power = .7 with a .01, two-tailed test?

*9. Suppose that the population correlation for the quantitative and verbal portions of the SAT equals .4.
   a. A sample of 80 psychology majors is found to have a quantitative/verbal correlation of .5. Is this correlation significantly different from the population at the .05 level, two-tailed?
   b. A sample of 120 English majors has a quantitative/verbal correlation of .3. Is the correlation for the psychology majors significantly different from the correlation for the English majors (alpha = .05, two-tailed)?

10. For a random sample of 50 adults the correlation between two well-known IQ tests is .8.
   a. Find the limits of the 95% confidence interval for the population correlation coefficient.
   b. If a sample of 200 participants had the same correlation, what would be the limits of the 95% CI for this sample?
   c. Compare the widths of the CIs in parts a and b. What can you say about the effects of sample size on the width of the confidence interval for the population correlation?

KEY FORMULAS

Pearson's product-moment correlation coefficient (definitional form; not convenient for calculating, unless z scores are readily available):

$$r = \frac{\sum z_x z_y}{N}$$

Formula 9.1

Pearson's product-moment correlation coefficient (convenient when the *biased* sample standard deviations have already been calculated):

$$r = \frac{\dfrac{\sum XY}{N} - \mu_x \mu_y}{\sigma_x \sigma_y}$$

Formula 9.2

Pearson's product-moment correlation coefficient (convenient when the *unbiased* sample standard deviations have already been calculated):

$$r = \frac{\dfrac{1}{N-1}\left(\sum XY = N\overline{X}\,\overline{Y}\right)}{s_x s_y}$$

Formula 9.3

Pearson's product-moment correlation coefficient (not a convenient formula for calculation; it is included here only because it appears in various other texts):

$$r = \frac{\sum (X - \bar{X})(Y - \bar{Y})}{\sqrt{SS_x SS_y}} = \frac{SP}{\sqrt{SS_x SS_y}}$$

Formula 9.4

Pearson's product-moment correlation coefficient, raw-score version:

$$r = \frac{N\sum XY - (\sum X)(\sum Y)}{\sqrt{[N\sum X^2 - (\sum X)^2][N\sum Y^2 - (\sum Y)^2]}}$$

Formula 9.5

A t value that can be used to test the statistical significance of Pearson's r against the H_0 that $\rho = 0$ (as an alternative, the critical r can be found directly from Table A.5):

$$t = \frac{r\sqrt{N - 2}}{\sqrt{1 - r^2}}$$

Formula 9.6

Delta, to be used in the determination of power for a particular hypothesized value of ρ and a given sample size:

$$\delta = \rho_A \sqrt{N - 1}$$

Formula 9.7

The required sample size to attain a given level of power (in terms of delta) for a particular hypothesized value of ρ:

$$N = \left(\frac{\delta}{\rho_A}\right)^2 + 1$$

Formula 9.8

Confidence interval for the population correlation coefficient (ρ) in terms of Fisher's Z transformation; the limits found by this formula must be transformed back into ordinary rs using the Fisher transformation in reverse:

$$\text{Upper } Z_\rho = Z_r + z_{\text{crit}} \frac{1}{\sqrt{N - 3}}$$

Formula 9.9

$$\text{Lower } Z_\rho = Z_r - z_{\text{crit}} \frac{1}{\sqrt{N - 3}}$$

z test to determine the significance of a sample r against a specific null hypothesis (other than $\rho_0 = 0$):

$$z = \frac{Z_r - Z_\rho}{\sqrt{\dfrac{1}{N - 3}}}$$

Formula 9.10

z test to determine whether two independent sample rs are significantly different from each other (the null hypothesis is that the samples were drawn from populations with the same ρ):

$$z = \frac{Z_{r_1} - Z_{r_2}}{\sqrt{\dfrac{1}{N_1 - 3} + \dfrac{1}{N_2 - 3}}}$$

Formula 9.11

LINEAR REGRESSION

You will need to use the following from previous chapters:

Symbols
μ: Mean of a population
$\overline{X}$: Mean of a sample
σ: Standard deviation of a population
s: Unbiased standard deviation of a sample
r: Pearson's product-moment correlation coefficient

Formulas
Formula 4.1: The z score
Formula 6.3: The t test for one sample
Formula 9.5: Pearson correlation (raw-score version)

Concepts
The normal distribution
Linear transformations

Chapter

CONCEPTUAL FOUNDATION

In the previous chapter I demonstrated that two variables could be perfectly correlated, even if each participant did not attain the same value on both variables. For instance, height and weight could be perfectly correlated for a particular group of people, although the units for the two variables are very different. The most useful property of perfect correlation is the perfect predictability that it entails. For a group in which height and weight are perfectly correlated, you can use a particular person's height to predict, without error, that person's weight (with a simple formula). Of course, the predictability would be just as perfect for two variables that were negatively correlated, as long as the negative correlation were perfect. It should also come as no surprise that when correlation is nearly perfect, prediction is nearly perfect as well and therefore very useful. Unfortunately, when predictability is needed most in real-life situations (e.g., trying to predict job performance based on a test), correlation is not likely to be extremely high and certainly not perfect. Fortunately, however, even when correlation is not very high, the predictability may be of some practical use. In this chapter you will learn how to use the linear relationship between two variables to make predictions about either one.

Perfect Predictions

The prediction rule that is used when linear correlation is perfect could not be simpler, especially when it is expressed in terms of z scores (as calculated by Formula 4.1). The rule is that the z score you predict for the Y variable is the same as the z score for the X variable. (It is conventional to plot the variable you wish to predict on the Y axis of a scatterplot and the variable you are predicting *from* on the X axis.) As a formula, this rule would be written as $z_{Y'} = z_X$, where the mark following the subscript Y signifies that it is the Y value that is being predicted. (Y' is pronounced "Y prime"; other symbols are sometimes used to indicate that Y is being predicted, but Y' is certainly one of the most popular.) For perfect negative correlation, the formula is the same except for the minus sign: $z_{Y'} = -z_X$. For nearly perfect correlation it might seem reasonable to follow the same rule, but whenever the correlation is less than perfect, the preceding rule does not give the best predictions. As the correlation becomes smaller in magnitude, there is a greater need for a modified prediction rule, as shown in the following.

Predicting with z Scores

Imagine trying to predict a student's math SAT score from his or her verbal SAT score. If we use the preceding simple prediction rule, a student two deviations above the mean in the verbal SAT score ($z = +2$) would be predicted to have a math SAT score two standard deviations above the mean as well. However, because the correlation between these two variables is always far from perfect, there is plenty of room for error, so predicting $z = +2$ for the math SAT would be going out on a limb. The usual procedure is to "hedge your bet," knowing that the lower the correlation, the greater room there is for error. A modified rule has been devised to minimize the error of predictions when the correlation is less than perfect. The rule is given by Formula 10.1:

$$z_{Y'} = rz_X$$

Formula 10.1

where r is the Pearson correlation coefficient described in Chapter 9.

Note that when r is $+1$ or -1, the formula reverts to the simple rule for perfect correlation that was discussed in the preceding. Note also that when there is no linear correlation between the two variables (i.e., $r = 0$), the prediction is always that the z score is zero, which implies that we are predicting that the Y variable will be at its mean. This strategy makes sense. If you have to predict the weight of each person in a group, and you have no information at all, error is minimized by predicting the mean weight in each case. Why this minimizes the error has to do with how we measure the error of predictions, which will be discussed later in this section. When correlation is less than perfect but greater than zero (in magnitude), Formula 10.1 represents a compromise between predicting the same z score as the first variable and predicting the mean of the second variable. As the correlation becomes lower, there is less of a tendency to expect an extreme score on one variable to be associated with an equally extreme score on the other. On the other hand, as long as the correlation is not zero, the first variable is taken into account in the prediction of the second variable; the first variable clearly has an influence, but that influence lessens as the correlation between the variables is reduced.

Calculating an Example

To make the discussion more concrete, let us apply Formula 10.1 to the prediction of math SAT scores from verbal SAT scores. If we assume that the correlation between these two variables is $+.4$, Formula 10.1 becomes $z_{Y'} = .4z_X$. A person with a verbal SAT z score of $+1.0$ would be predicted to have a math SAT z score of $+.4$. A person with a verbal SAT of $z = -2.0$ would be predicted to have $z = -.8$ in math. Of course, in most applications the data will not be in the form of z scores; to use Formula 10.1 you would first have to convert each score into a z score. Rather than converting to z scores, it is easier to use a prediction formula that is designed to deal with the original scores. The derivation of this formula will be presented shortly.

Regression toward the Mean

I have been calling Formula 10.1 a prediction formula, which it is, but it is more commonly called a *regression formula*. The term comes from the work of Sir Francis Galton, who, among other projects, studied the relation between the height of an individual and the heights of the individual's parents. He found that Formula 10.1 applied to his data with a correlation coefficient of about .67. In fact, Karl Pearson's derivation of the formula we now use for the correlation coefficient was motivated by and applied to Galton's work (Cowles, 1989). Galton noted the tendency for unusually tall people to have children shorter than themselves; the children were usually taller than average, of course, but not as tall as the parents. Similarly, unusually short people generally had children closer to the mean than themselves. Galton referred to this tendency as "regression toward mediocrity," but it is now referred to as *regression toward the mean*. At one point, this phenomenon was seen as some sort of natural pressure pushing toward mediocrity, but now scientists recognize that regression toward the mean is just a consequence of the laws of probability when correlation is less than perfect. Looking at Formula 10.1, you can see that as the correlation gets lower, the prediction gets closer to $z = 0$; that is, it regresses toward the mean. Because it is based on r, which measures linear correlation, Formula 10.1 is a formula for *linear* regression. Other forms of regression (e.g., polynomial) have been devised to handle more complex relationships among variables, but only linear (and multiple linear) regression will be covered in detail in this text.

Graphing Regression in Terms of z Scores

The use of Formula 10.1 can be clarified by means of a scatterplot. When a scatterplot is used in conjunction with z scores, the zero point of the graph (called the *origin*) is in the middle so that the horizontal and vertical axes can extend in the negative direction (to the left and down, respectively; see Figure 10.1). In terms of z scores, the scatterplot for perfect positive correlation is just a diagonal line (at an angle of 45 degrees) that passes through the origin,

Figure 10.1

Scatterplot for Perfect Correlation in Terms of z Scores

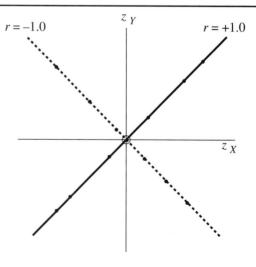

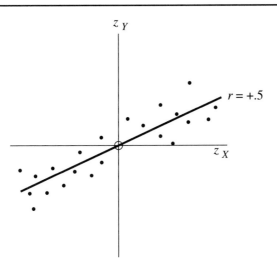

Figure 10.2

Scatterplot for $r = .5$ in Terms of z Scores

as in Figure 10.1. Notice that each point along the line corresponds to the same z score on both the X and the Y axis. Perfect negative correlation is represented by a similar diagonal line slanted in the opposite direction.

If correlation is not perfect, for example, if $r = +.5$, Formula 10.1 becomes $z_{Y'} = .5z_X$, which corresponds to the prediction line shown in Figure 10.2. In keeping with the traditional terminology, the prediction line is called the regression line. For any z_X, you can use this line to find the best prediction for z_Y (in this case, half of z_X). Notice that the regression line for $r = +.5$ makes a smaller angle with the X axis than the line for $r = +1$. (Compare Figure 10.2 to Figure 10.1.)

The angle of the regression line (as with any straight line on a graph) is called the *slope* of the regression line. The slope can be measured in degrees, but it is usually measured as the change in the Y axis divided by the change in the X axis. When $r = +1$, the line goes *up* one unit (change in Y axis) for each unit it moves to the *right* (change in X axis). However, when $r = +.5$, the line goes up only one-half unit each time it moves one unit to the right, so the slope is .5. The slope of the regression line, plotted in terms of z scores, always equals the correlation coefficient. Figure 10.2 illustrates the possible scatter of the data points around the regression line when $r = .5$. When correlation is perfect, all of the data points fall on the regression line, but as the correlation coefficient gets lower, the data points are more widely scattered around the regression line (and the slope of the regression line gets smaller, too). Regression lines are particularly easy to draw when dealing with z scores, but this is not the way regression is commonly done. However, if we want to deal directly with the original scores, we need to modify the regression formula as shown in the following.

The Raw-Score Regression Formula

To transform Formula 10.1 into a formula that can accommodate raw scores, Formula 4.1 for the z score must be substituted for z_X and $z_{Y'}$. (This is very similar to the way Formula 9.1 was transformed into Formula 9.2 in the previous chapter.) The formula that results must then be solved for Y' to be useful.

$$z_{Y'} = rz_X \qquad \text{so} \qquad \frac{Y' - \mu_Y}{\sigma_Y} = r\left(\frac{X - \mu_X}{\sigma_X}\right)$$

$$\text{thus, } Y' - \mu_Y = r\left(\frac{X - \mu_X}{\sigma_X}\right)\sigma_y \qquad \text{so} \qquad Y' - \mu_Y = \frac{\sigma_Y}{\sigma_X}r(X - \mu_X)$$

$$\text{finally,} \qquad Y' = \frac{\sigma_Y}{\sigma_X}r(X - \mu_X) + \mu_Y \qquad\qquad \textbf{Formula 10.2}$$

Note that Formula 10.2 is expressed in terms of population parameters (i.e., μ and σ). This is particularly appropriate when regression is being used for descriptive purposes only. In that case, whatever scores you have are treated as a population; it is assumed you have no interest in making inferences about a larger, more inclusive set of scores. In Section B, I will express these two formulas in terms of sample statistics and take up the matter of inference, but the formulas will change very little. In the meantime, Formula 10.2 can be put into an even simpler form with a bit more algebra. First I will create Formula 10.3A and define a new symbol, as follows:

$$b_{YX} = \frac{\sigma_Y}{\sigma_X}r \qquad\qquad \textbf{Formula 10.3A}$$

Formula 10.2 can be rewritten in terms of this new symbol:

$$Y' = b_{YX}(X - \mu_X) + \mu_Y$$

Multiplying to get rid of the parentheses yields

$$Y' = b_{YX}X - b_{YX}\mu_X + \mu_Y$$

One final simplification can be made by defining one more symbol, using Formula 10.4A:

$$a_{YX} = \mu_Y - b_{YX}\mu_X \qquad\qquad \textbf{Formula 10.4A}$$

If you realize that $-a_{YX} = b_{YX}\mu_X - \mu_Y$, you can see that Formula 10.2 can be written in the following form, which will be designated Formula 10.5:

$$Y' = b_{YX}X + a_{YX} \qquad\qquad \textbf{Formula 10.5}$$

The Slope and the Y Intercept

Formula 10.5 is a very convenient formula for making predictions. You start with an X value to predict from, you multiply it by b_{YX} and then you add a_{YX}; the result is your prediction for the Y value. You may also recognize Formula 10.5 as the formula for *any* straight line. (Usually in mathematics the formula for a straight line is given by the equation $y = mx + b$, but unfortunately the letters used for the constants conflict with the notation conventionally used for regression, so I will stick with the regression notation.) The term b_{YX} is the slope of the regression line using raw scores ($b_{YX} = r$ when regression is calculated in terms of z scores). If the line goes through the origin, all you need is the slope to describe the line. However, to describe lines that do not pass through the origin, you need to indicate at

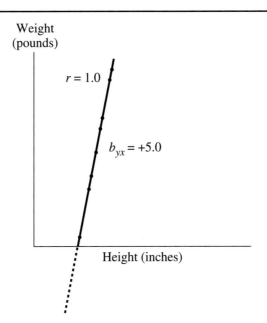

Figure 10.3

Regression Line: Perfect
Correlation

what point the line hits the Y axis (i.e., the value of Y when X is zero). This is called the *Y intercept*, and it is represented by a_{YX} in Formula 10.5. Again, these descriptions can be made more concrete by drawing a graph. Imagine that height and weight are perfectly correlated for a group of people. The scatterplot would form a straight line, as shown in Figure 10.3.

Predictions Based on Raw Scores

When correlation is perfect, as in Figure 10.3, the slope of the line is $b_{YX} = \sigma_Y/\sigma_X$ (or $b_{YX} = -\sigma_Y/\sigma_X$ for perfect negative correlation). In this example, the mean height is 69 inches; the standard deviation for height is about 3 inches. The mean weight is 155 pounds; the standard deviation for weight is about 15 pounds for this group. (σ_Y for this group is unusually small because all the people have the same "build"; it is only height that is producing the difference in weight.) Therefore, the slope for this graph is 15/3 = 5. This means that whenever two people differ by 1 inch in height, we know they will differ by 5 pounds in weight. Even though the correlation is perfect, you would not expect a slope of 1.0 because you would not expect a 1-inch change in height to be associated with a 1-*pound* gain in weight. The slope is always 1.0 for perfect correlation when the data are plotted in terms of z scores because you would expect a change of one *standard deviation* in height to be associated with a change of one *standard deviation* in weight.

It is important to point out that if the regression line in Figure 10.3 were extended, it would not pass through the origin. That is why we cannot predict weight in pounds by taking height in inches and multiplying by 5. If it were extended (as shown by the dotted portion of the line), the regression line would hit the Y axis at -190 pounds, as determined by Formula 10.4A:

$$a_{YX} = \mu_Y - b_{YX}\mu_X = 155 - 5(69) = 155 - 345 = -190$$

Given that a_{YX} is -190 pounds, the full regression equation is:

$$Y' = 5X - 190$$

Within this group, any person's weight in pounds can be predicted exactly: Just multiply their height in inches by 5 and subtract 190. For example, someone 6 feet tall would be predicted to weigh $5 \times 72 - 190 = 360 - 190 = 170$ pounds. Of course, there is no need to predict the weight of anyone in this group because we already know both the height *and* the weight for these particular people—that is how we were able to find the regression equation in the first place. However, the regression equation can be a very useful way to describe the relationship between two variables and can be applied to cases in which prediction is really needed. But first I must describe the regression line when correlation is not perfect.

In the previous example, the slope of the regression line was 5. However, if the correlation were not perfect, but rather $+.5$, the slope would be 2.5 $[b_{YX} = (\sigma_Y/\sigma_X)r = (15/3)(.5) = 5(.5) = 2.5]$. The Y intercept would be -17.5 $[a_{YX} = \mu_Y - b_{YX}\mu_X = 155 - 2.5(69) = 155 - 172.5 = -17.5]$. Therefore, the regression equation would be $Y' = 2.5X - 17.5$. A person 6 feet tall would be predicted to weigh $2.5 \times 72 - 17.5 = 180 - 17.5 = 162.5$ pounds. Notice that this prediction is midway between the prediction of 170 pounds when the correlation is perfect and the mean weight of 155 pounds. The prediction is exactly in the middle because the correlation in this example is midway between perfect and zero.

Interpreting the Y Intercept

The Y intercept for the above example does not make sense; weight in pounds cannot take on negative values. The problem is that the height–weight relationship is not linear for the entire range of weights down to zero. For many regression examples, either it does not make sense to extend the regression line down to zero or the relationship does not remain linear all the way down to zero. For instance, performance on a mental task may correlate highly with IQ over a wide range, but trying to predict performance as IQ approaches zero would not be meaningful. On the other hand, if vocabulary size were being predicted from the number of years of formal education, it would be meaningful to estimate vocabulary size for people with no formal schooling. As another example, you would expect a negative correlation between heart rate and average amount of time spent in aerobic exercise each week. When predicting heart rate from exercise time, the Y intercept would meaningfully represent the heart rate for individuals who do not exercise at all. If, instead, heart rate were being used as the predictor of, say, reaction time in a mental task, the Y intercept would not make sense because it would be the reaction time associated with a zero heart rate.

Quantifying the Errors around the Regression Line

When correlation is less than perfect, the data points do not all fall on the same straight line, and predictions based on any one straight line will often be in error. The regression equation (Formula 10.5) gives us the straight line that minimizes the error involved in making predictions. This will be easier to understand after I explain how error around the regression line is measured. Suppose we label as Y the variable that is to be predicted. (The equations can be reversed to predict the X variable from the Y variable, as I will show in Section B.) Then we can measure the difference between an actual Y value and the Y value predicted for it by the regression equation. This difference, $Y - Y'$, is called a *residual* because it is the amount of the original value that is left over after the prediction is subtracted out. The use of residuals to measure error from the regression line can be illustrated in a graph.

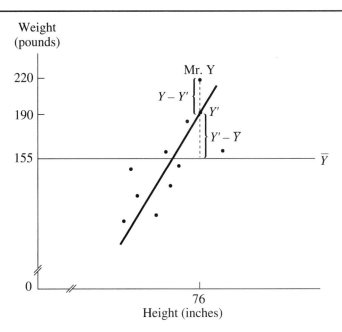

Figure 10.4

Regression Line:
Correlation Not Perfect

Figure 10.4 is a graph of height versus weight for a group of 12 men. Notice in particular the point on the graph labeled Mr. Y. This point represents a man who is quite tall (6 feet 4 inches) and also very heavy (220 pounds). The fact that this point is *above* the regression line indicates that this man is even heavier than would be expected for such a tall man. Using the regression line, you would predict that any man who is 76 inches tall would weigh 190 pounds (this point is labeled Y' on the graph). The difference between Mr. Y's actual weight (220) and his predicted weight (190), his residual weight, is $Y - Y'$, which equals 30. This is the amount of error for this one data point. To find the total amount of error, we must calculate $Y - Y'$ for each data point. However, if all of these errors were simply added, the result would be zero, because the data points below the regression line produce negative errors, which balance out the positive errors produced by data points above the line. The regression line functions like an "average," in that the total amount of error above the line always equals the total amount of error below it.

The Variance of the Estimate

To quantify the total amount of error in the predictions, all of the residuals (i.e., $Y - Y'$) are squared and then added together and divided by the total number, as indicated in Formula 10.6:

$$\sigma^2_{\text{est } Y} = \frac{\sum(Y - Y')^2}{N} \qquad \textbf{Formula 10.6}$$

If this formula looks like a variance, it should. It is the variance of the data points around the regression line, called the *variance of the estimate*, or the *residual variance* (literally, the variance of the residuals). The closer the data points are to the regression line, the smaller will be the error involved in the predictions, and the smaller $\sigma^2_{\text{est } Y}$ will be. Lower correlations are associated with more error in the predictions and therefore a larger $\sigma^2_{\text{est } Y}$. When correlation is at its lowest possible value (i.e., zero), $\sigma^2_{\text{est } Y}$ is at its maximum. What happens to $\sigma^2_{\text{est } Y}$ when $r = 0$ is particularly interesting.

When $r = 0$, the regression line becomes horizontal; its slope is therefore zero. The same prediction is therefore being made for all X values; for any X value the prediction for Y is the mean of Y. Because Y' is always $\overline{Y}$ (when $r = 0$), Formula 10.6 in this special case becomes

$$\sigma^2_{\text{est } Y} = \frac{\sum (Y - \overline{Y})^2}{N}$$

This means that the variance of the predictions around the regression line is just the ordinary variance of the Y values. Thus the regression line isn't helping us at all. Without the regression line, there is a certain amount of variability in the Y variable (i.e., σ^2_Y). If the X variable is not correlated with the Y variable, the regression line will be flat, and the variance around the regression line ($\sigma^2_{\text{est } Y}$) will be the same as the variance around the mean (σ^2_Y). However, for any correlation greater than zero, $\sigma^2_{\text{est } Y}$ will be less than σ^2_Y, and that represents the advantage of performing regression.

Explained and Unexplained Variance

The difference between the variance of the estimate and the total variance is the amount of variance "explained" by the regression equation. To understand the concept of *explained variance*, it will help to look again at Figure 10.4. Mr. Y's total deviation from the mean weight $(Y - \overline{Y})$ is $220 - 155 = 65$ pounds. This total deviation can be broken into two pieces. One piece is Mr. Y's deviation from the regression line $(Y - Y')$, which equals 30, and the other piece is the difference between the prediction and the mean $(Y' - \overline{Y})$, which equals 35. The two pieces, 30 and 35, add up to the total deviation. In general, $(Y - Y') + (Y' - \overline{Y}) = (Y - \overline{Y})$. The first part, $Y - Y'$, which I have been referring to as the error of the prediction, is sometimes thought of as the "unexplained" part of the variance, in contrast to the second part, $(Y' - \overline{Y})$, which is the "explained" part. In terms of Mr. Y's unusually high weight, part of his weight is "explained" (or predicted) by his height—he is expected to be above average in weight. But he is even heavier than someone his height is expected to be; this extra weight is not explained by his height, so in this context it is "unexplained."

If all the "unexplained" pieces were squared and added up $[\Sigma(Y - Y')^2]$, we would get the unexplained sum of squares, or unexplained SS; dividing the unexplained SS by N yields the *unexplained variance*, which is the same as the variance of the estimate discussed in the preceding. Similarly, we could find the explained SS based on the squared $Y' - \overline{Y}$ pieces and divide by N to find the amount of variance "explained." Together the explained and unexplained variances would add up to the total variance. The important concept is that whenever r is not zero, the unexplained variance is less than the total variance, so error or uncertainty has been reduced. We can guess a person's weight more accurately if we know his or her height than if we know nothing about the person at all. In terms of a scatterplot, when there is a linear trend to the data, the points tend to get higher (or lower, in the case of negative correlation) as you move to the right. The regression line slopes to follow the points, and thus it leads to better predictions and less error than a horizontal line (i.e., predicting the mean for everybody), which doesn't slope at all.

The Coefficient of Determination

If you want to know how well your regression line is doing in terms of predicting one variable from the other, you can divide the explained variance by the total variance. This ratio is called the *coefficient of determination* because it represents the proportion of the total variance that is explained (or determined) by the predictor variable. You might think that it would take a good deal of calculation to find the variances that form this ratio—and it would. Fortunately, the coefficient of determination can be found much more easily; it is always equal to r^2. If $r = .5$, the coefficient of determination is $.5^2 = .25$. If this were the value of the coefficient of determination in the case of height predicting weight, it would mean that 25% of the variation in weight is being accounted for by variations in height. It is common to say that r^2 gives the "proportion of variance accounted for." Because of the squaring involved in finding this proportion, small correlations account for less variance than you might expect. For instance, a low correlation of .1 accounts for only $.1^2 = .01$, or just 1%, of the variance.

The Coefficient of Nondetermination

Should you want to know the proportion of variance not accounted for, you can find this by dividing the unexplained variance (i.e., $\sigma^2_{est\,Y}$) by the total variance. Not surprisingly, this ratio is called the *coefficient of nondetermination*, and it is simply equal to $1 - r^2$, as shown in Formula 10.7A:

$$\frac{\sigma^2_{est\,Y}}{\sigma^2_Y} = 1 - r^2 \qquad \textbf{Formula 10.7A}$$

The coefficient of nondetermination is sometimes symbolized as k^2. For most regression problems, we'd like k^2 to be as small as possible and r^2 to be as large as possible. When $r = +1$ or -1, $k^2 = 1 - 1 = 0$, which is the most desirable situation. In the worst case, $r = 0$ and $k^2 = 1$, implying that the variance of the regression is just as large as the ordinary variance. For the preceding example in which $r = +.5$, $k^2 = 1 - .5^2 = 1 - .25 = .75$, indicating that the variance around the regression line is 75% of the total amount of variance. Because $k^2 = 1 - r^2$, the coefficient of determination added to the coefficient of nondetermination will always equal 1.0 for a particular regression problem.

Calculating the Variance of the Estimate

By rearranging Formula 10.7A, we can obtain a convenient formula for calculating the variance of the estimate. If we multiply both sides of Formula 10.7A by the population variance of Y, the result is Formula 10.7B:

$$\sigma^2_{est\,Y} = \sigma^2_Y(1 - r2) \qquad \textbf{Formula 10.7B}$$

This formula is a much easier alternative to Formula 10.6, once you have calculated the population variance of Y and the correlation between the two variables. (It is likely that you would want to calculate these two statistics anyway, before proceeding with the regression analysis.)

Bear in mind that I am only describing linear regression in this text; if two variables have a curvilinear relationship, other forms of regression will account for even more of the variance. The assumptions and limitations of linear regression are related to those of linear correlation and will be discussed in greater detail in the next section.

SUMMARY

1. For perfect linear positive correlation, the z score predicted for Y is the same as the z score for X. For perfect *negative* correlation, the z score predicted for Y is the same *in magnitude* as the z score for X, but opposite in sign.
2. If correlation is less than perfect, the z score predicted for Y is just r (i.e., Pearson's correlation coefficient) times the z score for X. If $r = 0$, the prediction for Y is always the mean of Y, regardless of the value of X.

3. The regression equation in terms of raw scores is $Y' = b_{YX}X + a_{YX}$, in which b_{YX} is the *slope* of the line and a_{YX} is the *Y intercept*. For regression in terms of z scores, the slope is just r, and the Y intercept is always zero.

4. For raw scores the slope is r times the ratio of the two standard deviations. The Y intercept is just the mean of Y minus the slope times the mean of X. The Y intercept is not always meaningful; it may not make sense to extend the regression line to values of X near zero.

5. The variance around the regression line is called the *variance of the estimate*, or the *residual variance*, symbolized by $\sigma^2_{\text{est } Y}$. When correlation is perfect (i.e., $r = +1$ or -1), $\sigma^2_{\text{est } Y} = 0$, and there is no error involved in the predictions. When $r = 0$, $\sigma^2_{\text{est } Y}$ equals σ^2_Y, which is the total variance of the Y scores.

6. The total variance can be divided into two portions: the unexplained variance (the variance of the estimate) and the explained variance. The ratio of the explained variance to the total variance is called the *coefficient of determination*, and it is symbolized by r^2.

7. The ratio of the variance of the estimate to the total variance is called the *coefficient of nondetermination* and is symbolized by k^2. In terms of Pearson's r, $k^2 = 1 - r^2$.

EXERCISES

1. Consider a math exam for which the highest score is 100 points. There will be a perfect negative correlation between a student's score on the exam and the number of points the student loses because of errors.
   a. If the number of points student A loses is half a standard deviation below the mean (i.e., he has $z = -.5$ for points lost), what z score would correspond to student A's score on the exam?
   b. If student B attains a z score of -1.8 on the exam, what z score would correspond to the number of points student B lost on the exam?

*2. Suppose that the Pearson correlation between a measure of shyness and a measure of trait anxiety is $+.4$.
   a. If a person is one and a half standard deviations above the mean on shyness, what would be the best prediction of that person's z score on trait anxiety?
   b. What would be the predicted z score for trait anxiety if a person's z score for shyness were $-.9$?

3. In Exercise 2,
   a. What proportion of the variance in shyness is accounted for by trait anxiety?
   b. If the variance for shyness is 29, what would be the variance of the estimate when shyness is predicted by trait anxiety?

*4. For the data in Exercise 9A6, write the regression equation to predict a z-score on exam 1 from a z-score on phobia rating. What proportion of the variance in exam scores is accounted for by phobia ratings?

5. On a particular regression line predicting heart rate in beats per minute (bpm) from number of milligrams of caffeine ingested, heart rate goes up 2 bpm for every 25 milligrams of caffeine. If heart rate is predicted to be 70 bpm with no caffeine, what is the raw score equation for this regression line?

6. For the data in Exercise 9A8, write the raw-score regression equation to predict GPA from verbal SAT. What is the value for the coefficient of nondetermination for these data?

*7. For a hypothetical population of men, waist size is positively correlated with height, such that Pearson's $r = +.6$. The mean height (μ_X) for this group is 69 inches with $\sigma_X = 3$; mean waist measurement (μ_Y) is 32 inches with $\sigma_Y = 4$.
   a. What is the slope of the regression line predicting waist size from height?
   b. What is the value of the Y intercept?
   c. Does the value found in part b above make any sense?
   d. Write the raw-score regression equation predicting waist size from height.

*8. Based on the regression equation found in Exercise 7,
   a. What waist size would you predict for a man who is 6 feet tall?
   b. What waist size would you predict for a man who is 62 inches tall?
   c. How tall would a man have to be for his predicted waist size to be 34 inches?
 9. a. In Exercise 7, what is the value of the coefficient of determination?
   b. How large is the coefficient of nondetermination?
   c. How large is the variance of the estimate (residual variance)?

*10. What is the magnitude of Pearson's r when the amount of unexplained variance is equal to the amount of explained variance?
 11. Describe a regression example in which the Y intercept has a meaningful interpretation.
*12. In a hypothetical example, the slope of the regression line predicting Y from X is -12. This means that
   a. A calculation error must have been made.
   b. The correlation coefficient must be negative.
   c. The magnitude of the correlation coefficient must be large.
   d. The coefficient of determination equals 144.
   e. None of the above.

Whenever two variables are correlated, one of them can be used to predict the other. However, if the correlation is very low, it is not likely that these predictions will be very useful; a low correlation means that there will be a good deal of error in the predictions. For instance, if $r = .1$, the variance of the data from the predictions (i.e., around the regression line) is 99% as large as it would be if you simply used the mean as the prediction in all cases. Even if the correlation is high, there may be no purpose served by making predictions. It might be of great theoretical interest to find a high correlation between the amount of repressed anger a person has, as measured by a projective test, and the amount of depression the person experiences, as measured by a self-report questionnaire, but it is not likely that anyone will want to make predictions about either variable from the other. On the other hand, a high correlation between scores on an aptitude test and actual job performance can lead to very useful predictions. I will use the following example concerning the prediction of life expectancy to illustrate how linear regression can be used to make useful predictions.

B

BASIC STATISTICAL PROCEDURES

Life Insurance Rates

To decide on their rates, life insurance companies must use statistical information to estimate how long individuals will live. The rates are usually based on life expectancies averaged over large groups of people, but the rates can be adjusted for subgroups (e.g., women live longer than men, nonsmokers live longer than smokers, etc.). Imagine that in an attempt to individualize its rates, an insurance company has devised a lifestyle questionnaire (LQ) that can help predict a person's life expectancy based on his or her health habits (amount of exercise, typical levels of stress, alcohol consumption, smoking, etc.). LQ scores can range from 0 to 100, with 100 representing the healthiest lifestyle possible. A long-term study is conducted in which each person fills out the LQ on his or her fiftieth birthday, and eventually his or her age at death is recorded. Because we want to predict the total number of years people will live based on their LQ scores, number of years will be the Y variable, and LQ score will be the X variable.

To streamline the procedures in this section, I will assume that the means and standard deviations for both variables, as well as the Pearson correlation coefficient, have already been calculated. Table 10.1 shows these values for a hypothetical sample of 40 subjects.

Table 10.1

LQ Score (X)		Number of Years Lived (Y)	
$\bar{X}$	36	$\bar{Y}$	74
s_X	14	s_Y	10
$r = +.6$			
$N = 40$			

Regression in Terms of Sample Statistics

Note that Table 10.1 gives the *unbiased* standard deviations. The regression formulas in Section A were given in terms of population parameters, as though we would have no desire to go beyond the actual data already collected. Actually, it is more common to try to extend the results of a regression analysis beyond the data given. In this section, the regression formula will be recast in terms of sample statistics, which can serve as unbiased estimators of population parameters. Formula 10.5 does not change, but if you are dealing with sample statistics, the slope is expressed as follows:

$$b_{YX} = \frac{s_Y}{s_X} r \qquad\qquad \textbf{Formula 10.3B}$$

and the Y intercept becomes:

$$a_{YX} = \overline{Y} - b_{YX}\overline{X} \qquad\qquad \textbf{Formula 10.4B}$$

You must continue to use the subscript *YX* on the variable for both the slope and the intercept because the subscript indicates that *Y* is being regressed on *X*, which is another way of saying that *Y* is being *predicted from X*. The slope and intercept will usually be different when *X* is being regressed on *Y*, as I will show later. We need not use the subscript on Pearson's *r* because there is no distinction between *X* correlated with *Y* and *Y* correlated with *X*.

Finding the Regression Equation

Once the means, standard deviations, and Pearson's *r* have been found, the next step in the regression analysis is to calculate the slope and the *Y* intercept of the regression line. We will use Formulas 10.3B and 10.4B, because it is the unbiased SDs that we have available (see Table 10.1).

$$b_{YX} = \frac{s_Y}{s_X} r = \frac{10}{14}(.6) = .714(.6) = .4$$

$$a_{YX} = \overline{Y} - b_{YX}\overline{X} = 74 = .43(36) = 74 - 15.43 = 58.6$$

Finally, we insert the values for b_{YX} and a_{YX} into Formula 10.5:

$$Y' = b_{YX}X + a_{YX} = .43X + 58.6$$

Making Predictions

The regression formula can now be used to predict a person's life expectancy based on his or her LQ score. For instance, the regression formula would give the following prediction for someone with an average LQ score (see Table 10.1):

$$Y' = .43X + 58.6 = .43(36) = 58.6 = 15.48 + 58.6 = 74.1$$

It is not surprising that someone with the average LQ score is predicted to have the average life expectancy. (The slight error is due to rounding off the slope to only two digits.) This will always be the case. Thinking in terms of z scores, the average of *X* corresponds to $z_X = 0$, which leads to a prediction of $z_Y = 0$, regardless of *r*.

Consider the life expectancy predicted for the person with the healthiest possible lifestyle (i.e., LQ = 100):

$$Y' = .43(100) + 58.6 = 43 + 58.6 = 101.6$$

On the other hand, the prediction for the least healthy lifestyle (LQ = 0) has already been found; it is the Y intercept (the point where the regression line hits the Y axis when LQ = 0):

$$Y' = .43(0) + 58.6 = 58.6$$

The process is the same for any value in between. If LQ equals 50, the prediction is

$$Y' = .43(50) + 58.6 = 21.5 + 58.6 = 80.1$$

For **LQ** = 10,

$$Y' = .43(10) + 58.6 = 4.3 + 58.6 = 62.9$$

These predictions can have very practical implications for the life insurance company. Although a correlation of .6 leaves plenty of room for error in the predictions, it would be possible to lower insurance rates somewhat for individuals with high LQ scores, while raising the rates proportionally for low LQ scorers. A correlation of .6 means that $.6^2$, or 36%, of the variance in life expectancy can be accounted for by the LQ scores; much of the remaining variance would be connected to, for instance, genetic factors. Nonetheless, the life insurance company could increase its profits by drawing in people with a healthy lifestyle with the promise of lower rates. (Of course, the company would not lower its rates too much; it would want to leave plenty of room for bad luck—that is, it would want to charge rates high enough to cover instances when, for example, a person with a high LQ score was killed by lightning.)

Using Sample Statistics to Estimate the Variance of the Estimate

When correlation is less than perfect, every prediction has a margin for error. With high correlation, it is unlikely that any of the predictions will be way off, so the margin is relatively small. However, the margin for error increases as correlation decreases. The margin for error is based on the degree to which the data points are scattered around the regression line, and this scatter is measured by the variance of the estimate, as described in the previous section. It would be useful for the insurance company to be able not only to generate life-expectancy predictions but also to specify the margin for error around each prediction in terms of a confidence interval. Certain assumptions are required if these confidence intervals are to be valid; the most important of these is *homoscedasticity*. This term means that the variance around the regression line in the population is the same at every part of the line. If your data exhibit homoscedasticity, the variance you calculate in Y for any particular X value will indicate about the same amount of spread no matter which X value you choose; this condition is illustrated in Figure 10.5. This assumption justifies our using the same value for the variance of the estimate as our margin for error, regardless of which part of the regression line we are looking at.

Figure 10.5

Scatterplot Depicting
Homoscedasticity

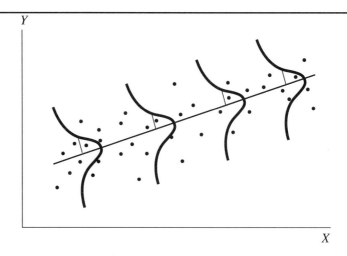

In most situations in which we would like to make predictions, it is not
realistic to use a variance-of-estimate formula that assumes we have data
for the entire population. The variance of the estimate for the entire population, $\sigma^2_{\text{est } Y}$, must be estimated from the available sample data according to
the following formula:

$$s^2_{\text{est } Y} = \frac{\sum (Y - Y')^2}{N - 2}$$

If you compare this formula to Formula 10.6, you will notice that the only
difference is that the denominator is $N - 2$ instead of N. Dividing by N
would result in a biased estimate of $\sigma^2_{\text{est } y}$. (As mentioned in Chapter 9, the degrees of freedom for correlational problems are $N - 2$ rather than $N - 1$.)
    There is a much easier way to calculate $s^2_{\text{est } Y}$, if you have already found s^2_Y
and r; you can use Formula 10.8A, which is very similar to Formula 10.7B:

$$s^2_{\text{est } Y} = \left(\frac{N - 1}{N - 2} \right) s^2_Y (1 - r^2)$$
                                        Formula 10.8

    Even though we are using the unbiased sample variance, we need the
factor of $(N - 1)/(N - 2)$ to ensure that we have an unbiased estimator of
the variance of the estimate.

Standard Error of the Estimate

Because confidence intervals are based on standard deviations, rather than
variances, we need to introduce one new term: the square root of the vari-
ance of the estimate, $\sigma_{\text{est } Y}$, which is called the *standard error of the estimate*.
This is the standard deviation of the points—in this case, in the vertical direction—from the regression line, and gives an idea of how scattered points
are from the line (about two thirds of the points should be within one standard error above or below the line). To find the standard error of estimate in
the population, we take the square root of both sides of Formula 10.7B:

$$\sigma_{\text{est } Y} = \sigma_Y \sqrt{1 - r^2}$$
                                        Formula 10.9

    To estimate this population value from sample data, we take the square
root of both sides of Formula 10.8A to create Formula 10.8B:

$$s_{\text{est } Y} = s_Y \sqrt{\frac{N-1}{N-2}(1-r^2)}$$ **Formula 10.8B**

Confidence Intervals for Predictions

To find a confidence interval for the *true* life expectancy of an individual with a particular LQ score, we first use the regression equation to make a prediction (Y') for life expectancy. Then that prediction is used as the center of the interval. Based on the principles described in Chapter 6, you would have good reason to expect the confidence interval to take the following form:

$$Y' \pm t_{\text{crit}} s_{\text{est } Y}$$

where t_{crit} depends on the level of confidence and $s_{\text{est } Y}$ is the estimated standard error of estimate. (For a 95% CI, t_{crit} corresponds to alpha = .05, two-tailed.) Unfortunately there is one more complication. The actual formula for the confidence interval contains an additional factor and therefore looks like this:

$$Y' \pm t_{\text{crit}} s_{\text{est } Y} \sqrt{1 + \frac{1}{N} + \frac{(X - \bar{X})^2}{(N-1)s_X^2}}$$ **Formula 10.10**

The reason for the additional factor is that the regression line, which is usually based on a relatively small sample, is probably wrong—that is, it does not perfectly represent the actual relationship in the population. So there are really two very different sources of error involved in our prediction. One source of error is the fact that our correlation is less than perfect; in the present example, unknown factors account for 64% of the variability in life expectancy. The second source (which necessitates the additional factor in the confidence interval formula) is the fact that different samples would lead to different regression lines and therefore different predictions. To demonstrate the use of this complex-looking formula, let us find a 95% confidence interval for a particular prediction.

An Example of a Confidence Interval

Earlier in this section, I used the regression equation to show that a person with an LQ score of 50 would be predicted to live 80.1 years. Now I will use interval estimation to supplement that prediction. The t_{crit} for the 95% CI with 38 degrees of freedom is about 2.02. The standard error of estimate for Y is found by using Formula 10.8B:

$$s_{\text{est } Y} = s_Y \sqrt{\frac{N-1}{N-2}(1-r_2)} = 10\sqrt{\frac{39}{38}(1-.6^2)} = 10\sqrt{.657} = 10(.81) = 8.1$$

Inserting the appropriate values into Formula 10.10, we get:

$$Y' \pm 2.02(8.1)\sqrt{1 + \frac{1}{40} + \frac{(50-36)^2}{(39)(196)}} = 80.1 \pm 16.36\sqrt{1 + .025 + .0256}$$

$$= 80.1 \pm 16.36\sqrt{1.0506} = 80.1 \pm 16.36(1.025) = 80.1 \pm 16.77$$

Our confidence interval predicts that a person with LQ = 50 will live somewhere between 63.33 and 96.87 years. Of course, 5% of these 95% CIs will turn out to be wrong, but that is a reasonable risk to take. If Pearson's r were higher, $s_{\text{est } Y}$ would get smaller (because $1 - r^2$ gets smaller), and therefore the 95% CI would get narrower. Increasing the sample size also tends to make the CI smaller by reducing both t_{crit} and the additional factor (in which N is in the denominator), but this influence is limited. With an extremely large N, t_{crit} approaches z_{crit}, and the additional factor approaches 1, but $s_{est\ Y}$ approaches $s_Y(1 - r^2)$, which is *not* affected by sample size. If Pearson's r is small, $s_{est\ Y}$ will not be much smaller than s_Y, no matter how large the sample size is. The bottom line is that unless the correlation is high, there is plenty of room for error in the prediction. The confidence interval just found may not look very precise, but without the LQ score, our prediction for the same individual would have been the mean (74), and the 95% confidence interval would have been from about 54 to 94 (about two standard deviations in either direction). Given the LQ information, the insurance company can be rather confident that the individual will not die before the age of 63, instead of having to use 54 as the lower limit.

Assumptions Underlying Linear Regression

As with linear correlation, linear regression can be used purely for descriptive purposes—to show the relationship between two variables for a particular group of people. In that case the assumptions described in the following need not be made. However, it is much more common to want to generalize your regression analysis to a larger group (i.e., a population) and to scores that were not found in your original sample but might be found in future samples. The confidence intervals discussed in the preceding subsection are valid only if certain assumptions are actually true.

Independent Random Sampling

This is the same assumption described for Pearson's r in Chapter 9, namely, that each case (i.e., pair of scores) should be independent of the others and should have an equal chance of being selected.

Linearity

The results of a linear regression analysis will be misleading if the two variables have a curvilinear relationship in the population.

Normal Distribution

At each possible value of the X variable, the Y variable must follow a normal distribution in the population.

Homoscedasticity

For each possible value of the X variable, the Y variable has the same population variance. This property is analogous to homogeneity of variance in the two-group t test.

Regressing X on Y

What happens if you wish to predict the X variable from scores on the Y variable, instead of the other way around? The obvious answer is that you can switch the way the two variables are labeled and use the equations already presented in this chapter. However, if you would like to try the regression analysis in both directions for the same problem (e.g., predict an individual's maximum running speed on a treadmill from his or her reaction time in a laboratory *and* predict reaction time from running speed), it would be confusing to relabel the variables. It is easier to switch the X and Y subscripts in the formulas to create a new set of formulas for regressing X on Y. For instance, the regression formula becomes

$$X' = b_{XY}Y + a_{XY}$$

where $b_{XY} = \dfrac{s_X}{s_Y}r$ and $a_{XY} = \overline{X} - b_{XY}\overline{Y}$

Unless the two variables have the same standard deviation, the two slopes will be different. If b_{XY} is less than 1, b_{YX} will be greater than 1, and vice versa. Moreover, it is only when the X and Y variables have both the same SDs *and* the same means that a_{XY} will be equalt to a_{YX}.

Raw Score Formulas

The formula that I recommend for the calculation of the regression slope $[b_{YX} = (s_y/s_x)r]$ is simple, but it does require that both standard deviations, as well as Pearson's r, be calculated first. Because it is hard to imagine a situation in which those latter statistics would not be computed for other reasons anyway, the formula for b_{YX} does not really require extra work. However, it is also true that b_{YX} can be calculated directly from the raw scores, without finding any other statistics. This tends to reduce the errors involved in rounding off intermediate statistics. By taking the raw-score formula for Pearson's r (Formula 9.5) and multiplying by the raw-score formula for the ratio of the two standard deviations, we derive Formula 10.11:

$$b_{YX} = \frac{N(\sum XY) - (\sum X)(\sum Y)}{N\sum X^2 - (\sum X)^2} \qquad \text{Formula 10.11}$$

You may recognize that the numerator in Formula 10.11 is the same as the numerator for calculating Pearson's r (Formula 9.5); it is the biased covariance. The denominator is just the biased variance of X (i.e., σ_X^2). So another way to view b_{YX} is that it is the covariance of X and Y divided by the variance of X. Although a similar raw-score formula could be derived for a_{YX}, it would not be efficient to use it. Once b_{YX} has already been calculated, Formula 10.4(A or B) is the most sensible way to calculate a_{YX}. You may encounter other formulas for b_{YX} and a_{YX} in terms of sums of squares and sums of products (SS and SP), but these are algebraically equivalent to the formulas presented here and therefore produce the same answers (SS divided by N yields a variance, whereas SP divided by N results in a covariance).

When to Use Linear Regression

Prediction

The most obvious application of linear regression is in predicting the future performance of something or someone—for example, a person's performance on a job or in college based on some kind of aptitude test. Hiring or admissions decisions can be made by finding a cutoff score on the aptitude test, above which an individual can be expected to perform adequately on the performance measure (e.g., college grades or job skill evaluation). The example used thus far in this section is another appropriate application of regression. Insurance rates can be individualized based on life-expectancy predictions. In addition to its practical applications, regression can also be used in testing theories. A regression equation could be devised based on a theoretical model, and the model's predictions could be confirmed or disconfirmed by the results of an experiment.

Statistical Control

Regression analysis can be used to adjust statistically for the effects of a confounding variable. For instance, if you are studying factors that affect vocabulary size (VS) in school-aged children, age can be a *confounding* variable. Across ages 6 through 14, for instance, VS will be highly correlated with height (and anything else that changes steadily with age). You can try to keep age constant within your study, but if VS varies linearly with age, you can control for the effects of age by using it to predict VS. An age-adjusted VS can be created by subtracting each child's VS from the VS predicted for his or her age. The result is a residual that should not be correlated with age; it should indicate whether a child's VS is high or low for his or her age. This procedure is referred to as "partialing out" the effects of age, and it will be discussed in greater detail in the context of multiple regression in Chapter 17 and again in the context of the analysis of covariance in Chapter 18.

Regression with Manipulated Variables

So far I have been describing regression in the situation in which neither of the variables is being controlled by the experimenter. Although it is much less common, regression can also be used to analyze the results of a genuine experiment in which participants are randomly assigned to different values of the X variable. For instance, prospective psychotherapy patients could be randomly assigned to one, two, three, four, or five sessions per week (X variable), and then after a year of such therapy the patients could be tested for severity of neurotic symptoms (Y variable). If the regression equation had a slope significantly different from zero, there would be evidence that the

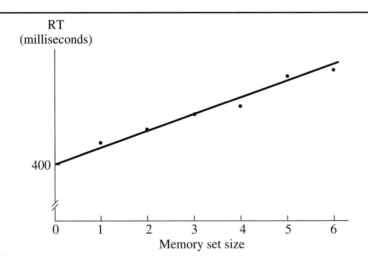

Figure 10.6

Regression Line for Predicting Reaction Time from Size of Memory Set number of sessions per week makes a difference in the effectiveness of psychotherapy. However, you might want to check for curvilinear effects, as well as linear ones; this can be handled fairly easily in terms of trend components, which are explained in Chapter 13, Section C.

For an experimental example in which the use of linear regression is especially appropriate, I turn to a classic experiment in cognitive psychology conducted by Saul Sternberg (1966). The mental task he assigned to his participants was to keep a small list of digits in mind and then answer as quickly as possible whether a test digit was included in that list. For instance, if one of the memory lists was 1, 3, 7, 8 and the test item were 7, the answer would be Yes, but if the test item were 4, the answer would be No. For half the test items the correct answer was No. The memory lists ranged in length from one to six digits, and, as you might expect, reaction time (RT) was longer for the longer lists because participants had to scan more digits in their minds. In fact, if you plot the average RT for each length list for the No trials (as Sternberg did), you can see that the relation is linear (see Figure 10.6, which is derived from Sternberg's published data). The slope of the line is about 38, which means that increasing the memory list by one digit lengthens RT by 38 milliseconds (ms). The slope of the regression line provides an estimate of how long it takes to scan each digit in the memory list. The Y intercept of the regression line is also meaningful. It can be thought of as a kind of "overhead"; it depends on the speed of motor commands, how quickly the test item can be recognized, and other fixed amounts that are not related to the size of the memory list.

The Sternberg regression model has practical applications in the exploration of cognitive functioning. For instance, one study found that the intake of marijuana (as compared with a placebo) increased the Y intercept of the regression line but did not alter the slope. There seemed to be a general slowing of the motor response but no slowing of the rate at which the digits were being mentally scanned. On the other hand, developmentally disabled subjects exhibited a steeper regression slope (indicating a slower rate of mental scanning) without much difference in the Y intercept. The regression approach allows the researcher to separate factors that change with the manipulated variable from factors that remain relatively constant.

1. If you are working with sample statistics, you can find the slope of the regression line by dividing the unbiased standard deviation for the Y variable by the unbiased standard deviation for the X variable and then multiplying by the correlation coefficient (you'll get the same value you would get if you had used the biased standard deviations); you can also substitute your sample means for population means in the formula for the Y intercept.

2. The formula for the variance around the regression line (which is the population variance of Y times $1 - r^2$) is biased if you are using sample data and are interested in extrapolating to the population. Even when using the unbiased variance of Y, there is a remaining bias that should be corrected by the following factor: $(N - 1)/(N - 2)$. The standard error of the estimate is just the square root of the variance around the regression line.

3. The width of the confidence interval around your prediction is based on the product of the standard error of the estimate and the appropriate critical value, but an additional factor is required to account for the

B

SUMMARY

fact that smaller samples give less accurate values for the regression line. The additional factor is reduced, and approaches 1.0, as the sample size increases.

4. The confidence interval calculated for a prediction is accurate only when the following assumptions are met:
    a. Random sampling
    b. A linear relationship between the two variables in the population
    c. Normal distributions for Y at each value of X
    d.   Homoscedasticity (i.e., the same variance of Y at each value of X).

5. The most common uses for linear regression are
    a.   Predicting future performance on some variable from the score on a different variable measured previously.
    b. Statistically removing the effects of a confounding or unwanted variable.
    c. Evaluating the linear relationship between the quantitative levels of a truly independent (i.e., manipulated) variable and a continuous dependent variable.
    d. Testing a theoretical model that predicts values for the slope and Y intercept of the regression line.

EXERCISES

*1. The data from Exercise 9A3, with the outlier eliminated, are reproduced in the following table.

Years (X)	Annual Salary (Y)
5	24
8	40
3	20
6	30
9	40
7	35
10	50
2	22

a. Find the regression equation for predicting an employee's annual salary from his or her number of years with the company.
b. What salary would you predict for someone who's been working at the same company for 4 years?
c. How many years would you have to work at the same company to have a predicted salary of $60,000 per year?

2. A statistics professor has devised a diagnostic quiz, which, when given during the first class, can accurately predict a student's performance on the final exam. So far data are available for 10 students, as follows:

Student	Quiz Score	Final Exam Score
1	5	82
2	8	80
3	3	75
4	1	60
5	10	92
6	6	85
7	7	86
8	4	70
9	2	42
10	6	78

a. Find the regression equation for predicting the final exam score from the quiz.
b. Find the (unbiased) standard error of the estimate.
c. What final exam score would be predicted for a student who scored a 9 on the quiz?
d. Find the 95% confidence interval for your prediction in part c.

*3. Refer to Exercise 9B5.
a. What proportion of the variance in orientation scores is accounted for by years of hospitalization?
b. Find the regression equation for predicting orientation scores from years of hospitalization.
c. What is the value of the Y intercept for the regression line? What is the meaning of the Y intercept in this problem?

d. How many years does someone have to be hospitalized before he or she is predicted to have an orientation score as low as 10?

4. For the data in Exercise 9A8,
   a. Find the standard error of the estimate for the population.
   b. Find the standard error of the estimate as based on data from a sample. Looking at the scatterplot you drew for the original exercise, why could this value (as well as the one in part a) be considered misleading (i.e., what regression assumption seems to be violated by these data)?
   c. What verbal SAT score would predict the average GPA? What general principle is being illustrated by this question?

*5. For the data in Exercise 9A6,
   a. Write the raw-score regression equation for predicting exam scores from phobia ratings.
   b. Given the limits of the phobia rating scale, what is the lowest exam score that can be predicted?
   c. What phobia rating would be required for a perfect exam score (i.e., 100) to be predicted? Is this possible? Explain.

6. A cognitive psychologist is interested in the relationship between spatial ability (e.g., ability to rotate objects mentally) and mathematical ability, so she measures 12 participants on both variables. The data appear in the following table:

Participant	Spatial Ability Score	Math Score
1	13	19
2	32	25
3	41	31
4	26	18
5	28	37
6	12	16
7	19	14
8	33	28
9	24	20
10	46	39
11	22	21
12	17	15

   a. Find the regression equation for predicting the math score from the spatial ability score.
   b. Find the regression equation for predicting the spatial ability score from the math score.
   c. According to your answer to part a, what math score is predicted from a spatial ability score of 20?

   d. According to your answer to part b, what spatial ability score is predicted from a math score of 20?

*7. For the data in Exercise 9B9,
   a. Use the correlation coefficient you calculated in part a of Exercise 9B9 to find the regression equation for predicting serum cholesterol levels from the number of hours exercised each week.
   b. Find the regression slope using the raw-score formula. Did you obtain the same value for b_{YX} as in part a?
   c. What cholesterol level would you predict for someone who does not exercise at all?
   d. What cholesterol level would you predict for someone who exercises 14 hours per week?
   e. Find the 99% confidence interval for your prediction in part d.

8. For the data in Exercise 9B10,
   a. Use the correlation coefficient calculated in Exercise 9B10 to find the regression equation for predicting reaction time from the number of letters in the string.
   b. How many milliseconds are added to the reaction time for each letter added to the string?
   c. What reaction time would you predict for a string with seven letters?
   d. Find the 95% confidence interval for your prediction in part c.

*9. If you calculate the correlation between shoe size and reading level in a group of elementary school children, the correlation will turn out to be quite large, provided that you have a large range of ages in your sample. The fact that each variable is correlated with age means that they will be somewhat correlated with each other. The following table illustrates this point. Shoe size is measured in inches, for this example, reading level is by grade (4.0 is average for the fourth grade), and age is measured in years.

Child	Shoe Size	Reading Level	Age
1	5.2	1.7	5
2	4.7	1.5	6
3	7.0	2.7	7
4	5.8	3.1	8
5	7.2	3.9	9
6	6.9	4.5	10
7	7.7	5.1	11
8	8.0	7.4	12

   a. Find the regression equation for predicting shoe size from age.

b. Find the regression equation for predicting reading level from age.

c. Use the equations from parts a and b to make shoe size and reading level predictions for each child. Subtract each prediction from its actual value to find the residual.

*10.a. Calculate Pearson's *r* for shoe size and reading level using the data from Exercise 9.

b. Calculate Pearson's *r* for the two sets of residuals you found in part c of Exercise 9.

c. Compare your answer in part b with your answer to part a. The correlation in part b is the partial correlation between shoe size and reading level after the confounding effect of age has been removed from each variable (see Chapter 17 for a much easier way to obtain partial correlations).

The Point-Biserial Correlation Coefficient

It is certainly not obvious that a two-group experiment can be analyzed by linear regression; however, not only is it possible, it is instructive to see how it works. As an example of a two-group experiment, suppose that participants in one group receive a caffeine pill, whereas participants in the other receive a placebo. All participants are then measured on a simulated truck-driving task. The results of this experiment can be displayed on a scatterplot, although it is a rather strange scatterplot because the X variable (drug condition) has only two values (see Figure 10.7). Drawing the regression line in this case is easy. The regression line passes through the mean of the Y values at each level of the X variable; that is, the regression line connects the means of the two groups, as shown in Figure 10.7. The group means serve as the predictions.

Remember that a prediction based on the regression line is better than predicting the mean of Y for everyone. The same is true for our two-group example. If we didn't know which group each participant was in, we would have to use the overall mean (i.e., the mean of all participants combined, regardless of group—the mean of the two group means when the groups are the same size) as the prediction for each participant. Therefore, we can divide each participant's score into an explained (i.e., predicted) and an unexplained (deviation from prediction) portion. We focus on Ms. Y, who has the highest score in the caffeine group. The part labeled $Y - Y'$ in Figure 10.7 is unexplained; we do not know why she scored so much better

---

Figure 10.7

Regression in the Case of Two Distinct Groups

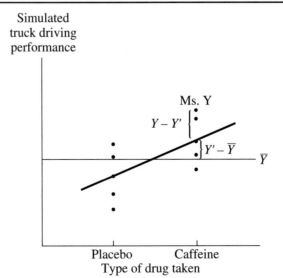

than the other participants in her group, all of whom had the same dosage of caffeine. The part labeled $Y' - \overline{Y}$ is the difference between using her own group mean to predict Ms. Y's score and using the mean of all participants to predict it. This part is explained; we could expect Ms. Y to perform better than the overall mean because she had taken a caffeine pill. Together $Y - Y'$ and $Y' - \overline{Y}$ add up to $Y - \overline{Y}$, the total deviation.

If all the unexplained pieces are squared and averaged, we get the variance of the estimate, but in this case the variance of the estimate is based on the variance around each group mean. In the two-group example, $s^2_{\text{est } y}$ is just the average of s^2_{caffeine} and s^2_{placebo}. (It would have to be a weighted average—a pooled variance—if the groups were not equal in size.) Clearly, the variances around each group mean are smaller than the variance of all participants around the overall mean (i.e., the total variance) would be. This reduction in variance is equal to the variance of the explained pieces. From Section A you know that the ratio of the explained variance to the total variance is equal to r^2. But what is r for this strange type of regression? If you divide the explained variance by the total variance and then take the square root in a two-group problem, the result is a correlation coefficient that is often called the *point-biserial r* (symbolized as r_{pb}).

Calculating r_{pb}

Strange as it may seem, we could have calculated Pearson's r directly for our two-group example. It is strange because to calculate r, each participant must have both an X and a Y value. Each participant has a Y value—his or her simulated truck-driving score—but what is the X value for each participant? The solution is to assign the same arbitrary number to all of the participants in the caffeine group and a different number to all of the placebo participants. The simplest way is to use the numbers 0 and 1 (or 1 and 2); the X value for all participants in the first group is 0, and for the second group it is 1. (The designations "first" and "second" are totally arbitrary.) Once each participant has both an X and a Y value, any of the formulas for Pearson's r can be used. However, the sign of r_{pb} is not important, because it depends on which group has the higher X value, and that is chosen arbitrarily. Therefore, r_{pb} can always be reported as a positive number as long as the means are given, or it is otherwise made clear in which direction the effect is going.

As a more concrete example, imagine that you have a sample of 10 men and 10 women, and you have measured everyone's height in inches. If you assign a 1 to the women and a 2 to the men and then calculate r for the 20 pairs of numbers (e.g., 1, 66; 1, 62; 2, 68; 2, 71 . . .), the resulting positive correlation would tell you that as the gender number goes up (from 1 to 2), height tends to go up, as well. Of course, if you had assigned a 1 to the men and a 2 to the women, your correlation would be the same size, but negative. However, whereas the sign of the correlation is not meaningful, the size of the correlation tells us something about how separated men and women are with respect to height. I will return to this example near the end of this section.

The repetition of the X values in the two-group case results in a simplified computational formula for r_{pb}, but it has become so easy to calculate a correlation with an electronic calculator or computer that I will not bother to show that formula here. Moreover, the specialized formula tends to obscure the fact that r_{pb} is like any other Pearson's r; the name *point-biserial r* is only to remind you that the correlation is being used in a special situation, in that one of the variables has only two values, chosen arbitrarily. Probably the most useful function of the point-biserial r is to provide supplementary information after a two-group t test has been conducted. I turn to this point next.

Deriving r_{pb} from a t Value

For a two-group experiment, you could calculate r_{pb} as an alternative to the t value and then test r_{pb} for statistical significance. To test r_{pb} you could use Table A.5 in the Appendix or Formula 9.6, which gives a t value for testing Pearson's r. One variation of Formula 9.6 is shown next, because it will provide a convenient stepping stone to a formula for r_{pb}:

$$t = \frac{r}{\sqrt{\dfrac{1 - r^2}{N - 2}}}$$

In the case of the two-group experiment, N is the total number of individuals in both groups: $n_1 + n_2$. Because $n_1 + n_2 - 2$ is the df for the two-group case, the preceding formula can be rewritten for r_{pb} as follows:

$$t = \frac{r_{pb}}{\sqrt{\dfrac{1 - r_{pb}^2}{df}}} \qquad \text{\textbf{Formula 10.12}}$$

The important connection is that the t value for testing the significance of r_{pb} is the same t value you would get by using the pooled-variance t test formula. So, instead of using the pooled-variance t test formula, you could arrive at the same result by calculating r_{pb} and then finding the t value by Formula 10.12 to test it for significance. However, this would probably not be any easier. What is more interesting is that you can find the pooled-variance t test first, plug it into the preceding formula, and solve for r_{pb}. You would probably want to find the t value anyway to test for significance, and r_{pb} is easily found using that t value. For convenience the preceding formula can be solved for r_{pb} ahead of time, yielding the very instructive and useful Formula 10.13:

$$r_{pb} = \sqrt{\frac{t^2}{t^2 + df}} \qquad \text{\textbf{Formula 10.13}}$$

Interpreting r_{pb}

The reason for spending this whole section on the point-biserial r is that this statistic provides information that the t value alone does not. Unfortunately, the size of the t value by itself tells you nothing about the effect size in your samples. As Chapters 7 and 8 emphasized, a large t value for a two-group experiment does not imply that the difference of the means was large or even that this difference was large compared to the standard deviations of the groups. There could be almost a total overlap between the two samples, and yet the t value could be very large and highly significant—as long as the df were very large. But r_{pb} *does* tell you something about the overlap of the two samples in your experiment; r_{pb}^2 tells you the proportion by which the variance is reduced by knowing which group each person is in (as compared to the variance around the overall mean). The point-biserial r tells you the *strength of association* between group membership and the dependent variable—that is, the strength of the tendency for scores in one group to be consistently higher than scores in the other group. If $r_{pb} = .9$, you know that the scores of the two groups are well separated; if $r_{pb} = .1$, you can expect a good deal of overlap. What r_{pb} does not tell you is anything about statistical significance. Even an r_{pb} as high as .9 would fail to reach significance if there are only two people in each group, and $r_{pb} = .1$ *can* be significant, but only if there are at least about 200 people per group. That is why the t value is also needed—to test for statistical significance.

Looking at Formula 10.13, you can see the relationship between t and r_{pb}. Suppose that the t value for a two-group experiment is 4 and the df is 16:

$$r_{pb} = \sqrt{\frac{4^2}{4^2 + 16}} = \sqrt{\frac{16}{32}} = \sqrt{.5} = .71$$

The point-biserial r is quite high, indicating a strong differentiation of scores between the two groups. On the other hand, suppose that t is once again 4, but for a much larger experiment in which df = 84:

$$r_{pb} = \sqrt{\frac{4^2}{4^2 + 84}} = \sqrt{\frac{16}{100}} = \sqrt{.16} = .4$$

The strength of association is still moderately high but considerably less than in the previous experiment. Finally, if $t = 4$ and df = 1,000,

$$r_{pb} = \sqrt{\frac{4^2}{4^2 + 1,000}} = \sqrt{\frac{16}{1,016}} = \sqrt{.016} = .125$$

Now the strength of association is rather low and unimpressive. A t of 4.0 indicates a strong effect in a small experiment, but not in a very large one.

Strength of Association in the Population (Omega Squared)

In its relationship to the t value, r_{pb} may remind you of g, the effect size in the sample (discussed in Chapter 8). There is a connection. Instead of r_{pb}, you could use $(\overline{X}_1 - \overline{X}_2)/s_p$ to describe the strength of association. However, sometimes a correlational measure is preferred, especially because r_{pb}^2 indicates directly the proportion of variance explained by group membership. Because they are alternative measures of effect size, it should not be surprising that there is a direct relationship between r_{pb} and g (or, more simply, between r_{pb}^2 and g^2), as shown in the following formula:

$$r_{pb}^2 = \frac{g^2}{g^2 + \dfrac{N_T(N_T - 2)}{n_1 n_2}}$$

where N_T equals $n_1 + n_2$. It is not obvious from looking at the preceding formula, but the term being added to g^2 in the denominator approaches 4.0 as the sample sizes get large if the two sample sizes are not very different. Fortunately, you would never use this formula for calcualtion, as it is so much easier to find r_{pb}^2 by dividing t^2 by the sum of t^2 and df. The preceding formula becomes more instructive to look at when the two samples are the same size.

$$r_{pb}^2 = \frac{g^2}{g^2 + 4\left(\dfrac{n - 1}{n}\right)}$$

Formula 10.14

where n is the size of each sample. It is easy to see that when the sample sizes are equal and fairly large, the denominator of the formula for r_{pb}^2 approaches $g^2 + 4$, and the actual sample size then has little effect on the relationship between r_{pb} and g.

When you look at r_{pb}^2 in the population, the term depending on sample size disappears entirely (assuming, of course, that the poulations representing the two conditions being compared are the same size—if they are not assumed to be infinite). Although you would have good reason to expect r_{pb}^2 to be symbolized as ρ_{pb}^2 when computed for an entire population, the term *omega squared* (symbolized as ω^2) has become popular instead to stand for the same quantity. As expected, g^2 becomes $\mathbf{d}^2$, as N_T approaches infinity, and the proportion of variance accounted for in the population (i.e., ω^2) bears a very simple relation to $\mathbf{d}^2$ (i.e., the population effect size squared), as shown in Formula 10.15:

$$\omega^2 = \frac{\mathbf{d}^2}{\mathbf{d}^2 + 4} \qquad \qquad \textbf{Formula 10.15}$$

Note that $\mathbf{d}$ can take on any value, but ω^2, like r_{pb}^2, is a proportion that cannot exceed 1.0. An unusually high $\mathbf{d}$ of 4 corresponds to

$$\omega^2 = \frac{4^2}{4^2 + 4} = \frac{16}{20} = .8$$

This means that the r_{pb} for a particular sample should come out to be somewhere near $\sqrt{.8}$, which is about .9 when $\mathbf{d}$ in the population is 4. A small effect size, $\mathbf{d} = .2$, corresponds to

$$\omega^2 = \frac{.2^2}{.2^2 + 4} = \frac{.04}{4.04} = .0099$$

Only about 1% of the variance in the population is accounted for by group membership, and the r_{pb} for a particular sample would be expected to be around $\sqrt{.01} = 1$.

If you wished to estimate ω^2 from r_{pb}, you could square r_{pb}, but the estimate would be somewhat biased. A better, though not perfect, estimate of ω^2 uses the t value corresponding to r_{pb}, and is given by Formula 10.16:

$$\text{est } \omega^2 = \frac{t^2 - 1}{t^2 + df + 1} \qquad \qquad \textbf{Formula 10.16}$$

Notice that Formula 10.16 is just a slight modification of the formula you would get by squaring Formula 10.13.

The quantity known as ω^2 gives us an alternative to $\mathbf{d}$ for describing population differences. For instance, in Chapter 8, I suggested that $\mathbf{d}$ for the difference in height between men and women was about $(69 - 65)/3 = 4/3 = 1.33$. For this $\mathbf{d}$, the corresponding ω^2 (using Formula 10.15) is

$$\omega^2 = \frac{1.33^2}{1.33^2 + 4} = \frac{1.78}{4.78} = .37$$

Therefore, we can say that about 37% of the variance in height among adults is accounted for by gender. (The square root of .37, which is about .61, gives us a rough estimate of how large r_{pb} is likely to be for equal-sized samples of men and women.) One important advantage of ω^2 is that whereas $\mathbf{d}$ is a measure that is specific to a two-group comparison, ω^2 can be used in many situations, including correlations and multigroup comparisons (the latter use of ω^2 will be described in Chapter 12).

Biserial *r*

The point-biserial *r* should only be used when the two groups represent a true dichotomy, such as male and female. On the other hand, in some studies two groups are formed rather arbitrarily according to differences on some continuous variable. For instance, a group of tall people and a group of short people can be compared on some dependent variable, such as self-esteem. If the variable used for dividing the two groups follows something like a normal distribution, as height does, assigning 0s and 1s to the two groups (usually according to a median split) and applying the Pearson formula will lead to a misleading result. It is more accurate to use a similar, but adjusted, correlation coefficient, known simply as the *biserial r*. Because it is so rarely used, I will not complicate matters any further by presenting the formula for the biserial *r* in this text. Please note that you generally lose a good deal of information by using a continuous variable as the basis for creating two distinct groups. It is usually more powerful to compute correlations with the original continuous variable than the grouping variable that is arbitrarily formed from it.

SUMMARY

1. If the results of a two-group experiment are described in the form of linear regression, the square root of the ratio of the explained variance to the total variance yields a correlation coefficient called the *point-biserial r*, symbolized as r_{pb}. The point-biserial *r* can be calculated directly, using any formula for Pearson's *r*, by assigning *X* values to all the participants, such that the participants in one group are all assigned the same value (e.g., 0) and the participants of the other group are all assigned some different value (e.g., 1).

2. The point-biserial *r* can also be found from the *t* value for a pooled-variance test. For a given *t* value, the larger the degrees of freedom, the smaller is r_{pb}. If the df stay the same, r_{pb} goes up as the *t* value increases.

3. The point-biserial *r* provides an alternative to *g* (the effect size in a sample) for assessing the strength of association between the independent and dependent variables in a two-group experiment.

4. A slight modification of the formula for r_{pb} squared can be used to estimate omega squared (ω^2), the proportion of variance accounted for in the dependent variable by group membership in the population. There is a simple relationship between ω^2 and the square of **d**, the population effect size.

5. The point-biserial *r* should only be used when dealing with two groups that represent a true dichotomy (e.g., male and female). If the two groups represent opposite sides of an underlying normal distribution (e.g., tall people and short people), a different correlation coefficient, the biserial *r*, should be used. However, it is usually better to work with the original continuous variable than to discard a considerable amount of information by forming two arbitrary groups.

EXERCISES

*1. a. Calculate r_{pb} for the data in Exercise 7B7, using any of the formulas for Pearson's *r*.

(*Note*: It is convenient to assign the values 0 and 1 to the two groups.)

 b. Use Formula 9.6 to find the *t* value to test the correlation coefficient you found in part a.

 c. Compare the *t* value you calculated in part b to the *t* value found in Exercise 7B7, part a.

2. a. Based on your answer to part a of Exercise 1, determine the proportion of variance in minutes waited that is accounted for by type of instructor.

b. Estimate the proportion of variance accounted for in the population (i.e., ω^2).

*3. If t for an experiment with two equal-sized groups equals 10, what proportion of variance in the dependent variable is accounted for
  a. When the sample size is 20?
  b. When the sample size is 200?
  c. What are your estimates for omega squared for parts a and b?

4. Suppose that you are planning a two-group experiment. How high would your t value have to be for half the variance of your dependent variable to be accounted for
  a. If you use 25 participants per group?
  b. If you use 81 participants per group?

*5. According to the guidelines suggested by Cohen (1988), $d = .8$ is a large effect size; any effect size much larger would probably be too obvious to require an experiment.

  a. What proportion of population variance is accounted for when **d** reaches this value?
  b. What proportion of population variance is accounted for when **d** is moderate in size, i.e., **d** = .5?
  c. How high does **d** have to be for half of the population variance to be accounted for?

6. a. Calculate g for the data in Exercise 7B3 using Formula 8.5 and the t value you found for that exercise. Then, use Formula 10.14 to calculate r_{pb}^2 for that problem. Next, find r_{pb}^2 directly from the t value using the square of Formula 10.13. Did you get the same proportion? Finally, estimate the proportion of variance accounted for in the population.
  b. Repeat part a for the data in Exercise 7B2.

KEY FORMULAS

Regression equation for predicting Y from X (in terms of z scores):

$$z_Y = r z_X$$

Formula 10.1

Regression equation for predicting Y from X (in terms of population parameters):

$$Y' = \frac{\sigma_Y}{\sigma_X} r(X - \mu_X) + \mu_Y$$

Formula 10.2

The slope of the regression line for predicting Y from X in the population:

$$b_{YX} = \frac{\sigma_Y}{\sigma_X} r$$

Formula 10.3A

The slope of the regression line, based on sample data:

$$b_{YX} = \frac{s_Y}{s_X} r$$

Formula 10.3B

The Y intercept of the regression line for predicting Y from X in the population:

$$a_{YX} = \mu_Y - b_{YX} \mu_X$$

Formula 10.4A

The Y intercept of the regression line, based on sample data:

$$a_{YX} = \overline{Y} - b_{YX} \overline{X}$$

Formula 10.4B

The regression equation for predicting Y from X, as a function of the slope and Y intercept of the regression line:

$$Y' = b_{YX} X + a_{YX}$$

Formula 10.5

The population variance of the estimate for Y (definitional formula, not convenient for calculating):

$$\sigma^2_{\text{est } Y} = \frac{\sum(Y - Y')^2}{N}$$

Formula 10.6

The coefficient of nondetermination:

$$k^2 = \frac{\sigma^2_{\text{est } Y}}{\sigma^2_Y} = 1 - r^2$$

Formula 10.7A

The population variance of the estimate for Y; convenient for calculating. (*Note*: This formula is a rearrangement of Formula 10.7A):

$$\sigma^2_{\text{est } Y} = \sigma^2_Y(1 - r^2)$$

Formula 10.7B

Variance of the estimate for Y (based on the *unbiased* estimate of total variance):

$$s^2_{\text{est } Y} = \left(\frac{N-1}{N-2}\right)s^2_Y(1 - r^2)$$

Formula 10.8A

Sample standard error of the estimate for Y (This formula is just the square root of Formula 10.8A):

$$s_{\text{est } Y} = s_Y\sqrt{\frac{N-1}{N-2}(1 - r^2)}$$

Formula 10.8B

The standard error of the estimate for Y in the population (it is the square root of Formula 10.7B):

$$\sigma_{\text{est } Y} = \sigma_Y\sqrt{1 - r^2}$$

Formula 10.9

The confidence interval for a prediction based on sample data:

$$Y' \pm t_{\text{crit}}s_{\text{est } Y}\sqrt{1 + \frac{1}{N} + \frac{(X - \overline{X})^2}{(N-1)s^2_X}}$$

Formula 10.10

The slope of the regression line for predicting Y from X (raw-score version; not recommended for calculations because errors are difficult to locate):

$$b_{YX} = \frac{N(\sum XY) - (\sum X)(\sum Y)}{N\sum X^2 - (\sum X)^2}$$

Formula 10.11

The t test for determining the significance of r_{pb} (equivalent to Formula 9.6):

$$t = \frac{r_{\text{pb}}}{\sqrt{\dfrac{1 - r^2_{\text{pb}}}{\text{df}}}}$$

Formula 10.12

The point-biserial correlation, based on the two-group (pooled-variances) t value and the degrees of freedom:

$$r_{\text{pb}} = \sqrt{\frac{t^2}{t^2 + \text{df}}}$$

Formula 10.13

Point-biserial r squared as a function of g^2 and n:

$$r^2_{pb} = \frac{g^2}{g^2 + 4\left(\dfrac{n-1}{n}\right)}$$

Formula 10.14

Omega squared (the proportion of variance accounted for in the population) expressed in terms of **d** (the population effect size):

$$\omega^2 = \frac{\mathbf{d}^2}{\mathbf{d}^2 + 4}$$

Formula 10.15

Estimate of omega squared, in terms of the two-group t value and the degrees of freedom:

$$\text{est } \omega^2 = \frac{t^2 - 1}{t^2 + df + 1}$$

Formula 10.16

THE MATCHED t TEST

You will need to use the following from previous chapters:

Symbols:
$\bar{X}$: mean of a sample
s: unbiased standard deviation of a sample
r: Pearson's correlation coefficient

Formulas:
Formula 3.15B: Computational formula for s
Formula 6.3: The t test for one sample
Formula 7.8: The t test for two equal-sized samples

Concepts:
The t distribution
Linear correlation

11
Chapter

$\mathcal{A}$

CONCEPTUAL FOUNDATION

The previous two chapters dealt with the situation in which each individual has been measured on two variables. In the simplest case, all participants are measured twice by the same instrument with some time in-between to assess test-retest reliability. A large positive correlation coefficient indicates that whatever is being measured about the person is relatively stable over time (i.e., high scorers tend to remain high scorers, and low scorers are still low scorers). However, the correlation would be just as high if all those involved tended to score a few points higher the second time (or all tended to score lower). This would make the mean score for the second testing higher (or lower), and this difference in means can be interesting in itself. The passage of time alone can make a difference—for example, people may tend to become happier (or less happy) as they get older—or the experimenter may apply some treatment between the two measurements—for example, patients may exhibit lower depression scores after brief psychotherapy. If the focus is on the difference in means between the two testings, a t test, rather than a correlation coefficient, is needed to assess statistical significance. But the t test for independent groups (described in Chapter 7) would not be the optimal choice. The correlation between the two sets of scores can be used to advantage by a special kind of t test, called the *repeated-measures* or *matched t test*, which is the topic of this chapter.

Before-After Design

Imagine that you have read about a new method for losing weight that involves no diet or real exercise—only periods of vividly imagining strenuous exercise. Thinking this is too good to be true, you get five friends to agree to follow a rigorous regimen of imagined exercise for 2 weeks. They agree to be weighed both before and after the 2-week "treatment." The data for this imaginary experiment are shown in Table 11.1.

Although not much weight was lost, the mean weight after the program was 1.6 pounds less than before, which suggests that the program was at least a little bit effective. However, if we try to publish these imaginary results, we shall have to deal with our imaginary adversary, the skeptical Dr. Null. There are quite a few important objections he could raise about the design of this particular experiment (e.g., the sample wasn't random),

Table 11.1	
Before	**After**
230	228
110	109
170	168
130	129
150	148
$\bar{X}_b = 158$	$\bar{X}_a = 156.4$
$s_b = 46.043$	
$s_a = 45.632$	

but even if it were perfectly designed, he could always suggest that our treatment is totally ineffective. In other words, he could bring up his famous null hypothesis and propose that we were just lucky this time that all of the participants were a bit lighter after the 2 weeks than before. After all, the average person's weight fluctuates a bit over time, and there's no reason that all five of our participants could not fluctuate in the same direction (and about the same amount) at the same time.

Having not read this chapter yet, you might be tempted to answer Dr. Null with the methods of Chapter 7 by conducting a *t* test for two independent groups, hoping to show that $\overline{X}_{\text{after}}$ is significantly less than $\overline{X}_{\text{before}}$. If you were to conduct an independent-groups *t* test on the data in Table 11.1, it would be easiest to use Formula 7.8 because the sample size before is the same as the sample size after. I will apply Formula 7.8 to the data in the table to show what happens:

$$t = \frac{(\overline{X}_1 - \overline{X}_2)}{\sqrt{\dfrac{s_1^2 + s_2^2}{n}}} = \frac{158 - 156.4}{\sqrt{\dfrac{46.04^2 + 45.63^2}{5}}} + \frac{1.6}{\sqrt{840.5}} = \frac{1.6}{29} = .055$$

We don't have to look up the critical *t* value to know that *t* = .055 is much too small to reject the null hypothesis, regardless of the alpha level. It is also easy to see why the calculated *t* value is so low. The average amount of weight loss is very small compared to the variation from person to person.

If you have the feeling that the preceding *t* test does not do justice to the data, you are right. When Dr. Null says that the before-after differences are random, you might be inclined to point out that although the weight losses are small, they are remarkably consistent; all five participants lost about the same amount of weight. This is really not very likely to happen by accident. To demonstrate the consistency in our data, we need a different kind of *t* test—one that is based on the before-after difference scores and is sensitive to their similarities. That is what the matched *t* test is all about.

The Direct-Difference Method

The procedure described in this section requires the addition of a column of difference scores to Table 11.1, as shown in Table 11.2.

The simplest way to calculate the matched *t* test is to deal only with the difference scores, using a procedure called the *direct-difference method*. To understand the logic of the direct-difference method, it is important to understand the null hypothesis that applies to this case. If the weight-loss program were totally ineffective, what could we expect of the difference scores? Because there would be no reason to expect the difference scores to be more positive than negative, over the long run the negative differences would be expected to balance out the positive differences. Thus, the null hypothesis would predict the mean of the differences to be zero. To reject this null hypothesis, the *t* test must show that the mean of the difference scores $(\overline{D})$ for our sample is so far from zero that the probability of beating $\overline{D}$ when the null hypothesis is true is too low (i.e., less than alpha) to worry about. To test $\overline{D}$ against zero requires nothing more than the one-group *t* test you learned about in Chapter 6.

To review the one-group *t* test, consider again Formula 6.3:

$$t = \frac{\overline{X} - \mu_0}{\dfrac{s}{\sqrt{n}}}$$

Formula 6.3

Table 11.2

Before	After	Difference
230	228	2
110	109	1
170	168	2
130	129	1
150	148	2

$\overline{D} = 1.6$

$s_D = .5477$

In the present case, $\overline{X}$ (the sample mean) is the mean of the difference scores, and we can therefore relabel it as $\overline{D}$. For the matched t test, the population mean predicted by the null hypothesis, μ_0, is the expected mean of the difference scores, so it can be called μ_D. Although it is possible to hypothesize some value other than zero for the mean of the difference scores, it is so rarely done that I will only consider situations in which $\mu_D = 0$. The unbiased standard deviation of the sample, s, is now the standard deviation of the difference scores, which can therefore be symbolized by adding a subscript as follows: s_D. Finally, Formula 6.3 can be rewritten in a form that is convenient to use for the matched t test; this revised form will be referred to as Formula 11.1:

$$t = \frac{\overline{D}}{\frac{s_D}{\sqrt{n}}}$$

Formula 11.1

where n is the number of difference scores. (I'm switching to a lower-case n to remind you that this is also the number of scores in *each* condition.) Let us apply this formula to our weight-loss problem using $\overline{D}$ and s_D from Table 11.2 and compare the result to that of our independent-groups t test:

$$t = \frac{1.6}{\frac{.5477}{\sqrt{5}}} = \frac{1.6}{\frac{.5477}{2.236}} = \frac{1.6}{.245} = 6.53$$

Again, we don't need to look up the critical t value. If you are somewhat familiar with Table A.2, you know that $t = 6.53$ is significant at the .05 level (two-tailed); the only time a calculated t value over 6 is not significant at the .05 level is in the extremely unusual case in which there is only one degree of freedom (i.e., two difference scores). Notice that the matched t test value is over 100 times larger than the t value from the independent groups test; this is because I have set up an extreme case to make a point. However, when the matched t test is appropriate, it almost always yields a value higher than would an independent-groups test; this will be easier to see when I present an alternative to the preceding direct-differences formula. Notice also that the numerators for both t tests are the same, 1.6. This is not a coincidence. The mean of the differences, $\overline{D}$, will always equal the difference of the means, $\overline{X}_1 - \overline{X}_2$ because it doesn't matter whether you take differences first and then average, or average first and then take the difference. It is in the denominator that the two types of t tests differ.

Because the matched t value in the previous example is statistically significant, we can reject the null hypothesis that the before and after measurements are really the same in the entire population; it seems that the imagined-exercise treatment may have some effectiveness. However, your intuition may tell you that it is difficult in this situation to be sure that it is the imagined exercise that is responsible for the before-after difference and not some confounding factor. The drawbacks of the before-after design will be discussed later in this section.

The Matched t Test as a Function of Linear Correlation

The degree to which the matched t value exceeds the independent-groups t value for the same data depends on how highly correlated the two samples (e.g., before and after) are. The reason the matched t value was more than 100 times larger than the independent-groups t value for the weight-loss example is that the before and after values are very highly correlated; in fact, $r = .99997$. To see how the correlation coefficient affects the value of the matched *t* test, let's look at a matched *t* test formula that gives the same answer as the direct-difference method (Formula 11.1), but is calculated in terms of Pearson's *r*. That formula will be designated Formula 11.2:

$$t = \frac{(\overline{X}_1 - \overline{X}_2)}{\sqrt{\dfrac{s_1^2 + s_2^2}{n} - \dfrac{2rs_1s_2}{n}}}$$ **Formula 11.2**

Note the resemblance between Formula 11.2 and Formula 7.8; the difference is that a term involving Pearson's *r* is subtracted in the denominator of the matched *t* test formula. However, when the two samples are *not* correlated (i.e., the groups are independent), $r = 0$, and the entire subtracted term in the denominator of Formula 11.2 becomes zero. So when $r = 0$, Formula 11.2 is identical to Formula 7.8, as it should be because the groups are not really matched in that case. As *r* becomes more positive, a larger amount is being subtracted in the denominator of the *t* test, making the denominator smaller. The smaller the denominator gets, the larger the *t* value gets, which means that if all other values remain equal, increasing the correlation (i.e., the matching) of the samples will increase the *t* value. On the other hand, if *r* turned out to be negative between the two samples, variance would actually be added to the denominator (but fortunately, when an attempt has been made to match the samples, there is very little danger of obtaining a negative correlation).

Comparing Formulas 11.1 and 11.2, you can see the relationship between the correlation of the two sets of scores and the variability of the difference scores. Because the numerators of the two formulas, $\overline{D}$ and $\overline{X}_1 - \overline{X}_2$, will always be equal (as mentioned above), it follows that the denominators of the two formulas must also be equal (because both formulas always produce the same *t* value). If the two denominators are equal, increasing the correlation (which decreases the denominator of Formula 11.2) must decrease the variability of the difference scores. This fits with the concept of correlation. If the same constant is added to (or subtracted from) all of the scores in the first set to get the scores in the second set (e.g., everyone loses the same amount of weight), the correlation will be perfect, and the variability of the difference scores will be zero. This would make the *t* value infinitely high, but this is extremely unlikely to happen in a real experiment. However, to the extent that participants tend to stay in the same relative position in the second set of scores as they did in the first, the correlation will be high, and the variability of the difference scores will be low.

The term being subtracted in the denominator of Formula 11.2 is a variance, and it can be thought of as an extraneous variance, which you are happy to eliminate. It is extraneous in the sense that it is the variance that the matched scores share—that they have in common due to general factors not related to the experimental treatment. In our example, most of the variance in the before scores reappears as variance in the after scores because heavy people tend to stay heavy, and light people tend to remain light. When this common variance is subtracted, the variance that remains in the denominator (the variance of the difference scores) is completely unrelated to the fact that people have very different weights at the start of the experiment; it is due to the fact that people react a bit differently to the treatment (i.e., everyone does not lose the same amount of weight).

Reduction in Degrees of Freedom

It may seem that any amount of positive correlation between two sets of scores will make the matched t test superior to the independent-samples t test. However, there is a relatively minor disadvantage to the matched t test that must be mentioned: The number of degrees of freedom is only half as large as for the independent-groups t test. For the independent t test for the weight-loss example, the df we would have used to look for the critical t was $n_1 + n_2 - 2 = 5 + 5 - 2 = 10 - 2 = 8$. For the matched t test the df is equal to $n - 1$, where n is the number of difference scores (this is also the same as the number of pairs of scores or the number of scores in each group or condition). The critical t would have been based on df $= n - 1 = 5 - 1 = 4$, which is only half as large as the df for the independent test. Because critical t gets higher when the df are reduced, the critical t is higher for the matched t test and therefore harder to beat. This disadvantage is usually more than offset by the increase in the calculated t due to the correlation of the two samples, unless the sample size is rather small *and* the correlation is fairly low. When the sample size is very large, the critical t will not increase noticeably even if the df are cut in half, and even a rather small positive correlation between the two sets of scores can be helpful.

You may have noticed that I have not mentioned using a z score formula for the matched design. I didn't bother because matched designs are rarely performed with very large samples. Of course, if you were dealing with hundreds of pairs of scores, you could simply substitute z for t in Formula 11.1 or 11.2. On the other hand, you are always correct using the t distribution regardless of sample size because the t distribution is approximated by the normal distribution ever more closely as the sample size increases.

Drawback of the Before-After Design

Although the results of a before-after t test are often statistically significant, you may feel uneasy about drawing conclusions from such a design. If you feel that something is missing, you are right. What is missing is a control group. Even though the weight loss in our example could not easily be attributed to chance factors, it is quite possible that the same effect could have occurred without the imagined exercise. Using a bogus diet pill that is really a placebo might have had just as large an effect. A well-designed experiment would include an appropriate control group, and sometimes more than one type of control group (e.g., one control group gets a placebo pill, and another is just measured twice without any experimental manipulation). However, even in a well-designed experiment, a before-after matched t test might be performed on each group. A significant t for the experimental group would not be conclusive in itself without comparison to a control group, but a *lack* of significance for the experimental group would indeed be disappointing. Moreover, a significant difference for the control group would alert you to the possibility of extraneous factors affecting your dependent variable.

Other Repeated-Measures Designs

The before-after experiment is just one version of a repeated-measures (or within-subject) design. It is possible, for instance, to measure each person twice on the same variable under different conditions within the same experimental session (practically simultaneously). Suppose that a researcher wants to test the hypothesis that people recall emotional words (i.e., words that tend to evoke affect, like *funeral* or *birthday*) more easily than they recall neutral words. Each participant is given a list of words to memorize, in which emotional and neutral words are mixed randomly. Based on a recall test, each participant receives two scores: one for emotional word recall and one for neutral word recall. Given that each participant has two scores, the researcher can find difference scores just as in the before-after example and can calculate the matched *t* test.

The researcher can expect that the emotional recall scores will be positively correlated with the neutral recall scores. That is, people who recall more emotional words than their fellow participants are also more likely to recall more neutral words; some people just have better memories, in general, than other people. This correlation will tend to increase the matched *t* value by decreasing the denominator. In addition to a reasonably high correlation, however, it is also important that recall scores be generally higher (or lower, if that is the direction hypothesized) for the emotional than for the neutral words. This latter effect is reflected in the numerator of the matched *t* test. As with any *t* test, too small a difference in the numerator can prevent a *t* value from being large enough to attain statistical significance, even if the denominator has been greatly reduced by matching.

The simultaneous repeated-measures design just described does not require a control group. However, there is always the potential danger of confounding variables. Before the researcher can confidently conclude that it is the emotional content of the words that leads to better recall (assuming the matched *t* test is significant and in the predicted direction), he or she must be sure that the emotional and neutral words do not differ from each other in other ways, such as familiarity, concreteness, imageability, and so forth. In general, simultaneous repeated measures are desirable as an efficient way to gather data. Unfortunately, there are many experimental situations in which it would not be possible to present different conditions simultaneously (or mixed together randomly in a single session). For instance, you might wish to compare performance on a tracking task in a cold room with performance in a hot room. In this case the conditions would have to be presented *successively*. However, it would not be fair if the cold condition (or the hot condition) were always presented first. Therefore, a *counterbalanced* design is used; this design is discussed further in Section B.

Matched-Pairs Design

Sometimes a researcher wants to compare two conditions, but it is not possible to test both conditions on the same person. For instance, it may be of interest to compare two very different methods for teaching long division to children to see which method results in better performance after a fixed amount of time. However, after a child has been taught by one method for 6 months, it would be misleading at best to then try to teach the child the same skill by a different method, especially because the child may have already learned the skill quite well. It would seem logical for the researcher to give up on the matched *t* test and its advantages and to resign herself to the fact that the matched *t* test is not appropriate for this type of situation. Although it is true that a repeated-measures design does not seem appropriate for this example, the advantage of a matched *t* test can still be gained by using a *matched-pairs* design. In fact, the term *matched t test* is derived from its use in analyzing this type of design.

The strategy of the matched-pairs design can be thought of in the following way. If it is not appropriate to use the same person twice, the next best thing is to find two people who are as similar as possible. Then each member of the pair is randomly assigned to one of the two different conditions. This is done with all of the pairs. After all of the measurements are made, the differences can be found for each pair, and Formula 11.1 can be used to find the matched t value. The similarities within each pair make it likely that there will be a high correlation for the two sets of scores. For some experiments, the ideal pairs are sets of identical twins because they have the same genetic makeup. This, of course, is not the easiest participant pool to find, and fortunately the use of twins is not critical for most psychology experiments. In the long-division experiment, the students can be matched into pairs based on their previous performance on arithmetic exams. It is not crucial that the two students in a pair be similar in every way as long as they are similar on whatever characteristic is relevant to the variable being measured in the study. The researcher hopes that the two students in each pair will attain similar performance on long division, ensuring a high correlation between the two teaching methods. However, it is also very important that one method consistently produce better performance. For instance, if method A works better than method B, it would be helpful (in terms of maximizing the matched t value) if the A member of each pair were to perform at least a little better than the B member and even more helpful if these differences were about the same for each pair.

Correlated or Dependent Samples

The t test based on Formula 11.1 (or Formula 11.2) can be called a *repeated-measures* t test, or a *matched* t test, because it is the appropriate statistical procedure for either design. For purposes of calculation it doesn't matter whether the pairs of scores represent two measurements of the same participant or measurements of two similar participants. In the preceding example, the sample of long-division performance scores for students taught with method B will likely be correlated with the sample of scores for students taught with method A because each score in one sample corresponds to a fairly similar score (the other member of the pair) in the other sample. Therefore, the matched t test is often called a t test for correlated samples (or sometimes just *related samples*). Another way of saying that two samples are correlated is to say that they are dependent. So the t test in this chapter is frequently referred to as a t test for two dependent samples. Because it is the means of the two dependent samples that are being compared, this t test can also be called a t test for the difference of two dependent means—in contrast to the t test for the difference of two independent means (or samples), which was the subject of Chapter 7.

When Not to Use the Matched t Test

When matching works—that is, when it produces a reasonably high positive correlation between the two samples—it can greatly raise the t value (compared to what would be obtained with independent samples) without the need to increase the number of participants in the study. Matching (or using the same participants twice, when appropriate) can be an economical way to attain a reasonable chance of statistical significance (i.e., to attain adequate power) while using a fairly small sample. However, in some situations in which it would be impossible to use the same participant twice, it is also not possible to match the participants. In such a situation, you must give up the increased power of the matched t test and settle for the t test for independent samples. An example of just such a situation follows.

A researcher wants to know whether young adults will more quickly come to the aid of a child or an adult when all they hear is a voice crying for help. To test this hypothesis, each participant is asked to sit in a waiting room while the "experimental" room is prepared. The experiment actually takes place in the waiting room because the researcher plays a recorded voice of someone crying for help so that the participant can hear it. The dependent variable is the amount of time that elapses between the start of the recording and the moment the participant opens the door from behind which the voice seems to be coming (the independent variable is the type of person on tape: adult or child). It should be obvious that having tested a particular participant with either the child or adult voice, the researcher could not run the same participant again using the other voice.

If you wished to obtain the power of a matched *t* test with this experiment, you would have to find some basis for matching participants into pairs; then one member of the pair would hear the child's voice, and the other would hear the adult's voice. But what characteristic could we use as a basis for matching? If we had access to personality questionnaires that the participants had previously filled out, we might match together individuals who were similar in traits of altruism, heroism, tendency to get involved with others, and so forth. However, it is not likely that we would have such information, nor could we be confident in using that information for matching unless there were reports of previous similar experiments to guide us. For the type of experiment just described, it would be very reasonable to give up on the matched design and simply perform a *t* test for independent groups. In the next section, I present an example for which the matched design is appropriate and demonstrate the use of the direct-difference method for calculating the matched *t* value.

SUMMARY

1. In an experiment in which each participant is measured twice (e.g., before and after some treatment) the variability of the *difference scores* will probably be less than the variability of either the before scores, or the after scores.
2. Because of this reduced variability, a one-group *t* test comparing the mean of the difference scores to zero will generally yield a higher *t* value than an independent *t* test comparing the means of the two separate sets of scores.
3. The *t* test on the difference scores, called the matched *t* test or the repeated-measures *t* test, as appropriate, relies on the consistency of the difference scores in its attempt to show that the mean of the difference scores is not likely to be merely a chance fluctuation from a mean of zero.
4. The matched *t* test can be expressed as a formula that contains a term that is subtracted in the denominator. The size of the subtracted term depends on the correlation between the two sets of scores: The higher (i.e., more positive) the correlation, the more that is subtracted (so the denominator gets smaller) and the larger the *t* value gets (all else staying equal).
5. A disadvantage of the matched *t* test is that the degrees of freedom are reduced by half, resulting in a higher critical *t* value to beat. However, if the matching is reasonably good, the increase in the calculated *t* will easily outweigh the increase in the critical *t*.
6. The before-after design is often inconclusive if there is no control group for comparison. However, there are other *repeated-measures* designs that involve simultaneous measurement or counterbalancing, in which a control group may not be needed (see Section B).

7. The difference scores in a matched t test can come from two measures on the same participant or from a *matched-pairs* design, in which participants matched for some particular similarity undergo different treatments. In either case, the differences can be analyzed in exactly the same way.

8. Another way of referring to the matched t test is as a t test for correlated (or dependent) samples (or means).

EXERCISES

1. For each of the following experimental designs, how many degrees of freedom are there, and how large is the appropriate critical t ($\alpha = .05$, two-tailed)?
   a. Twenty-five participants measured before and after some treatment.
   b. Two independent groups of 13 participants each.
   c. Two groups of 30 participants each, matched so that every participant in one group has a "twin" in the other group.
   d. Seventeen brother–sister pairs of participants.

*2. Can the depression of psychotherapy patients be reduced by treating them in a room painted in bright primary colors, as compared to a room with a more conservative look with wood paneling? Ten patients answered depression questionnaires after receiving therapy in a primary-colored room, and ten patients answered the same questionnaire after receiving therapy in a traditional room. Mean depression was lower in the colored room ($\overline{X}_{color} = 35$) than the traditional room ($\overline{X}_{trad} = 39$); the standard deviations were $s_{color} = 7$ and $s_{trad} = 5$, respectively.
   a. Calculate the t value for the test of two independent means (Formula 7.8).
   b. Is this t value significant at the .05 (two-tailed) level? (Make sure you base your critical t value on the appropriate degrees of freedom for this test.)

*3. Suppose that the patients in Exercise 2 had been matched in pairs, based on general depression level, before being assigned to groups.
   a. If the correlation were only .1, how high would the matched t value be?
   b. Is this matched t value significant at the .05 (two-tailed) level? Explain any discrepancy between this result and the decision you made in part b of Exercise 2.
   c. How high would the matched t value be if the correlation were .3?
   d. If the correlation were .5?

4. Exercise 7B3 described an experiment in which 12 students arbitrarily labeled "gifted" obtained a grade average of 87.2 with $s = 5.3$, as compared to 12 other students not so labeled, who had an average of 82.9, with $s = 4.4$. Suppose now that each gifted student was matched with a particular student in the other group and that the correlation between the two sets of scores was .4.
   a. Calculate the matched t and compare it to the t value you found in Exercise 7B3.
   b. Calculate the matched t if the correlation were .8 and compare that with the matched t you found in part a.

*5. Calculate the mean and unbiased standard deviation for each of the following sets of difference scores:
   a. $-6, +2, +3, 0, -1, -7, +3, -4, +2, +8$
   b. $+5, -11, +1, +9, +6, -2, 0, -2, +7$

6. Redo Exercise 7B2a as a matched t test, assuming that the correlation between the matched visual recall scores equals $+.2$.

7. a. Design an experiment for which it would be reasonable for the researcher to match the participants into pairs.
   b. Design an experiment in which it would be difficult to match participants into pairs.

*8. Suppose that the matched t value for a before-after experiment turns out to be 15.2. Which of the following can be concluded?
   a. The before and after scores must be highly correlated.
   b. A large number of participants must have been involved.
   c. The before and after means must be quite different (as compared to the standard deviation of the difference scores).
   d. The null hypothesis can be rejected at the .05 level.
   e. No conclusion is possible without more information.

B

To make the weight-loss example in Section A as simple as possible, I arranged the numbers so that all of the differences were in the same direction (i.e., everybody lost weight). In reality, the treatments we are usually interested in testing do not work on every individual, so some of the differences could go in the opposite direction. This makes calculation of the direct-difference formula a little trickier, as I will demonstrate in this section. For the following example I will use a matched-pairs design.

Let us suppose that the progressive Sunny Day elementary school wants to conduct its own experiment to compare two very different methods for teaching children to read. I will refer to these two methods as visual and phonic. The simplest way to conduct such a study would be to select two random samples of children who are ready to start learning to read, teach each sample with a different method, measure the reading ability of each child after 1 year, and conduct a *t* test for independent means to see if the groups differ significantly. However, as you learned in the previous section, we have a better chance of attaining statistical significance with a matched *t* test. Of course, a repeated-measures design is out of the question if we are interested in the initial acquisition of reading skills. So we are left to consider the matched-pairs design. It seems reasonable to suppose that a battery of tests measuring various prereading skills could be administered to all of the students and that we could find some composite measure for matching children that would tend to predict their ability to learn reading by either method. I will illustrate how the six steps of null hypothesis testing can be applied to this situation.

Step 1. State the Hypotheses

The research hypothesis that motivates this study is that the two methods differ in their effect on reading acquisition. The appropriate null hypothesis is that the population means representing the two methods are equal when measured on some appropriate variable (e.g., reading comprehension), $H_0: \mu_V = \mu_P$; or $H_0: \mu_V - \mu_P = 0$. In terms of difference scores, $H_0: \mu_D = 0$. The appropriate alternative hypothesis in this case is two-tailed. Although the researcher may expect a particular method to be superior, there would be no justification for ignoring results in the unexpected direction. The alternative hypothesis is expressed as $H_A: \mu_V \neq \mu_P$; or $H_A: \mu_V - \mu_P \neq 0$. In terms of difference scores, $H_A: \mu_D \neq 0$. A one-tailed version would be $H_A: \mu_D > 0$; or $H_A: \mu_D < 0$.

Step 2. Select the Statistical Test and the Significance Level

Because each child in one group is matched with a child in the other group, the appropriate test is the *t* test for correlated or dependent samples (i.e., the matched *t* test). There is no justification for using a significance level that is larger or smaller than .05, so we will use the conventional .05 level.

Step 3. Select the Samples and Collect the Data

First, one large sample is selected at random (or as randomly as possible within the usual practical constraints). Then participants are matched into pairs according to their similarity on relevant variables (for this example, the composite score on the battery of prereading skill tests). Finally, in this example, each child in a pair is randomly assigned to either the visual or the phonic group. (For instance, the flip of a fair coin could determine which member of the pair is assigned to the visual method and which to the phonic method). Because effective matching can help lead to a high t value, researchers tend to use smaller samples for a matched design than for an independent-groups design (i.e., matching increases power without having to increase sample size—see Section C). So, it is not entirely unreasonable that this example will be based on a total of 20 children, which means only 10 pairs, so $n = 10$. The reading levels after 1 year in the experimental program are recorded in Table 11.3 for each student. A reading level of 2.0 is considered average for a second grader and would be the level expected for these students had they not been in this experiment.

Pair	Visual	Phonic	D	D²	Table 11.3
1	2.3	2.5	−.2	.04	
2	2.0	1.9	+.1	.01	
3	2.1	2.6	−.5	.25	
4	2.4	2.2	+.2	.04	
5	1.9	2.1	−.2	.04	
6	2.2	2.5	−.3	.09	
7	1.8	2.2	−.4	.16	
8	2.4	2.7	−.3	.09	
9	1.6	1.9	−.3	.09	
10	1.7	1.6	+.1	.01	
Σ			−1.8	.82	

Step 4. Find the Region of Rejection

If the number of pairs were quite large, especially if n were greater than 100, we could safely use the normal distribution to test our null hypothesis. However, because n is small, and we do not know the standard deviation for either population (or for the population of difference scores), we must use the appropriate t distribution. The number of degrees of freedom is $n − 1$ (where n is the number of *pairs*), so for this example, df = $n − 1$ = $10 − 1 = 9$. Looking at Table A.2, in the .05 (two-tailed) column, we find that the critical t for df = 9 is 2.262. If our calculated matched t value is greater than +2.262 or less than −2.262, we can reject the null hypothesis (see Figure 11.1).

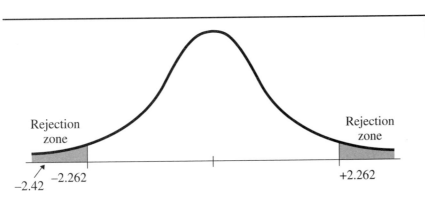

Figure 11.1

Rejection Zones for t Distribution (df = 9, Alpha = .05, Two-Tailed)

Rejection zone

Rejection zone

−2.42
−2.262

+2.262

Step 5. Calculate the Test Statistic

I will use Formula 11.1 to calculate the matched *t* value. The first step is to calculate the mean of the difference scores, $\overline{D}$, which equals $\Sigma D/n$. It is important when finding the sum of the difference scores to keep track of which differences are positive and which are negative (you will probably want to add the positive differences to get one sum and the negative differences to get a separate sum and then subtract the two sums). In our example, $\Sigma D = -1.8$, $\overline{D} = -1.8/10 = -1.8$. We can check this result by comparing it to $\overline{X}_V - \overline{X}_P$ to which it should be equal ($\overline{X}_V - \overline{X}_P = 2.04 - 2.22 = -.18$).

The next step is to find the unbiased standard deviation of the difference scores. Table 11.3 includes a column for the squared difference scores, so we can use Formula 3.15B (expressed in terms of difference scores) as follows:

$$s_D = \sqrt{\frac{1}{n-1}\left[\sum D^2 - \frac{(\sum D)^2}{n}\right]} = \sqrt{\frac{1}{9}\left[.82 - \frac{(-1.8)^2}{10}\right]}$$

$$= \sqrt{\frac{1}{9}(.496)} = \sqrt{.0551} = .235$$

(A perfectly reasonable alternative would be to enter the difference scores into a calculator that has the standard deviation as a built-in function, but then you must make sure that the sign of any negative difference score has been changed to a minus before it is entered.)

Now we have the values needed for Formula 11.1:

$$t = \frac{\overline{D}}{\frac{s_D}{\sqrt{n}}} = \frac{-.18}{\frac{.235}{\sqrt{10}}} = \frac{-.18}{.0743} = -2.42$$

Step 6. Make the Statistical Decision

Because the calculated *t* (-2.42) is less (i.e., more negative) than the critical *t* of -2.262, we can reject the null hypothesis (see Figure 11.1) and conclude that μ_V does not equal μ_P—that is, the two teaching methods do not have the same effect on reading acquisition. We can say that the phonic method is better than the visual method and that the difference between the two methods is statistically significant at the .05 level. The fact that our calculated *t* value was negative is just a consequence of subtracting the phonic score from the visual score instead of the other way around. It would have been perfectly acceptable to have reversed the order of subtraction to reduce the number of negative differences, in which case our calculated *t* would have been positive—and just as statistically significant.

Using the Correlation Formula for the Matched *t* Test

To see just how well matched our pairs of participants were, we could calculate the Pearson correlation coefficient for the visual and phonic scores. As a preliminary step, I have computed all of the cross products and the relevant summary statistics, as shown in Table 11.4.

Because we have already calculated the *unbiased* standard deviations, we use Formula 9.3 to calculate Pearson's *r*:

Table 11.4

Visual (X)	Phonic (Y)	XY
2.3	2.5	5.75
2.0	1.9	3.8
2.1	2.6	5.46
2.4	2.2	5.28
1.9	2.1	3.99
2.2	2.5	5.5
1.8	2.2	3.96
2.4	2.7	6.48
1.6	1.9	3.04
1.7	1.6	2.72
20.4	22.2	45.98
$s_x = .2875$	$s_y = .3553$	

$$r = \frac{\frac{1}{n-1}(\sum XY - N\overline{X}\overline{Y})}{s_x s_y} = \frac{\frac{1}{9}[45.98 - 10(2.04)(2.22)]}{(.2875)(.3553)}$$

$$= \frac{.0769}{.1021} = .753$$

Having calculated $r = .753$, we can see that the pairs of children were indeed well matched in their second-grade reading levels, suggesting that our matching criterion (the composite score of prereading skills) was appropriate. Now that we know Pearson's r, it is easy to demonstrate that Formula 11.2 gives exactly the same t value (unless there is some discrepancy due to rounding) as the direct-difference method:

$$t = \frac{(\overline{X}_1 - \overline{X}_2)}{\sqrt{\frac{s_1^2 + s_2^2}{n} - \frac{2rs_1s_2}{n}}}$$

$$= \frac{2.04 - 2.22}{\sqrt{\frac{(.2875)^2 + (.3553)^2}{10} - \frac{2(.753)(.2875)(.3553)}{10}}}$$

$$= \frac{-.18}{\sqrt{.0209 - .0154}} = \frac{-.18}{\sqrt{.0055}} = \frac{-.18}{.0743} = -2.42$$

Notice that both the numerator and denominator are the same in the preceding calculations as they were when we used Formula 11.1—but the denominator is found in a very different way. With Formula 11.2 it is easy to see that a large portion of the variance in the denominator is being "subtracted out" and that if Pearson's r were even higher, even more would have been subtracted out. If Pearson's r had turned out to be zero, or if an independent-groups test had been conducted instead, the denominator would have been larger, resulting in a smaller t value as follows:

$$t = \frac{(\overline{X}_1 - \overline{X}_2)}{\sqrt{\frac{s_1^2 + s_2^2}{n}}} = \frac{2.04 - 2.22}{\sqrt{\frac{(.2875)^2 + (.3553)^2}{10}}} = \frac{-.18}{\sqrt{.0209}} = \frac{-.18}{.1445} = -1.245$$

Without the matching, the t value would have been only about half as large, and the null hypothesis could not have been rejected.

Raw-Score Formula for the Matched t Test

Chapter 9 presented a formula for Pearson's r (Formula 9.5) that was based directly on sums and sums of squared values and did not require the intermediate steps of calculating means or standard deviations. There is also a raw-score formula for the matched t test. Although this raw-score formula (11.3) reduces the amount of calculation by leading directly from the raw scores to the matched t value, the benefit is slight and may not be worth the decreased chance of noticing a calculation error during the process. Formulas 11.1 and 11.2 have an advantage in that if any of the intermediate statistics they are based on (such as the mean of the difference scores) is drastically wrong, you will probably notice immediately. On the other hand, these formulas involve an increased risk of rounding-off error (compared to the raw-score formula) if not enough digits are retained in the intermediate steps. The worst thing about Formula 11.3 is that it looks confusing; its separate pieces do not correspond to commonly used statistics. However, to be comprehensive, and because the reader may encounter this formula elsewhere, I present the raw-score formula here:

$$t = \frac{\sum D}{\sqrt{\dfrac{1}{n-1}[n \sum D^2 - (\sum D)^2]}}$$

Formula 11.3

If you insert $\sum D$ and $\sum D^2$ from Table 11.3 into this formula, you will see that Formula 11.3 produces the same *t* value as Formula 11.1 or 11.2, except for any error due to rounding off. This will be the case, of course, for any set of data.

The Confidence Interval for the Difference of Two Population Means

The *t* value that is calculated for the matched *t* test, as with any *t* test, does not usually tell us all we want to know about the results of our experiment. For the example in this section, the *t* value tells us that we can reject the null hypothesis. Although the means tell us that the phonic method is better, the *t* value by itself does not help to tell us how much better.

From the results of the preceding example, our best guess is that the entire population learning by phonics would be .18 points better than the population learning by the visual method (that is the difference between the two sample means). Before a particular school system would go through the trouble and expense of changing to a new method for teaching reading (let's assume that they are using the visual method), they would have to judge whether a .18 point difference is worth the effort. However, they would also need to know how much confidence they could have that the population difference is really .18 points. After all, that is just an estimate based on 10 pairs of students. To know how much error is involved in our estimate of the population difference, we need to construct a confidence interval, as we did in Chapters 6 and 7.

Although two populations are involved, working with the difference scores allows us to find the confidence interval for a single population mean (i.e., the mean of the difference scores, which, of course, is the same as the difference of the two population means). I just need to modify Formula 6.5 slightly to indicate that difference scores are being used; the result is Formula 11.4:

$$\mu_{\text{lower}} = \overline{D} - t_{\text{crit}} s_{\overline{D}}$$

Formula 11.4

$$\mu_{\text{upper}} = \overline{D} + t_{\text{crit}} s_{\overline{D}}$$

in which $s_{\overline{D}}$ is just the denominator of the matched *t* test, $s_D/\sqrt{n}$. If we want to find the 95% confidence interval for the example in this section, t_{crit} is the .05, two-tailed *t* that we used for the significance test, 2.262. Inserting that value in Formula 11.4, we get the following confidence interval:

$$\mu_{\text{lower}} = .18 - (2.262)(.0743) = .18 - .168 = .012$$

$$\mu_{\text{upper}} = .18 + (2.262)(.0743) = .18 + .168 = .348$$

To have 95% confidence in our estimate, the interval must extend from +.012 to +.348. Knowing that there is a reasonable chance that the phonic method produces reading scores that are only about one hundredth of a point better than the visual method, few school systems would be inspired to switch. Note that zero is very close to being in the interval; this is consistent with the fact that the calculated t was only slightly larger than the critical t. Zero would be in the 99% CI, which tells us that the calculated t in this example would not be statistically significant at the .01 level.

Assumptions of the Matched t Test

Because the matched t test can be viewed as a one-sample hypothesis test of the difference scores, the assumptions behind it can be framed in terms of the population of difference scores. In addition to the requirement that the dependent variable be measured on an interval/ratio scale, there are just two major assumptions:

1. *Normality.* The population of differences scores should follow a normal distribution. As with other t tests, the normality assumption is not critical for relatively large samples. However, if the sample size is small (less than about 30), and the distribution seems to be very far from normal in shape, nonparametric statistics such as the Sign test or Wilcoxon's signed ranks test may be more valid (see Part VII).
2. *Independent random sampling.* Although there is a relationship between the two members of a pair of scores, each pair should be independent from all other pairs and, ideally, should be selected at random from all possible pairs. Of course, in psychology experiments with human participants the sample is more likely to be one of convenience than one that is at all random. This can limit the degree to which the results can be generalized to the larger population. However, a truly critical assumption for the matched-pairs design in particular is that the members of each pair are assigned randomly between the two conditions. If the assignment to conditions were made according to which member of the pair arrived earlier for the experimental session, for example, a systematic difference would be introduced between the two conditions other than the intended one, and you would have a confounding variable that would threaten the validity of your conclusions (e.g., it may be that individuals who are punctual are different in ways that affect your dependent variable and therefore contribute to the difference between your two conditions).

The Varieties of Designs Calling for the Matched t Test

The matched t test procedure described in this chapter can be used with a variety of repeated-measures and matched-pairs designs. I will consider the most common of these designs.

Repeated Measures

The two basic types of repeated-measures design are *simultaneous* and *successive*.

1. The experiment described in Section A, involving the recall of a mixed list of emotional and neutral words, is an example of the simultaneous design. It is not important that the participant literally see an emotional word and a neutral word simultaneously. The fact that the two types of words are randomly mixed throughout the session ensures that neither type has the temporal advantage of coming when the participant has, for instance, more energy or more practice.
2. The successive repeated-measures experiment has two major subtypes: the *before-after design* and the *counterbalanced design*. The before-after

design has already been discussed in terms of the weight-loss experiment. Recall that whereas this design can establish a significant change in scores from before to after, a control group is often required to rule out such alternative explanations as a practice or a placebo effect.

As an example of a counterbalanced design, imagine that a researcher proposes that people can solve more arithmetic problems while listening to happy music than to sad music. A simultaneous design is not feasible. But if participants always hear the same type of music first, their problem-solving scores while listening to the second type may reflect the advantage of a practice effect (or possibly the disadvantage of a fatigue effect). To eliminate *order* effects, half the participants would get the happy music first, and the other half would get the sad music first; the order effects should then cancel out when the data from all participants are considered. This is known as *counterbalancing* and will be dealt with again in Chapter 15 when more than two repeated measures are discussed. Although counterbalancing eliminates the bias in sample means that can be created by simple order effects, these order effects can still inflate the denominator of the *t* test (participants with different orders may have different difference scores, increasing the total variability of the difference scores). In Chapter 16, I will show you how to eliminate the extra variability produced by order effects.

Although counterbalancing eliminates the *simple* order effects described in the preceding, it cannot get rid of all types of order effects. If we use strong stimuli (e.g., selected movie segments) to make our participants sad or happy, the emotion induced first may linger on and affect the induction of the second emotion, producing what are called *carryover effects*. If these carryover effects are asymmetric (e.g., a prior sad mood interferes to a greater extent with the happiness induction than the other way around), they will not be eliminated by counterbalancing (in this case, they are often called *differential carryover effects*). A sufficient period of time between conditions, or the interposition of a neutral task, may help to avoid carryover effects. However, the two experimental conditions being compared may be such that one involves special instructions or hints (e.g., use visual imagery to remember items) that should not be used in the other condition. If participants are not able to forget or ignore those instructions in the second condition, a repeated-measures design is not recommended. A matched-pairs design may be the best solution.

Matched-Pairs

The matched-pairs design has two main subtypes, *experimental* and *natural*.

1. In the first type, the experimenter creates the pairs based either on a relevant pretest or on other available data (e.g., gender, age, IQ). The experiment described in Section B, comparing the two methods for teaching reading, is an example of an experimental matched-pairs design. Another way to create pairs is to have two different participants act as judges and rate the same stimulus or set of stimuli. For instance, the judges may be paired at random, with each member of the pair rating the same yearbook photo, but one member is told that the student in the photo is intelligent, whereas the other is told the opposite. The members of each pair are now matched in that they are exposed to the same stimulus (i.e., photo), and difference scores can be taken to see if the judge who is told the person in the photo is intelligent gives consistently higher (or lower) attractiveness ratings than the other member of the pair.

2. In contrast, in a natural matched-pairs design, the pairs occur naturally and are just selected by the experimenter. For example, husbands and wives can be compared with respect to their economic aspirations; daughters can be compared to their mothers to see if the former have higher educational goals. Because the pairs are not determined by the experiment, researchers must exercise extra caution in drawing conclusions from the results.

In general, if a repeated-measures design is appropriate (e.g., there are no differential carryover effects), it is usually the best choice because it yields more power (i.e., a better chance of statistical significance if the null hypothesis is not true) with fewer participants. When a repeated-measures design is not appropriate, a matched-pairs design offers much of the same advantage, as long as there is some reasonable basis for matching (e.g., a pretest or relevant demographic or background data on each participant). If an independent t test must be performed, it is helpful to use the principles described in Chapter 8 to estimate the sample size required for adequate power.

Publishing the Results of a Matched t Test

Reporting the results of a matched t test is virtually the same as reporting on an independent-groups t test. To report the results of the hypothetical reading acquisition experiment, we might use the following sentence: "A t test for matched samples revealed that the phonic method produced significantly better reading performance ($M = 2.22$) than the visual method ($M = 2.04$) when the pupils were tested at the end of 6 months, $t(9) = 2.42$, $p < .05$ (two-tailed)." Of course, the df reported in parentheses after the t value is the df that is appropriate for this matched t test. Note that I did not include the minus sign on the t value. Reporting the two means, as I did, is the clearest way to convey the direction of the results.

An Excerpt from Psychological Literature

I will illustrate the use of the matched t test in the psychological literature with an excerpt from an article by Kaye and Bower (1994). They report the results of a study of 12 newborns (less than 2 days old) that demonstrate their ability to match the shape of a pacifier in their mouths with a shape that they see on a screen. One of the main findings was that "the mean first-look duration at the image of the pacifier-in-mouth was 10.24 s ($SD = 8.20$). The mean first-look duration at the image of the other pacifier was 4.66 s ($SD = 4.02$). The difference was significant ($t = 3.25$, df = 11, $p < .01$). . . . There was no order effect." (p. 287). Note that the SDs given in parentheses are the standard deviations for each set of scores separately; the standard deviation of the difference scores is not given but can easily be found by multiplying the difference in means by the square root of n (i.e., df + 1) and then dividing by the given t value. (This is the same as solving Formula 11.1 for s_D.) As an exercise, use the information in this paragraph to calculate Pearson's r for the matched scores in the Kaye and Bower experiment (the answer is approximately .73).

1. The easiest procedure for performing a matched t test is the direct-difference method. The calculations must be done carefully, however, because of the nearly inevitable mixture of positive and negative numbers. The following steps are involved:
   a. Find the difference for each pair of scores, being careful to subtract always in the same order and to not lose track of any minus signs.

SUMMARY

b. Find the mean and unbiased standard deviation of the difference scores. If you are entering the difference scores into a calculator with a standard deviation function, don't forget to change the sign to minus before entering a negative difference score.

c. Insert the mean and SD into the appropriate t test formula, and compare the result to the critical value. Don't forget that the degrees of freedom equal one less than the number of difference scores.

2. An alternative method for calculating the matched t involves the correlation between the two sets of matched scores. The formula resembles the t test for two equal-sized independent groups, except that in the denominator (under the square root sign) a term is subtracted that is equal to 2 times the correlation coefficient multiplied by the standard deviations for the two sets of scores and divided by the number of pairs. The better the matching of the scores, the higher the correlation becomes, which results in a larger term being subtracted in the denominator and a larger value for the t test. Another alternative formula, called the raw-score formula, calculates t directly from the sum of the difference scores and the sum of the squared difference scores. This formula is confusing to look at but has the advantage of minimizing errors due to rounding off intermediate results before arriving at the value for t.

3. Knowing that the difference between two groups of scores is statistically significant is often not enough. In many cases, we would like an estimate of just how large the difference would be in the entire population. A confidence interval (CI), based on the standard error of the difference scores, can be used to provide the boundaries of such an estimate. A CI can be used to determine whether the difference between two means is large enough to have practical applications.

4. Assumptions:
   a. The entire population of difference scores is normally distributed (not a problem when dealing with a large sample).
   b. The pairs have been sampled independently of each other and at random from the population (not a problem, if the assignment to conditions is random for each pair).

5. The two basic types of repeated-measures design are: *simultaneous* and *successive*. The successive repeated-measures experiment has two major subtypes: the *before-after design* and the *counterbalanced design*. The before-after design usually requires a control group to draw valid conclusions. The problem with the counterbalanced design is that it does not always eliminate effects due to the order of conditions; when there are differential *carryover effects*, repeated measures should not be used.

6. The two main subtypes of matched-pairs design are: *experimental* and *natural*. The experimenter may create pairs based either on a relevant pretest or on other available data. Another way to create pairs is to have two participants rate or judge the same stimulus. Rather than creating pairs, the experimenter may use naturally occurring pairs, such as a father and his eldest son.

EXERCISES

1. The stress levels of 30 unemployed laborers were measured by a questionnaire before and after a real job interview. The stress level rose from a mean of 63 points to a mean of 71 points. The (unbiased) standard deviation of the difference scores was 18.
   a. What is the appropriate null hypothesis for this example?

b. What is the critical value of t for a .05, two-tailed test?

c. What is the observed (i.e., calculated) value of t?

d. What is your statistical decision with respect to the null hypothesis?

e. Given your conclusion in part d, could you be making a Type I or Type II error?

*2. In Exercise 7B9, participants in an individual motivation condition were compared to others in a group motivation condition in terms of task performance. Now assume that the participants had been matched in pairs based on some pretest. The data from Exercise 7B9 follow, showing the pairing of the participants.

Individual	11	17	14	10	11	15	10	8	12	15
Group	10	15	14	8	9	14	6	7	11	13

a. Perform a matched t test on these data (α = .05, two-tailed).

b. Compare the matched t with the independent t that you found for Exercise 7B9.

3. a. Using the data from Exercise 9B6, which follows, determine whether there is a significant tendency for verbal GRE scores to improve on the second testing. Calculate the matched t in terms of the Pearson correlation coefficient already calculated for that exercise.

b. Recalculate the matched t test according to the direct-difference method and compare the result to your answer for part a.

c. What would be the drawback to using the raw-score formula (11.3) to find the matched t for this problem?

Verbal GRE (1)	Verbal GRE (2)
540	570
510	520
580	600
550	530
520	520

*4. An educator has invented a new way to teach geometry to high school students. To test this new teaching method, 16 tenth-graders are matched into eight pairs based on their grades in previous math courses. Then the students in each pair are randomly assigned to either the new method or the traditional method. At the end of a full semester of geometry training, all students take the same standard high school geometry test. The scores for each student in this hypothetical experiment are as follows:

Traditional	New
65	67
73	79
70	83
85	80
93	99
88	95
72	80
69	100

a. Perform a matched t test for this experiment (α = .01, two-tailed). Is there a significant difference between the two teaching methods?

b. Find the 99% confidence interval for the population difference of the two teaching methods.

*5. In Exercise 9B8, a male and a female judge rated the same cartoon segments for violent content. Using the data from that exercise, which follows, perform a matched t test to determine whether there is a significant tendency for one of the judges to give higher ratings (use whichever formula and significance level you prefer).

Segment No.	Male Rater	Female Rater
1	2	4
2	1	3
3	8	7
4	0	1
5	2	5
6	7	9

*6. Do teenage boys tend to date teenage girls who have a lower IQ than they do? To try to answer this question, 10 teenage couples (i.e., who are dating regularly) are randomly selected, and each member of each couple is given an IQ test. The results are given in the following table (each column represents a different couple):

Boy	110	100	120	90	108	115	122	110	127	118
Girl	105	108	110	95	105	125	118	116	118	126

Perform a one-tailed matched t test (α = .05) to determine whether the boys have higher IQs than their girlfriends. What can you conclude?

7. A neuropsychologist believes that right-handed people will recognize objects placed

in their right hands more quickly than objects placed in their left hands when they are blindfolded. The following scores represent how many objects each participant could identify in 2 minutes with each hand.

Participant No.	Left	Right
1	8	10
2	5	9
3	11	14
4	9	7
5	7	10
6	8	5
7	10	15
8	7	7
9	12	11
10	6	12
11	11	11
12	9	10

a. Use Formula 11.1 to test the null hypothesis of no difference between the two hands (α = .05, two-tailed).
b. Recalculate the matched *t* using Formula 11.3, and compare the result to the *t* value you obtained in part a.

*8. A cognitive psychologist is testing the theory that short-term memory is mediated by subvocal rehearsal. This theory can be tested by reading aloud a string of letters to a participant, who must repeat the string correctly after a brief delay. If the theory is correct, there will be more errors when the list contains letters that sound alike (e.g., G and T) than when the list contains letters that look alike (e.g., P and R). Each participant gets both types of letter strings, which are randomly mixed in the same experimental session. The number of errors for each type of letter string for each participant are shown in the following table:

Participant No.	Letters That Sound Alike	Letters That Look Alike
1	8	4
2	5	5
3	6	3
4	10	11
5	3	2
6	4	6
7	7	4
8	11	6
9	9	7

a. Perform a matched *t* test (α = .05, one-tailed) on the data above and state your conclusions.
b. Find the 95% confidence interval for the population difference for the two types of letters.

9. For the data in Exercise 10B4:
a. Calculate the matched *t* value to test whether there is a significant difference (α = .05, two-tailed) between the spatial ability and math scores. Use the correlation coefficient you calculated to find the regression slope in Exercise 10B4.
b. Explain how the Pearson *r* for paired data can be very high and statistically significant, while the matched *t* test for the same data fails to attain significance.

10. For the data in Exercise 9B7:
a. Calculate the matched *t* value to test whether there is a significant difference (α = .05, *one*-tailed) between the odd and even items of that questionnaire. Use the correlation coefficient you calculated in Exercise 9B7.
b. Explain how the Pearson *r* for paired data can fail to come even close to being significant, while the matched *t* test for the same data does attain significance.

OPTIONAL MATERIAL

In Section A, I demonstrated that the matched *t* test will produce a much higher *t* value than the independent-groups test when the two sets of scores are well matched. All other things being equal, the better the matching (i.e., the higher the correlation), the higher the *t* value. You may recall from Chapter 8 that a higher expected *t* value (i.e., δ) means higher power. Thus matching can increase power without the need to increase the sample size. To illustrate how this works I will start with Formula 11.2, and replace each sample statistic with its expected value in the population (note that if we assume homogeneity of variance, σ^2 is the expected value for both s_1^2 and s_2^2):

$$\delta = \frac{\mu_1 - \mu_2}{\sqrt{\dfrac{\sigma^2 + \sigma^2 - 2\rho\sigma^2}{n}}} = \frac{\mu_1 - \mu_2}{\sqrt{\dfrac{2\sigma^2 - 2\rho\sigma^2}{n}}} = \frac{\mu_1 - \mu_2}{\sqrt{\dfrac{2(1-\rho)\sigma^2}{n}}}$$

The denominator can then be separated and part of it flipped over and put, effectively, in the numerator, as follows:

$$\delta = \frac{\mu_1 - \mu_2}{\sigma\sqrt{\dfrac{2(1-\rho)}{n}}} = \frac{\mu_1 - \mu_2}{\sigma}\sqrt{\frac{n}{2(1-\rho)}}$$

Finally, we can separate the term involving n from the term involving ρ to produce Formula 11.5, which finds δ for a matched t test:

$$\delta_{matched} = \frac{\mu_1 - \mu_2}{\sigma}\sqrt{\frac{n}{2}}\sqrt{\frac{1}{1-\rho}} \qquad \textbf{Formula 11.5}$$

Power of the Matched t Test

This new δ can then be looked up in Table A.3 to find power, as discussed in Chapter 8. Compare Formula 11.5 to Formula 8.1:

$$\delta = \frac{\mu_1 - \mu_2}{\sigma}\sqrt{\frac{n}{2}} \qquad \textbf{Formula 8.1}$$

The expected matched t ($\delta_{matched}$) is the same as the expected independent t (δ_{ind}) except for the term $\sqrt{1/(1-\rho)}$. This relationship can be expressed in the following manner:

$$\delta_{matched} = \delta_{ind}\sqrt{\frac{1}{1-\rho}} \qquad \textbf{Formula 11.6}$$

For instance a correlation of .5 would cause δ_{ind} to be multiplied by:

$$\sqrt{\frac{1}{1-\rho}} = \sqrt{\frac{1}{1-.5}} = \sqrt{\frac{1}{.5}} = \sqrt{2} = 1.41$$

A correlation of .9 would result in a multiplication factor of:

$$\sqrt{\frac{1}{1-.9}} = \sqrt{\frac{1}{.1}} = \sqrt{10} = 3.16$$

Suppose that, according to a particular alternative hypothesis, δ for an independent t test is only 2.0. This corresponds to a power of .52 (from Table A.3 with $\alpha = .05$, two-tailed). If the experiment could be changed to a matched design for which a correlation (ρ) of .5 could be expected, the new δ would be 1.41 times the old δ (see preceding calculation for $\rho = .5$), or 2.82. This new δ corresponds to a much more reasonable level of power—about .8. To get the same increase in power *without* matching, the number of subjects in each group would have to be *doubled* in this particular case.

The trickiest part of a power analysis for a matched-pairs or repeated-measures design is that, in addition to estimating the effect size (**d**) as in an independent-groups test, you must estimate ρ as well. As usual, you are often forced to rely on data from relevant studies already published. The relationship between changes in ρ and changes in power is not a simple one in mathematical terms, but generally speaking, increasing ρ will increase power. Of course, it is not always possible to increase ρ; there are limits due to the variability of individuals (and within individuals over time). Nevertheless, it can be worth the trouble to match experimental participants as closely as possible (or to keep conditions as constant as possible for each participant being measured twice).

There is also a potential danger involved in increasing the power by increasing ρ; it is the same danger inherent in using large samples, as discussed in Chapter 8. Like increasing the sample size, increasing ρ yields greater power without changing the relative separation of the two population means. Effect sizes that are otherwise trivial have a better chance of becoming statistically significant. From a practical point of view, we would be better off ignoring some of these tiny effects. For instance, a weight-loss program that results in every participant losing either 3 or 4 *ounces* of weight over a 6-month period will lead to a highly significant matched *t* value. Although the numerator of the *t* ratio will be small (the difference in means before and after will be about 3.5 ounces), the denominator will be even smaller (the standard deviation of the difference scores will be about half an ounce, which is then divided by the square root of *n*). The high *t* value, and corresponding low *p* value, are not due to the effectiveness of the weight-loss program in terms of helping people to lose a lot of weight, but rather they are due to the *consistency* of the weight-loss program (in that everyone loses about the same amount of weight). This is why it is a good idea to supplement the reporting of *t* values with confidence intervals and/or estimates of effect size.

Effect Size for the Matched *t* Test

I should mention here that an effect-size measure can be defined specifically for a matched design in a way that is analogous to the **d** that forms the basis of power analysis for two independent groups. To distinguish this new effect-size measure from **d**, I will call it d_{matched}, and define it as: $d_{\text{matched}} = (\mu_1 - \mu_2)/\sigma_{\text{diff}}$, where σ_{diff} is the standard deviation of the difference scores in the entire population. As you would expect, it is estimated by $g_{\text{matched}} = (\overline{X}_1 - \overline{X}_2/s_D) = \overline{D}/s_D$. If you look at Formula 11.1, you will see that the matched *t* value can be written as $g_{\text{matched}}\sqrt{n}$. Note the difference from the independent-groups case, in which *t* equals $g\sqrt{(n/2)}$. The factor of 2 is absent in the formula relating the matched *t* to g_{matched}, because in the latter case you are really dealing with only one population: the population of difference scores.

Because σ_{diff} equals $\sigma\sqrt{2(1 - \rho)}$, we can express d_{matched} as a function of **d** and ρ as follows:

$$d_{\text{matched}} = d\sqrt{\frac{1}{2(1 - \rho)}} \qquad \text{Formula 11.7}$$

Notice that when ρ drops below .5, d_{matched} is actually less than **d**, but in the more usual case when repeated measures or matching leads to a correlation larger than .5, d_{matched} is greater than **d**. Similarly, g_{matched} can be expressed as a function of *g* and *r*. However, g_{matched} (or d_{matched}) has a different interpretation from *g* (or **d**). Whereas *g* gives you an idea of the overall effectiveness of a treatment on a group-to-group basis, g_{matched} relates to the consistency of individual change that can be expected as a result of the treatment. Moreover, as Morris and DeShon (2002) point out, g_{matched} is much less commonly reported, and therefore much less familiar than *g* to psychological researchers, so it is probably best confined to meta-analyses in which most of the studies involve correlated samples, and the focus is on individual change.

Another choice, analogous to the one just described, arises with respect to displaying data from a matched design. In recent years, it seems that the most common way to depict a comparison of two independent groups has been with a bar graph, with each vertical bar representing one of the two conditions. Usually a thin line is drawn upward from the middle of the top surface of the bar and capped with a "T"; this is called an *error bar*. The error bar most often extends a distance equal to the standard error of the mean for that condition (though sometimes it represents the ordinary standard deviation), and is usually mirrored by an error bar drawn downward the same distance (though the downward portion my be hard to see if the main bar is shaded darkly). If the standard errors of the two conditions are comparable, one can see at a glance if the result is likely to be statisticaly significant; a separation in the heights of the two bars that is equal to nearly three standard errors is usually significant. (Note that when the two standard errors are equal, the standard error of the difference—i.e., the denominator of the t test—is equal to the common standard error multiplied by the square root of 2.)

The choice with respect to the matched design is whether to diplay the same error bars as in the independent-groups case, even though these error bars are not reduced by the correlation between the two sets of scores and therefore do not indicate the standard error of the difference for the matched t test, or to display error bars that *do* reflect the degree of matching, and are thus consistent with the statistical significance of the difference in the heights of the bars. Although it is the ordinary standard errors that usually form the basis of the error bars when psychologists display the results of a matched (or repeated measures) design, Loftus and Masson (1994) make a good case for the display of an appropriately reduced error bar. Their proposal amounts to using $s_{\bar{D}}/\sqrt{2}$ as the length of the error bar for both conditions. The advantage of using error bars of this length is that, as in the independent-group case, a separation in the heights of the bars equal to nearly three times the length of either error bar would indicate statistical significance. Thus, a bar graph for a matched design could be visually interpreted inferentially in the same way as a bar graph representing independent groups. For more detail on the proper use of confidence intervals for each condition in a repeated-measures or matched design, see Blouin and Riopelle (2005).

In general, higher t values are good because we can be more confident in rejecting the null hypothesis—more confident that our treatment is not totally ineffective. However, being more sure that the effect is not zero does not imply that the effect is at all large; as in the weight-loss example, the effect may be statistically significant but not large enough to be of any practical importance. On the other hand, the increased power of matching can help psychologists establish the direction of effects that are tiny but interesting theoretically. Just remember that matching, like increasing the sample size, helps us to detect small differences in population means and thus can be a useful tool or a nuisance, depending on the situation.

SUMMARY

1. Compared to an independent samples t test, and assuming the means and standard deviations remain the same, the matched t test is associated with a higher expected t (δ)—as long as a positive correlation can be expected between the two sets of scores.
2. Increasing the matching (i.e., the correlation) between the two sets of scores increases δ (all else remaining equal), and therefore the power, without the need to increase the number of participants.
3. In addition to estimating **d**, as in the independent-samples t test, the power analysis for a matched design requires estimating the population correlation (ρ) for the two sets of scores. Previous experimental results can provide a guideline.

4. It must be kept in mind that a well-matched design, like a study with very large sample sizes, can produce a high t value (and a correspondingly low p value), even though the separation of the means (relative to the standard deviation) is not very large and in fact may be too small to be of any practical interest. On the other hand, the increased power of the matched design can enable psychologists to answer interesting theoretical questions that may hinge on the directions of tiny effects.

5. The reduced error term due to successful matching can be used as the basis for expressing the increased reliability of the matched design in terms of $d_{matched}$, or its sample estimate, $g_{matched}$. The reduced error term can also be used as the basis of drawing bar graphs to display the results of a matched design.

EXERCISES

*1. Imagine that an experiment is being planned in which there are two groups, each containing 25 participants. The effect size (**d**) is estimated to be about .4.
  a. If the groups are to be matched, and the correlation is expected to be .5, what is the power of the matched t test being planned, with alpha = .05 and a two-tailed test?
  b. If the correlation in the preceding example were .7, and all else remained the same, what would the power be?
  c. Recalculate the power for part b above for alpha = .01 (two-tailed).

2. If a before-after t test is planned with 35 participants who are undergoing a new experimental treatment, and the after scores are expected to be one half of a standard deviation higher than the before scores, how high would the correlation need to be to have power = .8, with alpha = .05, two-tailed?

*3. A matched t test is being planned to evaluate a new method for learning foreign languages. From previous research, an effect size of .3, and a correlation of .6 are expected.
  a. How many participants would be needed in each matched group to have power = .75, with a two-tailed test at alpha = .05?
  b. What would your answer to part a be if alpha were changed to .01?

4. The correlation expected for a particular matched t test is .5.
  a. If it is considered pointless to have more than .7 power to detect a difference in population means as small as one tenth of a standard deviation, with an alpha = .05, two-tailed test, what is the largest sample size that should ever be used?
  b. What would your answer to part a be if the expected correlation were only .3?

*5. Which of the following can decrease the Type II error rate associated with a matched t test?
  a. Increasing the sample size.
  b. Increasing the correlation between the two sets of scores (i.e., improving the matching).
  c. Using a larger alpha level (e.g., .1 instead of .05).
  d. All of the above.
  e. None of the above.

6. For the data in Exercise 7B3, assume that δ_{ind} equals the t value that you calculated. Find $\delta_{matched}$ and $d_{matched}$ if this had been a matched-pairs design with ρ equal to:
  a. 0.1
  b. 0.5
  c. 0.8

KEY FORMULAS

The matched t test (difference-score method):

$$t = \frac{\overline{D}}{\dfrac{s_D}{\sqrt{n}}}$$

Formula 11.1

The matched t test, in terms of Pearson's correlation coefficient (definitional form):

$$t = \frac{(\overline{X}_1 - \overline{X}_2)}{\sqrt{\dfrac{s_2^1 + s_2^2}{n} - \dfrac{2rs_1s_2}{n}}}$$ Formula 11.2

The matched t test (raw-score form):

$$t = \frac{\sum D}{\sqrt{\dfrac{1}{n-1}[n \sum D^2 - (\sum D)^2]}}$$ Formula 11.3

The confidence interval for the difference of two population means, when dealing with matched pairs (or repeated measures):

$$\mu_{\text{lower}} = \overline{D} - t_{\text{crit}}s_{\overline{D}}$$ Formula 11.4
$$\mu_{\text{upper}} = \overline{D} + t_{\text{crit}}s_{\overline{D}}$$

Delta for the matched t test (power can then be found in Table A.3):

$$\delta_{\text{matched}} = \frac{\mu_1 - \mu_2}{\sigma} \sqrt{\frac{n}{2}} \sqrt{\frac{1}{1 - \rho}}$$ Formula 11.5

Delta for the matched t test as a function of delta for the independent-groups t test and ρ:

$$\delta_{\text{matched}} = \delta_{\text{ind}} \sqrt{\frac{1}{1 - \rho}}$$ Formula 11.6

The effect size for two matched populations in terms of Cohen's $\mathbf{d}$ and ρ:

$$\mathbf{d}_{\text{matched}} = \mathbf{d} \sqrt{\frac{1}{2(1 - \rho)}}$$ Formula 11.7

ONE-WAY INDEPENDENT ANOVA

Chapter

You will need to use the following from previous chapters:

Symbols
$\overline{X}$: Mean of a sample
s: Unbiased standard deviation of a sample
s^2: Unbiased variance of a sample
SS: Sum of squared deviations from the mean
MS: Mean of squared deviations from the mean (same as variance)

Formulas
Formula 3.12: Computational formula for SS
Formula 4.5: The standard error of the mean
Formula 7.5B: The pooled-variance t test

Concepts
Homogeneity of variance
The pooled variance, s_p^2
Null hypothesis distribution
One- and Two-tailed tests

CONCEPTUAL FOUNDATION

In Chapter 7, I described a hypothetical experiment in which the effect of vitamin C on sick days taken off from work was compared to the effect of a placebo. Suppose the researcher was also interested in testing the claims of some vitamin enthusiasts who predict that vitamin C combined with B complex and other related vitamins is much more effective against illness than vitamin C alone. A third group of subjects[1], a multivitamin group, could be added to the previous experiment. But how can we test for statistically significant differences when there are three groups? The simplest answer is to perform t tests for independent groups, taking two groups at a time. With three groups only three different t tests are possible. First, the mean of the vitamin C group can be tested against the mean of the placebo group to determine whether vitamin C alone makes a significant difference in sick days. Next, the multivitamin group can be tested against vitamin C alone to determine whether the multivitamin approach really adds to the effectiveness of vitamin C. Finally, it may be of interest to test the multivitamin group against the placebo group, especially if vitamin C alone does not significantly differ from placebo.

There is a better procedure for testing differences among three or more independent means; it is called the *one-way analysis of variance*, and by the end of this chapter and the next one, you will understand its advantages as well as the details of its computation. The term *analysis of variance* is usually abbreviated as ANOVA. The term *one-way* refers to an experimental design in which there is only one independent variable. In the vitamin experiment described above, there are three experimental conditions, but they are considered to be three different *levels* of the same independent variable, which can be called "type of vitamin." In the context of ANOVA, an independent variable

---

[1]*Note:* Although the term *participant* is now strongly preferred to *subject* in the field of psychology, the term *subject* is so deeply embedded in the traditional terminology used to express the concepts and formulas for ANOVA that I decided for the sake of clarity and simplicity to retain the term *subject* in this and the following chapters.

is called a *factor*; this terminology will be more useful when we encounter, in later chapters, ANOVAs with more than one factor. In this chapter I will deal only with the one-way ANOVA of *independent* samples. If the samples are matched, you must use the procedures described in Chapter 15.

Before proceeding, I should point out that when an experiment involves only three groups, performing all of the possible *t* tests (i.e., testing all three pairs), although not optimal, usually leads to the same conclusions that would be derived from the one-way analysis of variance. The drawbacks of performing multiple *t* tests become more apparent, however, as the number of groups increases. Consider an experiment involving seven different groups of subjects. For example, a psychologist may be exploring the effects of culture on emotional expressivity by comparing the means of seven samples of subjects, each group drawn from a different cultural community. In such an experiment there are 21 possible two-group *t* tests. (In the next chapter you will learn how to calculate the number of possible pairs given the number of groups.) If the .05 level is used for each *t* test, the chances are better than 50% that at least one of these *t* tests will attain significance even if all of the cultural populations have identical means on the variable being measured. Moreover, with so many groups, the psychologist's initial focus would likely be to see whether there are any differences at all among the groups rather than to ask whether any one particular group mean is different from another group mean. In either case, Dr. Null would say that all seven cultural groups are identical to each other (or, at least, that they share the same population mean for the measure of emotional expression being used). To test this null hypothesis in the most valid and powerful way, we need to know how much seven sample means are likely to vary from each other when all seven samples are drawn from the same population (or different populations that all have the same mean). That is why we need to develop the formula for the one-way ANOVA.

Transforming the *t* Test into ANOVA

I will begin with the *t* test for two independent groups and show that it can be modified to accommodate any number of groups. As a starting point, I will use the *t* test formula as expressed in terms of the pooled variance (Formula 7.5B):

$$t = \frac{\overline{X}_1 - \overline{X}_2}{\sqrt{s_p^2\left(\frac{1}{n_1} + \frac{1}{n_2}\right)}}$$

Formula 7.5B

We can simplify this formula by dealing with the case in which $n_1 = n_2$. The formula becomes:

$$t = \frac{\overline{X}_1 - \overline{X}_2}{\sqrt{s_p^2\left(\frac{1}{n} + \frac{1}{n}\right)}} = \frac{\overline{X}_1 - \overline{X}_2}{\sqrt{s_p^2\left(\frac{2}{n}\right)}}$$

which leads to Formula 12.1:

$$t = \frac{\overline{X}_1 - \overline{X}_2}{\sqrt{\frac{2s_p^2}{n}}}$$

Formula 12.1

When $n_1 = n_2$, the pooled variance is just the ordinary average of the two variances, and Formula 7.8 (the t test for equal-sized groups) is easier to use than Formula 12.1. But it is informative to consider the formula in terms of the pooled variance even when all of the ns are equal. The next step is to square Formula 12.1 so that we are dealing more directly with the variances rather than with standard deviations (or standard errors). After squaring, we multiply both the numerator and the denominator by the sample size (n) and then divide both by 2 to achieve the following expression:

$$t^2 = \frac{(\bar{X}_1 - \bar{X}_2)^2}{\dfrac{2s_p^2}{n}} = \frac{n(\bar{X}_1 - \bar{X}_2)^2}{2s_p^2} = \frac{\dfrac{n(\bar{X}_1 - \bar{X}_2)^2}{2}}{s_p^2}$$

Moving n in front of the fraction in the numerator yields Formula 12.2:

$$t^2 = \frac{n\dfrac{(\bar{X}_1 - \bar{X}_2)^2}{2}}{s_p^2}$$ **Formula 12.2**

Bear in mind that as strange as Formula 12.2 looks, it can still be used to perform a t test when the two groups are the same size, as long as you remember to take the square root at the end. If you do not take the square root, you have performed an analysis of variance on the two groups, and the result is called an F ratio (for reasons that will soon be made clear). Although Formula 12.2 works well for the two-group experiment, it must be modified to accommodate three or more groups to create a general formula for the one-way analysis of variance.

Expanding the Denominator

Changing the denominator of Formula 12.2 to handle more than two sample variances is actually very easy. The procedure for pooling three or more sample variances is a natural extension of the procedure described in Chapter 7 for pooling two variances. However, it is customary to change the terminology for referring to the pooled variance when dealing with the analysis of variance. (I introduced this terminology in Chapter 3 to prepare you, when I showed that the variance, s^2, equals SS/df, where SS stands for the sum of squares, i.e., the sum of squared deviations.) Because the variance is really a mean of squares, it can be symbolized as MS. Therefore, the pooled variance can be referred to as MS. Specifically, the pooled variance is based on the variability *within* each group in the experiment, so the pooled variance is often referred to as the *mean-square-within*, or MS_W. When all of the samples are the same size, MS_W is just the ordinary average of all the sample variances.

Expanding the Numerator

The numerator of the two-group t test is simply the difference between the two sample means. If, however, you are dealing with *three* sample means and you wish to know how far apart they are from each other (i.e., how spread out they are), there is no simple difference score that you can take. Is there some way to measure how far apart three or more numbers are from each other? The answer is that the ordinary variance will serve this purpose nicely. (So would the standard deviation, but it will be easier to deal only with variances.) Although it is certainly not obvious from looking at the numerator of Formula 12.2, the term that follows n in the numerator is equal to the (unbiased) variance of the two group means. To accommodate three or more sample means, the numerator of Formula 12.2 must be modified so that it equals n times the variance of all the sample means.

For example, if the average heart rates of subjects in three groups taking different medications were 68, 70, and 72 bpm, the (unbiased) variance of the three sample means would be 4. (Taking the square root would give you a standard deviation of 2, which you could guess just from looking at the numbers.) If the sample means were more spread out (e.g., 66, 70, and 74), their variance would be greater (in this case, 16). To produce the numerator of Formula 12.2 for this example, the variance of these group means must be multiplied by the sample size. (The procedure gets a little more complicated when the sample sizes are not all equal.) Therefore, if each sample had eight subjects, the numerator would equal 32 in the first case (when the variance of the means was 4) and 128 in the second case (when the sample means were more spread out and the variance was 16).

Like the denominator of the formula for ANOVA, the numerator also involves a variance, so it too is referred to as MS. In this case, the variance is between groups, so the numerator is often called the *mean-square-between*, or $MS_{\text{between-groups}}$ (MS_{bet}, for short).

The F Ratio

When MS_{bet}, as described above, is divided by MS_W, the ratio that results is called the *F ratio*, as shown in Formula 12.3:

$$F = \frac{MS_{\text{bet}}}{MS_W}$$ **Formula 12.3**

When the null hypothesis is true, the F ratio will follow a well-known probability distribution, called the *F distribution*, in honor of R.A. Fisher (mentioned in previous chapters), who created the analysis of variance procedure in the 1920s and first applied it to research in agriculture. The letter F is used to represent a test statistic that follows the F distribution. As with the t value from a two-group test, the F ratio gets larger as the separation of the group means gets larger relative to the variability within groups. A researcher hopes that treatments will make MS_{bet} much larger than MS_W and thus produce an F ratio larger than that which would easily occur by chance. However, before we can make any judgment about when F is considered large, we need to see what happens to the F ratio when Dr. Null is right—that is, when the treatments have no effect at all.

The F Ratio As a Ratio of Two Population Variance Estimates

To gain a deeper understanding of the F distribution and why it is the appropriate null hypothesis distribution for ANOVA, you need to look further at the structure of the F ratio and to consider what the parts represent. A ratio will follow the F distribution when both the numerator and the denominator are independent estimates of the same population variance. It should be relatively easy to see how the denominator of the F ratio, MS_W, is an estimate of the population variance, so I will begin my explanation of the F ratio at the bottom.

Pooling the sample variances in a t test is based on the assumption that the two populations have the same variance (i.e., homogeneity of variance), and the same assumption is usually made about all of the populations that are sampled in a one-way ANOVA. Under this assumption there is just one population variance, σ^2, and pooling all the sample variances, MS_W, gives the best estimate of it. If the null hypothesis is true, the numerator of the F ratio also serves as an estimate of σ^2, but why this is so is not obvious.

The numerator of the F ratio consists of the variance of the group means, multiplied by the sample size. (For simplicity I will continue to assume that all the samples are the same size.) To understand the general relationship between the variance of group means, $\sigma_{\bar{X}}^2$, and the variance of individuals in the population, σ^2, you have to go back to Formula 4.5:

$$\sigma_{\bar{X}} = \frac{\sigma}{\sqrt{n}}$$

Formula 4.5

and square both sides of the formula:

$$\sigma_{\bar{X}}^2 = \frac{\sigma^2}{n}$$

If we know the population variance and the size of the samples, we can use the preceding formula to determine the variance of the group means. On the other hand, if we have calculated the variance of the group means directly, as we do in the one-way ANOVA, and then multiplied this variance by n, the result should equal the population variance (i.e., $n\sigma_{\bar{X}}^2 = \sigma^2$). This is why the numerator of the F ratio in an ANOVA also serves as an estimate of σ^2. Calling the numerator of the F ratio $MS_{\text{between-groups}}$ can be a bit misleading if it suggests that it is just the variance among group means. MS_{bet} is actually an estimate of σ^2 based on multiplying the variance of group means by the sample size.

MS_{bet} is an estimate of σ^2 only when the null hypothesis is true; if the null hypothesis is false, the size of MS_{bet} reflects not only the population variance, but whatever treatment we are using to make the groups different. But, as usual, our focus is on null hypothesis testing, which means we are interested in drawing a "map" of what can happen when the null hypothesis *is* true. When the null hypothesis is true, both MS_{bet} and MS_W are estimates of σ^2. Moreover, they are independent estimates of σ^2. MS_W estimates the population variance directly by pooling the sample variances. MS_{bet} provides an estimate of σ^2 that depends on the variance of the group means and is not affected by the size of the sample variances in that particular experiment. Either estimate can be larger, and the F ratio can be considerably greater or less than 1 (but never less than zero). If we want to know which values are common for our F ratio and which are unusual when the null hypothesis is true, we need to look at the appropriate F distribution, which will serve as our null hypothesis distribution.

Degrees of Freedom and the *F* Distribution

Like the t distribution, the F distribution is really a family of distributions. Because the F distribution is actually the ratio of two distributions, each of which changes shape according to its degrees of freedom, the F distribution changes shape depending on the number of groups, as well as the total number of subjects. Therefore, two df components must be calculated. The df associated with MS_{bet} is 1 less than the number of groups; because the letter k is often used to symbolize the number of groups in a one-way ANOVA, we can say that:

$$df_{\text{bet}} = k - 1$$

Formula 12.4A

The df associated with MS_W is equal to the total number of subjects (all groups combined) minus 1 for each group. Thus:

$$df_W = N_T - k \qquad\qquad \textbf{Formula 12.4B}$$

When there are only two groups, $df_{bet} = 1$ and $df_W = N_T - 2$. Notice that in this case, df_W is equal to the df for a two-group t test ($n_1 + n_2 - 2$). When df_{bet} is 1, the F ratio is just the t value squared.

The Shape of the *F* Distribution

Let us look at a typical F distribution; the one in Figure 12.1 corresponds to $df_{bet} = 2$ and $df_W = 27$. This means that there are three groups of 10 subjects each, for a total of 30 subjects. The most obvious feature of the F distribution is that it is positively skewed. Whereas there is no limit to how high the F ratio can get, it cannot get lower than zero. F can never be negative because it is the ratio of two variances, and variances can never be negative. The mean of the distribution is near 1.0; to be more exact, it is $df_W/(df_W - 2)$, which gets very close to 1.0 when the sample sizes are large. Also, as the sample sizes get large, the F distribution becomes less skewed; as df_W approaches infinity, the F distribution becomes indistinguishable from a normal distribution with a mean of 1.0.

Notice that the upper 5% of the F distribution shown in Figure 12.1 is beyond $F = 3.35$; this is the critical value for this distribution when alpha equals .05. I will discuss in the following how to find these critical values in the F table, but first I must explain why ANOVA always involves a one-tailed test using the upper tail of the F distribution.

ANOVA As a One-Tailed Test

In the case of the t distribution, each tail represents a situation in which one of the two sample means is larger. But for the F distribution, one tail represents the situation in which all of the sample means are close together, whereas the other tail represents a spreading of the sample means. An F ratio larger than 1.0 indicates that the sample means are further apart than what we could expect (on average) as a result of chance factors. However, the size of F tells us nothing about *which* sample means are higher or lower than the others (we cannot rely on the sign of the F ratio to tell us anything because the F ratio is always positive). On the other hand, an F ratio smaller than 1.0 indicates that the sample means are closer together than could be expected by chance. Because the null hypothesis is that the population means are all the same (i.e., $\mu_1 = \mu_2 = \mu_3 = \ldots$), a small F, even an F that is unusually small and close to zero, can only suggest that the null is really true.

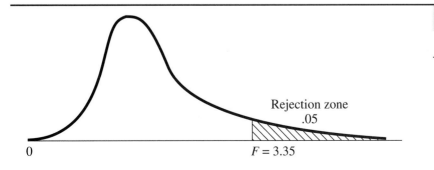

Figure 12.1

F Distribution (2, 27 df)

Rejection zone
.05

0

$F = 3.35$

Unfortunately, it can be confusing to refer to ANOVA as a one-tailed test because from dealing with t tests you get in the habit of thinking of a one-tailed test as a case in which you can change the critical value by specifying in advance which mean you expect to be larger. However, when you are dealing with more than two groups and performing the simple ANOVA described in this chapter, it makes no difference whether you make predictions about the relative sizes of the means. Thus some texts use the analogy of the t test and refer to the ANOVA F test as "two-tailed" because you are not predicting the order of the means. Whichever way you look at it, you do not have a one- or two-tail choice with ANOVA, as you do with the t test. The entire .05 area (or whatever alpha you are using) is always placed in the positive tail. Note that when there are only two groups, the usual F test is equivalent to a two-tailed t test at the same alpha (this makes sense because regardless of which mean is larger, a big difference between the two means will lead to a large positive F). If, in the two-group case, you can justify performing a one-tailed test, you should perform a one-tailed t test rather than an ANOVA.

Using Tables of *F* Values

To look up a critical F value, we must know the df associated with the numerator (df_{bet}) as well as the denominator (df_W). If you look at Tables A.7, A.8, and A.9 in Appendix A, you will see that the numerator df determines which column to look in, and the denominator df determines how far down to look. In the t table (Table A.2), the columns represent different alpha levels. In an F table, both the columns and rows represent the df, so the entire F table usually corresponds to just one alpha level. Appendix A includes three F tables, one for each of the following alpha levels: .05, .025, and .01. Each table represents a one-tailed test; the alpha for the table tells you how much area in the positive tail lies above (to the right of) the critical value. The .05 and .01 tables are used for ANOVAs that are tested at those alpha levels. The .025 table applies when using both tails of the F distribution (even if the tail near zero is quite short), which is appropriate for testing homogeneity of variance (see Section B).

An Example with Three Equal-Sized Groups

To demonstrate how easy it can be to perform an analysis of variance, I will present a simple example based on the experiment described at the beginning of this chapter. The independent variable is a type of vitamin treatment, and it has three levels: vitamin C, multivitamins, and placebo. The dependent variable is the number of sick days taken off from work during the experimental period. Three samples of 10 subjects each are selected at random, and each sample receives a different level of the vitamin treatment. At the end of the experiment, the means and standard deviations are as shown in Table 12.1.

In the special case when all of the samples are the same size, the formula for the one-way ANOVA becomes very simple. The numerator of the F ratio (MS_{bet}) is just the size of each sample (n) times the variance of the sample means:

$$MS_{bet} = n \frac{\sum (\overline{X}_i - \overline{X}_G)^2}{k - 1} = n s_{\overline{x}}^2 \qquad \textbf{Formula 12.5A}$$

where k is the number of groups in the experiment and $\overline{X}_G$ is the *grand mean* (i.e., the mean of all of the subjects in the whole experiment, regardless of group).

	Placebo	Vitamin C	Multivitamin	Table 12.1
$\overline{X}$	9	7	5.5	
s	3.5	3	2.5	

The denominator of the F ratio is just the average of all of the sample variances and can be calculated with Formula 12.5B: $MS_W = \Sigma s^2/k$. Combining this expression with the preceding one gives us Formula 12.5, which can only be used when all of the samples are the same size:

$$F = \frac{ns_{\overline{x}}^2}{\dfrac{\Sigma s^2}{k}}$$ **Formula 12.5**

Calculating a Simple ANOVA

We will begin by calculating MS_{bet} for the vitamin example, using Formula 12.5A. If you don't have a calculator handy that calculates SDs automatically, you will have to calculate the variance of the means the long way, as shown next. First, find $\overline{X}_G$, the grand mean. Because all the samples are the same size, the grand mean is just the average of the group means (otherwise we would have to take a weighted average): $\overline{X}_G = (7 + 5.5 + 9)/3 = 21.5/3 = 7.167$. There are three groups in this example, so $k = 3$. Now we can calculate the variance of the sample means:

$$\frac{\Sigma(\overline{X}_i - \overline{X}_G)^2}{k - 1} = \frac{(7 - 7.167)^2 + (5.5 - 7.167)^2 + (9 - 7.167)^2}{3 - 1}$$

$$= \frac{(-.167)^2 + (-1.667)^2 + (1.833)^2}{2}$$

$$= \frac{.0279 + 2.779 + 3.36}{2} = \frac{6.167}{2} = 3.083$$

The variance of the sample means (7, 5.5, and 9) is 3.083. The faster and more accurate alternative is to enter the three numbers into any calculator that has the unbiased standard deviation as a built-in function. The s for these three numbers [indicated as σ_{N-1} or s_{n-1} on most calculators] is 1.756, which when squared is 3.083.) Finally, because there are 10 subjects per group, $n = 10$ and $MS_{bet} = 10 \times 3.083 = 30.83$.

When all the samples are the same size, MS_W is just the average of the three variances, as indicated by Formula 12.5B: $(3^2 + 2.5^2 + 3.5^2)/3 = (9 + 6.25 + 12.25)/3 = 27.5/3 = 9.167$. (If you are given standard deviations, don't forget to square them.) To complete the calculation of Formula 12.5, we form a ratio of the results from Formulas 12.5A and 12.5B: F (i.e., MS_{bet}/MS_W) $= 30.83/9.167 = 3.36$.

Our calculated (or obtained) F ratio is well above 1.0, but to find out whether it is large enough to reject the null hypothesis, we need to look up the appropriate critical F in Table A.7, assuming that .05 is the alpha set for this ANOVA. To do this, we need to find the df components. The df for the numerator is df_{bet}, which equals $k - 1$. For our example, $k - 1 = 3 - 1 = 2$. The df for the denominator is df_W, which equals $N_T - k = 30 - 3 = 27$. Therefore, we go to the second column of the .05 table and then down the column to the entry corresponding to 27, which is 3.35. Because our calculated F (3.36) is larger than the critical F (just barely), our result falls in the rejection zone of the F distribution (see Figure 12.1), and we can reject the null hypothesis that all three population means are equal.

Interpreting the *F* Ratio

The denominator of the *F* ratio reflects the variability within each group of scores in the experiment. The variability within a group can be due to individual differences, errors in the measurement of the dependent variable, or fluctuations in the conditions under which subjects are measured. Regardless of its origin, all of this variability is unexplained, and it is generally labeled error variance. For this reason, the denominator of the *F* ratio is often referred to as MS_{error} rather than MS_W, and in either case it is said to be the *error term* of the ANOVA. According to the logic of ANOVA, subjects in different groups should have different scores because they have been treated differently (i.e., given different experimental conditions) or because they belong to different populations, but subjects within the same group ought to have the same score. The variability within groups is not produced by the experiment, and that is why it is considered error variance.

The variability in the numerator is produced by the experimental manipulations (or preexisting population differences), but it is also increased by the error variance. Bear in mind that even if the experiment is totally ineffective and all populations really have the same mean, we do not expect all the *sample* means to be identical; we expect some variability among the sample means simply as a result of error variance. Thus there are two very different sources of variability in the numerator, only one of which affects the denominator. This concept can be summarized by the following equation:

$$F = \frac{\text{estimate of treatment effect} + \text{between-group estimate of error variance}}{\text{within-group estimate of error variance}}$$

If the null hypothesis is true, the estimate of the treatment effect in the numerator will usually be close to zero, and both the top and the bottom of the *F* ratio will consist of error variance. You would expect the *F* ratio to equal 1.0 in this case—and on the average, it does equal about 1.0, but because the error variance in the numerator is estimated differently from the error variance in the denominator, either one can be larger. Therefore, the *F* ratio will fluctuate above and below 1.0 (according to the *F* distribution) when the null hypothesis is true.

When the null hypothesis is *not* true, we expect the *F* ratio to be greater than 1.0; but even when there is some treatment effect, the *F* can turn out to be less than 1.0 through bad luck. (The effect may be rather weak and not work well with the particular subjects selected, whereas the within-group variability can come out unusually high.) On the other hand, because we know the *F* ratio can be greater than 1.0 when there is no treatment effect at all, we must be cautious and reject the null only when our obtained *F* ratio is so large that only 5% (or whatever alpha we set) of the *F* distribution (i.e., the null hypothesis distribution) produces even larger *F* ratios. When we reject the null hypothesis, as we did for the vitamin example above, we are asserting our belief that the treatment effect is not zero—that the population means (in this case, placebo, vitamin C, and multivitamin) are not all equal.

We can picture the null hypothesis in the one-way ANOVA in terms of the population distributions; if three groups are involved, the null hypothesis states that the three population means are identical, as shown in Figure 12.2a. (We also assume that all the populations are normally distributed with the same variance, so the three distributions should overlap perfectly.)

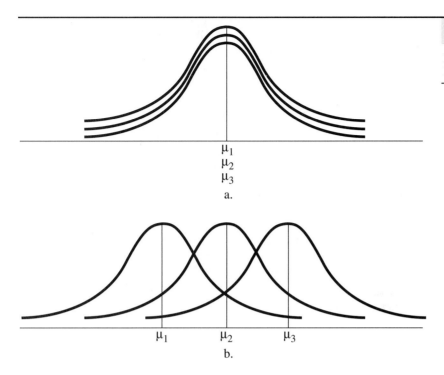

Figure 12.2

Relative Spread of Three
Population Distributions

The alternative hypothesis states that the three population means are not the same but says nothing about the relative separation of the distributions; one possibility is shown in Figure 12.2b. The larger the separation of the population means relative to the spread of the distributions, the larger the treatment effect and the higher F tends to be. Near the end of Section B, I will describe a way to quantify the relative separation of the population means (i.e., the effect size of the experiment).

Advantages of the One-Way ANOVA

Having rejected the null hypothesis for an experiment containing more than two groups, we would probably want to test each pair of group means to see more specifically where the significant differences are. (For instance, is the multivitamin condition significantly different from the vitamin C condition?) Why not go directly to performing the three t tests and just skip the ANOVA entirely? There are two main reasons. One reason was mentioned at the beginning of this chapter: The chance of making a Type I error increases as more pairs of means are tested. The problem is minor when there are only three groups, but it becomes quite serious as groups are added to the design. The second reason is that the ANOVA can find a significant difference among several group means even when no two of the means are significantly different from each other. These aspects of the one-way ANOVA will be illuminated when I discuss procedures for following the ANOVA with t tests in the next chapter.

Had our three groups not contained the same number of subjects, we could not have used Formula 12.5, and we might have been concerned about whether all three populations had the same variance. The general formula that can be used even when the groups are unequal in size is given in Section B, along with a numerical example.

SUMMARY

1. A *one-way* ANOVA has only one independent variable, which can have any number of *levels*. Each group in the experiment represents a different level of the independent variable.
2. The significance of the ANOVA is tested with an *F ratio*. The *F* ratio follows the *F distribution* when the null hypothesis is true. The null hypothesis is that all of the population means are equal (H_0: $\mu_1 = \mu_2 = \mu_3 = \ldots$).
3. The denominator of the *F* ratio, usually called MS_W (or MS_{error}), is just the average of the variances of all the groups in the experiment. (A weighted average, like that used for the pooled variance of the two-group *t* test, must be used when the groups are not all the same size.)
4. The numerator of the *F* ratio, usually called MS_{bet}, is the variance of the group means multiplied by the sample size. (The formula gets a bit more complicated when the groups are not all the same size.)
5. The *F* distribution is a family of distributions that tend to be positively skewed and to have a mean that is near 1.0. The exact shape of the *F* distribution is determined by both the df for the numerator ($df_{bet} = k - 1$, where k = the number of groups) and the df for the denominator ($df_W = N_T - k$, where N_T is the total number of subjects in all groups combined). Knowing df_{bet} and df_W, as well as alpha, you can look up the critical *F* in Table A.7, A.8, or A.9. If the calculated *F* ratio is larger than the critical *F*, the null hypothesis can be rejected.
6. The *F* test for ANOVA is always one-tailed, in that only *large* calculated *F* ratios (i.e., those in the positive, or right-hand, tail of the *F* distribution) lead to statistical significance. *F* ratios less than 1, even if close to zero, only indicate that the sample means are unusually close together, which is not inconsistent with the null hypothesis that all the population means are equal.
7. A significant *F* tells us that there is a treatment effect (i.e., the population means are not all equal) but does not tell us which pairs of population means differ significantly. The procedures of Chapter 13 are needed to test pairs of sample means following a multigroup ANOVA.

EXERCISES

*1. Consider a one-way ANOVA with five samples that are all the same size, but whose standard deviations are different: $s_1 = 10$, $s_2 = 15$, $s_3 = 12$, $s_4 = 11$, and $s_5 = 10$. Can you calculate MS_W from the information given? If so, what would be the value for MS_W?

2. a. If 120 subjects are divided equally among three groups, what are the dfs that you need to find the critical *F*, and what is the critical *F* for a .05 test?
 b. What is the critical *F* for a .01 test?
 c. If 120 subjects are divided equally among six groups, what are the dfs that you need to find the critical *F*, and what is the critical *F* for a .05 test?
 d. What is the critical *F* for a .01 test?

 e. Compare the critical *F*s in parts a and c, and do the same for parts b and d. What is the effect on critical *F* of adding groups, if the total number of subjects remains the same?

*3. If $df_{bet} = 4$, $df_W = 80$, and all groups are the same size, how many groups are there in the experiment, and how many subjects are in each group?

4. In Exercise 7B2, a two-group *t* test was performed to compare 15 "more vivid" visual imagers with 15 "less vivid" visual imagers on color recall. For the more vivid group, $\overline{X}_1 = 12$ colors with $s_1 = 4$; for the less vivid group, $\overline{X}_2 = 8$ colors with $s_2 = 5$.
 a. Calculate the *F* ratio for these data.

b. How does your calculated F in part a compare to the t value you found for Exercise 7B2? What is the general rule relating t and F in the two-group case?

c. What is the appropriate critical F for testing your answer to part a? How does this value compare with the critical t value you used in Exercise 7B2?

d. Which statistical test, t or F, is more likely to lead to statistical significance when dealing with two equal-sized groups? Explain your answer.

*5. The 240 students in a large introductory psychology class are scored on an introversion scale that they filled out in class, and then they are divided equally into three groups according to whether they sit near the front, middle, or back of the lecture hall. The means and standard deviations of the introversion scores for each group are as follows:

	Front	Middle	Back
$\overline{X}$	28.7	34.3	37.2
s	11.2	12.0	13.5

Calculate the F ratio.

6. Suppose the standard deviations in Exercise 5 were twice as large, as follows:

	Front	Middle	Back
s	22.4	24.0	27.0

Calculate the F ratio and compare it to the F ratio you calculated for Exercise 5. What is the effect on the F ratio of doubling the standard deviations?

*7. A psychologist is studying the effects of various drugs on the speed of mental arithmetic. In an exploratory study, 32 subjects are divided equally into four drug conditions, and each subject solves as many problems as he or she can in 10 minutes. The mean number of problems solved follows for each drug group, along with the standard deviations:

	Marijuana	Amphetamine	Valium	Alcohol
$\overline{X}$	7	8	5	4
s	3.25	3.95	3.16	2.07

a. Calculate the F ratio.
b. Find the critical F ($\alpha = .05$).
c. What can you conclude with respect to the null hypothesis?

8. If the study in Exercise 7 were repeated with a total of 64 subjects:
a. What would be the new value for calculated F?
b. How does the F ratio calculated in part a compare to the F calculated in Exercise 7? What general rule relates changes in the F ratio to changes in sample size (when all samples are the same size and all else remains unchanged)?
c. What is the new critical F ($\alpha = .05$)?

*9. Suppose that the F ratio you have calculated for a particular one-way ANOVA is .04. Which of the following can you conclude?
a. A calculation error has probably been made.
b. The null hypothesis can be rejected because $F < .05$.
c. There must have been a great deal of within-group variability.
d. The null hypothesis cannot be rejected.
e. No conclusions can be drawn without knowing the df.

10. Suppose that the F ratio you have calculated for another one-way ANOVA is 23. Which of the following can you conclude?
a. A calculation error has probably been made (F values this high are too unlikely to arise in real life).
b. The null hypothesis can be rejected at the .05 level (as long as all groups contain at least two subjects).
c. The group means must have been spread far apart.
d. There must have been very little within-group variability.
e. The sample size must have been large.

Just as in the case of the t test, the levels of an independent variable in a one-way ANOVA can be created experimentally or can occur naturally. In the following ANOVA example, I will illustrate a situation in which the levels of the independent variable occur naturally, and the experimenter randomly selects subjects from these preexisting groups. At the same time, I will be illustrating the computation of a one-way ANOVA in which the samples are not all the same size.

B

BASIC STATISTICAL PROCEDURES

An ANOVA Example with Unequal Sample Sizes

A psychologist has hypothesized that the death of a parent, especially before a child is 12 years old, undermines the child's sense of optimism and that this deficit is carried into adult life. She further hypothesizes that the loss of both parents before the age of 12 amplifies this effect. To test her hypothesis, the psychologist has found four young adults who were orphaned before the age of 12, five more who lost one parent each before age 12, and, for comparison, six young adults with both parents still alive. Each subject was tested with an optimism questionnaire whose possible scores ranged from zero to 50. The research hypothesis that parental death during childhood affects optimism later in life can be tested following the six steps of hypothesis testing that you have already learned.

Step 1. State the Hypotheses

In the case of a one-way ANOVA, the null hypothesis is very simple. The null hypothesis always states that all of the population means are equal; in symbols, $\mu_1 = \mu_2 = \mu_3 = \mu_4 = \ldots$, etc. Because our example involves three groups, the null hypothesis is H_0: $\mu_1 = \mu_2 = \mu_3$.

For one-way ANOVA, the alternative hypothesis is simply that the null hypothesis is *not* true, (i.e., that the population means are not all the same). However, this is not simple to state symbolically. The temptation is to write H_A: $\mu_1 \neq \mu_2 \neq \mu_3$, but this is *not* correct. Even if $\mu_1 = \mu_2$, the null hypothesis could be false if μ_3 were not equal to μ_1 and μ_2. In fact, in the case of three groups, there are four ways H_0 could be false: $\mu_1 = \mu_2 \neq \mu_3$; $\mu_1 \neq \mu_2 = \mu_3$; $\mu_1 \neq \mu_3 = \mu_2$; or $\mu_1 \neq \mu_2 \neq \mu_3$. The alternative hypothesis does not state which of these will be true, only that one of them will be true. Of course, with more groups in the experiment there would be even more ways that H_0 could be false and H_A could be true. Therefore, we do not worry about stating H_A symbolically, other than to say that H_0 is not true.

Step 2. Select the Statistical Test and the Significance Level

Because we want to draw an inference about more than two population means simultaneously, the one-way analysis of variance is appropriate—assuming that our optimism score can be considered as arising from an interval or ratio scale. As usual, alpha = .05.

Step 3. Select the Samples and Collect the Data

When you are dealing with preexisting groups and therefore not randomly assigning subjects to different levels of the independent variable, it is especially important to select subjects as randomly as possible. In addition, it is helpful to select samples that are all the same size and rather large so that you need not worry about homogeneity of variance (see the discussion of assumptions later in this section) or low power. However, practical considerations may limit the size of your samples, and one group may be more limited than another. In the present example, it may have been difficult to find more than four young adults with the characteristics needed for the first group. Then, having found five subjects for the second group, the psychologist probably decided it would not be worth the loss of power to throw away the data for one subject in the second group just to have equal-sized groups. The same considerations apply to having six subjects in the third group. The optimism rating for each subject is shown in Table 12.2 along with the mean and SD of each group.

Table 12.2

Both Parents Deceased	One Parent Deceased	Both Parents Alive
29	30	35
35	37	38
26	29	33
22	32	41
	25	28
		40
$\overline{X} = 28$	$\overline{X} = 30.6$	$\overline{X} = 35.83$
$s = 5.477$	$s = 4.393$	$s = 4.875$

Step 4. Find the Region of Rejection

Because we are using the F ratio as our test statistic, and it is the ratio of two independent estimates of the same population variance, the appropriate null hypothesis distribution is one of the F distributions. To locate the critical F for our example we need to know the df for both the numerator and the denominator. For this example, k (the number of groups) = 3, and N_T (the total number of subjects) = 4 + 5 + 6 = 15. Therefore, $df_{bet} = k - 1 = 3 - 1 = 2$, and $df_W = N_T - k = 15 - 3 = 12$. Looking at the .05 table (Table A.7), we start at the column labeled 2 and move down to the row labeled 12. The critical F listed in the table is 3.89. As you can see from Figure 12.3, the region of rejection is the portion of the F distribution that is above (i.e., to the right of) the critical F. The test is one-tailed; the .05 area is entirely in the upper tail, which represents sample means that are more spread out than typically occurs through random sampling alone.

Step 5. Calculate the Test Statistic

To deal with the unequal sample sizes in our example, we must modify Formula 12.5. The denominator is easy to modify; instead of taking the simple average of the variances, we go back to the concept of the pooled variance and take a weighted average, as we did for the t test. We can expand Formula 7.6A to accommodate any number of groups. (Remember that s_p^2 is now being referred to as MS_W.) When a one-way ANOVA contains groups of different sizes, I will use a lowercase n with a subscript to represent the size

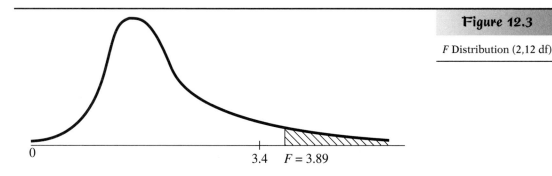

Figure 12.3

F Distribution (2,12 df)

of one particular group (the subscript i will be used to represent any group). As usual, an uppercase N with the subscript T will be used to represent total N. The formula for the denominator (MS_W) becomes:

$$MS_W = \frac{(n1 - 1)s_1^2 + (n2 - 1)s_2^2 + (n3 - 1)s_3^2 + \cdots + (n_k - 1)s_k^2}{N_T - k}$$

This expression can be rewritten more compactly to create Formula 12.6:

$$MS_W = \frac{\sum (n_i - 1)s_i^2}{df_W} \qquad \textbf{Formula 12.6}$$

The numerator of Formula 12.5 must also be modified to incorporate the different sample sizes; the squared difference of each sample mean from the grand mean must be weighted by the size of the sample, as shown next:

$$MS_{bet} = \frac{n_1(\overline{X}_1 + \overline{X}_G)^2 + n_2(\overline{X}_2 - \overline{X}_G)^2 + n_3(\overline{X}_3 - \overline{X}_G)^2 + \cdots + n_k(\overline{X}_k - \overline{X}_G)^2}{k - 1}$$

Written more compactly, this expression becomes Formula 12.7:

$$MS_{bet} = \frac{\sum n_i(\overline{X}_i - \overline{X}_G)2}{df_{bet}} \qquad \textbf{Formula 12.7}$$

The preceding equations represent the generally accepted approach to calculating the one-way ANOVA when the sample sizes are unequal. This approach is called the *analysis of weighted means*, and it is the only method I will discuss in this section. An alternative approach, the *analysis of unweighted means*, will be discussed briefly in Section C. Although the analysis of unweighted means is occasionally calculated for the one-way ANOVA (and sometimes recommended by statisticians), it is not available from the common statistical packages, which may serve partly as a cause, as well as an effect of, its lack of popularity.

I will apply these new equations to the data of the present example, beginning with MS_W, because that part is more straightforward. First, square the three (unbiased) standard deviations, as given in Table 12.1. The three variances are $s_1^2 = 30$; $s_2^2 = 19.3$, and $s_3^2 = 23.77$. Now that we have the three variances, it is easy to apply Formula 12.6:

$$MS_W = \frac{\sum (n_i - 1)s_i^2}{df_W} = \frac{3(30) + 4(19.3) + 5(23.8)}{12} = \frac{286.03}{12} = 23.84$$

Remember that MS_W is a weighted average of the sample variances, so it must be somewhere between the smallest and largest of your sample variances or you have made a calculation error. In this case, MS_W (= 23.84) is about in the middle of the three variances (30, 19.3, 23.77), as it should be.

Before we can apply Formula 12.7 to find MS_{bet}, we must find the grand mean, $\overline{X}_G$, which is simply the ordinary mean of all of your scores. The easiest way to find the grand mean is just to add all of the scores from all of the groups together and then divide by the total number of scores (N_T). However, if you had the group means but not the raw scores, you could take a weighted mean of the group means using Formula 3.10, which involves multiplying each group mean by its sample size, adding these products, and then dividing by the total N. I will illustrate the latter method:

$$\overline{X}_G = \frac{\sum n_i \overline{X}_i}{N_T} = \frac{4(28) + 5(30.6) + 6(35.83)}{15} = \frac{480}{15} = 32$$

Now we can calculate Formula 12.7:

$$MS_{bet} = \frac{\sum n_i(\overline{X}_i - \overline{X}_G)^2}{df_{bet}} = \frac{4(28 - 32)^2 + 5(30.6 - 32)^2 + 6(35.83 - 32)^2}{2}$$

$$= \frac{4(16) + 5(1.96) + 6(14.69)}{2}$$

$$= \frac{64 + 9.8 + 88.17}{2} = \frac{161.97}{2} = 80.98$$

Finally, we calculate the F ratio:

$$F = \frac{MS_{bet}}{MS_W} = \frac{80.98}{23.84} = 3.4$$

Step 6. Make the Statistical Decision

The calculated $F(3.4)$ is not as large as the critical $F(3.89)$, so our result does *not* land in the region of rejection (see Figure 12.3). We cannot reject the null hypothesis that all of the population means are equal. This is not surprising when you consider that with sample sizes as tiny as those in this experiment, the power of a one-way ANOVA is low for any effect size that is not very large. (The effect size for this example will be calculated later in this section.)

Interpreting Significant Results

Had we rejected the null, we could have concluded that the population means are not all the same, but we could not conclude that they are all different. We would need to conduct follow-up *t* tests to determine which pairs of population means are significantly different from each other (this will be explained in the next chapter). Moreover, because the three conditions were not created by the experimenter, rejecting the null would not mean that parental death *causes* changes in optimism. Alternative explanations are possible. For instance, pessimism may run in families, and pessimism may lead to poor health and early death. According to this hypothetical explanation, both the child's pessimism *and* the early parental death are caused by a third variable—that is, the parents' pessimism.

If the independent variable had involved an experimental manipulation (e.g., all subjects are given a "test" and then given randomly chosen feedback suggesting that their future job prospects are good, fair, or poor), significant differences in optimism between the (randomly assigned) groups could be attributed to the feedback manipulation. In this case, we might talk of the underlying population means being different, but we are not referring to actual populations; we are referring to theoretical populations (e.g., the mean of the population if everyone in the population had received positive feedback, the mean if everyone in the population had received negative feedback, etc.). To conclude that the population means are not all the same (i.e., reject the null hypothesis in a one-way ANOVA) in this case would be to say that the manipulation had some effect on optimism. To quantify the size of that effect, you will need to calculate an effect-size meaure like the one described at the end of this section, or the one described in Section C.

The Sums of Squares Approach

So far in this chapter, I have shown you how to calculate an ANOVA from the means and standard deviations of your samples. This is not the approach shown in most texts. Nearly all statistics texts emphasize the calculation of ANOVA directly from the raw scores and do not make it clear that the calculation of ANOVA from means and *SD*s is even possible. The approach I have taken here is useful in showing that ANOVA is not as different from the *t* test as it at first appears. Moreover, the mean/*SD* approach shows you how to calculate the ANOVA when the raw scores are not available; all you need is the descriptive table usually published for a multigroup experiment, which contains the $\overline{X}$, *SD*, and *n* for each group. The standard raw-score formulas used to have the advantage of decreasing computational steps. However, modern (and inexpensive) scientific calculators make it so easy to obtain means and standard deviations that the mean/*SD* approach is now the easier approach when using such a calculator. However, the raw-score approach still has an advantage in that it deals first with sums of squares (*SS*s), calculating variances only at the final step. Although the usual raw-score formulas are tedious and uninstructive (and generally used only by computer programs and students in statistics classes), dealing with the *SS*s as an intermediate step has several advantages.

First, let me make it clear what these *SS*s are. The numerator of Formula 12.7 is SS_{bet}, and the numerator of Formula 12.6 is SS_W. Therefore, $MS_{bet} = SS_{bet}/df_{bet}$, and $MS_W = SS_W/df_W$. These *SS*s have the convenient property that when you add them, you get the total *SS*—that is, the *SS* you would get from squaring every subject's deviation from the grand mean (regardless of group) and adding them all up. The *MS*s do not add up this way. MS_{bet} and MS_W do not add up to the total *MS* (i.e., the variance of all subjects from the grand mean). Consequently, it would make more sense to call ANOVA the analysis of sums of squares because it is the total *SS* that is being broken down into two additive components. The additive property of the *SS*s becomes especially useful when dealing with more complex ANOVA designs, as you will see in later chapters. Although it is not difficult to use the mean/*SD* approach (i.e., numerators of Formulas 12.6 and 12.7) to calculate SS_W and SS_{bet}, there is a good chance you will run into the standard raw-score formulas one place or another. The raw-score formulas also have the advantage of being far less susceptible to errors due to rounding off than the mean/*SD* approach, so I will present them in the following paragraphs.

Raw-Score Formulas

Within-Group Sum of Squares (SS_W)

One raw-score method for calculating SS_W involves calculating *SS* separately for each group using the computational formula (Formula 3.12) $SS = \sum X^2 - (\sum X)^2/n$. All of these within-group *SS*s are then summed to yield SS_W. (Compare the numerator of Formula 12.6 to Formulas 7.6A and 7.6B, and you'll see that Formula 12.6 also involves finding the *SS* for each group and then adding.) Fortunately, there is a shortcut formula that sums all of the within-group *SS*s in one step, but it requires some additional notation. To avoid using too many summation signs, I will define T_i as the sum of scores for one particular (i.e., the *i*th) group. Now the formula for SS_W can be written as:

$$SS_W = \sum X^2 - \sum \left(\frac{T_i^2}{n_i} \right)$$ **Formula 12.8**

where $\sum X^2$ is the sum of *all* the squared scores (from all of the groups). I will illustrate the use of this formula by applying it to the optimism example. To find $\sum X^2$ we must sum the squared scores for all 15 subjects. In this case, $\sum X^2 = 15{,}808$. Then we square the sum of each group, divide each squared sum by the appropriate group size, and add these up:

$$\frac{\sum T_i^2}{n_i} = \frac{112^2}{4} + \frac{153^2}{5} + \frac{215^2}{6} = \frac{12.544}{4} + \frac{23{,}409}{5} + \frac{46{,}225}{6}$$

$$= 3{,}136 + 4{,}681.8 + 7{,}704.17 = 15{,}521.97.$$

Finally, $SS_W = \sum X^2 - S\,T_i^2/n_i\; 15{,}808 - 15{,}521.97 = 286.03$. Note that this is exactly the same value we found for the numerator of Formula 12.6. We need to divide SS_W by df_W to yield MS_W. As before, $MS_W = 286.03/12 = 23.84$.

Between-Group Sum of Squares (SS_{bet})

Corresponding to the preceding computational formula for SS_W, there is a computational formula for SS_{bet}, but I need to introduce one new symbol to keep the notation simple. I will express the sum ($\sum X$) for all scores (regardless of group) as T_T—the total of all the group totals (T_T divided by N_T is the grand mean).

$$SS_{bet} = \sum \left(\frac{T_i^2}{n_i} \right) - \frac{T_T^2}{N_T}$$ **Formula 12.9**

Notice that the first term in the formula is exactly the same as the second term in Formula 12.8. The second term in the above formula is the sum of all scores being squared and divided by the total number of scores.

To complete the example, I will apply Formula 12.9 to the data in Table 12.1. First we need to find the total of all scores; $T_T = 480$. Next, notice that we do not have to recalculate the first term of Formula 12.9; we already found this value in the calculation of Formula 12.8: $\sum (T_i^2/n_i) = 15{,}521.97$. Inserting these values into Formula 12.9, we obtain:

$$SS_{bet} = \sum \left(\frac{T_i^2}{n_i} \right) - \frac{T_T^2}{N_T} = 15{,}521.97 - \frac{480^2}{15} = 15{,}521.97 - 15{,}360 = 161.97$$

The value for SS_{bet} must be divided by df_{bet} to yield MS_{bet}, which equals $161.97/2 = 80.98$—the same answer we obtained from Formula 12.7.

The Total Sum of Squares (SS_{tot})

If SS_{bet} and SS_W are added together, the result is SS_{tot}; I have expressed this as Formula 12.10:

$$SS_{total} = SS_{bet} + SS_W$$ **Formula 12.10**

The implication of this formula is that if you have calculated SS_{total}, it is not necessary to calculate both SS_{bet} and SS_W; you can calculate either one of them and then subtract it from SS_{total} to get the other one. It is quite simple to find SS_{total} directly, using the usual computational formula for any SS. In terms of the notation I introduced above, the computational formula for SS_{tot} is:

$$SS_{total} = \sum X^2 = \frac{T_T^2}{N_T}$$

Formula 12.11

Note that if you apply Formula 12.10 to the computational formulas for SS_W and SS_{bet}, you will get the computational formula for SS_{total} (this is easier to see if you add them in the reverse order of Formula 12.10, as follows):

$$SS_{total} = SS_W + SS_{bet} = \sum X^2 - \sum \left(\frac{T_i^2}{n_i}\right) + \sum \left(\frac{T_i^2}{n_i}\right) - \frac{T_T^2}{N_T} = \sum X^2 - \frac{T_T^2}{N_T}$$

Written this way, you can see that the two middle terms cancel each other out.

Let us apply Formula 12.11 to our example. We have already calculated the term $\sum X^2$ to find SS_W using Formula 12.8. The second term is also the second term of Formula 12.9, which we calculated in finding SS_{bet}. Therefore, Formula 12.11 gives us $SS_{total} = \sum X^2 - T_T^2/N_T = 15,808 - 15,360 = 448$.

Assumptions of the One-Way ANOVA for Independent Groups

The assumptions underlying the test described in this chapter are the same as the assumptions underlying the t test of two independent means. The assumptions are briefly reviewed in the following, but for greater detail you should consult Section B of Chapter 7.

Independent Random Sampling

Ideally, all of the samples in the experiment should be drawn randomly from the population(s) of interest. However, if you are not dealing with pre-existing populations, it is likely that you would collect just one convenient sample and then *randomly* assign subjects to the different experimental conditions, usually with the constraint that all groups turn out to be the same size. As mentioned in Chapter 7, the randomness of the assignment to conditions is critical to the validity of the experiment. It is also assumed that there is no connection between subjects in different groups. If all of the samples are in some way matched, as described for the matched t test, the procedures of this chapter are not appropriate. Methods for dealing with more than two matched samples are presented in Chapter 15.

Normal Distributions

It is assumed that all of the populations are normally distributed. As in the case of the t test, however, with large samples we need not worry about the shapes of our population distributions. In fact, even with fairly small samples, the F test for ANOVA is not very sensitive to departures from the normal distribution—in other words, it is robust, especially when all the distributions are symmetric or skewed in the same way. However, when dealing with small samples, and distributions that look extremely different from the normal distribution and from each other, you should consider using nonparametric tests, such as the Kruskal-Wallis H test (see Chapter 21) or data transformations (see Chapter 3, Section C). Especially if your data contain outliers, you should consider a robust form of ANOVA, as described in Section C of this chapter.

Homogeneity of Variance

It is assumed that all of the populations involved have the same variance. However, when the sample sizes are all equal, this assumption is routinely ignored. Even if the sample sizes and sample variances differ slightly, there is little cause for concern. Generally, if no sample variance is more than 4 times as large as another (i.e., no standard deviation is more than twice as large as another), and no sample is more than 1.5 times as large as another, you can proceed with the ordinary ANOVA procedure, using the critical F from the table with negligible error. It is when the sample sizes are considerably different (and not very large) and the sample variances are not very similar that there is some cause for concern. If you are concerned that the homogeneity of variance (HOV) assumption does not apply to your situation, you may want to consider testing the HOV assumption, as described next.

Testing Homogeneity of Variance

Two-Group Case

I mentioned in Chapter 7 (Section C) that the F ratio could be used to test homogeneity of variance. Now that you have learned something about the F distribution, I can go into greater detail. When performing an HOV test in the two-group case, whether to justify a pooled-variance t test or to determine whether an experimental treatment increases the variance of a population compared to a control procedure, the null hypothesis is that the two populations have the same variance ($H_0: \sigma_1^2 = \sigma_2^2$). Therefore, both sample variances are considered independent estimates of the same population variance. The ratio of two such estimates will follow the F distribution if the usual statistical assumptions are satisfied.

For example, suppose that the variance of a control group of 31 subjects is 135, and the variance of 21 experimental subjects is 315. Because our table of critical F values contains only Fs greater than 1.0, we take the larger of the two variances and divide by the smaller to obtain $F = 315/135 = 2.33$. To find the critical F, we need to know the df for both the numerator and denominator of the F ratio. Because we put the experimental group's variance in the numerator, $df_{num} = n_{num} - 1 = 21 - 1 = 20$. The $df_{denom} = n_{denom} - 1 = 31 - 1 = 30$ because the control group's smaller variance was used in the denominator. Finally, we must choose alpha. If we set $\alpha = .05$, however, we cannot use the F table for that alpha; we must use the F table for $\alpha/2 = .025$. The reason we must use $\alpha/2$ is that, in contrast to ANOVA, this time we are really performing a two-tailed test. The fact is that if the control group had had the larger variance, our F ratio would have been less than 1.0, and possibly in the left tail of the distribution. Of course, in that case, we would have turned the ratio upside down to make F greater than 1.0, but it would have been "cheating" not to allow for alpha in both tails because either variance could be the larger. (In ANOVA we would never turn the ratio upside down, so it is reasonable to perform a one-tailed test.)

Looking up $F(20, 30)$ at $\alpha = .025$, we find that the critical F is 2.20. Because our calculated F is higher than the critical F, the null hypothesis that the population variances are equal must be rejected. A pooled t test is not justified, and you can conclude that your experimental treatment alters the variance of the population. However, these conclusions are justified only if it is reasonable to make the required assumptions. The most important assumption in this case is that the two populations follow normal distributions. Whereas the t test and the ANOVA F test are robust with respect to violations of the normality assumption, the major drawback of the F test for homogeneity of variance is that it *is* strongly affected by deviations from normal distributions, especially when the samples are not large. This problem makes the F ratio suspect as an HOV test in most psychological applications, and therefore it is rarely used. Alternatives for the two-group case are discussed along with tests for the multigroup case next.

The Multigroup Case

The F test discussed in the preceding paragraph can be modified to test HOV for an experiment that includes more than two groups. An F ratio is formed by dividing the largest of the sample variances by the smallest; the resulting ratio is called F_{max} (Hartley, 1950), and it must be compared to a critical value from a special table. Hartley's F_{max} test is also sensitive to violations of the normality assumption, which is why it is rarely used these days. Alternative procedures, less sensitive to violations of the normality assumption, have been devised for testing HOV in the two-group as well as the multigroup case (e.g., Levene, 1960; O'Brien, 1981). For instance, Levene's (1960) test, which is the one provided by SPSS automatically when a t test is performed or when a homogeneity test is requested for an ANOVA, is based not on the raw scores but on their differences from their group mean, and it is not easily affected by the shape of the distribution. As these difference scores vary more in one group than another, indicating differences in variance, Levene's test produces a larger F ratio. If the p value corresponding to the F from Levene's test, is less than .05, you have a strong indication that you need to either transform your data or modify the one-way ANOVA.

On the other hand, a p only slightly greater than .05 for Levene's test cannot give you great confidence in the HOV assumption when you are dealing with fairly small sample sizes. Unfortunately, none of the homogeneity tests has much power when sample sizes are small, so you still must use some judgment in those cases. If it seems unreasonable to assume homogeneity of variance based on your data, and your sample sizes are not all equal, there are adjustments to the ANOVA procedure, which are analogous to the separate-variance t test, and should be considered. A discussion of two of these adjustments is presented in Section C.

Power and Effect Size for ANOVA

The concept of power as applied to the one-way independent ANOVA is essentially the same as the concept discussed in Chapter 8 with respect to the t test. When the null hypothesis is true but the ANOVA is repeated many times anyway, the F ratios will follow one of the ordinary (or "central") F distributions, which has an average, F_0, equal to $df_W/(df_W - 2)$. However, if the null hypothesis is not true, the F ratios will follow a similar distribution called the *noncentral F distribution*. The average, or expected, F of a noncentral F distribution (I'll call it F_A) is greater than F_0 and depends on a *noncentrality parameter* (ncp), just as in the case of the noncentral t distribution (it is also sometimes referred to as delta). To find the ncp for a particular alternative hypothesis, I will begin by using the same trick I used in the two-group case: I will take the formula for the F ratio when the sample sizes are equal and substitute the population value corresponding to each sample statistic. First, however, I will rewrite the F ratio in a form more convenient for my purposes. Recall that the numerator of the F ratio, MS_{bet}, can be written as $ns_{\bar{X}}^2$ (Formula 12.5A). Because MS is just another way to symbolize a variance, MS_W can be written as s_W^2, so the F ratio, when sample sizes are equal can be written as

$$F = \frac{ns_{\bar{X}}^2}{s_W^2}$$ **Formula 12.12**

Substituting population values for the sample statistics in the preceding formula gives me a quantity which is the square of something called ϕ (the Greek letter phi) and which is closely related to both the ncp and the expected value of F when the alternative hypothesis is true.

$$\phi^2 = \frac{n\sigma_{\bar{X}}^2}{\sigma^2}$$

I can use sigma squared in the denominator without a subscript, because I'm assuming that all the populations have the same variance. There is more than one way to define the ncp, but one convenient way is to use the symbol δ^2 (delta squared), so that $\delta^2 = k\phi^2$, where k is the number of groups (Hays, 1994, p. 409). Also bear in mind that neither δ^2 nor ϕ^2 is the expected F under the alternative hypothesis; $F_A = F_0 + k/(k-1) \times \phi^2 \times F_0$. However, it is the quantity ϕ (without being squared) that forms a convenient basis for power tables, so I will take the square root of both sides of the preceding equation to derive the following convenient formula:

$$\sigma = \sqrt{n}\,\frac{\sigma_{\bar{X}}}{\sigma}$$

Effect Size

It is ϕ that plays the same role for ANOVA that δ does for the t test; Table A.10 allows you to find the power that corresponds to a particular value of ϕ and vice versa. However, estimating ϕ is not an easy matter. As in the case of the t test, it is useful to separate the part that depends on sample size from the rest of the formula. The remainder of the formula is called $\mathbf{f}$ (the use of boldface signals that this is a population value), and it is a measure of effect size that is the multigroup analog of $\mathbf{d}$. This relation is shown in the next two formulas:

$$\mathbf{f} = \frac{\sigma_{\bar{X}}}{\sigma}$$ **Formula 12.13**

$$\phi = \mathbf{f}\sqrt{n}$$ **Formula 12.14**

When there are only two population means, you can take their difference and divide by their common standard deviation to obtain $\mathbf{d}$. With more than two populations, the spread of the populations is represented by the standard deviation of the population means and again divided by the common standard deviation to obtain $\mathbf{f}$. Because the (biased) standard deviation of two numbers is half of their difference, $\mathbf{f} = \mathbf{d}/2$ in the two-group case, but, of course, the concept is the same. Just as with $\mathbf{d}$, $\mathbf{f}$ must be estimated from previous research or by estimating values for the population means and standard deviation. My use of the symbol $\mathbf{f}$ comes from Cohen (1988), who offered the following guidelines for situations in which the estimation of $\mathbf{f}$ is difficult: $\mathbf{f} = .1$ is a small effect size, $\mathbf{f} = .25$ is medium, and $\mathbf{f} = .4$ is large.

Of course, the preceding formulas would not be so simple if the sample sizes were not all equal. However, even if slightly unequal sample sizes are fairly common, experiments are almost always planned in terms of equal sample sizes, so these formulas are probably all you will ever need for power analysis. The effect size formulas can be modified to accommodate unequal ns, along the lines of Formulas 12.6 and 12.7, but the concepts stay essentially the same. Therefore, I will confine my examples to the equal-n case.

Just as you can take the difference of the sample means from a two-group experiment and divide by s_{pooled} to get g (an estimate of **d**), you can divide MS_{bet} by n and then divide by MS_W to get a sample statistic I will call f (without the boldface). You can also find f from the F ratio you calculated for your experiment, according to the following relation:

$$f = \sqrt{\frac{F}{n}}$$

Formula 12.15

Whereas g is often used as an estimate of **d**, f is too biased as an estimator of **f** in the population to be used without correction. The correction factor is $(k - 1)/k$, as shown in the following formula for estimating **f** from your calculated F ratio:

$$\text{Estimated } \mathbf{f} = \sqrt{\left(\frac{k - 1}{k}\right)\frac{F}{n}}$$

Formula 12.16

The two denominators in the preceding formula, k and n, when multiplied yield N_T, so the formula is often written with $(k - 1)F/N_T$ under the square root sign. The version of the formula that I am using makes the correction factor stand out clearly.

I will illustrate the calculation of power for a one-way ANOVA in terms of the example in Section A. In that hypothetical experiment, k was equal to 3, n was equal to 10, and F was 3.36. We can estimate **f** from the sample values in that experiment using Formula 12.16:

$$\text{Estimated } \mathbf{f} = \sqrt{\left(\frac{3 - 1}{3}\right)\frac{3.36}{10}} = \sqrt{(.667)(.336)} = \sqrt{.224} = .473$$

If we take this estimate (.473 is a fairly large effect size) as the true value for **f**, and we plan to use 12 subjects in each of the three groups next time, ϕ for the next experiment (using Formula 12.14) equals $.473\sqrt{12} = .473(3.464) = 1.64$. We go to Table A.10 and look at the section for three ($k = 3$) groups. The column for ϕ that comes closest to our estimated value (1.64) is 1.6. In addition to finding that there is a different table for each number of groups (up to $k = 5$), you will notice another difference from the t test power table: You need to know the degrees of freedom for your error term (i.e., df_W). If there will be 12 subjects per group, $df_W = 36 - 3 = 33$. This is close to the entry for 30, so the power for the proposed experiment is about .65. [A better power estimate would be a little higher than this (about .68) because our df_W is a bit higher than 30, and ϕ is a bit higher than 1.6, but the power estimate is based on our guess that **f** will be related to f as in our last experiment, so there is often little point in trying to be very precise.]

The reason you did not need to use your df to look up power for the t test is that I was using the normal distribution as an approximation. The ANOVA power tables are more precise. The normal distribution approximation is included in each ANOVA table as the bottom row (infinite df). I have included a section for $k = 2$ in Table A.10 so that you can make more precise power estimates for t tests. However, you will notice a complication when you try to use the ANOVA tables backward to estimate the sample size required to attain a given level of power for a particular value of **f**. Suppose we wish to know what sample size to use if .7 power is sufficient, but we don't expect **f** to exceed .4. Solving for n in Formula 12.14, we get the following:

$$n = \left(\frac{\phi}{\mathbf{f}} \right)^2 \hspace{3cm} \textbf{Formula 12.17}$$

You can insert .4 for **f**, but ϕ must be looked up, which requires an estimate of df_W even though it is the sample size that you are trying to find. This is not as unreasonable as it sounds because your estimate of df_W can be quite crude; unless df_W is likely to be quite small, your estimate of df_W won't greatly affect the ϕ you look up. Suppose you think a df_W of 30 seems reasonable (about 10 subjects per group); for power = .7 ($\alpha = .05$), ϕ is about midway between 1.6 and 1.8. Thus, I will use 1.7 for ϕ and derive the following required sample size:

$$n = \left(\frac{1.7}{.4} \right)^2 = 4.25^2 = 18$$

Note that the above result suggests that you need 18 subjects in each of the three groups to have at least .7 power if **f** is no more than .4. If I had used $df_W = 20$, ϕ would have been about 1.71, and n would have been virtually unchanged. If I had used a very large value for df_W, ϕ would have been 1.6, yielding an n of 16, which is not very different from the preceding answer.

As you would expect, changing alpha to .01 reduces power—so unless you have a very good reason to be extra cautious about Type I errors, you would not want to use the .01 level (that is why I chose not to include values for alpha = .01 in Table A.10). To increase power without making alpha larger than .05, and without increasing your sample size, you would have to find a way to increase **f**. This can sometimes be done by using a more powerful experimental manipulation, thus spreading out the population means to a greater extent. When this is not feasible, one's focus can turn to minimizing subject-to-subject variability. A powerful way to reduce variability is to use repeated measures or match subjects, as described for the matched t test in Chapter 11; the application of matching to more than two treatments or conditions is described in Chapter 15. However, bear in mind that, as Formula 12.14 shows, power can be high even for extremely tiny values of **f**. If you are using very large sample sizes or other powerful procedures (e.g., repeated measures), you need to look carefully at the effect size estimate associated with any statistically significant result you find. Depending on the nature of your study, the estimated effect size may suggest that a particular significant result is not worth thinking about and that perhaps there was little point to using such powerful procedures in the first place.

Varieties of the One-Way ANOVA

I have already discussed that the levels of the independent variable in a one-way ANOVA can be experimentally created (e.g., vitamin treatments), or they can correspond to existing populations (e.g., number of living parents a subject has). The ANOVA is calculated the same way in both cases, but causal conclusions cannot be drawn in the latter case. There are two other notable variations concerning the determination of the levels of the independent variable in a one-way ANOVA; the first one described in the following does not affect the calculation of the ANOVA in the one-factor case, but the second one usually calls for an entirely different form of analysis.

Fixed versus Random Effects

The levels of an independent variable (IV) can be specific ones in which you are interested or just a sampling from many possible levels. For instance, for the vitamin experiment introduced in Section A, many different vitamins could have been tested. Vitamin C and the multivitamin were chosen specifically in that experiment because the researcher had a reason to be interested in their particular effects. Sometimes, however, there are so many possible treatments in which we are equally interested that we can only select a few levels at random to represent the entire set. For example, suppose a psychologist is creating a new questionnaire and is concerned that the order in which items are presented may affect the total score. There is a very large number of possible orders, and the psychologist may not be sure of which ones to focus on. Therefore, the psychologist may select a few of the orders at random to be the levels of a one-way ANOVA design. When the levels of the independent variable are selected *at random* from a larger set, the appropriate model for interpreting the data is a *random effects* ANOVA. The design that has been covered in this chapter thus far (in which the experimenter is specifically interested in the levels chosen and is not concerned with extrapolating the results to levels not actually included in the experiment) is called a *fixed effects* ANOVA. Effect size and power are estimated differently for the two models, and conclusions are drawn in a different manner. Fortunately, the calculation of the one-way ANOVA *F* ratio and its test for significance are the same for both fixed and random effects ANOVA (differences emerge only for more complex ANOVAs). Because the use of the random effects model is relatively rare in psychological research, and the concept behind it is somewhat difficult to explain, I will reserve further discussion of this topic to a chapter on advanced ANOVA design that I will post on my web site.

Qualitative versus Quantitative Levels

There is yet another way in which the levels of an independent variable in an ANOVA may be selected, and this distinction can lead to a very different procedure for analyzing the data. The levels can represent different values of either a *qualitative* (nominal, categorical) or a *quantitative* (interval/ratio) scale. The vitamin example of Section A involves a qualitative IV, and although the number of parents who die before a child is 12 is measured on a ratio scale, it is not likely that the levels (0, 1, or 2) would be treated as quantitative. As a clear example of a one-way ANOVA in which the independent variable has quantitative levels, consider the following experiment. A researcher wants to know if more therapy sessions per week will speed a patient's progress, as measured by a well-being score. Five groups are randomly formed and subjects in each group are assigned to either one, two, three, four, or five sessions per week for the 6-month experimental period. Because both the independent variable (i.e., number of sessions per week) and the dependent variable (i.e., well-being score) are quantitative, a linear regression approach may be advantageous. ANOVA can be considered a special case of regression, as shown in Section C, and in later chapters.

To get an idea of the advantage of linear regression over ANOVA when dealing with a quantitative independent variable, imagine that the mean well-being score is a perfect linear function of the number of sessions, as shown in Figure 12.4. Because all of the sample means fall on the same straight line, the SS explained by linear regression (called $SS_{regression}$, in this context) is the same as SS_{bet}, and the unexplained SS (called $SS_{residual}$) is equal to SS_W. The difference is in the degrees of freedom. With the linear regression approach, the numerator df is just 1, so unlike SS_{bet}, which would be divided by 4 for this example (i.e., $k - 1$), $SS_{regression}$ is only divided by 1. Thus the numerator for linear regression is 4 times as large as the numerator for the ANOVA. Because the denominator for regression will be a bit smaller than the denominator for ANOVA ($SS_{residual}$ is divided by $N - 1$, as compared to $N - k$ for SS_W), the F ratio for testing the linear regression will be somewhat more than 4 times the F ratio for ANOVA.

As the sample means deviate more and more from a straight line, $SS_{regression}$ becomes less than SS_{bet}, and the advantage of the linear regression approach decreases; but as long as the relation between the independent and dependent variables is somewhat linear, the regression approach will probably lead to a higher F than ANOVA. If the sample means fall on some kind of curve, a more sophisticated approach, such as multiple regression (Chapter 17) or the analysis of trend components, would probably have more power. The latter approach will be described in the next chapter. At this point I just want you to know that a study involving, for example, several dosage levels of the same drug or different amounts of training on some task or some other quantitative independent variable should probably *not* be analyzed with the methods in this chapter; some form of regression or trend analysis will usually yield more useful information and more power.

Publishing the Results of a One-Way ANOVA

Because the results of the parental death/optimism experiment were not statistically significant, they would probably be described for publication as follows: "Although there was a trend in the predicted direction, with

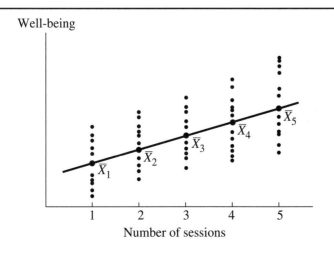

Figure 12.4

Linear Regression as an Alternative to ANOVA When the Independent Variable Has Quantitative Levels

parental loss associated with decreased optimism, the results of a one-way ANOVA failed to attain significance, $F(2, 12) = 3.4$, $MSE = 23.84$, $p > .05$." Had the F ratio been as large as, say 4.1, significant results could have been reported in the following way: "Early parental death had an impact on the levels of optimism in young adults. Subjects who had lost both parents before the age of 12 were least optimistic ($M = 28$), those who had lost only one parent were somewhat more optimistic ($M = 30.6$), and those whose parents were still alive were the most optimistic ($M = 35.83$). A one-way ANOVA demonstrated that these differences were statistically reliable, $F(2, 12) = 4.1$, $MSE = 23.84$, $p < .05$." The numbers in parentheses after F are df_{bet} followed by df_W. MSE stands for MS_{error}, which is another way of referring to MS_W in a one-way ANOVA. Reporting MSE is in accord with the recommendation of the APA style manual to include measures of variability along with means and the test statistic (in this case, F) so that the reader can reconstruct your analysis. For instance, with the preceding information, the reader could create a summary table. It is also becoming common to report, as an alternative measure of effect size, an estimate of the proportion of variance accounted for in the population by the independent variable. These types of effect size measures, such as omega squared, will be discussed in Section C.

Summary Table for One-Way ANOVA

In addition to presenting the means, the F ratio, and the significance level, the entire ANOVA is sometimes presented in the form of a *summary table*, including the value of each SS component. Summary tables are more likely to be presented when dealing with complex ANOVAs, but I will introduce the format here in Table 12.3, which shows the summary for the one-way

Table 12.3

Source	SS	df	MS	F	p
Between-groups	161.97	2	80.98	3.4	>.05
Within-groups	286.03	12	23.84		
Total	448.00	14			

ANOVA performed on the parental death/optimism data. The heading of the first column, "Source," refers to the fact that there is more than one source of variability if we look at all scores in the experiment. One source is "between-groups," sometimes called "treatment"; if the group means were further apart, this component would contain greater variability. The other source is "within-groups," also called "error"; increasing the variability within each group would lead to an increase in this component. These two sources are independent in that either SS can be increased without increasing the other; together they add up to the total sum of squares, as do the degrees of freedom. Notice that each SS in the table is divided by the appropriate df to yield MS, but the MSs do not sum to any meaningful quantity. (Some statisticians prefer not to use the term MS to stand for variance; they use s^2_{bet} and s^2_{within} instead.) Finally, the two MSs are divided to produce F. As usual, the p value can be given in terms of a significance level (for this example, $p > .05$), or it can be given exactly, which is the way it is given by most statistical packages (e.g., SPSS) that generate a summary table as part of the one-way ANOVA output.

Excerpt from the Psychological Literature

The one-way ANOVA for independent samples is commonly used in the psychological literature, most often with only three or four levels of the independent variable. The following excerpt illustrates the APA style for reporting the results of a one-way ANOVA. Winkler, et al. (1999) were investigating the notion that to learn a new language, you have to learn to recognize the sounds (i.e., phonemes) of that new language and distinguish them from other similar sounds. A particular brain wave phenomenon (based on averaged EEG) known as MMN (mismatch negativity) can be used to indicate when a subject hears a novel (i.e., different) sound in a string of familiar sounds. In their experiment, two similar vowel sounds from the Finnish language were used. These two vowel sounds make a difference in Finnish but not in the distantly related language of Hungarian. One group of subjects were native speakers of Finnish; another group were Hungarians who lived in Finland and had become fluent in Finnish; a third group were ("naive") Hungarians who did not speak Finnish. All subjects, after some practice in distinguishing the vowels, heard a series of vowel sounds, most of which were one of the two vowels, with a few of the other mixed in, and had to identify each. The main dependent variable (DV) was the MMN to the rare vowels. If you can barely hear the difference in the two vowels, your MMN will not be large in response to the relatively rare ones in the series. Otherwise, the rare ones come as a surprise and evoke a relatively large MMN. Consistent with their expectations, they found and reported the following results: "The MMN amplitude was significantly larger in the Finns and fluent Hungarians than in the naive Hungarians, one-way analysis of variance (ANOVA): $F(2, 27) = 8.03$, $p < .01$, with $t(27) = 3.33$ and 3.60, $p < .01$ both, for post hoc comparisons of the naive with the Finnish and fluent groups, respectively" (p. 640). Note that you can determine the total number of subjects in the experiment by adding together the two numbers in parentheses after F ($df_{bet} + df_W = df_{tot}$) and then adding 1: thus $2 + 27 + 1 = 30$, so there were a total of 30 subjects in the three groups (equally distributed in this case). The post hoc comparisons, about which you will learn in the next chapter, were also consistent with expectations. The "naive" Hungarians should differ significantly from both other groups, but the two fluent groups should not differ from each other (in fact, the authors report data to show that the two fluent groups performed in a very similar fashion).

SUMMARY

1. When the sample sizes are not equal, MS_{bet} is found by subtracting each group mean from the grand mean, multiplying each squared difference by the appropriate sample size, adding all of these terms, and then dividing the sum by 1 less than the number of groups (i.e., df_{bet}). MS_W is a weighted average of the within-group variances; multiply each group's variance by 1 less than its sample size, add these terms, and divide by the total N minus the number of groups (i.e., df_W).

2. It is customary to calculate the sums of squares (SSs) first in an ANOVA and then divide by the corresponding dfs to obtain the MSs. These intermediate results are sometimes displayed in a summary table. The SS components add up to SS_{total}, a property that becomes increasingly useful as the complexity of the ANOVA increases. The MSs do not add up to any meaningful quantity. The SSs can be found from raw-score formulas that are based on group sums, the total sum, and the sum of squared scores.

3. The assumptions of the ANOVA are essentially the same as for the *t* test: independent random sampling (or, at the least, the random assignment of subjects to conditions), normal distributions for the DV in each population sampled, and homogeneity of variance. The normality assumption becomes virtually unnecessary with large sample sizes, and homogeneity of variance is generally ignored when all of the samples are the same size. Of course, it is also assumed that the DV is measured on an interval or ratio scale.

4. An *F* ratio can be used to test HOV when there are two or more groups. Unfortunately, this test has been found to be overly sensitive to violations of the normality assumption, and is no longer routinely used. More robust alternatives have been devised for testing HOV (e.g., Levene's test), but all of these tests lack power when they are needed most (e.g., when the sample sizes are small), and so there is no universal agreement on the usefulness of HOV tests. One rule of thumb is that if your largest sample size isn't more than twice your smallest sample size, and your largest sample variance isn't more than 50% larger than your smallest sample variance, you needn't worry about the HOV assumption. If, however, your sample sizes *and* sample variances differ a great deal, you should consider one of the adjusted ANOVAs described in Section C.

5. The power of a one-way ANOVA depends on the alpha that is set and the *F* value that is expected. Power tables for ANOVA are often based on the square root of the expected *F*, which is a quantity called phi (ϕ). Phi is just the product of the effect size in the population (**f**), which equals the biased standard deviation of the population means divided by the common standard deviation, and the square root of the common sample size (in the equal-*n* case).

6. The sample effect size in the multigroup case is a measure called *f*, which is analogous to *g*; in the two-group case, $f = g/2$. In the equal-*n* case, *f* is just the square root of the calculated *F* ratio after it has been divided by *n*. However, if you want to use *f* to estimate **f** (its corresponding population value), you need to multiply it by a correction factor: the square root of $(k - 1)/k$.

7. The power of an ANOVA can be increased by: (a) increasing alpha (which increases the rate of Type I errors, and is therefore usually unacceptable; (b) increasing the sample sizes (which may be prevented by practical constraints); or (c) increasing the population effect size. The effect size can be made larger by increasing the spread of the population means (which may require unacceptably strong treatments), or reducing subject-to-subject variability (which can sometimes be accomplished through repeated measures or the matching of subjects across conditions).

8. The levels of the IV in a one-way ANOVA can be fixed (particular values or conditions in which you are interested) or random (a randomly selected subset from a large field of potential levels, all of which are equally interesting). The calculation is the same in either case. Also, the levels of the IV can be qualitative (i.e., measured on a categorical or nominal scale) or quantitative (i.e., measured on at least an ordinal scale and, preferably, either an interval or ratio scale). In the latter case, a linear regression (or trend analysis) approach can be considerably more powerful than ANOVA, provided that the sample means fall somewhat near a straight line (or simple curve) when the DV is plotted against the IV.

EXERCISES

*1. Are all antidepressant medications equally effective? To test this null hypothesis, a psychiatrist randomly assigns one of five different antidepressants to each of 15 depressed patients. At the end of the experiment, each patient's depression level is measured. Because some patients did not take their medication or dropped out of the experiment for other reasons, the final sample sizes are not equal. The means, standard deviations, and sample sizes are as follows: $\bar{X}_1 = 23$, $s_1 = 6.5$, $n_1 = 12$; $\bar{X}_2 = 30$, $s_2 = 7.2$, $n_2 = 15$; $\bar{X}_3 = 34$, $s_3 = 7$, $n_3 = 14$; $\bar{X}_4 = 29$, $s_4 = 5.8$, $n_4 = 12$; $\bar{X}_5 = 26$, $s_5 = 6$, $n_5 = 15$. Use the six-step one-way ANOVA procedure to test the null hypothesis at the .05 level.

2. Consider a one-way ANOVA with five samples that are not all the same size but whose standard deviations are the same: $s = 15$. Can you calculate MS_W from the information given? If so, what is the value of MS_W?

*3. A researcher suspects that schizophrenics have an abnormally low level of hormone X in their bloodstream. To test this hypothesis, the researcher measures the blood level of hormone X in five acute schizophrenics, six chronic schizophrenics, and seven normal control subjects. The measurements appear in the following table:

Acute	Chronic	Normal
86	75	49
23	42	28
47	35	68
51	56	52
63	70	63
	46	82
		36

a. Calculate the F ratio using the raw-score formulas (Formulas 12.9 and 12.11).
b. Recalculate the F ratio using the definitional formulas (Formulas 12.6 and 12.7).
c. Explain how an inspection of the means could have saved you a good deal of computational effort.

4. A social psychologist wants to know how long people will wait before responding to cries for help from an unknown person and whether the gender or age of the person in need of help makes any difference. One at a time, subjects sit in a room waiting to be called for an experiment. After a few minutes they hear cries for help from the next room, which are actually on a tape recording. The cries are in either an adult male's, an adult female's, or a child's voice; seven subjects are randomly assigned to each condition. The dependent variable is the number of seconds from the time the cries begin until the subject gets up to investigate or help.

Child's Voice	Adult Female Voice	Adult Male Voice
10	17	20
12	13	25
15	16	14
11	12	17
5	7	12
7	8	18
2	3	7

a. Calculate the F ratio.
b. Find the critical F ($\alpha = .05$).
c. What is your statistical conclusion?
d. Present the results of the ANOVA in a summary table.

*5. A psychologist is interested in the relationship between color of food and appetite. To explore this relationship, the researcher bakes small cookies with icing of one of three different colors (green, red, or blue). The researcher offers cookies to subjects while they are performing a boring task. Each subject is run individually under the same conditions, except for the color of the icing on the cookies that are available. Six subjects are randomly assigned to each color. The number of cookies consumed by each subject during the 30-minute session is shown in the following table:

Green	Red	Blue
3	3	2
7	4	0
1	5	4
0	6	6
9	4	4
2	6	1

a. Calculate the F ratio.
b. Find the critical $F(\alpha = .01)$.

c. What is your statistical decision with respect to the null hypothesis?

d. Present your results in the form of a summary table.

6. Suppose that the data in Exercise 5 had turned out differently. In particular, suppose that the number of cookies eaten by subjects in the green condition remains the same, but each subject in the red condition ate 10 more cookies than in the previous data set, and each subject in the blue condition ate 20 more. This modified data set follows:

Green	Red	Blue
3	13	22
7	14	20
1	15	24
0	16	26
9	14	24
2	16	21

a. Calculate the F ratio. Is the new F ratio significant at the .01 level?

b. Which part of the F ratio has changed from the previous exercise and which part has remained the same?

c. Put your results in a summary table to facilitate comparison with the results of Exercise 5.

7. A college is studying whether there are differences in job satisfaction among faculty members from different academic disciplines. Faculty members rate their own job satisfaction on a scale from 1 to 10. The data collected so far are listed in the following table by the academic area of the respondent:

Social Sciences	Natural Sciences	Humanities
6	8	7
7	10	5
10	7	9
8	8	4
8		
9		

a. State the null hypothesis in words and again in symbols.

b. Test the null hypothesis with a one-way ANOVA at the .05 level.

c. Based on your statistical conclusion in part b, what type of error (Type I or Type II) might you be making?

*8. A social psychologist is studying the effects of attitudes and persuasion on incidental and intentional memory. Subjects read a persuasive article containing 10 distinct arguments that either coincide or clash with their own opinion (as determined by a prior questionnaire). Half the subjects are told that they will be asked later to recall the arguments (intentional condition), and half are tested without warning (incidental condition). Five subjects are randomly assigned to each of the four conditions; the number of arguments recalled by each subject is shown in the following table:

Incidental-Agree	Incidental-Disagree	Intentional-Agree	Intentional-Disagree
8	2	6	7
7	3	8	9
7	2	9	8
9	4	5	5
4	4	8	7

a. Test the null hypothesis at the .05 level.

b. Test the null hypothesis at the .01 level.

c. Present the results of the ANOVA in a summary table.

*9. A psychologist is investigating cultural differences in emotional expression among small children. Three-year-olds from six distinct cultural backgrounds are subjected to the same minor stressor (e.g., their parent leaves the room), and their emotional reactions are rated on a scale from 0 to 20. The means and sample sizes are as follows: $\bar{X}_1 = 14.5$, $n_1 = 8$; $\bar{X}_2 = 12.2$, $n_2 = 7$; $\bar{X}_3 = 17.8$, $n_3 = 8$; $\bar{X}_4 = 15.1$, $n_4 = 6$; $\bar{X}_5 = 13.4$, $n_5 = 10$; $\bar{X}_6 = 12.0$, $n_6 = 7$.

a. Given that $\Sigma X^2 = 9,820$, use the raw-score formulas to calculate the F ratio. (*Hint:* $T_i^2/n_i = n_i\bar{X}_i^2$.)

b. What is the critical $F(\alpha = .05)$?

c. What is your statistical decision with respect to the null hypothesis?

10. A researcher is exploring the effects of strenuous physical exercise on the onset of puberty in girls. The age of menarche is determined for six young athletes who started serious training before the age of 8. For comparison, the same measure is taken on four girls involved in an equally serious and time-consuming training regimen starting at the same age but not involving such strenuous physical exercise—playing the violin. An additional six girls not involved in any such training are included as a control group. The data are as follows:

Controls	Athletes	Musicians
12	14	13
11	12	12
11	14	13
13	16	11
11	15	
12	13	
11		

a. Calculate the F ratio.
b. Find the critical $F(\alpha = .01)$.
c. What is your statistical decision?
d. What are the limitations to the conclusions you can draw from this study?

*11. a. What is the power associated with a five-group experiment that has seven subjects in each group and $f = .75$, if $\alpha = .05$?
b. What is the power for the same experiment if f is only .6 and $\alpha = .05$?

12. a. Approximately how many subjects per group are needed in a four-group experiment if f is expected to be .2 and power must be at least .77 for a .05 test? (*Hint*: Begin by assuming that df_{error} will be very large.)
b. How many subjects per group would be needed in part a if f were equal to .1? All

else being equal, what happens to the number of subjects required when f is cut in half?
c. If you have three groups of eight subjects each and you want power to be at least .80 for a .05 test, approximately, how large does f have to be?

*13. a. Redo the power calculation for Exercise 8B2a in terms of ϕ (*Hint*: Use the fact that $f = d/2$.)
b. Recalculate 8B5a; begin by assuming that df_{error} will be considerably larger than 60.

14. In Exercise 7B4, the standard deviation in diastolic blood pressure for 60 marathon runners was $s = 10$, and for the nonrunners it was $s = 8$. Use an F ratio to test the homogeneity of variance assumption for that study at the .05 level.

15. In Exercise 7B6, a sample of 12 anxiety neurotics had a standard deviation in enzyme concentration of .7, whereas the standard deviation for the 20 control subjects was .4.
a. Use an F ratio to test the homogeneity of variance assumption at the .05 level.
b. Would a pooled-variances t test be justified in this case?

Proportion of Variance Accounted For in ANOVA

OPTIONAL MATERIAL

In Chapter 10, to explain how a two-group experiment could be viewed in terms of linear regression and described by a correlation coefficient (i.e., r_{pb}), I graphed the results of an imaginary study involving the effects of caffeine and a placebo on a simulated truck driving task. It is a simple matter to add a third group to the study—for instance, a group that gets an even stronger stimulant, such as amphetamine. Figure 10.7 has been modified to include a third group and appears as Figure 12.5. If the three groups represented three levels of a quantitative independent variable (e.g., three dosage levels of caffeine), it would be appropriate to use a linear regression approach, as described briefly in Section B. In this section, however, I will show how a one-way ANOVA can be viewed in terms of regression even when the levels of the independent variable represent qualitative distinctions.

The breakdown of each score into separate components is exactly the same for three (or more) groups as it was for two groups. The grand mean of all the subjects in the experiment is shown in Figure 12.5, and the graph focuses on Mr. Y, the highest scoring subject in the amphetamine group. Mr. Y's deviation from the grand mean $(Y - \overline{Y}_G)$ can be divided into his deviation from the mean of the amphetamine group $(Y - \overline{Y}_A)$ and the deviation of the amphetamine group mean from the grand mean $(\overline{Y}_A - \overline{Y}_G)$. In Chapter 10, I called the deviations from the group mean "unexplained," and when they were all squared and added, the result was called the unexplained SS. The unexplained SS corresponds exactly to the quantity that I have been calling SS_W in this chapter. Similarly, the explained SS corresponds to SS_{bet}.

The explained SS divided by the total SS gives the proportion of explained variance—the percentage by which the variance is reduced by using the appropriate group mean to predict individual scores rather than using the grand mean for all subjects, regardless of group. In a correlational study, this proportion equals r_{XY}^2. In the two-group case, the same proportion is referred to as r_{pb}^2. In the multigroup case, this proportion is designated by a new term: η^2 (pronounced "eta squared"), as follows:

$$\eta^2 = \frac{SS_{bet}}{SS_{tot}} \qquad \textbf{Formula 12.18}$$

The term η^2, which is the proportion of variance accounted for in the dependent variable by the independent variable in the results of a particular one-way ANOVA, is interpreted in the same way as r_{pb}^2. (The use of η to denote a sample statistic is an exception to the common rule of using Greek letters to describe only population characteristics.)

If you are given an F ratio and the degrees of freedom, perhaps in a published article, but you do not have access to a summary table of SS or the raw data, you can still find η^2 by using Formula 12.19:

$$\eta^2 = \frac{df_{bet}F}{df_{bet}F + df_W} \qquad \textbf{Formula 12.19}$$

In the two-group case, $F = t^2$ and $df_{bet} = 1$, so Formula 12.19 becomes the same as the square of Formula 10.13 [i.e., $\eta^2 = r_{pb}^2 = t^2/(t^2 + df)$].

Whereas η^2 is an excellent statistic for describing your data and a valuable supplement to the information provided by the F ratio, the major weakness of η^2 is that, like r_{pb}^2, it gives a rather biased estimate of omega squared (ω^2), the proportion of variance accounted for in the population. Nearly all of the bias (i.e., *over*estimation) in η^2 can be corrected by using Formula 12.20 to estimate ω^2:

$$\text{Est. } \omega^2 = \frac{SS_{bet} - (k - 1)MS_W}{SS_{tot} + MS_W} \qquad \textbf{Formula 12.20}$$

I will apply these formulas to the ANOVA calculated in Section B. First, I will use η^2 (Formula 12.18) to find the proportion of variance in optimism that can be accounted for by early parental death in the *samples* of our hypothetical experiment:

$$\eta^2 = \frac{SS_{bet}}{SS_{tot}} = \frac{161.97}{448} = .36$$

(Note that you get the same answer from Formula 12.19: 6.8/18.8 = .36.)

Then, to estimate the true proportion of variance accounted for in the entire population, I will use Formula 12.20:

$$\text{Est. } \omega^2 = \frac{SS_{bet} - (k - 1)MS_W}{SS_{tot} + MS_W} = \frac{161.97 - 2(23.84)}{448 + 23.84} = \frac{114.29}{471.84} = .24$$

Eta squared shows us that 36% of the variance is accounted for in our data, which is really quite a bit compared to most psychological experiments. Normally, the F ratio would be significant with such a high η^2; it is because our sample sizes were so small (to simplify the calculations) that F fell short of significance for our ANOVA. Our estimate of ω^2 suggests that η^2 is overestimating—by 50%—the proportion of variance that would be accounted for in the population; however, $\omega^2 = .24$ is still a respectable proportion of explained variance in a three-group ANOVA. It is when ω^2 drops

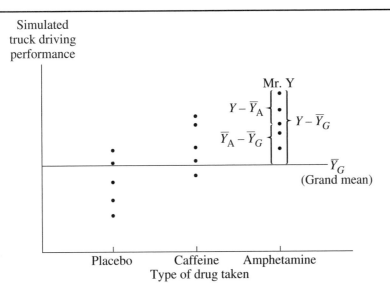

Figure 12.5

Regression in the Case of
Three Distinct Groups

below .01 that psychologists generally begin to question the value of a
study, although the results can be of some theoretical interest even when
less than 1% of the variance has been accounted for ($\omega^2 = .01$ corresponds
to the amount designated by J. Cohen, 1988, as a small effect size).

Just as the bias of Formula 12.18 can be corrected, as shown in Formula 12.20, so too can Formula 12.19 be corrected as follows:

$$\text{est. } \omega^2 = \frac{\text{df}_{\text{bet}}(F - 1)}{\text{df}_{\text{bet}}(F - 1) + N_T}$$

Formula 12.21

Although it certainly does not look obvious, Formula 12.21 will always yield
the same answer as Formula 12.20 (4.8 / 19.8 = .24). However, one advantage of Formula 12.21 is that this expression makes it clear that the estimate
of omega squared is not defined when F is less than 1.0 (ω^2 cannot be negative). The estimate for ω^2 comes out to zero when F equals 1.0, and by convention it is set to zero for any F below 1.0. Usually ω^2 is only estimated
when F is statistically significant, or was expected to be significant (or, perhaps, if one wants to make a point about how small it is), but very rarely
estimated when F is near 1.0 (recall that an F of 1.0 is telling you that the
variability of your sample means is just about what you would expect purely
from sampling error—without the contribution of any experimental effect).

Note that when there are only two groups, df_{bet} equals 1, and F equals
t^2, so Formula 12.21 reduces to the following:

$$\text{est. } \omega^2 = \frac{t^2 - 1}{t^2 - 1 + (\text{df}_W + 2)} = \frac{t^2 - 1}{t^2 + \text{df}_W + 1}$$

which is identical to Formula 10.16 (the unbiased estimate of omega-
squared associated with a two-group t test). Also, you may recall that in the
two-group case there is a simple relationship between ω^2 and **d** in the population, as given in Formula 10.15. There is a similar connection between ω^2
and **f** in the multigroup case, as shown in Formula 12.22:

$$\omega^2 = \frac{\mathbf{f}^2}{\mathbf{f}^2 + 1}$$

Formula 12.22

For example, an effect size (**f**) of .25 in the population, which is considered
to be of medium size, corresponds to a proportion of variance accounted

for in the population (ω^2) of $.25^2 / (.25^2 + 1) = .0625 / 1.0625 = .0588$—just a bit under 6%.

There is one more formula that I would like to show you, because it takes a different approach to estimating omega squared. This formula involves an adjustment of eta squared that is based on viewing eta squared as an R^2 obtained from multiple regression. Fortunately, you do not need to know what multiple regression is to understand the utility of the following formula:

$$\text{adj.}\eta^2 = \eta^2\left(1 - \frac{1}{F}\right)$$ **Formula 12.23**

I like the conceptual simplicity of this formula. Notice that η^2 is being adjusted by being multiplied by a correction factor that depends only on the F ratio for testing the ANOVA. As with Formula 12.21, you can see that the adjusted η^2 is zero when F equals 1, and that the adjustment is not valid for any F less than 1. You can also see that as F gets larger, the correction factor increases (producing *less* of an adjustment), eventually heading for its maximum value of 1.0 (i.e., no adjustment at all) as F becomes infinitely large. It does not matter if F is getting larger due to a larger effect size, or just larger sample sizes; larger Fs indicate that the effect in your samples is a more accurate reflection of the effect in the population.

However, it is important to note that, whereas Formula 12.23 will always yield an estimate of omega squared similar to the one from Formula 12.21, it is not algebraically equivalent to that formula, and will generally give a slightly different estimate. For instance, using the value for eta squared produced by Formula 12.18 for the optimism example (.36), and its associated F value (3.4), Formula 12.23 yields the following estimate:

$$\text{adj. }\eta^2 = .36\left(1 - \frac{1}{3.4}\right) = .36(.706) = .254$$

The difference between this estimate (.254) and the one produced by Formula 12.20 (.24) is not a simple artifact due to rounding off intermediate results. The latter estimate is generally preferred. However, if you use the General Linear Model/Univariate module of SPSS to perform a one-way ANOVA, SPSS will print a value labeled "*R* squared" immediately under the ANOVA output box. This value is the same as eta squared, as defined by either Formula 12.18 or Formula 12.19. In parentheses following that value is another one, which SPSS labels "Adjusted *R* Squared." This value corresponds to the adjusted η^2 as defined by Formula 12.23, and *not* the estimated omega squared, as defined by either Formula 12.20 or Formula 12.21. Formula 12.23 represents a perfectly reasonable way to estimate omega squared, and fortunately it will rarely differ by more than a tiny fraction from the more commonly reported value for the estimate of omega squared.

The Harmonic Mean Revisited

In Chapter 8, Section C, I presented a simplified formula for the harmonic mean that can only be used with two numbers (see Formula 8.8). Now that we are dealing with more than two groups at a time, we will need to be able to find the harmonic mean for any number of values (e.g., sample sizes). The general formula for the harmonic mean, n_h, is given next.

$$n_h = \frac{k}{\sum\dfrac{1}{n_i}}$$ **Formula 12.24**

For example, the harmonic mean of the numbers 10, 20, and 80 is not equal to their arithmetic mean (i.e., 37.5), but rather to a considerably lesser value, found from Formula 12.24 to be:

$$n_h = \frac{4}{\frac{1}{10} + \frac{1}{20} + \frac{1}{40} + \frac{1}{80}} = \frac{4}{.1 + .05 + .025 + .0125}$$

$$= \frac{4}{.1875} = 21.33$$

The Analysis of Unweighted Means for One-Way ANOVA

Formula 12.7 for MS_{bet} $[\sum n_i (X_i - X_G)^2/(k - 1)]$ weighs the squared difference of each group mean from the grand mean by the size of that group, and therefore forms the basis of what is sometimes called "the weighted-means" ANOVA. As an example, suppose we are comparing three patient groups on a psychiatric ward, with ns of 10, 15, and 25, and means of 7, 9, and 17, respectively. Then, the grand mean is $(10 \times 7 + 15 \times 9 + 25 \times 17)/N_T = (70 + 135 + 425)/50 = 630/50 = 12.6$. Thus, the ordinary MS_{bet} is $[10 \times (-5.6)^2 + 15 \times (-3.6)^2 + 25 \times (4.4)^2]/2 = (313.6 + 194.4 + 484)/2 = 992/2 = 496$. If the means of the largest and smallest groups were reversed, the grand mean would be reduced to 9.6 and MS_{bet} would become: $[10 \times (7.4)^2 + 15 \times (-.6)^2 + 25 \times (-2.6)^2]/2 = (547.6 + 5.4 + 169)/2 = 722/2 = 361$. Notice that when the most deviant mean (i.e., 17) is associated with the largest group (n = 25), MS_{bet} is considerably larger than when the most deviant mean is associated with the smallest group (496 versus 361, respectively).

By contrast, the *unweighted-means* formula for MS_{bet} (I'll refer to it as MS_{bet}') is identical to the formula for equal ns, (i.e., Formula 12.5A) except that the harmonic mean of the sample sizes (n_h) replaces "n": $MS_{bet}' = n_h s_{\bar{x}}^2$. Using Formula 12.24, n_h for 10, 15, and 25 is 14.52, and the unbiased variance of 7, 9, and 17 is 28, so $MS_{bet}' = 14.52 \times 28 = 406.6$. Note that this value for MS_{bet} falls between the two more extreme values for the weighted-means solution, and does not depend at all on the association between means and sample sizes (you can think of the unweighted-means formula as the *equally*-weighted formula).

The unweighted-means analysis makes the most sense when the differences in sample sizes are accidental, so that none of the samples actually represents a larger population, but as I mentioned earlier in this chapter, it is rarely used—which is why it is not included in major statistical packages like SPSS. Possibly, contributing to the lack of popularity of the method of unweighted means is the fact that the resulting F ratio may be slightly biased in the positive direction, increasing the Type I error rate above the alpha that is used to look up the critical F. So, you may rightly be wondering why I am bothering to mention this method at all. The reason is that I am, by contrast, trying to highlight an important property of the *weighted-means* solution for ANOVA. This will become helpful when I describe ways to adjust the ANOVA formula for a lack of homogeneity of variance (HOV) in the next subsection.

Adjusting the One-Way ANOVA for Heterogeneity of Variance

In Section B of this chapter, I mentioned that there was more than one way to adjust the one-way ANOVA when the sample sizes are not all equal *and* the sample variances are so different that an HOV test reaches statistical significance (some conservative statisticians recommend using an alpha of .1 or even .25 for the HOV test, especially when dealing with small samples). I will describe two such alternative procedures in this section, beginning with the one devised by Brown and Forsythe (1974). It should soon become clear that the Brown-Forsythe procedure could be reasonably referred to as "the separate-variances" ANOVA, as it is a natural extension of the separate-variances t test when dealing with more than two groups.

The Brown-Forsythe ANOVA

You may recall that the separate- and pooled-variances t test formulas share the same numerator. Similarly, the numerator of the Brown-Forsythe F (I'll refer to it as F') is the same MS_{bet} that you learned to calculate in Section B. However, the error term of F' (I'll refer to it as MS_W') resembles what you would get from dividing each sample variance by its corresponding sample size rather than pooling the variances. More precisely, MS_W' is found from the following formula:

$$MS_W' = \frac{\sum \left(1 - \frac{n_i}{N_T}\right) s_i^2}{df_{bet}} \qquad \textbf{Formula 12.25}$$

where the summation goes from 1 to k, which is the number of groups (i.e., levels of the IV). When all of the variances are equal, the top part of the formula reduces to $\sum(1 - n/N_T)s^2 = (k - 1)s^2$, which is why it must be divided by $k - 1$ (i.e., df_{bet}). A simple example involving three groups will help to demonstrate the difference between MS_W and MS_W'.

Let us return to the three patient groups, and posit that the groups of sizes 10, 15, and 25 are associated with variances of 3, 6, and 8, respectively. The ordinary MS_W would equal $(9 \times 3 + 14 \times 6 + 24 \times 8)/47 = 6.45$. However, MS_W' will be different:

$$MS_{W'} = \frac{\left(1 - \frac{10}{50}\right) \cdot 3 + \left(1 - \frac{15}{50}\right) \cdot 6 + \left(1 - \frac{25}{50}\right) \cdot 9}{2} = \frac{10.6}{2} = 5.3$$

In this example, as the group gets larger, so does its variance; this pattern always results in MS_W' being smaller than MS_W (e.g., $5.3 < 6.45$, in this comparison), which means that F' will be larger and more likely to attain significance than the usual F. Consequently, the usual F is conservative in this case, and whereas it usually has less power than F', the conservative statistician will not object to the use of the ordinary ANOVA in this situation. Let's reverse the variances of the largest and smallest groups, and see what happens. The ordinary MS_W is reduced to $(9 \times 8 + 14 \times 6 + 24 \times 3)/47 = 228/47 = 4.85$, but MS_W' increases to:

$$MS_{W'} = \frac{\left(1 - \frac{10}{50}\right) \cdot 8 + \left(1 - \frac{15}{50}\right) \cdot 6 + \left(1 - \frac{25}{50}\right) \cdot 3}{2} = \frac{12.1}{2} = 6.05$$

Now, MS_W' (6.05) is larger than MS_W (4.85), so F' is smaller than F. When the larger groups have the smaller variances, using the usual ANOVA has more power, but can result in a higher Type I error rate than the alpha you are using to make your statistical decisions. This possibility is unacceptable to the conservative researcher. However, if you want to use F' as an alternative, you will have to deal with the fact that even when the null hypothesis is true, F' does not follow an F distribution with the usual dfs. Fortunately, F' has been found to follow what is called a *quasi-F* distribution, which means that its distribution looks similar to an F distribution, but the df for the error term of the F' distribution is not df_W. If this problem sounds familiar, that's because it is closely related to the Behrens-Fisher problem you read about in Chapter 7. To solve this problem, Brown and Forsythe (1974) used a method originated by Satterthwaite (1946) to create a formula for the degrees of freedom for the error term of their F ratio.

The df-adjustment formula for F' in the general case is so complex that I felt there was little purpose to presenting it here. However, when all of the samples are the same size, the error df for F' (i.e., df_W') reduces to the following manageable formula:

$$df_W' = (n - 1)\frac{\left(\sum s_i^2\right)^2}{\sum s_i^4}$$ **Formula 12.26**

Again, the summations go from 1 to k. As you may have guessed, based on what you learned about the Welch-adjusted df in Chapter 7, df_W' ranges from a maximum of $k(n - 1)$, when all the sample variances are equal, down to a minimum of $n - 1$, as the variances maximally diverge. However, because F' equals the ordinary F when all the ns are equal, and discrepant population variances have relatively little impact on the distribution of F in the equal-n case, nobody bothers to adjust df_W in that case.

In the general case (ns not all equal), the df adjustment depends on the discrepancies among the w_is as defined in Chapter 7 (i.e., $w_i = s_i^2/n_i$). As in the case of two groups, there is little df adjustment when all of the samples have similar variance-to-size ratios, but the more the w_is diverge, the more that df_W' is reduced relative to the ordinary error df (and the more that F' is reduced below F).

The Welch ANOVA

When an ANOVA involves more than two groups, variations become possible that simply do not exist when comparing only two groups. One example, as mentioned earlier, is the weighted versus unweighted means solutions; these always produce the same answer for only two groups. An alternative to the Brown-Forsythe F, devised by Welch (1951), uses the weighted-means approach, but includes the variances, along with the sample sizes, in determining the weights. Thus, the Welch F (which I will label as F^*) adjusts, in effect, not only the denominator of the ordinary F ratio, but the numerator, as well. However, the way the Welch formula is presented, its numerator (which I will call W_{num}) and denominator do not correspond directly to MS_{bet} and MS_W, respectively. Nonetheless, the F ratios produced by the Welch formula do follow a quasi-F distribution that is similar (though not identical to) the distribution of F'. Because the denominator of F^* follows the same basic pattern as MS_W', I will concentrate on W_{num}, which has a novel element to it.

The formula for W_{num} looks a good deal like the ordinary formula for MS_{bet}, but with the weights changed from the sample sizes to a function of the w_is; W_{num} can be written as:

$$W_{num} = \frac{\sum \frac{1}{w_i}(\bar{X}_i - \bar{X}_{WG})^2}{k - 1}$$

It makes sense to use the reciprocal of w_i in the preceding formula, because in this case the weights are being applied to the numerator rather than the denominator (i.e., error term). Also, note that X_{WG} in the formula for W_{num} is not the usual grand mean, which can be found by weighing the various group means by their sample sizes, but rather a "Welch" grand mean that is found by using the w_is as the weighting factors.

What you can see from Formula 12.26 is that Welch's F^* can be seriously affected by whether the most discrepant means are associated with, for instance, large groups that have small variances, or small groups with large variances. The association of means and variances can even have an effect when all the ns are equal, so unlike F', F^* is not usually equal to the ordinary F even when all of the samples are the same size. Therefore, some statisticians recommend the use of F^* even when the ns are equal, if the variances are quite discrepant, but given the reputation of the ordinary F's robustness with respect to heterogeneity of variance when the ns are equal, this suggestion seems to have had little influence.

However, whereas Tomarkin and Serlin (1986) found that F^* seems to have greater power than F' for most combinations of means, variances, and sample sizes, Clinch and Keselman (1982) found that F' seems to maintain better control over the Type I error rate than does F^*, when the underlying distributions are skewed. Perhaps the fact that there is no clear way to decide when to use the Brown-Forsythe versus the Welch ANOVA explains, in part, why neither solution is commonly used, even though both are available as options when using the one-way ANOVA procedure from the Analyze/Compare Means menu of SPSS.

Bear in mind that, as simulation studies have shown, the various alternatives for F do not diverge dramatically until the variance of one group is at least several times the variance of another. A more serious problem occurs when your underlying distributions not only seem to be skewed, but they are skewed in different directions. However, if your samples exhibit extreme differences in variance and/or very different distribution shapes, it just does not seem sensible to test the null hypothesis that the population means are all equal. Obviously, whatever it is that distinguishes your groups is having some effect on your data, and it would seem incumbent upon you to explore your data further in an attempt to understand just how and why the distributions and/or their variances diverge so widely. If you are determined to test the equality of the population means, despite having messy data (e.g., extreme outliers), you should consider one of the robust ANOVA techniques described by Wilcox (2003). The complex computational methods presented by Wilcox (2003) include an application of the Welch ANOVA to trimmed means, and several bootstrap methods, which are based on the resampling procedure I briefly introduced in Chapter 6.

SUMMARY

1. A multigroup experiment can be analyzed as a form of regression in which each group mean serves as the prediction for the scores in that group. Squaring the deviation of each score from its own group mean and adding the squared values yields the unexplained SS, which is the SS_W of ANOVA. Squaring the deviation of each score from the *grand* mean and adding the squared values yields SS_{total}. The difference between SS_W and SS_{total} is SS_{bet}. Dividing SS_{bet} by SS_{total} gives you the proportion of variance in your dependent variable that is accounted for by your independent variable. This quantity is called eta squared (η^2) in the context of ANOVA.

2. In the two-group case, η^2 is equal to r^2_{pb}, and like r^2_{pb}, η^2 is a biased estimator of omega squared (ω^2). There are two different ways to correct this bias, but both lead to similar reductions in the original value of η^2. Note that both estimates come out to zero when the F ratio equals 1.0, and neither is defined for Fs less than 1. Either proportion-of-variance estimate, or the original η^2, can be reported along with an F ratio to give the reader a sense of how large an effect is being studied.

3. The harmonic mean of a set of numbers can be found by averaging the reciprocals of all of those numbers, and then taking the reciprocal of that average (e.g., to find n_h for 5 and 20, average .2 and .05 to get .125; and then find the reciprocal of .125, which is 8.

4. In the *unweighted-means* formula for ANOVA, MS_{bet} is found by multiplying the unbiased variance of the sample means by the harmonic mean of the sample sizes. However, the commonly used solution for the one-way ANOVA is the analysis of *weighted* means, as described in Section B.

5. The Brown-Forsythe F (F') is like an extension of the separate-variances t test; its numerator is always the same as in the ordinary ANOVA, and its denominator equals MS_w when all of the samples are the same size. When the larger groups are consistently associated with larger variances, F' tends to be larger than F, and the reduction of the error df is relatively slight (as compared to when the larger groups have the smaller variances).

6. The Welch F (F^*) is similar to F', and involves a similar adjustment in its error df. The major difference between F^* and either F' or the ordinary F ratio of ANOVA, is that F^* weights deviations of group means from the grand mean according to not just the size of the group, but the ratio of the group's size to its variance. Thus, a deviant sample mean will produce a larger increase in F^*, if it arises from a relatively large group with a relatively small variance. Large groups with relatively small variances are considered generally more reliable, and therefore more deserving of influence over the magnitude of F^*.

7. Although F^* differs from F even when all the ns are equal, and df_{error} can be adjusted for both F' and F* when the ns are equal, it is hard to imagine anyone using an adjusted ANOVA procedure in the equal-n case. In fact, it is only when there are considerable discrepancies among both the sample sizes and their variances that the Brown-Forsythe or Welch ANOVAs would be considered. As to the conditions under which one of these would be preferred to the other, there are no clear recommendations as of yet. However, if the differences among the sample variances seem to be due to outliers, or differing (or heavy-tailed) distribution shapes, a nonparametric version of ANOVA (see Chapter 21) or a robust ANOVA (e.g., an ANOVA based on trimmed means and/or a resampling procedure) should be considered.

*1. a. If $F = 5$ in a three-group experiment, what proportion of variance is accounted for (i.e., what is η^2) when the total number of subjects is 30? 60? 90?

b. If $F = 5$ in a six-group experiment, what proportion of variance is accounted for (i.e., what is η^2) when the total number of subjects is 30? 60? 90?

2. a. If $F = 10$ in a three-group experiment, what proportion of variance is accounted for (i.e., what is η^2) when the total number of subjects is 30? 60? 90?

b. If $F = 10$ in a six-group experiment, what proportion of variance is accounted for (i.e., what is η^2) when the total number of subjects is 30? 60? 90?

*3. a. For the data in Exercise 12B5, find η^2 (the proportion of variance in cookie consumption that is accounted for by color). Estimate ω^2 for the same data using Formula 12.20 (or explain why you cannot).

b. Answer part a for Exercise 12B6. Also, find the adjusted η^2 using Formula 12.23.

4. a. Find η^2 by Formula 12.19, and estimate ω^2 using Formula 12.21 for the data in Exercise 12B1. Also, find the adjusted η^2 using Formula 12.23. Are the values obtained from Formulas 12.21 and 12.23 the same? Explain.

b. Answer part a for Exercise 12B3.

*5. What proportion of variance is accounted for in the population, when f equals:
a. .1?   b. .25   c. .8?   d. 1.0?

6. How large does f have to be in order for ω^2 to equal:
a. .05?   b. .1?   c. .25?   d. .8?

*7. Recalculate the F ratio according to the unweighted-means solution for the data in:
a. Exercise 12B7.
b. Exercise 12B10.
c. For both part a and part b, explain the discrepancy between the F ratios from the weighted- and unweighted-means formulas.

8. Compare the Brown-Forsythe F ratio to the ordinary F for the data in:
a. Exercise 12B7.
b. Exercise 12B10.
c. For both part a and part b, explain the discrepancy between F and F'. In each case, determine whether F' would be statistically significant even with the most extreme df adjustment that is possible.

*9. a. Calculate the adjusted df for a Brown-Forsythe ANOVA on the data in Exercise 12B5.
b. Compare the adjusted df to the ordinary error df for ANOVA, and explain why there is such a large discrepancy.

10. a. Repeat part a of Exercise 9 for the data in Exercise 12B4.
b. Why is the df adjustment so much less severe than it was in the previous exercise?

KEY FORMULAS

The F ratio (in terms of two independent population variance estimates):

$$F = \frac{MS_{bet}}{MS_W}$$

Formula 12.3

The degrees of freedom associated with the numerator of the F ratio:

$$df_{bet} = k - 1$$

Formula 12.4A

The degrees of freedom associated with the denominator of the F ratio:

$$df_W = N_T - k$$

Formula 12.4B

Simplified formula for the mean square between-groups variance estimate, when all the groups are the same size:

$$MS_{\text{bet}} = ns_{\bar{x}}^2 \qquad\qquad \textbf{Formula 12.5A}$$

Simplified formula for the mean square within-groups variance estimate, when all the groups are the same size:

$$MS_W = \frac{\sum s^2}{k} \qquad\qquad \textbf{Formula 12.5B}$$

The *F* ratio when all the groups are the same size (means and variances have already been calculated):

$$F = \frac{ns_{\bar{x}}^2}{\dfrac{\sum s^2}{k}} \qquad\qquad \textbf{Formula 12.5}$$

Mean square within-groups variance estimate (unequal sample sizes):

$$MS_W = \frac{\sum (n_i - 1)s_i^2}{\text{df}_W} \qquad\qquad \textbf{Formula 12.6}$$

Mean square between-groups variance estimate (unequal sample sizes):

$$MS_{\text{bet}} = \frac{\sum n_i(\bar{X}_i - \bar{X}_G)^2}{\text{df}_{\text{bet}}} \qquad\qquad \textbf{Formula 12.7}$$

Sum of squares within groups—raw score form (T_i is the total for one particular group):

$$SS_W = \sum X^2 - \sum\left(\frac{T_i^2}{n_i}\right) \qquad\qquad \textbf{Formula 12.8}$$

Sum of squares between groups—raw score form (T_T is the total of all the observations):

$$SS_{\text{bet}} = \sum\left(\frac{T_i^2}{n_i}\right) - \frac{T_T^2}{N_T} \qquad\qquad \textbf{Formula 12.9}$$

The total sum of squares in terms of its two components:

$$SS_{\text{tot}} = SS_{\text{bet}} + SSW \qquad\qquad \textbf{Formula 12.10}$$

The total sum of squares (raw-score form):

$$SS_{\text{tot}} = \sum X^2 = \frac{T_T^2}{N_T} \qquad\qquad \textbf{Formula 12.11}$$

F for an equal-n ANOVA, in terms of the unbiased variance of the group means and the average within-group variance:

$$F = \frac{ns_{\bar{X}}^2}{s_w^2} \qquad\qquad \textbf{Formula 12.12}$$

Measure of effect size for ANOVA based on relative spread of population means (analogous to **d** in the two-group case):

$$\mathbf{f} = \frac{\sigma_{\bar{X}}}{\sigma}$$

Formula 12.13

Phi as a function of sample size and effect size (analogous to delta in the two-group case):

$$\phi = \mathbf{f}\sqrt{n}$$

Formula 12.14

Sample effect size as a function of the observed F value and the size of each sample:

$$\mathbf{f} = \sqrt{\frac{F}{n}}$$

Formula 12.15

Sample estimate of **f** based on calculated F ratio and common sample size, corrected for bias:

$$\text{Estimated } \mathbf{f} = \sqrt{\left(\frac{k-1}{k}\right)\frac{F}{n}}$$

Formula 12.16

Required size of each sample for desired power, for a given population effect size:

$$n = \left(\frac{\phi}{\mathbf{f}}\right)^2$$

Formula 12.17

Eta squared (the proportion of variance accounted for in your data):

$$\eta^2 = \frac{SS_{\text{bet}}}{SS_{\text{tot}}}$$

Formula 12.18

Eta squared (in terms of the calculated F ratio and the degrees of freedom):

$$\eta^2 = \frac{df_{\text{bet}}F}{df_{\text{bet}}F + df_W}$$

Formula 12.19

Estimate of omega squared (the proportion of variance accounted for in the population):

$$\text{est.}\omega^2 = \frac{SS_{\text{bet}} - (k-1)MS_W}{SS_{\text{tot}} + MS_W}$$

Formula 12.20

Estimate of omega squared (the proportion of variance accounted for in the population):

$$\text{est.}\omega^2 = \frac{df_{\text{bet}}(F-1)}{df_{\text{bet}}(F-1) + N_T}$$

Formula 12.21

Relation between two measures of effect size for ANOVA:

$$\omega^2 = \frac{\mathbf{f}^2}{\mathbf{f}^2 + 1}$$

Formula 12.22

An alternative approach to estimating the proportion of variance accounted for in the population:

$$\text{adj.}\eta^2 = \eta^2\left(1 - \frac{1}{F}\right)$$

Formula 12.23

The harmonic means of k numbers:

$$n_h = \frac{k}{\sum \frac{1}{n_i}}$$

Formula 12.24

The denominator (i.e., error term) of the Brown-Forsythe F ratio (F'):

$$MS_{W'} = \frac{\sum \left(1 - \frac{n_i}{N_T}\right) s_i^2}{df_{bet}}$$

Formula 12.25

Adjusted error df for the Brown-Forsythe ANOVA, when all of the ns are equal:

$$df_{w'} = (n-1)\frac{\left(\sum s_i^2\right)^2}{\sum s_i^4}$$

Formula 12.26

MULTIPLE COMPARISONS

Chapter

CONCEPTUAL FOUNDATION

You will need to use the following from previous chapters:

Symbols
$\bar{X}$: Mean of a sample
k: The number of groups in a one-way ANOVA
s_p^2: The pooled variance
MS_W: Mean square within-groups
(denominator from the formula for a one-way ANOVA)

Formulas
Formula 7.5B: The pooled-variance t test

Concepts
Homogeneity of variance
Type I and Type II errors

In Section A of Chapter 12, I described an experiment comparing the effects on illness of three treatments: a placebo, vitamin C, and a multivitamin supplement. In that example I rejected the null hypothesis that all three population means were equal, and I mentioned that additional tests would be required to discern which pairs of means were significantly different. A significant F in the one-way ANOVA does not tell us whether the multivitamin treatment is significantly different from vitamin C alone, or whether either vitamin treatment is significantly different from the placebo. The obvious next step would be to compare each pair of means with a t test, performing three t tests in all. This procedure would not be unreasonable, but it can be improved upon somewhat, as I will describe shortly. Performing all of the possible t tests becomes more problematic as the number of conditions or groups increases. The disadvantages of performing many t tests for one experiment, and particularly the various procedures that have been devised to modify those t tests, comprise the main topic of this chapter.

The Number of Possible t Tests

To understand the main drawback of performing multiple t tests, consider an example of a multigroup study in which the null hypothesis could reasonably be true. Imagine a fanciful researcher who believes that the IQ of an adult depends to some extent on the day of the week on which that person was born. To test this notion, the researcher measures the mean IQ of seven different groups: one group of people who were all born on a Sunday, another group of people who were all born on a Monday, and so forth. As mentioned in the previous chapter, the number of possible t tests when there are seven groups is 21. Let us see how that number can be found easily. When picking a pair of groups for a t test, there are seven possible choices for the first member of the pair (any one of the seven days of the week). For each of those seven choices, there are six possibilities for the second member of the pair, so there are $7 \times 6 = 42$ pairs in all. However, half of those pairs are the same as the other half but in reverse order. For example, picking Monday first and then Thursday gives the same pair for a t test as picking Thursday first and then Monday. Therefore there are

$42/2 = 21$ different t tests. The general formula for finding the number of possible t tests is as follows:

$$\frac{k(k-1)}{2}$$

Formula 13.1

where k is the number of groups.

Experimentwise Alpha

Now suppose that our researcher does not know about ANOVA and performs all 21 t tests, each at the .05 level. Assuming that all seven population means are indeed equal, the null hypothesis will be true for each of the 21 t tests. Therefore, if any of the t tests attains statistical significance (e.g., Monday turns out to be significantly different from Thursday), the researcher has made a Type I error. The researcher, not knowing that the null hypothesis is true, might try to publish the finding that, for example, people born on Mondays are smarter than those born on Thursdays—which, of course, would be a misleading false alarm. Even if only one of the 21 t tests leads to a Type I error, we can say that the experiment has produced a Type I error, and this is something researchers would like to prevent. The probability that an experiment will produce *any* Type I errors is called the *experimentwise alpha*. (Note that it is becoming increasingly popular to use the term *familywise alpha* because a family of tests can be more precisely defined. For the one-way ANOVA, however, this distinction is not important, so I will continue to use the term *experimentwise* throughout this chapter.) When t tests are performed freely in a multigroup experiment, the experimentwise alpha (α_{EW}) will be larger than the alpha used for each t test (the testwise α). Furthermore, α_{EW} will increase as the number of groups increases because of the increasing number of opportunities to make a Type I error.

You can get an idea of how large α_{EW} can become as the number of groups increases by considering a simple case. Suppose that a researcher repeats the same totally ineffective two-group experiment (i.e., $\mu_1 = \mu_2$) 21 times. What is the chance that the results will attain significance one or more times (i.e., what is the chance that the researcher will make at least one Type I error)? The question is very tedious to answer directly—we would have to find the probability of making one Type I error and then the probability of exactly two Type I errors, up to the probability of committing a total of 21 Type I errors. It is easier to find the probability of making *no* Type I errors; subtracting that probability from 1.0 gives us the probability of making one or more Type I errors. We begin by finding the probability of *not* making a Type I error for just one t test. If H_0 is true, and alpha = .05, the probability of not making a Type I error is just $1 - .05 = .95$. Now we have to find the probability of *not* making a Type I error 21 times in a row. If each of the 21 t tests is independent of all the others (i.e., the experiment is repeated with new random samples each time), the probabilities are multiplied (according to the multiplication rule described in Chapter 4, Section C). Thus the probability of *not* making a Type I error on 21 independent occasions is $.95 \times .95 \times .95 \ldots$ for a total of 21 times, or .95 raised to the 21st power ($.95^{21}$), which equals about .34. So the probability of making at least one Type I error among the 21 tests is $1 - .34 = .66$, or nearly two thirds. The general formula is:

$$\alpha_{EW} = 1 - (1 - \alpha)^j$$

Formula 13.2

where j is the number of independent tests. Formula 13.2 does not apply perfectly to multiple t tests within the same multigroup experiment because the t tests are not all mutually independent (see Section C), but it does give us an idea of how large α_{EW} can get when many t tests are performed.

Complex and Planned Comparisons

The calculation I just showed for α_{EW} should make it clear why it is not acceptable to perform multiple t tests without performing some other procedure to keep α_{EW} under control. The fanciful researcher in the preceding example would have had much more than a .05 chance of finding at least one significant difference in IQ between groups from different days of the week, even assuming that the null hypothesis is true. Before I can describe the various procedures that have been devised to keep α_{EW} at a reasonable level, I need to introduce some new terms. For instance, following an ANOVA with a t test between two of the sample means is an example of a *comparison*. When the comparison involves only two groups, it can be called a *pairwise comparison*, so this term is another way of referring to a two-sample t test. *Complex comparisons* involve more than two means. As an example, imagine that the fanciful researcher mentioned suggests that people born on the weekend are smarter than those born during the week. If the average of the Saturday and Sunday means is compared to the average of the remaining 5 days, the result is a complex comparison. This section will deal only with pairwise comparisons because they are more common and easier to describe. The methods for testing complex comparisons will be explained in Section B.

The α used for each test that follows an ANOVA can be called the *alpha per comparison*, or α_{pc}. (Because this term is more commonly used than the term *testwise* α and just as easy to remember, I will adopt it.) As you will see, adjusting α_{pc} is one way to control α_{EW}.

Another important distinction among comparisons is between those that are planned before running a multigroup experiment and those that are chosen after seeing the data. Comparisons that are planned in advance are called *a priori* comparisons and do not involve the same risk of a high α_{EW} as *a posteriori* comparisons, that is, comparisons a researcher decides on after inspecting the various sample means. Because a priori (planned) comparisons can get quite sophisticated, I will reserve the bulk of my discussion of them for the second half of Section B. Most of this section and the first part of Section B will be devoted to a posteriori comparisons, which are more often called *post hoc* (i.e., after the fact) comparisons.

Fisher's Protected t Tests

In the days-of-the-week example I described a researcher who didn't know about ANOVA to demonstrate what happens when multiple t tests are performed freely, without having obtained a significant ANOVA. A real researcher would know that something has to be done to keep α_{EW} from becoming too high. The simplest procedure for keeping down the experimentwise alpha is not to allow multiple t tests unless the F ratio of the one-way ANOVA is statistically significant. If this procedure were adopted, the days-of-the-week experiment would have only a .05 chance (assuming that alpha = .05) of producing a significant F; only about 5 out of 100 totally ineffective experiments would ever be analyzed with multiple t tests. It is true that once a researcher is "lucky" enough to produce a significant F with a totally ineffective experiment, there is a good chance at least one of the multiple t tests will also be significant, but demanding a significant F means that 95% (i.e., $1 - \alpha$) of the totally ineffective experiments will never be followed up with t tests at all.

The procedure of following only a significant ANOVA with t tests was invented by Fisher (1951), and therefore the follow-up t tests are called *Fisher's protected t tests*. The t tests are "protected" in that they are not often performed when the null hypothesis is actually true because a researcher must first obtain a significant F. Also, the t tests are calculated in a way that is a little different from ordinary t tests, and more powerful, as you will soon see. To explain the formula for calculating Fisher's protected t tests, I will begin with Formula 7.5B, which follows (without the $\mu_1 - \mu_2$ term in the numerator, which can be assumed to equal zero):

$$t = \frac{(\overline{X}_1 - \overline{X}_2)}{\sqrt{s_p^2 \left(\frac{1}{n_1} + \frac{1}{n_2} \right)}}$$

Formula 7.5B

The use of s_p^2 indicates that we are assuming that there is homogeneity of variance, so the pooling of variances is justified. If the assumption of homogeneity of variance is valid for the entire multigroup experiment (i.e., the variances of all the populations are equal), MS_W (also called MS_{error}) is the best estimate of the common variance and can be used in place of s_p^2 in Formula 7.5B. In particular, I will assume that for the days-of-the-week experiment, pooling all seven sample variances (i.e., MS_W) gives a better estimate of σ^2 than pooling the sample variances for only the two samples being compared in each t test that follows the ANOVA. Substituting MS_W for s_p^2 yields Formula 13.3:

$$t = \frac{(\overline{X}_i - \overline{X}_j)}{\sqrt{MS_W \left(\frac{1}{n_i} + \frac{1}{n_j} \right)}}$$

Formula 13.3

where the subscripts i and j indicate that any two sample means can be compared.

If homogeneity of variance cannot be assumed, there is no justification for using MS_W. If homogeneity of variance cannot be assumed for a particular pair of conditions *and* the sample sizes are not equal, some form of separate-variances t test must be performed for that pair (see Chapter 7, Section C). (Because matters can get rather complicated, in that case, I will deal only with analyses for which homogeneity of variance can be assumed for all pairs.) If all of the samples in the ANOVA are the same size, both n_i and n_j in Formula 13.3 can be replaced by n without a subscript, producing Formula 13.4:

$$t = \frac{(\overline{X}_i - \overline{X}_j)}{\sqrt{\frac{2MS_W}{n}}}$$

Formula 13.4

Note that the denominator of Formula 13.4 is always the same, regardless of which two groups are being compared. The constancy of the denominator when all sample sizes are equal leads to a simplified procedure, called *Fisher's least significant difference (LSD) test*, which will be described more fully in Section B.

The advantage of using Formula 13.3 (or Formula 13.4) for follow-up t tests, instead of Formula 7.5B, is that the critical t used is based on df_W, which is larger (leading to a smaller critical t) than the df for just the two groups involved in the t test. However, the whole procedure of using protected t tests has a severe limitation, which must be explained if you are to understand the various alternative procedures.

Complete versus Partial Null Hypotheses

The problem with Fisher's protected t tests is that the protection that comes from finding a significant F only applies fully to totally ineffective experiments, such as the days-of-the-week example. By "totally ineffective" I mean that the null hypothesis of the ANOVA—that all of the population means are equal—is actually true. The null hypothesis that involves the equality of *all* the population means represented in the experiment is referred to as the *complete null hypothesis*. Fisher's protected t test procedure keeps α_{EW} down to .05 (or whatever α is used for the ANOVA) only for experiments for which the complete null hypothesis is true. The protection does not work well if the null hypothesis is only partially true, in which case α_{EW} can easily become unreasonably large. To illustrate the limitations of Fisher's procedure, I will pose an extreme example.

The Partial Null Hypothesis

Imagine that a psychologist believes that all phobics can be identified by some physiological indicator, regardless of the type of phobia they suffer from, but that some phobias may exhibit the indicator more strongly than others. Six types of phobics (social phobics, animal phobics, agoraphobics, claustrophobics, acrophobics, and people who fear knives) were measured on some relevant physiological variable, as was a control group of nonphobic subjects. In such a study, we could test to see if all the phobics combined differ from the control group (a complex comparison), and we could also test for differences among the different types of phobics. Depending on the variable chosen, there are many ways these seven groups could actually differ, but I will consider one simple (and extreme) pattern to make a point. Suppose that, for the physiological variable chosen, the phobic population means are different from the control population mean, but that all six phobic population means are equal to each other (i.e., H_0: $\mu_1 = \mu_2 = \mu_3 = \mu_4 = \mu_5 = \mu_6 \neq \mu_7$). In this case, the complete null hypothesis is not true, but a *partial null hypothesis* is true. Next, suppose that the psychologist dutifully performs a one-way ANOVA. If the control population differs only very slightly from the phobic populations, the chance of attaining a significant ANOVA may be only a little greater than alpha. However, it is quite possible that the control population differs greatly from the phobics (even though the phobics do not differ from each other), so the ANOVA is likely to be significant.

If our psychologist finds her F ratio to be significant at the .05 level, and she adheres to the Fisher protected t test strategy, she will feel free to conduct all the possible pairwise comparisons, each with $\alpha_{pc} = .05$. This strategy includes testing all possible pairs among the six phobic groups which (using Formula 13.1) amounts to $6(6 - 1)/2 = 30/2 = 15$ pairwise comparisons, for which the null hypothesis is true in each case. If these 15 t tests were all mutually independent, the α_{EW} (using Formula 13.2) would become $1 - (1 - .05)^{15} = 1 - .95^{15} = 1 - .46 = .54$. Although these t tests are not totally independent, it should be clear that there is a high chance of committing a Type I error, once the decision to perform all the t tests has been made. (The remaining six t tests involve comparing the control group with each of the phobic groups and therefore cannot lead to any Type I errors in this example.) Note that without the control group, the complete null would be true, and the chance of attaining a significant ANOVA would only be alpha. There would be no drawback to using Fisher's procedure. Unfortunately, the addition of the control group can make it relatively easy to attain a significant ANOVA, thus removing the protection involved in Fisher's procedure and allowing α_{EW} to rise above the value that had been set for the overall ANOVA.

The Case of Three Groups

The one case for which Fisher's procedure gives adequate protection even if the complete null is *not* true is when there are only three groups. In that case the only kind of partial null you can have is one in which two population means are equal and a third is different. A significant ANOVA then leads to the testing of at most only *one* null hypothesis (i.e., the two population means that are equal), so there is no buildup of α_{EW}. However, Fisher's procedure allows a buildup of α_{EW} when there are more than three groups and the complete H_0 is not true—and the greater the number of groups, the larger the buildup in α_{EW}. For this reason, the protected t test has gotten such a bad reputation that researchers are reluctant to use it even in the common three-group case for which it is appropriate. This is unfortunate because the Fisher procedure has the most power of any post hoc comparison procedure in the three-group case, as will become clear as we analyze the various alternatives.

Tukey's HSD Test

To provide complete protection—that is, to keep α_{EW} at the value chosen regardless of the number of groups or whether the null is completely or only partially true—Tukey devised an alternative procedure for testing all possible pairs of means in a multigroup experiment. His procedure is known as *Tukey's honestly significant difference (HSD) procedure*, in contrast to Fisher's least significant difference (LSD) test. (The term *difference*, as used in HSD and LSD, will be clarified in the next section.) The implication is that Fisher's procedure involves some cheating because it provides protection only when used with experiments for which the complete null hypothesis is true. To understand the protection required when all possible pairs of means are being tested, imagine that you have conducted a multigroup experiment and you are looking at the different sample means. If you are hoping to find at least one pair of means that are significantly different, your best shot is to compare the largest sample mean with the smallest. It is helpful to understand that in terms of your chance of making at least one Type I error when the complete null is true, testing the smallest against the largest mean is the same as testing all possible pairs. After all, if the two sample means that differ most do not differ significantly, none of the other pairs will, either. If a procedure can provide protection against making a Type I error when comparing the smallest to the largest mean, you are protected against Type I errors when testing all possible pairs. This is the strategy behind Tukey's procedure. The test that Tukey devised is based on the distribution of a statistical measure called the *studentized range statistic*, which I will explain next.

The Studentized Range Statistic

The t distribution will arise whenever you draw two samples from populations with the same mean and find the difference between the two sample means (and then estimate σ^2 from the sample data). If you draw three or more samples under the same conditions and look at the difference between the smallest and largest means, this difference will tend to be greater than the difference you found when drawing only two samples. In fact, the more samples you draw, the larger the difference will tend to be between the largest and smallest means. These differences are due only to chance (because all the population means are equal). To protect ourselves from being fooled, we need a critical value that accounts for the larger differences that tend to be found when drawing more and more samples. Fortunately, the distribution of the *studentized range statistic* allows us to find the critical values we need, adjusted for the number of samples in the multigroup experiment. (The statistic is "studentized" in that, like the ordinary *t* value—which, as you may recall, is sometimes called *Student's t*—it relies on sample variances in its denominator, to estimate the population variances, which are usually unknown.) Moreover, these critical values assume that all of the samples are the same size, so Formula 13.5 for performing *t* tests according to Tukey's procedure looks very much like Formula 13.4 for Fisher's protected *t* test with equal *n*s, as follows:

$$q = \frac{(\overline{X}_i - \overline{X}_j)}{\sqrt{\dfrac{MS_W}{n}}}$$

Formula 13.5

The letter *q* stands for the critical value of the studentized range statistic; critical values are listed in Table A.11 in Appendix A. The use of this table will be described in Section B. You may have noticed that the number 2, which appears in the denominator of Formula 13.4, is missing in Formula 13.5. This does not represent a real difference in the structure of the test. For ease of computation Tukey decided to include the factor of 2 (actually $\sqrt{2}$, because the 2 appears under the square root sign) in making Table A.11; thus the original critical values of the studentized range statistic were each multiplied by the square root of 2 to produce the entries in Table A.11.

Advantages and Disadvantages of Tukey's Test

The advantage of Tukey's HSD procedure is that the alpha that is used to determine the critical value of *q* is the experimentwise alpha; no matter how many tests are performed, or what partial null is true, α_{EW} remains at the value set initially. If you choose to keep α_{EW} at .05, as is commonly done, α_{pc} (the alpha for each pairwise comparison) must be reduced below .05 so that the accumulated value of α_{EW} from all possible tests does not exceed .05. The more groups, the more possible tests that can be performed, and the more α_{pc} must be reduced. The user of the HSD test does not have to determine the appropriate α_{pc}, however, because the critical value can be found directly from Table A.11 (a larger critical value corresponds, of course, to a smaller α_{pc}). As you might guess, the critical values in Table A.11 increase with the number of groups in the study. I will illustrate the use of Table A.11, when I present an example of Tukey's test in Section B.

The disadvantage of the HSD test in the three-group case is that it results in a reduction in power, and therefore more Type II errors, as compared to the LSD test. When dealing with more than three groups, the LSD test remains more powerful than the HSD test for the same initial alpha, but this comparison is not entirely fair in that the LSD test derives most of its extra power (when *k* > 3) from allowing α_{EW} to rise above the initially set value. By now you should be familiar with the fact that you can always increase the power of a statistical test by increasing alpha, but in doing so you are decreasing the rate of Type II errors by allowing a larger percentage of Type I errors. Because most researchers consider it unacceptable to increase power in a multigroup experiment by allowing α_{EW} to rise, the HSD test is generally preferred to the LSD test for more than three groups.

Statisticians frequently say that the Tukey test is more *conservative* than the Fisher procedure because it is better at keeping the rate of Type I errors down. All other things being equal, statistical tests that are more conservative are less powerful; as the Type I error rate is reduced, the Type II error rate (β) increases, which in turn reduces power ($1 - \beta$). Tests that are more powerful because they allow α_{EW} to build up are referred to as too *liberal*. The Fisher protected t test procedure is the most liberal (and for that reason the most powerful) way to conduct post hoc comparisons (other than unprotected t tests). The Tukey HSD procedure is one of the most conservative of the post hoc tests.

Unlike Fisher's LSD test, the HSD test does not require that the overall ANOVA be tested for significance. Although it is unlikely, it is possible for the HSD test to find a pair of means significantly different when the overall ANOVA was not significant. Requiring a significant ANOVA before using the HSD test would reduce its power slightly—and unnecessarily, because it is already considered adequately conservative. On the other hand, it is also possible, but unlikely, to find no significant pairwise comparisons with the HSD or even the LSD test when following up a significant ANOVA. The only guarantee that follows from the significance of an ANOVA is that some comparison among the means will be significant, but the significant comparison could turn out to be a complex one. As mentioned previously, complex comparisons will be discussed further in Section B.

Finally, one minor disadvantage of Tukey's HSD test is that its accuracy depends on all of the samples being the same size. The small deviations in sample sizes that most often occur accidentally in experimental studies can be dealt with by calculating the harmonic mean of all the ns (see Formula 12.24). However, if there are large discrepancies in your sample sizes, some alternative post hoc comparison procedure should be used.

Other Procedures for Post Hoc Pairwise Comparisons

Most post hoc tests for pairwise comparisons fall somewhere between Fisher's protected t tests and Tukey's HSD on the liberal to conservative spectrum. For the most part, these tests differ according to the trade-off they make between the control of Type I errors and power; unfortunately, some of the more powerful tests gain their extra power through some laxity with regard to Type I errors (e.g., the first test described below). The following list of alternative pairwise tests is far from exhaustive, but they are among the best known.

The Newman-Keuls Test

In recent years, the leading competitor of Tukey's HSD test for following a simple one-way ANOVA with pairwise comparisons has been the procedure known as the *Newman-Keuls* (N-K) *test*, also called the *Student-Newman-Keuls test* because its critical values come from the studentized range statistic. The major advantage of this test is that it is usually somewhat more powerful than HSD but more conservative than LSD. Therefore, the N-K test was widely considered a good compromise between LSD and HSD, with adequate control over Type I errors. The chief disadvantage of the N-K test used to be that it was more complicated to apply. Rather than using the same critical value for each pairwise comparison, you must arrange the means in order and use the *range* between any two of them (two adjacent means in the order have a range of 2; if there is one mean between them, the range is 3; etc.) instead of the number of groups in the overall ANOVA to look up the critical value in Table A.11. Now that most computer packages supply the N-K test as an option after an ANOVA, the computational complexities are not important. However, a more serious drawback of the N-K test is that, unlike Tukey's HSD, it does not keep α_{EW} at the level used to determine the critical value of the studentized range statistic. For that reason, statisticians do not recommend it. Now that it is becoming more widely known that the N-K test's apparent edge over Tukey's HSD in power is due chiefly to an inflation of α_{EW} (which gets worse as the number of groups increases), the popularity of the N-K test seems to be declining in the psychological literature.

Dunnett's Test

Recall the example involving six different phobic groups and one nonphobic control group that I used to describe a partial null hypothesis. If the experimenter wished to compare each of the phobic groups to the control group, but did *not* wish to compare any of the phobic groups with each other, the best method for pairwise comparisons would be the one devised by Dunnett (1964). However, *Dunnett's test* requires the use of special tables of critical values, and the situation to which it applies is quite specific, so I will not describe it here. Be aware, though, that Dunnett's test is performed by some statistical software packages, and, when it applies, it is the most powerful test available that does not allow α_{EW} to rise above its preset value.

REGWQ Test

It appears that the REGWQ test accomplishes what researchers thought the N-K test could do (but doesn't)—it modifies Tukey's test to make it more powerful without allowing α_{EW} to creep above whatever value (usually .05) is set for the test. Like the N-K test, the REGWQ test is based on q, the studentized range statistic, and adjusts the critical value separately for each pair of means, depending on how many steps separate each pair when the means are put in order. The REGWQ test does this by adjusting the alpha corresponding to q for each pair; unfortunately, this leads to fractional values of alpha that don't appear in conventional tables. The lack of tables for this test explains why it hasn't been used until recently. However, now that major statistical packages provide this test as an option (e.g., SPSS, SAS), you can expect its use to increase. (The test is named for the people who contributed to its development: <u>R</u>yan, <u>E</u>inot, <u>G</u>abriel, and <u>W</u>elsch; the Q is for the studentized range statistic, upon which the test is based.)

The Modified LSD (Fisher-Hayter) Test

Tukey's HSD is easy to use and understand but it is more conservative than necessary. In comparing various multiple-comparison methods, Seaman, Levin, and Serlin (1991) used computer simulations to demonstrate that under typical data analytic conditions, Tukey's procedure tends to keep the experimentwise alpha between about .02 and .03 when you think you are setting the overall alpha at .05. Accordingly, Hayter (1986) devised a hybrid of the LSD and HSD tests in order to squeeze more power out of HSD without allowing α_{EW} to rise above .05. Hayter's new test employs the two-step process originated by Fisher, which is why it is called the Fisher-Hayter (F-H), as well as the modified LSD test. The first step of the modified LSD (modLSD) test is to evaluate the significance of the one-way ANOVA. If (and only if) the ANOVA is significant, you are allowed to proceed to the second step, which involves the calculation of HSD, but with an important modification: the critical value associated with HSD (i.e., q) is found in Table A.11 by setting the number of groups to $k - 1$, rather than k, resulting in a smaller value for q, and therefore a smaller difference of means that must be exceeded to attain significance (hence, greater power).

Seaman, Levin, and Serlin (1991) found the modLSD (or F-H) test to be nearly as powerful as the REGWQ test, and always acceptably conservative with respect to α_{EW}. Moreover, the modLSD test has the advantage of being much easier to explain, and requires only an ordinary calculator and a table of q values—not sophisticated statistical software. Note that the modLSD test reduces to the ordinary LSD test when only three samples are involved, but it is more powerful than HSD even when there are as few as four groups. I will demonstrate the use of the modLSD test in Section B.

The Advantage of Planning Ahead

There is an advantage to planning particular comparisons before collecting the data, as compared to performing all possible comparisons or selecting comparisons after seeing the data (which is essentially the same as performing all possible comparisons). This advantage is similar to the advantage involved in planning a one-tailed instead of a two-tailed test: You can use a smaller critical value. As in the case of the one-tailed test, planned comparisons are less appropriate when the research is of an exploratory rather than a confirmatory nature. Moreover, the validity of planned comparisons depends on a promise that the researcher truly planned the stated comparisons and no others; researchers are aware of how easy it is, after seeing the data, to make a comparison that appears to have been planned. Therefore, planned comparisons are almost always well grounded in theory and/or previous research and stated as reasonable hypotheses before the results section of an empirical journal article. A common way to test planned comparisons is with the Bonferroni test, described next.

Bonferroni t, or Dunn's Test

When I presented Formula 13.2, which shows how large α_{EW} becomes when testing a number of independent comparisons, I mentioned that this equation is not accurate when the tests are not all mutually independent—such as when you are performing all possible pairwise comparisons. However, there is an upper limit to α_{EW} that will never be exceeded. Based on the work of the mathematician Bonferroni, we can state that for a given number of comparisons (which will be symbolized as j), the experimentwise alpha will never be more than j times the alpha used for each comparison. This is one form of the *Bonferroni inequality*, and it can be symbolized as follows:

$$\alpha_{EW} \leq j\alpha_{pc}$$

The Bonferroni inequality provides a very simple procedure for adjusting the alpha used for each comparison. The logic is that if α/j is used to test each individual comparison, α_{EW} will be no more than j (α/j) = α.

Therefore, α is set to whatever value is desired for α_{EW} (usually .05), and then it is divided by j to find the proper alpha to use for each comparison, as shown in Formula 13.6:

$$\alpha_{pc} = \frac{\alpha_{EW}}{j} \qquad \textbf{Formula 13.6}$$

For instance, if $\alpha_{EW} = .05$ is desired, and five tests have been planned, α_{pc} would be set to .05/5 = .01. This procedure works, because when five tests are performed (H_0 being true for each test), each at the .01 level, the probability of making at least one Type I error among the five tests is not more than $5 \times .01 = .05$. The pairwise comparisons can be tested by ordinary t tests (using the formula for protected t tests), except that the critical t is based on α_{pc} as found by Formula 13.6. Therefore, the procedure is often called the *Bonferroni t*. When it was first introduced, the major difficulty involved in using the Bonferroni procedure was that the α that must be used for each comparison is often some odd value not commonly found in tables of the t distribution. For instance, if $\alpha_{EW} = .05$ is desired but four tests are planned, the α for each test will be .05/4 = .0125. How do you look up the critical t corresponding to $\alpha = .0125$? There are equations that can be used to approximate the critical t, but Dunn (1961) worked out tables to make this test easier to use, which is why the test is also called *Dunn's test* or the *Bonferroni–Dunn test*. Of course, computers now make the Bonferroni test easy to perform; some statistical programs offer the Bonferroni test as an option when running a one-way ANOVA.

A more serious drawback of the Bonferroni test is that it is very conservative, often keeping the actual α_{EW} *below* the level that was initially set. Recall that the test is based on an inequality and that the preset α_{EW} is an upper limit; especially when there are many tests planned, that upper limit will not be reached, even in the worst case. This is why the Bonferroni test is overly conservative and therefore not recommended when you are performing all possible pairwise comparisons. For instance, if you had run a five-group experiment and planned to perform all 10 possible pairwise comparisons, the Bonferroni test would set α_{pc} to .05/10 = .005, but the Tukey test effectively sets α_{pc} to about .0063 (a larger α_{pc} indicates greater power). However, if you can eliminate as few as three of the ten pairwise comparisons from consideration, the Bonferroni test becomes more powerful than Tukey's HSD (.05/7 = .00714). Some planning is required, however, because it is only legitimate to eliminate tests from consideration *before* you see the data. The Bonferroni test is therefore best used with planned comparisons and is only used as a post hoc test when other more powerful procedures are not applicable. Consequently, this test will be discussed further only in the context of planned comparisons.

SUMMARY

1. If there are k groups in an experiment, the number of different t tests that are possible is $k(k - 1)/2$.
2. If all of the possible t tests in an experiment are performed, the chances of making at least one Type I error (i.e., the experimentwise alpha, or α_{EW}) will be larger than the alpha used for each of the t tests (i.e., the alpha per comparison, or α_{pc}). The α_{EW} will depend on α_{pc}, the number of tests, and the extent to which they are mutually independent.
3. If all possible t tests are performed, or if t tests are selected after seeing the results of the experiment, α_{EW} can easily become unacceptably high. To keep α_{EW} down, a procedure for *post hoc* (or *a posteriori*) comparisons is needed. If you can plan the comparisons before seeing the data, *a priori* procedures can be used.

4. The simplest procedure for post hoc comparisons is to begin by performing the one-way ANOVA and proceed with t tests only if the ANOVA is significant. The MS_W term from the ANOVA is used to replace the sample variances in these t tests, which are generally referred to as Fisher's protected t tests. When all the groups are the same size, Fisher's least significant difference (LSD) can be calculated, thus streamlining the procedure.

5. If all of the population means represented in a multigroup experiment are actually equal, the *complete null hypothesis* is true. Fisher's procedure provides full protection against Type I errors only in this case. Protection is not adequate if a partial null is true (i.e., some, but not all, of the population means are equal); in this case Fisher's procedure allows α_{EW} to become unacceptably high when there are more than three groups.

6. Tukey devised a procedure—the *honestly significant difference* (HSD) test—that allows α_{EW} to be set before conducting any t tests and assures that α_{EW} will not rise above the initially set value no matter how many groups are involved in the experiment and no matter which partial null may be true. It is not necessary to obtain a significant ANOVA before proceeding. Tukey's test is based on the *studentized range statistic* (q).

7. The HSD test is more *conservative* than the LSD test because it does a better job of keeping Type I errors to an acceptably low level. A conservative test is less powerful than a more liberal test. The reduction of Type II errors associated with a liberal test generally comes at the expense of an increase in Type I errors.

8. The Newman-Keuls test is more powerful than Tukey's HSD, because it adjusts the critical value of q according to the number of steps by which a pair of means differ when all the means are put in order. However, now that it appears that the N-K test gains its extra power by letting α_{EW} rise above the preset value, this test is losing its popularity.

9. Dunnett's test is highly recommended in the special case when you want to compare one particular group mean (e.g., a control group) to each of the others in your study. The REGWQ test is a modification of Tukey's test that seems to provide greater power without reducing the control of Type I errors. The modified LSD test is a simpler alternative to the REGWQ test that has almost as much power, together with good control over α_{EW}.

10. All else being equal, comparisons that can be planned in advance will have greater power than post hoc comparisons. The Bonferroni test involves dividing the desired α_{EW} by the number of comparisons being planned. It is too conservative to be used for post hoc comparisons, but can be very powerful when relatively few tests are planned.

EXERCISES

*1. How many different pairwise comparisons can be tested for significance in an experiment involving
   a. Five groups?
   b. Eight groups?
   c. Ten groups?

2. If a two-group experiment were repeated with a different independent pair of samples

each time, and the null hypothesis were true in each case ($\alpha = .05$), what would be the probability of making at least one Type I error
   a. In five repetitions?
   b. In ten repetitions?

*3. In Exercise 12A5, the introversion means and standard deviations for students seated in three classroom locations ($n = 80$ per group) were as follows:

	Front	Middle	Back
$\overline{X}$	28.7	34.3	37.2
s	11.2	12.0	13.5

a. Use Formula 13.4 to calculate a t value for each pair of means.

b. Which of these t values exceed the critical t based on df_W, with alpha = .05?

4. Assume that the standard deviations from Exercise 3 were doubled.

a. Recalculate the t value for each pair of means.

b. Which of these t values now exceed the critical t?

c. What is the effect on the t value of doubling the standard deviations?

*5. a. Recalculate the t values of Exercise 3 for a sample size of $n = 20$.

b. What is the effect on the t value of dividing the sample size by 4?

6. Describe a five-group experiment for which a complex comparison would be interesting and meaningful. Indicate which means would be averaged together.

7. Describe a five-group experiment in which the complete null hypothesis is not true, but a partial null hypothesis *is* true. Indicate which population means are equal.

*8. What α_{pc} would you use if you had decided to perform the Bonferroni test with $\alpha_{EW} = .05$

a. To test a total of eight comparisons for one experiment?

b. To compare all possible pairs of means in a six-group experiment?

c. To compare a third of the possible pairs of means in a seven-group experiment?

9. Compared to Fisher's LSD test, Tukey's HSD test

a. Is more conservative

b. Is more powerful

c. Is less likely to keep α_{EW} from building up

d. Uses a smaller critical value

*10. What would be the implication for post hoc comparisons in a multigroup experiment if there were only two possibilities concerning the null hypothesis: Either the complete null hypothesis is true (all population means are equal) or all the population means are different from each other (no partial null is possible)?

a. Tukey's test would become more powerful than Fisher's.

b. Fisher's protected t tests would be sufficiently conservative.

c. Neither Fisher's nor Tukey's test would be recommended.

d. There would be no difference between the Fisher and Tukey procedures.

B

BASIC STATISTICAL PROCEDURES

In the three-group experiment to test the effects of vitamins on sick days described in the previous chapter, the F ratio was significant, allowing us to reject the null hypothesis that the three population means were equal. However, the designer of that study would not want to end the analysis with that result; he or she would probably want to ask several more specific questions involving two population means at a time, such as, is there a significant difference between the vitamin C group and the multivitamin group? Because the F ratio was significant, it is acceptable to answer such questions with Fisher's protected t tests.

Calculating Protected t Tests

Protected t tests can be calculated using Formula 13.3 or Formula 13.4, but first we need some pieces of information from the ANOVA. We need to know MS_W as well as the mean and size of each group. For the vitamin example, $MS_W = 9.17$, and all the groups are the same size: $n = 10$. The means are as follows: $\overline{X}_{Plac} = 9$, $\overline{X}_{VitC} = 7$, and $\overline{X}_{MVit} = 5.5$. Inspecting the means, you can see that the largest difference is between $\overline{X}_{Plac}$ and $\overline{X}_{MVit}$, so we test that difference first. When all of the groups are the same size, the largest difference gives us our best chance of attaining statistical significance. If the largest difference is not significant, we needn't bother testing any other pairs of means.

Because the samples being tested are the same size, we can use Formula 13.4 to find the t value that corresponds to $\overline{X}_{MVit}$ versus $\overline{X}_{Plac}$:

$$t = \frac{(\overline{X}_i - \overline{X}_j)}{\sqrt{\frac{2MS_W}{n}}} = \frac{9 - 5.5}{\sqrt{\frac{2(9.17)}{10}}} = \frac{3.5}{\sqrt{1.834}} = \frac{3.5}{1.354} = 2.58$$

The critical t for any protected t test is determined by the df corresponding to MS_W, that is, df_W (which is also called df_{error}). In this case, $df_W = N_T - k = (3 \times 10) - 3 = 30 - 3 = 27$. From Table A.2, using $\alpha = .05$, two-tailed, we find that $t_{crit} = 2.052$. Because the calculated t (2.58) is greater than the critical t, the difference between $\overline{X}_{MVit}$ and $\overline{X}_{Plac}$ is declared statistically significant; we can reject the hypothesis that the two population means represented by these groups are equal. The next largest difference is between $\overline{X}_{VitC}$ and $\overline{X}_{Plac}$, so we test it next.

$$t = \frac{9 - 7}{\sqrt{\frac{2(9.17)}{10}}} = \frac{2}{\sqrt{1.834}} = \frac{2}{1.354} = 1.477$$

The critical t is, of course, the same as in the test above (2.052), but this time the calculated t (1.477) is less than the critical t, and therefore the difference cannot be declared significant at the .05 level.

You may have noticed some redundancy in the preceding two tests. Not only is the critical t the same, but the denominator of the two t tests (1.354) is the same. As you can see from Formula 13.4, the denominator will be the same for all the protected t tests following a particular ANOVA if all of the samples are the same size. The fact that both the critical t and the denominator are the same for all of these tests suggests a simplified procedure. There has to be some difference between means that when divided by the constant denominator is exactly equal to the critical t. This difference is called *Fisher's least significant difference (LSD)*. This relationship is as follows:

$$t_{crit} = \frac{LSD}{\sqrt{\frac{2MS_W}{n}}}$$

Calculating Fisher's LSD

If the difference between any two means is less than LSD, it will correspond to a t value that is less than t_{crit}, and therefore the difference will not be significant. Any difference of means greater than LSD *will* be significant. To calculate LSD, it is convenient to solve for LSD in the preceding expression by multiplying both sides by the denominator, to produce Formula 13.7:

$$LSD = t_{crit}\sqrt{\frac{2MS_W}{n}} \qquad\qquad \textbf{Formula 13.7}$$

Note that LSD can be calculated only when all the sample sizes are the same, in which case the n in Formula 13.6 is the size of any one of the samples. Let us calculate LSD for the vitamin ANOVA:

$$LSD = 2.052\sqrt{\frac{2(9.17)}{10}} = 2.052\sqrt{1.834} = 2.052(1.354) = 2.78$$

Table 13.1	$\overline{X}_{\text{Vitc}}$	$\overline{X}_{\text{MVit}}$
$\overline{X}_{\text{Plac}}$	2	3.5*
$\overline{X}_{\text{VitC}}$		1.5

Once LSD has been calculated for a particular ANOVA, there is no need to calculate any t tests. All you need to do is calculate the difference between every pair of sample means and compare each difference to LSD. A simple way to display all the differences is to make a table like Table 13.1 for the vitamin experiment. The asterisk next to the difference between $\overline{X}_{\text{Plac}}$ and $\overline{X}_{\text{MVit}}$ indicates that this difference is larger than LSD (2.78) and is therefore significant, whereas the other differences are not. It is easy to see that in an experiment with many groups, the calculation of LSD greatly streamlines the process of determining which pairs of means are significantly different. However, if the groups are not all the same size, you will need to perform each protected t test separately, using Formula 13.3. If there are differences in sample size, and it is not reasonable to assume homogeneity of variance, you cannot use MS_W, and each protected t test must be performed as a separate-variances t test, as described in Chapter 7.

Calculating Tukey's HSD

For the vitamin experiment, you could calculate Tukey's HSD instead of LSD. I don't recommend using HSD when there are only three groups because in that case the procedure is unnecessarily conservative. However, for comparison purposes I will calculate HSD for the preceding example. We begin with Formula 13.5, replacing the difference of means with HSD and the value of q with q_{crit}, as follows:

$$q_{\text{crit}} = \frac{\text{HSD}}{\sqrt{\dfrac{MS_W}{n}}}$$

Next we solve for HSD to arrive at Formula 13.8:

$$\text{HSD} = q_{\text{crit}}\sqrt{\frac{MS_W}{n}} \qquad \text{\textbf{Formula 13.8}}$$

To calculate HSD for the vitamin example, we must first find q_{crit} from Table A.11 (assuming $\alpha = .05$). We look down the column labeled "3," because there are three groups in the experiment, and we look down to the rows labeled "24" and "30". Because df_W (or df_{error}) = 27, which is midway between the rows for 24 and 30, we take as our value for q_{crit} a value that is midway between the entries for 24 (3.53) and 30 (3.49)—so q_{crit} is about equal to 3.51. (This value is approximate because we are performing linear interpolation, and the change in q is not linear.) The values for MS_W and n are the same, of course, as for the calculation of LSD. Plugging these values into Formula 13.8, we get:

$$\text{HSD} = 3.51\sqrt{\frac{9.17}{10}} = 3.51(.958) = 3.36$$

Referring to Table 13.1, you can see that the difference between $\overline{X}_{\text{MVit}}$ and $\overline{X}_{\text{Plac}}$ (3.5) is greater than HSD (3.36) and is therefore significant, according to the Tukey procedure. The other two differences are less than HSD and therefore not significant. In this case, our conclusions about which pairs of population means are different do not change when switching from Fisher's to Tukey's method. This will often be the case. However, note that HSD (3.36) is larger than LSD (2.78) and that the difference of 3.5 between the multivitamin and the placebo groups, which easily exceeded LSD, only barely surpassed HSD. Had there been two sample means that differed by 3, these two means would have been declared significantly different by Fisher's LSD, but not significantly different by Tukey's HSD. As discussed in the previous section, the Tukey procedure is more conservative (too conservative in the three-group case), and using it makes it harder for a pairwise comparison to attain significance.

Interpreting the Results of Post Hoc Pairwise Comparisons

What are the implications of the statistical conclusions we drew from our pairwise comparisons in the vitamin example? First, we have some confidence in recommending multivitamins to reduce sick days. However, we cannot say with confidence that vitamin C alone reduces sick days. A naive researcher might be tempted to think that if multivitamins differ significantly from the placebo, but vitamin C alone does not, the multivitamins must be significantly better than vitamin C. However, we have seen that the difference between vitamin C alone and multivitamins is *not* statistically significant. It is not that we are asserting that vitamin C alone is no different from the placebo or that multivitamins are no different from vitamin C alone. But we must be cautious; the differences just mentioned are too small to rule out chance factors with confidence. It is only in the comparison between multivitamins and the placebo that we have sufficient confidence to declare that the population means are different. (*Note*: There are, of course, no actual populations being compared in this case. What we are really implying is that the population mean you would find if everyone were taking multivitamins would be different from the population mean you would get with everyone taking placebos. This is also the same as saying that multivitamins have some [not zero] effect on the population compared to placebos.)

Declaring that the difference between the placebo and multivitamin conditions is significant is, by itself, not very informative. Before deciding whether to bother taking multivitamins to reduce your annual sick days, you would probably want to see (if you have been studying this text) an interval estimate of the number of sick days likely to be reduced. How to create such an estimate will be discussed next.

Confidence Intervals for Post Hoc Pairwise Comparisons

Both the ordinary and modified LSD tests are considered examples of *sequential* comparison methods, because a decision made at one step can affect the results at the next step. In particular, a lack of significance of the overall ANOVA at the first step stops the procedure completely, preventing any significant results occurring at the next step (it is not very unusual for a one-way ANOVA to fail to reach significance, even though one or more ordinary t tests among the sample means would be significant). In contrast, Tukey's HSD procedure is considered a *simultaneous* comparison method, in that it consists of only one step. Simultaneous methods lend themselves more easily to the construction of valid confidence intervals (CIs) for all possible pairs of population means.

The 95% CI for the difference of any two population means represented in a multi-group study can be expressed in terms of Tukey's q criterion as:

$$\mu_i - \mu_j = \overline{X}_i - \overline{X}_j \pm q_{.05}\sqrt{\frac{MS_W}{n}} \qquad \textbf{Formula 13.9}$$

or more simply, as: $\mu_i - \mu_j = \overline{X}_i - \overline{X}_j \pm HSD$, where the i and j subscripts represent any two groups (i.e., levels of the independent variable). Note that because Tukey's HSD test keeps α_{EW} at (or below) .05 simultaneously for all possible pairwise tests, the 95% confidence level applies jointly to all of the possible two-group CIs. For example, in a five-group study, there are ten 95% CIs that can be constructed, and if these CIs are based on Formula 13.9, the long-run probability is .95 that *all ten* of these CIs will simultaneously capture their corresponding population mean differences.

Because we have already found HSD to be 3.36 for the vitamin example, we can easily find the 95% CI for the difference between the multivitamin and placebo means, as follows: $\mu_i - \mu_j = \overline{X}_i - \overline{X}_j \pm 3.36 = 9 - 5.5 \pm 3.36 = 3.5 \pm 3.36 = .14$ *to* 6.86. Notice that this interval does not include zero, which tells you that this difference is significant according to Tukey's test. On the other hand, the 95% CI for the Vitamin C/placebo difference is: 2 ± 3.36, which extends from -1.36 to 5.36, and *does* include zero. As I have mentioned before, Tukey's HSD is a bit overly conservative with only three groups, but the Bonferroni correction is even more conservative, as I will show next.

The Bonferroni correction expressed in Formula 13.8 can be used as the basis for constructing simultaneous CIs. As mentioned in Section A, the critical t for a five-group study equals $t_{.005}$, if the desired α_{EW} is .05. Therefore, the 95% Bonferroni CI looks like this:

$$\mu_i - \mu_j = \overline{X}_i - \overline{X}_j \pm t_{.005}\sqrt{\frac{2MS_W}{n}} = \overline{X}_i - \overline{X}_j \pm t_{.005}\sqrt{2}\sqrt{\frac{MS_W}{n}}.$$

Because $t_{.005}(60)\sqrt{2}$ equals $2.915*1.414 = 4.12$, which is slightly larger than $q_{.05}(5, 60) = 3.98$, you can see that the Bonferroni CIs are even more conservative. The Tukey CIs are conservative enough! As I mentioned before, the Bonferroni correction is not appropriate when you are performing all possible pairwise comparisons.

Tukey's HSD versus ANOVA

As I mentioned in Section A, it is possible for a pair of means to differ significantly according to Tukey's test, even in the context of a one-way ANOVA that falls short of significance itself. This situation is most likely to occur when all of the sample means are clumped together, except for the smallest and largest, like this: 10, 19, 20, 21, 30. Let us calculate MS_{bet} for these means, given that each group contains 13 observations: $MS_{bet} = ns_{\overline{x}}^2 = 13(50.5) = 656.5$. If MS_W happens to equal 280, the ANOVA F will be $656.5/280 = 2.345$, which is not significant at the .05 level, because $F_{.05}(4, 60) = 2.53$. However, $q_{.05}(5, 60) = 3.98$, and therefore:

$$HSD = 3.98\sqrt{\frac{280}{13}} = 3.98(4.641) = 18.47$$

The difference of the two extreme means (20) is greater than 18.47, and therefore significant (surprisingly) at the .05 level by Tukey's conservative test.

The reverse situation can occur when the means are mostly spread out, like this: 10, 11, 20, 29, 30. Now, $MS_{bet} = 13(90.5) = 1176.5$. Even with MS_W as large as 350, the ANOVA F, $1176.5/350 = 3.36$, is easily significant (the critical F is still 2.53). However, HSD has increased to:

$$HSD = 3.98\sqrt{\frac{350}{13}} = 3.98(5.19) = 20.65.$$

Although the ordinary ANOVA is significant at the .05 level, Tukey's test is so conservative that it does not find any of the pairs of means to differ significantly. It is this conservatism that is addressed by the modified LSD test, as shown next.

The Modified LSD (Fisher-Hayter) Test

Given the significance of the F ratio in the immediately preceding example, the modified LSD test would proceed by calculating HSD based on $q_{.05}$ $(k - 1, \text{df}_W) = q_{.05}(4, 60) = 3.74$, so modLSD would be equal to: 3.74 (5.19), which equals 19.41. Notice that, unlike the HSD test, the slightly more powerful modified LSD test allows the two extreme means, 10 and 30, to be declared significantly different at the .05 level. Ordinary LSD would be equal to only: $LSD = 2.00\sqrt{2*350/13} = 2.00(7.34) = 14.68$, allowing three more pairs to be considered significantly different (10 vs. 29; 11 vs. 30; 11 vs. 29). But ordinary LSD is considered too liberal to be applied to a five-group analysis.

Which Pairwise Comparison Procedure Should You Use?

The choice is clear when your study involves only three different groups, and the ANOVA is significant. In that case, neither the modified LSD nor the Newman-Keuls test provide any advantage over ordinary Fisher protected t tests (they produce identical results in the equal-n case), and Tukey's HSD is overly conservative. When trying to publish your results you may have to include a reference to the literature to justify using Fisher's procedure at all (e.g., Seaman, Levin, & Serlin, 1991), but there is no controversy about this choice. There is no difficulty performing protected t tests with different sample sizes, but if the sample variances are very different, you should consider seperate-variance t tests.

When you are dealing with four or more groups, you have more options. If the sample sizes are equal, or nearly equal (calculate n_h in the latter case), Tukey is always a reasonable choice—even if the ANOVA is not significant. For a little more power, you might consider the REGWQ test, or the modified LSD test (if the ANOVA *is* significant), but be prepared to reference the literature to justify your choice, because even though these tests are adequately conservative, they are fairly new and relatively unknown. Additional alternatives have been devised for pairwise comparisons, but they are even less known and rarely have any advantage over the ones mentioned here. If you happen to meet the assumptions of Dunnett's test (in particular, if you are comparing each of several conditions to the same control group), that test is always the best choice.

If your sample sizes differ considerably, and especially if the sample variances are quite different as well, the Games-Howell test (Games & Howell, 1976) is a well-respected option; it is a seperate-variances version of Tukey's test. You will sometimes see Scheffé's test being used for pairwise comparisons, but unless other, more complex comparisons were performed or considered during the same analysis, this test is usually much more conservative than necessary. I will discuss Scheffé's test in the context of complex comparisons in the next section. Finally, if you are performing no more than about half of all of the pairwise tests that are possible, a series of tests based on the Bonferroni adjustment can be a powerful and very flexible choice. However, there are more powerful versions of the Bonferroni test than the one I have already described; some of these will be described in the remainder of this section.

Complex Comparisons

I introduced the concept of a complex comparison in Section A in the context of the days-of-the-week experiment. A researcher who wants to compare IQs of people born on a weekend with those of people born during the week will want to calculate and test the difference between the Saturday/Sunday IQ average, and the average IQ for the other five days of the week. As usual, what the researcher really wants to know is the magnitude of this difference in the population. The population value for this complex comparison involves a comparison of the averages of population means, and is symbolized by Ψ (the Greek letter psi, pronounced "sigh"). For our example:

$$\Psi = \frac{\mu_{Sat} + \mu_{Sun}}{2} - \frac{\mu_{Mon} + \mu_{Tue} + \mu_{Wed} + \mu_{Thu} + \mu_{Fri}}{5}$$

This is written more conveniently by dividing each term by its denominator as follows:

$$\Psi = \frac{1}{2}\mu_{Sat} + \frac{1}{2}\mu_{Sun} - \frac{1}{5}\mu_{Mon} - \frac{1}{5}\mu_{Tue} - \frac{1}{5}\mu_{Wed} - \frac{1}{5}\mu_{Thu} - \frac{1}{5}\mu_{Fri}$$

This is a *linear* combination of population means (linear in the same sense as in linear transformation). The more general way of writing any linear combination of population means is:

$$\Psi = c_1\mu_1 + c_2\mu_2 + c_3\mu_3 + \ldots + c_k\mu_k = \sum_{i=1}^{k} c_i\mu_i \qquad \textbf{Formula 13.10}$$

where the c_is are the coefficients (or weights) by which the population means are multiplied. If the c_is add up to zero (this means that at least one of the c_is must be negative, but not all), the linear combination is called a *linear contrast*. Note that our example above has this property ($1/2 + 1/2 - 1/5 - 1/5 - 1/5 - 1/5 - 1/5 = 0$), so it qualifies as a linear contrast. To avoid dealing with coefficients that are fractions, you can multiply all of the coefficients in the example above by their lowest common denominator, which is 10, to yield the following expression:

$$\Psi = 5\mu_{Sat} + 5\mu_{Sun} - 2\mu_{Mon} - 2\mu_{Tue} - 2\mu_{Wed} - 2\mu_{Thu} - 2\mu_{Fri}$$

This, of course, also multiplies ψ by 10 as well, but there will be a way to compensate for this when we get to significance testing for linear contrasts.

Sample Estimates for Linear Contrasts

A linear contrast is just a difference score involving group means; it can be as simple as a pairwise comparison (in this case, $c_1 = +1$ and $c_2 = -1$, so $\psi = u_1 - u_2$) or as complex as the combination in the weekend versus weekday comparison. No matter how complex the expression that gives us the difference score, it can be tested for significance with a simple t test that places that difference score in the numerator. It is, however, a bit more convenient to use the ANOVA approach, but as you know, that only means squaring the t test. Although Ψ is the quantity whose value we really want to know, we have to estimate it with our sample means (or measure the entire population). The sample estimate of Ψ is often symbolized as a Ψ with some modification to it, but in keeping with my desire to use Roman letters for sample statistics, I will use a capital L (for linear contrast) to represent this estimate, as shown in the following formula:

$$L = \sum_{i=1}^{k} c_i \overline{X}_i \qquad \textbf{Formula 13.11}$$

In the simple case of the pairwise comparison, $L = \overline{X}_1 - \overline{X}_2$.

Calculating the *F* ratio to Test a Linear Contrast

To perform a one-way ANOVA for a contrast, L is used as the basis for calculating a term that resembles $SS_{between}$, which is called $SS_{contrast}$. If all the groups contributing to L are the same size, n, the formula is simply:

$$SS_{contrast} = \frac{nL^2}{\sum c_i^2}$$

Formula 13.12

Note that if you multiply all of the coefficients by some constant, both L^2 and $\sum c_i^2$ are multiplied by the square of the constant, which cancels out when you divide them. That is why we can multiply our coefficients to get rid of fractions without affecting the size of $SS_{contrast}$.

In keeping with what you have learned about post hoc comparisons, the error term for testing a contrast is just MS_W from the one-way ANOVA. Because a linear contrast amounts to a single difference score, it involves only one degree of freedom; therefore, $MS_{contrast} = SS_{contrast}$, always. Thus, the F ratio for testing the significance of a linear contrast, involving equal-sized groups is:

$$F = \frac{nL^2/\sum c_i^2}{MS_W}$$

Formula 13.13

Note that in the two-group case this turns out to be:

$$F = \frac{\dfrac{n(\overline{X}_1 - \overline{X}_2)^2}{2}}{MS_W}$$

If you look back to the beginning of Chapter 12, you will see that this is what you get when you square the formula for the t test and rearrange terms a bit.

If the groups are not all the same size, the formula for $SS_{contrast}$ must be modified as follows:

$$SS_{contrast} = \frac{L^2}{\sum(c_i^2/n_i)}$$

Formula 13.14

Note that each squared coefficient must be divided by the size of the group with which it is associated before being summed. Then $SS_{contrast}$, which is the same as $MS_{contrast}$, is divided by MS_W to create the F ratio.

As an example, let us return to the experiment analyzed in Chapter 12, Section B, but this time our aim is to test a linear contrast that compares the group of subjects who have not experienced a parental death with the average of the two groups who have. The first step is to find L. The way that the groups are ordered in Table 12.2, $L = 1/2\ \overline{X}_1 + 1/2\ \overline{X}_2 - \overline{X}_3$, or $L = \overline{X}_1 + \overline{X}_2 - 2\overline{X}_3$. Using the latter version, $L = 28 + 30.6 - 2(35.83) = 58.6 - 71.67 = -13.07$. The next step is to use Formula 13.13 to find $SS_{contrast}$:

$$SS_{contrast} = \frac{(-13.07)^2}{\dfrac{1^2}{4} + \dfrac{1^2}{5} + \dfrac{-2^2}{6}} = \frac{170.82}{1.117} = 152.93$$

The final step is to take $SS_{contrast}$ ($= MS_{contrast}$), and divide it by MS_W from the ANOVA:

$$F = \frac{152.93}{23.84} = 6.41$$

Testing the Significance of a Planned Contrast

Given that the numerator of this F ratio is associated with only one degree of freedom (e.g., we could take its square root and perform a t test), and the denominator (i.e., MS_W) has df_W degrees of freedom, the critical F for this test is $F_{.05}(1, 12) = 4.75$. Because $F_{contrast}$ (6.41) > 4.75, the contrast is statistically significant at the .05 level. In case you don't recall, the F ratio for the original ANOVA was 3.4, which was not significant. Also, note that the critical value for the original ANOVA was actually *smaller* than F_{crit} for the contrast (3.89 vs. 4.75), because the ANOVA has *two* df in the numerator; the contrast was significant despite having an even larger critical F to beat! So, why was the calculated contrast F so much larger than the ANOVA F? Because we chose the right contrast to test. You'll see shortly what happens if you choose the wrong one.

The Advantage of Planned Complex Comparisons

The significance test just performed assumes that the contrast was planned, based on theoretical formulations and/or previous results. For instance, previous research may have indicated that the death of one parent has such a profound negative effect on optimism, that a second parental death cannot reduce it much further. This could have led to a prediction with respect to the pattern (i.e., spacing) of the three sample means—specifically, that $\overline{X}_{one}$ and $\overline{X}_{both}$ would be close together with $\overline{X}_{none}$ (i.e., both parents still alive) considerably separated from the other two. If you expect this pattern, it is to your advantage to calculate a complex comparison between the average of the two means expected to be close together and the one mean expected to be different from the other two. In the original ANOVA, SS_{bet} was 161.97 (see Table 12.3), but, being based on the variance of three means, it had to be divided by 2 (i.e., $k - 1$). Therefore, MS_{bet} was 80.98. $SS_{contrast}$ was 152.93, but because it was based on a single df (i.e., the magnitude of L), $MS_{contrast}$ was also 152.93, and therefore much larger than MS_{bet}.

The more closely your contrast matches the actual pattern of the sample means, the larger the proportion of SS_{bet} that will be captured by $SS_{contrast}$. In this case, had $\overline{X}_{one}$ and $\overline{X}_{both}$ been equal to each other, but not to $\overline{X}_{none}$, $SS_{contrast}$ would have been equal to SS_{bet}, and therefore $MS_{contrast}$ would have been twice as large as MS_{bet}. On the other hand, when the sample means fall in a pattern that is counter to your expectations, your $F_{contrast}$ could actually come out smaller than the ANOVA F. As an example of betting on the wrong contrast, imagine that your theory was telling you that the loss of both parents before a certain age would be much more devastating to optimism than the loss of one parent, because in the latter case, the effect on optimism would be mitigated by the remaining parent. The corresponding contrast would involve averaging $\overline{X}_{none}$ and $\overline{X}_{one}$ and subtracting $\overline{X}_{both}$. Let us calculate $F_{contrast}$ for this case: $L = -2\overline{X}_{both} + \overline{X}_{one} + \overline{X}_{none} = -2(28) + 30.6 + 35.83 = 10.43$

$$SS_{contrast} = \frac{10.43^2}{\frac{-2^2}{4} + \frac{1^2}{5} + \frac{1^2}{6}} = \frac{108.785}{1 + .2 + .1667} = \frac{108.785}{1.3667} = 79.6 = MS_{contrast}$$

$F_{contrast} = 79.6/23.84 = 3.34$, which is actually slightly smaller than the ANOVA F of 3.4. Even if $F_{contrast}$ were slightly larger, we would have lost ground, because the critical F is considerably larger for the contrast than for the ANOVA. It would not be fair to gain power through choosing a planned contrast if there were no possibility of losing power by making the wrong choice. In that way, planned contrasts are like one-tailed tests in a two-group analysis; they are valid only when you predict the direction or pattern of the means *before* looking at your results.

Although one-tailed tests are rarely reported in the more selective journals, planned contrasts seem to be increasing in popularity. Perhaps, planned contrasts are viewed as more acceptable than one-tailed tests, because they involve more elaborate and sophisticated rationales, which cannot easily be devised to fit any pattern of results that you may obtain. Although it is considered acceptable to create a *post hoc complex comparison* to aid in the interpretation of your data, such a procedure is allowed only when the overall ANOVA attains statistical significance. Thus, it is not considered acceptable to use a post hoc complex comparison in order to attain significance when the ANOVA is *not* significant. To prevent the latter practice, researchers are expected to test their post hoc complex comparisons according to *Scheffé's test*, as described next.

Scheffé's Test

Scheffé (1953) understood that the best anyone can do when creating a complex contrast is to capture all of the SS_{bet} in a single-df comparison, so that SS_{bet} is divided by 1, instead of df_{bet}. Therefore, in the best-case scenario, $MS_{contrast}$ equals df_{bet} times MS_{bet}, and $F_{contrast}$ equals df_{bet} times F_{ANOVA}. It follows, then, that if F_{ANOVA} does not exceed its critical value, $F_{contrast}$ won't exceed df_{bet} times that critical value. This led Scheffé to a simple strategy to prevent any complex comparisons from attaining significance when the overall ANOVA does not. In fact, Scheffé (1953) discovered that you can test all the comparisons from a set of means that you want, complex or pairwise, and regardless of whether the complete or some partial null hypothesis is true, without α_{EW} exceeding .05, if you take the usual critical F from the overall ANOVA—often referred to as the *omnibus* ANOVA—and multiply it by df_{bet}. If we use F_S to represent Scheffé's stricter critical F, it is obtained from the following simple formula:

$$F_S = (k - 1)F_{crit}(k - 1, N_T - k) \qquad \textbf{Formula 13.15}$$

where N_T is the total number of subjects from all groups combined, and k is the number of groups. Whatever alpha you use to look up F_{crit}, that is the α_{EW} you get by using F_S.

If the contrast I tested in the previous subsection were the result of "data snooping," and had not been planned in advance, the F that I obtained (6.41) should be compared to F_S, which is two times ($k - 1 = 2$) the critical F from the ANOVA, or $2 \times 3.89 = 7.78$. In that case the contrast would not have been significant, which is consistent with the results of the omnibus ANOVA. When a contrast has been planned, it should be compared to a critical F based on 1 and $N_T - k$ degrees of freedom, which will always be larger than the critical F for the ANOVA, but considerably smaller than F_S.

Scheffé's test has a number of advantages. It is quite robust with respect to the usual ANOVA assumptions, and it requires no special tables (you use your usual table of the F distribution). It doesn't require the sample sizes to be equal, and it leads easily to the creation of confidence intervals. But you pay a price for all that versatility. Scheffé's test is very conservative. For instance, if your intention were to conduct pairwise comparisons only, and you used F_S as your critical value for each pair, you would have considerably less power than you would have with Tukey's HSD test (assuming that the sample sizes were equal or nearly equal). There is a good chance that pairs that differed only a bit more than HSD would fail to differ significantly by Scheffé's test. Although the desirable properties of Scheffé's test have made it quite popular in psychological research, it need not be used when you have no interest in any comparisons more complex than those involving just one pair of means, and there are no large differences in the sizes of your various groups. There is no reason to reduce your power when other methods for pairwise comparisons (e.g., Tukey's HSD) provide adequate control of Type I errors. Bear in mind that it is considered acceptable to test planned comparisons first, and then use an appropriate post hoc test to look for any additional comparisons that may be significant, according to the stricter criteria of your post hoc test.

Orthogonal Contrasts

Recall that when planning several (but not nearly all) of the possible pairwise comparisons for a multi-group study, it was recommended that you divide your desired α_{EW} by the number of comparisons planned (i.e., use a Bonferroni adjustment). When planning more than one complex comparison, there is a similar concern about the possible buildup of Type I errors. However, the accepted notion is that because the ordinary ANOVA involves $k - 1$ degrees of freedom in its numerator, a researcher ought to be allowed to spend those $k - 1$ df on complex comparisons as an alternative to performing an omnibus ANOVA. Following the usual definition of a complex linear contrast, these comparisons involve only one df each, so it should be allowable to plan $k - 1$ complex comparisons. However, in order to prevent various $SS_{contrast}$s from adding up to more than the original SS_{bet}, each of the comparisons being planned should be *orthogonal* to all of the others. To explain this concept, I will being with a case in which k is equal to 4.

Imagine a four-group experiment in which two groups receive different forms of psychotherapy, a third group reads books about psychotherapy, and the fourth group receives no treatment at all. The latter two groups can both be considered control groups. One very reasonable complex comparison would be to compare the average improvement in the control groups with the average improvement in the two therapy groups. Assuming all four groups are the same size, this comparison can be represented as $L = \frac{1}{2}\overline{X}_1 + \frac{1}{2}\overline{X}_2 - \frac{1}{2}\overline{X}_3 - \frac{1}{2}\overline{X}_4$. It is likely that the researcher would then be interested in comparing the two therapy groups with each other. Note that this second comparison is independent of the first; that is, you can increase the difference between the average score of the two control groups and the average score of the two therapy groups without increasing the difference between the two therapy groups, and vice versa (see Figure 13.1). The researcher might also have some interest in testing the difference between the two control groups (perhaps, to see if reading the books produces a placebo effect). If so, this third comparison would also be independent of the first as well as of the second. It would not be possible, however, to find a fourth comparison that would be independent of the other three (except for the rather trivial comparison in which the average of all four groups is compared to zero, which is the same as testing whether the grand mean differs from zero).

Comparisons that represent mutually independent pieces of information are called *orthogonal comparisons* or *orthogonal contrasts*. The maximum number of orthogonal contrasts in a set is related to the number of groups in the experiment; if there are k groups, it will be possible to create $k - 1$ mutually orthogonal contrasts in a set, but it won't be possible to construct a larger set. For our four-group example, there can be a set no larger than $4 - 1 = 3$ mutually independent comparisons. Bear in mind, however, that we could have constructed a different set of three mutually independent comparisons. (For example, Control 1 vs. the average of the other three; Control 2 vs. the average of the two therapies; Therapy 1 vs. Therapy 2.)

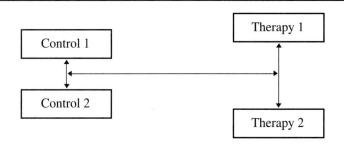

Note that when you calculate the $SS_{contrast}$ for each of $k - 1$ orthogonal contrasts, these SSs will sum to the $SS_{between}$ of the omnibus ANOVA. That is why it is considered acceptable, as an alternative to the one-way ANOVA, to test each of the comparisons in an orthogonal set using .05 as the alpha for each (provided, of course, that the comparisons have been specified in advance). In both cases, you are using the same number of degrees of freedom.

Fortunately, it is not difficult to determine whether two contrasts are orthogonal. There is a simple mathematical procedure that tells you whether the coefficients of two linear contrasts are uncorrelated when the samples are all the same size. Just sum the cross products of the coefficients of two contrasts (multiply corresponding coefficients and add); if the sum is zero, the two contrasts are orthogonal. As an example, take the contrast for comparing the two therapy groups with the two control groups, and check it against the contrast for comparing the two therapy groups to each other:

$$1/2\, \overline{X}_1 + 1/2\, \overline{X}_2 - 1/2\, \overline{X}_3 - 1/2\, \overline{X}_4$$

$$(0)\, \overline{X}_1 + (0)\, \overline{X}_2 + (1)\, \overline{X}_3 - (1)\, \overline{X}_4$$

Normally, the second contrast would be written as $\overline{X}_3 - \overline{X}_4$, with the coefficients understood, but filling in the coefficients makes it easy to sum the cross products as follows: $+1/2 \times 0 + 1/2 \times 0 + (-1/2) \times (+1) + (-1/2)(-1) = 0 + 0 - 1/2 + 1/2 = 0$. If you compare either of the two contrasts above with the comparison of the two control groups, you will also obtain a zero sum for the cross products, confirming that these contrasts are also orthogonal.

In order to demonstrate that the sum of the SSs for a test of orthogonal contrasts will add up to SS_{bet} from the ANOVA, I will use the simple vitamin example from Chapter 12, Section A.

The Additivity of the SSs from Orthogonal Contrasts

Orthogonality is a simpler matter when all of the samples are the same size, which is why I am now turning to the vitamin example, rather that the optimism example. With three groups there are only three possible complex comparisons: any one of the three means can be compared to the average of the other two. For the vitamin example (see Table 12.1), the logical choice it to compare the average of vitamin C and multivitamin means with the placebo group. First note that for this example, MS_{bet} was already calculated, so we can find SS_{bet} as $df_{bet} \times MS_{bet}$, which equals $2 \times 30.833 = 61.67$. For the complex contrast just mentioned, $L = 1 \times 9 + (-.5)7 +$

$(-.5)5.5 = 9 - 3.5 - 2.75 = 9 - 6.25 = 2.75$. Using Formula 12.12, $SS_{contrast}$ equals:

$$\frac{n \cdot L^2}{\Sigma c_i^2} = \frac{10(2.75)^2}{1^2 + (-.5)^2 + (-.5)^2} = \frac{10(7.5625)}{1 + .25 + .25} = \frac{75.625}{1.5} = 50.4167.$$

The remaining SS—i.e., $SS_{bet} - SS_{contrast} = 61.67 - 50.417 = 11.25$—should be contained in the only contrast that is orthogonal to the one just calculated: the comparison between the two vitamin conditions. Let us see. For that pairwise comparison, $L = 0 \times 9 + (1)7 + (-1)5.5 = 7 - 5.5 = 1.5$; therefore, $SS_{contrast}$ equals: $10(1.5)^2/(0 + 1^2 + 1^2) = 22.5/2 = 11.25$. (Note that any more rounding off in the SS_{bet} result would have obscured this comparison.) One of the most common uses of orthogonal contrasts is polynomial trend analysis, which is the main topic of Section C. However, before closing this section, I need to say something about planning a set of complex comparisons that are not mutually orthogonal.

Nonorthogonal Planned Comparisons

If your planned comparisons are not orthogonal, or you have planned more than $k - 1$ comparisons, the usual recommendation is to divide your desired experimentwise alpha by the number of planned comparisons and use that smaller alpha (i.e., the Bonferroni adjustment) to perform each test. However, it may seem unfair that with $k - 1$ orthogonal contrasts you can use .05 for each, but add one more, and you have to divide .05 by k. This was Keppel's (1991) sentiment when he proposed a modification to Bonferroni's test. Keppel reasoned as follows: If it is acceptable to perform $k - 1$ orthogonal comparisons, each at the predetermined alpha, it should be acceptable to set α_{EW} to $(k - 1)\alpha$, where α is the value you would normally use for each orthogonal comparison (usually .05). Then, if more than $k - 1$ comparisons are planned, α_{pc} is set to α_{EW}/j, which equals $(k - 1)\alpha/j$, where j is the number of planned comparisons. For instance, if five tests are planned for a five-group experiment, $\alpha_{pc} = (k - 1)\alpha/j = (5 - 1) \cdot .05/5 = .2/5 = .04$. Although the Bonferroni procedure almost invariably forces you to deal with unusual alpha levels, this is not a problem when you calculate your statistical tests with a major statistical software package that provides you with an exact p value for each comparison requested.

Modified Bonferroni Tests

In addition to Keppel's proposed modification of the Bonferroni test, a number of testing schemes have been devised to increase the power associated with the Bonferroni adjustment, while maintaining its flexibility and strict control over the experimentwise alpha. One of the simplest of these improves upon the inequality that forms the very basis of the Bonferroni test. Note that the procedures described next can be validly used whether the comparisons have been planned or not.

The Sharper Bonferroni: Sidak's Test

Bonferroni's inequality states that the α_{EW} will be less than or equal to the number of tests conducted (j) times the alpha used for each comparison (α_{pc}). So, in the case of a seven-group experiment, in which all of the 21 possible pairwise comparisons are to be tested, each at the .05 level, Bonferroni tells us that α_{EW} will be less than or equal to $j \times \alpha_{pc} = 21 \times .05 =$

1.05. Of course, we already knew that just by the way we define probability. As j gets larger, the Bonferroni inequality becomes progressively less informative. However, if we can assume that all of the tests we are conducting are mutually independent, we can use a sharper (i.e., more accurate) inequality, based on Formula 13.2. Solving that formula for alpha, we obtain an adjustment for α_{pc} that is somewhat less severe than the original Bonferroni correction, as follows:

$$\alpha_{pc} = 1 - (1 - \alpha_{EW})^{\frac{1}{j}} \qquad \text{Formula 13.16}$$

Although it is true that when you are performing several comparisons that are not orthogonal to each other your tests may not be mutually independent, Sidak (1967) showed that this lack of independence normally serves to keep the use of Formula 13.16 (i.e., Sidak's test) on the conservative side, while still adding power relative to the use of the usual Bonferroni formula (i.e., Formula 13.6). Unfortunately, Sidak's test leads to only a small power gain, and its greater complexity would act as a deterrent to its use, were it not for the easy availability of statistical software. As an example of the Sidak test's advantage, let us consider the case of performing all 21 of the possible pairwise comparisons following a seven-group experiment. Given that $j = 21$, and we want to keep α_{EW} at .05, Sidak's adjusted α_{pc} equals: $1 - (1 - .05)^{1/21} = 1 - .95^{1/21} = 1 - .99756 = .00244$. Note that this alpha is only slightly larger than the one you would obtain from Formula 13.6: $\alpha_{pc} = .05/21 = .002381$.

Typically, the Bonferroni and Sidak tests are used as simultaneous multiple-comparison tests, but they can be made considerably more powerful by incorporating their alpha adjustments into a sequential test. I will consider two such tests, but only in terms of Bonferroni's correction, because the Sidak formula is so messy to deal with computationally.

A Sequential Bonferroni Test

Holm (1979) demonstrated that you could add considerable power to your Bonferroni test with a simple step-down procedure. Holm's *sequentially rejective* test begins by checking whichever of your comparisons has the smallest p value. If your smallest p value is not less than the ordinary Bonferroni-corrected alpha, the testing stops right there; none of your results will be considered significant according to Holm's test. However, if your smallest p value is significant according to the ordinary Bonferroni criterion, Holm's test allows you to compare your next smallest p value to a slightly relaxed criterion—i.e., $\alpha_{EW}/(j - 1)$, instead of α_{EW}/j. If the second smallest p value is not significant, testing stops, and only the comparison with the smallest p value is declared to be significant. As you might now expect, if the second smallest p is significant, the third smallest is compared to $\alpha_{EW}/(j - 2)$, and testing continues only if this p is significant. If testing continues all the way to your largest p value, that p is compared to the desired α_{EW} without adjustment.

As a concrete example, imagine that you have ten p values, and they are (in order from smallest to largest) as follows: .002, .0054, .007, .008, .009, .0094, .012, .015, .028, .067. Because the smallest p is less than .005, it is significant. The second smallest, .0054, does not have to be smaller than .005 according to Holm's procedure, it only has to be less than .0056 (i.e., .05/9), and it is, so testing proceeds. Next, .007 is compared to .05/8 = .00625; because it is *not* less than this value, it is not declared significant, and testing stops at this point. It does not matter, for instance, that the sixth smallest $p(.0094)$ happens to be less than .05/5 = .01, and would have been significant if you had gotten that far. Once you hit a nonsignificant result in the sequence you cannot test any larger p values without ruining the Type I error control of this system. A clever improvement to Holm's procedure was later proposed by Shaffer (1986), but her modification is too complex to be described here.

An Even Sharper Bonferroni Test

More recently, Hochberg (1988) demonstrated that his reversed sequential Bonferroni procedure was even more powerful than Holm's test. Hochberg's *sequentially acceptive* test begins with a test of the *largest p* value in the set. If this p is less than .05, then all of the comparisons are declared significant (e.g., if all 10 of the ps are between .02 and .04, they are all declared significant at the .05 level, even though none of them would be significant with the usual Bonferroni adjustment). If the largest p is not less than .05, you continue by following the same kind of step-by-step procedure described for Holm's test, but in reverse. I will illustrate Hochberg's test using the same set of ten p values to which I applied Holm's procedure. First, the largest p value is compared to α_{EW}. Because .067 is not less than .05, this comparison is not declared significant—but testing does not stop. The second largest p is then compared to $.05/2 = .025$. Because .028 is not less than .025, this result is not significant either, but testing continues. The third largest p is compared to $.05/3 = .0167$, and because this time p (.015) *is* less than the Hochberg criterion value, the corresponding comparison is declared significant. Moreover, testing stops at this point, and all of the smaller p values (i.e., .012 down to .002) are automatically declared significant without further testing.

It is easy to see that Holm's test is not as powerful as Hochberg's test. The former test found only the ps of .002 and .0054 to be significant, whereas the latter found all of the p values except for the two largest (.028 and .067) to be significant. Of course, I deliberately devised an example to emphasize the difference between these two procedures. In most real-life cases, the conclusions from the two methods will not differ nearly so dramatically. Rom (1990) designed a modification of the critical values of Hochberg's test to further increase its power, but the improvement seems too slight and complex to be worth describing here.

SUMMARY

1. When you are performing protected t tests and all of your samples are the same size, the denominator will be the same for every pair of sample means. This naturally leads to a shortcut for performing your t tests. Multiplying the common denominator of all the t tests by the critical value of t for your chosen alpha yields LSD, the least significant difference of sample means. Any pair of means whose difference is larger than LSD will differ significantly.

2. HSD is calculated and used the same way as LSD except that MS_W is not multiplied by 2 (that factor has been absorbed into the table of values), and the critical value is found from the table for the studentized range statistic rather than the t distribution.

3. In addition to requiring independent random samples, normal distributions, and homogeneity of variance, Fisher's protected t tests require that the ANOVA be significant and, to keep alpha$_{EW}$ at .05, that the complete null hypothesis be true.

4. When the sample sizes are equal, LSD can be calculated as a convenience. However, Tukey's test *must* be calculated in the form of HSD, and this requires that the sample sizes be equal. Small random differ-

ences in sample sizes can be accommodated by calculating the harmonic mean of the sample sizes. Tukey's test is recommended when you want to construct CIs for the difference of each possible pair of population means, because it is a simultaneous, rather than a sequential, post hoc test. Tukey's test can actually find a significantly different pair of means when the ANOVA is *not* significant (especially if most of the sample means are clustered in the middle between two extreme means), or fail to find any significant pairs, even when the ANOVA *is* significant (especially if the sample means fall into two widely separated clusters).

5. Complex comparisons can be represented as a weighted combination of several of your group means which results in a single "difference" score, called L. To be considered a *linear contrast*, the weights, (i.e., coefficients) that are applied to your means must sum to zero. The better the match between your coefficients and the actual pattern of sample means you obtain, the larger L will be in magnitude (e.g., coefficients that result in averaging together those means that are similar and subtracting them from a mean that is quite different will tend to yield a large L).

6. A linear contrast can be tested for significance by squaring L, multiplying by n (this works only if the groups are all the same size), dividing by the sum of the squared coefficients, and then dividing this quantity (called $SS_{contrast}$, but also equal to $MS_{contrast}$ because it is based on one df) by MS_W to get an F ratio. This procedure must be modified a bit if the sample sizes are not all equal.

7. If your complex comparison was planned you can compare your F ratio to a critical value based on 1 and $N_T - k$ degrees of freedom. If the comparison was devised based on seeing the data, you should use Scheffé's (post hoc) test. This means your critical F should be $k - 1$ times the critical F from the omnibus ANOVA. Scheffé's test is very flexible, but it is too conservative to be used when you are conducting only pairwise comparisons.

8. Two linear contrasts that are not correlated with each other (the size of one does not affect the size of the other) are said to be *orthogonal* to each other. If you multiply the corresponding coefficients of two linear contrasts, and then take the sum of these cross products and find that the sum is zero, you know that the two contrasts are orthogonal.

9. In any experiment with k groups, you can create a set of $k - 1$ contrasts that are mutually orthogonal, but not a larger set. As an alternative to the one-way ANOVA, you can plan a set of orthogonal contrasts, testing each at the conventional alpha (usually .05). Orthogonal contrasts can have greater power than the overall ANOVA if you correctly predict the way the means are clustered.

10. A number of modifications to the Bonferroni test have been proposed. Keppel suggested multiplying your desired α_{EW} by $k - 1$ before dividing it among your planned comparisons. In Holm's sequentially rejective test, you arrange your tests in increasing order of their corresponding p values and compare the smallest p to α/j, the next to $\alpha/(j - 1)$, and so on, until a comparison is not significant or the largest p value is compared to alpha, where α is the desired α_{EW}, and j is the total number of comparisons being tested. Hochberg's sequentially acceptive test proceeds in the reverse manner: Your largest p value is compared to your desired α_{EW}. If significant, all of the smaller p values are declared significant as well. If not, the next largest p value is compared to $\alpha_{EW}/2$. If not significant, the third largest p value is compared to $\alpha_{EW}/3$, and so on.

1. What is the critical value (α = .05) of the studentized range statistic (q) for an experiment in which there are
   a. Four groups of six subjects each?
   b. Six groups of four subjects each?
   c. Eight groups of 16 subjects each?

*2. Use the results of the ANOVA you performed in Exercise 12B4 to calculate Fisher's LSD and Tukey's HSD at the .05 level.
   a. Which pairs of means exceed LSD?
   b. Which pairs of means exceed HSD?
   c. Which procedure seems to have more power in the three-group case?
   d. Would it be permissible to follow the ANOVA you performed in Exercise 12B5 with Fisher's protected t tests? Explain.

*3. In Exercise 12A7, the following means and standard deviations were given as the hypothetical results of an experiment involving the effects of four different drugs (n = 8 subjects per group).

	Marijuana	Amphetamine	Valium	Alcohol
$\overline{X}$	7	8	5	4
s	3.25	3.95	3.16	2.07

   a. Calculate Fisher's LSD (α = .05), whether or not it is permissible.
   b. Calculate Tukey's HSD (α = .05).
   c. Use HSD to construct 95% CIs for each pair of drug conditions.

4. Recalculate Fisher's LSD and Tukey's HSD for the data in Exercise 3, assuming that the number of subjects per group was 16.
   a. What effect does increasing the number of subjects have on the size of LSD and HSD?
   b. What conclusions can you draw from the LSD test?
   c. Does the HSD test lead to different conclusions?
   d. Which test is recommended in the four-group case and why?

*5. a. Calculate Fisher's protected t tests for each pair of groups in Exercise 12B10, using alpha = .05.
   b. What specific conclusions can you draw from the tests in part a? What do you know from these tests that you did not know just from rejecting the null hypothesis of the ANOVA?

6. According to Keppel's modified Bonferroni test,

   a. If you normally use α = .05, what would be the appropriate α_{EW} for a four-group experiment? For a six-group experiment?
   b. What α_{pc} should you use if you have planned six tests for a four-group experiment?
   c. What α_{pc} should you use if you have planned 10 tests for a six-group experiment?

*7. For the data in Exercise 12B9,
   a. Use the Bonferroni test (at the .05 level) to compare the sixth cultural group with each of the others; use Formula 13.3 and the appropriate α_{pc}. Which groups differ significantly from the sixth one?
   b. Use the modified LSD (α = .05) to make the same comparisons (use the harmonic mean of the sample sizes for n). Which groups differ significantly from the sixth one with this test?
   c. Which of the two tests seems to have more power when applied in this manner?

8. In Exercise 12B1, an experiment involving five different antidepressants yielded the following means and standard deviations: $\overline{X}_1$ = 23, s_1 = 6.5; $\overline{X}_2$ = 30, s_2 = 7.2; $\overline{X}_3$ = 34, s_3 = 7; $\overline{X}_4$ = 29, s_4 = 5.8; $\overline{X}_5$ = 26, s_5 = 6.
   a. Assuming that none of the original subjects were lost (i.e., n = 15), calculate Tukey's HSD for this experiment.
   b. Which pairs of means differ significantly?
   c. Calculate the modified LSD. Would using this test change your answer to part b?

*9. A clinical psychologist is studying the role of repression in eight different types of phobias: four animal phobias (rats, dogs, spiders, and snakes) and four nonanimal phobias (fear of meeting strangers at a party, fear of speaking to groups, claustrophobia, and acrophobia). Sixteen subjects were sampled from each of the eight phobic groups, and their mean repression scores were as follows: $\overline{X}_{rat}$ = 42.8; $\overline{X}_{dog}$ = 44.1; $\overline{X}_{spider}$ = 41.5; $\overline{X}_{snake}$ = 42.1; $\overline{X}_{party}$ = 28.0; $\overline{X}_{speak}$ = 29.9; $\overline{X}_{claus}$ = 36.4; $\overline{X}_{acro}$ = 38.2. Assume that MS_W = 18.7.
   a. Describe the most logical complex comparison that involves all of the means. Express this comparison as a linear contrast (L) in terms of the group means and appropriate coefficients.
   b. Calculate the value of L, and test the contrast for statistical significance as a post hoc test (i.e., use Scheffé's test).

c. Test all possible pairs of means with both Tukey's and Scheffé's test (Hint: Remember that L for a pairwise comparison is just the difference of sample means).

d. What value for L is just at the borderline of significance for Scheffé's test? How does this value for L compare to HSD? What does this tell you about the relative power of the two tests for pairwise comparisons?

10. Analyze the data in Exercise 12B4 in terms of two meaningful orthogonal contrasts. Test each of the contrasts for significance as a planned comparison.

*11. Analyze the data in Exercise 12B10 in terms of two meaningful orthogonal contrasts. test each of the contrasts for significance as a planned comparison.

12. Suppose that you performed eight statistical tests and obtained the following eight p values: .006, .04, .012, .008, .005, .011, .06, .02. Assuming that you want to keep α_{EW} down to .05, which of your p values would be considered significant according to:

a. The ordinary Bonferroni correction?

b. Holm's sequential Bonferroni test?

c. Hochberg's sequential Bonferroni test?

The Analysis of Trend Components

As I pointed out in Chapter 12, the levels of an independent variable can involve quantitative rather than qualitative differences. For instance, an experiment may involve groups of subjects given the same list of 50 words to memorize, but are given either 3, 6, 9, or 12 repetitions of the list. Different groups might be given either two, three, or four therapy sessions per week, or subjects could be selected for smoking either one, two, or three packs a day of cigarettes. When the levels of the independent variable are quantitative, Pearson's r can be calculated between the IV and the DV, and both the sign and magnitude of the correlation will make sense. It doesn't matter that the X variable has relatively few values, whereas the Y variable usually has a different value for every subject.

Using Linear Regression with a Quantitative IV

For instance, in the memorization example I just mentioned, each subject has 3, 6, 9, or 12 as her X value and some number from 0 to 50 (words recalled) as her Y value. The linear correlation between the IV and DV would be meaningful in this case, telling us something about how closely the number of repetitions is related to recall. However, you would want to test the Pearson's r for statistical significance before taking it too seriously. The important thing to note is that the significance test for r in this case is likely to be more interesting and powerful (in the sense of reducing Type II errors) than the results of an ANOVA performed on the four groups of subjects. Such is the case for the hypothetical data for the memorization example given in Table 13.2.

Table 13.3 is a summary table for an ordinary one-way ANOVA on these data. As you can see, the F ratio falls just short of significance at the .05 level. However, a scatterplot of the data (see Figure 13.2) shows a linear trend that is not captured by performing an ordinary ANOVA.

The Pearson's r between the IV and the DV is a statistically significant .654. (You should verify this as an exercise by calculating the r for the data in Table 13.2, as 16 pairs of numbers: 3, 17; 3, 9; 3, 7; 3, 17; 6, 11; 6, 25; and so on.) If we frame the analysis as a linear regression, the significance test for the regression (which is equivalent to testing Pearson's r) can be performed by dividing the total sum of squares into $SS_{regression}$ and $SS_{residual}$ (error). The results can be displayed in an ANOVA summary table, as shown in Table 13.4.

$SS_{regression}$ can be found by multiplying SS_{total} by r^2 ($748 \times .654^2 = 320$). Now the F ratio is significant (the p value for F is the p value associated with testing r for significance). Notice that $SS_{regression}$ is a bit less than $SS_{between}$; this is because the group means do not fall on one straight line. However, also notice that $SS_{regression}$ is only being divided by one df instead

Table 13.2			
NUMBER OF REPETITIONS			
3	6	9	12
17	11	20	30
9	25	24	18
7	15	29	19
15	21	15	29

Source	SS	df	MS	F	p
Between	336	3	112	3.26	.059
Within	412	12	34.33		
Total	748	15			

Table 13.3

Figure 13.2

Linear Trend over
Four Levels of a
Quantitative IV

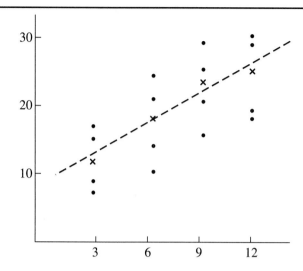

of three. Even if $SS_{\text{regression}}$ were as low as 135, the linear regression would still lead to a higher F ratio. It only takes one degree of freedom to predict all four group means with a single straight line (given that one df for the intercept of the line has already been subtracted from the total), but it takes three degrees of freedom to use the actual means of the four groups as your predictions (performing the ordinary ANOVA is like using a broken line to connect the group means; the broken line would consist of three line segments and therefore use up three df).

If you have a quantitative IV and your group means are fairly close to being a linear function of your IV, linear regression is a good alternative to the ordinary ANOVA. However, there is a more convenient approach within the context of ANOVA; it involves the use of trend components, as will be described shortly.

Testing a Curvilinear Trend

Many variables affecting physical or mental health have some optimal level (e.g., too much or too little sensory stimulation can be harmful to an infant's development); a plot of the IV versus the DV would suggest a curve more than a straight line. As an example, imagine that you want to investigate the relationship between alcohol consumption and blood levels of HDL (good) cholesterol. You randomly assign five subjects to each of six levels of alcohol consumption: 0, 1, 2, 3, 4, or 5 ounces per day and measure each subject's HDL cholesterol level after 6 months (for simplicity, I'll just assume that all

Table 13.4

Source	SS	df	MS	F	p
Regression	320	1	320	10.5	.006
Residual	428	14	30.57		
Total	748	15			

Table 13.5

	NUMBER OF OUNCES OF ALCOHOL PER DAY					
	0	**1**	**2**	**3**	**4**	**5**
	55	65	76	53	52	50
	53	64	73	60	38	37
	62	59	67	58	42	30
	57	71	70	62	45	43
	<u>48</u>	<u>66</u>	<u>64</u>	<u>67</u>	<u>48</u>	<u>40</u>
Mean	55	65	70	60	45	40

Table 13.6

Source	SS	df	MS	F	p
Regression	1,501.8	1	1,501.8	16.4	<.001
Residual	2,562.4	28	91.5		
Total	4,064.2	29			

subjects start out with the same HDL and prior history of alcohol consumption). The data are displayed in Table 13.5 and graphed in Figure 13.3. Table 13.6 gives the summary for a linear regression analysis of these data.

As in the memorization example, the linear results are significant. In fact, Pearson's r, which is negative in this case ($-.608$), is almost as large as in the previous example. And yet, the group means clearly form a curve, peaking at 2 ounces (see Figure 13.3). You can see that a good deal of between-group variance remains to be explained if you calculate the ordinary ANOVA: $SS_{between}$ equals 3,354.2, whereas the SS for linear regression is only 1,501.8. The difference is 1,852.4. (When you add this difference to the within-group error SS from the ANOVA, which equals 710, you get the residual SS in Table 13.6.) To find out if this remaining between-group SS is statistically significant, we must convert it to a variance by dividing it by the appropriate degrees of freedom, which are $df_{between} - 1 = 4$ (you lose one df for subtracting the linear SS). This variance, which can be called $MS_{nonlinear}$ is equal to $1,852.4/4 = 463.1$. To test this variance, divide it by MS_W from the omnibus ANOVA (710/24) to form an F ratio: $463.1/29.58 = 15.7$. This F ratio makes it obvious that there is a significant amount of nonlinear variance in addition to the significant linear variance. To further separate the nonlinear SS into additional orthogonal components that represent different degrees of "curviness," we will need contrast coefficients that have been designed for this purpose.

Polynomial Trend Components

One of the simplest and most obvious ways for a function to be nonlinear is for the height of the function to reverse direction (i.e., to be U-shaped, or resemble an inverted U) as you move along the horizontal axis. The function shown in Figure 13.3 is an example of a curve that exhibits one reversal. A complete quadratic function (e.g., $Y = X^2$) always includes a reversal, and can usually be used as a building block, along with a linear function and, perhaps, higher-order components, to match data that you have collected that follows a curve (I am assuming that both your independent and dependent variables have been measured on interval/ratio scales). In fact, a quadratic function can help you to model your data even if the curve does not actually reverse, but just levels off. However, if your data should exhibit more than one reversal, you can expect to need polynomial components higher than the quadratic to completely describe your data, though such higher-order components are rarely significant in psychological research.

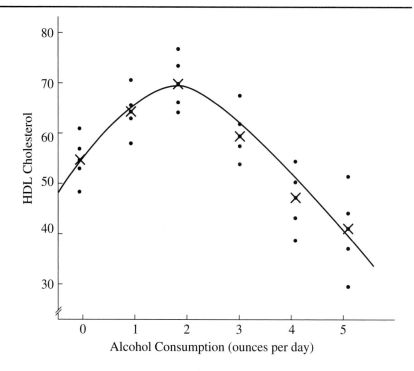

For example, when your IV has six levels, you can have, in addition to linear and quadratic components, cubic (X^3), quartic (X^4), and quintic (X^5) components. If the plot of your DV against your six-level IV happens to exhibit two reversals, a cubic component (at the least) will be required to account for all of the between-group variance. However, the converse is not true; even if the function has only one reversal, it may bend in such a way as to have a considerable cubic component. Three reversals demand a quartic component, and four reversals, a quintic. You cannot have more than four reversals with only six levels, so with $k = 6$, the highest component ever needed is quintic (see Figure 13.4). In general, the highest power needed as a predictor for k levels is $k - 1$, and the maximum number of reversals is $k - 2$ (e.g., when you have three groups, you can only have $3 - 2 = 1$ reversal).

Trend Components as Orthogonal Contrasts

A convenient way to separate the SS_{bet} from an ANOVA with k quantitative levels into $k - 1$ meaningful portions that add up to SS_{bet} is by calculating a set of orthogonal contrasts, such that each successive contrast represents a higher degree of curviness (e.g., more reversals). Each contrast should be a *linear contrast*, as defined in Section B—that is, the coefficients that define the contrast should sum to zero. The first (i.e., lowest) contrast for trend is also linear in the geometric sense, as well—that is, the coefficients progress linearly. For instance, for $k = 6$, the coefficents that are applied to the six sample means are: $-5, -3, -1, +1, +3, +5$, respectively. (Note that you cannot use $-3, -2, -1, +1, +2, +3$ for this purpose, because those coefficients are not geometrically linear; the distance between -1 and $+1$ is 2 units, while the other adjacent coefficients are only 1 unit apart.)

For $k = 6$, the quadratic trend has the following coefficents: $+5, -1, -4, -4, -1, +5$. If you were to plot these coefficents as Y values for the X values of 1 to 6, you would see a U- (i.e., parabolic) shape, which captures

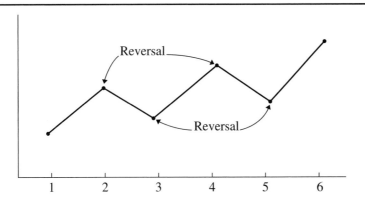

Figure 13.4

Maximum Number of
Reversals Possible with
Six Groups

the degree to which the curve reverses its direction once. You can check that the quadratic coefficients are orthogonal to the linear ones by adding the cross-products: $(-5)(+5) + (-3)(-1) + (-1)(-4) + (+1)(-4) + (+3)(-1) + (+5)(+5) = -25 + 3 + 4 + (-4) + (-3) + 25 = 0$. The full set of orthogonal polynomial contrasts for six groups is shown in Table 13.7.

Level	1	2	3	4	5	6
Linear	−5	−3	−1	1	3	5
Quadratic	5	−1	−4	−4	−1	5
Cubic	−5	7	4	−4	−7	5
Quartic	1	−3	2	2	−3	1
Quintic	−1	5	−10	10	−5	1

Table 13.7

The coefficients for other numbers of levels are shown in Table A.12 (for $k = 7$ and above, components higher than quintic can be computed, but such components so rarely have any interpretation that they are almost never computed and so have been left out of the table in Appendix A). The coefficients in Table A.12 are meant to be used with IV levels that are equally spaced. Appropriate values can be computed for levels not equally spaced, but such computations will not be discussed here. If you cross-multiply the coefficients in any pair of rows in Table 13.7, you will see that they are orthogonal. If you calculate the SSs for each of the five contrasts in the table, these SS components will sum to the $SS_{between}$ of the ordinary ANOVA.

Testing Trend Components for Significance

If the number of subjects at each level of the IV is the same, Formula 13.12 can be used to find the SS components from the means in Table 13.5 and the coefficients in Table 13.7 (otherwise the calculations are still doable, but more complicated). As an example, I will find the SS associated with the quadratic component for the alcohol experiment. The first step is to calculate L, the sum of means as weighted (i.e., multiplied) by the appropriate trend coefficients:

$$L = 55(5) + 65(-1) + 70(-4) + 60(-4) + 45(-1) + 40(5)$$

$$= 275 - 65 - 280 - 240 - 45 + 200 = -155.$$

The sum of the squared coefficients is $25 + 1 + 16 + 16 + 1 + 25 = 84$ (this sum is included in Table A.12), and the number, n, of scores at each level is 5. Plugging these values into Formula 13.12, we get

$$SS_{quadratic} = \frac{nL^2}{\sum c_i^2} = \frac{5(-155^2)}{84} = \frac{120,125}{84} = 1,430.1$$

Each trend component has only one df, so it is also true that $MS_{quadratic}$ equals 1,430.1. Because trend components are treated like any other set of orthogonal contrasts in a one-way ANOVA, each one is tested by forming an F ratio with MS_W in the denominator. (I stated earlier that, for these data, $SS_W/df_W = 710/24 = 29.58$, but you can also obtain that value by simply averaging the variances in the six columns of Table 13.5.) Therefore, $F_{quadratic}$ equals 1,430.1/29.58, which equals an easily significant value of 48.3.

SS_{linear} was already calculated for these data in terms of linear regression (see Table 13.6), and found to be 1,501.8. You can obtain this same value (and you should, as an exercise) by applying the relevant linear trend coefficients $(-5, -3, -1, +1, +3, +5)$ to the means in Table 13.5. If you subtract both SS_{linear} and $SS_{quadratic}$ from SS_{bet} (3,354.2) for the ANOVA, $SS_{residual}$ will include only the SS components above quadratic, which together add up to only 422.3. However, even when $SS_{residual}$ is divided by the three remaining df for between-group variation to form $MS_{residual}$, testing against MS_W reveals that there is a significant amount of trend variance that remains to be accounted for (see Table 13.8). If you calculate the cubic, quartic, and quintic components, you will find that they are 306.25, 111.6, and 4.46, respectively. The cubic component is significant at the .05 level, but the remaining components fall short.

Trend Components as Planned Comparisons

If your independent variable is being observed, rather than manipulated (i.e., it is not truly an IV), and is being measured precisely, it is likely that most of your subjects will have different values on the IV—you will have almost as many IV values as you do subjects. In such a case, you would likely begin with linear regression as described at the beginning of this section and then proceed to multiple polynomial regression for functions that are not linear and may even exhibit reversals (see Chapter 17). If you are assigning the levels of your quantitative IV randomly to your subjects, it is a good idea to have considerably fewer levels than subjects; you will probably want to have at least five to ten subjects for each level (if not more). It is in this type of situation that an ANOVA approach, including trend components, is likely to be appropriate.

If you are planning to look for trends before you design the experiment, you need to choose the levels of your IV carefully. Equally spaced intervals are simpler to deal with but may not be the best way to capture the function you are studying. For instance, if a child's vocabulary expands slowly from 18 to 24 months of age and then much more rapidly during the next 6 months, you may want to place your measurements at shorter intervals during the latter period. You would also want to cover the entire range of the IV that may be of interest. If it is possible that a substantial number of people are exercising to the point that they are harming their health (as measured by your DV), you may want to include high enough levels to capture that part of the trend.

Table 13.8	**Source**	**SS**	**df**	**MS**	**F**	**p**
	Linear	1,501.8	1	1,501.8	50.4	<.001
	Quadratic	1,430.1	1	1,430.1	48.3	<.001
	Higher	422.3	3	140.8	4.76	<.01
	Error	710.0	24	29.6		
	Total	4,064.2	29			

Of course, it generally makes the analysis simpler to sample the same number of subjects at each level, regardless of how the intervals are spaced.

Once you have obtained your data, if you have a theoretical model or predictions based on past research that suggest testing particular trend components, you can proceed directly with the relevant tests. However, even if you have clear expectations, it is a good idea to look at a plot of your DV against your IV. It is possible that an exponential or other trend will be a better fit to your data, in which case you should forgo the orthogonal polynomials in terms of a more appropriate function.

If you have decided to use orthogonal polynomials but do not have specific expectations concerning the trends you will find, and you don't have more than six levels, it is not entirely unreasonable to simply extract and test all of the possible trend components. With more than six levels, it is sensible to stop with the quintic trend; higher components are so unlikely to be significant and so difficult to interpret that they are rarely tested (that is why the higher components do not appear in most tables of trend coefficients). However, to avoid capitalizing on chance, it is frequently suggested that you proceed as I did in the alcohol example. That is, test the linear trend, then subtract SS_{linear} from $SS_{between}$, and proceed only if the remainder is significant when tested against MS_W from the ordinary ANOVA. Testing stops when what is left of $SS_{between}$ (i.e., all the higher trend SSs lumped together) is not significant when tested against MS_W, or the quintic component has been reached, whichever comes first (some may feel comfortable setting an even lower limit).

SUMMARY

1. If the levels of your IV are quantifiable, linear regression can be a more powerful alternative to the one-way ANOVA. The SS for regression will be less than $SS_{between}$ for the ANOVA (unless all the group means fall on one straight line), but to form the numerator of the F ratio, $SS_{regression}$ is divided by only one df (instead of $k - 1$ for $SS_{between}$) which usually leads to a higher F ratio than for the ordinary ANOVA (unless the trend is far from being linear).

2. If different groups of participants have been assigned to different levels of a quantitative factor, an analysis of trend components should be considered a potentially much more powerful alternative to the ordinary one-way ANOVA, or even linear regression. The MS for the linear trend will be the same as the MS for linear regression, but the ANOVA error term (MS_W) tends to be considerably smaller than the error variance around the regression line, generally offsetting the loss of df in calculating MS_W (df$_W$ will be smaller than the df around the regression line).

3. If a plot of your group means follows a curve, and especially if the curve involves a reversal (e.g., a U- or inverted U-shape), you will probably want to test for a quadratic component. The more reversals you see, the more likely it is that higher-order trend components will attain significance. When a quantitative IV has k levels, the SS_{bet} of the ANOVA can always be analyzed into k-1 mutually orthogonal trend components. For equally spaced levels of the IV, appropriate contrast coefficents for any number of groups from 3 to 10 can be found in Table A.12.

4. If the IV has many levels, it is customary to begin by testing the linear trend and then to test the remainder of $SS_{between}$ after subtracting the linear and each successively higher trend component. Extracting higher trends stops when the remaining $SS_{between}$ is not significant or the highest polynomial of interest has been tested (components above quintic are rarely tested). Of course, if you have theoretical predictions regarding the presence of particular trend components, you can test those components immediately.

5. When planning to use a quantitative IV, it is simpler to plan equally spaced intervals, but sometimes different spacings are needed to best capture changes in the IV/DV function. It is also important to cover the full range of IV levels that may be of interest. However, viewing the data may suggest that polynomial regression is not the most relevant approach to your data. Some other mathematical function, such as a logarithmic curve, may provide a better fit.

EXERCISES

*1. When I introduced the notion of performing linear regression as an alternative to the one-way ANOVA when your IV has quantitative levels, I used the example of five separate groups of subjects undergoing therapy either one, two, three, four, or five times per week. Suppose that the mean well-being score for each group after six months (along with its standard deviation) is as follows and that there are 15 subjects in each group.

Number of sessions/ week	1	2	3	4	5
Means	6.8	7.6	8.0	8.2	8.3
SDs	2.06	1.67	2.19	1.73	2.28

a. Calculate the ordinary one-way ANOVA for these data, and test for significance at the .05 level.
b. Test both the linear and quadratic trend components at the .05 level.
c. How does the test of the linear component in part b compare to your results in part a? What does this tell you about the advantage of planning to test trend components?
d. Plot the cell means, and describe the trend that you see. Given the number of groups in this problem, what is the largest number of reversals your plot could have had?

2. A medical researcher is testing the effects of exercise on HDL (good) cholesterol. Twenty subjects are assigned at random to each of six levels of exercise. The HDL and its standard deviation is given for each group.

Number of hours of exercise per week	3	6	9	12	15	18
Means	55	60	70	75	77	72
SDs		12.4	16.7	21.9	17.4	21.8	19.7

a. Calculate the ordinary one-way ANOVA for these data, and test for significance.
b. Test trend components until the residual is not significant. What is the highest trend component that is significant? What is the highest component that *could* be tested given these many groups?
c. Plot the cell means, and describe the trend that you see.

*3. In Exercise 12B6, the IV was the color of the cookies being eaten. In this exercise I will use the same data (following) but replace the color with sugar content in grams.

10 g	20 g	30 g
3	13	22
7	14	20
1	15	24
0	16	26
9	14	24
2	16	21

a. Analyze the data with simple linear regression.
b. Compare your results in part a with the one-way ANOVA results you obtained in Exercise 12B6.
c. Calculate the SS for the linear trend component for these data.
d. Calculate the SS for the quadratic trend component for these data. Sum the SSs for the linear and quadratic components. How does this sum compare to SS_{bet} from the results of Exercise 12B6? Explain the relationship.

4. A psychologist is studying the effects of caffeine on the performance of a video game that simulates driving a large truck at night. Six subjects are tested at each caffeine level; the number of driving errors committed by each subject are shown in the following table:

MILLIGRAMS OF CAFFEINE			
0	100	200	300
25	16	6	8
19	15	14	18
22	19	9	9
15	11	5	10
16	14	9	12
20	23	11	13

a. Calculate both the linear and quadratic trend components for these data, and test each for statistical significance.
b. After subtracting the linear and quadratic components from SS_{bet}, test the residual for significance.
c. What trend component corresponds to the residual found in part b? How could you justify testing this component given your results in part a?

*5. A cognitive psychologist is studying the amount of time required to solve anagrams as a function of the number of letters in the scrambled word. Subjects solve a set of ana-grams with either 5, 6, 7, 8, 9, or 10 letters. The average time for solution in minutes for each subject follows:

NUMBER OF LETTERS IN ANAGRAM					
5	6	7	8	9	10
1.3	1.4	4.8	5.2	6.6	6.9
1.2	3.0	3.4	6.3	6.1	6.3
2.4	1.8	4.5	6.4	6.3	6.9
1.8	2.2	3.7	5.8	7.1	7.5
2.3	2.6	4.1	5.3	6.9	6.9

a. Analyze the data with linear regression.
b. Test trend components until the residual is not significant. What is the highest trend component that is significant? What is the highest component that could be tested given these many groups?
c. Use a plot of the means to explain the results of this experiment. Based on your plot, which components would you expect to be significant? Explain.

The number of different t tests that are possible among k sample means:

$$\frac{k(k-1)}{2}$$

Formula 13.1

KEY FORMULAS

Experimentwise alpha for j independent comparisons (does not apply to the case when all possible pairwise comparisons are being performed because they cannot all be mutually independent):

$$\alpha_{EW} = 1 - (1 - \alpha)^j$$

Formula 13.2

Fisher's protected t test (it is only protected if the one-way ANOVA is significant):

$$t = \frac{(\overline{X}_i - \overline{X}_j)}{\sqrt{MS_W\left(\frac{1}{n_i} + \frac{1}{n_j}\right)}}$$

Formula 13.3

Fisher's protected t test (when both samples involved are the same size):

$$t = \frac{(\overline{X}_i - \overline{X}_j)}{\sqrt{\frac{2MS_W}{n}}}$$

Formula 13.4

Post hoc t test, according to Tukey's procedure (critical values for q are listed in Table A.11):

$$q = \frac{(\overline{X}_i - \overline{X}_j)}{\sqrt{\frac{MS_W}{n}}}$$

Formula 13.5

The Bonferroni-adjusted alpha for each comparison in terms of the desired experimentwise alpha and the number of comparisons planned:

$$\alpha_{pc} = \frac{a_{EW}}{j}$$

Formula 13.6

Fisher's LSD procedure (a simplified procedure for performing Fisher's protected t tests when all the samples are the same size):

$$LSD = t_{crit}\sqrt{\frac{2MS_W}{n}}$$

Formula 13.7

Tukey's HSD procedure (assumes all samples are the same size; if the sample sizes differ slightly, Formula 13.15 can be used to find n):

$$HSD = q_{crit}\sqrt{\frac{MS_W}{n}}$$

Formula 13.8

The 95% confidence interval for the difference between a pair of population means, based on the studentized range statistic:

$$\mu_i - \mu_j = \overline{X}_i - \overline{X}_j \pm q_{.05}\sqrt{\frac{MS_W}{n}}$$

Formula 13.9

The general formula for any linear combination of population means (if the coefficients sum to zero, it is a linear contrast):

$$\Psi = c_1\mu_1 + c_2\mu_2 + c_3\mu_3 + \ldots + c_k\mu_k = \sum_{i=1}^{k}c_i\mu_i$$

Formula 13.10

A linear combination of sample means (usually used to test a complex comparison for significance):

$$L = \sum_{i=1}^{k}C_i\overline{X}_i$$

Formula 13.11

The sum of squares associated with a linear contrast when all the samples are the same size:

$$SS_{contrast} = \frac{nL^2}{\sum c_i^2}$$

Formula 13.12

The F ratio for testing the significance of a linear contrast when all the samples are the same size:

$$F = \frac{nL^2/\sum c_i^2}{MS_W}$$

Formula 13.13

The sum of squares associated with a linear contrast when all the samples are not the same size:

$$SS_{contrast} = \frac{L^2}{\sum c_i^2/n_i}$$

Formula 13.14

The critical F for Scheffé's test:

$$F_s = (k - 1)F_{crit}(k - 1, N_T - k)$$

Formula 13.15

A sharper form of the Bonferroni adjustment, which forms the basis of Sidak's test:

$$\alpha_{pc} = 1 - (1 - \alpha_{EW})^{1/j}$$

Formula 13.16

TWO-WAY ANOVA

You will need to use the following from previous chapters:

Symbols
k: Number of groups in a one-way ANOVA
s^2: Unbiased variance of a sample
SS_{bet}: Sum of squared deviations based on group means
SS_W: Sum of squared deviations within groups
N_T: Total number of all observations in an experiment

Formulas
Formula 12.3: The F ratio
Formula 12.5: F ratio for equal-sized groups
Formula 12.18: Eta squared
Formula 12.20: Estimate of omega squared

Concepts
The F distribution
One-way ANOVA
Planned and post hoc comparisons

Chapter 14

A
CONCEPTUAL FOUNDATION

Calculating a two-way ANOVA is very similar to calculating two one-way ANOVAs, except for one extra sum of squares that almost always appears. This extra SS is the *interaction* of the two independent variables, and sometimes it is the most interesting and important part of a two-way ANOVA. Occasionally, it is a nuisance. In any case, the primary goal of this chapter is to help you understand the possible interactions of two independent variables as well as the simpler aspects of the two-way ANOVA.

Calculating a Simple One-Way ANOVA

To introduce the concept of a two-way ANOVA, I will start with a simple one-way ANOVA and show how it can be changed into a two-way ANOVA. For the one-way ANOVA, suppose we are conducting a three-group experiment to test the effects of a new hormone-related drug for relieving depression. Six depressed patients are chosen at random for each of the groups; one group receives a moderate dose of the drug, a second group receives a large dose, and a third group receives a placebo. The depression ratings for all three groups of patients after treatment appear in Table 14.1, along with their means and (unbiased) standard deviations. Because all of the groups are the same size, we can use Formula 12.5A to calculate MS_{bet}.

$$MS_{bet} = ns_{\bar{X}}^2 \qquad \textbf{Formula 12.5A}$$

As I mentioned in Chapter 12, Section A, the part of the formula that follows n is just the unbiased variance of the group means (33, 32.5, and 27.83), which is 8.12. (You can calculate this variance the long way by first finding the grand mean and then subtracting the grand mean from each group mean, squaring, and so forth, but if you have a handheld calculator with the standard deviation as a built-in function, the short way is just to enter the three group means, press the key for the unbiased standard deviation, and then square the result.) Given that MS_{bet} is simply n (the

Table 14.1		Placebo	Moderate Dose	Large Dose
		38	33	23
		35	32	26
		33	26	21
		33	34	34
		31	36	31
		28	34	32
	$\overline{X}$	33	32.5	27.83
	SD	3.406	3.450	5.269

number of subjects in each group) times the variance of the group means: $MS_{bet} = 6 \times 8.12 = 48.72$. Because the groups are all equal in size, MS_W is the simple average of the group variances: $MS_W = 1/3\ (11.6 + 11.9 + 27.76) = 1/3\ (51.26) = 17.087$.

The F ratio for the one-way ANOVA equals $MS_{bet}/MS_W = 48.72/17.087 = 2.85$. The critical F for $\alpha = .05$, $df_{bet} = 2$, and $df_W = 15$ is 3.68. Thus the calculated F is *not* greater than the critical F, and the null hypothesis cannot be rejected. The variability within the groups is too large compared to the differences among the group means to reach statistical significance. But as you may have guessed, this is not the end of the story.

Adding a Second Factor

What I neglected to mention in the initial description of the experiment is that within each group, half the patients are men and half are women. It will be easy to modify Table 14.1 to show this aspect of the experiment because the first three subjects in each group are women. Table 14.2 shows the depression ratings grouped by gender, along with the means and standard deviations for the new groups.

You may have already noticed the advantage of this new grouping: The variability within these new groups is generally smaller. In the one-way ANOVA, differences between men and women contributed to the variability within groups, but in Table 14.2, male-female differences now contribute to differences *between* groups. This fact can be illustrated dramatically by recalculating the one-way ANOVA with the six new groups. First, calculate

Table 14.2		Placebo	Moderate Dose	Large Dose
Women		38	33	23
		35	32	26
		33	26	21
Mean		35.33	30.33	23.33
SD		2.517	3.786	2.517
Men		33	34	34
		31	36	31
		28	34	32
Mean		30.67	34.67	32.33
SD		2.517	1.155	1.528

the variance of the six group means, and multiply it by 3, which is the number of subjects in each of these new groups. $MS_{bet} = 3 \times 18.66 = 56.0$. MS_W equals the average of the group variances $= 1/6\ (6.335 + 14.33 + 6.335 + $

6.335 + 1.334 + 2.335) = 1/6 (37.0) = 6.17. The new F ratio is 56.0/6.17 = 9.08, which is much higher than the F of 2.85 calculated without the gender separation and is also much higher than the critical F for that design. However, there is also a new critical F because of the gender division; df_{bet} is now 5 (one less than the number of groups), whereas $df_W = N_T - k = 18 - 6 = 12$, so the new critical F is 3.11 (a bit less than the old critical F). Because the new calculated F is larger than the new critical F, we can now reject the null hypothesis.

The new calculated F allows us to conclude that the six population means are not all the same, but how are we to interpret this information? The problem is that the six groups differ in *two ways*: on the basis of both drug treatment and gender. To sort out these differences we need to perform a *two-way* ANOVA.

Regrouping the Sums of Squares

Before proceeding, I want to point out something important about the SS components of the two one-way ANOVAs I just calculated. Multiplying each MS by its df will yield the corresponding SS. For the three-group ANOVA, $SS_{bet} = 2 \times 48.72 = 97.44$, and $SS_W = 15 \times 17.087 = 256.3$. Therefore, $SS_{total} = 97.44 + 256.3 = 353.74$. For the six-group ANOVA, $SS_{bet} = 5 \times 56.0 = 280.0$, and $SS_W = 12 \times 6.17 = 74.04$, so $SS_{total} = 280.0 + 74.04 = 354.04$. Except for rounding-off error, these two totals are the same (the drawback of using the mean/SD method of calculating ANOVA as compared to using group sums is that you must retain quite a few decimal places for your means and SDs, or the rounding off error can get unacceptably large). When you realize that SS_{total} depends on each score's deviation from the grand mean, and that the grand mean does not change just from regrouping the scores, it makes perfect sense that the total sum of squares would not change. What does change is how SS_{total} is divided into SS_{bet} and SS_W. We can see a drastic reduction in SS_W and a complementary increase in SS_{bet} because of the regrouping. The advantage of reducing SS_W should be obvious because it forms the basis for the denominator of the F ratio. The increase in SS_{bet} is a more complex matter; we will be dividing SS_{bet} into subcomponents, as you will soon see.

New Terminology

Before I describe the mechanics of the two-way ANOVA, I need to introduce some new terms. An ANOVA is referred to as a *two-way ANOVA* if the groups differ on *two* independent variables. In the preceding example, the two independent variables are drug treatment and gender. Of course, gender is not a truly independent variable—it is not a condition created by the experimenter—but as you saw for the t test and one-way ANOVA, grouping variables like gender can be treated as independent variables as long as you remember that you cannot conclude that the grouping variable is the *cause* of changes in the dependent variable.

An independent variable in an ANOVA is often called a *factor*. The simplest way to combine two or more factors in an experiment is to use what is called a *completely crossed factorial design*. In the case of two factors, a completely crossed design means that every level of one factor is combined with every level of the other factor such that every possible combination of levels corresponds to a group of subjects in the experiment. The preceding example is a completely crossed design because each combination of drug treatment and gender is represented by a group of subjects. If, for instance, there were no men in the placebo group, the design would not be completely crossed. The completely crossed factorial design is more commonly called a *factorial* design, for short. The two-way factorial design can always be represented as a matrix in which the rows represent the levels of one factor and the columns represent the levels of the other.

Our example could be represented by the matrix shown in Table 14.3. Each box in the matrix is called a *cell* and represents a different combina-

Table 14.3		**DRUG TREATMENT**	
	Placebo	Moderate Dose	Large Dose
Women			
Men			

tion of the levels of the two factors. Another way to define a completely crossed design is to say that none of the cells in the matrix is empty; there is at least one subject in each cell. If all of the cells have the same number of subjects, the design is said to be a *balanced design*. Factorial designs that are not balanced can be analyzed in more than one way; this complication will be discussed in Section C. In this section as well as in Section B, I will deal only with balanced factorial designs.

A common way of describing a factorial design is in terms of the number of levels of each factor. The design represented in Table 14.3 can be referred to as a 2 × 3 (pronounced "2 by 3") or a 3 × 2 ANOVA (the order is arbitrary). If the subjects in each cell are selected independently from the subjects in all other cells, the design is a two-way independent-groups ANOVA. This is the only kind of two-way design that will be considered in this chapter.

Calculating the Two-Way ANOVA

Now we are ready to calculate a two-way ANOVA for our drug treatment × gender design. A useful way to begin is to fill in the cells of Table 14.3 with the mean and standard deviation of each group, as shown in Table 14.4. The means within the cells are called, appropriately, *cell means*. The means below each column and to the right of each row are called the *marginal means*, and they are the means of each column and row, respectively. Because all the cells have the same n (for this example, $n = 3$), the mean of each row is just the ordinary average of the cell means in that row (similarly for the columns). The number in the lower right corner of Table 14.4 is the grand mean, and it can be found by averaging all the column means or all the row means (or all the cell means).

I will calculate the two-way ANOVA based on the cell means, cell variances, and cell sizes, rather than using sums of scores (and sums of squared scores). Although there are raw-score computational formulas for the two-

Table 14.4		Placebo	Moderate Dose	Large Dose	Row Means
	Women	35.33	30.33	23.33	29.67
		2.517	3.786	2.517	
	Men	30.67	34.67	32.33	32.56
		2.517	1.155	1.528	
	Column Means	33	32.5	27.83	31.11

way ANOVA, which are simple extensions of the raw-score formulas in Chapter 12, Section B, it is easy to lose sight of what you are measuring when you are using these formulas. I see no reason to use raw-score formulas, unless no computer is available and it is important to make accurate calculations (avoiding round-off error). In Section B, I will present a computational trick that makes it easy to calculate the two-way ANOVA without a computer if you have a handheld scientific (or statistical) calculator.

Calculating MS_W

For the two-way ANOVA, MS_W is based on the variances of each *cell*. In a balanced design all the cell sizes are equal, so MS_W is simply the ordinary average of all the cell variances. I calculated MS_W for the two-way ANOVA when I calculated MS_W for the six-group one-way ANOVA; each of the six groups corresponds to a cell of the two-way ANOVA. MS_W equals 6.17.

Calculating MS_{bet} for the Drug Treatment Factor

Next we need to calculate MS_{bet} for the drug treatment factor. This can be done by applying Formula 12.5A to the column means in Table 14.4 (each column has the same number of subjects). This is another quantity that I have already calculated; it is the same as MS_{bet} for the three-group ANOVA. (Notice that the column means are the same in Table 14.4 as in Table 14.1; there is no reason why grouping subjects as men and women would change these means.) However, the *F* ratio for testing the drug factor is different from what it was in the three-group ANOVA because MS_W has been reduced. The new *F* ratio is 48.72/6.17 = 7.90, which is much higher than the previous *F* of 2.85. The critical *F* we need to compare to is also higher than in the three-group ANOVA, but fortunately it is only slightly higher. The reason for the change in critical *F* is that whereas df_{bet} has not changed (it is 2 in both cases), df_W has been reduced. When MS_W was based on the variability within each of the three drug groups, the appropriate df_W was $18 - 3 = 15$. However, in the two-way ANOVA, MS_W is based on the variability within *cells*, and the appropriate df_W involves losing one df for each cell (N_T − number of cells), so $df_W = 18 - 6 = 12$. The critical *F* for $\alpha = .05$, $df_{bet} = 2$, and $df_W = 12$ is 3.89. The *F* ratio comparing the drug treatments (7.90) is now significant at the .05 level, and this time we can reject the null hypothesis that the population means corresponding to the three drug conditions are equal. The increase in the calculated *F* compared to the one-way ANOVA far outweighs the slight increase in the critical *F*. Separating the men from the women in this two-way ANOVA reduced the error term (i.e., MS_W) sufficiently to attain statistical significance.

Calculating MS_{bet} for the Gender Factor

The next step in the two-way ANOVA is to perform a one-way ANOVA on the other factor—in this case, gender. Because gender has only two levels, a *t* test could be used to test for statistical significance, but to be consistent, I will use Formula 12.5A again. The $\overline{X}_i$s are the means for the women and the men; *k* is the number of genders, which equals 2; and *n* is the number of subjects in each gender, which equals 9. Because there are only two means to deal with, I'll take this opportunity to demonstrate the calculation of Formula 12.5A the long way (i.e., without a statistical calculator to obtain the variance of the means automatically).

$$MS_{bet} = \frac{9\sum(\overline{X}_i - \overline{X}_G)^2}{2 - 1} = 9[(29.67 - 31.11)^2 + (32.56 - 31.11)^2]$$

$$= 9[(-1.44)^2 + 1.45^2] = 9(2.074 + 2.103) = 9(4.17) = 37.55$$

The F ratio for the gender difference uses the same error term (i.e., the average within-cell variance) as the F ratio for drug treatment; for gender, F is $37.55/6.17 = 6.09$. The critical F based on $\alpha = .05$, $df_{bet} = 1$, and $df_W = 12$ is 4.75, so this F ratio is also statistically significant. The women are, on the average, less depressed than the men in this experiment. However, if you look at the columns of Table 14.4, you'll see that women are actually *more* depressed than men in the placebo group. The row means are therefore somewhat misleading. This problem will be addressed shortly.

As you have seen, the two-way ANOVA begins with two one-way ANOVAs, and the numerator of each one-way ANOVA is found simply by ignoring the existence of the other factor. The denominator of both one-way ANOVAs is the same: the MS_W based on the variability within each cell. To the extent that the population variance within the cells is smaller than the variance within an entire row (or column) of the two-way table, these one-way ANOVAs are more powerful than ordinary one-way ANOVAs. But this reduction in MS_W does not apply to all two-way ANOVAs, nor is it the most interesting reason to perform a two-way ANOVA. It is the third MS that must be calculated for all two-way ANOVAs that is often the most interesting; it corresponds to the amount of *interaction* between the two independent variables. Though the amount of interaction is easy to quantify, it can be hard to understand using only mathematical formulas. Before I show you how to calculate the interaction part, I want you to be able to visualize the size of an interaction.

Graphing the Cell Means

You may have noticed that although the women are less depressed than the men for either drug dosage, the reverse is true for the placebo condition. It is these kinds of reversals that can lead to a significant amount of interaction. However, there are many ways that an interaction can combine with the row and column effects to produce a pattern of cell means; any particular pattern can be seen by drawing a graph of the cell means. A graph of the depression example will make it easier to explain the concept of interaction. The dependent variable (depression scores) is always placed on the Y axis. For the X axis we choose whichever of the two independent variables makes it easiest to interpret the graph. If one of the two factors has more levels, it is usually convenient to place that one on the X axis. For this example, I will place drug treatment on the X axis. Now, the cell means for each level of the second factor are plotted separately. Each level of the second factor becomes a different line on the graph when the cell means are connected; see Figure 14.1.

For the present example, the cell means for the male groups form one line and the cell means for the female groups form a second line. When the two lines are perfectly parallel, SS for interaction will always be zero, as will the F ratio for testing the interaction. When the lines diverge or converge strongly, or even cross each other, as in Figure 14.1, the F for interaction can be quite large, but whether it will be statistically significant depends on the relative size of MS_W.

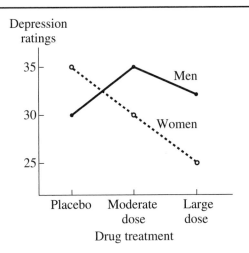

Depression
ratings

Figure 14.1

Graph of Cell Means for
Data in Table 14.4

The Case of Zero Interaction

To understand better what interactions involve, it will be helpful to start with the simplest case: parallel lines, zero interaction. Suppose the cell means for our example were as shown in Table 14.5. The graph of these cell means would consist of two parallel lines; see Figure 14.2. Notice that because the lines are parallel, the male-female difference is always the same (10 points), regardless of the drug condition. Similarly, the difference between any two drug conditions is the same for men as it is for women.

	Placebo	Moderate Dose	Large Dose	Row Means
Women	35	30	25	30
Men	25	20	15	20
Column Means	30	25	20	25

Table 14.5

The Additive Model of ANOVA

Another way to describe the lack of interaction in Table 14.5 is in terms of separate column and row effects relative to the grand mean. As you can see from the marginal means of Table 14.5, the effect of being a woman is to add 5 points to the depression score (as compared to the grand mean of 25), whereas the effect of being a man is to subtract 5 points. The effect of the placebo is to add 5 points, and the effect of the large dose is to subtract 5 points compared to the grand mean. (This doesn't mean that the placebo makes you more depressed, but rather that depression in the placebo group is higher than in the groups that get a dose of the drug.) The total lack of interaction between the factors assures that the mean for any cell will equal the grand mean plus the sum of the row and column effects for that cell. For example, to find the cell mean for women in the placebo group, we start with 25 (the grand mean) and add 5 points for being a woman and 5 more points for taking the placebo: $25 + 5 + 5 = 35$. We can write this as a general equation: Cell mean = Grand mean + Row effect + Column effect. This way of describing an ANOVA is sometimes referred to as the *additive model*, for the obvious reason that the components are added to yield the final result.

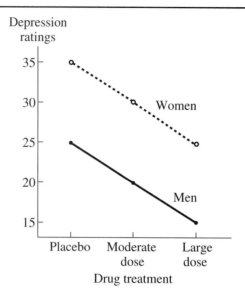

Figure 14.2

Graph of Cell Means
Depicting Zero
Interaction

General Linear Model

The preceding equation only applies when there is zero interaction, and it is one way of defining a total lack of interaction. The more general equation for a two-way ANOVA allows for interaction: Cell mean = Grand mean + Row effect + Column effect + Interaction effect. Any particular score can be expressed in terms of its cell mean plus an error term, so if we rearrange this equation to solve for a particular score, we have Score = Grand mean + Row effect + Column effect + Interaction effect + Error. (The difference between a subject's score and his or her cell mean is considered error because it is totally unpredictable—it is not related to any manipulation in the experiment.)

Mathematically, the preceding equation is linear (e.g., no terms are being squared) and shows that ANOVA can be viewed in terms of a system for understanding data called the *general linear model* (GLM). The equation for simple linear regression presented in Chapter 10 is probably the simplest case of the GLM. By expressing ANOVA in terms of the GLM, it can be seen that any ANOVA is just a special case of multiple regression (this will be made clear in Chapter 18). Because it is convenient to work in terms of deviations from the grand mean, I will subtract the grand mean from both sides of the previous equation to create the following equation: Score − Grand mean = Row effect + Column effect + Interaction effect + Error. If these quantities are squared and summed for all subjects, the result is a very useful way to analyze a two-way design:

$$SS_{total} = SS_{row} + SS_{column} + SS_{interaction} + SS_{error} \qquad \textbf{Formula 14.1}$$

You are already familiar with SS_{total} from the one-way ANOVA, and SS_{error} is just another name for SS_W. SS_{row} is the SS_{bet} obtained from the row means, and SS_{column} is the SS_{bet} obtained from the column means. Together the SS_{bet} for the two factors (i.e., SS_{row} and SS_{column}) and $SS_{interaction}$ will add up to the SS_{bet} you would get by treating each cell as a separate group in a one-way ANOVA (like our six-group analysis). I will use the abbreviations SS_R, SS_C, and SS_{inter} to refer to SS_{row}, SS_{column}, and $SS_{interaction}$, respectively.

Some text authors prefer to use SS_A and SS_B to refer to the two factors because this notation easily generalizes to any number of factors (which can be referred to as C, D, etc.). For my present purposes, I prefer the concreteness of the row and column notation.

Calculating the Variability Due to Interaction

Now it is time to calculate the variability due to interaction in the drug-gender example. To do this I need to work with the SS components; the MS components do not add up in any convenient way. Because I calculated the MSs directly, I now have to multiply each MS by its df to obtain the corresponding SS. For instance, SS_R, which in the above example is the SS due to the difference in gender means, equals $1 \times (MS_{gender}) = 1 \times 37.55 = 37.55$. (There are only two levels of this factor and therefore only one degree of freedom.) To find SS_C for the same example, we need to multiply MS_C (the MS_{bet} for the drug factor, which is 48.72) by the appropriate df (which is 2 because the drug factor has three levels). So $SS_C = 2 \times 48.72 = 97.44$. The key to finding SS_{inter} rests on the fact that the SS_{bet} that is calculated as though all the cells were separate groups in a one-way ANOVA (i.e., the SS_{bet} from the six-group analysis) contains variability due to the differences in row means (SS_R), the differences in column means (SS_C), and the degree of interaction (SS_{inter}). This relation can be summarized as $SS_{bet} = SS_R + SS_C + SS_{inter}$. Therefore,

$$SS_{inter} = SS_{bet} - SS_R - SS_C \hspace{3cm} \textbf{Formula 14.2}$$

For the present example, SS_{inter} equals $280 - 37.55 - 97.44 = 145.0$.

Next we need to divide SS_{inter} by df_{inter}. The df for the interaction is always equal to the df_{bet} for one factor multiplied by the df_{bet} for the other factor. (Symbolically, $df_{inter} = df_{R \times C} = df_R \times df_C$.) In this case, $df_{inter} = 2 \times 1 = 2$. Therefore, $MS_{inter} = 145.0/2 = 72.50$. Finally, to find the F ratio for the interaction, we use MS_W once again as the denominator (i.e., error term), so $F = 72.50/6.17 = 11.75$. Because the df are the same, the critical F for the interaction is the same as the critical F for the drug treatment factor (3.89), so we can conclude that the interaction is statistically significant. However, that conclusion does not tell us anything about the pattern of the cell means. We would need to refer to a graph of the cell means to see how all of the effects in the experiment look when they are combined.

Types of Interactions

The row and column effects of the two-way ANOVA are called the *main effects*. When there is no interaction, the two-way ANOVA breaks into two one-way ANOVAs with nothing left over. All you have are the two main effects. However, in addition to the two main effects, normally you have some degree of interaction. To describe the different ways that the main effects can combine with the interaction to form a pattern of cell means, I will concentrate on the simplest type of two-way ANOVA, the 2×2 design. I'll begin with an example in which the interaction is more important than either of the main effects.

Suppose a psychiatrist has noticed that a certain antidepressant drug works well for some patients but hardly at all for others. She suspects that the patients for whom the drug is not working are bipolar depressives (i.e., manic-depressives) who are presently in the depressed part of their cycle. She decides to form two equal groups of patients: unipolar depressives and

Figure 14.3

Ordinal Interaction

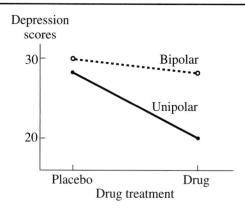

bipolar depressives. Half of the subjects in each group get a placebo, and the other half get the drug in question. If the psychiatrist is right, a graph of the results (in terms of depression scores) might look like the one in Figure 14.3. The divergence of the lines suggests the presence of an interaction. The main effects are usually easier to grasp in a table that includes the marginal means. Table 14.6 corresponds to Figure 14.3. As you can see from

Table 14.6

	Placebo	Drug	Row Means
Bipolar	30	28	29
Unipolar	28	20	24
Column Means	29	24	26.5

the marginal means, the two main effects are equally large and might be statistically significant. However, the psychiatrist would not be concerned with the significance of either main effect in this example. If the drug had been proven effective many times before, the main effect of drug versus placebo would not be new information. The main effect of bipolar depression versus unipolar depression would be meaningless if the psychiatrist deliberately chose subjects to make the groups comparable in initial depression scores. On the other hand, a significant interaction can imply that the psychiatrist found what she expected: The drug works for unipolar depressives but not bipolar depressives.

The practical application of this result should be obvious. Furthermore, the large amount of interaction tells us to use some caution in interpreting the main effects. In this example, the main effect of drug versus placebo is somewhat misleading because it is an average of a strong drug effect for unipolar and a weak effect for bipolar depressives. The main effect of unipolar depression versus bipolar depression is similarly misleading because the difference is large only under the drug and not the placebo condition. On the other hand, the main effects are not *entirely* misleading because the *direction* of the drug effect is the same for both depressive groups; it is the amount of effect that differs considerably. (Similarly, the direction of the bipolar–unipolar difference is the same for both drug conditions.) This type of interaction, in which the direction (i.e., the order) of the effects is consistent, is called an *ordinal interaction*.

As an illustration of an interaction that strongly obscures the main effects, consider the following hypothetical experiment. Researchers have

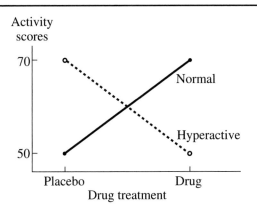

Figure 14.4

Disordinal Interaction

found that amphetamine drugs, which act as strong stimulants to most people, can have a paradoxically calming effect on hyperactive children. Imagine that equal-sized groups of hyperactive and normal children are each divided in half, and the subjects are given placebos or amphetamines. A graph of the results (in terms of activity scores) could look like Figure 14.4. The nonparallel lines again indicate the presence of an interaction. The *crossing* of the lines indicates a *disordinal interaction*, that is, an interaction in which the direction of the effects reverses for different subgroups. Note, however, that the lines do not have to form a perfect X and completely obliterate the main effects to reveal a disordinal interaction; it is sufficient if the lines cross at all in the graph.

Figure 14.4 represents the most extreme case of a disordinal interaction. To see how such an interaction can completely obliterate the main effects, look at Table 14.7, which corresponds to Figure 14.4, and pay particular attention to the row and column (i.e., marginal) means. Because the marginal means are the same for both the columns and rows, the *SS* components for the numerators of the two main effects will be zero, as will the corresponding *F* ratios—so of course, neither of the main effects can be significant. This lack of significance can be quite misleading if it is taken to imply either that the drug is simply ineffective or that hyperactive children do not differ from normal children on a measure of activity. A significant interaction tells you not to take the main effects at face value. An interaction anywhere near statistical significance tells you that it is a good idea to graph the cell means and look at the nature of the interaction (but see Separating Interactions from Cell Means, which follows).

To complicate matters a bit more, there are always two ways to graph the cell means of a two-way ANOVA, depending on which of the two factors is placed on the *X* axis. It is possible for the interaction to be disordinal in one graph and ordinal in the other. (If the lines cross, but slant in the same direction, the other graph will have an ordinal interaction; if the lines don't cross, but slant in different directions—that is, one slants up, while the other slants down—the other graph will exhibit a disordinal interaction.) If a significant interaction is "doubly" ordinal (lines slant in the same direc-

	Placebo	Drug	Row Means	**Table 14.7**
Hyperactive	70	50	60	
Normal	50	70	60	
Column Means	60	60	60	

tion and do *not* cross in either graph as in Figure 14.3), both main effects should still be interpretable; if a significant interaction is doubly disordinal (lines slant in opposite directions, and the lines *do* cross in either graph, as in Figure 14.4), both main effects are likely to be misleading. If a significant interaction is mixed (lines cross in one graph but not in the other), look at the graph in which the lines cross; the main effect whose levels are laid out along the horizontal axis of the graph should be interpretable, but the main effect whose levels are represented by the crossing lines should be interpreted with caution, if at all. In fact, even when an interaction falls short of significance, the presence of a disordinal interaction suggests that you should use some caution in interpreting the results of the main effects involved. Remember, however, that no matter how extreme an interaction looks, you can't be sure it will be statistically significant until you test it (it is always possible that MS_W will be relatively large). A major advantage of the two-way ANOVA is the opportunity to test the size of an interaction for statistical significance.

The interaction of two independent variables is called a *two-way interaction*. To test whether three variables mutually interact, a three-way ANOVA would be required as described in Section C. Similarly, a four-way ANOVA is required to test a four-way interaction, and so on. ANOVA designs involving more than two factors are called *higher-order designs*; interactions involving more than two factors are called *higher-order interactions*.

Another Definition of Interaction

A useful way to define an interaction in a two-way ANOVA is to say that an interaction is present when the effects of one of the independent variables change with different levels of the other independent variable. For instance, in one of the preceding examples, the effects of an antidepressant drug changed depending on whether the patients were unipolar or bipolar depressives. For the unipolar level of the grouping variable, the drug treatment had a dramatic effect, but for the bipolar level the drug effect was very small. Of course, we can always look at a two-way interaction from the reverse perspective. We can say that the unipolar-bipolar effect (i.e., difference) depends on the level of the drug treatment. With the placebo, there is only a very small unipolar-bipolar difference, but with the drug this difference is large. The preferred way to view an interaction depends on which perspective makes the most sense for that experiment.

Separating Interactions from Cell Means

The subsection labeled "Types of Interactions" should probably have been labeled "Patterns of Cell Means" instead. As Rosnow and Rosenthal (1989, 1991) point out, when you are looking at a graph of the cell means, you are not looking just at the interaction, but at a combination of both main effects *with* the interaction. If you want to look at just the interaction by itself, you would need to subtract the appropriate row and column effects from each cell mean, and graph the remainders (i.e., the *residuals*). A graph of a 2 × 2 interaction *with main effects removed* will always look like Figure 14.4. However, as Rosnow and Rosenthal (1991) concede: "We rarely advise plotting interactions because they are seldom of theoretical interest to researchers, but we almost always advise plotting cell means because they are frequently of great interest" (p. 575). Rosnow and Rosenthal are arguing against the strong habit of psychological researchers to say they are interpreting "the interaction" when they are really looking at a graph of cell means, which therefore includes the main effects. It is not clear, however, how this confusion in terminology translates to any confusion on a practical level concerning the findings and applications of a study.

In my opinion, a more important criticism of the way psychologists typically analyze and interpret the data from a 2×2 factorial design is that the breakdown of the data into two main effects and an interaction is not always the most appropriate and powerful way to analyze such data (this criticism has also been made by Rosnow and Rosenthal, 1996). Consider again the bipolar/unipolar example. If the two groups had been well-matched for depression initially, and the drug had virtually no effect on bipolars, three of the cell means would have been about the same, with only the drug/unipolar cell being considerably different. If that were the pattern expected, it would have been quite reasonable to dispense with the two-way ANOVA approach entirely and test the cell means with a planned contrast that compares the drug/unipolar cell mean to the average of the other three. I suspect that this approach is not used as often as it could be because of all the variability and paradoxical results that frequently arise in many areas of psychological research; researchers may sometimes lack the confidence to plan specific contrasts even when there is a good theoretical basis for them. I will have more to say about planning comparisons for a two-way ANOVA in Section C.

The *F* Ratio in a Two-Way ANOVA

The structure of each of the three *F* ratios in a two-way ANOVA is the same as that of the *F* ratio in a one-way ANOVA. For each *F* ratio in a two-way ANOVA, the denominator is an estimate of the population variance based on the sample variances (all the populations are assumed to have the same variance). Each numerator is also an estimate of the population variance when the appropriate null hypothesis is true, but these estimates are based on differences among row means, column means, or the cell means (after row and column mean differences have been subtracted out). These three numerators are statistically independent; that is, the size of any one numerator is totally unrelated to the size of the other two numerators. Moreover, the significance of the interaction in a two-way ANOVA implies nothing about whether either of the main effects is significant, and vice versa. The significance of any one of the three *F* ratios that are tested in a two-way ANOVA does not depend on the significance of any other *F* ratio. For instance, the interaction can be significant when both main effects are not, and both main effects can be significant when the interaction is not. There are eight possible combinations of the three *F* ratios with respect to which are significant ($2 \times 2 \times 2 = 8$); these are illustrated in Figure 14.5. (In Figure 14.5, assume that any effect that looks significant *is* significant, and any effect that does not look significant, is not.)

The significance of the interaction *does* have implications for interpreting the main effects, which is why it makes sense to test the *F* for interaction first. If the interaction is *not* significant, you can test the main effects and interpret them as two separate one-way ANOVAs. But if the interaction *is* significant, you must be cautious in interpreting the significance of the main effects. It is quite possible that the main effects are misleading. The way you follow up a two-way ANOVA with more specific (e.g., pairwise) comparisons differs depending on the significance of the interaction, as you will see at the end of Section B.

Figure 14.5

All Possible
Combinations of
Significant and
Nonsignificant Effects in
a Two-Way ANOVA

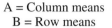

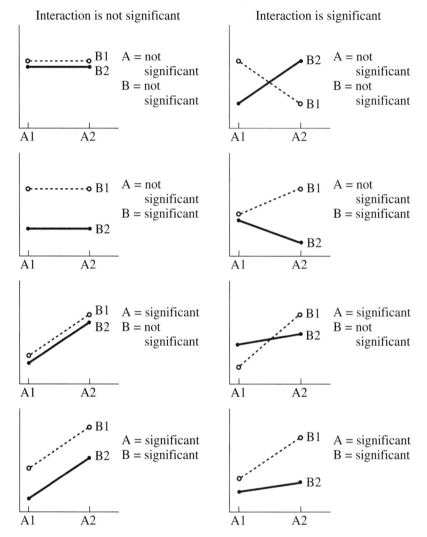

Advantages of the Two-Way Design

If your study consists of an independent variable that involves an experimental manipulation (e.g., type of drug given), and you add a grouping
variable for the chief purpose of reducing MS_W, the two-way design that
results is often called a *treatment X blocks design*. The subjects at each level
of the grouping variable—which can be a categorical variable (e.g., marital
status) or a continuous one that has been divided into ranges (e.g., age,
classified as twenties, thirties, etc.)—are considered to comprise a separate
"block." The main advantage of this design is lost, however, if the grouping
variable is not relevant to the dependent variable being measured. If, in our
drug-gender example, there were virtually no male-female differences in
depression under any conditions, there would be very little reduction in
MS_W. In such a case, the two-way ANOVA can actually produce a disadvan-

tage because df_W is reduced (the more cells you create, the more df you lose), which causes an increase in the critical *F*. If your grouping variable is continuous, and has a linear relationship with your dependent variable, you should probably not break it into distinct categories, but rather perform an analysis of covariance instead (see Chapter 18).

Another important reason to add a grouping variable to create a two-way ANOVA is the opportunity to observe the interaction of the grouping factor with the experimental factor. A significant interaction serves to increase our understanding of how the experimental manipulation works with different groups of people; an almost complete lack of an interaction increases the generality of the experimental effect. When both factors involve experimental manipulations, the lack of an interaction implies that we have conducted two one-way ANOVAs (i.e., the two main effects) in an efficient and economical way. On the other hand, many two-way ANOVAs are performed with the primary goal of discovering whether there is any interaction between two experimental factors. An example of such an experiment is described in Section B.

SUMMARY

1. The groups in a *two-way* ANOVA design vary according to two independent variables (often referred to as *factors*), each having at least two *levels*. In a *completely crossed factorial* ANOVA (usually referred to as just a *factorial* ANOVA), each level of one factor is combined with each level of the other factor to form a *cell*, and none of the cells is empty of subjects. The two-way ANOVA is often described by the number of levels of each factor (e.g., a 3×4 design has two factors, one with three levels and the other with four).

2. In a *balanced* design all of the cells contain the same number of subjects (observations, measurements, scores, etc.). If the design is not balanced, there is more than one legitimate way to analyze it (see Section C).

3. The cells can be arranged in a matrix so that the rows are the levels of one factor and the columns are the levels of the other factor. The means of the rows and columns are called *marginal means*, and they show the effect of each factor, averaging across the other.

4. The two-way ANOVA yields three *F* ratios, each of which is compared to a critical *F* to test for statistical significance. The denominator of each ratio (called the error term) is MS_W, which is the simple average of all the cell variances in a balanced design.

5. As in the one-way ANOVA, SS_{tot} can be divided into SS_W and SS_{bet}, with the latter referring to between-cell variability in the two-way ANOVA. The SS_{bet} can be further subdivided into three components: SS_R (based on the variability of row means), SS_C (based on the variability of column means), and SS_{inter} (the variability left over after SS_R and SS_C are subtracted from SS_{bet}). Dividing by the corresponding df yields MS_R, MS_C, and MS_{inter}.

6. The *F* ratios based on MS_R and MS_C are like the *F*s from two one-way ANOVAs, each ignoring (averaging over) the other factor, except that MS_W for the two-way ANOVA may be smaller than in either one-way ANOVA because it is based on variability within cells rather than entire rows or columns. These two embedded one-way ANOVAs are called the *main effects* of the two-way ANOVA.

7. The *F* ratio based on MS_{inter} is used to test for the significance of the *interaction* between the two factors. If the amount of interaction is zero, a graph of the cell means will produce parallel lines. If the

amount of interaction is relatively small and not close to statistical significance, the two-way ANOVA can be viewed as an economical way to perform two one-way ANOVAs (i.e., the main effects can be taken at face value).

8. If the amount of interaction is relatively large and statistically significant, the main effects may be misleading. If the lines on a graph of the cell means cross or slant in opposite directions, the interaction is at least partially *disordinal*, and the results of the main effects should not be taken at face value. Even if the interaction is doubly *ordinal*, a large amount of interaction means that the main effects should be interpreted cautiously.

EXERCISES

*1. How many cells are there and what is the value of df_W in the following ANOVA designs? (Assume $n = 6$ subjects per cell.)
   a. 2 × 5
   b. 4 × 4
   c. 6 × 3

2. a. Graph the cell means in the following table, and find the marginal means.
   b. Which effects (i.e., *F* ratios) might be significant, and which cannot be significant?

	Factor A, level 1	Factor A, level 2
Factor B, level 1	75	70
Factor B, level 2	60	65

*3. a. Graph the cell means in the following table, and find the marginal means.
   b. Which effects might be significant, and which cannot be significant?

	Factor A, level 1	Factor A, level 2
Factor B, level 1	70	70
Factor B, level 2	60	65

4. a. Graph the cell means in the following table, and find the marginal means.
   b. Which effects might be significant, and which cannot be significant?

	Factor A, level 1	Factor A, level 2
Factor B, level 1	75	70
Factor B, level 2	75	70

*5. a. Graph the cell means in the following table, and find the marginal means.

b. Which effects might be significant, and which cannot be significant?

	Factor A, level 1	Factor A, level 2
Factor B, level 1	75	70
Factor B, level 2	70	75

6. A researcher is studying the effects of both regular exercise and a vegetarian diet on resting heart rate. A 2 × 2 matrix was created to cross these two factors (exercisers versus nonexercisers, and vegetarians versus nonvegetarians), and 10 subjects were found for each cell. The mean heart rates and standard deviations for each cell are as follows:

	Exercisers	Nonexercisers
Vegetarians	$\bar{X} = 60$	$\bar{X} = 70$
	$s = 15$	$s = 18$
Nonvegetarians	$\bar{X} = 65$	$\bar{X} = 75$
	$s = 16$	$s = 19$

a. What is the value of MS_W?
b. Calculate the three *F* ratios. (*Hint*: Check to see if there is an interaction. If there is none, the calculation is simplified.) State your conclusions.
c. How large would these *F* ratios be if there were 40 subjects per cell? Compare these values to the ones you calculated for part b. What can you say about the effect on the *F* ratio of increasing the sample size? (What is the exact mathematical relationship?)

d. What conclusions can you draw based on the *F* ratios found in part c? What are the limitations on these conclusions (in terms of causation)?

*7. The interaction for a particular two-way ANOVA is statistically significant. This implies that
a. Neither of the main effects is significant.
b. Both of the main effects are significant.
c. The lines on a graph of cell means will intersect.
d. The lines on a graph of cell means will not be parallel.

8. When there is no interaction among the population means in a two-way design, the numerator of the *F* ratio for interaction in the data from an experiment is expected to be
a. About the same as MS_W
b. Zero
c. About the same as the sum of the numerators for the two main effects
d. Dependent on the number of cells in the design

The drug treatment–gender example in Section A represents a common situation for which the two-way ANOVA is appropriate: adding a grouping variable to an experimental factor. As I mentioned, an important advantage of this design is that the grouping variable can reduce the within-group variability, as represented by MS_W. However, when both factors involve actual experimental manipulations, adding one factor to the other is not expected to reduce the error term. In this case, you are either exploring two independent variables in an efficient manner or looking for an interaction between them. The next example falls under the latter category.

Dr. Sue Pine, a cognitive psychologist, is studying the effects of sleep deprivation and compensating stimulation on the performance of complex motor tasks. The dependent variable is the subject's score on a video game that simulates driving a large truck at night. There are four different levels of sleep deprivation. Each subject spends 4 days in the sleep lab, but subjects are either: (1) allowed to sleep on their own schedule (control group–no sleep deprivation), (2) allowed to sleep their usual *amount* but not allowed to sleep at all during the time of day (e.g., 11 P.M. to 7 A.M.) they usually sleep (jet lag), (3) allowed to sleep their usual amount per day but not more than 2 hours at one time (kept awake for at least 1 hour after any stretch of sleep that lasts 2 hours)—I'll call this the "interrupted" condition, or (4) deprived of sleep entirely (total deprivation). Subjects in each sleep group are randomly assigned to one or another of three stimulation conditions: (1) They are given a pill they are told is caffeine but is really a *placebo*, (2) they are given a *caffeine* pill and told what it is, (3) they are given mild electric shocks for mistakes during the game and monetarily rewarded for good performance (reward condition). The resulting design is a 4×3 factorial, with a total of 12 cells. Its analysis can be described according to the usual six-step procedure.

B

BASIC STATISTICAL PROCEDURES

Step 1. State the Null Hypothesis

In a two-way ANOVA there are three null hypotheses, one corresponding to each main effect and one for the interaction. The null hypothesis for each main effect is the same as it would be for the corresponding one-way ANOVA. For sleep level, H_0: $\mu_{control} = \mu_{jetlag} = \mu_{interrupt} = \mu_{total}$; for stimulation, H_0: $\mu_{placebo} = \mu_{caffeine} = \mu_{reward}$. As for the null hypothesis concerning the interaction, there is no simple way to state it symbolically. We can say that this H_0 implies that the effects of the two factors will be additive or that the effects of one factor do not depend on the levels of the other factor.

When dealing with ANOVA, it is easiest to state the alternative hypothesis simply as the null hypothesis not being true. However, as ANOVA designs become increasingly complex, it is common to state specific research hypotheses, which correspond to just a subset of the possible alternative hypotheses. These will be discussed further in Section C, as planned comparisons.

Step 2. Select the Statistical Test and the Significance Level

We are comparing population means along two dimensions, or factors, so the two-way ANOVA is appropriate (assuming that the dependent variable is measured on an interval/ratio scale). The same alpha level is used for testing all three F ratios, usually .05. We will use .05 for the present example.

Step 3. Select the Samples and Collect the Data

As is virtually always the case when a design has two experimental factors, we will plan to have an equal number of subjects in each cell. If we consider five subjects per cell a good balance between power and economy, we will need a total of sixty subjects. Ideally, we would select 12 independent random samples from the population, each containing five subjects. In reality, we would find a convenient sample of 60 subjects and assign subjects as randomly as possible to the 12 cells, making sure the design is balanced. Table 14.8 displays the video game scores for each subject, along with the means and standard deviations for each cell, and the marginal means.

Step 4. Find the Regions of Rejection

Because there are three null hypotheses, we need to find three critical values. The appropriate distribution in each case is the F distribution, so we need to know the appropriate df for the numerator and denominator corresponding to each F ratio tested. Our task is simplified by the fact that the df for the denominator is df_W for all three tests. To find df_W we can use the fact that the total number of subjects (N_T) is equal to the number of cells (12) times the size of each cell (5), so $N_T = 12 \times 5 = 60$. Then df_W is $N_T -$ the number of cells $= 60 - 12 = 48$.

The df for the numerator for the F test of the sleep factor is 1 less than the number of its levels (just as in a one-way ANOVA, where $df_{bet} = k - 1$), so $df_R = 4 - 1 = 3$. The df for the stimulation factor (df_C) is 2 because there are three stimulation levels. Finally, these two df are multiplied together to give the df for the numerator for the F ratio that tests the interaction ($df_{inter} = df_R \times df_C$), so $df_{inter} = 3 \times 2 = 6$.

Now we can find the critical Fs to test each of the three null hypotheses. For the sleep main effect, we need to look up $F_{.05}$ (3, 48). However, because the value of 48 does not appear in the table under df for the denominator, I will be slightly conservative and use 40 as an approximation in each case. The critical value for the sleep effect is approximated by 2.84. For the stimulation effect, we need $F_{.05}$ (2, 48), which is approximately 3.23. For the interaction we need $F_{.05}$ (6, 48), which is approximately 2.34.

Step 5. Calculate the Test Statistics

For a two-way ANOVA there are three different F ratios to calculate. Because they all have the same denominator (i.e., MS_W), I will calculate

	Placebo	Caffeine	Reward	Row Means	Table 14.8
Control	24	26	28		
	20	22	23		
	29	20	24		
	20	30	30		
	28	27	33		
Cell Means	24.2	25.0	27.6		
(*SD*)	(4.266)	(4.0)	(4.159)	25.6	
Jet Lag	22	25	26		
	18	31	20		
	16	24	32		
	25	27	23		
	27	21	30		
Cell Means	21.6	25.6	26.2		
(*SD*)	(4.615)	(3.715)	(4.919)	24.47	
Interrupt	16	23	16		
	20	28	13		
	11	26	12		
	19	17	18		
	14	19	19		
Cell Means	16.0	22.6	15.6		
(*SD*)	(3.674)	(4.615)	(3.05)	18.07	
Total	14	23	15		
	17	16	11		
	12	26	19		
	18	18	11		
	10	24	17		
Cell Means	14.2	21.4	14.6		
(*SD*)	(3.347)	(4.219)	(3.578)	16.73	
Column Means	19.0	23.65	21.0	21.217	

that term first. Because we are dealing with a balanced design, I recommend that you calculate MS_W by taking the ordinary average of the 12 cell variances. The traditional way to calculate MS_W, however, is to first calculate SS (the numerator of the variance) for each cell (usually with a raw-score formula), add the SSs to get SS_W, and then divide by df_W. My recommendation is based on the fact that, for descriptive purposes, you will want to find the standard deviation for each cell anyway. These standard deviations are almost always included when publishing a table of cell means (and they are easy to obtain with handheld calculators). Once you have the (unbiased) cell standard deviations, you need only square each one to produce the cell variances, which are then averaged to obtain MS_W. Then if you need SS_W you can always multiply MS_W by df_W. Another advantage of dealing with cell standard deviations is that if you make a gross mistake with a standard deviation, you are likely to notice that it is an inappropriate value for your data (by now I hope you can "eyeball" a standard deviation, at least very roughly), whereas it is not easy to look at an SS and know if it is reasonable.

Squaring and averaging the cell standard deviations from Table 14.8 we get: MS_W = 1/12 (18.2 + 16 + 17.3 + 21.3 + 13.82 + 24.24 + 13.5 + 21.3 + 9.3 + 11.2 + 17.8 + 12.8) = 1/12(196.7) = 16.39.

As you saw in Section A, because of the way the SS components add up, it makes sense to find all of the components of $SS_{between}$ and then divide each by its df to get the corresponding *MS*. We begin by calculating $SS_{between}$, which, in the context of a two-way ANOVA, is often called $SS_{between-cells}$ to avoid confusion; as the name (ungrammatically) implies, this component gives us the total amount of variation among the cell means. This amount includes the *SS* due to each main effect, as well as the interaction, so after finding $SS_{between-cells}$ we will have to subtract from it both SS_R (variation among the row means) and SS_C (variation among the column means) to find $SS_{interaction}$. The traditional computational formula for finding $SS_{between-cells}$ in a two-way ANOVA is similar to Formula 12.9, but as I have mentioned before, the raw-score computational formulas look confusing and do not take advantage of the ease with which standard deviations and variances can be found with inexpensive calculators. Alternatively, we can use Formula 12.5A as we did in Section A, but then we still have the annoyance of having to convert an *MS* to an *SS*. Fortunately, I discovered a computational trick for finding the desired SSs directly that is even easier than using Formula 12.5A, as described next.

The shortcut I discovered for obtaining $SS_{between-cells}$ works by first finding the *biased* variance of the cell means and then multiplying by the total *N*. This computational procedure is convenient only if you are using a statistical calculator that has the *biased* standard deviation as a built-in function, so you can square the result to obtain the biased variance of the cell means. To indicate that I am going to calculate the biased variance of a set of numbers, I will use the following notation: $\sigma^2(X_1, X_2, X_3 \ldots)$. For instance, to indicate the biased variance of the cell means in Table 14.8, I would write: $\sigma^2(24.2, 25, 27.6, 21.6, 25.6, 26.2, 16, 22.6, 15.6, 14.2, 21.4, 14.6) = 21.82$. Therefore, $SS_{between-cells} = 21.82 \times N_T = 21.82 \times 60 = 1,309.2$. This computational shortcut is designed for the case when all of the cells are the same size, but you will find out how to modify this method for (2 × 2) unbalanced designs in Section C.

What makes the above calculation method so surprisingly convenient is that exactly the same formula can be used to find SS_R and SS_C, the *SS* components associated with the two main effects. To find SS_R (SS_{sleep}, for this example) calculate the biased variance of the row means and multiply by N_T: SS_R ($=SS_{sleep}$) $= \sigma^2(25.6, 24.47, 18.07, 16.73) \times 60 = 14.96 \times 60 = 897.6$. Similarly, SS_C ($=SS_{stim}$) is based on the column means, so $SS_C = \sigma^2(19, 23.65, 21) \times 60 = 3.627 \times 60 = 217.6$. Finally, SS_{inter} is found by subtracting SS_R and SS_C from $SS_{between-cells}$. $SS_{inter} = 1309.2 - 897.6 - 217.6 = 194$. The general formula for $SS_{between-cells}$, SS_R, or SS_C is

$$SS_{bet} = N_T \sigma^2 \text{ (means)}$$

Formula 14.3

where the means in parentheses can be group means (this formula also works for a one-way ANOVA with equal *n*s), cell means, row means, or column means (or even individual scores, as you will soon see).

To check your work, you can make use of the fact that the *SS* components must add up to the total *SS*, as expressed in Formula 14.1. For this example, $SS_{total} = 897.6 + 217.6 + 194 + 786.7 = 2,095.9$. This sum can then be compared to the value you get by calculating SS_{total} directly from the data, using the same method I used to get all of the *SS* components (except for SS_W). Just find the biased variance of all 60 scores (the individual scores function as the "means" in Formula 14.3), and multiply by 60 (if the method seemed strange for the other *SS* components, it should certainly make sense in this instance). You will find that, within error due to rounding off, the value you get for SS_{total} directly from the data agrees with the result from Formula 14.1 (if it doesn't, you know to look for a computational error). On the other hand, you can use the calculation of SS_{total}, which is very straightforward, to reduce rather than check your work. By subtracting $SS_{between}$ from SS_{total}, you can get SS_W without having to compute the standard deviation for each cell (although, as I mentioned previously, you will probably need to for descriptive purposes anyway). Another way to check your work as you are calculating the SS components is to check the mean whenever you calculate the biased variance for row, column, or cell means (or the individual scores). The mean of the means will be the grand mean in each case and so should be the same each time.

Now that you have calculated all of the SS components, you need to find their corresponding degrees of freedom (actually I already calculated the df components to find the critical Fs, but here I'll present the formulas in a more formal way). Using r to represent the number of rows, c for the number of columns, and n for the number of subjects in each cell, the formulas for the df components are as follows:

a. $df_{row} = r - 1$ **Formula 14.4**

b. $df_{col} = c - 1$

c. $df_{inter} = (r - 1)(c - 1)$

d. $df_W = rcn - rc$

e. $df_{total} = rcn - 1$

For this example:

$df_{sleep} = 4 - 1 = 3$

$df_{stim} = 3 - 1 = 2$

$df_{inter} = 3 \cdot 2 = 6$

$df_W = 60 - 12 = 48$

$df_{total} = 60 - 1 = 59$

Note that the four df components add up to df_{total}. This will always be the case if you don't make an error. One way to show how the total degrees of freedom are partitioned into components (corresponding to the sums of squares) is by drawing a "degrees of freedom tree." Figure 14.6 displays the df tree for both the one-way and two-way ANOVAs. Now we are ready to divide each SS by its df to obtain all of the MSs for the two-way ANOVA. The general formulas are as follows:

a. $MS_R = \dfrac{SS_R}{df_R}$ **Formula 14.5**

b. $MS_C = \dfrac{SS_c}{df_c}$

c. $MS_{inter} = \dfrac{SS_{inter}}{df_{inter}}$

d. $MS_W = \dfrac{SS_W}{df_W}$

Figure 14.6

The df Trees for the One-
and Two-Way ANOVA

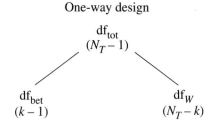

One-way design

$$df_{tot}$$
$$(N_T - 1)$$

$$df_{bet}$$
$$(k - 1)$$

$$df_W$$
$$(N_T - k)$$

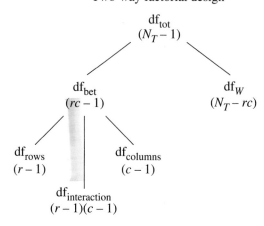

Two-way factorial design

$$df_{tot}$$
$$(N_T - 1)$$

$$df_{bet}$$
$$(rc - 1)$$

$$df_W$$
$$(N_T - rc)$$

$$df_{rows}$$
$$(r - 1)$$

$$df_{columns}$$
$$(c - 1)$$

$$df_{interaction}$$
$$(r - 1)(c - 1)$$

There is no point to calculating MS_{total}, because the other MS components will not add up to it. For this example,

$$MS_{sleep} = \frac{897.6}{3} = 299.2$$

$$MS_{stim} = \frac{217.6}{2} = 108.8$$

$$MS_{inter} = \frac{194}{6} = 32.33$$

$$MS_W = \frac{786.7}{48} = 16.39$$

Finally, the three F ratios are calculated using MS_W as the denominator (i.e., error term) for each:

a. $F_R = \dfrac{MS_R}{MS_W}$ 　　　　　　　　　　　　　　　　**Formula 14.6**

b. $F_C = \dfrac{MS_C}{MS_W}$

c. $F_{inter} = \dfrac{MS_{inter}}{MS_W}$

For this example,

$$F_{sleep} = \frac{MS_{sleep}}{MS_W} = \frac{299.2}{16.39} = 18.3$$

$$F_{stim} = \frac{MS_{stim}}{MS_W} = \frac{108.8}{16.39} = 6.64$$

$$F_{inter} = \frac{MS_{inter}}{MS_W} = \frac{32.33}{16.39} = 1.97$$

Step 6. Make the Statistical Decisions

There are three independent decisions to be made—one for each F ratio. The F ratio for the main effect of sleep deprivation is 18.3, which is greater than the corresponding critical F, 2.84, so this null hypothesis can be rejected. The main effect of stimulation is also significant because its calculated F, which is 6.64, is greater than the corresponding critical F, 3.23. Finally, the F for interaction, 1.97, is *not* significant because it is less than the appropriate critical F, 2.34.

The Summary Table for a Two-Way ANOVA

The summary table for a two-way ANOVA has the same column headings as the table for a one-way ANOVA (see Chapter 12, Section B), but it is more complicated because there are more sources of variation. Each SS component of the two-way ANOVA corresponds to a different source of variation in Table 14.9. Notice that the summary table shows the total SS

Source	SS	df	MS	F	p	
Between-Cells	1,309.2	11				**Table 14.9**
Sleep Deprivation	897.6	3	299.2	18.3	<.001	
Stimulation	217.6	2	108.8	6.64	<.01	
Interaction	194	6	32.33	1.97	>.05	
Within-Cells	786.7	48	16.39			
Total	2,095.9	59				

divided between $SS_{between-cells}$ and SS_W, and the $SS_{between-cells}$ further subdivided into its three components. Such a presentation parallels the actual procedure for analyzing the data and is the format used in the output of many statistical computer programs. If the summary table is published at all, it is common to leave out the first row ("Between Cells") because that information could be easily derived by adding the SS and df for the following three rows. Because I calculated MS_W directly from the cell variances, without finding SS_W first, I had to multiply MS_W by df_W ($16.39 \times 48 = 786.7$) to obtain SS_W for the summary table. (I could also have found SS_W by calculating SS_{total} directly from the data and substracting $SS_{between-cells}$.)

Interpreting the Results

The fact that the interaction in this case was not statistically significant allows us to focus our interpretation on the main effects. We could base our interpretation on the row and column means, but given that the amount of

Figure 14.7

Graph of Cell Means for
Data in Table 14.8

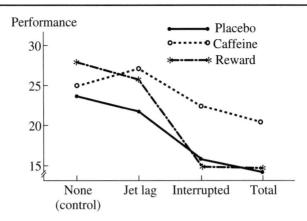

interaction was actually rather large, it is not a bad idea to look at the graph
of the cell means anyway. We may see a tendency toward interaction that
can be explored further in future experiments. Indeed, from the graph in
Figure 14.7, you can see that the different stimulation groups tend to show
a greater separation with more severe sleep deprivation. However, because
that tendency is not statistically significant, it would not be explored with
follow-up tests in this study.

Turning to the main effects, we can see that the significant effect of stimulation is due to caffeine producing the highest performance, and placebo,
the lowest (this is easiest to see not from Figure 14.7, but from the column
means in Table 14.8). The caffeine and reward means can each be tested
against the placebo mean, and against each other, using follow-up t tests that
are similar to those described in the previous chapter. Similarly, the significant sleep deprivation main effect suggests that you may want to perform
the six pairwise comparisons that are possible when a factor has four levels.

However, you may have noticed that these follow-up tests don't really
answer the questions that probably motivated the design of the study (e.g.,
does caffeine improve performance relative to placebo as much in the jet-
lag condition as it does in the interrupted condition? Does reward do so?).
Unfortunately, the failure of the interaction to reach significance in this
case tells us that we didn't get the results that would have been most interesting. Later, I'll describe the tests that would have followed a significant
interaction in this example, and outline some general plans for following
up on a significant interaction in any two-way ANOVA.

Post Hoc Comparisons for the Significant Main Effects

If the interaction between factors in an experiment is not significant, nor
large enough to obscure the main effects, the comparisons following the
ANOVA are very straightforward. The ANOVA is followed with pairwise or
complex comparisons among the levels of whichever main effect is significant *and* involves more than two levels (this may be both main effects as in
the sleep deprivation example). The appropriate post hoc comparison procedure should be used to keep α_{EW} from becoming too large, as described in
Chapter 13. You may have noticed that, in each of my two-way ANOVA
examples, *three F ratios* were tested, with no concern about the buildup of
α_{EW}. This is because the three F tests are generally viewed as planned comparisons. When performing post hoc comparisons, it is common to view
the tests involving each factor as a separate "family" (the interaction is a

third family) and to control α_{EW} within each family rather than across the entire experiment. That is why many statisticians prefer to speak about *familywise alpha* rates rather than the experimentwise alpha. (The distinction does not arise in a one-factor design, so I did not discuss this topic in the previous chapter.)

When a significant main effect has only three levels, as in the case of the stimulation factor in the current example, the LSD test is an appropriate (and powerful) choice for follow-up comparisons. The formula requires just a slight modification for use in a two-way ANOVA:

$$LSD = t_{crit} \sqrt{\frac{2MS_W}{N_j}}$$

The modification is that N_j has replaced n; N_j is the number of subjects contributing to each of the means being compared. In a balanced two-way ANOVA, N_j is equal to either N_{row} (the number of subjects in each row), or N_{col} (the number of subjects in each column), where $N_{row} = cn$, and $N_{col} = rn$. Each level of the simulation factor (i.e., each column in Table 14.8) involves 20 scores (i.e., 4×5), so in this case, N_{col} equals 20, and therefore LSD (for $\alpha = .05$) is equal to:

$$LSD = 2.01 \sqrt{\frac{2(16.39)}{20}} = 2.01\sqrt{1.639} = 2.01(1.28) = 2.57$$

According to this test, we can declare that the caffeine condition differs significantly from both the placebo and reward conditions (though just barely, in the latter case), but the placebo and reward conditions do not differ significantly from each other (the difference between 19.0 and 21.0 is smaller in magnitude than 2.57).

Because the sleep deprivation factor was also significant, it is appropriate to explore it further, but given that it has four levels, Tukey's HSD test is needed to control the possible accumulation of Type I errors. For a row effect, the HSD formula would look like this:

$$HSD = q_{crit} \sqrt{\frac{MS_W}{N_{row}}} = 3.77 \sqrt{\frac{16.39}{15}} = 3.77(1.045) = 3.94$$

Looking at the row means in Table 14.8, you can see that all of the pairs of means differ significantly, except for Control versus Jet Lag and Interrupt versus Total. You could also follow up this significant main effect with complex comparisons, such as the average of Control and Jet Lag to the average of Interrupt and Total, but as a post hoc comparison it would be important to use Scheffé's test. I will discuss the possibility of planning complex comparisons for the two-way ANOVA at the beginning of Section C.

Effect Sizes in the Two-Way ANOVA

The proportion of variance accounted for by each factor in a particular two-way design can be found by calculating η^2 (eta squared), much as you would for a one-way design. For the row effect, Formula 12.18 is modified slightly to become SS_R/SS_{total}. The corresponding formula for the column effect is, of course, SS_C/SS_{total}, and for the interaction it is SS_{inter}/SS_{total}.

However, this method of calculating eta squared actually becomes misleading when both of your factors involve experimental manipulations, as in the sleep-deprivation example. I will illustrate the problem by calculating eta squared for the stimulation effect: $\eta^2_{stim} = 217.6/2{,}095.9 = .104$. Note that variability due to the other experimental factor (in this case, type of sleep deprivation), along with any interaction between the two factors, is included in SS_{total} (the denominator of η^2), thus reducing the proportion of variance explained by the stimulation factor relative to what it would be in a one-way ANOVA by itself. The larger the SS for the other factor becomes, the smaller η^2_{stim} gets, even if SS_{stim} stays the same—and that can be misleading. The simple solution is to calculate what J. Cohen (1973) called partial eta squared, as expressed in the following formula:

$$\text{partial } \eta^2 = \frac{SS_{effect}}{SS_{effect} + SS_W} \qquad \textbf{Formula 14.7}$$

For the stimulation factor, partial η^2 (often symbolized as η^2_p) equals $217.6/(217.6 + 786.7) = 217.6/1{,}004.3 = .217$, which is a more realistic estimate of the effect that the stimulation factor may have in the context of some future experiment (except that its bias needs to be corrected, as will be shown shortly). (You can check, as an exercise, that η^2_p for the sleep-deprivation factor is .533.) If you don't have easy access to the SSs needed in Formula 14.7, but you have the F ratio and its associated dfs for testing a particular effect, you can use a variation of Formula 12.19, as shown next:

$$\text{partial } \eta^2 = \frac{df_{effect}F_{effect}}{df_{effect}F_{effect} + df_W} \qquad \textbf{Formula 14.8}$$

Given that F_{stim} was 6.64 with dfs of 2 and 48, η^2_p equals $13.28/(13.28 + 48) = 13.28/61.28 = .217$. Within rounding error, Formulas 14.7 and 14.8 will always yield the same answer. However, when one (or both) of your two factors does not involve an experimental manipulation, but rather the sampling of pre-existing groups, the rationale for a partial eta squared may not apply. This brings us back to the ordinary η^2.

For an example of a two-way ANOVA that includes both an experimental and an individual-differences factor, I will return to the drug/gender example introduced in Section A. If you recall that MS_W decreased when gender was added as a factor, it should not be surprising that this type of combination of factors has been referred to as following the *variance-reduction* model (Gillett, 2003). (Gillett [2003] refers to individual-differences factors as "stratified factors." Other authors [e.g., Olejnik & Algina, 2003] have referred to such factors as "measured factors" or "blocking factors.") When finding eta squared for an experimental factor in a two-way ANOVA in which the other factor is a grouping factor, it makes sense to calculate the ordinary, rather than the partial, η^2 (i.e., divide the SS_{bet} for the experimental factor by the entire value of SS_{total}).

For instance, for the drug factor in the Section A example, ordinary η^2 equals $SS_{drug}/SS_{total} = SS_{drug}/(SS_{bet-cells} + SS_W) = 97.44/(280 + 74) = 97.44/354 = .275$ (or 27.5%), which would be considered a good deal of variance accounted for. In this case, the other factor, gender, is not increasing the error variance at each level of the drug factor, because one would expect both genders to be included in the experiment. In fact, a partial η^2 would be misleadingly large, as it would reflect the drug effect size when only one gender is tested. (Note that you need to know the total SS in order to calculate ordinary η^2. There is no simple formula, like Formula 14.8, that can find ordinary η^2 from F ratios alone.) Although it would not be unreasonable to find the η^2_p for the gender factor in this example, it is not likely to be of much interest. In general, researchers are more interested in the effect sizes of factors they create, or factors involving unusual groups. You will usually be interested in a partial η^2 for each of your factors whenever both factors involve the random assignment of subjects to conditions. Gillett (2003) refers to this case as the *variance-preservation* model; adding one experimental factor to another will generally *not* reduce the error term of your *F* ratio.

Unbiased Estimates of Omega Squared

As in the one-way ANOVA, the value you calculate for eta squared from your data is a biased estimate of omega squared (i.e., the proportion of variance in your dependent variable that is accounted for by your factor in the entire population). The correction for the bias of an ordinary eta squared in a two-way ANOVA is the same as the correction in a one-way ANOVA, as was given in Formula 12.20. That formula is repeated next with a slight change in notation:

$$\text{est. } \omega^2 = \frac{SS_{\text{effect}} - df_{\text{effect}}MS_W}{SS_{\text{total}} + MS_W} \qquad \textbf{Formula 14.9}$$

For the drug factor in the Section A example, the estimate of ω^2 equals $(97.44 - 2 \times 6.17)/(354 + 6.17) = 85.1/360.17 = .236$, which is a relatively minor decrease from the ordinary eta squared of .275. However, to correct the bias in η_p^2 rather than η^2, you need a formula that estimates *partial omega squared* (ω_p^2), such as the one that follows:

$$\text{est. } \omega_p^2 = \frac{SS_{\text{effect}} - df_{\text{effect}}MS_W}{SS_{\text{effect}} + (N_T - df_{\text{effect}})MS_W} \qquad \textbf{Formula 14.10}$$

Applied to the stimulation factor for the sleep-deprivation experiment, Formula 14.10 shows us that an unbiased estimate of ω_p^2 is equal to $[217.6 - 2(16.39)]/[217.6 + 58(16.39)] = 184.82/1,168.22 = .158$, which is a considerable reduction from the value for η_p^2, which was .217 (but still much larger than the value of ordinary η^2, which was only .104).

Partial *d* and *g*

Just as eta squared comes in both ordinary and partial versions, so too can a distance effect-size measure (ESM), such as *g* or *f*. Given that distance ESMs seem to be preferred for meta-analysis, this distinction is worth considering. Gillett (2003) deals with a simplified case of a two-way ANOVA in which one of the factors has only two levels, thus allowing him to confine his discussion (as I will, as well) to *g* and "partial *g*." As in the case of eta squared, the two versions of *g* in a two-way ANOVA differ in terms of the inclusiveness of their denominators.

Recall that in the case of just two samples, the denominator of *g* is s_p, the square root of the pooled variance. The analogous measure in a two-way ANOVA would have the square root of MS_W in its denominator. Gillett (2003) points out that dividing the difference of the means for the two-level factor by $\sqrt{MS_W}$ yields a partial *g*, rather than an ordinary *g*, which is appropriate only for the variance-preservation (i.e., two experimentally created factors) model. For example, if you wanted an ESM for the caffeine versus placebo effect in the sleep-deprivation experiment, you could subtract one of those column means from the other, and divide by $\sqrt{MS_W}$, in which case you would get: $(23.65 - 19)/\sqrt{16.39} = 4.65/5.04 = 1.15$. It should be clear that this measure is appropriate, because $\sqrt{MS_W}$ is not affected by the presence or absence of the other factor (type of sleep deprivation).

On the other hand, it should also be clear why $\sqrt{MS_W}$ is not the right denominator to use when you want an ordinary g in a two-way ANOVA that follows the variance-reduction model. Suppose you want an ESM for the drug factor in the Section A example. Using $\sqrt{MS_W}$ would ignore the variation due to gender, although such variation would be expected in any similar experiment not artificially restricted to just one gender. To find ordinary g for comparing the placebo to the large-dose condition, you would calculate the variance for all placebo participants—not separated by gender—and do the same for the large-dose participants. Pooling these two variances, and then taking the square root of the result, would yield an appropriate denominator (Gillett calls it s_{level}) with which to divide the difference of the placebo and large-dose means (given the usual HOV assumption, it would also be reasonable to pool the variances of all three drug condtions to obtain this effect-size denominator). This is not hard to do if you have the raw data; just proceed as though you were calculating a one-way ANOVA for the drug condition, and forget about gender entirely. However, if you have only the three F ratios of the two-way ANOVA, and their corresponding dfs, you would have to obtain g by using a complex-looking formula presented by Gillett (2003, Equation 19, p. 427).

Post Hoc Comparisons for a Significant Interaction

When the F ratio for the interaction in a two-way ANOVA reaches statistical significance, or even just approaches significance, it will often be difficult to interpret the results for the main effects in a straightforward manner. Moreover, if the interaction is markedly disordinal, it will usually be pointless to perform comparisons on row or column means even if one or both of the main effects is significant. It is more likely that comparisons would focus ultimately on cell means, but there is more than one way to proceed in this situation. I will discuss two possible approaches, the first being the more frequently employed.

Simple Effects

To gain a deeper understanding of two-way interactions and how to analyze them, let us look again at the graph of the cell means for the example in Section B (reprinted with an added feature as Figure 14.8). Notice that the main effect of the sleep factor is a bit different at each level of stimulation (i.e., the lines are distinctly not parallel). Each line on the graph represents what is called a *simple main effect*; it is the effect of the different levels of sleep deprivation at only *one* level of the stimulation factor. Of course, we could redraw the graph with the stimulation levels along the horizontal axis, and then each line would represent a simple main effect of stimulation at *one* level of sleep deprivation. In general, a simple main effect is the effect of one factor while holding the other factor fixed at one level. The main effects that are tested in the two-way ANOVA are actually averages of the appropriate simple main effects.

In Figure 14.8, the heavy line labeled "average" is *the* main effect of sleep deprivation that was tested in the two-way ANOVA ($F = 18.3$); it is the average of the three simple main effects shown in the graph (note: the average line is based on the row means from Table 14.8). Averaging the simple main effects of a factor into a single main effect that is tested for significance and, perhaps, followed up with pairwise comparisons makes sense only if the simple main effects do not really differ from each other in the entire population. How can we know they are the same in the population?

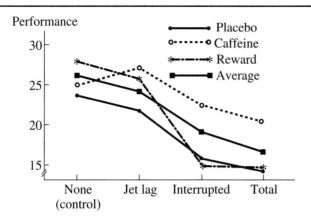

Performance

30

25

20

15

None   Jet lag   Interrupted   Total
(control)

Placebo
Caffeine
Reward
Average

Figure 14.8

Graph of Cell Means for
Data in Table 14.8
(Including Main Effect
of the Sleep Deprivation
Factor)

In practice we never really know, but when we test the interaction effect, we are testing the null hypothesis that the simple main effects are identical to each other (in the population) for each factor. When the simple main effects differ from each other sufficiently (relative to MS_W), we can reject this null hypothesis and declare that the interaction is statistically significant.

One approach to analyzing a significant interaction is to test each simple main effect (or simple effect, for short) as a one-way ANOVA, except that MS_W from the two-way ANOVA is used as the error term in each case (if the homogeneity of variance assumption seems reasonable). For instance, to test the simple effect of sleep deprivation at the caffeine level, first calculate SS_{bet} based on the four cell means that comprise this effect (as though the four cells were the four groups of a one-way ANOVA); $SS_{bet} = N \times \sigma^2(25.0, 25.6, 22.6, 21.4)$, where N is the number of subjects included in that simple effect (in this case, $N = 20$). $SS_{bet} = 20 \times 2.95 = 59.0$, and $MS_{bet} = 59/(k - 1) = 59/3 = 19.67$ (where k is the number of levels or groups included in the simple effect). Finally, $F = 19.67/16.39 = 1.2$. (If you are analyzing your data by computer, you can obtain the numerator MS by selecting data only for the group of interest and then calculating a one-way ANOVA—but save your MS_W from the two-way ANOVA output for the denominator MS.) The critical F is the same as for the main effect of sleep deprivation (i.e., 2.84), so this simple effect is not significant (caffeine seems to be good at eliminating the effects of different types of sleep deprivation). Were it significant, this simple effect could have been followed by pairwise (or complex) comparisons of the cells comprising that effect (e.g., the jet-lag/caffeine cell mean vs. the interrupted/caffeine cell mean); however, because there are six possible pairwise comparisons to follow up the simple effect I just tested, Tukey's HSD (or a similar test) should be used (if any complex comparisons are involved, Scheffé's test would be appropriate).

For an ANOVA in which both factors have quite a few levels it may not make sense to follow a significant interaction by testing every simple effect and then performing all possible pairwise comparisons on every significant simple effect. The design of the study usually makes some tests more interesting than others. For the present study, testing the simple effect of stimulation when there is no sleep deprivation may not be particularly interesting in itself. Often researchers will perform only the tests that are relevant to their theoretical questions (see the subsection on Planned Comparisons in Section C). On the other extreme, researchers sometimes bypass the analysis of simple effects and follow a significant interaction by testing all meaningful pairs of cell means. In our example, there are a total of 12 cells, so 66 pairwise tests are possible. Not all of these tests are useful, however. Comparing jet lag/caffeine to interrupted/placebo makes little sense because two factors have changed levels. Comparisons are interpretable only within the same row or column of the design. Thus, there are only 30 tests that are meaningful. A test like Tukey's HSD should be used, but modified to protect you for a total of 30 rather than 66 tests (Cicchetti, 1972). Of course, if a researcher has planned relatively few particular pairwise comparisons to test his or her hypotheses, it is likely that a Bonferroni adjustment will yield considerably more power than a post hoc method that adjusts for all possible tests or even all meaningful tests.

Interaction Contrasts

If the interaction in a 2×2 ANOVA is statistically significant, the only follow-up tests that are possible are the simple effects (which, in this case, are just pairwise comparisons). However, a 2×3 ANOVA can be separated into three 2×2 interactions. For instance, in the example of Section A, you can delete either the placebo, the moderate dose, or the large dose and calculate the interaction for each remaining 2×2 ANOVA. The MS for the overall interaction of the 2×3 ANOVA is equal to the average of the interaction MSs for the three possible 2×2 subsets. However, as you can see in Figure 14.1, the amount of interaction can differ greatly from one subset to another (e.g., in the Section A example, there is a good deal more interaction between the placebo and moderate dose than between the moderate and large doses). If the interaction in a 3×2 ANOVA is significant, it can be studied further by testing the three possible 2×2 interactions, rather than analyzing simple effects. These 2×2 subsets are called *interaction contrasts*.

Although the interaction failed to reach significance in the sleep-deprivation experiment, to illustrate the calculation of an interaction contrast, it will be useful to pretend that it was. If the interaction were significant, we might want to look at a graph of the cell means to see if the bulk of the interaction could be encapsulated in a single 2×2 contrast. Looking at Figure 14.8, it appears that a large portion of whatever interaction exists can be captured by comparing jet-lag and interrupted sleep conditions for just the caffeine and reward stimulation levels. This 2×2 comparison can be quantified with a single-df linear contrast, just like the ones you learned about in the previous chapter. The critical step is to find L.

One way to think about any 2×2 interaction is to recognize that it is the difference between the two simple main effects of either variable—that is, it is either the difference between the two rows in one column subtracted from the difference between the two rows in the other column, or it is the difference between the two columns in one row subtracted from the difference between the two columns in the other row (L will come out to the same value in either case). For the 2×2 contrast described in the previous paragraph, we can calculate L as the caffeine-reward difference for jet lag minus the caffeine-reward difference for interrupted sleep: $L = (25.6 - 26.2) - (22.6 - 15.6) = -.6 - 7.0 = -7.6$. To calculate the SS for this contrast, we can use Formula 13.12, where n is the size of each cell in the two-way ANOVA (5, in this case), and all of the cs have a value of $+1$ or -1 (in the preceding calculation of L, the cs are $+1$, -1, -1, $+1$, respectively, which, as necessary for a contrast, sum to zero). Therefore, $SS_{cont} = 5 \times (-7.6)^2/(1 + 1 + 1 + 1) = 5(57.76)/4 = 288/4 = 72.2$, and so MS_{cont} equals 72.2, as well. To test this contrast for significance, we can use the error term from the entire two-way ANOVA; thus, $F_{cont} = MS_{cont}/MS_W = 7.2/16.39 = 4.4$.

Scheffé's Test

If this contrast had been planned, the appropriate critical value to test against would be $F_{.05}(1, 48) = 4.04$ (approx.); the numerator has one df, as does the interaction in any 2×2 ANOVA, and the denominator has whatever df corresponds to the error term being used, regardless of the number of scores involved in the actual contrast. Because $F_{cont} > F_{crit}$, this contrast would have been significant if it had been planned before inspecting the data, and that would have entitled us to test appropriate pairs of cells in the 2×2 subset that was tested (e.g., caffeine versus reward for just the jet-lag condition). However, if this contrast were tested as part of an attempt to localize a significant interaction from the omnibus ANOVA, it would be considered a type of complex post hoc comparison and therefore would require Scheffé's test to control Type 1 errors. We need only modify the degrees of freedom appropriately to adapt Scheffé's test to interactions in the two-way design, as follows:

$$F_S = df_{inter} F_{crit}(df_{inter}, df_W) \qquad \text{Formula 14.11}$$

This is just the critical F for testing the interaction of the entire design multiplied by the df for the interaction from the entire design. For the example in Section B, the critical value you would use to test any 2×2 post hoc contrast at the .05 level is

$$F_S = 6 \cdot F_{.05} (6, 48) = 6 \cdot 2.34 = 14.04.$$

Effect Size

It should come as no surprise that this contrast would not be significant as a post hoc test, given that the omnibus interaction was not significant in the original two-way ANOVA. The use of Scheffé's test ensures that this will be the case. However, if the 2×2 contrast we tested were considered significant, we might want to look at its effect size. Because, like any of the linear contrasts I have described, the 2×2 contrast reduces to a single distance measure, L, we need only divide this measure by an appropriate standard deviation to create an effect-size measure akin to g. I would like to use g_c to represent the g for a contrast, but it is so often referred to as d_c (just as d_p, without the boldface, is used to represent a partial g) that I will go along with that notation. For a balanced two-way ANOVA, it is reasonable to use $\sqrt{\Sigma c_i^2 MS_W/2}$ as the denominator, so for the example in the preceding paragraph, d_c equals $L/\sqrt{(4 * MS_W/2)} = 7.6/\sqrt{32.78} = 7.6/5.7 = 1.33$—a rather large effect (note that the sign of L is ignored when calculating an ESM). Sometimes, it is more convenient to obtain d_c from the F ratio of the contrast, in which case the following simple formula can be used for any contrast:

$$d_c = \sqrt{\frac{2F_{cont}}{n}} \qquad \text{Formula 14.12}$$

where n is the size of each cell in the design (note that $\sqrt{(2 * 4.4/5)} = \sqrt{1.76} = 1.33$).

Partial Interactions

With a 3×3 or larger ANOVA, there are subsets larger than interaction contrasts that can serve as the initial steps for decomposing a significant interaction. One logical way to begin is to retain all of the levels for one of the factors but only two levels for the other factor. The interaction calculated for this kind of subset can be called a *partial interaction*. For instance, if you look at Figure 14.8, you'll see that most of the interaction occurs between the jet lag and interrupted levels of the sleep factor. A 2×3 subset of the 4×3 ANOVA can be created by dropping the control and total levels of the sleep factor but retaining all three levels of the stimulation factor. The 2-df interaction calculated for that subset is a partial interaction. A partial interaction can also involve a complex comparison. For example, imagine that you have retained all the levels of the sleep factor, but rather than dropping one of the stimulation levels to create a 4×2 design, you average the reward and caffeine levels together and compare this average to the placebo. This also creates a 4×2 design and a potentially interesting (complex) partial interaction.

Boik (1979) details a method for modifying Scheffé's test to determine the significance of a partial interaction while protecting the experimentwise alpha rate. A significant partial interaction can then be further analyzed into interaction contrasts that are subjected to the ordinary Scheffé test. For instance, a significant interaction between the stimulation factor and the jet lag and interrupted levels of the sleep factor could be followed by testing the three possible 2×2 interactions, dropping one of the three stimulation levels at a time. However, it is hard to imagine a researcher devising a partial interaction as a post hoc test (I have never seen it), so I will postpone any further discussion of partial interactions until I cover the topic of Planned Comparisons for a Two-Way ANOVA in Section C.

Assumptions of the Two-Way ANOVA

The assumptions for the two-way ANOVA for independent groups are the same as for the one-way ANOVA (see Chapter 12, Section B), so I will not bother to repeat them here. Bear in mind that when you are dealing with a balanced design (i.e., all cell sizes are equal), you do not have to worry about the homogeneity of variance assumption. If the design is not balanced, it may be appropriate to test for heterogeneity of variance and adjust the ANOVA according to procedures in advanced texts. However, even if homogeneity of variance can be assumed for a particular unbalanced design, there is still more than one reasonable procedure for conducting the ANOVA, and some judgment is involved in selecting the procedure that fits your purposes (see Section C). On the other hand, even in a balanced design, if the cell sizes are small *and* it seems that the underlying population distributions are *very* far from normal in shape and/or extremely different in variance, it may be appropriate to perform a data transformation or even to abandon the ANOVA formulas of this chapter entirely and use an advanced form of nonparametric statistics.

Advantages of the Two-Way ANOVA with Two Experimental Factors

Economy

The results of the study described in this section would have been more interesting in the presence of a significant interaction. Imagine, instead, a simpler study in which all of the subjects have been totally sleep deprived, and the two factors consist of four levels of caffeine and three levels of reward. You may not expect or want an interaction in this case. If the interaction is not significant, you have performed two one-way ANOVAs in an efficient, economical way (each subject would be serving in two experiments simultaneously—a caffeine experiment and a reward experiment). To have equal power with two separate one-way ANOVAs, you would have to use 60 subjects in each study—that is, twice as many subjects. You would probably want to use post hoc tests on any factor that is significant, but note that when a factor has quantitative levels, such as caffeine dosages or amounts of reward, your follow-up tests would probably be designed to look for linear effects of the factor or other trends, as were described in the previous chapter.

Exploration of Interactions

In an ANOVA with two experimental factors, the interaction can be more interesting than either of the main effects. For instance, in a study of anagram solving, one factor might be the degree of time pressure (higher rewards for solving the problems in shorter amounts of time), and the other factor might be the difficulty of the anagram. Researchers may expect that subjects will solve fewer problems as difficulty increases and that they will solve more problems, the greater the reward is to solve them quickly. However, what may be more interesting is the interaction of those two variables. With relatively easy anagrams, time pressure should result in more problems solved, but with difficult anagrams time pressure may be counterproductive and actually reduce the number of problems solved (see Figure 14.9). The calculation of interactions involving trend components will be described briefly in Section C.

Reduction in Error Term

In some cases, extraneous variables are not controlled in an experiment because they are thought to be totally irrelevant. If one of those variables is later found to be relevant, adding it as a factor in your ANOVA design can reduce MS_W. For instance, the subjects in a social psychology experiment may be run by different research assistants, some male and some female. If the gender of the assistant affects the subjects' performance, variability (in the error term) will be increased. In that case, adding "gender of assistant" as an experimental factor can reduce MS_W. You could also reduce MS_W by using assistants of only one gender, but this could limit the generalizability of the results. Another researcher might fail to replicate your results simply because of using assistants of the opposite gender.

Advantages of the Two-Way ANOVA with One Grouping Factor
Reduction in Error Term

Even when experimental conditions are controlled as much as possible, there can be a great deal of subject-to-subject variability because of individual differences. In some experiments, a significant amount of subject-to-subject variability can be attributed to a grouping variable, such as gender. (Remember that a grouping variable such as gender will not be relevant in all experiments; it depends on whether there are gender differences for the

Figure 14.9

An Example of
Interaction

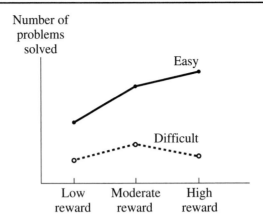

particular dependent variable being studied.) When a grouping variable contributes to variability, it should be added as a factor to reduce MS_W. The ultimate way to reduce subject-to-subject variability is to use the same subject in several conditions, or to match subjects carefully before assigning them randomly among the different conditions. This is an extension of the procedure described in Chapter 11 to the multigroup case, and it forms the focus of the next chapter.

Exploration of Interactions

The experimental variable (e.g., drug versus placebo) may operate differently for different levels of the grouping variable. Unless the grouping variable is included in the ANOVA, the effects of the experimental variable can be obscured in a misleading way (e.g., if the experimental variable had opposite effects on men and women, the variable might simply appear to be ineffective—unless the genders are later separated in the analysis). Analyzing the interaction can yield valuable information about the conditions under which the experimental variables are most effective. On the other hand, if the interaction is not near significance, you have increased your confidence about generalizing the experimental effect to each of the subgroups included in your study.

Advantages of the Two-Way ANOVA with Two Grouping Factors

Exploration of Interactions

When both factors in a two-way ANOVA are grouping variables, the focus is generally on the interaction. For instance, if achievement motivation is being measured in a two-way design in which one factor is job level (executive versus clerical staff) and the other is length of time at the job (newly hired versus experienced), the primary interest would be to see if the difference due to experience is the same for both job levels (or, conversely, if the difference between the two job levels remains the same with experience). Of course, the nature of the two independent variables precludes the making of any conclusive causal inferences, but the findings can support or disconfirm a theory and lead to further studies.

Publishing the Results of a Two-Way ANOVA

The results of the sleep deprivation experiment could be reported in the following manner: "The video-simulation performance scores were subjected to a 4×3 independent-groups ANOVA with type of sleep deprivation and type of stimulation as the two factors. Both main effects were significant ($MSE = 10.8$): for sleep deprivation, $F(3, 48) = 18.3$, $p < .001$, $\eta_p^2 = .533$, and for stimulation, $F(2, 48) = 6.64$, $p < .01$, $\eta_p^2 = .217$. The interaction, however, fell short of significance, $F(6, 48) = 1.97$, $p > .05$." (If an interaction had been predicted, or had been considered desirable, it is likely that the last result would have taken advantage of the fact that the p value for the interaction was about .09 and would have been reported as: "The interaction approached significance, $F(6, 48) = 1.97$, $p < .1$.") Because the mean square for error (MSE—another name for MS_W) is the same for all of the F ratios in a two-way independent groups design, it need not be repeated. An alternative to reporting eta squared for each of your significant F ratios is to report an estimate of omega squared; for instance, for stimulation, $F(2, 48) = 6.64$, $p < .01$, est. $\omega^2 = .158$.

An Excerpt from the Psychological Literature

As an example of the typical use of two-way ANOVAs in the psychological literature, I have chosen an article entitled "Evidence of codependency in women with an alcoholic parent: Helping Out Mr. Wrong" (Lyon & Greenberg, 1991). One of the independent variables in this study was a grouping factor: Half the subjects "came from families in which one parent was alcohol dependent" (the codependent group), whereas the other half (the control group) did not (i.e., neither parent was alcohol dependent). The second independent variable involved an experimental manipulation. In one condition, the subject heard positive feedback about the experimenter in his absence (that the experimenter was a nurturing individual) from a "confederate" (an assistant to the experimenter, posing as another subject). In the other condition, the subject heard negative feedback (that the experimenter was an exploitive individual). Half of the subjects with alcoholic parents were randomly assigned to each condition, as were half of the control subjects. In each case, the experimenter returned to the room after the feedback and asked the subject to volunteer to help him as a research assistant. The amount of time volunteered (between 0 and 180 minutes for each subject) served as the dependent variable. The principal results were reported as follows.

> We conducted a 2×2 ANOVA on the amount of time volunteered. A significant main effect was found for group membership, $F(1, 44) = 9.89$, $p < .003$, so that codependents were generally more helpful ($M = 97.9$ min) than were the members of the control group ($M = 31.3$ min). We also found a main effect for condition, $F(1, 44) = 4.99$, $p < .03$; subjects in the exploitive experimenter condition volunteered significantly more time overall ($M = 72.04$ min) than subjects in the nurturant experimenter condition ($M = 36.00$ min). The main effects were qualified, however, by a significant two-way interaction, $F(1, 44) = 43.64$, $p < .0001$, that strongly supported the primary prediction of the study; . . . codependents were significantly more helpful to the exploitive experimenter than to the nurturant experimenter, $t(23) = 6.38$, $p < .001$. In contrast, the control group was more helpful to the nurturant experimenter than to the exploitive one, $t(24) = 2.47$, $p < .05$.

The interaction is very large (F is over 40!) and disordinal, which tells us not to take the significant main effects at face value. Although one main effect suggests that codependents are "generally more helpful," this was only true when the experimenter was perceived as an exploitive individual (one who presumably resembled an alcoholic parent). The significant interaction justifies the subsequent *t* tests, in which individual cells were compared. (Sometimes, follow-up pairwise comparisons are presented as *F* ratios instead of *t* values, but, of course, the *p* values will be the same.) The exact *p* values (e.g., $p < .03$) reported by these researchers are undoubtedly a result of the availability of computer analysis.

For a 2×2 ANOVA, it is often convenient to present the marginal means in the text of the report, as in the preceding excerpt. For more complex designs, such as the sleep deprivation example, it is more common to report the marginal means as part of a larger table of cell means and standard deviations (a table like Table 14.8, but without the individual scores). To display the cell means when an interaction is present, a bar graph or other figure might be preferred.

B SUMMARY

1. The easiest and fastest way to calculate a two-way ANOVA without a computer is:
   a. Calculate the following four components: SS_{total}, $SS_{between-cells}$, SS_R, and SS_C. Each one is calculated the same way: Find the biased variance of the appropriate means, and multiply by the total N. (For SS_{total}, the "means" are the individual scores; for $SS_{between-cells}$, the means are the cell means; for SS_R and SS_C, the means are the row means and column means, respectively.)
   b. SS_W can then be found by subtracting $SS_{between-cells}$ from SS_{total}, and SS_{inter} can be found by subtracting SS_R and SS_C from $SS_{between-cells}$. (If you want to check your work at this point, you can calculate SS_W independently by averaging the cell variances, multiplying by df_w, and seeing if this value agrees with the one you obtained by subtraction.)
   c. SS_R, SS_C, SS_{inter}, and SS_W are each divided by the appropriate df to produce MS_R, MS_C, MS_{inter}, and MS_W respectively.
   d. Finally, MS_R, MS_C, and MS_{inter} are each divided by MS_W to form the three *F* ratios, which are then compared to their corresponding critical values to test for significance.

2. If a main effect is significant and has more than two levels, it can be followed by post hoc comparisons, pairwise or complex, among its various levels—as long as you use the appropriate method to control Type I errors. When using a formula like the ones for LSD or HSD, the *n* is not the cell size but rather the number of subjects at each level of the factor being tested (assuming a balanced design). If the interaction is statistically significant, and especially if it is large and/or disordinal, the focus usually shifts to tests of simple main effects or interaction contrasts.

3. Each individual row or column in a two-way design represents a *simple effect*. When an interaction is significant, it means that the simple effects, which are averaged together to create the main effects, differ significantly. In this case, it makes sense to test the simple effects for significance and then follow up on significant simple effects with pairwise (or complex) comparisons. However, it is possible to skip this step, and to proceed by performing pairwise (or complex) comparisons only on cell means within the same column or row, adjusting α_{pc} accordingly. For all such comparisons, MS_W is used as the error term, unless there are serious concerns about homogeneity of variance (which is very rare for a balanced design).

4. An alternative strategy for analyzing a significant interaction in a 2×3 or larger ANOVA, which is particularly relevant when the interaction is much stronger in some parts of the design than others, is to focus on the interactions in various 2×2 subsets of the overall design, called interaction contrasts. In a 3×3 or larger ANOVA, interaction contrasts may or may not be preceded by the testing of partial interactions. In their simple form, partial interactions involve two levels of one factor crossed with all of the levels of the other factor. Both partial interactions and interaction contrasts can involve complex comparisons.

5. Effect size can be estimated for each factor in a two-way ANOVA the same way as in a one-way ANOVA: The SS associated with each main effect is divided by the total variability. However, if both factors involve experimental manipulations, both factors contribute to the total variability, which reduces eta squared for each factor compared to what it would be in a one-factor experiment. To correct for this, the SS due to each main effect is divided not by the total SS but by the numerator SS plus the SS for error, thus creating a partial eta squared. Similarly, a partial g can be calculated, and there is a modified formula for calculating an estimate of partial omega squared.

6. Types of Two-Way ANOVA
   a. *Two experimental factors*: Allows the exploration of the interaction of two variables. When the interaction is not significant, the principle advantage is economy.
   b. *One experimental factor, one grouping factor*: Adding the grouping factor can increase the power involved in testing the experimental factor, by reducing the error term. This design also determines whether the experimental factor functions similarly for different subgroups of the population.
   c. *Two grouping factors*: In most cases, the possible interaction is of greater interest than either of the main effects. However, none of the significant effects in this design can demonstrate causation.

EXERCISES

1. For a 3×5 ANOVA with nine subjects in each cell,
   a. Find the value of each df component, and show that they sum to df_{total}.
   b. Find the critical value of F for each main effect and the interaction.

*2. A cognitive psychologist knows that concrete words, which easily evoke images (e.g., *sunset*, *truck*), are easier to remember than abstract words (e.g., *theory*, *integrity*), but she would like to know if those who frequently use visual imagery (*visualizers*) differ from those with little visual imagery (*nonvisualizers*) when trying to memorize these two types of words. She conducts a 2×2 experiment with five subjects in each cell and measures the number of words each subject recalls. The data appear in the following table.

	Visualizers	Nonvisualizers
Concrete	17	18
	20	19
	18	17
	21	17
	20	20
Abstract	14	18
	15	18
	15	17
	17	17
	16	19

a. Perform a two-way ANOVA and create a summary table.
b. Draw a graph of the cell means and describe the interaction, if any.
c. Report the results in part a in the form of a paragraph written in APA style.
d. Calculate the appropriate version of eta squared for each main effect.

3. A neuropsychologist is studying brain lateralization—the degree to which the control of various cognitive functions is localized more in one cerebral hemisphere than the other. This researcher is aware that for men, being right-handed seems to be associated with a greater degree of lateralization than does being left-handed but that this relationship may not apply to women, who seem to be less lateralized in general. Various tasks are used in combination to derive a lateralization score for each subject as shown in the following table.

	Left-Handed	Right-Handed
Men	9	14
	12	25
	8	15
	9	17
	10	21
	11	20
Women	13	10
	10	8
	16	11
	19	13
	22	9
	12	10

a. Perform a two-way ANOVA and create a summary table.
b. Graph the cell means and describe the interaction, if any.
c. Calculate the MS for interaction as a 2×2 contrast, and show that you get the same result as in part a.
d. If both main effects were significant would you want to find a partial eta squared for either one? Explain.

*4. The following summary table corresponds to a hypothetical experiment in which subjects must complete a mental task at five levels of difficulty and with three different levels of reward. Much of the table has deliberately been left blank, but all of the missing entries can be found by using the information in the previous sentence and the information in the table. (Assume the design is balanced.)

Source	SS	df	MS	F	p
Difficulty	100				
Reward	150				
Interaction			5		
Within-Cells		90			
Total	1,190				

a. Complete the summary table.
b. How many subjects are in each cell of the two-way design?
c. Calculate η_p^2 for each of the three effects, using the corresponding F ratios.

5. A social studies teacher is exploring new ways to teach history to high school students, including the use of videotapes and computers. He also wants to know if these new techniques will have the same impact on average students as on gifted students. Twelve average students are randomly divided into three equal-sized groups: one instructed by the traditional method, one by videotapes, and one by an interactive computer program. In addition, 12 gifted students are divided in a similar fashion, resulting in a 3×2 design. At the end of the semester all students take the same final exam. The data appear in the following table.

	Traditional	Videotape	Computer
Average	72	69	63
	83	66	72
	96	78	78
	79	64	59
Gifted	83	96	89
	95	87	93
	89	93	86
	98	86	95

a. Perform a two-way ANOVA and report the results as a paragraph, using the appropriate APA style.
b. Draw a graph of the cell means. Regardless of whether the F ratio for the interaction was statistically significant, test the simple main effects of method for both the average and gifted students.
c. Calculate the ordinary η^2 and an estimate of ordinary ω^2 for the main effect of method. Why are these measures justified?

*6. A clinical psychologist is trying to find the most effective form of treatment for panic attacks. Knowing that antidepressant drugs have been surprisingly helpful in reducing the number and severity of attacks, this researcher decides to test three forms of psychological treatment, both with and without the use of antidepressants as an adjunct. The dependent variable—the total number of panic attacks during the final 30 days of treatment—is shown for each subject in the following table.
a. Perform a two-way ANOVA. Which effects were significant at the .05 level? At the .01 level?

b. Calculate and test for significance each of the three possible 2 × 2 interaction contrasts.
c. Calculate an estimate of partial omega squared for each of the main effects.

	Psychoanalysis	Group Therapy	Behavior Modification
Therapy Alone	4	6	1
	3	5	3
	2	8	0
	5	6	2
	4	8	4
	5	7	4
	5	6	2
	4	6	2
	2	7	1
	6	8	2
Therapy Plus Drug	1	3	0
	1	4	1
	3	3	3
	1	4	2
	2	2	0
	1	2	0
	3	5	1
	4	3	2
	2	2	1
	3	5	0

7. A college is conducting a study of its students' expectations of employment upon graduation. Students are sampled by class and major area of study and are given a score from 0 to 35 according to their responses to a questionnaire concerning their job preparedness, goal orientation, and so forth. The data appear in the following table.

	Humanities	Sciences	Business
Freshmen	2	5	7
	4	6	8
	3	9	7
	7	10	12
Sophomores	3	10	20
	4	12	13
	6	16	16
	5	14	15
Juniors	7	14	20
	8	15	25
	7	13	22
	7	12	21
Seniors	10	16	30
	12	18	33
	9	16	34
	13	19	29

a. Perform a two-way ANOVA and create a summary table.
b. Draw a graph of the cell means. Does the interaction obscure the interpretation of the main effects?
c. Use Tukey's HSD to determine which pairs of class years differ significantly.
d. For just the freshmen and seniors, calculate the three possible interaction contrasts. Which, if any, would be significant according to Scheffé's test?

*8. The data from Exercise 12B8 for a four-group experiment on attitudes and memory are reproduced below. Considering the relationships among the four experimental conditions, it should be obvious that it makes sense to analyze these data with a two-way ANOVA.

Incidental-Agree	Incidental-Disagree	Intentional-Agree	Intentional-Disagree
8	2	6	7
7	3	8	9
7	2	9	8
9	4	5	5
4	4	8	7

a. Perform a two-way ANOVA and create a summary table of your results. (*Note:* You can use the summary table from Exercise 12B8 as the basis for a new table.)
b. Compare your summary table to the one you produced for Exercise 12B8.
c. What conclusions can you draw from the two-way ANOVA?

9. Suppose that the table of means in Exercise 13C2 represents only half of an experiment. What was not mentioned in the previous exercise is that those participants were following a strict diet along with their assigned number of exercise hours. In the table below, I have added the means for a second group, also consisting of 20 participants assigned to each of the six exercise levels, who maintained their usual eating habits. The table does not include SDs, but for convenience we'll use an error term close to the one in the previous exercise; assume that $MS_W = 350$.

hours/week	3	6	9	12	15	18
Exercise + Diet	55	60	70	75	77	72
Exercise Only	60	62	66	70	71	64

a. Perform a two-way ANOVA on these data, and report the results in paragraph form, using APA style.
b. Conduct pairwise comparisons for the main effect of amount of exercise. Which pairs of levels differ significantly?
c. Calculate the simple main effects of exercise hours for each diet group, and test each for significance at the .05 level.
d. Calculate partial eta squared, partial g, and an estimate of partial omega squared for the diet factor.

*10. For Exercise 13C1, I invented an experiment in which the five levels of the independent variable were determined by the number of weekly sessions of therapy given to the participants. A better experiment would include a control condition—a fake form of therapy—administered at the same five levels of frequency. The group means from Exercise 13C1 are reprinted below as the therapy level of a second factor whose other level is labeled "control." Consistent with the previous exercise, there are 15 participants in

each cell. Rather than giving you the SDs for all ten cells, I'll just tell you that $MS_W = 4.0$ for the entire two-way ANOVA.

# of weekly sessions	1	2	3	4	5
Therapy	6.8	7.6	8.0	8.2	8.3
Control	7.0	6.2	5.3	6.4	6.7

a. Perform a two-way ANOVA on these data, and organize the results into a summary table.
b. Calculate partial eta squared, partial g, and an estimate of partial omega squared for the therapy (vs. control) factor.
c. Find the 2×2 interaction contrast that yields the largest L, and test it for significance as a planned contrast at the .05 level. Calculate the g for that contrast (i.e., d_c).
d. Would the contrast you tested in part c be significant by Scheffé's test? Could you have answered this question solely from your results in part a? Explain.

Planned Comparisons for a Two-Way ANOVA

Research hypotheses often come in a form that predicts a 2×2 interaction. For instance, the proximity (arm's length or across the room) of food will affect eating for people who are overweight but not for those near their ideal weight. Or, depressed people will remember more negative than positive words from a memorization list, whereas the reverse will be true for people who are not depressed. In a larger ANOVA such as the 4×3 study we have been using as an example, it is often the case that smaller 2×2 subsets of the study correspond to particular research hypotheses. For instance, in the Section B example, the 2×2 interaction of jet lag versus interrupted as one factor and caffeine versus reward as the other answers the question of whether different types of partial sleep interference are best compensated for, by different forms of stimulation (see Figure 14.8). A test for interaction in one or more of these 2×2 subsets, is often planned before conducting a factorial experiment. However, in a 4×3 ANOVA there are 18 possible 2×2 subsets, and it is not likely that you would plan to test all of these or that all would be interesting. In fact, it would be reasonable to plan and test six of these interactions at the .05 level because that is the number of degrees of freedom for the interaction in a 4×3 design. The most defensible scheme for testing six interaction contrasts, each at the .05 level, is to plan a set of six mutually orthogonal 2×2 contrasts—an example of which is shown in Figure 14.10.

Quite often when you test a main effect for significance, you do not merely expect differences among its levels; you have more specific predictions concerning the order and relative spacing of the group means. For instance, in the sleep experiment you would not only expect a main effect of type of sleep deprivation, but more specifically, you might expect the control group to have the highest scores, the "total" group to have the low-

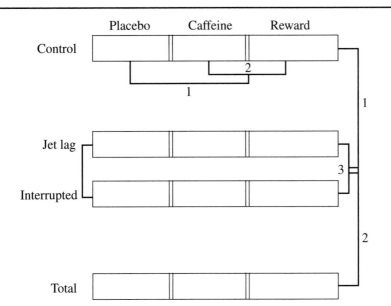

Figure 14.10

A Set of Orthogonal
Contrasts to Decompose
the Interaction in the
Sleep Deprivation
Experiment

Each of the three sleep-level comparisons is crossed
with each of the two stimulation-level comparisons to
create six orthogonal interaction contrasts.

est scores, and the other two groups to be somewhere between. You could, for instance, plan a complex comparison in which the average of the two partial-sleep groups is tested against the control group. In this way, you could analyze your main effects with complex rather than pairwise comparisons. However, if the comparison just described is used as one two-level factor, and the other factor is placebo versus caffeine (dropping the reward group), you have created a 2 × 2 complex interaction contrast. If you do *not* drop the reward group, you have planned a 2 × 3 partial interaction, as described in Section B.

The possibilities for such complex contrasts are quite numerous, but each planned contrast should be predictable from a theory or from previous results. Given that these interaction contrasts are planned, it is legitimate to test each at the .05 level, as long as the number of planned tests does not exceed df_{inter} for the full design ($r − 1$ times $c − 1$). For some clever ways to plan a single-df interaction contrast that encompasses an entire 3 × 3 (or larger) ANOVA design, see Abelson and Prentice (1997). These authors emphasize that if you are going to reduce your entire table of cell means to a single linear contrast, it is important to also test the residual SS: $MS_{residual} = (SS_{bet\text{-}cells} − SS_{contrast})/(df_{bet\text{-}cells} − 1)$; $F_{residual} = MS_{residual}/MS_W$, in order to determine whether there is a significant amount of between-cell variation that remains to be explained.

Interaction of Trend Components

When one of the factors in a two-way ANOVA has quantitative levels, it is likely that the experimenter would be interested in testing trend components not only for the main effect of the quantitative factor, but for the interaction of the two factors as well. For example, suppose that two methods for weight-lifting training are being compared (traditional and new), each at four levels of practice time (4, 8, 12, or 16 hours per week). An equal number of participants would be randomly assigned to each of the eight cells in the design. The main effect of weekly practice time would be of little concern, as it is averaged over the two different methods that the experimenter wants to compare. The main effect of weight-lifting method would be of interest, but the study has been designed to obtain a more detailed comparison. The interaction has three degrees of freedom, and these can be separated into three orthogonal contrasts: the interactions of method with the linear, quadratic, and cubic trend components of the practice-time factor. The calculation of these interaction contrasts in a balanced two-way ANOVA is straightforward. For example, testing the interaction of method with linear trend implies testing the difference in the linear trends of the two methods. Calculate L separately for each weight-lifting method, using the appropriate coefficients from Table A.12, and subtract the two Ls to find L_{diff}. L_{diff} is then tested like any other linear contrast, using Formula 13.13, where n is the size of each cell, and Σc_i^2 includes two sets of linear trend coefficients. In a similar fashion, one can test the interactions for the quadratic and cubic trends. Two-way ANOVAs in which both factors are quantitative can get quite complex, but they are too rare to merit space in this text.

The Two-Way ANOVA for Unbalanced Designs

For the two-way design, Formula 14.2 is based on the assumption that the between-cells SS can be divided into three components (i.e., SS_R, SS_C, and SS_{inter}) that do not overlap—that is, that these three components are mutually orthogonal and therefore add up to $SS_{between-cells}$. That is why we could find SS_{inter} by subtraction. The between-cell SS components of the two-way ANOVA will only be orthogonal, however, if the cell sizes are equal. When the design is not balanced, the SS components are not mutually independent, they do not add up to $SS_{between-cells}$, and SS_{inter} cannot be found by subtraction. The problem with an unbalanced design is that if you perform an ordinary two-way ANOVA on the data, one main effect can actually influence the size of the other main effect, producing a misleading result. Consider the following example.

All of your subjects have dyslexia and you are studying a treatment for dyslexia that involves improving communication between the left and right sides of the brain. Because this treatment may work better for left-handers than right-handers, you decide to create a two-way design crossing condition (dyslexia treatment vs. control treatment) with handedness (left vs. right). The dependent variable consists of subjects' scores on a speeded reading test after 3 months of treatment. Suppose that after running 30 left- and 30 right-handers in the treatment, you do not have many left-handed subjects remaining in your subject pool (which is not so surprising given the relative scarcity of left-handers), so, although you can run 30 right-handers in the control condition, you are forced to run only 10 left-handers in that condition. Finally, let us suppose that there is a main effect of handedness—left-handers on average are better than right-handers on this task—but that your dyslexia treatment has absolutely no effect at all. The cell means could come out in the pattern shown in Table 14.10.

From the cell means you can see that left-handers have a 10-point advantage, but for each subgroup of subjects, the treatment and control means are exactly the same. The problem appears in the row means. Because there are equal numbers of left- and right-handers for the treatment, the average of the two cells is just the average of 60 and 50, which is 55. However, in the "control" row there are only 10 left-handers averaging 60,

	Left-Handed	Right-Handed	Row Means	
Treatment	60	50	55	
Control	60	50	52.5	
Column Means	60	50		

Table 14.10

who must be combined with the 30 right-handers who average 50. The sum of 10×60 is added to 30×50, which equals $600 + 1{,}500$, which is then divided by the total number of control subjects: $2{,}100/40 = 52.5$. This is equivalent to a weighted average of the two cells. This method of dealing with unequal group sizes is called the *analysis of weighted means*, and it is the method we used to analyze the one-way ANOVA with groups of different sizes.

Because the left-handers are better on the task, and the control group is deficient in left-handers, it appears that the treatment is producing some small increase in reading speed (which could be significant with a small enough error term), even though it is not. Analyzing the row means in Table 14.10 as part of an ordinary ANOVA (i.e., weighted means approach) is clearly not an acceptable solution. However, there is more than one acceptable way to analyze an unbalanced design. I will describe only the simplest and most logical of these methods. Major statistical packages make it easy to analyze unbalanced designs in several ways; some of these will be described in Section A of Chapter 18.

To some extent, the procedure chosen to analyze an unbalanced design should depend on why the cell sizes are different. Probably the most common cause of unbalanced designs is the unexpected loss of a few subjects from an experiment that had been planned with equal cell sizes. There are many reasons for eliminating a subject's data: The subject may not have completed the experiment, or you may find out later that the subject misunderstood the instructions or knew too much about the purposes of the experiment. Sometimes subjects can be replaced, but there are situations in which this is not feasible. For instance, if a study is tracking the progress of subjects in psychotherapy over a 6-month period and a subject drops out of the study after 5 months, it may not be practical to replace the subject.

Before computers made the analysis of unbalanced designs rather easy, researchers would often replace the data from a missing subject with average values from that subject's cell. However, as long as the loss of subjects is random and unrelated to the experimental conditions, there should be no difficulty in analyzing the results from an unbalanced design and drawing valid conclusions. On the other hand, if the loss of subjects is systematically related to the experimental conditions, you are dealing with a confounding variable that may make it impossible to draw valid conclusions, regardless of the statistical procedure you use.

As a simple example of the nonrandom loss of subjects, consider an experiment comparing a psychotherapy group with a control group (involving some mock therapy) over a 6-month period. Suppose that the patients with the most severe psychological problems become frustrated with the control group and drop out, whereas similar patients in the real therapy group experience slight progress and stay in the experiment. At the end of 6 months, it may seem that the control group has even less pathology than the therapy group because those with the highest pathology scores have left the control group. This kind of problem can affect any of the experimental designs covered in this text. Although truly random samples are rare in psychological experiments, the random assignment of subjects to conditions is critical. If subjects are lost for reasons that are not random but are related to the experimental conditions, the random assignment is compromised, and no method of statistical analysis can guarantee valid conclusions.

For unbalanced designs in which the cell sizes are not very different, and the differences are due mainly to chance, the ordinary two-way ANOVA procedure can be performed with only minor modifications. First, MS_W is calculated as it would be for a one-way ANOVA with different-sized groups; a weighted average of the cell variances is calculated. (This is the same as adding the within-cell SS from all of the cells to get SS_W and then dividing by df_W.) Second, when using the computational shortcut introduced in Section B to find the $SS_{between-cells}$ components, the value for the total N (N_T) must be modified. You need to calculate the harmonic mean of all the cell sizes using Formula 12.24 (k in that formula becomes the number of cells, rc) and then multiply that value by the number of cells. This process will reduce N_T only slightly if the cell sizes are similar but can produce a very noticeable reduction when the cell sizes are widely discrepant (in which case this procedure is not recommended). Third, each row and column mean is found by taking the simple average of the cell means in that row or column, which may differ from the average of all the individual scores in that row or column.

The method just described for dealing with unbalanced designs is an extension of the *analysis of unweighted means*, introduced in Chapter 12, Section C. When applied to the two-way ANOVA, you don't take a weighted average of cell means to find the marginal means (as I did in Table 14.10); instead, you treat the cell means equally (i.e., unweighted) even though they may be of different sizes. For a step-by-step computational example of this procedure, see Cohen (2002). An alternative, but related, method for dealing with unbalanced factorial designs is known simply as the "regression approach" (or the *analysis of unique sources*). The regression approach (labeled "Type III SS" in SPSS) is the default method for factorial ANOVA in several major statistical packages; it yields identical results to the analysis of unweighted means when applaied to an unbalanced 2 × 2 design, but is preferred for larger designs. The regression approach will be discussed in detail in Chapter 18, along with alternative methods that may be preferred when the lack of balance in the design is not accidental (e.g., when the cell sizes are proportional to the different-sized subpopulations they represent).

The Concepts of the Three-Way Factorial ANOVA

At the end of Section B, I reported the results of a published study, which was based on a 2 × 2 ANOVA. In that study one factor contrasted subjects who had an alcohol-dependent parent with those who did not. I'll call this the *alcohol* factor and its two levels, *at risk* (of codependency) and *control*. The other factor (the *experimenter* factor) also had two levels; in one level subjects were told that the experimenter was an *exploitive* person, and in the other level the experimenter was described as a *nurturing* person. All of the subjects were women. If we imagine that the experiment was replicated using equal-sized groups of men and women, the original two-way design becomes a three-way design with gender as the third factor. For simplicity, we will assume that all eight cells of the 2 × 2 × 2 design contain the same number of subjects. The cell means for a three-factor experiment are often displayed in published articles in the form of a table, such as Table 14.11.

Graphing Three Factors

The easiest way to see the effects of this experiment is to graph the cell means. However, putting all of the cell means on a single graph would not

		Nurturing	Exploitive	Row Mean	**Table 14.11**
Control:	Men	40	28	34	
	Women	30	22	26	
	Mean	35	25	30	
At risk:	Men	36	48	42	
	Women	40	88	64	
	Mean	38	68	53	
	Column mean	36.5	46.5	41.5	

be an easy way to look at the three-way interaction. It is often clearer to use two graphs side by side, as shown in Figure 14.11. With a two-way design one has to decide which factor is to be placed along the horizontal axis, leaving the other to be represented by different lines on the graph. With a three-way design one chooses both the factor to be placed along the horizontal axis *and* the factor to be represented by different lines, leaving the third factor to be represented by different graphs. These decisions result in six different ways that the cell means of a three-way design can be presented.

Let us look again at Figure 14.11. The graph for the women shows the two-way interaction you would expect from the study on which it is based. The graph for the men shows the same kind of interaction, but to a considerably lesser extent (the lines for the men are closer to being parallel). This difference in amount of two-way interaction for men and women constitutes a three-way interaction. If the two graphs had looked exactly the same, the *F* ratio for the three-way interaction would have been zero. However, that is not a necessary condition. A main effect of gender could raise the lines on one graph relative to the other without contributing to a three-way interaction. Moreover, an interaction of gender with the experimenter factor could rotate the lines on one graph relative to the other, again without contributing to the three-way interaction. As long as the difference in slopes (i.e., the amount of two-way interaction) is the same in both graphs, the three-way interaction will be zero.

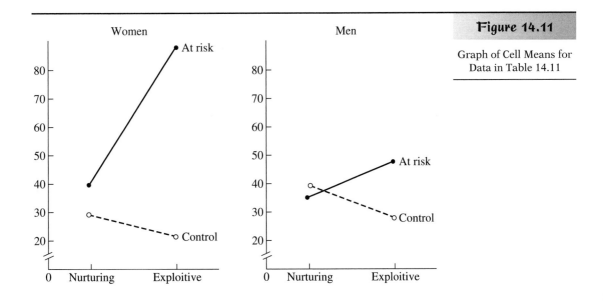

Figure 14.11

Graph of Cell Means for Data in Table 14.11

Simple Interaction Effects

A three-way interaction can be defined in terms of simple effects in a way that is analogous to the definition of a two-way interaction. A two-way interaction is a difference in the simple main effects of one of the variables as you change levels of the other variable (if you look at just the graph of the women in Figure 14.11, each line is a simple main effect). In Figure 14.11 each of the two graphs can be considered a simple effect of the three-way design—more specifically, a simple interaction effect. Each graph depicts the two-way interaction of alcohol and experimenter at one level of the gender factor. The three-way interaction can be defined as the difference between these two simple interaction effects. If the simple interaction effects differ significantly, the three-way interaction will be significant. Of course, it doesn't matter which of the three variables is chosen as the one whose different levels are represented as different graphs—if the three-way interaction is statistically significant, there will be significant differences in the simple interaction effects in each case.

Varieties of Three-Way Interactions

Just as there are many patterns of cell means that lead to two-way interactions (e.g., one line is flat while the other goes up or down, the two lines go in opposite directions, or the lines go in the same direction but with different slopes), there are even more distinct patterns in a three-way design. Perhaps the simplest is when all of the means are about the same, except for one, which is distinctly different. For instance, in our present example the results might have shown no effect for the men (all cell means about 40), no difference for the control women (both means about 40), and a mean of 40 for at-risk women exposed to the nice experimenter. Then, if the mean for at-risk women with the exploitive experimenter were well above 40, there would be a strong three-way interaction. This is a situation in which all three variables must be at the right level simultaneously to see the effect—in this variation of our example the subject must be female *and* raised by an alcohol-dependent parent *and* exposed to the exploitive experimenter to attain a high score. Not only might the three-way interaction be significant, but one cell mean might be significantly different from all of the other cell means, making an even stronger case that all three variables must be combined properly to see any effect (if you were sure that this pattern were going to occur, you could test a contrast comparing the average of seven cell means to the one you expect to be different and not bother with the ANOVA at all).

More often the results are not so clear-cut, but there is one cell mean that is considerably higher than the others (as in Figure 14.11). This kind of pattern is analogous to the ordinal interaction in the two-way case and tends to cause all of the effects to be significant. On the other hand, a three-way interaction could arise because the two-way interaction reverses its pattern when changing levels of the third variable (e.g., imagine that in Figure 14.11 the labels of the two lines were reversed for the graph of men but not for the women). This is analogous to the disordinal interaction in the two-way case. Or, the two-way interaction could be strong at one level of the third variable and much weaker (or nonexistent) at another level. Of course, there are many other possible variations. And consider how much more complicated the three-way interaction can get when each factor has more than two levels.

Fortunately, three-way (between-subjects) ANOVAs with many levels for each factor are not common. One reason is a practical one: the number of subjects required. Even a design as simple as a $2 \times 3 \times 4$ has 24 cells (to find the number of cells, you just multiply the numbers of levels). If you want to have at least 5 subjects per cell, 120 subjects are required. This is not an impractical study, but you can see how quickly the addition of more levels would result in a required sample size that could be prohibitive.

Main Effects

In addition to the three-way interaction there are three main effects to look at, one for each factor. To look at the gender main effect, for instance, just take the average of the scores for all of the men and compare it to the average of all of the women. If you have the cell means handy and the design is balanced, you can average all of the cell means involving men and then all of the cell means involving women. In Table 14.11, you can average the four cell means for the men (40, 28, 36, 48) to get 38 (alternatively, you could use the row means in the extreme right column and average 34 and 42 to get the same result). The average for the women (30, 22, 40, 88) is 45. The means for the other main effects have already been included in Table 14.11. Looking at the bottom row you can see that the mean for the nurturing experimenter is 36.5 as compared to 46.5 for the exploitive one. In the extreme right column you'll find that the mean for the control subjects is 30, as compared to 53 for the at-risk subjects.

Two-Way Interactions in Three-Way ANOVAs

Further complicating the three-way ANOVA is that, in addition to the three-way interaction and the three main effects, there are three two-way interactions to consider. In terms of our example there are the gender by experimenter, gender by alcohol, and experimenter by alcohol interactions. We will look at the last of these first. Before graphing a two-way interaction in a three-factor design, you have to "collapse" (i.e., average) your scores over the variable that is not involved in the two-way interaction. To graph the alcohol by experimenter ($A \times B$) interaction you need to average the men with the women for each combination of alcohol and experimenter levels (i.e., each cell of the $A \times B$ matrix). These means have also been included in Table 14.11.

The graph of these four means is shown in Figure 14.12. If you compare this overall two-way interaction with the two-way interactions for the men and women separately (see Figure 14.11), you will see that the overall interaction looks like an average of the two separate interactions; the amount of

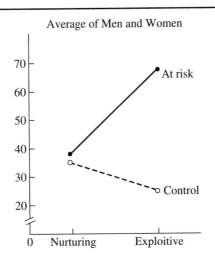

Average of Men and Women

Figure 14.12

Graph of Cell Means in Table 14.11 after Averaging Across Gender

Figure 14.13a

Graph of Cell Means in Table 14.11 Using the "Alcohol" Factor to Distinguish the Panels

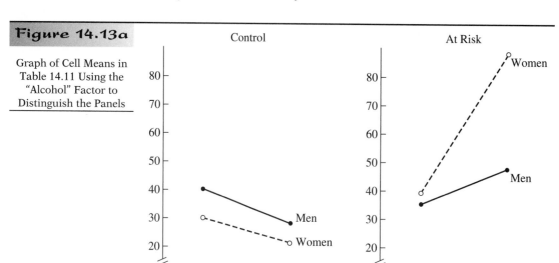

Figure 14.13b

Graph of Cell Means in Table 14.11 after Averaging Across the "Alcohol" Factor

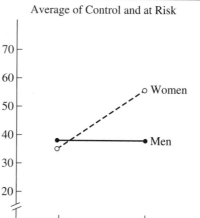

interaction seen in Figure 14.12 is midway between the amount of interaction for the men and that amount for the women. Does it make sense to average the interactions for the two genders into one overall interaction? It does if they are not very different. How different is too different? The size of the three-way interaction tells us how different these two two-way interactions are. A statistically significant three-way interaction suggests that we should be cautious in interpreting any of the two-way interactions. Just as a significant two-way interaction tells us to look carefully at, and possible test, the simple main effects (rather than the overall main effects), a significant three-way interaction suggests that we focus on the simple interaction effects—the two-way interactions at each level of the third variable (which of the three independent variables is treated as the third variable is a matter of convenience). Even if the three-way interaction falls somewhat short of significance, I would recommend caution in interpreting the two-way interactions and the main effects, as well, whenever the simple interaction effects look completely different and, perhaps, show opposite patterns.

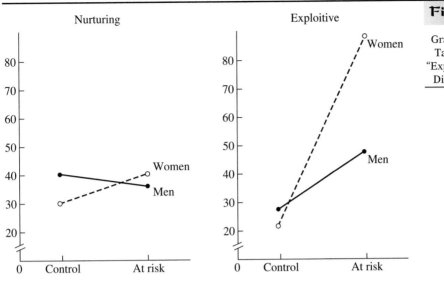

Figure 14.14a

Graph of Cell Means in Table 14.11 Using the "Experimenter" Factor to Distinguish the Panels

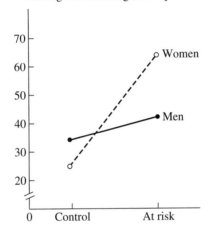

Figure 14.14b

Graph of Cell Means in Table 14.11 after Averaging Across the "Experimenter" Factor

So far I have been focusing on the two-way interaction of alcohol and experimenter in our example, but this choice is somewhat arbitrary. The two genders are populations that we are likely to have theories about, so it is often meaningful to compare them. However, I can just as easily graph the three-way interaction using alcohol as the third factor, as I have done in Figure 14.13a. To graph the overall two-way interaction of gender and experimenter, you can go back to Table 14.11 and average across the alcohol factor. For instance, the mean for men in the nurturing condition is found by averaging the mean for control group men in the nurturing condition (40) with the mean for at-risk men in the nurturing condition (36), which is 38. The overall two-way interaction of gender and experimenter is shown in Figure 14.13b. Note that once again the two-way interaction is a compromise. (Actually, the two two-way interactions are not as different as they look; in both cases the slope of the line for the women is more positive—or at least less negative). For completeness, I have graphed the three-way interaction using experimenter as the third variable, and the overall two-way interaction of gender and alcohol in Figures 14.14a and 14.14b, respectively.

Figure 14.15a

Rearranging the Cell
Means of Table 14.11 to
Depict a Disordinal
3-Way Interaction

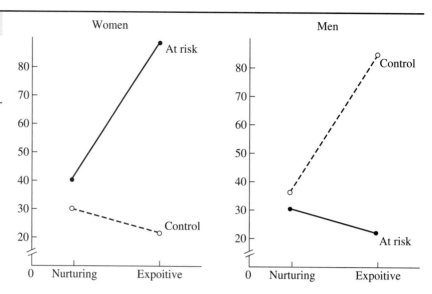

An Example of a Disordinal Three-Way Interaction

In the three-factor example I have been describing, it looks like all three
main effects and all three two-way interactions, as well as the three-way
interaction, could easily be statistically significant. However, it is important
to note that in a balanced design all seven of these effects are independent;
the seven F ratios do share the same error term (i.e., denominator), but the
sizes of the numerators are entirely independent. It is quite possible to have
a large three-way interaction while all of the other effects are quite small. By
changing the means only for the men in our example, I will illustrate a large,
disordinal interaction that obliterates two of the two-way interactions and
two of the main effects. You can see in Figure 14.15a that this new three-way
interaction is caused by a reversal of the alcohol by experimenter interaction
from one gender to the other. In Figure 14.15b, you can see that the overall
interaction of alcohol by gender is now zero (the lines are parallel); the gender by experimenter interaction is also zero (not shown). On the other hand,
the large gender by alcohol interaction very nearly obliterates the main

Figure 14.15b

Regraphing Figure
14.15a after Averaging
Across Gender

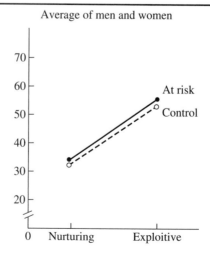

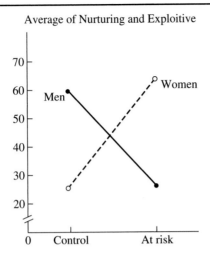

Average of Nurturing and Exploitive

Figure 14.15c

Regraphing Figure 14.15a
after Averaging Across
the "Experimenter"
Factor

effects of both gender and alcohol (see Figure 14.15c). The main effect of experimenter is, however, large, as can be seen in Figure 14.15b.

An Example in which the Three-Way Interaction Equals Zero

Finally, I will change the means for the men once more to create an example in which the three-way interaction is zero, even though the graphs for the two genders do not look the same. In Figure 14.16, I created the means for the men by starting out with the women's means and subtracting 10 from each (this creates a main effect of gender); then I added 30 only to the men's means that involved the nurturing condition. The latter change creates a two-way interaction between experimenter and gender, but because it affects both the men/nurturing means equally, it does not produce any three-way interaction. One way to see that the three-way interaction is zero in Figure 14.16 is to subtract the slopes of the two lines for each gender. For the women the slope of the at-risk line is positive: $88 - 40 = 48$. The slope of the control line is negative: $22 - 30 = -8$. The difference of the slopes is

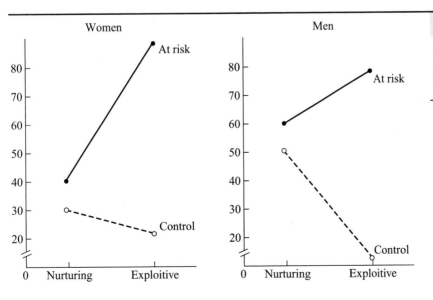

Figure 14.16

Rearranging the Cell
Means of Table 14.11 to
Depict a Zero Amount of
Three-Way Interaction

$48 - (-8) = 56$. If we do the same for the men, we get slopes of 18 and -38, whose difference is also 56. You may recall that a 2×2 interaction has only one df, and can be summarized by a single number, L, that forms the basis of a simple linear contrast. The same is true for a $2 \times 2 \times 2$ interaction or any higher-order interaction in which all of the factors have two levels. Of course, quantifying a three-way interaction gets considerably more complicated when the factors have more than two levels, but it is safe to say that if the two (or more) graphs are exactly the same, there will be no three-way interaction (they will continue to be identical, even if a different factor is chosen to distinguish the graphs). Bear in mind, however, that even if the graphs do not *look* the same, the three-way interaction will be zero if the amount of two-way interaction is the same for every graph.

Calculating the Three-Way ANOVA

Calculating a three-way independent-groups ANOVA is a simple extension of the method for a two-way independent-groups ANOVA, using the same basic formulas. In particular, there is really nothing new about calculating MS_W (the error term for all the F ratios); it is just the ordinary average of the cell variances when the design is balanced. (It is hard to imagine that anyone would calculate an unbalanced three-way ANOVA with a calculator rather than a computer, so I will not consider that possibility. The analysis of unbalanced designs is described in general in Chapter 18, Section A.) Rather than give you all of the cell standard deviations or variances for the example in Table 14.11, I'll just tell you that SS_W equals 6,400; later I'll divide this by df_W to obtain MS_W. (If you had all of the raw scores, you would also have the option of obtaining SS_W by calculating SS_{total} and subtracting $SS_{between-cells}$ as defined in the following.)

Main Effects

The calculation of the main effects is also the same as in the two-way ANOVA; the SS for a main effect is just the biased variance of the relevant group means multiplied by the total N. Let us say that each of the eight cells in our example contains five subjects, so N_T equals 40. Then the SS for the experimenter factor (SS_{exper}) is 40 times the biased variance of 36.5 and 46.5 (the nurturing and exploitive means from Table 14.11), which equals 40(25) = 1000 (the shortcut for finding the biased variance of two numbers is to take the square of the difference between them and then divide by 4). Similarly, $SS_{alcohol} = 40(132.25) = 5290$, and $SS_{gender} = 40(12.25) = 490$.

The Two-Way Interactions

When calculating the two-way ANOVA, the SS for the two-way interaction is found by subtraction; it is the amount of the $SS_{between-cells}$ that is left after subtracting the SSs for the main effects. Similarly, the three-way interaction SS is the amount left over after subtracting the SSs for the main effects and the SSs for all the two-way interactions from the overall $SS_{between-cells}$. However, finding the SSs for the two-way interactions in a three-way design gets a little tricky. In addition to the overall $SS_{between-cells}$, we must also calculate some intermediate "two-way" $SS_{between}$ terms.

To keep track of these I will have to introduce some new subscripts. The overall $SS_{between-cells}$ is based on the variance of all the cell means, so no factors are "collapsed," or averaged over. Representing gender as G, alcohol as A, and experimenter as E, the overall $SS_{between-cells}$ will be written as SS_{GAE}. We will also need to calculate an $SS_{between}$ after averaging over gender. This is based on the four means (included in Table 14.11) I used to graph the alcohol by experimenter interaction and will be represented by SS_{AE}. Because the design is balanced, you can take the simple average of the appropriate male cell mean and female cell mean in each case. Note that SS_{AE} is not the SS for the alcohol by experimenter interaction because it also includes the main effects of those two factors. In similar fashion, we need to find SS_{GA} from the means you get after averaging over the experimenter factor and SS_{GE} by averaging over the alcohol factor. Once we have calculated these four $SS_{between}$ terms, all of the SSs we need for the three-way ANOVA can be found by subtraction.

Let's begin with the calculation of SS_{GAE}; the biased variance of the eight cell means is 366.75, so $SS_{GAE} = 40(366.75) = 14{,}670$. The means for SS_{AE} are 35, 25, 38, 68, and their biased variance equals 257.75, so $SS_{AE} = 10{,}290$. SS_{GA} is based on the following means: 34, 26, 42, 64, so $SS_{GA} = 40(200.75) = 8{,}030$. Finally, SS_{GE}, based on means of 38, 38, 35, 55, equals 2,490.

Next we find the SSs for each two-way interaction:

$$SS_{A \times E} = SS_{AE} - SS_{alcohol} - SS_{exper} = 10{,}290 - 5{,}290 - 1{,}000 = 4{,}000$$
$$SS_{G \times A} = SS_{GA} - SS_{gender} - SS_{alcohol} = 8{,}030 - 490 - 5{,}290 = 2{,}250$$
$$SS_{G \times E} = SS_{GE} - SS_{gender} - SS_{exper} = 2{,}490 - 490 - 1{,}000 = 1{,}000$$

Finally, the SS for the three-way interaction ($SS_{G \times A \times E}$) equals:

$$SS_{GAE} - SS_{A \times E} - SS_{G \times A} - SS_{G \times E} - SS_{gender} - SS_{alcohol} - SS_{exper}$$
$$= 14{,}670 - 4{,}000 - 2{,}250 - 1{,}000 - 490 - 5{,}290 - 1{,}000 = 640$$

Formulas for the General Case

It is traditional to assign the letters A, B, and, C to the three independent variables in the general case; variables D, E, and so forth, can then be added to represent a four-way, five-way, or higher ANOVA. I'll assume that the following components have already been calculated using Formula 14.3 applied to the appropriate means: SS_A, SS_B, SS_C, SS_{AB}, SS_{AC}, SS_{BC}, SS_{ABC}. In addition, I'll assume that SS_W has also been calculated, either by averaging the cell variances and multiplying by df_W or by subtracting SS_{ABC} from SS_{total}. The remaining SS components are found by Formula 14.13:

Formula 14.13

a. $SS_{A \times B} = SS_{AB} - SS_A - SS_B$
b. $SS_{A \times C} = SS_{AC} - SS_A - SS_C$
c. $SS_{B \times C} = SS_{BC} - SS_B - SS_C$
d. $SS_{A \times B \times C} = SS_{ABC} - SS_{A \times B} - SS_{B \times C} - SS_{A \times C} - SS_A - SS_B - SS_C$

At the end of the analysis, SS_{total} (whether or not it has been calculated separately) has been divided into eight components: SS_A, SS_B, SS_C, the four interactions listed in Formula 14.13, and SS_W. Each of these is divided by its corresponding df to form a variance estimate, MS. Using a to represent the number of levels of the A factor, b for the B factor, c for the C factor, and n for the number of subjects in each cell, the formulas for the df components are as follows:

Formula 14.14

a. $df_A = a - 1$
b. $df_B = b - 1$
c. $df_C = c - 1$
d. $df_{A \times B} = (a - 1)(b - 1)$
e. $df_{A \times C} = (a - 1)(c - 1)$
f. $df_{B \times C} = (b - 1)(c - 1)$
g. $df_{A \times B \times C} = (a - 1)(b - 1)(c - 1)$
h. $df_W = abc(n - 1)$

Completing the Analysis for the Example

Because each factor in the example has only two levels, all of the numerator df's are equal to 1, which means that all of the MS terms are equal to their corresponding SS terms—except, of course, for the error term. The df for the error term (i.e., df_W) equals the number of cells (abc) times one less than the number of subjects per cell (this gives the same value as N_T minus the number of cells); in this case $df_W = 8(4) = 32$. $MS_W = SS_W/df_W$; therefore, $MS_W = 6400/32 = 200$. (*Reminder:* I gave the value of SS_W to you to reduce the amount of calculation.)

Now we can complete the three-way ANOVA by calculating all of the possible F ratios and testing each for statistical significance:

$$F_{gender} = \frac{MS_{gender}}{MS_W} = \frac{490}{200} = 2.45$$

$$F_{alcohol} = \frac{MS_{alcohol}}{MS_W} = \frac{5,290}{200} = 26.45$$

$$F_{exper} = \frac{MS_{exper}}{MS_W} = \frac{1,000}{200} = 5$$

$$F_{A \times E} = \frac{MS_{A \times E}}{MS_W} = \frac{4,000}{200} = 20$$

$$F_{G \times A} = \frac{MS_{G \times A}}{MS_W} = \frac{2,250}{200} = 11.35$$

$$F_{G \times E} = \frac{MS_{G \times E}}{MS_W} = \frac{1000}{200} = 5$$

$$F_{G \times A \times E} = \frac{MS_{G \times A \times E}}{MS_W} = \frac{640}{200} = 3.2$$

Because the df happens to be 1 for all of the numerator terms, the critical F for all seven tests is $F_{.05}(1,32)$, which is equal (approximately) to 4.15. Except for the main effect of gender, and the three-way interaction, all of the F ratios exceed the critical value (4.15) and are therefore significant at the .05 level.

Follow-Up Tests for the Three-Way ANOVA

Decisions concerning follow-up comparisons for a factorial ANOVA are made in a top-down fashion. First, one checks the highest-order interaction for significance; in a three-way ANOVA it is the three-way interaction. (Two-way interactions are the simplest possible interactions and are called *first-order* interactions; three-way interactions are known as *second-order* interactions, etc.) If the highest interaction is significant, the post hoc tests focus on the various simple effects or interaction contrasts, followed by appropriate cell-to-cell comparisons. In a three-way ANOVA in which the three-way interaction is not significant, as in the present example, attention turns to the three two-way interactions. Although all of the two-way interactions are significant in our example, the alcohol by experimenter interaction is the easiest to interpret because it replicates previous results.

It would be appropriate to follow up the significant alcohol by experimenter interaction with four t tests (e.g., one of the relevant t tests would determine whether at-risk subjects differ significantly from controls in the exploitive condition). Given the disordinal nature of the interaction (see

Figure 14.12), it is likely that the main effects would simply be ignored. A similar approach would be taken to the two other significant two-way interactions. Thus, all three main effects would be regarded with caution. Note that because all of the factors are dichotomous, there would be no follow-up tests to perform on significant main effects, even if none of the interactions were significant. With more than two levels for some or all of the factors, it becomes possible to test partial interactions, and significant main effects for factors not involved in significant interactions can be followed by pairwise or complex comparisons, as described for the two-way ANOVA.

$2 \times 2 \times 2$ Contrasts

If at least one of the three factors has more than two levels, a significant three-way interaction can be followed up with more specific $2 \times 2 \times 2$ contrasts, which are a logical extension of the 2×2 contrasts that may follow a multi-level two-way ANOVA. Just as the L for a 2×2 contrast is a difference of differences (e.g., the difference between the gender difference in a drug condition and the gender difference in a placebo condition), a $2 \times 2 \times 2$ contrast is a difference between the Ls for two 2×2 contrasts. For example, from Table 14.11, you can see that the difference in gender differences between the Nurturing and Exploitive conditions, just for the Control subjects, is $L_{Cont} = (40 - 30) - (28 - 22) = 10 - 6 = 4$, whereas the L for the At Risk subjects is $L_{Risk} = (36 - 40) - (48 - 88) = -4 - (-40) = -4 + 40 = 36$. Therefore, the L for the $2 \times 2 \times 2$ contrast, which happens to be the entire three-way interaction in this simple case, is equal to $L_{Risk} - L_{Cont} = 36 - 4 = 32$. (Note that I performed the final subtraction in the order that yields a positive result, but this is not critical, given that the final L is going to be squared anyway. However, you must be careful to be consistent in the order you use for subtraction in obtaining the two 2×2 Ls, and to keep track of the sign of each difference you calculate.) Bearing in mind that the coefficients for all 8 cells are either $+1$ or -1, we can now calculate the SS for the $2 \times 2 \times 2$ contrast with Formula 13.12:

$$SS_{Contrast} = \frac{nL^2}{\sum c_i^2} = \frac{5 \times 32^2}{8} = \frac{5120}{8} = 640$$

Notice that this is the same value we just calculated for $SS_{G \times A \times E}$ by subtraction. Another benefit of all the factors having only two levels is that the unweighted-means solution described earlier in this section would yield the same F ratios as the regression approach (the default procedure in SPSS, labeled Type III SS), even if the design were not balanced (this would also be true of a $2 \times 2 \times 2 \times 2$ design, and so on).

Types of Three-Way Designs

Let us look at some of the typical situations in which three-way designs arise. In one situation all three factors involve experimental manipulations. For instance, subjects perform a repetitive task in one of two conditions: They are told that their performance is being measured or that it is not. In each condition half of the subjects are told that performance on the task is related to intelligence, and the other half are told that it is not. Finally, within each of the four groups just described, half the subjects are treated respectfully and half are treated rudely. The work output of each subject can then be analyzed by a $2 \times 2 \times 2$ ANOVA.

Another possibility involves three grouping variables, each of which involves selecting subjects whose group is already determined. For instance, a group of people who exercise regularly and an equal-sized group of those who don't are divided into those high and those relatively low on self-esteem (by a median split). If there are equal numbers of men and women in each of the four cells, we have a balanced $2 \times 2 \times 2$ design. More commonly one or two of the variables involve experimental manipulations and two or one involve grouping variables. The example calculated earlier in this section involved two grouping variables (gender and having an alcohol-dependent parent or not) and one experimental variable (nurturing vs. exploitive experimenter). (For a description of partial effect sizes in various three-way designs, see Gillett [2003]).

Higher-Order ANOVA

This text will not cover factorial designs of higher order than the three-way ANOVA. Although higher-order ANOVAs can be difficult to interpret, no new principles are introduced. The four-way ANOVA produces 15 different F ratios to test: four main effects, 6 two-way interactions, 4 three-way interactions, and 1 four-way interaction. Testing each of these 15 effects at the .05 level raises serious concerns about the increased risk of Type I errors. Usually, all of the F ratios are not tested; specific hypotheses should guide the selection of particular effects to test. Of course, the potential for an inflated rate of Type I errors only increases as factors are added. In general, an N-way ANOVA produces $2^N - 1$ F ratios that can be tested for significance. For more complex varieties of three-way ANOVA, look for my chapter on complex ANOVA designs on my stats web page.

SUMMARY

1. Each interaction contrast involves one between-cell degree of freedom; therefore it falls under the category of a single-df comparison. These contrasts often correspond to specific research hypotheses, and several can be conducted as planned comparisons in one study. No adjustment of α_{pc} is necessary if the number of planned interaction contrasts is no more than df_{inter} from the overall design. A particularly powerful way to analyze a two-way ANOVA, in which a significant interaction is expected, is with a complete set of mutually orthogonal interaction contrasts. If one of the factors has quantitative levels, finding the interactions of trend components is often desirable.

2. When a two-way ANOVA is not balanced, the SS for interaction is not independent from SS_R and SS_C; the three components do *not* have to add up to the SS for between-cell variability (i.e., SS_{bet}). This complicates the analysis because there is more than one legitimate way to partition the total sum of squares. The simplest way to analyze an unbalanced design is to use the analysis of *unweighted means*. This method applies best when the cell sizes differ only slightly and by accident.

3. The analysis of unweighted means can be calculated with the same computational shortcut that I presented for the two-way ANOVA of a balanced design, except that
   a. MS_W is calculated as it is in a one-way ANOVA with groups of unequal size (i.e., take a weighted average of the cell variances).
   b. The total N is adjusted by finding the *harmonic mean* of all the cell sizes and multiplying by the number of cells.
   c. Each row and column mean is found by taking the simple average of the cell means in that row or column.

4. When analyzing an unbalanced design with statistical software, the preferred method is the "regression" method. This method is indicated in the same circumstances for which the analysis of unweighted means is appropriate and tends to give similar results. (The two methods yield

identical results when none of the ANOVA factors has more than two levels.) If the cell sizes differ considerably and systematically (e.g., the cell sizes reflect the relative sizes of subgroups in the population), more sophisticated methods are called for. If the loss of subjects from different cells is related to the experimental conditions, the validity of the experimental conclusions becomes questionable, regardless of the statistical procedure you use to analyze the data.

5. To display the cell means of a three-way factorial design, it is convenient to create two-way graphs for each level of the third variable and place these graphs side by side (you have to decide which of the three variables will distinguish the graphs and which of the two remaining variables will be placed along the X axis of each graph). Each two-way graph depicts a simple interaction effect; if the simple interaction effects are significantly different from each other, the three-way interaction will be significant.

6. Three-way interactions can occur in a variety of ways. The interaction of two of the factors can be strong at one level of the third factor and close to zero at a different level (or even stronger at a different level). The direction of the two-way interaction can reverse from one level of the third variable to another. Also, a three-way interaction can arise when all of the cell means are similar except for one.

7. The main effects of the three-way ANOVA are based on the means at each level of one of the factors, averaging across the other two. A two-way interaction is the average of the separate two-way interactions (simple interaction effects) at each level of the third factor. A two-way interaction is based on a two-way table of means created by averaging across the third factor. The error term for the three-way ANOVA, in a balanced design, is the simple average of all of the cell variances. There are seven F ratios that can be tested for significance: the three main effects, three two-way interactions, and the three-way interaction.

8. Averaging simple interaction effects together to create a two-way interaction is reasonable only if these effects do not differ significantly. If they do differ, follow-up tests usually focus on the simple interaction effects themselves or particular 2×2 interaction contrasts. If the three-way interaction is not significant, but a two-way interaction is, the significant two-way interaction is explored as in a two-way ANOVA—with simple main effects or interaction contrasts. Also, when the three-way interaction is not significant, any significant main effect can be followed up in the usual way if that variable is not involved in a significant two-way interaction.

EXERCISES

1. Describe a complete set of mutually orthogonal interaction contrasts which could be used to decompose the overall interaction in Exercise 14B7

2. This problem is based on the data in Exercise 14B10.
   a. For the main effect of the number of weekly therapy sessions, test the significance of the linear and quadratic trends.

   Do the higher-order trends account for a significant amount of variability?
   b. Test the significance of the interaction between the therapy factor and the linear trend of the sessions factor. Calculate the effect size (i.e., d_c) for this interaction.
   c. Repeat part b for the interaction with the quadratic trend of the sessions factor.

*3. This problem is based on the data in Exercise 14B9 (note that $MS_W = 350$).

   a. Test the significance of the interaction between the exercise/diet factor and the linear trend of the number-of-hours factor. Calculate the effect size (i.e, d_c) for this interaction.

   b. Calculate the SS for the quadratic and cubic trends for the Exercise Only group.

   c. Repeat part b for the Exercise + Diet group, testing each SS for significance at the .05 level.

4. A psychologist suspects that emotions affect people's eating habits, but that the eating habits of obese people are affected differently. Ten obese subjects are randomly divided into two equal-sized groups, one of which views a horror movie while the other watches a comedy. Ten normal-weight control subjects are similarly divided. The number of ounces of popcorn eaten by each subject during the movie is recorded. Unfortunately, one obese subject who had watched the horror movie admitted later that it was his favorite movie, and therefore that subject's data had to be eliminated (after it was too late to run additional subjects). The data for the remaining subjects are as follows:

	Horror	Comedy
Control	5	8
	2	6
	3	5
	7	11
	6	9

	Horror	Comedy
Obese	15	13
	19	12
	17	9
	16	16
		15

   a. Conduct a two-way ANOVA using the unweighted means solution.

   b. What can you say about the effects of emotion on eating, and how does this statement change depending on whether obese or nonobese people are considered?

*5. A cognitive psychologist has spent 3 months teaching an artificial language to 32 children, half of whom are bilingual and half of whom are monolingual. Half the children in each group are 5 years old and half are 8 years old. The following table shows the number of errors made by each child on one of the comprehension tests for the new language. Unfortunately, as you can see by the cell sizes, some of the children had to drop out so late in the study that they could not be replaced.

   a. Conduct a two-way ANOVA using the unweighted means solution; create a summary table of your results.

   b. Report your results in paragraph form, including an estimate of ordinary omega squared for each main effect.

	5-Year-Olds	8-Year-Olds
Monolingual	18	8
	19	10
	18	12
	16	9
	11	6
	17	12
		11
Bilingual	8	5
	11	8
	11	6
	10	10
	13	4
	18	5
		6
		4

*6. Imagine an experiment in which each subject is required to use his or her memories to create one emotion: either happiness, sadness, anger, or fear. Within each emotion group, half of the subjects participate in a relaxation exercise just before the emotion condition, and half do not. Finally, half the subjects in each emotion/relaxation condition are run in a dark, sound-proof chamber, and the other half are run in a normally lit room. The dependent variable is the subject's systolic blood pressure when the subject signals that the emotion is fully present. The design is balanced, with a total of 128 subjects. The results of the three-way ANOVA for this hypothetical experiment are as follows: $SS_{emotion} = 223.1$, $SS_{relax} = 64.4$, $SS_{dark} = 31.6$, $SS_{emo \times rel} = 167.3$, $SS_{emo \times dark} = 51.5$; $SS_{rel \times dark} = 127.3$, and

$SS_{emo \times rel \times dark} = 77.2$. The total sum of squares is 2,344.

   a. Calculate the seven F ratios, and test each for significance.

   b. Calculate partial eta squared for each of the three main effects. Are any of these effects at least moderate in size?

7. In this exercise there are 20 subjects in each cell of a $3 \times 3 \times 2$ design. The levels of the first factor (location) are urban, suburban, and rural. The levels of the second factor are no siblings, one or two siblings, and more than two siblings. The third factor has only two levels: presently married and not presently married. The dependent variable is the number of close friends that each subject reports having. The cell means are as follows:

	Urban	Suburban	Rural
No Siblings			
Married	1.9	3.1	2.0
Not Married	4.7	5.7	3.5
1 or 2 Siblings			
Married	2.3	3.0	3.3
Not Married	4.5	5.3	4.6
2 or more Siblings			
Married	3.2	4.5	2.9
Not Married	3.9	6.2	4.6

   a. Given that SS_W equals 1,094, complete the three-way ANOVA, and present your results in a summary table.

   b. Draw a graph of the means for Location × Number of Siblings (averaging across marital status). Describe the nature of the interaction.

   c. Using the means from part b, test the simple effect of number of siblings at each location.

*8. Seventy-two patients with agoraphobia are randomly assigned to one of four drug conditions: SSRI (e.g., Prozac), tricyclic antidepressant (e.g., Elavil), antianxiety (e.g., Xanax), or a placebo (offered as a new drug for agoraphobia). Within each drug condition, a third of the patients are randomly assigned to each of three types of psychotherapy: psychodynamic, cognitive/behavioral, and group. The subjects are assigned so that half the subjects in each drug/therapy group are also depressed, and half are not. After 6 months of treatment, the severity of agoraphobia is mea-sured for each subject (30 is the maximum possible phobia score); the cell means ($n = 3$) are as follows:

	SSRI	Tricyclic	Antianxiety	Placebo
Psychodynamic				
Not Depressed	10	11.5	19.0	22.0
Depressed	8.7	8.7	14.5	19.0
Cog/Behav				
Not Depressed	9.5	11.0	12.0	17.0
Depressed	10.3	14.0	10.0	16.5
Group				
Not Depressed	11.6	12.6	19.3	13.0
Depressed	9.7	12.0	17.0	11.0

   a. Given that SS_W equals 131, complete the three-way ANOVA, and present your results in a summary table.

   b. Draw a graph of the cell means, with separate panels for depressed and not depressed. Describe the nature of the therapy × drug interaction in each panel. Does there appear to be a three-way interaction? Explain.

   c. Given your results in part a, describe a set of follow-up tests that would be justifiable.

   d. Test the $2 \times 2 \times 2$ interaction contrast that results from deleting Group therapy and the SSRI and placebo conditions from the analysis.

9. An industrial psychologist is studying the relation between motivation and productivity. Subjects are told to perform as many repetitions of a given clerical task as they can in a 1-hour period. The dependent variable is the number of tasks correctly performed. Sixteen subjects participated in the experiment for credit toward a requirement of their introductory psychology course (credit group). Another 16 subjects were recruited from other classes and paid $10 for the hour (money group). All subjects performed a small set of similar clerical tasks as practice before the main study; in each group (credit or money) half the subjects (selected randomly) were told they had performed unusually well on the practice trials (positive feedback), and half were told they had performed poorly (negative feedback). Finally, within each of the four groups created by the manipulations just

described, half of the subjects (at random) were told that performing the tasks quickly and accurately was correlated with other important job skills (self motivation), whereas the other half were told that good performance would help the experiment (other motivation). The data appear in the following table:

| | CREDIT SUBJECTS | | PAID SUBJECTS | |
	Positive Feedback	Negative Feedback	Positive Feedback	Negative Feed back
Self	22	12	21	25
	25	15	17	23
	26	12	15	30
	30	10	21	26
Other	11	20	33	21
	18	23	29	22
	12	21	35	19
	14	26	29	17

a. Perform a three-way ANOVA on the data. Test all seven F ratios for significance, and present your results in a summary table.
b. Use graphs of the cell means to help you describe the pattern underlying each effect that was significant in part a.
c. Based on the results in part a, what post hoc tests would be justified?

10. How many F ratios can be tested in a 5-way ANOVA? How many of these represent two-way interactions (you can use Formula 13.1 to calculate this)?

KEY FORMULAS

The total sum of squares in terms of its components (when the design is balanced):

$$SS_{total} = SS_{row} + SS_{column} + SS_{interaction} + SS_{error}$$ **Formula 14.1**

SS for interaction in terms of previously computed SS components (if the design is balanced):

$$SS_{inter} = SS_{bet} - SS_R - SS_C$$ **Formula 14.2**

General formula for any between-cells SS component in a balanced design (can also be used for SS_{bet} in a one-way ANOVA):

$$SS_{bet} = N_T \sigma^2 \text{ (means)}$$ **Formula 14.3**

The df components of a two-way ANOVA (r = no. of rows, c = no. of columns, n = no. of subjects in each cell):

a.　$df_{row} = r - 1$ **Formula 14.4**

b.　$df_{col} = c - 1$

c.　$df_{inter} = (r - 1)(c - 1)$

d.　$df_W = rcn - rc$

e.　$df_{total} = rcn - 1$

The variance estimates of a two-way ANOVA:

a.　$MS_R = \dfrac{SS_R}{df_R}$ **Formula 14.5**

b.　$MS_C = \dfrac{SS_C}{df_C}$

c. $MS_{\text{inter}} = \dfrac{SS_{\text{inter}}}{df_{\text{inter}}}$

d. $MS_W = \dfrac{SS_W}{df_W}$

The three F ratios of a two-way ANOVA:

a. $\quad F_R = \dfrac{MS_R}{MS_W}$ $\qquad\qquad\qquad$ **Formula 14.6**

b. $\quad F_C = \dfrac{MS_C}{MS_W}$

c. $\quad F_{\text{inter}} = \dfrac{MS_{\text{inter}}}{MS_W}$

Formula for eta squared (proportion of variance accounted for in sample data) for one component of a two-way ANOVA, as though that component were tested in a one-way design:

$$\text{Partial } \eta^2 = \dfrac{SS_{\text{effect}}}{SS_{\text{effect}} + SS_W} \qquad\qquad \textbf{Formula 14.7}$$

Partial eta squared for an effect as a function of the F ratio for that effect, and the corresponding dfs:

$$\text{Partial } \eta^2 = \dfrac{df_{\text{effect}}F_{\text{effect}}}{df_{\text{effect}}F_{\text{effect}} + df_W} \qquad\qquad \textbf{Formula 14.8}$$

General formula for estimating omega squared (variance accounted for in the population) for one component of a two-way ANOVA:

$$\text{Est. } \omega^2 = \dfrac{SS_{\text{effect}} - df_{\text{effect}}MS_W}{SS_{\text{total}} + MS_W} \qquad\qquad \textbf{Formula 14.9}$$

An (almost) unbiased estimate of partial omega squared:

$$\text{Est. } \omega_P^2 = \dfrac{SS_{\text{effect}} - df_{\text{effect}}MS_W}{SS_{\text{effect}} + (N_T - df_{\text{effect}})MS_W} \qquad\qquad \textbf{Formula 14.10}$$

The critical F for Scheffé's test, for testing interaction comparisons:

$$F_S = df_{\text{inter}}\, F_{\text{crit}}(df_{\text{inter}}, df_W) \qquad\qquad \textbf{Formula 14.11}$$

The sample effect size (could be called g_c) for a single-df contrast, in terms of the F ratio for testing that contrast

$$d_c = \sqrt{\dfrac{2F_{\text{cont}}}{n}} \qquad\qquad \textbf{Formula 14.12}$$

The SS components for the interaction effects of the three-way ANOVA with independent groups.

a. $SS_{A \times B} = SS_{AB} - SS_A - SS_B$ $\qquad\qquad$ **Formula 14.13**

b. $SS_{A \times C} = SS_{AC} - SS_A - SS_C$

c. $SS_{B \times C} = SS_{BC} - SS_B - SS_C$

d. $SS_{A \times B \times C} = SS_{ABC} - SS_{A \times B} - SS_{B \times C} - SS_{A \times C} - SS_A - SS_B - SS_C$

The df components for the three-way ANOVA with independent groups:

a. $df_A = a - 1$ **Formula 14.14**

b. $df_B = b - 1$

c. $df_C = c - 1$

d. $df_{A \times B} = (a - 1)(b - 1)$

e. $df_{A \times C} = (a - 1)(c - 1)$

f. $df_{B \times C} = (b - 1)(c - 1)$

g. $df_{A \times B \times C} = (a - 1)(b - 1)(c - 1)$

h. $df_W = abc\,(n - 1)$

REPEATED MEASURES ANOVA

You will need to use the following from previous chapters:

Symbols
σ^2: Biased variance of a sample
c: Number of columns in the two-way ANOVA table
N_T: Total number of observations in an experiment

Formulas
Formula 3.11: SS in terms of squared scores and the mean
Formula 14.2: SS_{inter} (by subtraction)
Formula 14.3: SS_{bet} or one of its components

Concepts
Matched t test
SS components of the one-way ANOVA
SS components of the two-way ANOVA
Interaction of factors in a two-way ANOVA

Chapter

**CONCEPTUAL
FOUNDATION**

In describing various experimental designs for which the matched t test would be appropriate (see Chapter 11, Section A), I mentioned a hypothetical experiment in which the number of emotion-evoking words (e.g., "funeral," "vacation") a subject could recall was to be compared with the number of neutral words recalled. Suppose, however, that the researcher is concerned that the number of words evoking positive emotions that a subject recalls might be different from the number evoking negative emotions. The researcher would like to design the experiment with three types of words: positive, negative, and neutral. Of course, three matched t tests could be performed to test each possible pair of word types, but, as you should remember from Chapter 12, this approach does not control Type I errors, and the potential for increasing α_{EW} would get worse if more specific types of words were added (e.g., words evoking sadness or anger). What is needed is a statistical procedure that expands the matched t test to accommodate any number of conditions, just as the one-way ANOVA expands the t test for independent groups. There is such a procedure, and it is called the repeated measures (RM) ANOVA. This chapter will deal only with the *one-way repeated measures ANOVA* (i.e., one factor or independent variable). As with the matched t test, the analysis is the same whether the measures are repeated on the same subjects or the subjects are matched across conditions.

Suppose that the previously mentioned experiment was performed with eight subjects, each of whom was presented with all three types of words mixed together. Table 15.1 shows the number of words each subject correctly recalled from each category.

Calculation of an Independent-Groups ANOVA

To show the advantage of the RM ANOVA, I will first calculate the ordinary one-way ANOVA as though the scores were not connected across categories—that is, as though three separate groups of subjects were presented with only one type of word each. I begin by calculating SS_{total}. If you have the appropriate calculator, just enter all 24 numbers from Table 15.1, get the biased standard deviation, square it, and multiply by 24 (you are using

Table 15.1	Subject No.	Neutral	Positive	Negative
	1	18	16	22
	2	6	11	12
	3	14	17	20
	4	5	4	13
	5	12	13	15
	6	15	16	19
	7	6	7	12
	8	<u>20</u>	<u>20</u>	<u>23</u>
	Column Means	12	13	17

the principle behind Formula 14.3). If you can only obtain the unbiased standard deviation, square it and multiply it by 23 (i.e., n – 1). If you do not have a calculator at all, you can use Formula 3.11 to find SS_{total} as you would find any other SS. To use Formula 3.11, you must first square every individual score in Table 15.1 and add up the squared scores. The result, ΣX^2, is 5,398. The μ in Formula 3.11 should be replaced by the grand mean of all the scores in Table 15.1, which is the mean of the column means (because there are the same number of scores for each condition, as will generally be the case for an RM ANOVA). $\overline{X}_G = (12 + 13 + 17)/3 = 42/3 = 14$. Inserting these values into Formula 3.11, we obtain

$$SS_{total} = \sum X^2 - N\overline{X}_G^2 = 5,398 - 24(14^2) = 5,398 - 4,704 = 694$$

Next, I will calculate SS_{bet} using the all-purpose SS_{bet} formula (Formula 14.3), in which the means are the column means from Table 15.1:

$$SS_{bet} = N_T\sigma^2(means) = 24 \cdot \sigma^2(12, 13, 17) = 24(4.67) = 112$$

If you do not have a statistical calculator, you can use Formula 3.13A to find the biased variance of the column means. Now, I can find SS_W by subtraction: $SS_W = SS_{total} - SS_{bet} = 694 - 112 = 582$. (You could instead skip the calculation of SS_{total} and SS_W by calculating MS_W directly from the unbiased variances of the three columns in Table 15.1). The calculations of the MSs are as follows:

$$MS_{bet} = \frac{SS_{bet}}{df_{bet}} = \frac{112}{2} = 56$$

$$MS_W = \frac{SS_W}{df_W} = \frac{582}{21} = 27.7$$

Finally, the F ratio equals $MS_{bet}/MS_W = 56/27.7 = 2.02$.

The critical F for 2, 21 df ($\alpha = .05$) is 3.47, so the calculated F is not even near statistical significance. The null hypothesis that the population means for the three types of words are all equal cannot be rejected. This should not be surprising considering how small the groups are and how small the differences among groups are compared to the variability within each group.

The One-Way RM ANOVA
as a Two-Way Independent ANOVA

When you realize that the scores in each row of Table 15.1 are all from the same subject, you can notice a kind of consistency. For instance, the score for negative words is the highest of the three conditions for all of the subjects. In the two-group case (Chapter 11) we were able to take advantage of such consistency by focusing on the difference scores, thus converting a two-group problem to a design involving only one group of scores. There is also a simple way to take advantage of this consistency within a one-way RM ANOVA, but in this case we take a one-factor (RM) design and treat it as though it were a two-way (independent-groups) ANOVA. The levels of the second factor are the different subjects. Table 15.2 is a modification of Table 15.1 that highlights the use of the (eight) different subjects as levels of a second factor. Notice that the table looks like it corresponds to a two-way completely crossed factorial design with one measure per cell (so it is also balanced).

Subject No.	Neutral	Positive	Negative	Row Means
1	18	16	22	18.67
2	6	11	12	9.67
3	14	17	20	17
4	5	4	13	7.33
5	12	13	15	13.33
6	15	16	19	16.67
7	6	7	12	8.33
8	20	20	23	21
Column Means	12	13	17	14

Table 15.2

After studying the main example of Section A of Chapter 14, the idea of using subjects as different levels of a factor should not seem strange. It made sense in that example to separate men and women as two levels of a second factor because the two genders differed consistently on the dependent variable. When each subject receives the same set of related treatments, individual differences also tend to have a type of consistency that can be exploited. A subject who scores highly at one level of the treatment tends to score highly at the other levels, and vice versa. For instance, in Table 15.1, subject 8 scores highly at all three levels, whereas subject 4 has consistently lower scores. It is not surprising that some subjects have better recall ability than others and that this ability applies to a variety of conditions; subjects who recall more positive words than other subjects are likely to be above average in recalling negative and neutral words as well.

Stated another way, the correlation should be high for any pair of treatment levels. You may remember that higher positive correlations meant higher t values in the matched t test. The same principle applies here. With three treatment levels, however, the situation is more complicated. There are three possible pairs of levels and therefore three correlation coefficients to consider. Moreover, if you would like to use the direct-difference method of Chapter 11, three sets of difference scores could be created. And with even more treatment levels, the complexity grows quickly. Fortunately, the two-way ANOVA approach makes it all rather simple, as described in the following.

Calculating the SS Components of the RM ANOVA

To perform a one-way RM ANOVA on the data of Table 15.2, we can begin by performing a two-way ANOVA (as described in the previous chapter) to obtain the SS components. The first thing you should notice when trying to perform the two-way ANOVA with the data above is that you can calculate neither SS_W nor MS_W because there is only one measure per cell, and hence no within-cell variance. This would be a problem if we were planning to complete the two-way ANOVA procedure, but we are going to use the SS components in a different way. Instead of being a problem, the lack of an MS_W term just simplifies the calculations. Unable to calculate SS_W, you might begin the two-way ANOVA by calculating $SS_{\text{beween-cells}}$ using Formula 14.3. But then you should notice that the cell means are just the individual scores, (i.e., $n = 1$). Therefore, the calculation for $SS_{\text{between-cells}}$ becomes exactly the same as the calculation for SS_{total}, which I already performed. Thus, we know that $SS_{\text{between-cells}} = 694$.

Next, we can calculate the SS for rows and columns using Formula 14.3 on the row and column means. SS_R corresponds to the variability among subject means, so in the RM ANOVA it is called SS_{subject} (or SS_{sub}, for short). SS_C corresponds to the variability among treatment means, which is exactly the same variability labeled SS_{bet} in the one-way independent ANOVA. Because this component corresponds to the variability due to the repeated-measures factor, I will refer to it as SS_{RM}. We already know that SS_{RM} equals 112, so the only new component that we need to calculate is SS_{sub}:

$$SS_{\text{sub}} = N_T\sigma^2(\text{row means}) = 24 \cdot 22.85 = 548.4$$

Now we can find SS_{inter} by subtraction, using Formula 14.2 (rewritten in terms of the new notation for the RM design):

$$SS_{\text{inter}} = SS_{\text{bet}} - SS_{\text{sub}} - SS_{\text{RM}} = 694 - 548.4 - 112 = 33.6$$

To calculate the MSs we need to know the df components. The df for rows (df_R or df_{sub}) is one less than the number of rows (i.e., the number of subjects), which equals $8 - 1 = 7$; df_C (or df_{RM}) is one less than the number of columns (i.e., the number of treatment levels), which equals $3 - 1 = 2$. The df for interaction (df_{inter}) equals df_{sub} multiplied by $df_{\text{RM}} = 7 \times 2 = 14$.

The MS for subjects would be $SS_{\text{sub}}/df_{\text{sub}}$, but this MS plays no role in the RM ANOVA, so it is almost never calculated. The rationale for ignoring MS_{sub} will be discussed shortly. The MS for the independent (i.e., treatment) variable, MS_{RM}, equals $SS_{\text{RM}}/df_{\text{RM}} = 112/2 = 56$ (the same as MS_{bet} from the one-way independent ANOVA). The MS for the interaction equals $SS_{\text{inter}}/df_{\text{inter}} = 33.6/14 = 2.4$.

The F Ratio for the RM ANOVA

We are finally ready to calculate the F ratio for the RM ANOVA. The only F ratio we will calculate is the F ratio that tests differences among the treatment levels. The numerator for this F ratio, therefore, is MS_{RM}. Because there is no MS_W, it may not be obvious which MS should be used for the denominator. Had I not just mentioned that MS_{sub} is generally not calculated, you might have guessed that this MS would be used as the error term. For reasons to be explained shortly, the MS used for the error term is MS_{inter}. Thus the F ratio for the one-way RM ANOVA is $MS_{\text{RM}}/MS_{\text{inter}}$. For this example, $F = 56/2.4 = 23.33$.

Comparing the Independent ANOVA with the RM ANOVA

The new calculated F ratio is considerably higher than the critical F for the one-way independent ANOVA, but the degrees of freedom have changed, requiring us to find a new critical F as well. The numerator df remains the same for both types of ANOVA (in this case, $df_{num} = 2$), but for the RM ANOVA the denominator df becomes smaller. In this example, df_{denom} (i.e., df_{inter}) is 14, whereas df_{denom} for the independent ANOVA (i.e., df_W) was 21. The new critical F is 3.74, which is *higher* than the previous critical $F(3.47)$ because of the reduced df for the error term. However, whereas the critical F is slightly higher for the RM ANOVA, the calculated F is much higher (23.33) than it was for the independent ANOVA (2.02). Consequently, the RM ANOVA is significant—and would be even if an alpha much smaller than .05 were used, implying that the null hypothesis concerning the treatment levels can be rejected easily. As is often the case, the increase in the calculated F that results from matching across conditions far outweighs the relatively small increase in the critical F. The advantage of the RM design can be seen more clearly by comparing the SS components for this design with those from the independent ANOVA.

The total SS for the one-way independent ANOVA is divided into only two pieces: SS_{bet} and SS_W. The total SS for the one-way RM ANOVA is divided into three pieces: SS_{sub}, SS_{RM}, and SS_{inter}. First, it should not be surprising that SS_{total} is the same for both designs; regrouping the scores does not change the total SS. It should also make sense that $SS_{bet} = SS_{RM}$. Therefore, the comparison becomes quite simple:

Independent ANOVA: $SS_{total} = SS_{bet} + SS_W$

RM ANOVA: $SS_{total} = SS_{RM} + SS_{sub} + SS_{inter}$

Because SS_{total} is the same in both designs and $SS_{bet} = SS_{RM}$, it must be that the sum of SS_{sub} and SS_{inter} will always equal SS_W. This is true for the present example, in which $SS_{sub} = 548.4$ and $SS_{inter} = 33.6$; adding these two components, we get 582, which equals SS_W.

The Advantage of the RM ANOVA

Now it is easy to see the possible advantage of the *repeated measures (RM) design*. The SS_W that would be calculated for an RM design if independent groups were assumed (i.e., if the matching were ignored) is divided into two pieces when it is recognized that the scores are matched. One of these pieces, SS_{sub}, is ignored. The error term is based on SS_{inter}; as you will see, the better the matching, the smaller SS_{inter} becomes, and the larger SS_{sub} becomes. When the matching is totally ineffective (i.e., zero correlation between any pair of treatment levels), SS_W is divided into SS_{sub} and SS_{inter} in such a way that, after dividing by the appropriate df, MS_W and MS_{inter} become equal. To the extent that the matching gets any better than random, MS_{inter} becomes smaller than MS_W. MS_{inter} actually gets larger than MS_W in the unusual case when the matching works in reverse (i.e., the correlation is negative between pairs of treatment levels). If the correlation is positive but very low, the slight increase in the calculated F ratio may not outweigh the loss in degrees of freedom and the resulting increase in the critical value for F. As with the matched t test, poor matching combined with fairly small sample sizes can actually mean that the results will be closer to significance with the independent-groups test. Conceptually, matching works the same way in the RM ANOVA as it does for the matched t test, but it does get more complicated when there are more than two treatment levels. A graph can make the effect of matching more understandable.

Figure 15.1

Graph of the Data in
Table 15.1

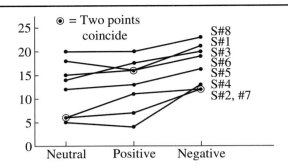

Picturing the Subject by Treatment Interaction

If we are going to calculate the RM ANOVA as though it were a two-way ANOVA, it can be helpful to extend the analogy by graphing the cell means (which are just the individual scores, in this case). Although either factor could be placed on either axis, the graph will be easier to look at if we place the treatment levels along the X axis and draw separate lines for each subject, as in Figure 15.1 (which uses the data from Table 15.1). By noting the degree to which the lines are parallel, you can estimate whether SS_{inter} will be relatively large or small. (If the lines were all parallel, SS_{inter} would equal zero.) In Figure 15.1 you can see that the lines are fairly close to being parallel. The more the lines tend to be parallel, the smaller the SS for the interaction, and therefore the smaller the denominator of the F ratio for the RM ANOVA. Of course, a smaller denominator means a larger calculated F ratio and a greater chance of statistical significance.

Interactions are usually referred to in terms of the two factors involved; in the RM design our focus is on the subject by treatment (or subject $\times$ treatment) interaction. A relatively small subject $\times$ treatment interaction indicates a consistent pattern from subject to subject. That is, different subjects are responding in the same general way to the different treatment conditions. In Figure 15.1, with a couple of exceptions, the number of words subjects recall goes up a little from neutral to positive words and then increases to a greater degree from positive to negative words. This consistency is what the RM ANOVA capitalizes on—just as the matched t test was based on the similarity of the difference scores. Note that the overall level of each subject can be very different (compare subjects 4 and 8 in Figure 15.1) without affecting the interaction term. Differences in overall level between subjects (i.e., average recall over the three word types) contribute to the size of SS_{sub}, but fortunately this component does not contribute to the error term. Again, you can draw a comparison to the matched t test, in which subjects' absolute scores are irrelevant, and only the consistency of the difference scores is important.

Comparing the RM ANOVA to a Matched t Test

The similarity between the RM ANOVA and a matched t test may be more obvious if you realize that in the case of only two treatment levels, either procedure can be applied (just as either the two-group t test or the one-way ANOVA can be applied to two independent groups). As an illustration, I will calculate the RM ANOVA for the data from the matched t test in Exercise 11B4. I begin by finding the row and column means, as shown in Table 15.3.

Table 15.3

Subject No.	Traditional	New Method	Row Means
1	65	67	66
2	73	79	76
3	70	83	76.5
4	85	80	82.5
5	93	99	96
6	88	95	91.5
7	72	80	76
8	69	100	84.5
Column Means	76.875	85.375	81.125

The SS components are found using Formula 14.3 on the individual scores as well as on the row and column means:

$$SS_{total} = N_T\sigma^2(\text{scores}) = 16(122.61) = 1{,}961.8$$

$$SS_{RM} = N_T\sigma^2(\text{column means}) = 16 \cdot \sigma^2(76.875, 85.375)$$

$$= 16(18.0625) = 289$$

$$SS_{sub} = N_T\sigma^2(\text{row means}) = 16(80.61) = 1{,}289.8$$

$$SS_{inter} = 1{,}961.8 - 289 - 1{,}289.8 = 383$$

The relevant MS values are:

$$MS_{RM} = \frac{SS_{RM}}{df_{RM}} = \frac{289}{1} = 289$$

$$MS_{inter} = \frac{SS_{inter}}{df_{inter}} = \frac{383}{7} = 54.7$$

Finally, the F ratio $= MS_{RM}/MS_{inter} = 289/54.7 = 5.28$.

The matched t value calculated for the same data was 2.30, which when squared is equal (within rounding error) to the F ratio calculated in the preceding, as will always be the case. The standard deviation of the difference scores (s_D) was 10.46 in Exercise 11B4, which when squared equals 109.4—exactly twice the value of MS_{inter} as just found (this will always be true). Thus in the two-group case, the variability of the difference scores is a direct function of the amount of subject × treatment interaction.

This relationship can also be shown graphically. Figure 15.2 graphs the data from the preceding example, and you can see a fair amount of subject by treatment interaction. This interaction helps to explain why the matched t was not significant. (The matched t will always be associated with the same p value as the F from the corresponding RM ANOVA.) The slope of each line represents the size of a difference score (negative slopes indicate negative difference scores); more variability in difference scores means more variability in the slopes of the lines, which in turn means a larger interaction term. When the differences are all the same, there is no variability in difference scores, and the lines are all parallel, indicating no interaction. When there are more than two treatment levels, there is more than one set of difference scores. The MS for interaction is a function of the average of the variances for each set of difference scores. For instance, if your RM fac-

Figure 15.2

Graph of the Data in
Table 15.3

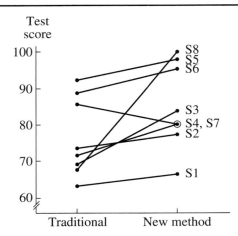

tor has four levels, there are six possible pairs of levels for which difference scores can be calculated. Average the variances of these six sets of difference scores and the result will always be exactly twice as large as MS_{inter} for the overall RM analysis.

The issues to consider in deciding when the repeated measures design is appropriate are the same as those discussed for the matched t test, except that some matters become more complicated when there are more than two treatment levels. Perhaps the least problematic repeated-measures design is one in which the different treatment levels are presented simultaneously or mixed randomly together, as in the preceding word recall example. Including more than two levels does not automatically create any problems. On the other hand, successive repeated measures designs can get quite complicated.

Dealing with Order Effects

In describing the successive repeated measures design with two levels (see Chapter 11), I pointed out the danger of *order effects*. For example, if the same subject is to calculate math problems during both happy and sad pieces of music, whichever type of music is played second will have an unfair advantage (practice) or disadvantage (fatigue) for the subject, depending on the details of the study. The solution in Chapter 11 was to use *counterbalancing*: Half the subjects get the happy music first and the other half get the sad music first. Counterbalancing becomes more complicated (although no less necessary) when there are more than two treatment levels.

If you wish to study the effects of classical music, jazz, and heavy metal rock music on math calculations, six different orders are possible (three choices for the first piece of music times two choices for the second times one choice for the third). If 48 subjects were in the study, 8 would be assigned to each of the six orders. However, if a fourth type of music were added (e.g., new age music), there would be 24 possible orders, and only two of the 48 subjects could be assigned to each order. For any study with five or more levels to be administered successively, complete counterbalancing, as described above, becomes impractical. Fortunately, there are clever counterbalancing schemes, such as the *Latin square design*, that do not require all possible orders to be presented and yet do eliminate simple order effects. Because it greatly simplifies experimental procedures, the Latin square design is commonly employed even when there are only three or four treatment levels. Such designs are discussed in detail in Section C.

Differential Carryover Effects

Whereas counterbalancing schemes can eliminate the confounding effects of practice, fatigue, and other phenomena that result from the order of the treatment levels, such schemes cannot be counted upon to eliminate *carryover effects*, which can differ for each particular order of treatments. For instance, if the rock music is played loudly enough, it might affect the subject's ability to appreciate (or even to hear) the next piece of music played, thus attenuating the effect of that music. This effect would not be symmetrical (e.g., hearing classical music would not likely affect a subject's ability to hear a piece of rock music right after) and would therefore not be balanced out by counterbalancing.

With more than two treatment levels, the possibility of complex, asymmetrical carryover effects increases. In some cases, leaving more time between the presentation of treatment levels or imposing some neutral, distracting task between the experimental conditions can eliminate carryover effects. In other cases, carryover cannot be eliminated. For instance, a cognitive psychologist may want to know which of three different types of hints is the most helpful to subjects who are solving a particular problem. The psychologist cannot expect a subject to forget a previous hint and use only the present one; different subjects must be used to test different hints.

The Randomized-Blocks Design

When carryover effects make it inappropriate to use repeated measures, the best alternative is to match subjects as closely as possible. Suppose you wanted to compare three methods for teaching algebra to 14-year-olds. The students could be matched into groups of three so that all three would be as similar as possible in mathematical ability (based on previous math tests and grades). The three matched students in each group would then be randomly assigned to the three different teaching methods. (This is the same procedure I described for the matched *t* test, in which each group consisted of a pair of subjects.) At the end of a semester all students would be given the same algebra test; the scores would be analyzed using an RM ANOVA, as though the three students in each group were really the same subject measured under three different conditions. Of course, the matching procedure would be the same for four teaching methods or any other number of treatment levels. In this way, repeated measures can be viewed as the closest possible form of matching; the same subject measured under three conditions will usually produce scores that are more similar than three matched subjects (although a particular subject will vary in performance somewhat from time to time). Fortunately, when repeated measures are not appropriate, good matching can yield much of the same advantage.

I have been talking about the goal of good matching in terms of producing similar scores for the different conditions, but it is important to remember that we also want the scores to differ between the conditions, with some conditions associated with consistently higher scores than others. We expect the scores from the same subject (or matched subjects) to be more similar than scores from entirely different subjects under the same conditions, but it is also important that the treatment means differ (i.e., parallel lines on a graph of the scores but not flat, or horizontal, lines).

In the context of ANOVA, a group of subjects who are matched together on some relevant variable is called a *block*. A design in which members of a block are randomly assigned to different experimental conditions is known as a *randomized-blocks (RB) design*; the matched pairs design described in Chapter 11 is the simplest form of randomized-blocks design. The ANOVA performed on data from this type of design is called a randomized-blocks ANOVA (RB ANOVA). When the number of subjects in a block is the same as the number of treatment levels, the RB ANOVA is calculated in exactly the same way as the RM ANOVA (just as the matched *t* test is calculated exactly like a repeated-measures *t* test). Therefore, I will use the term RM ANOVA, regardless of whether it is applied to an RM design or a simple RB design. It is possible to create more complicated RB designs in which the number of subjects in a block is some multiple of the number of treatment levels, but such designs are not common, and I will not consider their analyses in this chapter.

SUMMARY

1. The one-way RM ANOVA expands the matched *t* test to accommodate any number of levels of the independent (i.e., treatment) variable simultaneously.

2. The one-way RM ANOVA can be viewed as a two-way ANOVA in which the subjects are the different levels of the second factor and there is only one subject per cell. Although you cannot calculate SS_W, you can find the other SS components of a two-way ANOVA.

3. Using the two-way ANOVA approach, there is no MS_W to act as the denominator of the *F* ratio. Instead, MS_{RM} (which corresponds to MS_{bet} in a one-way ANOVA or MS_{col} in a two-way ANOVA) is divided by $MS_{interaction}$. In this context, the latter term is often called MS_{error}.

4. Compared to the corresponding one-way independent-groups ANOVA, the advantage of the RM ANOVA is that the error term is almost always smaller (i.e., MS_{inter} will be smaller than MS_W unless the matching is no better than random). The better the matching, the smaller the error term gets, and the larger the *F* ratio becomes for the RM ANOVA.

5. One disadvantage of the RM ANOVA is the reduced df in the error term, which leads to a larger critical *F* than for the independent ANOVA. However, the increase in the calculated *F* for RM ANOVA that results from matching usually outweighs the increase in the critical *F*.

6. When there are only two treatment levels, either a matched *t* test or RM ANOVA can be performed. The variance of the difference scores in the matched *t* test is always twice as large as the MS_{inter} for the RM ANOVA on the same data. The matched *t* value is equal to the square root of the *F* ratio from the RM ANOVA.

7. If the treatment levels in an RM design are presented in succession and always in the same order, the effects of order (practice, fatigue, etc.) may contribute to the treatment effects. Counterbalancing provides a system for averaging out the order effects.

8. If the effects of particular treatment levels interfere (i.e., carry over) with subsequent levels, it is unlikely that these carryover effects will be compensated for by counterbalancing. In this case, matching subjects into blocks will eliminate the problem. The randomized-blocks (RB) design is analyzed in exactly the same way as the RM design as long as the number of subjects in each block equals the number of treatment levels. If the matching is effective, the RB design will have much of the power associated with the RM design. (The RM design usually represents the best possible matching because the same subject at different times is usually more similar than well-matched but different subjects.)

EXERCISES

1. The following data were originally presented in Exercise 9B8 and were reproduced in Exercise 11B5, for which you were asked to perform a matched t test. Now perform an RM ANOVA on these data, and compare the calculated F ratio with the t value you found in Exercise 11B5.

Segment	Male Rater	Female Rater
1	2	4
2	1	3
3	8	7
4	0	1
5	2	5
6	7	9

*2. The data in the following table are from an experiment on short-term memory involving three types of stimuli: digits, letters, and a mixture of digits and letters. Draw a graph of these data and describe the degree of interaction between the various pairs of levels.

Subject	Digit	Letter	Mixed
1	6	5	6
2	8	7	5
3	7	7	4
4	8	5	8
5	6	4	7
6	7	6	5

*3. The progress of 20 patients is measured every month for 6 months. An RM ANOVA on these measurements produced the following sums of squares: $SS_{total} = 375$; $SS_{RM} = 40$; $SS_{subject} = 185$.
   a. Complete the analysis and find the F ratio for the RM ANOVA.
   b. Calculate the F ratio that would be obtained for an independent-groups ANOVA.

4. In Exercise 12A7, eight subjects were tested for problem-solving performance in each of four drug conditions, yielding the following means and standard deviations.
   If the same eight subjects were tested in all four conditions, and if SS_{sub} were equal to 190.08, how large would the F ratio for the RM ANOVA be?

	Marijuana	Amphetamine	Valium	Alcohol
$\overline{X}$	7	8	5	4
s	3.25	3.95	3.16	2.07

5. In Exercise 13C4, independent groups of subjects performed a video game after being given one of four possible doses of caffeine (including zero). In this exercise, we will imagine that the subjects had been matched into blocks of four before being randomly assigned to one dosage level or another. The data from Exercise 13C4 are reprinted next, with the addition of a column that indicates that there are six blocks of subjects (the scores are the number of errors committed, so lower scores demonstrate better performance).

	Amount of caffeine			
Block #	zero mg	100 mg	200 mg	300 mg
1	25	16	6	8
2	19	15	14	18
3	22	19	9	9
4	15	11	5	10
5	16	14	9	12
6	20	23	11	13

   a. Graph these data, with caffeine amount on the X axis and the blocks represented by separate lines. Describe the general trend of the data with respect to caffeine dosage. Does the amount of the subject-by-dosage-level interaction look relatively large or small?
   b. Calculate the F ratio for a one-way RM ANOVA on these data. Can the null hypothesis be rejected at the .05 level? At the .01 level?

*6. A psychotherapist is studying his patients' progress over time by having each patient fill out a depression questionnaire at the end of each month of therapy for a total of 5 months. The monthly depression scores for each of eight patients are displayed in the following table.

Patient #	Month 1	Month 2	Month 3	Month 4	Month 5
1	22	16	13	12	15
2	12	12	11	15	19
3	29	28	21	18	22
4	35	25	24	21	18
5	19	23	18	19	17
6	16	11	17	14	20
7	25	17	24	20	16
8	30	35	23	26	27

a. Graph these data, with month on the X axis and the blocks represented by separate lines. Describe the general trend of the data over time.

b. Calculate the F ratio for a one-way RM ANOVA on these data. Can the null hypothesis be rejected at the .05 level? Given your decision, what type of error could you be making, Type I or Type II?

7. Which of the following tends to increase the size of the F ratio in an RM ANOVA?

a. Larger subject-to-subject differences (averaging across treatment levels)

b. Better matching of the scores across treatments

c. Larger subject by treatment interaction

d. Smaller differences between treatment means

*8. An RM ANOVA on a certain set of data did not attain statistical significance. What advantage could be derived from performing an independent-groups ANOVA on the same data?

a. A larger alpha level would normally be used.

b. The numerator of the F ratio would probably be larger.

c. The critical F would probably be smaller.

d. The error term would probably be larger.

9. The chief advantage of the randomized blocks design (compared to the RM design) is that

a. Differential carryover effects are eliminated.

b. The matching of scores is usually better.

c. The critical F is reduced.

d. All of the above.

*10. The chief advantage of counterbalancing is that

a. Subject by treatment interaction is reduced.

b. Simple order effects are prevented from contributing to differences among treatment means.

c. Differential carryover effects are eliminated.

d. All of the above.

BASIC STATISTICAL PROCEDURES

In the previous section I dealt with a repeated-measures design in which it was reasonable to mix together items from different treatment levels (i.e., the emotional content of the words). In this section I will consider a design in which such a mixture would be detrimental to performance. For quite some time, cognitive psychologists have been interested in how subjects solve transitive inference problems, such as the following: Bill is younger than Nancy. Tom is older than Nancy. Who is the youngest? There is some debate about the roles of various cognitive mechanisms (e.g., visual imagery) in solving such problems. It appears that the way the problem is presented may affect the choice of mental strategy, which in turn may affect performance. Consider a study comparing three modes of presentation: visual-simultaneous (both premises and the question appear simultaneously on a computer screen until the subject responds), visual-successive (only one of the three lines appears on the screen at a time), and auditory (the problem is read aloud to the subject—which, of course, is a successive mode of presentation). The subject is given the same amount of time to answer each problem, regardless of presentation mode. The dependent variable is the number of problems correctly answered out of the 20 presented in each mode.

Changing the presentation mode randomly for each problem would be disorienting to the subject and probably would not be an optimal test of the research hypothesis. A reasonable approach would be to present three sets of transitive inference problems to each subject, each set being presented in a different mode. However, if the three sets were presented in the same order to all the subjects, this alone could determine which condition would lead to the best performance. The solution is to counterbalance. In a completely counterbalanced design, all six of the possible orders would be used; see Table 15.4. As usual, I will minimize the calculations by creating the simplest possible design: one subject for each of the six orders.

Subject No.	First Block	Second Block	Third Block	
1	Visual-sim	Visual-succ	Auditory	**Table 15.4**
2	Visual-sim	Auditory	Visual-succ	
3	Auditory	Visual-sim	Visual-succ	
4	Auditory	Visual-succ	Visual-sim	
5	Visual-succ	Visual-sim	Auditory	
6	Visual-succ	Auditory	Visual-sim	

The research hypothesis is that the mode of presentation will affect the subjects' performance on transitive inference problems. (It would also be a good idea to collect some data on the subjects' uses of various mental strategies, but for simplicity, I will ignore that aspect of the study.) This research hypothesis can be tested by using the usual six-step procedure.

Step 1. State the Hypotheses

The null hypothesis for the one-way RM ANOVA is the same as for the one-way independent-groups ANOVA. With three treatment levels, H_0: $\mu_1 = \mu_2 = \mu_3$. The alternative hypothesis is also the same; H_A is best stated indirectly by claiming that H_0 is not true.

Step 2. Select the Statistical Test and the Significance Level

Because each subject is observed under all three conditions, and we expect a good degree of matching across conditions, the RM ANOVA is the appropriate procedure (assuming that the dependent variable has been measured on an interval/ratio scale). The conventions with respect to alpha are no different for this design, so we will once again set $\alpha = .05$.

Step 3. Select the Samples and Collect the Data

Because this is an RM design, there is only one sample to select, which ideally should be a random sample from the population of interest but more often is one of convenience. Because we are considering the *same* subject's performance at each treatment level, the means for the treatment levels cannot be attributed to differences in samples of subjects. The data for this hypothetical study appear in Table 15.5.

Step 4. Find the Region of Rejection

If the null hypothesis is true, the F value calculated for the RM ANOVA will represent the ratio of two independent estimates of the population variance

Subject No.	Vis-sim	Vis-succ	Auditory	Row Means	
1	15	13	12	13.33	**Table 15.5**
2	14	16	15	15	
3	20	17	10	15.67	
4	17	12	11	13.33	
5	12	7	5	8	
6	18	8	7	11	
Column Means	16	12.17	10	12.72	

and will therefore follow one of the F distributions. To determine which F distribution is appropriate, we need to know the degrees of freedom for both the numerator and the denominator. The df for the numerator is the same as in the independent ANOVA; it is one less than the number of treatments, which for this example is $3 - 1 = 2$. The df for the denominator is the df for the subject $\times$ treatment interaction, which is one less than the number of subjects times one less than the number of treatments; in this case, $df_{denom} = (6 - 1)(3 - 1) = 5 \times 2 = 10$. The critical F with $(2, 10)$ df and $\alpha = .05$ is 4.10. Thus the region of rejection is the tail of the $F(2, 10)$ distribution above the critical value of 4.10.

Step 5. Calculate the Test Statistic

First, the SS components are found using Formula 14.3 on the individual scores as well as on the row and column means:

$$SS_{total} = N_T \sigma^2 (scores) = 18(16.645) = 299.6$$

$$SS_{RM} = N_T \sigma^2 (column\ means) = 18 \cdot \sigma^2(16,\ 12.17,\ 10)$$

$$= 18(6.153) = 110.8$$

$$SS_{sub} = N_T \sigma^2 (row\ means) = 18 \cdot \sigma^2 (13.33,\ 15,\ 15.67,\ 13.33,\ 8,\ 11)$$

$$= 18(6.646) = 119.6$$

The subject by treatment interaction (SS_{inter}) can now be found by subtraction, using a relabeled version of Formula 14.2, which will become Formula 15.1:

$$SS_{inter} = SS_{total} - SS_{RM} - SS_{sub} \qquad \textbf{Formula 15.1}$$

For the present example, $SS_{inter} = 299.6 - 110.8 - 119.6 = 69.2$.

The next step is to divide the total number of degrees of freedom into components that match the SSs. As usual, $df_{total} = N_T - 1$. I will define n as the number of different subjects (or blocks in an RB design) and use c (the number of columns in a two-way design) for the number of levels of the repeated measures factor. Thus, N_T, the total number of observations, or scores, equals c times n. For this example, df_T equals $18 - 1 = 17$. The other components are found using the following formulas, which, collectively, are Formula 15.2:

a. $df_{sub} = n - 1$ **Formula 15.2**

b. $df_{RM} = c - 1$

c. $df_{inter} = (n - 1)(c - 1)$

In this example, $n = 6$ and $c = 3$, so

$$df_{sub} = 6 - 1 = 5$$

$$df_{RM} = 3 - 1 = 2$$

$$df_{inter} = 5 \cdot 2 = 10$$

Note that the df components add up to 17, which equals df_{total}.

The two MSs of interest can be found by dividing the appropriate SS and df components, as follows:

a. $MS_{RM} = \dfrac{SS_{RM}}{df_{RM}}$ **Formula 15.3**

b. $MS_{inter} = \dfrac{SS_{inter}}{df_{inter}}$

For this example, the MSs are

$$MS_{RM} = \frac{110.8}{2} = 55.4$$

$$MS_{inter} = \frac{69.2}{10} = 6.92$$

Finally, the F ratio for the RM ANOVA is found by Formula 15.4:

$$F = \frac{MS_{RM}}{MS_{inter}}$$ **Formula 15.4**

The F ratio for the transitive inference experiment is $F = 55.4/6.92 = 8.0$.

Step 6. Make the Statistical Decision

Because the observed F ratio (8.0) is larger than the critical F ratio (4.10), our experimental result falls in the region of rejection for the null hypothesis—we can say that the result is statistically significant at the .05 level. The research hypothesis that the population means corresponding to the different presentation modes are not the same has been supported. However, as with any one-way ANOVA with more than two groups, rejecting the complete null hypothesis does not tell us which pairs of population means differ significantly.

Interpreting the Results

From the column means in Table 15.5, it is clear that the simultaneous visual condition renders the transitive inferences easiest to make, and you can be reasonably sure that this condition will differ significantly from the auditory condition. As with independent samples, post hoc comparisons are usually conducted to locate population differences more specifically following a significant RM ANOVA, and these can include both pairwise and

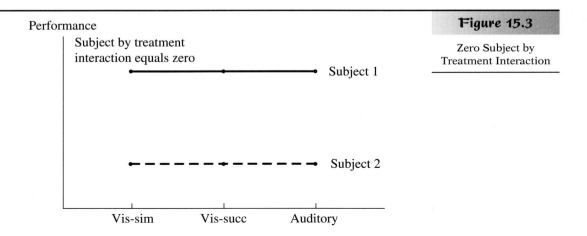

Performance

Subject by treatment interaction equals zero

Subject 1

Subject 2

Vis-sim   Vis-succ   Auditory

Figure 15.3

Zero Subject by Treatment Interaction

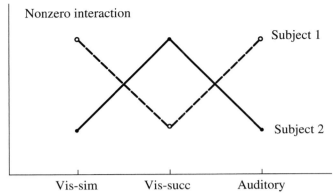

complex comparisons. If you are conducting only pairwise comparisons, Fisher's LSD for a three-level experiment and Tukey's HSD for four or more levels can be applied, substituting MS_{inter} for MS_W. Scheffé's test should be performed whenever complex comparisons are involved. However, just as the use of MS_W for multiple comparisons requires the assumption of homogeneity of variance for an independent-samples ANOVA, using MS_{inter} as the error term for comparisons following a significant RM ANOVA requires a corresponding homogeneity assumption. I will return to the topic of post hoc comparisons after discussing the assumptions that underlie the RM ANOVA.

The Residual Component

When you are dealing with repeated measures, issues arise that you don't have to think about when using independent groups. For instance, consider the null hypothesis for the preceding example: The three population means are all equal. There are two different ways this can occur in an RM design: with or without interaction. Consider just two subjects. In Figure 15.3, the two subjects differ in overall problem-solving ability, but the three treatment means are equal—and there is no interaction. Now look at Figure 15.4. The treatment means are *still* equal (they are the same as they were before), but now there is less of a difference on average between the two subjects—and there *is* an interaction. Expanding this description to populations of subjects, we can say that three population means can be equal because each subject has the same score under all three conditions, or they can be equal because subjects differ from treatment to treatment, but in such a way that these differences cancel out.

The latter possibility is considered a more realistic version of the null hypothesis (e.g., some subjects benefit when problems are presented in the auditory mode, some are hindered by it, with a net effect of zero). If we adopt this way of viewing the RM design (i.e., we do *not* assume the subject × treatment interaction to be zero), there is, fortunately, no change in the way the F ratio for the treatment effect is calculated. However, there is a difference in the way we view the structure of the F ratio. For the independent ANOVA, I pointed out that F represents the following ratio:

$$F = \frac{\text{est. treatment effect} + \text{between-group est. error variance}}{\text{within-group est. of error variance}}$$

In the case of the RM ANOVA, the denominator of the F ratio is based on MS_{inter}. If MS_{inter} is assumed to be zero in the population, the size of the denominator could be attributed entirely to error. However, if MS_{inter} is *not* assumed to be zero, the denominator would be the sum of the actual MS_{inter} in the population plus error. The amount of MS_{inter} would also show up in the numerator along with the amount of error because both components can affect the variability of the treatment means in a particular experiment. So, for the RM ANOVA, the F ratio is usually viewed in the following manner:

$$F = \frac{\text{est. treatment effect} + \text{est. interaction} + \text{est. error}}{\text{est. interaction} + \text{est. error}}$$

Notice that according to the null hypothesis, the treatment effect is expected to be zero, and therefore the F ratio is expected to be around 1.0.

Because the denominator of the F ratio is generally considered to be a sum of both the true amount of subject × treatment interaction and the error, the denominator is commonly referred to as $MS_{residual}$ (or MS_{res}, for short) rather than MS_{inter}. The term *residual* makes sense when you consider that SS_{inter} was found by subtracting both SS_{sub} and SS_{RM} from SS_{total}; what is left over is usually thought to contain a mixture of interaction and error components and is therefore referred to in a more neutral way as $SS_{residual}$. Because there is no separate estimate of error (i.e., no SS_W), we cannot separate the contributions of interaction and error to $SS_{residual}$, so this term is often preferred to the term SS_{inter} when dealing with the RM design. However, it is still common to refer to $SS_{residual}$ as the error term, or SS_{error}, so you can use either one when referring to SS_{inter} in the RM ANOVA. I will continue to use the term SS_{inter} for its mnemonic value with respect to the calculation of that term.

The Effect Size of an RM ANOVA

The effect-size measure (ESM) I described in Section C of Chapter 11, $g_{matched}$, could be extended to the multi-group case by a measure I will call f_{RM}. I will discuss this measure in the context of power analysis, but I have never seen it used for descriptive purposes. To describe the effect size for an RM ANOVA, a form of eta squared, or an estimate of omega squared, is recommended. A measure analogous to $g_{matched}$ would be η_{RM}^2, as defined in the following formula:

$$\eta_{RM}^2 = \frac{SS_{RM}}{SS_{RM} + SS_{inter}} \qquad\qquad \textbf{Formula 15.5}$$

This measure suffers from the same failing as $g_{matched}$; η_{RM}^2 tends to overestimate the impact that an independent variable would have when tested with independent groups, or considerably less effective matching. A more familiar and generalizable ESM would be the ordinary eta squared (Olejnik & Algina, 2003), calculated as though the groups were independent—that is, η^2 equals SS_{RM}/SS_{Total}, which equals $SS_{RM}/(SS_{RM} + SS_{Sub} + SS_{inter})$. Usually, η^2 will not be as large as η_{RM}^2, but it will provide a more stable basis for planning future experiments. The difference between these two measures can be illustrated by applying them to the data in Table 15.5:

$$\eta_{RM}^2 = \frac{110.8}{110.8 + 69.2} = \frac{110.8}{180} = .616$$

$$\eta^2 = \frac{SS_{RM}}{SS_{RM} + SS_{Sub} + SS_{inter}} \quad \frac{110.8}{110.8 + 119.6 + 69.2} = \frac{110.8}{299.6} = .370$$

If you use an appropriately modified version of Formula 12.19 to obtain the proportion of variance accounted for you will get η^2_{RM} (except for error due to rounding off the value of F), rather than η^2, as shown next.

$$\eta^2_{RM} = \frac{df_{RM}F}{df_{RM}\,F + df_{inter}} = \frac{2(8.0)}{2(8.0) + 10} = \frac{16}{26} = .615$$

As in the case of partial η^2, described in the previous chapter, when you obtain η^2 from the F ratio for the effect of interest, rather than from the SSs, F is based on a specific error term—not on SS_{Total}. Because you need to determine SS_{Total} in order to calculate ordinary η^2, and this is rarely possible from a published article, you will usually be able to calculate ordinary η^2 only for your own data.

The value you get from η^2_{RM} is almost always more dramatic than ordinary eta squared, but it is rather specific to your experimental design. If another researcher decides to avoid the possibility of carry-over effects, and therefore run your study with matched blocks of subjects instead of repeated measures, η^2_{RM} is likely to be a substantial overestimation of the effect in the future study. Ordinary η^2 would provide a more neutral basis for planning the next study; an estimate of the correlation expected from matching subjects could then be used to increase that projected value (see further discussion in this chapter under the heading "Power of the RM ANOVA").

Of course, even ordinary η^2 is an overestimate of its corresponding value in the population, but the bias can be (almost completely) corrected by estimating omega squared (ω^2), with a slight change in the notation of Formula 12.20, as follows:

$$\text{est. } \omega^2 = \frac{SS_{RM} - (c - 1)MS_W}{SS_{total} + MS_W} \qquad \textbf{Formula 15.6}$$

where MS_W is the error term you would calculate if you ignored the matching of the scores. (If you have already separated your SSs for the RM ANOVA, you can find MS_W by adding SS_{sub} and SS_{inter} together to get SS_W, and then dividing SS_W by the sum of df_{sub} and df_{inter}—i.e., df_W.)

I have never seen a measure for estimating what might be labeled ω^2_{RM}, either in theory or in practice, so I am comfortable not including one here. With respect to the construction of confidence intervals, the same issues apply here as were discussed in Chapter 11.

Power of RM ANOVA

In the case of the independent-groups ANOVA, power depends on four basic factors: alpha, the size of the samples, the separation of the population means, and the subject-to-subject variability within the populations. For an RM ANOVA, the first three of these factors apply, but the fourth does not. Instead of depending on subject-to-subject variability, the power of the RM ANOVA depends on the amount of interaction that exists between subjects and factor levels—i.e., the extent to which the profile of any one subject across IV levels is parallel to all of the other subjects. As the lines on a graph of all subjects in the population get closer to being parallel the interaction gets smaller, and power increases.

You may recall that when an RM factor has only two levels, the amount of interaction depends inversely on the size of the correlation between the two sets of scores. If, in the multi-level case, the correlation were the same for every possible pair of levels, the interaction of the RM ANOVA would depend on the size of that common correlation coefficient in the population, ρ. The population effect size for an RM ANOVA, f_{RM}, could then be expressed in terms of the ordinary f (ignoring the matching of the scores), like this:

$$f_{RM} = \frac{f}{\sqrt{(1 - \rho)}} \qquad \qquad \text{Formula 15.7}$$

Note the resemblance to Formula 11.7.

When the ρs are the same for every pair of levels for your RM factor, we say that the population exhibits *compound symmetry*. This is certainly a convenient property to assume when planning an RM ANOVA; for instance, you can see just how the effect size increases for different degrees of matching. However, this assumption is considered too restricting, as I will explain in the section on Assumptions of the RM ANOVA. Nonetheless, if you expect all the ρs to be about the same, you can expect your effect size to be larger than the effect size of a corresponding independent groups ANOVA by a factor of roughly $1/\sqrt{(1 - \rho_{avg})}$, where ρ_{avg} is the average of the pairwise correlations that you expect to exist in the population (e.g., if you expect the ρs to center around .7, you can expect your independent-groups f to be multiplied by a factor of about 2).

You can estimate f_{RM} directly by using Formula 12.16, using for F the calculated value from a similar, previously performed RM ANOVA. If instead you have an estimate of ω^2 from a relevant independent-groups ANOVA (or a value of ω^2 that you consider reasonable to shoot for), you can convert it to an estimate of ordinary f by reversing Formula 12.22 as follows:

$$f = \frac{\omega}{\sqrt{(1 - \omega^2)}} \qquad \qquad \text{Formula 15.8}$$

This estimate of f can then be combined with an estimate of ρ, using Formula 15.7, to estimate f_{RM}. Once you have settled on an estimate of f_{RM}, you can multiply it by the square root of n (the size of each group you are planning to use), in order to find the value of φ with which to enter Table A.10. Alternatively, you can look up the value for φ needed for a desired level of power, divide it by your estimate of f_{RM}, and square the result to get an estimate of the sample size you will need, as described in Chapter 12.

Assumptions of the RM ANOVA

The assumptions underlying the hypothesis test conducted in this section are essentially the same as those for the independent ANOVA, except for an important additional assumption.

Independent Random Sampling

In the RM design, only one sample is drawn, and ideally it should be selected at random from the population of interest, with each subject being selected independently of all others. More commonly, however, the sample is one of convenience. This can limit the generalizability of the results but does not threaten the validity of the experiment, in that differences between treatments cannot be due to differences in the samples of subjects at different levels of the IV (the subjects are the same at each level). However, order effects can threaten the validity of the experiment (see "Counterbalancing," in Section C). In the RB design, subjects are matched together in small groups (usually the size of a block is equal to the number of treatment levels), and then the subjects within each block are *randomly assigned* to the different treatment conditions. Although ideally the subjects should be chosen at random from the general population, it is the random assignment of subjects to treatment conditions that is vital if the results of the RB design are to be considered valid.

Normal Distributions

As usual, it is assumed that the dependent variable follows a normal distribution in the population for each treatment level. In addition, the joint distribution that includes all levels of the independent variable is assumed to be a multivariate normal distribution. I described the bivariate normal distribution in Chapter 11. This is the distribution that is assumed in the two-group case of RM ANOVA (or the matched t test). *Multivariate normal distribution* is the generic term for a joint normal distribution that can contain any number of variables. Unfortunately, when there are more than two variables (or more than two conditions in an RM ANOVA), the multivariate normal distribution that applies is impossible to depict in a two-dimensional drawing. Fortunately, the RM ANOVA is not very sensitive to departures from the multivariate normal distribution, so this assumption is rarely a cause for concern. If, however, this assumption is severely violated and the samples are fairly small, you should consider a data transformation or a nonparametric test, such as the Friedman test (see Chapter 21, Section C).

Homogeneity of Variance

In the RM or RB design, there is always the same number of observations at each treatment level, so except for the most extreme cases of heterogeneity of variance, it is common to ignore this assumption.

Homogeneity of Covariance

The last assumption, *homogeneity of covariance*, does not apply when the groups are independent or when an RM or RB design has only two treatment levels. However, when a matched (or repeated) design has more than two levels, the covariance (as defined in Chapter 9) can be calculated for each pair of levels. Homogeneity of covariance exists in the population only when all pairs of treatment levels have the same amount of covariance.

The implications of this last assumption can be difficult to understand, and there is some debate about what to do if it is violated. First, I should point out that if the third and fourth assumptions are both true, the population displays a condition I mentioned earlier called *compound symmetry*. When compound symmetry exists, the population correlation (ρ) between any pair of treatment levels is the same as between any other pair. Compound symmetry is certainly a desirable condition; when it is true for an RM ANOVA (along with the first two assumptions), you can find critical Fs, as described earlier in this section, without worrying about your Type I error rate. However, compound symmetry is a stricter assumption than is necessary. The third and fourth assumptions can be relaxed as long as the variances and covariances follow a pattern that is called *sphericity*, or *circularity*.

Sphericity is defined mathematically in terms of the matrix of variances and covariances that apply to the various treatment levels and pairs of levels (see Chapter 16, Section C), but an easy way to understand this assumption is in terms of the amount of interaction (in the population) between any two levels of the independent variable: Sphericity implies that all of these pairwise interactions will be equally large. (This is the same as requiring that the variance of the difference scores will be the same no matter which pair of treatment levels you look at.) By graphing the results of your experiment for each subject, you can see whether the amount of inter-action seems to be about the same for any pair of levels. A relative lack of sphericity is illustrated by the hypothetical results depicted in Figure 15.5. Notice that there is very little interaction between condition A and condi-tion B in Figure 15.5, but a great deal between condition B and condition C (and therefore between condition A and condition C, as well). The total amount of subject × treatment interaction for the experiment would be moderately large—an average of the three pairwise amounts of interaction. But that average would be misleading.

Recall that in the independent ANOVA the error term is MS_W, which is an average of the sample variances. The use of this average was justified by assuming homogeneity of variance. Similarly, MS_{inter} is the average of the pairwise interactions (or the variances of the difference scores), and its use as the error term in the RM ANOVA is justified by assuming that sphericity exists in the population. However, it is important to note that although a population is somewhat more likely to show sphericity than compound symmetry, for many types of RM designs, especially those involving repeated measurements over time, the sphericity assumption will rarely be true. This represents a serious problem for a combination of two reasons.

Consequences of Violating Sphericity

First, the RM ANOVA is *not* robust with respect to violations of sphericity, as it is with respect to violations of the normal distribution or homogeneity of variance assumptions; when sphericity does not apply, your null hypoth-esis will not conform to the F distribution that you think it will. Second, the effect of this violation is to inflate your calculated F; the F ratio is said to be positively biased in this case, which means that the actual Type I error rate will be larger than the alpha you use to determine critical F. This problem is more serious than, for instance, the violation of homogeneity of variance in the case of two independent samples because it is more likely to occur and more likely to produce a large increase in the Type I error rate.

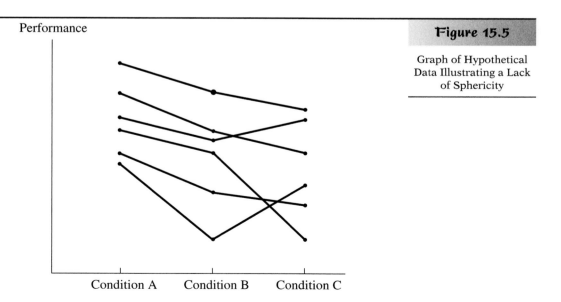

Figure 15.5

Graph of Hypothetical Data Illustrating a Lack of Sphericity

Dealing with a Lack of Sphericity

The problem just described is further complicated by the fact that it can be difficult to decide whether it is reasonable to assume sphericity in the population. Mauchly (1940) devised a way of using the variances and covariances in your sample to calculate a statistic, W, that allows you to make an inference about sphericity in the population (a test of this statistic is supplied automatically when an RM ANOVA is performed by some of the major statistical packages, such as SPSS). Unfortunately, like most statistical tests, this one does not have much power when small samples are used, so when that is the case, you will probably end up accepting the null hypothesis for Mauchly's W test, falsely concluding that sphericity is present in the population—unless the departure from sphericity is particularly severe. However, given how common it is for RM designs to violate the sphericity assumption, it makes sense to estimate the degree of sphericity exhibited in your data, regardless of the p value associated with Mauchly's test, before proceeding with the ordinary RM ANOVA.

Estimating Sphericity in the Population

The various pairwise interactions in your experimental data can be used to calculate a factor called epsilon (ϵ), which is an estimate of the degree of sphericity in the population. Epsilon ranges from a maximum of 1.0 when sphericity holds, down to a minimum of $1/(c - 1)$ when your data suggest a total lack of sphericity. The ordinary df components from the RM ANOVA are multiplied by ϵ, and then the adjusted df components are used to find the critical value of F. Calculating ϵ is not trivial, and there are at least two reasonable methods that lead to different answers—one developed by Greenhouse and Geisser (1959) and a somewhat less conservative adjustment devised by Huynh and Feldt (1976), which is considerably more powerful when ϵ is not very far from 1.0. Moreover, the adjusted df components are not likely to appear in any standard table of the F distribution. Fortunately, most computer statistical packages now calculate ϵ whenever a repeated factor has more than two levels and give an exact p value for the observed F, regardless of how the degrees of freedom come out. In fact, some packages calculate ϵ by both methods mentioned in the preceding and give p values for both. However, there is one simple case for which you will not need a computer—that is, when your F ratio is significant with the least possible number of degrees of freedom for a RM ANOVA, as discussed next.

Lower-Bound Epsilon

Geisser and Greenhouse (1958) demonstrated that the lowest value that is ever required for ϵ (when there is a total lack of sphericity) is $1/(c - 1)$. Because the ordinary df for an RM ANOVA are $c - 1$ and $(c - 1)(n - 1)$, multiplying by this lower-bound for ϵ always yields $(c - 1)/(c - 1)$ and $(c - 1)/[(n - 1)(c - 1)]$, which equals 1 and $n - 1$. Therefore, if your F is significant with the conventional df for an RM ANOVA, it makes sense to look up the critical F for 1 and $n - 1$ df. If your F exceeds this larger critical F, you can declare your results to be statistically significant without concern about Type I error inflation due to a lack of sphericity. On the other hand, if your F is not significant with the conventional df, there is also no danger of Type I error inflation, and nothing left to do. There is no df adjustment that will make it easier to attain significance in this situation.

When your observed F falls somewhere between the conventional F_{crit} and the worst-case F_{crit} (the latter F_{crit} is actually considered overly conservative), the use of statistical software to obtain a more exact solution is recommended. The whole procedure of adjusting the critical F as necessary to accommodate an apparent lack of sphericity in the population is known as the *modified univariate approach,* in contrast to the multivariate approach (i.e., MANOVA)—an entirely different method that I will not be prepared to discuss until Section C of Chapter 18. As an example of the modified univariate approach, let us test the F ratio for the transitive inference experiment, which was conventionally significant, against the Geisser-Greenhouse conservatively adjusted F: $F_{.05}(1, 5) = 6.61$. Because our observed F (8.0) is larger than this worst-case critical F, we can declare our results to be significant without the need to calculate ϵ more precisely.

Post Hoc Comparisons

The optimal method to use for post hoc comparisons depends on whether it is reasonable to assume sphericity. If it is reasonable to make this assumption (e.g., your sample size is large and Mauchly's test for sphericity was not significant), you can proceed with multiple comparisons by any of the methods described in Chapter 13, substituting MS_{inter} for MS_W. On the other hand, if it is not reasonable to assume sphericity, the use of the error term from the overall analysis is not justified for each post hoc comparison. The problem of inflated experimentwise Type I error rates that occurs when the ordinary RM ANOVA is used to analyze studies for which sphericity does not exist in the population becomes especially severe when MS_{inter} is then used for follow-up tests. The problem with post hoc tests is so serious that even when a reasonable judgment is made to use the ordinary RM ANOVA without a sphericity correction, MS_{inter} is not recommended for the follow-up tests unless there is very compelling evidence that the sphericity assumption has been satisfied.

The simplest and safest procedure for pairwise comparisons when sphericity cannot be assumed is to perform the ordinary matched t test for each pair of conditions using a Bonferroni adjustment to control experimentwise alpha. The error term for each test depends only on the two conditions involved in that test, so sphericity is not an issue. Moreover, the amount of power compares favorably to that of other procedures (Lewis, 1993). It is not even necessary to perform an overall RM ANOVA if particular pairwise comparisons have been planned or your post hoc tests are properly protected.

To illustrate the use of post hoc comparisons for an RM ANOVA when sphericity is a safe assumption to make, I will apply an ordinary multiple-comparison method to the example of this section. Because the ANOVA was significant, and has only three levels, it is appropriate to calculate LSD to conduct the post hoc tests (t_{crit} will be based on $\alpha = .05$, and df = df_{inter} = 10):

$$LSD = t_{crit} \sqrt{\frac{2\,MS_{inter}}{n}} = 2.228 \sqrt{\frac{2(6.922)}{6}} = 2.228 \sqrt{2.3073}$$

$$= 2.228(1.519) = 3.38$$

We can see from Table 15.5 that the mean for Vis-sim differs significantly from the mean for Vis-succ (i.e., $16.0 - 12.17 = 3.83 > 3.38$), and from the mean for the Auditory condition, as well (i.e., $16.0 - 10.0 > 3.38$). However, Vis-succ does not differ significantly from Auditory (i.e., $12.17 - 10.0 < 3.38$). If we were to take the more conservative approach of performing the three possible matched t tests, we would find that only the Vis-sim/Auditory comparison would yield a p value less than .05, and even that p value (.022) would not be significant against a Bonferroni-adjusted criterion (.05/3 = .0167).

Varieties of Repeated-Measures and Randomized-Blocks Designs

The types of RM and RB designs are basically just extensions of the two-level designs already discussed in Chapter 11, with some added complications, as described in the following.

Simultaneous RM Design

Sometimes several aspects of a stimulus can be manipulated simultaneously (e.g., the subject sees a slide and is then asked about the number of objects, their colors, their locations, etc.); more commonly, discrete trials of different types can be randomly interspersed (e.g., a subject views a series of slides, some pleasant, some unpleasant, some neutral). In either case, simple order effects cannot occur, but there is still the possibility of complex carryover effects if some of the stimuli are distracting (simultaneous case) or evoke strong reactions (interspersed case).

Successive RM Design

In some types of research, the manipulations underlying the levels of the RM factor are such that it is not practical to repeat the levels more than once each (e.g., testing a subject's blood pressure response to three different drugs). In this case, the order of levels for each subject becomes important, and some form of counterbalancing is necessary. It is also important to avoid carryover effects. More detail concerning counterbalancing will be given in Section C. A special case of successive design, in which counterbalancing is not possible, is the design in which time is the independent variable. This design is discussed next.

Repeated Measures over Time

A common design in some areas of psychological research involves repeated measures of the same dependent variable over time, often at fixed intervals (e.g., each month during treatment). A significant RM ANOVA of this type is often followed by an analysis of trend components to determine whether changes over time are linear, quadratic, and so forth (see Section C). This design is an extension of the before-after design and also usually requires a control group to pin down the cause of any changes observed over time (RM designs with more than one independent group will be discussed in the next chapter).

Repeated Measures with Quantitative Levels

If the levels of your IV are quantifiable, such as different stimulus intensities (e.g., the DV might be changes in blood pressure), some form of regression approach may prove to be a more powerful way to analyze the data than RM ANOVA. In any case, it is important to either randomly intersperse the levels or counterbalance the order of levels to avoid the confounding of order effects and to separate conditions sufficiently to eliminate carryover effects. A significant RM ANOVA can be followed by an analysis of trend components as in the RM over-time design because the levels have been determined quantitatively.

Randomized Blocks

If you cannot avoid serious carryover effects in an RM design, and you have a relevant basis on which to match your subjects before beginning your experiment, the RB design can be a powerful alternative to using independent groups. Although the RB design does not have the same economy as an RM design in reducing the number of subjects required (you need different subjects for each condition), the increased power of the RB design means that it requires fewer subjects than a corresponding independent-samples design. A variation of the RB design, the treatment by blocks design, will be considered in the next chapter. Another possible alternative is the analysis of covariance (see Chapter 18, Section B). Unfortunately, because of the difficulty and expense involved in matching blocks of subjects, the experimental RB design is not very common, especially when the independent variable consists of more than three levels. However, given that the RB design usually has more power than using independent groups and avoids both order and carryover effects completely, it should probably be considered more often.

Naturally Occurring Blocks

Sometimes matched blocks of subjects are created by nature. An example would be a study of families containing three children that explored the effect of birth order on some personality variable. The children are "matched" in that they have all been raised by the same set of parents. Other natural (i.e., not created by the experimenter) blocks are basketball teams, sets of coworkers, and so forth. The disadvantage of using natural blocks is that you cannot infer that the change in the dependent variable is *caused* by whatever distinguishes the members of the blocks (this is the usual problem when dealing with preexisting groups). On the other hand, natural blocks may be relatively easy to use, and their results may suggest interesting experiments.

Publishing the Results of an RM ANOVA

The results of a one-way RM ANOVA are reported in the same way as those for a one-way independent ANOVA. For the example in this section, the results could be presented in the following manner: "Simultaneous visual presentation produced the most accurate performance ($M = 16$), followed by auditory ($M = 12.17$), and then visual-successive ($M = 10$). A repeated-measures ANOVA determined that these means were significantly different, $F(2, 10) = 8.0$, $p = <.05$." It is also likely that this analysis would be referred to as a "within-subjects" ANOVA.

The Summary Table for RM ANOVA

The components of a repeated-measures analysis are sometimes displayed in a summary table that follows the general format described for the independent ANOVA, but with a different way of categorizing the sources of variation. In the independent ANOVA, the total sum of squares is initially divided into between-groups and within-groups components; for the two-way ANOVA, the between-groups SS is then divided further. However, in the RM ANOVA there is only one group of subjects, so the SS is initially divided into a *between-subjects* and a *within-subjects* component. For the one-way RM ANOVA, the between-subjects *SS* consists only of the component I have been referring to as SS_{sub}, and it is not of interest in this design.

On the other hand, the within-subjects *SS* can be divided into two *SS* components. Within each subject there are several scores (one for each treatment level), and these will usually vary. There are two reasons for this variation. One reason is the fact that each treatment level may have a different mean (as the researcher usually hopes), and the other reason is interaction (each subject may have his or her own reactions to the treatment levels that do not conform exactly with the differences in means). Thus the within-subjects *SS* is divided into SS_{RM} and SS_{inter} (or SS_{res}). The successive division of the variation in a one-way RM design can be displayed in the form of a df tree, as shown in Figure 15.6. This division of variation is also evident in the usual format of the summary table for the one-way RM design. The summary table for the transitive inference example is given in Table 15.6.

An Excerpt from the Psychological Literature

The one-way RM ANOVA is not uncommon in the psychological literature (though not as common as more complicated designs). The following description of a published study, along with an excerpt from a report of its results, provides a typical example of the use of the one-way RM ANOVA in psychological research.

Harte and Eifert (1995) measured urinary hormone concentrations for each of 10 marathon runners under three different running conditions (outdoors, on an indoor treadmill while listening to environmental sounds, and on an indoor treadmill while listening to amplified sounds of their own breathing) as well as in baseline (resting) and control (reading sports magazines) conditions. The main findings were as follows:

> One-way repeated measures ANOVA . . . confirmed significant treatment effects for adrenaline ($F[4, 36] = 21.11, p < .0001, \epsilon = 0.6628$), noradrenaline ($F[4, 36] = 32.06, p < .0001, \epsilon = 0.3983$), and cortisol ($F[4, 36] = 41.08, p < .0001, \epsilon = .5309$). Newman-Keuls tests show that levels of adrenaline, noradrenaline, and cortisol were significantly higher after all three running conditions as compared with baseline levels or the control activity. However, Newman-Keuls tests indicate that there were no differences between hormonal concentrations at baseline and following the control session. There were also no hormonal differences among any of the three running conditions except that levels of noradrenaline and cortisol were significantly higher after the indoor run with internal attention focus [listening to breathing] than after the outdoor run (pp. 52–53).

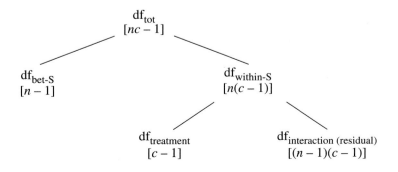

Figure 15.6

df Tree for the One-Way RM Design

Source	SS	df	MS	F	p	Table 15.6
Between-Subjects	119.6	5				
Within-Subjects						
Between-Treatments	110.8	2	55.4	8.0	<.01	
Interaction (Residual)	69.2	10	6.92			
Total	299.6	17				

Note that the epsilon (ϵ) values following the p values are based on estimates of the degree to which the sphericity assumption has been violated, as I described in an earlier subsection. The extremely large F ratios are due to the dramatic differences in the levels of hormones secreted while running as compared to the levels at rest. This experimental design is one in which a partial null hypothesis seems quite possible—that is, the three running conditions are equal, the baseline and control conditions are equal, but these two subsets of conditions are not equal to each other. This is not a good situation in which to use the Newman-Keuls test; if a partial null hypothesis is true, the experimentwise alpha is likely to be greater than the level set for the Newman-Keuls test. Given the low values for epsilon, separate matched t tests using Bonferroni-corrected alpha levels would be more appropriate (but with such dramatic results, it is not likely that any of the conclusions would change).

1. A simple way to compute a one-way RM ANOVA is first to calculate the SS components. SS_{total}, SS_{RM}, and SS_{sub} are calculated by finding the biased variance of the individual scores, the column (i.e., treatment) means, and the row (i.e., subject) means, respectively, and multiplying each by the total number of scores. SS_{RM} and SS_{sub} are then subtracted from SS_{total} to obtain SS_{inter} (also called $SS_{residual}$ or SS_{error}). MS_{RM} and MS_{inter} are then found and formed into an F ratio to test the null hypothesis that the population means at each level of the RM factor are identical to each other.

2. There are effect-size measures that take advantage of the reduced error term of the RM ANOVA (e.g., η^2_{RM}, f_{RM}), but these can be misleadingly large when planning future studies that may not include matching or repeated measures. An estimate of ordinary omega squared is recommended, as it is less specific to the experimental design used. The power of an RM ANOVA increases as the correlations between pairs of RM levels increase. If all of the correlations were the same, Cohens f for an RM ANOVA would be larger than the ordinary f by a factor of one divided by the square root of one minus the common correlation coefficient.

3. The validity of the F ratio previously described depends on the following assumptions:
   a. *Independent random sampling.* In an RM design, subjects should all be selected independently. In an RB design, subjects in different blocks should be independent, but subjects *within* a block should be matched.
   b. *Multivariate normal population distribution.*
   c. *Homogeneity of variance.*
   d. *Homogeneity of covariance.* When this assumption and assumption c are true simultaneously, the population exhibits compound symmetry, which is desirable, but not necessary for the RM ANOVA to

B

SUMMARY

be valid. Assumptions c and d can be relaxed as long as a condition called sphericity (or circularity) can be assumed. Sphericity is said to exist when the variability of the difference scores between any two levels of the independent variable in the population is the same as the variability of the difference scores for any other pair of levels.

4. When the null hypothesis is true but the population does *not* exhibit sphericity, the F ratios tend to be larger than normally expected (there is a positive bias to the test), and therefore the Type I error rate tends to be larger than the alpha used to determine the critical F. Unfortunately, sphericity is likely to be violated in an RM design, and tests of the sphericity assumption have little power to reveal this when sample sizes are small.

5. You need not worry about the sphericity assumption if your F ratio is less than the usual critical F or greater than a worst-case conservatively adjusted F (the Geisser-Greenhouse F test). If the F ratio lands between these extremes, and you suspect a violation of sphericity, a more precise adjustment of the degrees of freedom based on a coefficient called epsilon is called for; there are two common ways to calculate epsilon, both of which can be obtained easily using statistical software.

6. When sphericity can be assumed, post hoc comparisons are based on MS_{inter} in place of MS_W in the usual post hoc procedures. When this assumption is not reasonable, the safe approach is to conduct separate matched t tests for each pair of conditions, adjusting alpha according to the Bonferroni test.

7. If it is feasible, the RM design is the most desirable because it generally has more power than independent groups or randomized blocks. In a simultaneous design the different conditions are presented as part of the same stimulus or as randomly mixed trials. A successive design usually requires counterbalancing to avoid simple order effects and is not valid if there are differential carryover effects.

8. The RB design has much of the power of the RM design and avoids the possible order and carryover effects of a successive RM design. In the simplest experimental RB design, the number of subjects per block is the same as the number of experimental conditions, and the subjects in each block are assigned to the conditions at random. An RB design can also be based on naturally occurring (i.e., preexisting) blocks. The drawback to this design is that you cannot conclude that your independent variable *caused* the differences in your dependent variable.

EXERCISES

*1. This exercise is based on data from a hypothetical repeated-measures experiment in which the independent variable is the level of noise: ordinary background noise, moderately loud popular music, and very loud heavy metal music. The number of tasks completed by each of five subjects under all three noise conditions is shown in the table at the right:

a. Perform an RM ANOVA on the data and present the results in a summary table. Is the F ratio significant at the .05 level? At the .01 level?

b. Would the results be significant at the .05 level if you were to assume a total lack of sphericity? Show how you arrived at your answer.

Subject	Background Noise	Popular Music	Heavy Metal Music
1	10	12	8
2	7	9	4
3	13	15	9
4	18	12	6
5	6	8	3

2. To illustrate the effect of increasing subject-to-subject variance on the *F* ratio of the RM ANOVA, the data from Exercise 1 have been modified to produce the following table. Ten points were added to each of the scores for subject 1, 20 points for subject 3, and 30 points for subject 4.

Subject	Background Noise	Popular Music	Heavy Metal Music
1	20	22	18
2	7	9	4
3	33	35	29
4	48	42	36
5	6	8	3

a. Perform an RM ANOVA on the data, and present the results in a summary table.
b. Compare the summary table for this exercise with the one you made for Exercise 1. Which of the *SS* components reflects the change in subject-to-subject variability? What effect does an increase in subject-to-subject variability have on the *F* ratio?
c. Calculate η^2_{RM} for this exercise and for the previous exercise, and compare the values. What general principle is being illustrated?
d. Calculate an estimate of ordinary ω^2 for both this exercise and the previous one. Explain the difference between the two values.

*3. A psychophysiologist wishes to explore the effects of public speaking on the systolic blood pressure of young adults. Three conditions are tested. The subject must vividly imagine delivering a speech to one person, to a small class of 20 persons, or to a large audience consisting of hundreds of fellow students. Each subject has his or her blood pressure measured (mm Hg) under all three

conditions. Two subjects are randomly assigned to each of the six possible treatment orders. The data appear in the following table:

Subject No.	One Person	Twenty People	Large Audience
1	131	130	135
2	109	124	126
3	115	110	108
4	110	108	122
5	107	115	111
6	111	117	121
7	100	102	107
8	115	120	132
9	130	119	128
10	118	122	130
11	125	118	133
12	135	130	135

a. Perform an RM ANOVA on the blood pressure data and write the results in words, as they would appear in a journal article. Does the size of the audience have a significant effect on blood pressure at the .05 level? (*Hint*: Subtract 100 from every entry in the preceding table before computing any of the SSs. This will make your work easier without changing any of the *SS* components or *F* ratios.)
b. What might you do to minimize the possibility of carryover effects?
c. Calculate η^2_{RM} from the *F* ratio you calculated in part a. Does this look like a large effect? How could this effect size be misleading in planning future experiments?
d. Test all the pairs of means with protected *t* tests using the error term from the RM ANOVA. Which pairs differ significantly at the .05 level?

*4. A statistics professor wants to know if it really matters which textbook she uses to teach her course. She selects four textbooks that differ in approach and then matches her 36 students into blocks of four based on their similarity in math background and aptitude. Each student in each block is randomly assigned to a different text. At some point in the course, the professor gives a surprise 20-question quiz. The number of questions each student answers correctly appears in the following table:

Block No.	Text A	Text B	Text C	Text D
1	17	15	20	18
2	8	6	11	7
3	6	5	10	6
4	12	10	14	13
5	19	20	20	18
6	14	13	15	15
7	10	7	14	10
8	7	7	11	6
9	12	11	15	13

a. Perform an RM ANOVA on the data, and present the results of your ANOVA in a summary table. Does it make a difference which textbook the professor uses?

b. Considering your answer to part a, what type of error could you be making (Type I or Type II)?

c. Would your F ratio from part a be significant at the .01 level if you were to assume a maximum violation of the sphericity assumption? Explain.

d. Test all the pairs of means with Tukey's HSD, using the error term from the RM ANOVA. Which pairs differ significantly at the .05 level?

5. a. Perform a one-way independent-groups ANOVA on the data from Exercise 4. (*Note*: You do not have to calculate any *SS* components from scratch; you can use the *SS*s from your summary table for Exercise 4.)

b. Does choice of text make a significant difference when the groups of subjects are considered to be independent (i.e., the matching is ignored)?

c. Comparing your solution to this exercise with your solution to Exercise 4, which part of the F ratio remains unchanged? What can you say about the advantages of matching in this case?

*6. A neuropsychologist is exploring short-term memory deficits in people who have suffered damage to the left cerebral hemisphere. He suspects that memory for some types of material will be more affected than memory for other types. To test this hypothesis he presented six brain-damaged subjects with stimuli consisting of strings of digits, strings of letters, and strings of digits and letters mixed. The longest string that each subject in each stimulus condition could repeat correctly is presented in the following table. (One subject was run in each of the six possible orders.)

Subject	Digit	Letter	Mixed
1	6	5	6
2	8	7	5
3	7	7	4
4	8	5	8
5	6	4	7
6	7	6	5

a. Perform an RM ANOVA. Is your calculated F value significant at the .05 level?

b. Would your conclusion in part a change if you could not assume that sphericity exists in the population underlying this experiment? Explain.

c. Based on the graph you drew of these data for Exercise 15A2, would you say that the RM ANOVA is appropriate for these data? Explain.

d. Test all the possible pairs of means with separate matched t tests (or two-group RM ANOVAs) at the .05 level.

*7. A school psychologist is interested in determining the effectiveness of an antidrug film on the attitudes of eighth-grade students. Each student's antidrug attitude is measured on a scale from 0 to 20 (20 representing the strongest opposition to drug use) four times: the day before the film, as students are sitting in class waiting for the film to start, immediately after the film, and the next day in class. The data for eight subjects appear in the following table:

Subject	Day Before	Prior to Film	After Film	Day After
1	14	15	18	17
2	10	13	19	17
3	15	15	18	18
4	13	16	20	18
5	7	9	15	16
6	10	9	14	11
7	16	17	19	19
8	8	10	16	15

a. Perform an RM ANOVA on these data to determine if there is a difference in attitude over time at the .01 level.

b. Calculate an estimate of ordinary ω^2 for these data. Does the film appear to be having a large impact on the students' attitudes? Explain.

c. Test all the pairs of means that are adjacent in time with separate matched t tests (or two-group RM ANOVAs) at the .05

level. According to the Bonferroni test, what alpha should be used for each pairwise comparison?

8. A clinical psychologist wants to test the effects of exercise and meditation on moderate depression. She matched 30 of her patients into blocks of three based on the severity of their depression and their demographic characteristics. Patients were randomly assigned from each block to one or another of the following three treatments: aerobic exercise, meditation, and reading inspirational books (control). After two daily 30-minute sessions of the assigned treatment over the course of a month, each subject's level of depression was measured with a standard questionnaire. The depression scores for each condition appear in the following table:

Subject	Aerobic Exercise	Meditation	Control
1	22	25	28
2	13	13	12
3	24	31	34
4	11	14	13
5	18	19	26
6	16	19	23
7	15	15	18
8	10	13	14
9	15	14	18
10	9	10	16

a. Perform an RM ANOVA on these data. Is your obtained F significant at the .05 level?

b. Assuming sphericity, which pairs of treatments differ significantly at the .05 level?

*9. a. Calculate f corresponding to the F ratio you found in Exercise 5 in this section. Then calculate f_{RM} for Exercise 4 (use Formula 12.16 in both cases). Plug these two effect-size estimates into Formula 15.7, and solve for ρ. What is your estimate for the average correlation between pairs of levels in the data of Exercise 4?

b. What can you say about the degree of matching achieved in Exercise 4?

10. a. For Exercise 7 in this section, use Formula 15.8 to convert your estimate of ω^2 to an estimate of f. Given this value for ordinary f, how many subjects would be needed in each of four independent groups to have power equal to .7 for a .05, two-tailed test?

b. Given the value of f you found in part a, and the number of subjects used in Exercise 7, how large would ρ_{avg} have to be to have power equal to .7 for a .05, two-tailed test?

Counterbalancing

As I have mentioned before, counterbalancing can eliminate simple order effects but not complex order effects such as differential carryover effects. To see the type of order effects you have in your data, if any, you can plot your treatment levels as a function of serial position. Of course, you must have kept track of the treatment order given to each subject. This information is found in Table 15.4 for the inference experiment. Notice that any particular condition (e.g., vis-sim) occurs twice in the first place, twice as the second treatment, and twice as the last treatment. To plot position and treatment on the same graph, begin by marking off the three positions along the X axis. Then plot a line for each treatment level in the following manner. To find the value for vis-sim in position 1, average together the vis-sim scores for the two subjects who received the vis-sim condition first (i.e., the first two subjects, according to the orders listed in Table 15.4). As another example, you can find the value for auditory in position 2 by averaging together the auditory scores of subjects 2 and 6. Do this for all of the other positions and the other treatments. The resulting position by treatment graph is shown in Figure 15.7.

Note the almost total lack of interaction for the two visual conditions. The nearly parallel lines indicate that if only these two conditions were considered, we could conclude that only simple order effects are present (if

OPTIONAL MATERIAL

Figure 15.7	
Graph of the Data in Table 15.5, Averaging across Serial Position	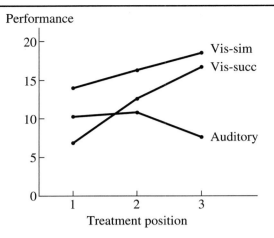

there were no order effects of any kind, all of the lines would be horizontal). The positive slopes of the two lines indicate a practice effect that is about the same for both conditions. In the ideal case of simple order effects, the practice (or fatigue) effect, and therefore the slope, would be exactly the same for all three conditions, and counterbalancing would completely balance out the order effects. However, the auditory condition is not parallel to the two visual conditions, producing an overall interaction in the graph. This interaction indicates that the order effects are not entirely simple and cannot be completely eliminated through counterbalancing (the auditory condition is not benefiting from practice like the other conditions, and therefore its overall mean will be misleadingly low compared to the visual conditions). The degree of position by treatment interaction can be quantified and tested for significance in the context of a mixed-design ANOVA, as will be described in the next chapter.

Whereas counterbalancing succeeds in averaging out simple order effects, so that no treatment level attains an unfair advantage by virtue of its ordinal position, even simple order effects lead to an increase in the error variance. Different subjects receive a particular treatment level in different positions in the order, so some of the variability in scores at a given treatment level is due to the different order effects at different positions. It takes a mixed-design ANOVA (see next chapter) to separate this added variance and remove it from the error term.

Latin Square Designs

Complete counterbalancing becomes a nuisance when there are four levels, and becomes impractical when there are five or more levels. (With 5 levels, for instance, the number of orders is given by the mathematical function 5!—pronounced 5 factorial—which equals 120. The factorial function is described in Chapter 19, Section C.) Fortunately, complete counterbalancing is not necessary for averaging out simple order effects. One of various schemes for partial counterbalancing, such as the *Latin square design*, will suffice. For the transitive inference experiment, only three orders are needed:

> vis-sim, vis-succ, auditory
> vis-succ, auditory, vis-sim
> auditory, vis-sim, vis-succ

Notice that each treatment level appears once in each of the three ordinal positions.

With four treatment levels, four orders are needed (in the Latin square design, the number of orders always equals the number of treatment levels), but there are a variety of schemes that will balance out order effects. For example, if the treatment levels are labeled a, b, c, and d, the following four orders can be used:

> a, b, c, d
> b, c, d, a
> c, d, a, b
> d, a, b, c

Although each treatment level appears in each ordinal position only once, this set of orders is not considered the most desirable. The problem is that a particular level is always preceded by the same other level (e.g., b is always preceded by a, except, of course, when b comes first). A more desirable set of orders is the following:

> a, b, c, d
> c, a, d, b
> b, d, a, c
> d, c, b, a

Now each treatment level appears in each ordinal position only once, *and* each level is preceded by each other level once (e.g., b is preceded by a in the first order, then by d in the second, and finally by c in the fourth). This design is said to be *digram balanced*. When there is an even number of treatments (c), a digram-balanced Latin square with c orders can be created. When c is odd, however, two Latin squares of c orders each must be constructed to achieve digram balancing. In the case of three conditions, the only way to have digram balancing is to use all six possible orders. Because there is little advantage to using three orders instead of six, complete counterbalancing is generally recommended for $c = 3$. For five conditions, a total of 10 orders is required for digram balancing, but this is still an enormous reduction from the 120 orders that are possible.

Random Orders

In some experimental designs, the number of possible orders is extremely large. For instance, there are over three million possible orders in which you can present a list of 10 words for memorization. When balancing the orders seems impractical, but you want the generalizability created by giving different orders to different subjects, you can select one order randomly (out of the many available) for each subject. Whereas the random selection of orders is not likely to represent any kind of perfect balancing of the orders, it is also not likely to represent any particular kind of bias.

Trend Analysis with Repeated Measures

When the factor in an RM ANOVA has quantitative levels, it is likely that a test of trend components will have much greater power than the ordinary one-way RM ANOVA (this applies equally to the RB design), just as we have seen for independent-groups ANOVA. However, the calculation procedure is quite different for the RM design; the test of an RM trend actually requires a fancy version of the matched *t* test. The coefficients from Table A.12 are applied to individual subjects (or blocks) rather than treatment means. This process creates a single trend score for each individual; the mean of the trend scores is tested against zero just as in a matched *t* test.

The method will be illustrated by testing the linear trend in the data of Exercise 15B7, reproduced below.

Table 15.7	Subject	Day Before	Prior to Film	After Film	Day After	L_{linear}
	1	14	15	18	17	+12
	2	10	13	19	17	+27
	3	15	15	18	18	+12
	4	13	16	20	18	+19
	5	7	9	15	16	+33
	6	10	9	14	11	+8
	7	16	17	19	19	+11
	8	8	10	16	15	+27

The rightmost column (L_{linear}) was created by applying the following linear trend coefficients (from Table A.12) to the four scores in each row: $-3, -1, +1, +3$ (e.g., $-3 \cdot 14 - 15 + 18 + 3 \cdot 17 = -42 - 15 + 18 + 51 = +12$). These L scores are then treated exactly like the difference scores in a matched t test. (Note: Although all of the Ls in Table 15.7 have the same sign, this will not always be the case.) I'll just change the Ds to Ls in Formula 11.1, to create a formula for the L scores.

$$t_{trend} = \frac{\overline{L}_{trend}}{\frac{s_L}{\sqrt{n}}}$$

Formula 15.9

$$t_{trend} = \frac{18.625}{\frac{9.3034}{\sqrt{8}}} = \frac{18.625}{3.289} = 5.66$$

If you want to report the corresponding F ratio, just square the t value. In this example, $F_{linear} = 5.66^2 = 32.04$, which happens to be only slightly larger than F_{RM} (30.7). Although the data rise fairly linearly over time, there are considerable quadratic and cubic components, as well (e.g., the data drop off slightly, but rather consistently, from "After Film" to "Day After").

Trend components are often more interesting when they interact with a second factor, as described in the previous chapter. In particular, the experiment just described is sorely in need of a control group—perhaps, a group that is shown an irrelevant film of the same length. This would create a second factor: type of film. If the same subjects were shown both films (weeks apart) in counterbalanced order, or subjects could be matched in pairs and randomly assigned to one film group or the other, one could test for differences in trend between the two film groups—i.e., an interaction with trend components. This is a special case of a two-way repeated-measures ANOVA. First, I will describe the calculation of the two-way RM ANOVA with two qualitative factors, and then consider the case in which one of the factors has quantitative levels.

The Two-Way RM ANOVA

Given the power advantage of repeated measures or matching, it should not be surprising that researchers try to use RM and RB designs whenever possible—including for two-way (and higher) factorial ANOVAs. The two-way RM design is used commonly in cognitive psychology, for instance, in which it is possible to mix many trials of different types randomly, so that all of the conditions are presented essentially simultaneously. Imagine a

3×4 memory experiment, in which an equal number of very concrete, moderately concrete, or only slightly concrete words are presented in a very long study list either one, two, three, or four times each. Later, some aspect of recall or recognition of the words is measured. Although each of the words on the recall/recognition list can be categorized as belonging to one or another of the 12 cells in the two-way design, order effects are not an issue. However, when conditions must be presented successively, counter-balancing quickly becomes problematic for a two-way RM ANOVA with multiple levels for both factors. In such cases, it is likely that the levels will be repeated for only one of the factors, while matched or even independent groups are used at each level of the other factor. Mixing an RM with an independent factor in a two-way ANOVA is the topic of the next chapter. However, when one factor is repeated and the other involves matching, the calculations are exactly the same as for a two-way RM ANOVA. Therefore, in the cause of realism, I will use just such an experimental design in this section to demonstrate the computation of the two-way RM ANOVA.

Imagine that the data in Table 15.5 represent just half of an experiment: those subjects who did not receive visual imagery instructions for transitive inference problem solving. For each of the six subjects whose scores appear in Table 15.5, there is a matched subject (based on a problem-solving pretest) who received imagery instructions before solving the same problems in the same order. The data from Table 15.5 appear in the following table in the columns designated by a "U" for uninstructed; the matched subjects are in the "I" (for instructed) columns. The "M" columns contain the means of the scores from each pair of matched subjects. The Type means are the means of each subject's scores across the three presentation *types*, and the Block means are the averages of the Type means across the two subjects matched in each block. (Alternatively, you can think of all six scores—three presentation types at two levels of instruction—as coming from a single subject, or a single block of six matched subjects. It would not affect the computations in this section.)

As you learned in Section B, a one-way RM ANOVA can be computed as though it were a two-way ANOVA with only one score per cell. Similarly, a two-way RM ANOVA can be computed as though it were a *three*-way ANOVA with one score per cell, using the procedure of Section C in the previous chapter. For instance, the data in Table 15.8 can be viewed as coming from a 3 (presentation types) $\times$ 2 (instruction types) $\times$ 6 (blocks) ANOVA. (When using the formulas for the three-way ANOVA, I recommend changing the letter designation of the third variable from C to S, for the subject factor.) Just as the MS for the interaction of the treatment and subject factors is used as the error term in a one-way RM ANOVA, the MS for the $A \times S$ interaction serves as the error term for testing the main effect of the A factor, the $B \times S$ interaction is used as the error term for the B factor, and the $A \times B \times S$ interaction serves as the error term for the $A \times B$ interaction.

To make the analysis of the data in Table 15.8 as concrete as possible, I will begin by testing the main effects for both the presentation- and instruction-type factors. To test the main effect of presentation type, we average the scores across the two instruction types, and perform a one-way RM ANOVA on those averages—exactly as in Section B of this chapter. Because those means are already in Table 15.8, we can just apply the procedures of Section B (see Step 5) to the data in the M columns (bear in mind that N_T is 36 for all of the calculations for this two-way RM ANOVA). The biased variance of the 18 scores in the M columns is 11.278, so $SS_{total} = 36 \cdot 11.278 = 406$. Based on the means of those columns, $SS_{RM} = N_T \cdot \sigma^2$ (15.33, 11.67, 11.5) $= 36 \cdot 3.12957 = 112.66$, and based on the block means, $SS_{sub} = 36 \cdot \sigma^2$ (13.167, 14.333, 15.5, 13.833, 9.0, 11.167) $= 36 \cdot 4.6574 = 167.67$. So, SS_{inter}

= 406 − 112.66 − 167.67 = 125.67. (In the notation of the three-way ANOVA, $SS_{total} = SS_{AS}$; $SS_{RM} = SS_A$; $SS_{sub} = SS_S$; and $SS_{inter} = SS_{A \times S}$.) The dfs are the same as in the Section B example, so $MS_{RM} = 112.66/2 = 56.33$, and $MS_{inter} = 125.67/10 = 125.7$. Therefore, the F ratio for the main effect of the presentation factor is $56.33/12.57 = 4.48$, which is larger than the critical F at the .05 level (4.10), and therefore statistically significant—if sphericity is assumed. This calculated F, however, does not exceed the most conservatively adjusted F_{crit} (i.e., $F_{.05}$ (1, 5) = 6.61), so the more cautious approach would be to use statistical software to find the p value, after the dfs have been adjusted according to Greenhouse and Geisser's (1959) formula for epsilon.

In exactly the same manner, we can analyze the 12 values that appear in the two columns labeled "Type Means" to find the main effect of instruction type. The SS_{total} for these 12 values is 178.33. Based on the column means, $SS_{RM} = 36 \cdot \sigma^2$ (12.72, 12.94) = .44. SS_{sub} is based on the same block means in the previous computation, so again $SS_{sub} = 167.67$. So, $SS_{inter} = 178.33 − .44 − 167.67 = 10.22$. (In the notation of the three-way ANOVA, $SS_{total} = SS_{BS}$; $SS_{RM} = SS_B$; $SS_{sub} = SS_S$; and $SS_{inter} = SS_{B \times S}$.) Because $df_{RM} = 1$, $MS_{RM} = .44$. $MS_{inter} = 10.22/5 = 2.044$, so the F ratio for the main effect of the instruction factor is $.44/2.044 = .22$, which is obviously not significant.

To calculate the F ratio for the interaction of the two RM factors involves a bit more tedium, but not as much as you might think. First, we need to calculate $SS_{between-cells}$, which is based on the cell means for the six combinations of presentation types and instruction types; these means appear in the bottom row of Table 15.8. $SS_{bet-cells} = 36 \cdot \sigma^2$ (16, 12.17, 10.0, 14.67, 11.17, 13) = $36 \cdot 4.11 = 148$. (In the notation of the three-way ANOVA, $SS_{bet-cells} = SS_{AB}$.) The SS for the interaction of the two factors (i.e., $SS_{A \times B}$) is found by subtracting the SSs for the two main effects from $SS_{bet-cells}$: $SS_{A \times B} = 148 − 112.66 − .44 = 34.9$, so $MS_{A \times B} = 34.9/2 = 17.45$.

Finally, the error term for the interaction of the two factors is based on the three-way interaction of the subject factor with the two experimental factors (i.e., $SS_{A \times B \times S}$); it can be found by subtracting the SSs for the three main effects (i.e., the two SS_{RM}s we calculated, and SS_{sub}), and the three interaction SSs (i.e., the two SS_{inter}s we calculated, and $SS_{A \times B}$), from the total SS for all 36 observations in the analysis: "total" SS_{total} (i.e., SS_{ABS}) = $36 \cdot 12.8056 = 461$. Therefore, $SS_{A \times B \times S} = 461 − 112.66 − .44 − 167.67 − 125.67 − 10.22 − 34.9 = 9.44$. The df for this $3 \times 2 \times 6$ interaction is $2 \times 1 \times 5 = 10$, so $MS_{A \times B \times S} = .944$, and $F_{A \times B} = 17.45/.944 = 18.48$, thus completing the analysis. The interaction is significant, even with the dfs adjusted for a total lack of sphericity (i.e., $18.48 > 6.61$). It is easy to see the source of the interaction by inspecting a graph of the cell means, as shown in Figure 15.8.

Given that the interaction is significant, you might want to follow up your ANOVA by testing the simple main effects of the presentation factor separately for each level of the instruction factor. In fact, the one-way RM

Table 15.8

Block No.	Vis-sim (U)	Vis-succ (U)	Auditory (U)	Type Means (U)	Vis-sim (I)	Vis-succ (I)	Auditory (I)	Type Means (I)	Vis-sim (M)	Vis-succ (M)	Auditory (M)	Block Means
1	15	13	12	13.33	12	12	15	13	13.5	12.5	13.5	13.17
2	14	16	15	15	13	14	14	13.67	13.5	15	14.5	14.33
3	20	17	10	15.67	18	15	13	15.33	19	16	11.5	15.5
4	17	12	11	13.33	15	12	16	14.33	16	12	13.5	13.83
5	12	7	5	8	13	8	9	10	12.5	7.5	7	9
6	18	8	7	11	17	6	11	11.33	17.5	7	9	11.17
Col. Means	16	12.17	10	12.72	14.67	11.17	13	12.94	15.33	11.7	11.5	12.833

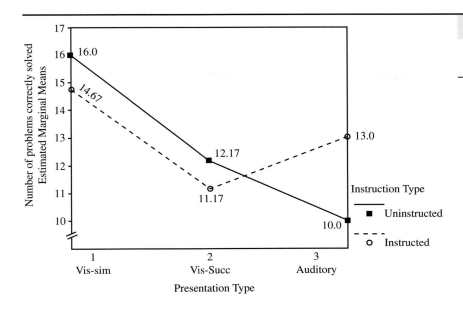

Figure 15.8

Cell Means from
Table 15.8

ANOVA in Section B is one of these simple main effects, and we already know that it is significant. Of course, the Section B ANOVA does not use the presentation-factor error term from the two-way RM ANOVA (which includes data from the instructed group), and therefore loses some potential power. However, the more conservative approach of basing your error term just on the subset of data in the simple effect you are testing, and not on omnibus error terms, seems generally preferred, anyway, in order to avoid making an assumption about sphericity. In the same vein, significant simple effects can be further explored by testing pairs of levels with ordinary matched t tests, using a Bonferroni-adjusted alpha. (Had the interaction not been significant, the significant main effect of the presentation factor would have justified the performance of matched t tests for pairs of presentation types, using data from both of the instruction levels.) On the other hand, considering the significant interaction and the pattern of cell means in Figure 15.8, it would make sense to test the 2×2 interaction contrast formed by ignoring the data from the first level of the presentation factor (i.e., vis-sim).

As you might guess, a three-way RM ANOVA is calculated like a four-way independent-groups ANOVA, and so forth. However, a more common three-way factorial design would arise by adding an independent-groups factor to the two-way RM ANOVA we just calculated (e.g., select half the subjects for their high scores on a test of visual imagery or spatial ability, and the other half for their low scores on the same test). I will deal with such complex ANOVA designs in an extra chapter to be posted on the web and linked to my statistics home page. I will end this chapter, as promised, by considering the case of the two-way RM ANOVA with one quantitative factor.

Interaction with Trend Components

Imagine that a comparison group was added to the experiment described in Exercise 15B7; a second data table, similar to Table 15.7, could be constructed for the data from the comparison group. If the comparison subjects had been matched with the experimental subjects before randomly assigning them (or they are the *same* subjects run at a different time), the two columns of L scores would also be matched into eight pairs of linear-trend scores. Then, a simple matched t test could be performed on the matched L scores to determine whether there is a significant difference between the two film groups in the linear slope of the attiude scores over time. Quite often, looking at how scores change over time is much more informative than taking a snapshot of the data at just one, arbitrarily selected, point in time.

Of course, it would be just as easy to test for differences in quadratic trends between the two groups. In fact, the L scores could represent any linear contrast of interest. For example, for the data in Table 15.8, L scores could be created by subtracting the Auditory score from the average of the two visual conditions for each individual, and then tested to see if they differ significantly between the two instruction groups. What if there were three matched instruction groups (e.g., visual imagery, verbal rehearsal, and no instructions)? Instead of a matched t test between two sets of L scores, you could perform a one-way RM ANOVA on three (or more) matched sets of L scores, likely followed by pairwise comparisons (i.e., matched t tests). Thus, any two-way RM ANOVA can be transformed into a one-way RM ANOVA, by combining the levels of one of the factors into a single linear-contrast. When matching or repeated measures is not convenient for one of the factors (e.g., you may want to show the anti-drug and the neutral films at the same time to two different classes, and can find no relevant basis on which to match the students between the two classes), the second factor can be a between-groups factor, creating a mixed-design ANOVA. It is not difficult to test for differences in trend components between independent groups, but it will require methods to be discussed in the next chapter.

SUMMARY

1. If only simple order effects are present in a successive RM design (i.e., there is no interaction between position and treatment), counterbalancing will average them out, so no treatment level has an unfair advantage as a result of its ordinal position. Moreover, a mixed-design ANOVA can be used to remove from the error term the extra amount of variability due to the order effects, thus increasing the power of the RM ANOVA (see the next chapter). However, when a graph of the treatments as a function of ordinal position reveals a strong interaction, counterbalancing will not do a good job of removing order effects from your treatment effects. A special ANOVA procedure can test the amount of position by treatment interaction but cannot remove the possible confounding due to carryover effects.

2. When there are only two or three treatment levels, complete counterbalancing is a practical solution. When there are four or more levels, a partial counterbalancing design such as the *Latin square* is almost always more convenient. The number of orders required by the Latin square design is equal to the number of treatment levels. If digram balancing is desired and there is an odd number of treatments, a combination of two Latin squares is required.

3. When an RM factor has quantitative levels (e.g., different numbers of repetitions of a subliminal stimulus), it usually makes sense to analyze it with polynomial trend components. Instead of applying the appropriate trend coefficients to the means of the different levels of the RM factor, the coefficients are applied to the data for each individual (or block) to create L scores. These L scores (say, for the linear trend) are then tested for significance exactly like the difference scores of a matched t test (i.e., a one-sample test is conducted against a null hypothesis of zero).

4. A two-way RM ANOVA is appropriate when both factors involve either repeated measures or matching of subjects into blocks. The calculation of

the two-way RM ANOVA is exactly the same as a three-way ANOVA on the same data, viewing the subjects as the different levels of the third factor (leaving only one observation in each cell, and therefore no MS_W term). The error term for each main effect is based on the interaction of the subject factor with that effect, after averaging across the other factor; the interaction of the two RM factors is tested against the three-way interaction of the subject factor with the other two factors (the magnitude of the three-way interaction depends on the degree to which different subjects each exhibit the same two-way interaction for the two RM factors).

5. A linear contrast performed on one of two RM factors reduces the analysis to a one-way RM ANOVA on the L scores created by applying that linear contrast to the data from each subject. If one of the factors has quantitative levels, it is common to create polynomial trend scores for that factor, and then perform a one-way RM ANOVA to look for differences in trends among the levels of the other factor. The F ratio from that one-way RM ANOVA provides a test of the interaction between the qualitative factor, and the trends of the quantitative factor.

EXERCISES

*1. Referring to Exercise 3 in the previous section, if you wanted to use a Latin square design for counterbalancing, instead of using all possible orders
   a. How many different orders would you have to use?
   b. Write out all of the treatment orders you would use.
   c. Given the sample size for that exercise, how many subjects would be assigned to each order?

2. In Exercise 4 of Section A, each subject solved problems under four different drug conditions: marijuana, amphetamine, valium, and alcohol. Write out all the orders of the drug conditions that would be used in a Latin square design that balanced for preceding condition as well as ordinal position.

*3. Calculate, and test for significance, the linear, quadratic, and cubic trends for the data in Exercise 5 of Section A.

4. Calculate, and test for significance, the linear and quadratic trends for the data in Exercise 6 of Section A. Test the residual SS for significance. Are tests of higher-order trends warranted? Explain.

*5. Imagine that subjects are matched in blocks of three based on height, weight, and other physical characteristics; six blocks are formed in this way. Then the subjects in each block are randomly assigned to three different weight-loss programs. Subjects are measured before the diet, at the end of the diet program, 3 months later, and 6 months later. The

results of the two-way RM ANOVA for this hypothetical experiment are given in terms of the SS components, as follows: $SS_{diet} = 403.1$, $SS_{time} = 316.8$, $SS_{diet \times time} = 52$, $SS_{diet \times S} = 295.7$, $SS_{time \times S} = 174.1$, and $SS_{diet \times time \times S} = 230$.
   a. Calculate the three F ratios, and test each for significance.
   b. Find the conservatively adjusted critical F for each test. Will any of your conclusions be affected if you do not assume that sphericity exists in the population?

6. A psychologist wants to know how both the affective valence (happy vs. sad vs. neutral) and the imageability (low, medium, high) of words affect their recall. A list of 90 words is prepared with 10 words from each combination of factors (e.g., happy, low imagery: promotion; sad, high imagery: cemetery) randomly mixed together. The number of words recalled in each category by each of the six subjects in the study is given in the following table:
   a. Perform a two-way RM ANOVA on the data. Test the three F ratios for significance, and present your results in a summary table.
   b. Find the conservatively adjusted critical F for each test. Will any of your conclusions be affected if you do not assume that sphericity exists in the population?
   c. Draw a graph of the cell means, and describe any trend toward an interaction that you can see.

Subject No.	Low	SAD Medium	High	Low	NEUTRAL Medium	High	Low	HAPPY Medium	High
1	5	6	9	2	5	6	3	4	8
2	2	5	7	3	6	6	5	5	6
3	5	7	5	2	4	5	4	3	7
4	3	6	5	3	5	6	4	4	5
5	4	9	8	4	7	7	4	5	9
6	3	5	7	4	5	6	6	4	4

   d. Based on the variables in this exercise, and the results in part a, what post hoc tests would be justified and meaningful?

*7. Suppose that the data in Exercise 5 of Section A represent only half of that experiment: subjects who had been deprived of sleep for 24 hours. Matched to those six blocks of subjects were six other blocks of subjects who were well rested when tested on the same videogame with the same doses of caffeine. The data for both (matched) sets of subjects appear in the following table

8. Suppose that the data in Exercise 6 of Section A represent only half of that experiment: patients treated with psychodynamic therapy (PDT). Matched to these eight patients were eight other patients with similar symptoms and depression ratings, who were treated by a different therapist, employing cognitive/behavioral therapy (CBT) techniques. The monthly depression

Well-Rested	0 mg	100 mg	200 mg	300 mg	Sleep-Deprived	0 mg	100 mg	200 mg	300 mg
1	19	11	10	6	1	25	16	6	8
2	15	10	5	14	2	19	15	14	18
3	13	14	14	11	3	22	19	9	9
4	10	6	8	9	4	15	11	5	10
5	18	9	15	7	5	16	14	9	12
6	16	18	10	5	6	20	23	11	13

(i.e., it is as though there were only a total of six different subjects in the entire study, each run under eight different dosage/sleep conditions).

   a. Perform a two-way RM ANOVA on these data, testing each of the F ratios at the .05 level. Which effects would remain significant if a total lack of sphericity is assumed?

   b. Test the linear, quadratic, and cubic trends for the main effect of caffeine dosage (i.e., average across the two sleep levels).

   c. Test the differences (i.e., interactions) between the two sleep groups for the linear, quadratic, and cubic trends over the

ratings for the matched sets of patients appear in the following table.

   a. Perform a two-way RM ANOVA on these data, testing each of the F ratios at the .05 level. Which effects would remain significant if a total lack of sphericity is assumed?

   b. Test the differences (i.e., interactions) between the two therapy groups for both the linear and the quadratic trends over time. Use a graph of the data to help explain your results.

   c. For whichever trend component interacts significantly with the therapy-type factor, test that trend separately at each therapy level.

PDT	Mon 1	Mon 2	Mon 3	Mon 4	Mon 5	CBT	Mon 1	Mon 2	Mon 3	Mon 4	Mon 5
1	20	19	17	18	12	1	22	16	13	12	15
2	15	12	11	10	8	2	12	12	11	15	19
3	25	28	23	24	18	3	29	28	21	18	22
4	31	29	28	25	17	4	35	25	24	21	18
5	27	23	24	20	19	5	19	23	18	19	17
6	16	18	17	17	12	6	16	11	17	14	20
7	25	27	24	26	20	7	25	17	24	20	16
8	32	35	29	26	21	8	30	35	23	26	27

The SS for the subject-by-treatment interaction (found by subtraction):

$$SS_{inter} = SS_{tot} - SS_{sub} - SS_{treat}$$ **Formula 15.1**

The degrees of freedom for the components of the one-way RM ANOVA:

a. $df_{sub} = n - 1$ **Formula 15.2**

b. $df_{RM} = c - 1$

c. $df_{inter} = (n - 1)(c - 1)$

The variance estimates (MS) for the one-way RM ANOVA:

a. $MS_{RM} = \dfrac{SS_{RM}}{df_{RM}}$ **Formula 15.3**

b. $MS_{inter} = \dfrac{SS_{inter}}{df_{inter}}$

The F ratio for a one-way RM ANOVA:

$$F = \frac{MS_{RM}}{MS_{inter}}$$ **Formula 15.4**

Proportion of *within-subject* variance accounted for in an RM design (this measure is likely to be misleadingly large, if you are trying to predict the size of the effect in a replication with independent groups):

$$\eta^2_{RM} = \frac{SS_{RM}}{SS_{RM} + SS_{inter}}$$ **Formula 15.5**

Estimate of ordinary omega squared in an RM ANOVA (i.e., this measure is not inflated by the degree to which the scores are matched):

$$\text{Est. } \omega^2 = \frac{SS_{RM} - (c - 1)MS_W}{SS_{total} + MS_W}$$ **Formula 15.6**

Cohen's f measure of population effect size for an RM ANOVA, as a function of the ordinary (i.e., corresponding independent-groups) f, and the population correlation coefficient:

$$f_{RM} = \frac{f}{\sqrt{(1 - \rho)}}$$ **Formula 15.7**

Cohen's f measure of population effect size as a function of omega squared (the proportion of variance accounted for in the population):

$$f = \frac{\omega}{\sqrt{(1 - \omega^2)}}$$ **Formula 15.8**

The t test for a trend component involving repeated measures:

$$t_{trend} = \frac{\overline{L}_{trend}}{\dfrac{s_L}{\sqrt{n}}}$$ **Formula 15.9**

TWO-WAY MIXED DESIGN ANOVA

Chapter 16

You will need to use the following from previous chapters:

Symbols
k: Number of independent groups in a one-way ANOVA
c: Number of levels (i.e., conditions) of an RM factor
n: Number of subjects in each cell of a factorial ANOVA
N_T: Total number of observations in an experiment
ϵ: coefficient to estimate the degree of sphericity in the population

Formulas
Formula 14.2: SS_{inter} (by subtraction)
Formula 14.3: SS_{bet} or one of its components

Concepts
Advantages and disadvantages of the RM ANOVA
SS components of the one-way RM ANOVA
SS components of the two-way ANOVA
Interaction of factors in a two-way ANOVA

A

**CONCEPTUAL
FOUNDATION**

If you want the economy of a two-way factorial design, or its ability to detect the interaction of two independent variables, and at the same time you want the added power of repeated measures, you may be able to use a two-way repeated-measures (and/or randomized-blocks) design. The two-way RM ANOVA parallels the analysis of a three-way factorial design, just as the calculation of a one-way RM ANOVA resembles the two-way ANOVA for independent groups (as described in Section C of the previous chapter). However, at least as common as the two-way RM design is a two-way factorial design in which one of the factors involves repeated measures (or matched subjects) and the other factor involves independent groups of subjects. For obvious reasons, this design is often called a *mixed design*, although this designation is not universal. Mixed designs are sometimes called *split-plot* designs, a description that arises from their early use in agricultural research. A potential source of confusion is that the term *mixed design ANOVA* sounds similar to the term *mixed model ANOVA*, which denotes a mixing of random and fixed effects in the same design. However, random effects are rarely used in psychological research (except for subjects, but the subject factor is almost never tested for significance), so there are few occasions for confusion. In this text, all of the ANOVA effects tested in earlier chapters were fixed effects, and that will continue to be the case for the remainder of the text (but see supplemental material on my Web site).

Because mixed designs can have any number of *within-subjects factors* and *between-subjects factors*, it is usually necessary to specify the number of each type of factor. For instance, a three-way mixed design can have either two RM factors and one between-groups factor or one RM factor and two between-groups factors. In this chapter you will encounter only the simplest mixed design: the two-way mixed design. So long as the appropriate assumptions can be made, the data from a mixed design experiment can be analyzed with the mixed design ANOVA described in this chapter.

The One-Way RM ANOVA Revisited

To demonstrate how the mixed-design ANOVA works, I will return to the experiment described at the beginning of Chapter 15, in which each of six subjects must try to recall three types of words: positive, negative, and neutral. I will change the data, however, to suit the purposes of this chapter; see Table 16.1.

First I will conduct a one-way repeated measures ANOVA for these new data. As usual, I will begin by finding SS_{total}. I can use Formula 14.3, and find that $SS_{total} = N_T\sigma^2(\text{scores}) = 18(25.89) = 466$. Alternatively, I can obtain the same answer with Formula 3.11:

$$SS_{total} = \sum X^2 - N\bar{X}_G^2 = 3{,}508 - 18(13^2) = 3{,}508 - 3{,}042 = 466$$

Next I will find SS_{RM}, which involves the means for each treatment (the column means in Table 16.1):

$$SS_{RM} = N_T\sigma^2(\text{column means}) = 18 \cdot \sigma^2(14.33, 12.5, 12.17)$$
$$= 18(.903) = 16.25$$

(Remember that you can use Formula 3.13A or an equivalent formula to calculate the biased variance of the column means if you do not have a statistical or scientific calculator.) Now I need to calculate SS_{sub} so it can be subtracted from the total. I will use the means for each subject (the row means in Table 16.1):

$$SS_{sub} = N_T\sigma^2(\text{row means}) = 18 \cdot \sigma^2(19.33, 15, 16.67, 12.67, 8, 6.33)$$
$$= 18(21.19) = 381.4$$

Finally, I can find $SS_{residual}$ (i.e., SS_{inter}) by subtraction:

$$SS_{resid} = 466 - 16.25 - 381.4 = 68.35$$

To find the F ratio, first I find that $MS_{RM} = SS_{RM}/df_{RM} = 16.25/2 = 8.13$, and $MS_{resid} = SS_{resid}/df_{resid} = 68.35/10 = 6.84$. Therefore, $F = 8.13/6.84 = 1.19$. The results of the RM ANOVA are summarized in Table 16.2. Because

Subject No.	Neutral	Positive	Negative	Row Means	
1	20	21	17	19.33	**Table 16.1**
2	16	18	11	15	
3	17	15	18	16.67	
4	15	10	13	12.67	
5	10	4	10	8	
6	8	7	4	6.33	
Column Means	14.33	12.5	12.17	13	

Source	SS	df	MS	F	
Between-Subjects	381.4	5			**Table 16.2**
Within-Subjects	84.6	12			
Treatment	16.25	2	8.13	1.19	
Residual	68.35	10	6.84		
Total	466	17			

Figure 16.1

Graph of the Data in
Table 16.1

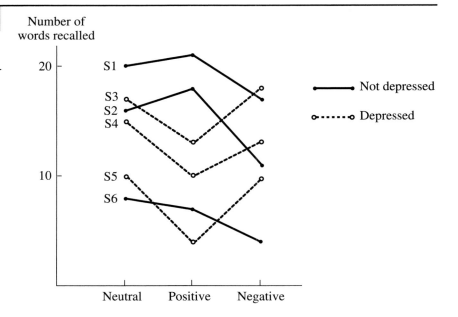

the critical $F_{.05}(2, 10)$ equals 4.1, we cannot reject the null hypothesis in this case.

As you may have guessed, this is not the end of the story for this experiment. What I didn't tell you is that three of the subjects were selected because of high scores on a depression inventory, whereas the remaining three showed no signs of depression. The depressed subjects are graphed as dashed lines in Figure 16.1; solid lines represent the nondepressed subjects.

Converting the One-Way RM ANOVA to a Mixed Design ANOVA

The one-way RM ANOVA just performed can be transformed into a two-way mixed ANOVA by adding a between-groups factor with two levels: depressed and nondepressed. However, the advantage of this design may not be obvious from looking at Figure 16.1. On average, the two groups do not differ much in overall recall ability, nor does the variability within each group seem less than the total variability. The difference between the groups becomes apparent, though, when you focus on the subject × treatment interaction.

In Figure 16.2, the two groups are graphed in separate panels to make it obvious that the subject × treatment ($S \times T$) interaction within each group is much smaller than the total amount of interaction when all subjects are considered together. The calculation of the mixed ANOVA can take advantage of this smaller $S \times T$ interaction by analyzing the SS components of the RM ANOVA further.

Analyzing the Between-Subjects Variability

Even though we would normally know about the depressed and nondepressed subgroups before analyzing our data, it would be reasonable to begin the mixed ANOVA by ignoring this distinction and calculating the

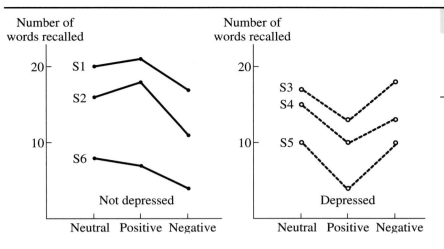

Number of words recalled

S1
S2

20

10

S6

Not depressed

Neutral   Positive   Negative

Number of words recalled

20

S3
S4

10

S5

Depressed

Neutral   Positive   Negative

Figure 16.2

Graph of the Data in Table 16.1, Separating Depressed from Not Depressed Subjects

RM ANOVA as we just did. The next step is to analyze further the between-subjects SS component shown in Table 16.2. This is the component we happily throw away in the one-way RM design because we do not care about subject-to-subject variability. However, in the mixed design some of the subject-to-subject variability is due to the difference in the means of the two (or more) groups. This SS component is like the SS_{total} for a one-way independent ANOVA; it can be divided into SS_W (variability within each group) and SS_{bet} (variability due to the group means). To avoid confusion I will refer to this SS_{bet} as SS_{groups} in discussing the mixed design.

To obtain SS_{groups} first find the means for the two subgroups. The mean for the depressed subjects (subjects 3, 4, and 5) is 12.44, and the mean for the nondepressed subjects (subjects 1, 2, and 6) is 13.55. Now we can use Formula 14.3:

$$SS_{groups} = N_T\sigma^2(\text{group means}) = 18 \cdot \sigma^2(12.44, 13.55) = 18(.308) = 5.54$$

To find SS_W, we need only subtract SS_{groups} from the total SS for this part of the analysis, which is 381.4. (The total SS for this part of the analysis is the between-subjects SS from the one-way RM analysis, also called SS_{sub}.) So, $SS_W = 381.4 - 5.54 = 375.86$.

These SS components can now be converted to variance estimates so they can be put into an F ratio. What I am actually doing is a one-way independent ANOVA on the subject-to-subject variability. Thus, I will use the notation of Chapter 12, except that the term *groups* is substituted for *between*:

$$MS_{groups} = \frac{SS_{groups}}{k-1} = \frac{5.54}{1} = 5.54$$

$$MS_W = \frac{SS_W}{N-k} = \frac{375.86}{4} = 94.0$$

Note that N in the MS_W formula refers to the total number of subjects (not observations or scores, which is N_T), which in this case is 6, and k is the number of different groups (2). I will make the notation less confusing in the next section. Here I want to emphasize the connections with the analyses in previous chapters. Finally:

$$F = \frac{MS_{groups}}{MS_W} = \frac{5.54}{94} = .059$$

This F value (.059) allows us to test whether the means for the two groups differ significantly.

Because we are conducting a factorial ANOVA, we can say that this F ratio tests the *main effect* of depression. Clearly, we do not have to look up a critical F to know that the null hypothesis cannot be rejected. The fact that our observed F is much less than 1.0 (indeed, unusually so) is of no interest to us. This just tells us that the means for the depressed and nondepressed subjects are surprisingly close together, given all of the variability within each group.

Analyzing the Within-Subjects Variability

Although we already calculated the F for the main effect of word type when we performed the one-way RM ANOVA, this F ratio must be recalculated for the mixed design to take into account the separation of subjects into subgroups. As usual, it is not the numerator of the F ratio that will change; this numerator depends on the separation of the means for the three word types, and it does not change just because we have regrouped the subjects within conditions. On the other hand, the denominator (the error term) does change, for reasons that can be seen by comparing Figure 16.2 to Figure 16.1. Notice that the subject × treatment interaction is fairly small within each group (Figure 16.2), but it looks rather large when all subjects are considered together (Figure 16.1). We can say that most of the $S \times T$ interaction is really due to a group by word type interaction, which should be removed from the total $S \times T$ interaction. This is the same as taking the two much smaller interactions of the subgroups and averaging them. Mathematically, what we need to do is further analyze the $SS_{residual}$ from the RM ANOVA into smaller components.

To subtract the *group* by word type interaction from the *subject* by word type interaction, we must calculate the former. This is done as it is for any two-way ANOVA and involves calculating the $SS_{between-cells}$ component. For this step we need to know the mean for each combination of group and word type. Table 16.3 shows the data from Table 16.1 rearranged into cells, with the cell means.

Table 16.3	**Group**	**Neutral**	**Positive**	**Negative**
	Nondepressed	20	21	17
		16	18	11
		8	7	4
	Cell Means	14.67	15.33	10.67
	Depressed	17	15	18
		15	10	13
		10	4	10
	Cell Means	14	9.67	13.67

The cell means from this table can then be inserted into Formula 14.3, as follows:

$$SS_{between-cells} = N_T \sigma^2_{(cell\ means)}$$

$$= 18 \cdot \sigma^2(14.67, 15.33, 10.67, 14, 9.67, 13.67)$$

$$= 18(4.364) = 78.55$$

We have already calculated the SS for word type (i.e., SS_{RM}) and the SS for groups, so we are ready to find the SS for the group by treatment interaction ($SS_{G \times RM}$) by subtraction:

$$SS_{G \times RM} = 78.55 - 16.25 - 5.54 = 56.76$$

Now we can go back to the original SS_{resid} from the one-way RM ANOVA and subtract $SS_{G \times RM}$ to get the new smaller SS_{resid} for the mixed design (the sum of the subject by treatment interactions *within each group*). This new error component can be referred to simply as SS_{resid} or, more specifically, as the SS for the interaction of the subject factor with the repeated-measures factor, $SS_{S \times RM}$.

$$SS_{S \times RM} = 68.35 - 56.76 = 11.59$$

Now we have calculated all the SS components needed to complete the mixed design analysis. To summarize: The one-way RM ANOVA gave us three components, SS_{RM}, SS_{sub}, and SS_{resid}. SS_{sub} was then further divided to yield SS_{groups} and SS_{W}. SS_{resid} was also divided into two components: $SS_{G \times RM}$ and $SS_{S \times RM}$. SS_{RM} was left alone. We have already tested the main effect of groups by forming an F ratio from MS_{groups} and MS_{W}. Now we can recalculate the F ratio for the repeated factor. The numerator MS is the same as in the one-way analysis: 8.13. But the new error term is based on $SS_{S \times RM}$ (11.59) divided by $df_{S \times RM}$ (8). So $MS_{S \times RM} = 11.59/8 = 1.45$. Thus, the new F for testing the main effect of word type is:

$$F = \frac{MS_{RM}}{MS_{S \times RM}} = \frac{8.13}{1.45} = 5.61$$

The critical F is based on two and eight degrees of freedom; $F_{.05}(2, 8) = 4.46$. Because of the smaller error term in the mixed design, the main effect of word type is now significant at the .05 level.

Two-Way Interaction in the Mixed Design ANOVA

Like any other two-way ANOVA, the mixed design ANOVA can give us one more F ratio: a test of the interaction of the two factors. We find $MS_{G \times RM}$ by dividing $SS_{G \times RM}$ by $df_{G \times RM}$. We have already found that $SS_{G \times RM} = 56.76$, and you will have to take on faith for the moment that $df_{G \times RM} = 2$. So $MS_{G \times RM} = 56.76/2 = 28.38$. (Formulas for the mixed-design ANOVA, including those for the df, will be presented more formally in Section B.) To complete the F ratio, however, we need to know which MS to use as the error term. So far both MS_{W} and $MS_{S \times RM}$ have been used as error terms. Can we use one of these error terms, or is there some third error term to use? The answer is that $MS_{S \times RM}$ is the appropriate error term, for reasons that I will make clear shortly. To test whether depression and word type interact significantly, we form the following F ratio:

$$F = \frac{MS_{G \times RM}}{MS_{S \times RM}} = \frac{28.38}{1.45} = 19.57$$

You should not have to look up a critical F to know that such a large observed F ratio must be significant at the .05 level (except in the ridiculous case when some of your groups contain only one subject).

To understand why $MS_{S \times RM}$ is the appropriate error term for the interaction of the two factors, it may help to compare a graph of the cell means (see Figure 16.3) to a graph of the data from individual subjects (see Figure

Figure 16.3

Graph of Cell Means for
the Data in Table 16.3

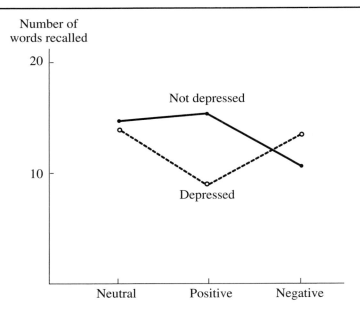

16.2). Notice that each line in Figure 16.3 represents the average number of words recalled by the subjects in that group. To the extent that the individual subjects in a group are not parallel to the group average, or to each other, $MS_{S \times RM}$ increases. As $MS_{S \times RM}$ increases, so does the chance of finding an accidental interaction between the groups (just as increased variation within groups leads to a greater chance of finding a large difference among group means in an independent ANOVA). Thus, $MS_{G \times RM}$ contains error arising from $MS_{S \times RM}$ in addition to any real (i.e., population) interaction between the factors. When the null hypothesis is true, there is no real interaction contributing to $MS_{G \times RM}$, so the F ratio—$MS_{G \times RM}/MS_{S \times RM}$—should equal about 1.0.

Summarizing the Mixed Design ANOVA

The results of our mixed ANOVA are summarized in Table 16.4. The structure of the table tells us a great deal about the structure of our analysis. First, the table is broken into two distinct sections. The upper section, "Between-Subjects," arises from the variation of the row means in Table 16.1; if all the row (i.e., subject means) were equal, the SS components for this section would all be zero. The differences that do exist (on average) between subjects can be divided into the difference between the means of the subgroups (SS_{groups}) and the differences between subjects within each group (SS_W); the latter forms the basis of the error term for this part of the table. The number of degrees of freedom for between-subjects variation is one less than the total number of different subjects, which equals $6 - 1 = 5$. These five df break down into one df for group differences ($k - 1 = 2 - 1 = 1$) and four df for differences within groups (number of subjects $-$ number of groups $= 6 - 2 = 4$).

The lower section of Table 16.4, "Within-Subjects," deals with the variation of scores within each *row* of Table 16.1; if each subject produced the same recall score for all word types, the SS components in this section of the table would all be zero (even though subjects might differ from each other). The differences that do exist between levels of the repeated factor

Source	SS	Df	MS	F	p	
Between-Subjects	381.4	5				*Table 16.4*
Groups	5.54	1	5.54	.06	>.05	
Within-Groups	375.86	4	94.0			
Within-Subjects	84.6	12				
Treatment	16.25	2	8.13	5.61	<.05	
Group × Treatment	56.76	2	28.38	19.57	<.01	
Residual	11.59	8	1.45			
Total	466	17				

(in this case, word type) can be divided into average differences among the word types (SS_{RM}), the interaction of groups with word type ($SS_{G \times RM}$), and the interaction of individual subjects with word type ($SS_{S \times RM}$). For instance, if a particular subject recalls more positive than negative words, it may be because subjects tend to do this in general, or because those in this subject's particular subgroup tend to do this, or because this particular subject has an individual tendency to do this. The last of these sums of squares ($SS_{S \times RM}$) forms the basis of the error term for the lower section of Table 16.4. The number of degrees of freedom for within-subjects variation equals the total number of subjects (six) times one less than the number of repeated measures ($c - 1 = 3 - 1 = 2$), which equals 12. These 12 df break down into 2 df for treatment differences ($c - 1$), 2 df for the group by treatment interaction [$(k - 1)(c - 1)$], and the remainder, 8 df, for the residual, which is the number of groups times one less than the number of repeated conditions times one less than the number of subjects in each subgroup. I will present df formulas for this design in a more formal way in Section B.

Interpreting the Results

The between-groups factor in our example was not found to be significant, whereas the within-subjects factor was. This pattern is probably more common than the reverse pattern because the test of the repeated factor is likely to have greater power, for reasons discussed in Chapter 15. However, we should interpret the results of the main effects cautiously because of the significance of the interaction. The effect of the repeated factor is different for the two subgroups, and further analysis would be appropriate to localize these differences. As with other two-way ANOVAs, a graph of the cell means can help you understand how the main effects combine with the interaction to produce the pattern of results. You can see from Figure 16.3 that the depressed subjects have relatively poor recall for positive words, whereas it is the recall of negative words that is weak for nondepressed subjects. The testing of follow-up comparisons for a mixed design will be discussed in Section B.

The Varieties of Mixed Designs
The Between-Groups Factor Is a Grouping Variable

In the example above, the RM factor involves an experimental manipulation (i.e., type of word), whereas the between-groups factor is based on preexisting individual differences in depression. This is a common form of the mixed design. The reduction in SS_{resid} caused by separating the depressed and nondepressed subjects is reminiscent of the reduction in SS_W that followed the separation of men and women in the example of a depression drug in Chapter 14. This reduction will occur whenever there is an interaction between the experimental and grouping factor. However, adding a grouping variable causes a reduction in df for the error term (from 10 to 8 in the example above), which makes the critical F larger. This also means that SS_{resid} will be divided by a smaller number, and that tends to increase your error term. Therefore, if the interaction is quite small, adding the grouping variable can actually lower your F ratio. The lesson is that grouping variables should not be added casually.

Often, the reason a researcher adds a grouping variable is that he or she expects the between-groups factor to interact with the RM factor; in such cases, there is little interest in the main effect of the between-groups factor. The chief purpose of such a design is to determine whether the effects of the RM factor are the same for various subgroups in the population. In the preceding example, it is of interest that the type of word remembered most easily depends on the subjects' level of depression.

The Between-Groups Factor Is an Experimental Variable

Another type of mixed design arises when the between-groups factor involves experimental manipulation that does not lend itself to repeated measures. For instance, consider an experiment in which each subject completes a series of tasks of varying difficulty. In one condition, the subjects are told that the tasks come from a test of intelligence and that their performance will give an indication of their IQ. In another condition, subjects are given monetary rewards for good performance, and in a third condition, subjects are simply asked to work as hard as they can. It should be clear that once the researcher has run a subject in the IQ condition, running the same subject again in a different condition (using similar tasks) would yield misleading results. Nor could you always run the IQ condition last without confounding your results with an order effect. A reasonable solution is the mixed design, with the three motivational conditions as the levels of a between-groups factor, and task difficulty as a within-subjects factor.

One purpose of the preceding experiment could be to explore the interaction between the motivational condition and task difficulty. (For instance, are performance differences between difficulty levels the same for different types of motivation?) However, if there were no interaction, it might still be interesting to examine the main effect of motivational condition. (The main effect of difficulty would merely confirm that the difficulty manipulation worked.) In general, when there is no interaction between the factors of a mixed design, the analysis reduces to two one-way ANOVAs: an independent-groups ANOVA and an RM ANOVA.

In one common form of the mixed design, the repeated-measures factor is merely the passage of time; this type of RM factor is rarely used unless the between-groups factor involves an experimental manipulation (e.g., measurements can be taken before, during, and/or after some treatment, with different groups assigned to different treatments). An example of this type of design will be used in Section B to illustrate the systematic calculation of a two-way mixed design.

The RM Factor Can Be Based on Repeated Measures or Randomized Blocks

The levels of the RM factor in a mixed design can be administered in several ways, as discussed for the one-way RM design. The levels can be interspersed for a presentation that is virtually simultaneous, or the levels can be presented successively. Successive presentations usually produce order effects that can be averaged out by counterbalancing, but nonetheless inflate the error term and decrease power. A mixed-design ANOVA can be employed to remove the influence of order effects from the error term, as I will describe in Section C. In this context, methods can be used to quantify the magnitude of differential carryover effects. If carryover effects are severe and unavoidable, the experiment should be designed to match subjects into blocks; randomized blocks can provide much of the benefit of repeated measures with no order effects at all. Finally, if carryover cannot be avoided, and no basis for matching can be found, both factors would have to consist of independent samples, and the two-way ANOVA described in Chapter 14 would be appropriate.

SUMMARY

1. The two-way mixed design (also called a *split-plot* design) includes one between-subjects factor and one within-subjects factor; the latter involves either repeated measures or matched blocks of subjects. The total variability in a two-way mixed design can be initially divided into between-subjects and within-subjects variation.
2. The between-subjects variation can be further subdivided into a portion that depends on the separation of the group means and a portion that depends on subject-to-subject variation within each group. The within-group variability is used as the error term for testing the main effect of the grouping factor.
3. The within-subjects variation can be divided into three components: one that reflects variation among the means of the repeated conditions, one that reflects the interaction of the two factors, and one that reflects the interaction of subjects with the RM factor within each group. The last of these components is used as the error term for testing both the main effect of the within-subjects factor and the interaction of the two factors.
4. Adding a between-subjects factor to a one-way RM design is likely to reduce the error term of the RM factor if the between-subjects factor is based on preexisting groups and there is some interaction between the two factors. Adding the grouping factor also allows you to test the interaction of the two factors and thereby determine whether the effect of the RM factor is similar for different subgroups of the population. However, adding the grouping factor reduces the degrees of freedom for the error term, so it can be counterproductive if the grouping variable is not related to your dependent variable.
5. A mixed design commonly arises by adding the passage of time (RM factor) to a between-subjects factor that is based on an experimental manipulation. In such a design, it is usually only the interaction of the two factors that is interesting. Another common type of mixed design is one in which one factor lends itself easily to repeated measures (trials at different levels of difficulty, randomly interspersed), but the other factor does not (telling some subjects that the task tests intelligence, and others that it does not).
6. The within-subjects (RM) factor in a mixed design has the same advantages and disadvantages as the within-subjects factor in a one-way RM design. This factor is likely to have greater power than the between-subjects factor because subject-to-subject variability is ignored; however, if the repeated measures are not simultaneous, you may need to counterbalance, and if differential carryover effects are likely, you may need to match subjects in blocks as an alternative to repeated measures.

1. a. Devise a mixed design experiment in which the between-subject variable is quasi-independent (i.e., based on pre-existing groups).
   b. Devise a mixed design experiment in which the between-subjects variable is manipulated by the experimenter.
   c. Devise a mixed design experiment in which the within-subjects variable involves matched subjects rather than repeated measures.
*2. A researcher tested two groups of subjects—six alcohol abusers and six moderate social drinkers—on a reaction time task. Each subject was measured twice: before and after drinking 4 ounces of vodka. A mixed-design ANOVA produced the following SS components: $SS_{groups} = 88$, $SS_W = 1380$, $SS_{RM} = 550$, $SS_{G \times RM} = 2.0$, and $SS_{S \times RM} = 134$. Complete the analysis and present the results in a summary table.
3. Exercise 15B4 described a randomized-blocks experiment involving four different textbooks and nine blocks of subjects. The RM ANOVA produced the following SS components: SS_{treat} (SS_{RM}) $= 76.75$, $SS_{subject} = 612.5$, and $SS_{resid} = 27.5$. Now suppose that the nine blocks of subjects can be separated into three subgroups on the basis of overall ability, and that the mixed design ANOVA yields $SS_{groups} = 450$ and $SS_{G \times RM} = 8.5$. Complete the analysis and present the results in a summary table.
*4. The following table shows the number of ounces of popcorn consumed by each subject while viewing two emotion-evoking films,

one evoking happiness and one evoking fear. Half the subjects ate a meal just before the film (preload condition), whereas the others did not (no load condition). Graph the data for all of the subjects on one graph.

	Happiness	Fear
Preload	10	12
	13	16
	8	11
	16	17
No Load	26	20
	19	14
	27	20
	20	15

a. Does there appear to be about the same amount of subject × treatment interaction in each group?
b. Does there appear to be a considerable amount of group × repeated-measure interaction?
5. If you calculate an RM ANOVA and then assign the subjects to subgroups to create a mixed design, the observed F ratio for the RM factor may get considerably larger. Under which of the following conditions is this likely?
a. The degrees of freedom associated with the error term are reduced considerably.
b. There is a good deal of subject × RM treatment interaction.
c. There is a good deal of (sub)group × RM treatment interaction.
d. There is a good deal of subject to subject variability.

B

BASIC STATISTICAL PROCEDURES

In Chapter 11, I pointed out the weakness of the simple before-after design. Even if the before-after difference turns out to be statistically significant for some treatment, without a control group it is difficult to specify the cause of the difference. Was the treatment really necessary to produce the difference, or would just the act of participating in an experiment be sufficient? When you add a control group and continue to measure your variable twice in both groups, you have created a mixed design. For example, suppose that you have devised a new treatment for people afraid of public speaking. To show that the effects of your new treatment are greater than a placebo effect, half the subjects (all of whom have this phobia) are randomly assigned to a control group; their treatment consists of hearing inspirational talks about the joys of public speaking. Suppose further that you wish to demonstrate that the beneficial effects of your treatment last beyond the end of the treatment period. Consequently, you measure the degree of each subject's phobia not only before and after the treatment period, but also 6 months after the end of treatment (follow-up). I will apply the usual six-step hypothesis testing procedure to this experiment.

Step 1. State the Hypotheses

The design in this example consists of two factors that have been completely crossed (i.e., a two-way factorial design). As such, the design involves three independent null hypotheses that can be tested. The H_0 for the main effect of treatment (the between-subjects factor) is $\mu_{exp} = \mu_{con}$. The H_0 for the main effect of time (the within-subjects, or repeated, factor) is $\mu_{bef} = \mu_{aft} = \mu_{fol}$. The H_0 for the interaction of the two factors is a statement that the experimental-control difference will be the same at each point in time (before, after, and at follow-up) or, more simply, that the effects of one factor are independent of the effects of the other factor. The alternative hypothesis for the main effect of treatment can be stated either as one-tailed (e.g., $\mu_{exp} > \mu_{con}$) or as two-tailed ($\mu_{exp} \neq \mu_{con}$). I will take the more conservative approach and use the two-tailed H_A. For the main effect of time there are three levels, so the only simple way to state H_A is to state that H_0 is not true. Also, for simplicity the H_A for the interaction is a statement that the corresponding H_0 is not true.

Step 2. Select the Statistical Test and the Significance Level

The time factor involves three measures on each subject, but the treatment factor involves different subjects. Because our purpose is to detect differences in population means along these two different dimensions, a mixed design ANOVA is appropriate. The conventional approach is to use .05 as alpha for all three null hypothesis tests.

Step 3. Select the Samples and Collect the Data

To minimize the calculations I will assume that only eight phobic subjects are available for the experiment and that four are selected at random for each treatment group. The dependent variable will be a 10-point rating scale of phobic intensity with respect to public speaking (from 0 = relaxed when speaking in front of a large audience to 10 = incapable of making a speech in front of more than one person). Because each subject is measured three times, there will be a total of $8 \times 3 = 24$ ratings or observations, as shown in Table 16.5.

Group	Subject No.	Before	After	Follow-up	Row Means	
Phobia	1	8	4	6	6	**Table 16.5**
Treatment	2	9	6	5	6.67	
	3	6	3	5	4.67	
	4	7	5	4	5.33	
	Cell Means	7.5	4.5	5.0	5.67	
Control	5	9	8	7	8	
Treatment	6	7	7	8	7.33	
	7	7	6	7	6.67	
	8	6	4	7	5.67	
	Cell Means	7.25	6.25	7.25	6.92	
	Column Means	7.375	5.375	6.125	6.292	

Step 4. Find the Regions of Rejection

Given that certain assumptions have been met (these will be discussed shortly), it is appropriate to use the F distribution to find a critical value for each null hypothesis. However, we need to know the degrees of freedom that apply in each case. The breakdown of the df can get complicated for a mixed design, so a df tree can be especially helpful when dealing with this type of design (see Figure 16.4). In Figure 16.4, I use k to represent the number of different groups (i.e., the number of levels for the between-groups factor), as I did for the one-way independent ANOVA, and c to represent the number of treatment conditions presented to each subject (i.e., the number of levels for the within-subjects factor), as I did for the one-way RM ANOVA. Thus $k \times c$ (or just kc) is the number of cells in the two-way mixed design. I will deal only with the case in which each group has the same number of subjects (or blocks), so n can be used to represent the number of subjects assigned to each level of the between-groups factor (i.e., the number of subjects per group); thus kn is the total number of different subjects in the experiment (I called this N without a subscript, in Section A). However, there are c measurements, or scores, for each subject, so nkc is the total number of observations in the experiment (i.e., N_T). Now we can express the total df as:

$$df_{tot} = nkc - 1 \quad \text{or} \quad (N_T - 1)$$

The total df are divided into the df associated with variation between subjects ($df_{between\text{-}S}$) and the df associated with variation within subjects ($df_{within\text{-}S}$). The $df_{between\text{-}S}$ are simply the total number of different subjects minus 1; the $df_{within\text{-}S}$ are equal to the total number of *observations* minus the total number of different subjects. These relationships are expressed in the first two parts of Formula 16.1:

a. $df_{between\text{-}S} = nk - 1$ **Formula 16.1**

b. $df_{within\text{-}S} = nk(c - 1) \quad \text{or} \quad nkc - nk$

As you can see from Figure 16.4, the df are further subdivided in each branch. The $df_{between\text{-}S}$ are divided into the following two components:

c. $df_{groups} = k - 1$ **Formula 16.1**

d. $df_W = k(n - 1) \quad \text{or} \quad nk - k$ **(cont.)**

Figure 16.4

Degrees of Freedom Tree for Two-Way Mixed Design

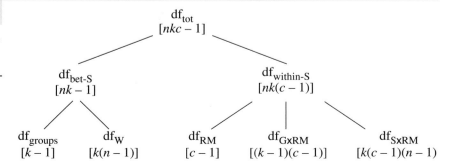

where df_W stands for the df associated with subject-to-subject variation within each group. The $df_{within\text{-}S}$ are divided into three components:

e. $df_{RM} = c - 1$ **Formula 16.1**
(cont.)

f. $df_{G \times RM} = (k - 1)(c - 1)$

g. $df_{S \times RM} = k(c - 1)(n - 1)$

The df for the interaction of the two factors, $df_{G \times RM}$, is equal to the df for groups times the df for the repeated measures factor. The $df_{S \times RM}$ component is the df corresponding to the sum of the df for the subject by repeated measure interactions for each group.

For the present example, $k = 2$, $c = 3$, and $n = 4$, so $df_T = 4 \times 2 \times 3 - 1 = 24 - 1 = 23$, which can be broken down as follows:

$df_{between\text{-}S} = (4 \cdot 2) - 1 = 8 - 1 = 7$

$df_{within\text{-}S} = 4 \cdot 2(3 - 1) = 8 \cdot 2 = 16$

Then, $df_{bet\text{-}S}$ can be divided as follows:

$df_{groups} = 2 - 1 = 1$

$df_W = 2(4 - 1) = 6$

Similarly, $df_{within\text{-}S}$ can be divided into these components:

$df_{RM} = 3 - 1 = 2$

$df_{G \times RM} = 1 \cdot 2 = 2$

$df_{S \times RM} = 2(3 - 1)(4 - 1) = 2 \cdot 2 \cdot 3 = 12$

Now we can find the critical F value for each of our three null hypothesis tests. For the main effect of phobia treatment, the df are df_{groups} (1) and df_W (6); $F_{.05}(1, 6) = 5.99$. For the main effect of time, the df are df_{RM} (2) and $df_{S \times RM}$ (12); $F_{.05}(2, 12) = 3.89$. And for the interaction of the two factors, the df are $df_{G \times RM}$ (2) and $df_{S \times RM}$ (12); therefore, the critical F for this test is also 3.89.

Step 5. Calculate the Test Statistics

For each of the df components delineated in the preceding step, there is a corresponding SS component. As usual, these SS components will add up to SS_{total} which, for this example, equals $N_T \sigma^2(scores) = 24(2.457) = 58.97$. A convenient next step is to calculate $SS_{between\text{-}S}$, which depends on the means for each subject (the eight row means in Table 16.5). I will be calculating all of the SS components with Formula 14.3, except for those which are more conveniently found by subtraction.

$SS_{between\text{-}S} = 24 \cdot \sigma^2(6.0, 6.67, 4.67, 5.33, 8.0, 7.33, 6.67, 5.67)$

$= 24(1.0386) = 24.93$

Now we find the SS for the grouping factor, which depends on the means for each group (these are also the row means if you think of the rows as consisting of cell means, rather than individual scores):

$$SS_{\text{groups}} = N_T \sigma^2(\text{group means}) = 24 \cdot \sigma^2(5.67, 6.92) = 24(.391) = 9.38$$

By subtracting SS_{groups} from $SS_{\text{between-S}}$, we obtain SS_W (Formula 16.2):

$$SS_W = SS_{\text{between-S}} - SS_{\text{groups}} \qquad \textbf{Formula 16.2}$$

For this example:

$$SS_W = 24.93 - 9.38 = 15.55$$

One branch of the total SS has now been analyzed into its components. To find the total of the other branch we subtract $SS_{\text{between-S}}$ from SS_{total} to find $SS_{\text{within-S}}$:

$$SS_{\text{within-S}} = SS_{\text{total}} - SS_{\text{between-S}} \qquad \textbf{Formula 16.3}$$

In this case:

$$SS_{\text{within-S}} = 58.97 - 24.93 = 34.04$$

We now turn our attention to analyzing $SS_{\text{within-S}}$ into its components, starting with the SS for the repeated measures factor (SS_{RM}), which is based on the means of the columns in Table 16.5:

$$SS_{\text{RM}} = N_T \sigma^2(\text{column means}) = 24 \cdot \sigma^2(7.375, 5.375, 6.125)$$
$$= 24(.6806) = 16.3$$

The SS for the interaction of the two factors ($SS_{G \times \text{RM}}$), as in any two-way ANOVA, requires that we calculate $SS_{\text{between-cells}}$ and then subtract the SS components for the two main effects. Using the cell means from Table 16.5, we obtain:

$$SS_{\text{between-cells}} = N_T \sigma^2(\text{cell means}) = 24 \cdot \sigma^2(7.5, 4.5, 5.0, 7.25, 6.25, 7.25)$$
$$= 24(1.363) = 32.71$$

Now we can use Formula 16.4 to obtain $SS_{G \times \text{RM}}$:

$$SS_{G \times \text{RM}} = SS_{\text{between-cells}} - SS_{\text{groups}} - SS_{\text{RM}} \qquad \textbf{Formula 16.4}$$

For these data:

$$SS_{G \times \text{RM}} = 32.71 - 9.38 - 16.33 = 7.0$$

The third component of $SS_{\text{within-S}}$ is also found by subtraction, according to Formula 16.5:

$$SS_{S \times \text{RM}} = SS_{\text{within-S}} - SS_{\text{RM}} - SS_{G \times \text{RM}} \qquad \textbf{Formula 16.5}$$

Therefore:

$$SS_{S \times \text{RM}} = 34.04 - 16.33 - 7.0 = 10.71$$

To obtain the variance estimates (i.e., *MS*) that we will need to form our three *F* ratios, we must divide each of the five final *SS* components we found above by their corresponding dfs as shown in Formula 16.6:

$$MS_{\text{groups}} = \frac{SS_{\text{groups}}}{df_{\text{groups}}}$$ **Formula 16.6**

$$MS_W = \frac{SS_W}{df_W}$$

$$MS_{\text{RM}} = \frac{SS_{\text{RM}}}{df_{\text{RM}}}$$

$$MS_{G \times \text{RM}} = \frac{SS_{G \times \text{RM}}}{df_{G \times \text{RM}}}$$

$$MS_{S \times \text{RM}} = \frac{SS_{S \times \text{RM}}}{df_{S \times \text{RM}}}$$

Inserting the values for this example yields:

$$MS_{\text{groups}} = \frac{9.38}{1} = 9.38$$

$$MS_W = \frac{15.55}{6} = 2.59$$

$$MS_{\text{RM}} = \frac{16.33}{2} = 8.17$$

$$MS_{G \times \text{RM}} = \frac{7}{2} = 3.5$$

$$MS_{S \times \text{RM}} = \frac{10.71}{12} = .89$$

Finally, we can form the following *F* ratios to test each of the three null hypotheses stated in Step 1:

a. $$F_{\text{groups}} = \frac{MS_{\text{groups}}}{MS_W}$$ **Formula 16.7**

b. $$F_{\text{RM}} = \frac{MS_{\text{RM}}}{MS_{S \times \text{RM}}}$$

c. $$F_{G \times \text{RM}} = \frac{MS_{G \times \text{RM}}}{MS_{S \times \text{RM}}}$$

To test the main effect of groups, the *F* ratio comes out to:

$$F_{\text{groups}} = \frac{9.38}{2.59} = 3.62$$

To test the main effect of the within-subject factor, $MS_{S \times \text{RM}}$ is used as the error term:

$$F_{RM} = \frac{8.17}{.89} = 9.19$$

The same error term is used again in the F ratio to test the two-way interaction:

$$F_{G \times RM} = \frac{3.5}{.89} = 3.93$$

Step 6. Make the Statistical Decisions

The observed F for the main effect of the phobia treatment is 3.62, which is less than the critical $F(5.99)$, so we cannot reject the null hypothesis for this factor. On the other hand, the F ratio for the time factor (before versus after versus follow-up) is 9.19, which is well above the critical $F(3.89)$, so the null hypothesis for this effect can be rejected. For the interaction, the observed F ratio (3.93) is only slightly above the critical $F(3.89)$, but that is all that is required to reject this null hypothesis, as well.

Interpreting the Results

At first, the lack of statistical significance for the main effect of treatment group may seem discouraging; it seems to imply that the phobia treatment didn't work, or that at best it was no more effective than the control procedure. However, the significant interaction should remind you to graph the cell means before trying to interpret the results of the main effects. You can see from Figure 16.5 that the two groups are very similar in phobic intensity before the treatment (which is to be expected with random assignment) but diverge considerably after treatment. Despite the similarity of the groups before treatment, the two later measurements might have caused the main effect of group to be significant had not the samples been so small. The F ratio for groups is sensitive to the variability from subject to subject, and with small samples power is low unless the between-group difference is relatively large.

Figure 16.5

Graph of Cell Means for the Data in Table 16.5

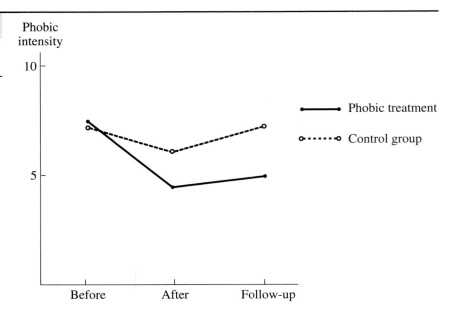

On the other hand, subject-to-subject variability does not affect the F ratio for the time factor. As long as the subjects exhibit similar patterns over time within each group, $MS_{S \times RM}$ will tend to be small and F_{RM} will tend to be large, as is the case in this example. However, the significance of the time factor must also be interpreted cautiously, given that the interaction is significant. The significance of $F_{G \times RM}$ tells you that the effect of time is different for the two groups. From Figure 16.5 you can see that the before-after reduction is much larger for the experimental group, and the increase from after to follow-up is somewhat greater for the control group.

It is likely that after obtaining these results a researcher would think of some more specific hypotheses to test—for instance, are the two groups significantly different just after the treatment? Or, is the before-after phobia reduction significant for the control group alone? Unless these specific questions have been planned in advance, the researcher should use procedures for post hoc comparisons. The choice of procedure depends on whether comparisons are being made among a series of repeated measures or among different groups and on whether the interaction is significant (see the subsection on "Post Hoc Comparisons").

Publishing the Results of a Mixed ANOVA

The results of the preceding phobia treatment experiment could be reported in a journal article in this manner: "The phobia intensity ratings were submitted to a 2 × 3 mixed design ANOVA, in which treatment group (experimental versus placebo control) served as the between-subjects variable, and time (before versus after versus follow-up) served as the within-subjects variable. The main effect of treatment group did not attain significance, $F(1, 6) = 3.61$, $MSE = 2.6$, $p > .05$, but the main effect of time did reach significance, $F(2, 12) = 9.19$, $MSE = .89$, $p < .05$. The results of the main effects are qualified, however, by a significant group by time interaction, $F(2, 12) = 3.94$, $MSE = .89$, $p < .05$. The cell means reveal that the before-after decrease in phobic intensity was greater, as predicted, for the phobia treatment group and that this group difference was maintained at follow-up. In fact, at follow-up, the control group's phobic intensity had nearly returned to its level at the beginning of the experiment."

The preceding paragraph would very likely be accompanied by a table or graph of the cell means and followed by a report of more specific comparisons, such as testing for a significant group difference just at the follow-up point. Although it is not likely that an ANOVA summary table would be included in a journal article (due to space constraints), such tables are produced by most statistical software packages, and because they are instructive, I am including a summary table (Table 16.6) for the phobia treatment example.

Source	SS	df	MS	F	p	
Between-Subjects	24.96	7				**Table 16.6**
Groups	9.38	1	9.38	3.61	>.05	
Within-Groups	15.58	6	2.6			
Within-Subjects	34	16				
Time	16.33	2	8.2	9.19	<.05	
Group × Time	7.0	2	3.5	3.94	<.05	
Residual (S × Time)	10.67	12	.89			
Total	58.96	23				

Assumptions of the Mixed Design ANOVA

I didn't say anything about how phobia was measured in the preceding experiment, other than to say that a 10-point scale was used, but a basic assumption of all the parametric statistical tests in this text (i.e., all the inferential tests covered thus far) is that the measurement scale used has the interval property, as defined in the first chapter. Nonetheless, as I have mentioned earlier, it is common in psychological research to perform parametric tests on data arising from subjective rating scales without demonstrating that the scale possesses the interval property. You will occasionally see the matter debated, but I will remain neutral on this matter. The other usual ANOVA assumptions apply as well to the mixed design; this includes independent random sampling (but random assignment is usually substituted), normal distributions, and homogeneity of variance.

Homogeneity of Variance

When the design is balanced, there is usually little concern about this assumption. However, in a mixed design there are two very different ways in which the design can be unbalanced. One way involves the between-subjects factor. It is not uncommon to end up with different numbers of subjects in each independent group, either because you are dealing with intact groups (e.g., there may be different numbers of patients available for each diagnosis represented in the study) or because some subjects have to be deleted from the data after the study has been concluded. Because unequal group sizes complicate the analysis, especially if the within-group variances are quite different from one group to another, I will not be covering such cases in this chapter, but the analysis is fairly routine and similar to the unbalanced two-way independent ANOVA.

The other way the mixed design can become unbalanced involves one or more subjects missing data for one or more of the RM levels. This can occur when a subject drops out of an experiment before the final session or leaves out some responses on a questionnaire or provides data that is just not usable for some of the conditions. Missing RM data can produce such statistical complications that, in the past, most researchers either just replaced the subject entirely or filled in the missing values by using some technique to estimate the most likely value for each missing data point. Modern statistical methods now exist for dealing with missing RM data, but are well beyond the scope of this text. When there is no missing data, the homogeneity of variance assumption with respect to the RM levels is not important if the sphericity assumption is satisfied, as described in the previous chapter.

Homogeneity of Covariance Across Groups

An assumption unique to mixed designs is that the covariance structure among the RM levels must be the same for all of the independent groups. In terms of the preceding example, the phobia group should have the same amount of before-after interaction (in the population) as the control group, and this should hold for the other two pairs of RM levels, as well. This form of homogeneity can be tested in terms of a statistic known as Box's M criterion (Huynh & Mandeville, 1979), which will be discussed further in Section C.

Dealing with Sphericity in Mixed Designs

If a test of Box's M is not significant, the interactions for pairs of RM levels can be calculated by averaging across the different independent groups before testing for sphericity with Mauchly's W statistic. If the W statistic is far from being significant, one can proceed with the calculations of the mixed design ANOVA as demonstrated in this chapter. However, if W is significant, you should consider a df adjustment as described in the next paragraph. If Box's M statistic is significant, the methods of this chapter are not justified, and you may have to proceed by performing separate one-way RM ANOVAs for each of your independent groups, followed by cell-to-cell comparisons (see Section C).

As described in the previous chapter, a lack of sphericity increases the likelihood of a significant F ratio, even if all of the population means for the repeated treatments are equal. The same conservative df adjustment discussed previously can be applied to the RM factor of a mixed design, as well. Of course, if F_{RM} does not meet the usual criterion for significance, you need do nothing further. However, a significant F_{RM} can be tested against a conservatively adjusted critical F, with $df_{num} = 1$ and $df_{denom} = df_W = k(n - 1)$. If the F ratio for the RM factor (F_{RM}) surpasses this larger critical F, it can be considered statistically significant without assuming sphericity. If F_{RM} falls between the usual critical F and the conservative F, you will want to use statistical software to estimate the degree of sphericity (i.e., epsilon) in the population and adjust the degrees of freedom accordingly.

The F ratio for the interaction of the two factors ($F_{G \times RM}$) uses the same error term as F_{RM} and can therefore be similarly biased when sphericity is violated. A similar df adjustment can be used for $F_{G \times RM}$, except that the degrees of freedom for the numerator of the adjusted critical F are $df_{num} = df_{groups} = k - 1$ (df_{denom} still equals df_W from the mixed-design ANOVA). This modified univariate approach can be replaced, however, by a multivariate analysis of both the RM factor and the interaction of the two factors as will be described in Section C of Chapter 18.

A Special Case: The Before-After Mixed Design

The simplest possible mixed design is one in which an experimental group and a control group are measured before and after some treatment. Some statisticians have argued that a two-way ANOVA is usually unnecessary in this case and may even prove misleading (Huck & McLean, 1975). The main effect of group is misleading because the group difference *before* the treatment (which is expected to be very close to zero) is being averaged with the group difference *after* the treatment. The main effect of time is equally misleading because the before-after difference for the experimental group is being averaged with the before-after difference for the control group. The only effect worth testing is the interaction, which tells us whether the before-after differences for the experimental group are different from the before-after differences for the control group. We do not need ANOVA to test this interaction—we need only find the before-after difference for each subject and then conduct a t test of two independent samples (experimental versus control group) on these difference scores. Squaring this t value will give the F that would be calculated for the interaction in the mixed design ANOVA. If before and after measurements have been taken on more than two groups, a one-way independent-groups ANOVA on the difference scores yields the same F ratio as a test of the interaction in the mixed design. Of course, you would probably want to follow a significant F with pairwise tests on the difference scores to determine which groups differ significantly. Note that when the RM factor has only two levels, the sphericity assumption does not apply, and homogeneity of covariance across groups becomes the ordinary homogeneity of variance for the difference scores.

Although testing the interaction of the mixed design, or testing the difference scores, as just described, seem to be the most common ways of evaluating group differences in a before-after design, they are not the most powerful. A more sensitive test of group differences is based on a procedure known as the *analysis of covariance* (*ANCOVA*). Instead of simply subtracting the before score from the after score, you can use linear regression to predict the after score from the before score. The analysis proceeds in terms of the residual scores (after score minus predicted after score) rather than on the difference scores. The residual scores always have less variance (unless the regression slope happens to be 0 or 1.0) and therefore tend to yield a higher t or F. The logic and mechanics of ANCOVA will be explained further in Chapter 18, Section B.

Post Hoc Comparisons

When the Two Factors Do Not Interact

If the interaction of the two factors is not statistically significant (nor large and disordinal), whichever main effects are significant can be explored in a straightforward manner. If the RM factor is significant and has more than two levels, a post hoc comparison method (e.g., Tukey's HSD) can be used to test each pair of RM levels for significance (see Figure 16.6). If you feel strongly that all of the assumptions of the mixed design ANOVA have been met, $MS_{S \times RM}$ can be used as the error term for testing each pair of levels. However, if there is any doubt about the sphericity assumption, the safe thing to do is to base your error term on the variability of the difference scores for only the two levels being compared, for reasons discussed in the previous chapter. (The subject by treatment interaction for the two levels would be calculated separately for each group and then pooled, assuming that these interactions were similar from group to group.) Of course, complex comparisons can also be conducted among the RM levels using Scheffé's test (for the preceding example, "beffore" vs. the average of "after" and "follow-up" would be a likely contrast followed by a test between "after" and "follow-up"). Again, it is safer to base your error term on only the RM levels involved in the comparison.

Figure 16.6	

Pairwise Comparisons in a Mixed Design (When the Interaction Is Not Significant)

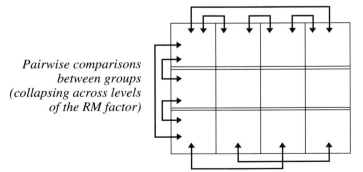

Pairwise comparisons between groups (collapsing across levels of the RM factor)

Pairwise comparisons between levels of the RM factor (collapsing across groups)

Pairwise comparisons following a mixed design with significant main effects, but no significant interaction between the factors

Similarly, a significant between-groups factor with more than two levels would be followed with pairwise or complex comparisons among the group means (averaging across the repeated measures), using the between-groups error term, MS_W. With the same number of subjects in each group, homogeneity of variance across the groups is not a serious concern. However, if the groups have different sizes and considerably different variances, the equivalent of separate-variance t tests should be used.

The alpha level is usually set at .05 for each "family" of comparisons; in this case, each main effect represents a separate family. If the interaction turns out to be significant, or specific comparisons have been planned, the analysis of the data would proceed differently, as described next.

When the Interaction of the Two Factors is Significant

If $F_{G \times RM}$ is statistically significant, the main effects are usually not explored further, and post hoc comparisons are focused instead on the cell means, along the lines already described in Chapter 14. The most common approach is an analysis of simple main effects. You can test the effects of the RM factor separately for each group or test the effects of the between-groups factor separately at each level of the RM factor or conduct tests in both ways (see Figure 16.7). Unfortunately, the interaction of the two factors in a mixed design complicates the choice of an error term for each of the simple effects, as I will explain.

The simplest and safest way to test simple effects involving the RM factor is to conduct a separate one-way RM ANOVA for each group, as though the other groups do not exist. If there is homogeneity among the groups in terms of the subject by treatment interaction, more power can be gained by using $MS_{S \times RM}$ from the overall analysis rather than the $MS_{S \times RM}$ for just the group being tested. However, the more conservative approach, in light of a significant interaction between the factors, is to sacrifice the small amount of extra power from the pooled error term, in favor of not increasing the risk of a Type I error, in case there actually is no homogeneity among the groups in the population.

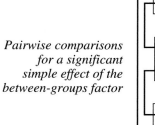

*Pairwise comparisons
for a significant
simple effect of the
between-groups factor*

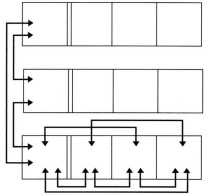

Figure 16.7

Pairwise Comparisons in
a Mixed Design (When
the Interaction Is
Significant)

*Pairwise comparisons for a significant
simple effect of the RM factor*

Pairwise comparisons following a mixed design with a significant
interaction

To follow up a significant simple effect, you would probably want to conduct pairwise comparisons among the different RM levels (e.g., before versus after for the control group; after versus follow-up for the experimental group). If you had a strong reason to assume sphericity within the significant simple effect, you could use the $MS_{S \times RM}$ for that group as the error term for each pairwise test. However, because the sphericity assumption is generally considered quite risky, especially for pairwise comparisons, it is strongly recommended that you base your error term only on the two levels being tested. (This is equivalent to performing a simple matched t test between a pair of RM levels for one of the groups.) The Type I error rate can be controlled by using the Bonferroni test, setting alpha at .05 for the entire family of possible RM comparisons.

Simple effects can also be tested by comparing the different groups at each level of the RM factor. However, the proper error term for each of these effects is not MS_W, as calculated for the mixed design. You may have noticed that what I'm calling MS_W in the mixed design is not the same as MS_W in a two-way independent ANOVA. In the independent ANOVA, variances are calculated for each *cell* and then averaged (I'll call this $MS_{\text{within-cell}}$). In the mixed design, variances are not calculated separately for each cell; scores are averaged over the RM levels, and then variances are calculated within each group (for now, I'll call MS_W from the mixed design $MS_{\text{within-group}}$). In fact, if you calculated $SS_{\text{within-cell}}$ for a mixed design, it would equal the sum of $SS_{\text{within-group}}$ and $SS_{S \times RM}$.

The relation just described provides another way to understand the error terms in the mixed design. You can begin by calculating $SS_{\text{within-cell}}$ as for an independent ANOVA. Then this SS is divided into two pieces: one ignores subjects' differing reactions to different RM levels by averaging across those levels ($SS_{\text{within-group}}$), whereas the other piece is insensitive to overall differences between subjects and measures only the extent to which subjects react similarly to the differing RM levels within each group ($SS_{S \times RM}$). If we are performing post hoc comparisons in which we are looking at differences among the groups at one level of the RM factor at a time, it is not appropriate to use an error term that averages across RM levels (i.e., $MS_{\text{within-group}}$), especially given that the grouping variable interacts significantly with the RM factor. The appropriate error term is $MS_{\text{within-cell}}$.

Unfortunately, the homogeneity assumption required to justify the use of $MS_{\text{within-cell}}$ as the error term for post hoc comparisons is likely to be violated, leading to a biased F ratio. For large sample sizes this bias will be negligible, but for samples smaller than 30 an adjustment of the degrees of freedom (similar to the adjustment in the separate-variances t test) is recommended before you determine the critical F. The problem can be avoided, with some loss of power, by pooling error terms for only the cells involved in the analysis. For instance, if you are comparing only "after" measurements for two experimental groups and a control group, the error term can be based on pooling the MS_W for just those cells, rather than being based on $MS_{\text{within-cell}}$ from the entire mixed design. This localized error term can then be used to test pairs of groups within a particular level of the RM factor, whenever the simple effect of groups is significant at that level of the RM factor. (Of course, if there are only two groups in the design, there are no follow-up tests to be done on that factor.)

As an alternative to simple effects analysis, 2×2 interaction contrasts can be performed on a mixed design using Scheffé's test. These interaction contrasts can incorporate complex comparisons. For example, in the phobia experiment, "before" and the average of "after" and "follow-up" could be crossed with the two treatment groups to create an interaction contrast.

The error term could be the same as for the omnibus ANOVA, but it can be argued that the safer error term would be based on the contrast scores themselves (as with trend interactions—see Section C).

In general, the simplest solution to the error term problem is to calculate each interaction contrast as a separate mixed design ANOVA without attempting to use the error term from the larger analysis. Some power is sacrificed in favor of increased caution regarding the experimentwise alpha rate.

Effect Sizes for a Mixed Design

In Chapter 14, we saw that an effect size measure (ESM) can be reduced by the presence of a second, experimentally manipulated variable, and this can be misleading when comparing effects with other studies (or planning future studies) that do not include the same second factor. The remedy was to use a partial ESM. On the other hand, we saw in Chapter 15 that the effect of an experimental variable can be misleadingly inflated by the reduction that occurs in the error term of a design with repeated measures. The remedy was to include between-subject variability in the denominator of our ESM. In the mixed-design ANOVA we may have to deal with both of these influences, when trying to portray the effect size of one or the other of our two factors.

The Between-Groups Factor Has Been Manipulated

Let us first consider the case in which the between-groups (BG) factor is an experimental one. In calculating eta squared for the BG factor, we do not want to include the variability due to the RM factor, which by its nature is inevitably an experimental factor itself (imagine trying to repeat an individual-difference factor), nor its interaction with the grouping factor. At the same time we do not want to use a measure of error which has been reduced by matching, which is often the case even for SS_W (and therefore MS_W), when calculated as part of a mixed design. The solution is to include in the denominator both between-group and RM sources of error, but not SS_{RM} or $SS_{G \times RM}$, to create a measure that Olejnik and Algina (2003) labeled *generalized eta squared*, because it can be easily compared even among studies that differ in experimental design.

$$\eta^2_{\exp(RM)} = \frac{SS_{group}}{SS_{group} + SS_W + SS_{S \times RM}}$$

The subscript "exp(RM)" attached to η^2 is to remind you that this generalized eta squared measure corresponds to an experimental between-groups factor in a mixed design. Given that the sum of SS_W and $SS_{S \times RM}$ is $SS_{within-cell}$, the preceding version of eta squared can be written as:

$$\eta^2_{\exp(RM)} = \frac{SS_{group}}{SS_{group} + SS_{within-cell}} \qquad \textbf{Formula 16.8}$$

To obtain the ordinary (i.e., generalized) eta squared for either the RM or Group × RM interaction effects, just substitute SS_{RM} or $SS_{G \times RM}$, respectively, for SS_{group} in both the numerator and denominator of Formula 16.8.

As usual, the formula for $\eta^2_{\exp(RM)}$ needs to be adjusted to create a relatively unbiased estimate of what Olejnik and Algina (2003) call *generalized omega squared*, as shown in Formula 16.9.

$$\text{est. } \omega^2_{\exp(RM)} = \frac{SS_{group} - df_{group} MS_W}{SS_{group} - df_{group} MS_W + N_T MS_{within-cell}} \qquad \textbf{Formula 16.9}$$

The formulas for estimating generalized omega squared for the RM and interaction effects are even more complicated, as you can see from Table 4 in Olejnik and Algina (2003).

The Between-Groups Factor Has Been Measured (i.e., Selected For)

Another way to express generalized eta squared is in terms of which SS components need to be subtracted from SS_{total} in the denominator, so Formula 16.8 can be expressed alternatively as:

$$\eta^2_{\text{exp(RM)}} = \frac{SS_{\text{group}}}{SS_{\text{total}} - SS_{\text{RM}} - SS_{G \times \text{RM}}}$$

If SS_{group} arises from a pre-existing grouping factor, then we do not want to subtract its interaction with the RM factor from the denominator of eta squared, but we do want to subtract variability due to the main effect of the RM factor. Therefore, the formula for generalized eta squared for a *measured* between-groups factor is:

$$\eta^2_{\text{meas(RM)}} = \frac{SS_{\text{group}}}{SS_{\text{total}} - SS_{\text{RM}}} \qquad \textbf{Formula 16.10}$$

So, put another way, the denominator of Formula 16.10 is equal to the denominator of Formula 16.8 plus $SS_{G \times \text{RM}}$ (or, the sum of SS_{group}, $SS_{G \times \text{RM}}$, and $SS_{\text{within-cell}}$). To better understand this denominator, it may help to think of it as something I will call $SS_{\text{within-RM}}$; this is what you would get if you calculated all of the subject-to-subject SS separately for each RM condition (ignoring the grouping factor), and then added these SSs together. To obtain eta squared for the group by RM interaction you would merely substitute $SS_{G \times \text{RM}}$ for the numerator in Formula 16.10, and leave the denominator as is. Finally, for the RM factor, ordinary eta squared (unmodified) can be used—that is, $SS_{\text{RM}}/SS_{\text{total}}$. You do not want a partial eta squared because the between-groups factor has not been manipulated, and you do not want η^2_{RM} (i.e., you *do* want to include subject-to-subject variability in the denominator), if you want an ESM that is easy to generalize from.

The adjustment to $\eta^2_{\text{meas(RM)}}$ that you need if you want to estimate $\omega^2_{\text{meas(RM)}}$ is as follows:

$$\text{est. } \omega^2_{\text{meas(RM)}} = \frac{SS_{\text{group}} - \text{df}_{\text{group}} MS_W}{SS_{\text{total}} - SS_{\text{RM}} + MS_W + \text{df}_{\text{RM}} MS_{S \times \text{RM}}} \qquad \textbf{Formula 16.11}$$

For the RM factor, the ordinary eta squared can be adjusted to provide an estimate of ordinary (generalized) omega squared, like this:

$$\text{est. } \omega^2 = \frac{SS_{\text{RM}} - \text{df}_{\text{RM}} MS_{S \times \text{RM}}}{SS_{\text{total}} + MS_W} \qquad \textbf{Formula 16.12}$$

To estimate ω^2 for the interaction of the two factors in this design, the reader is again referred to Table 4 in Olejnik and Algina (2003).

An Excerpt from the Psychological Literature

The following example of a mixed design ANOVA in the psychological literature comes from an article entitled "Affective Valence and Memory in Depression: Dissociation of Recall and Fragment Completion" (Denny & Hunt, 1992). This study contains several ANOVAs, but the one I have chosen resembles the example in Section A, except that it does not include neu-

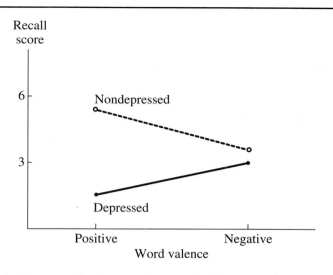

Recall score

Figure 16.8

Graph of Cell Means for Data in Denny and Hunt (1992) Study

tral words. Thus word valence refers to whether a word tended to evoke positive or negative affect. The results were reported as follows.

> Recall data were subjected to an analysis of variance (ANOVA) with group as a between-subjects variable and word valence as a within-subjects variable. The results revealed a significant main effect for group: Recall level was higher for the nondepressed group than for the depressed group, $F(1, 30) = 30.21$, $MS_e = 2.05$, $p < .0001$. This effect was qualified, however, by a highly significant Group $\times$ Word Valence interaction, $F(1, 30) = 30.29$, $MS_e = 1.45$, $p < .0001$. The results of t-tests indicated that, as predicted, the depressed group recalled more negative than positive words, $t(15) = 4.45$, $SE_{difference} = .393$, $p < .001$. Within the nondepressed group, the opposite pattern was observed. Recall of positive words was significantly higher than that of negative words, $t(15) = 3.42$, $SE_{difference} = .456$, $p < .01$. Finally, comparisons revealed a significant between-groups difference in recall of positive words, $t(30) = 6.99$, $SE_M = .517$, $p < .001$, but not negative words, $t(30) = .75$, $SE_M = .411$, $p < .20$.

Note that the article gives the error term (MS_e) for each F ratio after stating that ratio; this is the recommended practice so that a reader is equipped to perform his or her own follow-up analyses. Similarly, the denominator is given for the two-sample independent t test (SE_M) and for the matched t test ($SE_{difference}$). The authors refer to the interaction as being "highly significant." Be aware that some statistical purists abhor this expression; they argue that a result is either significant (i.e., p is less than the predetermined alpha) or it is not. What the authors meant in this case is that the p value for the interaction was very small; the interaction would have been significant even if a very small alpha (e.g., .0001) had been set.

Figure 16.8 is a graph of the cell means reported by Denny and Hunt. The figure illustrates the nature of the significant interaction and should make it obvious why the main effect of group was significant but the main effect of word valence was not.

1. The order for calculating the *SS* components for the mixed design, as given in step 5 of this section, is meant to be instructive and reveal the concepts behind the analysis. A simpler, but equivalent, description follows. All *SS* components not found by subtraction are found by multiplying the biased variance of the appropriate means by the total

B

SUMMARY

number of observations (i.e., the number of different subjects or blocks times the number of levels of the repeated measures factor).

   a. First, calculate the basic SS components of an independent two-way ANOVA (as in Chapter 14): SS_{total}, as usual, is based on all of the individual scores, and $SS_{between\text{-}cells}$ is based on the cell means (i.e., the mean for each group at each level of the RM factor); $SS_{columns}$ is called SS_{RM}, and SS_{rows} is called $SS_{between\text{-}S}$ (assuming the row and column arrangement of Chapter 15). An additional component, SS_{groups}, is found from the means of the groups (averaging across all subjects in the group and all levels of the RM factor). The remaining SS components can be found by subtraction.

   b. SS_W is found by subtracting SS_{groups} from $SS_{between\text{-}S}$. $SS_{G \times RM}$ is found by subtracting both SS_{groups} and SS_{RM} from $SS_{between\text{-}cells}$. $SS_{S \times RM}$ can be found by subtracting SS_{RM}, $SS_{between\text{-}S}$, and $SS_{G \times RM}$ from SS_{total}.

   c. Divide SS_{RM}, SS_{groups}, $SS_{G \times RM}$, SS_W, and $SS_{S \times RM}$ by their respective df to create the needed MSs.

   d. Divide MS_{groups} by MS_W, and both MS_{RM} and $MS_{G \times RM}$ by $MS_{S \times RM}$ to complete the analysis.

2. If the different independent groups have different numbers of subjects, the mixed-design ANOVA becomes more complicated (and susceptible to violations of homogeneity of variance), but the analysis is still routine. However, if any subjects are missing data for any of the RM conditions, advanced statistical techniques are required to analyze the data appropriately unless the subjects with missing data are deleted completely or the missing data points are replaced with reasonable estimates.

3. The mixed-design ANOVA rests on the assumption that the interaction between any two RM levels is the same for all the independent groups (i.e., for every level of the between-subjects factor). This assumption can be tested with Box's M criterion. If M is significant, the mixed-design ANOVA, as presented in this chapter, is questionable. If M is not significant, a sphericity test should be performed after pooling the pairwise interactions across groups.

4. If the sphericity test is significant, the df should be adjusted for both the RM factor and the two-way interaction before finding the critical F in each case. Even if the sphericity test is not significant, an adjustment should be considered when your sample sizes are small.

5. If two independent groups are measured before and after some treatment whose impact differs on the two groups, it is usually only the interaction of the mixed design that is of interest, and this F ratio can also be found by performing an independent t test on the before-after difference scores of the two groups and squaring the result. Similarly, if there are more than two independent groups, a one-way ANOVA can be performed on the difference scores. However, an analysis of covariance of the "after" scores, using the "before" scores as the covariate is almost always the more powerful way to analyze this design.

6. When the two-way interaction is not significant, either main effect can be followed with post hoc comparisons in a routine way if it is significant. However, although there is little concern about using MS_W from the overall analysis to test pairs of groups, $MS_{S \times RM}$ should only be used for pairs of RM levels if there is a strong reason to believe that sphericity exists in the population. When in doubt, your error term should be based on only the two RM levels being compared.

7. A significant interaction is often followed by an analysis of simple (main) effects. Simple effects of the RM factor are most safely

explored by conducting separate one-way RM ANOVAs for each group; $MS_{S \times RM}$ from the overall analysis should be used as the error term for simple effects only when there is a strong basis to assume that all of the homogeneity assumptions of the mixed design have been met. Similarly, significant simple RM effects can be followed by pairwise tests for pairs of levels, but the error term for the pairwise tests should be based on the two levels involved, rather than the pooled error term from the simple effect, unless sphericity can be assumed for that simple effect.

8. Simple effects can also be tested among the groups at each level of the RM factor. For these tests, the appropriate error term is $MS_{\text{within-cell}}$, the average variance of the cells, as in a two-way independent ANOVA (in a mixed design, $SS_{\text{within-cell}}$ is subdivided into $SS_{S \times RM}$ and a portion I've been calling SS_W to form the basis of the two error terms). For small samples, and when the homogeneity of cell variances is in doubt, the safer approach is to pool the variances only for the cells involved in the simple effect tested. This more limited pooled error term can then be used to test pairs of groups for a simple effect that reached significance.

9. As with any two-way ANOVA, a significant interaction can be followed by a test of interaction contrasts to further localize the effect. However, sphericity concerns argue against using $MS_{S \times RM}$ from the overall analysis as the error term. The simple solution is to perform a separate mixed-design ANOVA on the desired subset of the full design (e.g., interaction contrast), using the error term from just that subset.

10. Although the RM factor is always considered a manipulated factor, the between-groups (BG) factor can be either manipulated or selected for, which determines whether a partial eta squared is appropriate. At the same time, it can be misleading to use an error term that is reduced by using repeated measures or matching. Combining these two considerations, the generalized eta squared for an experimental BG factor is the SS for that factor divided by the total SS minus the SSs for both the RM and group by RM interaction effects. However, if the BG factor is a measured one (i.e., based on pre-existing groups), only SS_{RM} is subtracted in the denominator of the generalized eta squared formula.

EXERCISES

*1. Imagine a study conducted to compare the effects of three types of training on the acquisition of a motor skill. Thirty-six subjects are divided equally into three groups, each receiving a different type of training. The performance of each subject is measured at five points during the training process.
   a. Construct a df tree (like the one in Figure 16.4) for the mixed design ANOVA that would be used to analyze this experiment.

   b. Find the critical values of F for each of the F ratios that would be calculated.

2. In Exercise 14B2, a group of visualizers and a group of nonvisualizers were each divided in half, and all subjects were presented with either concrete or abstract words to recall. A more powerful way to conduct that experiment would be to give each subject the same list containing a mixture of concrete and abstract words. The data from Exercise 14B2 follow, rearranged to make it easier to

see the number of concrete and abstract words each subject recalled.

	Concrete	Abstract
Visualizers	17	14
	20	15
	18	15
	21	17
	20	16
Nonvisualizers	18	18
	19	18
	17	17
	17	17
	20	19

a. Perform a two-way mixed design ANOVA on the data above (*Note*: You can save time by using some of the SSs that you calculated for Exercise 14B2.)

b. Present your results in a summary table and compare it to the summary table you created for Exercise 14B2.

c. Calculate the appropriate value for generalized eta squared for both main effects. Also, calculate the corresponding estimate of omega squared in each case.

*3. A psychologist is studying the relationship between emotion and eating. All of his subjects view the same two film segments. One segment evokes happiness and one segment evokes fear; the order in which subjects view the film segments is counterbalanced. Half of the subjects are randomly assigned to a condition that requires them to eat a full meal just before viewing the film segments (preload condition); the remaining half are not permitted to eat during the 4 hours preceding the experiment (no load condition). The subjects are offered an unlimited amount of popcorn while viewing the film segments. The amount of popcorn (in ounces) consumed by each subject in each condition appears in the following table.

	Happiness	Fear
Preload	10	12
	13	16
	8	11
	16	17
No Load	26	20
	19	14
	27	20
	20	15

a. Perform a mixed design ANOVA.

b. Draw a graph of the cell means. (Note that you already graphed the data for the individual subjects in Exercise 16A4.) Describe the nature of each significant effect.

c. Calculate the happiness-fear difference score for each subject, and then perform a two-group independent *t* test on these difference scores. Which of the *F* ratios calculated in part a is related to the *t* value you just found? What is the relationship?

4. In Exercise 15B1, subjects performed a clerical task under three noise conditions. Now suppose a new group of subjects is added to study the effects of the same three conditions on the performance of a simpler, more mechanical task. The data from Exercise 15B1 follow, along with the data for the mechanical task.

a. Perform a mixed design ANOVA, and display the results in a summary table.

b. Calculate generalized eta squared for the main effect of the type-of-task factor, and the corresponding estimate of generalized omega squared. Does this look like a large effect size? Explain.

	Background	Popular	Heavy Metal
Clerical Task	10	12	8
	7	9	4
	13	15	9
	18	12	6
	6	8	3
Mechanical Task	15	18	20
	19	22	23
	8	12	15
	10	10	14
	16	19	19

*5. Dr. Jones is investigating various conditions that affect mental effort—which, in this experiment, involves solving anagrams. Subjects were randomly assigned to one of three experimental conditions. Subjects in the first group were told that they would not be getting feedback on their performance. Subjects in the second and third groups were told they *would* get feedback, but only subjects in the third group were told (erroneously) that anagram solving was highly correlated with intelligence and creativity (Dr. Jones hoped this information would produce ego involvement). The list of anagrams given to each subject contained a random mix of problems at four levels of difficulty determined by the number of letters presented (five, six, seven, or eight). The number of anagrams correctly solved by each subject in each condition and at each level of difficulty is given in the following table:

a. Draw a degrees of freedom tree for this experiment.
b. Perform a mixed analysis of variance, and display the results in a summary table. Would any of your conclusions change if you do not assume sphericity? Explain.
c. Perform post hoc pairwise comparisons for both main effects, using the appropriate error term from part b in each case. Explain why these follow-up tests are appropriate given your results in part b.

	Five	Six	Seven	Eight
No Feedback	9	6	4	2
	10	7	4	3
	12	9	7	5
Feedback	19	16	15	12
	19	15	11	11
	22	20	17	14
Feedback + Ego	30	25	22	21
	31	30	27	23
	34	32	28	24

6. A psychiatrist is comparing various treatments for chronic headache: biofeedback, an experimental drug, self-hypnosis, and a control condition in which subjects are asked to relax as much as possible. Subjects are matched in blocks of four on headache frequency and severity and then randomly assigned to the treatment conditions. To check for any gender differences among the treatments, four of the blocks contain only women and the other four blocks contain only men. The data appear in the following table:

Blocks	Control	Biofeedback	Drug	Self-Hypnosis
Men	14	5	2	11
	11	8	0	4
	20	9	7	13
	15	12	3	10
Women	15	7	1	10
	10	8	1	4
	14	8	8	12
	17	14	4	11

a. Before conducting the ANOVA, graph the cell means and guess which of the three F ratios is (are) likely to be significant and which is (are) not.
b. Test the significance of each of the three F ratios at the .05 level. Would any of your conclusions change if you do not assume sphericity? Explain.
c. Perform post hoc pairwise comparisons on the RM factor, assuming sphericity.
d. Calculate ordinary eta squared for the RM factor. Does the effect look large? Explain.

*7. A market researcher is comparing three types of commercials to determine which will have the largest positive effect on the typical consumer. One type of commercial is purely informative, one features a celebrity endorsement, and the third emphasizes the glamour and style of the product. Six different subjects are randomly assigned to watch each of the three types of commercial. Subjects rate their likelihood of buying the product on a scale from 0 (very unlikely to buy) to 10 (very likely to buy) both before and after viewing the assigned commercials. The data appear in the following table:

INFORMATIVE		CELEBRITY		GLAMOUR/STYLE	
Before	After	Before	After	Before	After
3	5	6	8	5	8
6	6	6	9	7	8
5	7	4	4	5	5
7	8	5	6	5	7
4	6	7	8	6	7
6	5	2	4	3	6

a. Perform a mixed design ANOVA and test the three F ratios at the .01 level.
b. Would any of your conclusions in part a change if you do not assume sphericity? Explain.
c. Calculate the before-after difference score for each subject, and then perform a one-way independent groups ANOVA on these difference scores. Which of the F ratios calculated in part a is the same as the F ratio you just found? Explain the connection.

8. Exercise 15B6 described a neuropsychologist studying subjects with brain damage to the left cerebral hemisphere. Such a study would probably include a group of subjects with damage to the right hemisphere and a group of control subjects without brain damage. The data from Exercise 15B6 (the number of digit or letter strings each subject

recalled) follow, along with data for the two comparison groups just mentioned.

a. Perform a mixed design ANOVA and test the three F ratios at the .05 level. What can you conclude about the effects of brain damage on short-term recall for these types of stimuli?

b. Draw a graph of these data, subject by subject. Do the assumptions of the mixed design ANOVA seem reasonable in this case? Explain.

c. Perform post hoc pairwise comparisons for both main effects. Do *not* assume sphericity for the RM factor.

	Digit	Letter	Mixed
Left Brain Damage	6	5	6
	8	7	5
	7	7	4
	8	5	8
	6	4	7
	7	6	5
Right Brain Damage	9	8	6
	8	8	7
	9	7	8
	7	8	8
	7	6	7
	9	8	9
Control	8	8	7
	10	9	9
	9	10	8
	9	7	9
	8	8	8
	10	10	9

*9. Treat the data for the sleep-deprived condition in Exercise 15C7 as though they come from a group of subjects that is completely independent from a separate group of subjects that produced the "well-rested" data.

a. Perform a mixed-design ANOVA on the data from Exercise 15C7, with sleep condition as the between-groups variable.

b. Explain any differences from the results you obtained in Exercise 15C7. Refer to a graph of the cell means to describe whatever interaction you see, regardless of whether it is statistically significant.

10. Treat the data for the CBT therapy condition in Exercise 15C8 as though they come from a group of subjects that is completely independent from a separate group of subjects that was given psychodynamic therapy.

a. Perform a mixed-design ANOVA on the data from Exercise 15C8, with type of therapy as the between-groups variable.

b. Explain any differences from the results you obtained in Exercise 15C8. Refer to a graph of the cell means to describe whatever interaction you see, regardless of whether it is statistically significant.

The Variance-Covariance Matrix for ANOVAs with an RM (or RB) Factor

In introducing the concept of sphericity in the previous chapter, I mentioned that it can be defined in terms of the matrix of variances and covariances that correspond to the levels of an RM design. The variance-covariance matrix (VCM) for a data set involving three or more conditions tells an important story about the relations among the various treatment levels; the structure of the VCM determines the most appropriate way to analyze the data. Here, I will describe only a few simple examples of VCM structures, including, of course, a description of a VCM that exhibits sphericity. Suppose your experiment consists of measuring your participants before and after a 3-month treatment, and then again 6 months later (follow-up). The VCM for the population could be represented as shown in Table 16.7.

The entries on the diagonal (i.e., σ^2_B, σ^2_A, σ^2_F) are the population variances for the three treatment levels: before, after, and follow-up, respectively (these are subject-to-subject variances for each level, ignoring the fact that there are repeated measures). To represent homogeneity of variance among the levels, the three entries on the (downward sloping) diagonal

	Before	After	Follow-up
Before	σ^2_B	σ^2_{BA}	σ^2_{BF}
After	σ^2_{AB}	σ^2_A	σ^2_{AF}
Follow-up	σ^2_{FB}	σ^2_{FA}	σ^2_F

Table 16.7

Variance-Covariance
Matrix for Three
Treatment Levels

would all be replaced by the variance symbol (σ^2) without a subscript. The symbol σ^2_{BA}, for example, represents the covariance between the before and after measures. Using this notation, the Pearson correlation between the before and after measures (in the population) can be expressed in terms of the following ratio: $\rho_{BA} = \sigma^2_{BA}/\sigma_B\sigma_A$. Therefore, $\sigma^2_{BA} = \rho_{BA}\sigma_B\sigma_A$. Note that just as ρ_{BA} always equals ρ_{AB}, σ^2_{BA} equals σ^2_{AB}, so the three entries above (and to the right) of the diagonal are redundant with the three below the diagonal.

The VC Matrix for Compound Symmetry

When there is homogeneity of *covariance,* all six of the entries off the diagonal will be equal to each other. When you have both homogeneity of variance and homogeneity of covariance in the same data set, all of the covariances can be written as $\rho\sigma\sigma$ (without subscripts), which equals $\rho\sigma^2$. As I mentioned in Section B, this special condition is called *compound symmetry.* Because all of the variances are equal to each other, and all of the covariances are equal to each other (but different from the variances, unless correlation is always perfect), the correlation between any pair of levels will be the same as for any other pair. This leads to a very simple way of expressing the VCM for compound symmetry, as shown in Table 16.8.

Compound symmetry is especially unlikely to apply to measurements taken over time. For most DVs you could imagine, the correlation between two measures that are close together in time will be larger than the correlation for two measures further apart in time. In the example just described, the "before" measure would be expected to correlate more highly with the "after" measure than the one at follow up. One simple model for how correlations will decline over time is called the first-order autoregressive model. The term "first-order" refers to the situation in which a measure is correlated directly only with the immediately preceding measure. Applied to our example, "before" is directly correlated to "after" (let's say by the amount ρ), and "after" to "follow up" (also by ρ), but "before" is correlated with "follow up" only indirectly through each of their correlations with "after." Therefore, the before/follow-up correlation would be $\rho \times \rho = \rho^2$, leading to a dramatic drop-off in covariance as measures are separated further in time (unless ρ is perfect, squaring it will make it smaller). These relations are shown in the VC matrix in Table 16.9.

	Before	After	Follow-up
Before	σ^2	$\rho\sigma^2$	$\rho\sigma^2$
After	$\rho\sigma^2$	σ^2	$\rho\sigma^2$
Follow-up	$\rho\sigma^2$	$\rho\sigma^2$	σ^2

Table 16.8

Compound Symmetry

	Before	After	Follow-up
Before	σ^2	$\rho\sigma^2$	$\rho^2\sigma^2$
After	$\rho\sigma^2$	σ^2	$\rho\sigma^2$
Follow-up	$\rho^2\sigma^2$	$\rho\sigma^2$	σ^2

Table 16.9

First-order
Autoregressive Model

Of course, there are other more complex models that are sometimes needed to account for changes in covariance over time; some of these fall under the topic of time-series analysis, and there is a growing body of modern methods that deal with the *analysis of change* (e.g., Collins & Sayer, 2001; Singer & Willet, 2003). In any case, it rarely makes sense to apply the ordinary RM ANOVA to a series of measures taken over time. Perhaps, the simplest method for analyzing repeated measurements over time is the analysis of trend components, as described in the previous chapter.

The VC Matrix for Sphericity

Sphericity can be defined as a condition in the population in which every possible pair of levels for the RM factor exhibits the same amount of subject-by-treatment interaction as every other pair. The difference between sphericity and homogeneity of covariance can be made clearer by expressing the MS for the interaction between two RM levels in terms of the correlation between the scores at those two levels. I will derive a formula for this relationship by beginning with the equality of two very different ways to find the error term of the RM t test; let us compare the denominators of Formulas 11.1 and 11.2

$$\frac{s_D}{\sqrt{n}} = \sqrt{\frac{s_1^2 + s_2^2}{n} - \frac{2r_{s_1 s_2}}{n}}$$

Squaring and then multiplying both sides by n, we see how the variance of the difference scores can be expressed as a function of the variances of the two sets of scores and their correlation (the subscripts i and j are used so that the formula can accommodate any pair of levels from a multilevel study):

$$s_D^2 = s_i^2 + s_j^2 - 2 r_{ij} s_i s_j$$

As I mentioned in the previous chapter, MS_{resid} (the error term for an RM ANOVA) for two levels is equal to half of the variance of the difference scores between those two levels; therefore, dividing both sides of the equation by 2, and substituting the corresponding population parameters, yields the following equation for the population:

$$\sigma_{\text{resid}}^2 = \frac{\sigma_i^2 + \sigma_j^2}{2} - \rho_{ij} \sigma_i \sigma_j$$

Note that the first term on the right side of the equation is the pooled variance for the two levels (it is a simple average in this case because the two sets of scores are paired and so must have the same n), and the term being subtracted from it is the covariance of those two levels; using the notation I introduced earlier in this section, the latter term can be written as σ_{ij}. Solving for the covariance in the preceding equation yields the following covariance formula:

$$\sigma_{ij}^2 = \frac{\sigma_i^2 + \sigma_j^2}{2} - \sigma_{\text{resid}}^2 \qquad \textbf{Formula 16.13}$$

Recall that sphericity is defined as a condition in which σ_{resid}^2 is always the same, regardless of which pair of levels you are considering, so σ_{resid}^2 can be replaced by a constant (C) in Formula 16.13, when sphericity applies. Applying Formula 16.13 to our example, the off-diagonal entries in Table 16.7 (i.e., the covariances) can be replaced by the corresponding entries shown in Table 16.10 under the sphericity condition. Note that the variances for the B, A, and F levels need not be equal to each other in Table 16.10, and if they are not all the same (i.e., homogene-

	Before	After	Follow-up	**Table 16.10**
Before	σ_B^2	$\dfrac{\sigma_B^2 + \sigma_A^2}{2} - C$	$\dfrac{\sigma_B^2 + \sigma_F^2}{2} - C$	Sphericity
After	$\dfrac{\sigma_A^2 + \sigma_B^2}{2} - C$	σ_A^2	$\dfrac{\sigma_A^2 + \sigma_F^2}{2} - C$	
Follow-up	$\dfrac{\sigma_F^2 + \sigma_B^2}{2} - C$	$\dfrac{\sigma_F^2 + \sigma_A^2}{2} - C$	σ_F^2	

ity of variance does not apply), the off-diagonal entries of Table 16.10 will not all be equal to the same value (i.e., homogeneity of covariance will not apply, either). However, sphericity applies in Table 16.10, because σ_{resid}^2 is the same for all pairs of levels. (The equation immediately preceding Formula 16.13 shows that σ_{resid}^2 for any two levels is equal to the difference between the pooled variance of the two levels and their covariance; if that difference is the same in the population (i.e., C) for all pairs, the sphericity assumption is valid.) If you were to require HOV in Table 16.10, then you would also have homogeneity of covariance in that table, in which case Table 16.10 could be replaced by Table 16.8.

Multi-Group Sphericity

Now that VC matrices have been defined, it is easy to describe the multi-group sphericity assumption of a mixed-design ANOVA. To validly use the F ratios outlined in Section B, one must assume that the VCM for one level of the between-group variable is identical to the VCM at any other level (i.e., they are all the same). It is this assumption that allows the VCMs from all of the groups in your data to be averaged together into a single VCM that can then be tested for sphericity with Mauchly's W (i.e., the composite VCM is tested to see if it differs significantly from the structure of Table 16.10). The multi-group sphericity assumption (i.e., the equality of the VCM matrices) can be tested with Box's M criterion (Huynh & Mandeville, 1979), a multivariate test. (SPSS will give you "Box's Test of Equality of Covariance Matrices" whenever you run a mixed-design ANOVA with GLM/Repeated Measures, and request Homogeneity tests as an Option.) Not surprisingly, Box's M test has little power when it is needed most—when you are dealing with small samples. Therefore, a lack of significance for this test when your ns are small cannot give you much confidence in the accuracy of your p values from a mixed-design ANOVA. However, if Box's test is significant, you might want to consider skipping the omnibus tests of the mixed-design ANOVA entirely, and proceed directly to post hoc comparisons (e.g., simple effects and cell-to-cell comparisons), using a Bonferroni adjustment to reduce your alpha for each comparison. Even when Box's M test is not significant, there are reasons to consider using a multivariate ANOVA test as an alternative to the mixed-design ANOVA described in this chapter (see Chapter 18, Section C).

Planned Comparisons for a Mixed-Design ANOVA: Trend Interactions

When planning a mixed-design experiment, you should consider the extra power that can be gained by planning pairwise or complex comparisons among the levels of either the between-group factor, or the RM factor, or both. If you are expecting some interaction between the factors, your test of the interaction can be considerably more powerful, as with any two-way ANOVA, if you plan to test some particular interaction contrasts. When the RM factor has quantitative levels (e.g., measurements over time, different difficulty levels of the same task, different dosages of the same drug), the most obvious contrasts to test involve polynomial trend components. (The case in which it is the between-groups factor that has the quantitative levels, and *not* the RM factor, is so rare that I will not take the space to discuss it here.) In fact, the use of trend components can reduce a design with many repeated measurements to a design that is as simple as the before-after case, as I will show next.

Reproduced in Table 16.11 are the data from Exercise 16B5. The values in the added column, labeled "Linear," were found by applying the same set of (decreasing) linear trend coefficients $(+3, +1, -1, -3)$ to the data from each of the nine subjects in the table (e.g., $3 \times 9 + 1 \times 6 - 1 \times 4 - 3 \times 2 = 27 + 6 - 4 - 6 = 23$). (Note that I used a set of decreasing linear trend coefficients to avoid dealing with minus signs; you would obtain the same test statistic with the standard, increasing coefficients.) If you have already computed the appropriate ANOVA for this exercise, then you know that the interaction yields an F ratio of merely 1.0—just what you would expect from chance factors alone. However, you can often gain greater power with this type of design by testing the interactions of various trend components of the quantitative RM (e.g., number-of-letters) factor with the grouping factor (e.g., type of feedback). Moreover, it is computationally easy to do. For example, to test the interaction of the linear trend of the RM factor with the grouping factor, I need only perform a one-way ANOVA on the linear trend scores, using type of feedback as the independent variable (just as you might do with a set of after-before difference scores).

A quick way to perform this one-way ANOVA is to find the total SS of the nine linear trend scores ($SS_{total} = 122.22$), and then SS_{bet}, based on the means of the linear scores for the three feedback groups: $SS_{bet} = N_T \sigma^2$ (23.33, 25.67, 30.33) = $9 \times 8.466 = 76.19$. Therefore, $SS_W = 122.22 - 76.19 = 46.03$; $MS_{bet} = 76.19/2 = 38.095$, and $MS_W = 46.03/6 = 7.672$. Finally, the F ratio for testing the interaction of group by linear trend is $38.095/7.672 = 4.97$. Because the critical F is $F_{.05}(2, 6) = 5.14$, this result is very nearly significant, which is rather surprising when you recall the the F for the omnibus interaction was only 1.0. In fact, you might think that it would be impossible for $F_{Group \times Linear}$ to be more than three times larger than the omnibus F_{inter} in this example, because the best you can do with the numerator of the F ratio is to stuff all of the SS_{inter} into 2 instead of 6 df. That is true, but it is also important to note that I'm not dividing $MS_{Group \times Linear}$ by the omnibus (i.e., $MS_{S \times RM}$) error term from the mixed-design ANOVA. The error term I am using to test the trend interaction is based only on the subject-to-subject variability in linear trend scores, which could turn out to be larger or smaller than $MS_{S \times RM}$ (in this example, it happens to be consid-

Feedback Group	Five	Six	Seven	Eight	Linear
No Feedback	9	6	4	2	23
No Feedback	10	7	4	3	24
No Feedback	12	9	7	5	23
Feedback	19	16	15	12	22
Feedback	19	15	11	11	28
Feedback	22	20	17	14	27
Feedback + Ego	30	25	22	21	30
Feedback + Ego	31	30	27	23	27
Feedback + Ego	34	32	28	24	34

Table 16.11

Linear Trends in a Mixed Design

erably smaller). Using the more specific error term is generally preferred, because to use the omnibus error term you would need to assume that the subject-to-subject variability would be the same in the population for the quadratic and cubic trends, as it is for the linear trend.

Removing Error Variance from Counterbalanced Designs

In a one-way repeated-measures experiment, counterbalancing prevents the presence of simple order effects from systematically affecting the numerator of the *F* ratio, but order effects will inflate the denominator, which in turn reduces the power of the test. The extra variance produced by order effects can and, in most cases, should be removed from your error term before you obtain your *F* ratio and test it for significance. The easiest way to do this is to convert your one-way RM ANOVA into a two-way mixed ANOVA in which each group consists of subjects who were given the RM levels in the same order. For instance, if your IV had four levels and you used a Latin square design, your mixed design would have four groups, each representing a different order (your total number of subjects would have to be divisible by 4). (I am describing the *single Latin-square design with replications*. Instead of having just four orders, you could have a different Latin square for each set of four subjects. Such a design is considerably more tedious to run and to analyze; therefore, it is rarely used, even though it provides a more thorough balancing of possible complex order effects.) I will refer to the added between-groups factor as the "order" factor and to the RM factor as the "treatment" factor.

In the mixed design just described, both simple order effects and asymmetric carryover effects contribute to the order × treatment interaction. If it is not obvious that simple order effects would contribute to that interaction, it will help if you imagine a memory study in which each subject's recall is tested in two rooms, one painted red and one painted green. Imagine also that room color has no effect on memory, so each subject's line on a graph—with red and green on the horizontal axis—should be flat. However, if the red and green conditions are counterbalanced and there is a practice effect that gives a boost to whichever condition is second, the lines of the "red-first" subjects will slant one way (i.e., green higher than red), and the "green-first" subjects will slant the opposite way, creating a subject by treatment interaction—(i.e., the error term for the one-way analysis). Adding the order factor separates the order × treatment interaction from the subject × treatment interaction, so the latter is no longer influenced by order effects (the *SS* for the subject × treatment interaction is calculated separately for each order and then summed).

Performing the mixed design in the manner just described tends to increase the *F* ratio for your treatment effect (relative to a simple one-way RM ANOVA) when there are order effects embedded in your ANOVA. However, you can lose a considerable number of degrees of freedom in your error term by adding order as a factor, and this makes the error term larger ($MS_{error} = SS_{error}/df_{error}$), so unless you actually have considerable order effects, adding order can hurt more than help. If you perform the mixed-design ANOVA and find that both the order main effect and its interaction with the treatment are very small, it is recommended that you then drop order as a factor and proceed with your analyses in the usual manner. If your original study is already a mixed design (e.g., three different cognitive tasks, with half the subjects told that the tasks measure intelligence and half that they do not), adding task order as a second between-subjects factor creates a three-way ANOVA with one RM factor.

If the order × treatment interaction is large in your mixed ANOVA, you know that you have large order effects (of one type or another) that have been removed from your error term. That sounds like a problem solved, but, unfortunately, both differential carryover effects and simple order effects can contribute to the order × treatment interaction. You may recall from the previous chapter that differential carryover effects are a type of order effect that can represent a confounding variable in your study, spuriously increasing or decreasing the separation of your treatment means. Before you accept the conclusions of your mixed design, you would want to rule out the possibility of significant carryover effects. (Note that a large and/or significant main effect for the order effect is a bad sign, which suggests the presence of carry-over effects.) You can check for carryover effects by drawing a graph of your treatment means versus serial position as in Figure 15.8. The further the lines on your graph deviate from being parallel, the larger the carryover effects. To quantify the magnitude of your carryover effects, you can use the following analysis.

Analyzing a Latin-Square Design

The analysis I'm about to describe is usually referred to as the analysis of a Latin-square design, although it would work as well for a design that is completely counterbalanced (the latter design is rarely used when there are more than three treatment levels). The Latin-square analysis begins with the mixed design I described in the preceding subsection, but further refines it by dividing the order × treatment effect into two pieces. I mentioned that one problem with the mixed design approach is that the order × treatment interaction contains both simple order and carryover effects. To separate these two components, it is necessary to quantify the simple order effect by calculating the SS for the main effect of serial position (SS_P). This is found by calculating the mean of your dependent variable at each serial position (e.g., to find the mean for position 1, you have to average every subject's first condition together even though you will be averaging together different experimental conditions). For example, using the orders in Table 15.4, I have rearranged the data in Table 15.5 in terms of serial position instead of the actual experimental conditions and created Table 16.12. You can see by the column means in Table 16.12 that there is a practice effect such that tasks performed later have an advantage.

SS_P is calculated as for any main effect: $N_T \times \sigma^2$(position means). Then SS_P is subtracted from the order by treatment interaction; the leftover portion is the position by treatment interaction ($SS_{P \times T}$), which is a measure of the variability due to differential carryover effects. $SS_{P \times T}$ can be tested for significance by first dividing it by its df, which equals $(c - 1)(c - 2)$, to create $MS_{P \times T}$ and then dividing by the $MS_{S \times RM}$ error term from the mixed design ANOVA. This is one F ratio that you do not want to be significant or

Table 16.12	Subject No.	Position 1	Position 2	Position 3
	1	15	13	12
	2	14	15	16
	3	10	20	17
	4	11	12	17
	5	7	12	5
	6	8	7	18
	Column Means	10.833	13.167	14.167

even close to significance. A significant position by treatment interaction, together with a graph showing noticeable carryover effects, suggests that you should reanalyze your data using only the first treatment condition for each subject. This converts your RM design to a between-subjects design, which means a large loss of data and power, but any significant results you *do* get cannot be attributed to carryover effects. Finding that the results of the between-group analysis are similar to the RM analysis is often used as justification for retaining the RM analysis, but any serious discrepancy argues for using the between-group results.

You can also test MS_P (divide SS_P by $c - 1$) for significance, using the same error term as for $MS_{P \times T}$. The significance of this effect won't affect your interpretation of your main treatment effect, but it can be interesting in its own right. A significant main effect of position may demonstrate the importance of a practice effect, a fatigue effect, or both (the means may be highest for the middle positions). If you are wondering how to do the Latin-square analysis with one of the major statistical packages, there is no easy way at present. Probably the simplest choice is to use the computer to do the mixed-design analysis and then calculate SS_P with a handheld calculator and complete the analysis yourself. You can get the computer to calculate SS_P, but, unless your data set is huge, it hardly seems worth the trouble because you would have to either reenter the data by position instead of condition or create new position variables with program statements or a series of menu operations.

Analyzing the Latin-Square Design without Replications

The Latin-square analysis just described assumes that you have more than one subject assigned to each order condition; otherwise you could not perform a mixed design. But even if each subject in your experiment receives a different order of experimental conditions (e.g., there are eight conditions and eight subjects run in a Latin-square design), it is still possible to remove order effects from your error term and test the main effect of position (but no longer possible to test the position by treatment interaction).

The example in Section B of the previous chapter is not a Latin square (it is completely counterbalanced), but it involves only one subject per order. To extract order effects from that design, begin with the usual RM analysis, as performed in the previous chapter. To calculate SS_P, it is helpful to rearrange the scores according to their serial positions, as I did in Table 16.12. Then, $SS_P = 18 \times \sigma^2(10.83, 13.167, 14,167) = 18 \times 1.951 = 35.12$. When there is only one subject per order, the order $\times$ treatment interaction is identical with the subject $\times$ treatment interaction, so SS_P is subtracted from $SS_{S \times RM}$ to create an adjusted error term: $SS_{adj.\ error} = 69.2 - 35.12 = 34.08$. $SS_{adj.\ error}$ must be divided by the proper df to create $MS_{adj.\ error}$. The original df_{error} was $2 \times 5 = 10$. Extracting SS_P, however, results in the loss of two df (the number of df lost is always one less than the number of levels or serial positions), so $df_{adj.\ error} = 8$. $MS_{adj.\ error} = SS_{adj.\ error}/8 = 34.08/8 = 4.26$. MS_{RM} is still 55.4, so the new F ratio is $55.4/4.26 = 10.53$, which is considerably larger than the original F ratio of 8.0. The reduction in the SS for the error term outweighed the reduction in df_{error}, which is generally the case when the order effects are fairly large.

You can also test the main effect of serial position in the design just described. If you divide SS_P by df_P, you get MS_P, which equals $35.12/2 = 17.56$. Then divide MS_P by the same adjusted error term I just calculated: $F_P = 17.56/4.26 = 4.12$. The critical $F_{.05}(2,8) = 4.46$, so the serial position effect falls short of significance. This is no surprise with such a small sample size, and it seems reasonable to make the error adjustment even though the order effect fails to reach significance. The adjusted error term is a measure of the position by treatment interaction, but it cannot itself be tested for significance (there is no additional error term to test it against). Therefore, with one subject per order, one must look carefully for carryover effects before making conclusions. Also, if your design is a Latin square with one subject per order, the loss in degrees of freedom due to extracting SS_P can reduce your power to a ridiculously low level. If you are going to run such a design, the analysis just described is not recommended unless you have at least five conditions/subjects.

Relative Efficiency

The advantage of extracting SS_P can be expressed in terms of the *relative efficiency* (RE) of that analysis as compared to the ordinary one-way RM ANOVA. (I first introduced the RE concept in Chapter 6, Section C, in the context of estimators and their properties.) The RE for two corresponding analyses can be expressed as a ratio of the error terms of the two analyses, adjusted for sample size:

$$\text{RE} = \left(\frac{MS_{\text{error2}}}{MS_{\text{error1}}} \right)\left(\frac{df_{\text{error1}} + 1}{df_{\text{error1}} + 3} \right)\left(\frac{df_{\text{error2}} + 3}{df_{\text{error2}} + 1} \right) \qquad \textbf{Formula 16.14}$$

If the analysis subtracting SS_P is considered analysis 1, and the original analysis is considered analysis 2, the relative efficiency of the two analyses can be found by inserting the relevant values into the preceding formula:

$$\text{RE} = \left(\frac{6.92}{4.26} \right)\left(\frac{8 + 1}{8 + 3} \right)\left(\frac{10 + 3}{10 + 1} \right) = (1.624)(.818)(1.18) = 1.57$$

Expressed as a percentage, the relative efficiency of the SS_P analysis is 157% as compared to the ordinary RM ANOVA, which makes it well worth the trouble. Another interesting use of the RE formula is to quantify the advantage of an RM ANOVA relative to its between-groups counterpart (an interesting exercise would be to do this for the preceding analysis). And if you compare the SS_P analysis directly to the analysis that doesn't use repeated measures at all, the RE should be quite impressive (it is more than 270% for the preceding example).

SUMMARY

1. The variance-covariance matrix (VCM) for an RM ANOVA contains the covariances for every pair of RM levels, and on the downward-sloping diagonal, the variances for each RM level (its covariance with itself). Under the assumption of compound symmetry, the population variances in the VCM are all equal to each other, and the covariances are all equal to that common population variance multiplied by the common population correlation coefficient for any two RM levels. On the other hand, sphericity does not require any two values in the population VCM to be the same; however, the covariance of two RM levels must differ from the pooled variance of those same two levels by some fixed amount (C), which is the same for all possible pairs of RM levels (C is equal to the variance attributable to the subject by treatment interaction for those two levels).

2. The multi-group sphericity assumption for the mixed-design ANOVA requires that the population VCM for any one level of the between-group factor is identical to the population VCM for any other level. This is the assumption that is tested by Box's M criterion, but unfortu-

nately, this test has too little power in most common cases to be of much use in deciding whether or not to use the mixed-design ANOVA.

3. When the RM factor in a mixed design has quantitative levels, it is often of interest to test the interaction of the linear, and perhaps higher-order, polynomial trend components. Just apply (i.e., cross-multiply) the appropriate trend coefficients from Table A.12 to the data from each individual subject to obtain single trend scores for all of the subjects. Then, perform an ordinary one-way ANOVA using the between-groups variable as your factor and the trend scores as your dependent variable. Testing the resulting F ratio in the normal way yields a test for the significance of the interaction between the grouping factor and the trend component being calculated. A significant test allows you to conclude that the trends over the RM levels differ among the subgroups of the study. Bear in mind that this test is not based on the omnibus error term (i.e., $MS_{S \times RM}$) from the mixed-design ANOVA, but rather on an error term involving only the particular trend component being tested.

4. In a counterbalanced design, order effects contribute to the subject by treatment interaction and therefore make the error term for the RM factor larger. Separating subjects according to the order in which they received the treatments and adding "order" as a between-groups factor removes order effects from the subject by treatment interaction and places them in a group by treatment interaction, where they can be tested for significance. Removing order effects often increases the F ratio of the RM factor, but adding order as a factor always reduces the df of the error term, which tends to reduce the F ratio. If order effects are small, the increase in F due to removing order effects won't offset the decrease in F due to the decrease in df, so the order factor should not be added in such cases.

5. If your order by treatment interaction is significant and/or very large and disordinal, you may want to subtract the simple order effects from the more complex effects, and test the latter separately, because asymmetric order effects can threaten the validity of your results. First, the mean is calculated for each ordinal position (across treatments). Then, SS_p can be calculated from these means and subtracted from the SS for the order $\times$ treatment interaction. If the order $\times$ treatment interaction is still significant, you should inspect a graph of treatment by position to look for differential carryover effects that may be confounded with your main RM effect. If strong complex order effects are evident, you may have to convert your RM design to a between-groups design by including only the first treatment given to each subject.

6. The preceding analysis is usually performed on the results of a single Latin-square design, in which the number of orders equals the number of treatments, and the number of subjects is some multiple of the number of treatments (it can also be applied to a case of complete counterbalancing). If your design is a Latin square (or completely counterbalanced design) with only one subject per order (i.e., without replication), you can still calculate the SS for the main effect of position and subtract order effects from the error term, but you cannot separate simple from complex order effects or test the latter separately. You must rely instead on a position by treatment graph to determine if your order effects are simple or potentially confounding.

7. Adding order as a factor to an RM design creates two changes that have opposite effects on the F ratio and hence on power. The error term tends to go down, but so do the df for the error term. These two

opposing effects are incorporated in a measure called the *relative efficiency* (RE) of two analyses. If the two analyses differ only in their error terms and corresponding dfs, RE gives us a measure of the advantage one design or analysis has over the other. The RE can give you some idea of whether matching subjects was worth the trouble for a particular study or whether adding order as a factor makes a large difference.

1. Create variance-covariance matrices from the data in Exercise 8 of the previous section (one VCM for each of the three brain-damage groups). Which of the three VCMs seems to exhibit the most homogeneity of variance? Which has the most homogeneity of covariance? Does it look likely that multi-group sphericity would exist in the population?

*2. In Exercise 9 of the previous section, you were asked to calculate a mixed-design ANOVA on the data of Exercise 15C7, as though the data from the two sleep conditions were from entirely independent groups. Viewing the data of Exercise 15C7 as coming from a mixed design again:
   a. Test (at the .05 level) the interactions of the linear, quadratic, and cubic trend components over caffeine dosages with the sleep-condition factor.
   b. How does the sum of the SSs for the three interaction components computed in part a compare to the SS for the interaction in Exercise 15C7? Use this comparison to explain the potential increased power gained by planning interaction contrasts.

3. In Exercise 10 of the previous section, you were asked to calculate a mixed-design ANOVA on the data of Exercise 15C8, as though the data from the two therapy conditions were from entirely independent groups. Viewing the data of Exercise 15C8 in that way again:
   a. Average across the therapy conditions, and test the significance of the linear and quadratic trends of the main effect of time. Compare these results to the test of the main effect of time you preformed in Exercise 15C8, and explain any discrepancies.
   b. Test (at the .05 level) the interactions of the linear and quadratic trend components over time with the therary-type factor.

*4. Suppose that the experiment described in Exercise 15B3 was not completely counterbalanced and that the first four subjects followed the order "One," "Twenty," "Large,"; the second four followed "Twenty," "Large," "One,"; and the last four, "Large," "One," "Twenty."
   a. Add order as a between-groups factor, and complete the mixed design ANOVA. Compare the F ratio for the RM factor in the mixed design with the F ratio from the original one-way RM ANOVA.
   b. Draw a graph of the treatments as a function of serial position. Can you see evidence of complex order effects?
   c. Complete the Latin-square analysis. Test the main effect of position and the position by treatment interaction for significance.

5. In Exercise 15B6, the orders in which the subjects received the Digit (D), Letter (L), and Mixed (M) tasks were as follows: 1. DLM; 2. DML; 3. LMD; 4. LDM; 5. MDL; 6. MLD.
   a. Complete the Latin-square analysis, and draw a graph of the treatments as a function of serial position. Is the main effect of position significant?
   b. Compare the F ratio for the RM factor in the Latin-square design, with the F ratio from the original one-way RM ANOVA.

*6. a. Calculate the relative efficiency (RE) of the mixed design in Exercise 4 in comparison with the original one-way RM ANOVA.
   b. Calculate the (RE) of the Latin-square design in Exercise 5 in comparison with the original one-way RM ANOVA.

The df components for the mixed design:

a. $df_{\text{between-S}} = nk - 1$ **Formula 16.1**

b. $df_{\text{within-S}} = nk(c - 1)$ or $nkc - nk$

c. $df_{\text{groups}} = k - 1$

d. $df_W = k(n - 1)$ or $nk - k$

e. $df_{\text{RM}} = c - 1$

f. $df_{G \times \text{RM}} = (k - 1)(c - 1)$

g. $df_{S \times \text{RM}} = k(c - 1)(n - 1)$

The sum of squares due to subject-to-subject variability within groups, found by subtraction:

$$SS_W = SS_{\text{between-S}} - SS_{\text{groups}} \qquad \text{\textbf{Formula 16.2}}$$

The sum of squares due to variation among the several measurements within each subject (this component is further divided into subcomponents corresponding to mean differences between levels of the RM factor, the interaction of the two factors, and the interaction of the subjects with the RM factor within each group):

$$SS_{\text{within-S}} = SS_{\text{total}} - SS_{\text{between-S}} \qquad \text{\textbf{Formula 16.3}}$$

The sum of squares due to the interaction of the two factors, found by subtraction:

$$SS_{G \times \text{RM}} = SS_{\text{between-cells}} - SS_{\text{groups}} - SS_{\text{RM}} \qquad \text{\textbf{Formula 16.4}}$$

The sum of squares due to the interaction of subjects with the repeated factor within each group, found by subtraction:

$$SS_{S \times \text{RM}} = SS_{\text{within-S}} - SS_{\text{RM}} - SS_{G \times \text{RM}} \qquad \text{\textbf{Formula 16.5}}$$

The mean-square components (variance estimates) of the two-way mixed design:

$$MS_{\text{groups}} = \frac{SS_{\text{groups}}}{df_{\text{groups}}} \qquad \text{\textbf{Formula 16.6}}$$

$$MS_W = \frac{SS_W}{df_W}$$

$$MS_{\text{RM}} = \frac{SS_{\text{RM}}}{df_{\text{RM}}}$$

$$MS_{G \times \text{RM}} = \frac{SS_{G \times \text{RM}}}{df_{G \times \text{RM}}}$$

$$MS_{S \times \text{RM}} = \frac{SS_{S \times \text{RM}}}{df_{S \times \text{RM}}}$$

The F ratios for testing the three effects in a two-way mixed design:

a. $F_{\text{groups}} = \dfrac{MS_{\text{groups}}}{MS_W}$ **Formula 16.7**

b. $F_{\text{RM}} = \dfrac{MS_{\text{RM}}}{MS_{S \times \text{RM}}}$

c. $F_{G \times \text{RM}} = \dfrac{MS_{G \times \text{RM}}}{MS_{S \times \text{RM}}}$

Generalized eta squared for an experimental between-groups factor in a mixed design (where $SS_{\text{within-cell}}$ is the sum of SS_W and $SS_{S \times \text{RM}}$):

$$\eta^2_{\exp(\text{RM})} = \frac{SS_{\text{group}}}{SS_{\text{group}} + SS_{\text{within-cell}}}$$ **Formula 16.8**

Estimate of generalized omega squared for an experimental between-groups factor in a mixed design:

$$\text{est. } \omega^2_{\exp(\text{RM})} = \frac{SS_{\text{group}} - \text{df}_{\text{group}} MS_W}{SS_{\text{group}} - \text{df}_{\text{group}} MS_W + N_T MS_{\text{within-cell}}}$$ **Formula 16.9**

Generalized eta squared for a measured (i.e., selected for) between-groups factor in a mixed design:

$$\eta^2_{\text{meas}(\text{RM})} = \frac{SS_{\text{group}}}{SS_{\text{total}} - SS_{\text{RM}}}$$ **Formula 16.10**

Estimate of generalized omega squared for a measured between-groups factor in a mixed design:

$$\text{est. } \omega^2_{\text{meas}(\text{RM})} = \frac{SS_{\text{group}} - \text{df}_{\text{group}} MS_W}{SS_{\text{total}} - SS_{\text{RM}} + MS_W + \text{df}_{\text{RM}} MS_{S \times \text{RM}}}$$ **Formula 16.11**

Estimate of generalized omega squared for the RM factor in a mixed design with a measured between-groups factor:

$$\text{est. } \omega^2 = \frac{SS_{\text{RM}} - \text{df}_{\text{RM}} MS_{S \times \text{RM}}}{SS_{\text{total}} + MS_W}$$ **Formula 16.12**

The population covariance of two RM levels expressed as the difference between their pooled variance and the variance of the subject by treatment interaction between them:

$$\sigma^2_{ij} = \frac{\sigma^2_i + \sigma^2_j}{2} - \sigma^2_{\text{resid}}$$ **Formula 16.13**

The relative efficiency of two analyses in terms of their error terms and error df:

$$\text{RE} = \left(\frac{MS_{\text{error2}}}{MS_{\text{error1}}} \right) \left(\frac{\text{df}_{\text{error1}} + 1}{\text{df}_{\text{error1}} + 3} \right) \left(\frac{\text{df}_{\text{error2}} + 3}{\text{df}_{\text{error2}} + 1} \right)$$ **Formula 16.14**

We will need to use the following from previous chapters:

Symbols:
s: unbiased standard deviation of a sample
r: Pearson's product-moment correlation coefficient
$z_{Y'}$: predicted z score for Y
b: (raw-score) slope of a regression line

Formulas:
Formula 9.6: The t test for a correlation coefficient
Formula 10.1: The regression line for standardized scores
Formula 10.5: The raw-score regression line

Concepts:
The limitations of linear correlation
The assumptions of linear regression

Chapter

CONCEPTUAL FOUNDATION

In Chapter 10, I described how one variable can be used to predict another when the two variables have a reasonably good linear relationship. However, in many practical situations your predictions can be improved by combining several variables rather than using just one. At first, the example in Section B of Chapter 10, in which the score from a lifestyle questionnaire (LQ) was used to predict longevity, looks like an example in which just one variable can be used to make an excellent prediction of another. But saying that LQ is just one variable is cheating a bit. It is really a composite of several very different variables, including amount of exercise, stress, smoking, and so forth.

The lifestyle questionnaire combines these very different variables into a single measure that can be used like any other variable. But there is more than one way that the component variables of LQ can be combined. Should alcohol consumption be given the same weight as cigarette consumption? (And how are we to determine how many beers it takes per month to shorten your life as much as a given number of cigarettes?) Should we include every variable we can think of that may affect life expectancy? Is there any way to decide which variables are really helping the prediction and which are not?

All of these questions can be dealt with by one form or another of a statistical procedure called *multiple linear regression*, or just *multiple regression*, for short. It is a natural extension of simple linear regression that results in an equation you can use to make predictions about one particular variable of interest, usually called the *criterion* variable. But there can be several predictor variables in the equation, each with a different multiplier (or "weight"), and determining those multipliers, and which predictors to include, can get quite complicated. However, if you have only two variables available as predictors, multiple regression does not get *too* complicated, and you can actually do all of the calculations quickly with a basic handheld calculator. Because many of the principles of multiple regression are easier to explain in the two-predictor case, I will discuss that case thoroughly in this section. In Section B, I will deal with the conceptual complications of having many potential predictors, but I will leave most of the calculation for commonly available statistical software packages. In

Section C I will take up some more advanced complications, such as what happens when the variables interact with each other or have nonlinear relationships with the criterion.

Before we go any further, I don't want you to form the impression that the major use of multiple regression by psychologists is for making predictions. In psychological research, multiple regression is more often used to provide evidence that the role of one variable is more primary or direct in affecting the criterion than another. Multiple regression has also been used frequently to determine the relative importance of various variables in affecting psychological outcomes, but this application has become rather controversial (the problems with this application of multiple regression will be discussed in detail in Section B). However, the use of multiple regression for predictions is more mechanical and therefore easier to explain. So I will begin with an emphasis on prediction and only later describe the role of multiple regression in forming and confirming psychological theories.

Uncorrelated Predictors

If you want to predict a student's performance in high school (as measured, perhaps, by an average grade over all areas of study), a reasonable place to start is by obtaining a measure of the student's general inclination to learn school material quickly and easily with some form of aptitude test. Let us say that you have done this for a sample of students, and the correlation between aptitude and high school grades (HSG) is .4. This is a pretty good start. You have accounted for $.4^2 = .16$, or 16%, of the variance in HSG with your aptitude measure. However, 84% of the variance in HSG remains to be explained.

As with most cases of human performance, aptitude is not the only important factor; usually some measure of effort or motivation is useful in accounting for actual (rather than just potential) performance. Let us say then that we also have a reliable estimate of how many hours each student in our sample has spent studying during the semester for which we have grades. For our example, the correlation between study hours (SH) and HSG is .3. Not as good as aptitude, but SH accounts for another $.3^2 = .09$, or 9%, of the variance in HSG. But does this 9% represent a different part of the variance from the 16% accounted for by aptitude? Can we simply add the percentages and say that together aptitude and SH account for $9 + 16 = 25\%$ of the variance in HSG? We can if the correlation between aptitude and SH is zero.

If the total variance in HSG is represented by a rectangle, aptitude and SH can be represented as separate circles within the rectangle, with the area of each circle proportional to the percentage of variance accounted for by the predictor it represents, as shown in Figure 17.1. Such a figure, which can represent various sets that do or do not overlap, is called a *Venn diagram*. In the Venn diagram of Figure 17.1, the fact that the circles do not overlap tells us that the predictors they represent are independent of each other (i.e., they have a Pearson's *r* of zero with each other). The areas of the two circles can be added to determine how much of the criterion variance can be accounted for by combining the two predictors. If we use the symbol R^2 to represent the total amount of variance accounted for, we can see that in this case $R^2 = r_{apt}^2 + r_{SH}^2 = .4^2 + .3^2 = .25$. Just as r^2 was called the coefficient of determination, R^2 is called the *coefficient of multiple determination*.

Without the squaring, R is called the *multiple correlation coefficient*; it is the correlation between your predictions for some criterion (based on two or more predictors) and the actual values for that criterion. For this

Figure 17.1

Two Uncorrelated
Predictors

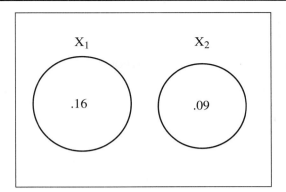

example, $R = .5$, which is less than adding the two rs, but more than either one of them (this will always be the case for two positive, independent rs).

The Standardized Regression Equation

You may recall from Chapter 10 that the regression equation becomes very simple when the variables are expressed as z scores. For one predictor: $z_{y'} = rz_x$. This is called the *standardized regression equation*, and it is especially helpful when dealing with multiple regression. For the two-predictor example, I will label aptitude as X_1, SH as X_2, and HSG, the criterion, as Y. If you had only aptitude as a predictor of HSG, the standardized regression equation would be $z_{y'} = .4z_{x_1}$, and if you had only SH, it would be $z_{y'} = .3z_{x_2}$. However, if you know both aptitude and the number of hours studied for each subject, you can make a better prediction about high school grades than using either predictor variable by itself. The way multiple regression works is by taking a weighted combination of the two predictor variables to make the best possible prediction. When the two predictors are uncorrelated with each other, the optimal weights are just the ordinary Pearson rs of each predictor separately with the criterion:

$$z_{y'} = r_{yx_1}z_{x_1} + r_{yx_2}z_{x_2}$$

Formula 17.1

For this example, $z_{y'} = .4z_{x_1} + .3z_{x_2}$. Using this notation, the formula for R when you have two uncorrelated predictors is:

$$R = \sqrt{r_{yx_1}^2 + r_{yx_2}^2}$$

Formula 17.2

The correlation of each predictor with the criterion is called a *validity*. As this formula shows, you cannot add the validities of the predictors you are combining to find R. You must add the squared rs and then take the square root. Note also that there is a limit to the validities of uncorrelated predictors. If one predictor correlates .7 with the criterion, a second predictor can only correlate very slightly more than .7 with the same criterion if it is uncorrelated with the first ($.7^2 + .72^2$ is greater than 1.0).

More Than Two Mutually Uncorrelated Predictors

The case of three predictor variables in which $r = 0$ for each pair is a simple extension of the two-predictor case just described. The multiple regression equation would be:

$$z_{y'} = r_{yx_1}z_{x_1} + r_{yx_2}z_{x_2} + r_{yx_3}z_{x_3}$$

Both Formulas 17.1 and 17.2 easily generalize to any number of mutually uncorrelated predictors.

However, although it is not very common to find two virtually uncorrelated predictors with high validities, finding three mutually uncorrelated predictors, each having a meaningful correlation with the criterion, is extremely rare unless the variables have been specially created to have this property (e.g., one could create subscales of a questionnaire to have this property, usually using a method called factor analysis). Unfortunately, correlations among predictors can greatly complicate the process of finding the multiple regression equation. Even in the two-predictor case, a variety of complications arise depending on the correlation between the two predictors, as you will see shortly.

The Sign of Correlations

To simplify matters in this chapter, I will deal only with the case in which every predictor variable has a positive correlation with the criterion. This is not really a restriction because any negatively correlated variable can be measured in a reverse way so that its validity will be positive. For instance, in golf a lower score means better performance (fewer strokes to get the ball in the cup), so we might expect a negative correlation between subjects' golf scores (GS) and a measure of their eye-hand coordination (EHC). However, if we simply transform the original golf scores to $160 - GS$, the new scores will have a positive correlation with EHC and nothing important will have changed. (I'm subtracting from 160 because few golf scores are higher, and we would want to avoid negative scores. Also, transformed scores over 99 will be very rare, so we could avoid using three digits.)

On the other hand, if we use the original golf score to predict EHC, the correlation between the predictions and the actual EHC scores will be positive, even though the GS/EHC correlation is negative. That's because the prediction equation uses the sign of the correlation to make its predictions (e.g., a low GS leads to a high prediction for EH), so if the GS/EHC correlation is $-.4$, the correlation between the actual EHC scores and the predictions based on GS will be $+.4$. Because the multiple correlation (R) is the correlation between predicted and actual scores, it is always positive, even if we don't have positive validities.

Two Correlated Predictors

Given positive validities, the most common situation when dealing with two predictors is not the $r = 0$ case just described but rather a moderate degree of positive correlation between the two predictors. Returning to our example, let us assume a slight tendency for students with higher aptitude scores to spend more time studying, so the correlation between aptitude and SH is .2. I can write this as $r_{x_1x_2} = .2$, but it will be neater from this point on to write just 1 instead of X_1, and so forth, wherever possible, so I will write $r_{12} = .2$ (also $r_{1y} = .4$ and $r_{2y} = .3$). The correlation between the two predictors is depicted as an area of overlap between the two circles in a Venn diagram, like the one in Figure 17.2; part of the variance accounted for by one predictor is also accounted for by the other. (Note that the areas in Figure 17.2 are not drawn to scale in order to enhance the interesting parts).

We want to predict HSG using the weighted combination of both predictors that will produce the highest possible R. (When our "weights" give

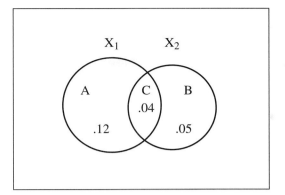

Figure 17.2

Two Positively
Correlated Predictors

us the highest possible R, we have also minimized the sum of the squared errors from our predictions, so the resulting regression equation is known as a *least-squares* solution.) In Figure 17.2, R^2 is the total area of the two circles combined; because of the overlap, R^2 is less than simply adding the sizes of the two circles as we did last time. You would be counting the overlap area (area C) twice. The total area is actually areas A, B, and C added together (.12 + .05 + .04), so $R^2 = .21$. We can't use the regression equation $z_{y'} = .4z_{x_1} + .3z_{x_2}$ from the uncorrelated example because it implies that $R^2 = .25$, which is more predictability than we have. Rather than using the validities as our regression weights, we have to reduce the weights a bit to account for the overlap and the reduced R^2.

The Beta Weights

The reduced weights are often symbolized by the Greek letter beta (there is no relation to the use of beta for the Type II error rate) when referring to their true standardized population values and are therefore called *beta weights*. When referring to estimates of these betas calculated from a sample, it is common to use the uppercase letter B and still call them beta weights. Unfortunately, this usage is not universal, and some authors use the Greek letter even for the sample statistic (other variations on the letter B are used as well). I will stick to the principle of using Greek letters for population values and Roman letters for sample statistics.

In the two-predictor case, the beta weights are a simple function of the correlations involved:

$$B_1 = \frac{r_{1y} - r_{2y}r_{12}}{1 - r_{12}^2} \qquad B_2 = \frac{r_{2y} - r_{1y}r_{12}}{1 - r_{12}^2} \qquad \text{\textbf{Formula 17.3}}$$

The standardized multiple regression equation is expressed in terms of the beta weights as follows:

$$z_{y'} = B_1 z_{x_1} + B_2 z_{x_2} \qquad \text{\textbf{Formula 17.4}}$$

Note that if $r_{1y} = r_{2y}$, $B_1 = B_2$, which means that both predictors will have equal weight in the regression equation, which makes sense. Also note that when the predictors are not correlated (i.e., $r_{12} = 0$), $B_1 = r_{1y}$ and $B_2 = r_{2y}$, and Formula 17.4 becomes Formula 17.1. In any case, the beta weights are chosen in such a way that the predictions they produce have the highest possible correlation with the actual values of Y; no other weights would produce a higher value for R. Given that $r_{12} = .2$, $r_{1y} = .4$, and $r_{2y} = .3$:

$$B_1 = \frac{.4 - .06}{1 - .04} = \frac{.34}{.96} = .354; \quad B_2 = \frac{.3 - .08}{1 - .04} = \frac{.22}{.96} = .229$$

and $z_{y'} = .354 z_{x_1} = .229 z_{x_2}$

With two correlated predictors the multiple R becomes:

$$R = \sqrt{B_1 r_{1y} + B_2 r_{2y}} \qquad \qquad \textbf{Formula 17.5}$$

This formula can be extended easily to accommodate any number of variables. For the preceding example:

$$R = \sqrt{.354 \cdot .4 + .229 \cdot .3} = \sqrt{.21} = .459$$

and $R^2 = .21$. Once you have calculated R^2, you can subtract this value from $r_{1y}^2 + r_{2y}^2$ to find the size of the overlap between the two predictors (area C in Figure 17.2) and then subtract the overlap from each r^2 to find areas A and B. To calculate R^2 directly you can use Formula 17.5 without the square root, or you can skip the calculation of the beta weights and calculate R^2 in terms of the three correlations with Formula 17.6:

$$R^2 = \frac{r_{1y}^2 + r_{2y}^2 - 2r_{1y}r_{2y}r_{12}}{1 - r_{12}^2} \qquad \qquad \textbf{Formula 17.6}$$

If the correlation between aptitude and SH were even higher (with the validities remaining the same), their overlap would be greater, with more of a reduction in R^2. For instance, if r_{12} were equal to .5, the overlap would be as shown in Figure 17.3. The beta weights would be smaller than in the previous example:

$$B_1 = \frac{.4 - .15}{1 - .25} = \frac{.25}{.75} = .333 \quad B_2 = \frac{.3 - .2}{1 - .25} = \frac{.1}{.75} = .133$$

According to Formula 17.5, $R^2 = .333(.4) + .133(.3) = .133 + .04 = .173$, and $R = .416$. As an exercise, you should insert the three rs into Formula 17.6 and demonstrate for yourself that you get the same value for R^2. Notice what happens in the numerator of Formula 17.6 if r_{12} is negative. If the two predictors are negatively correlated with each other, the numerator actually gets larger instead of smaller (the sign does not affect the denomi-

Figure 17.3

Two Highly
Correlated Predictors

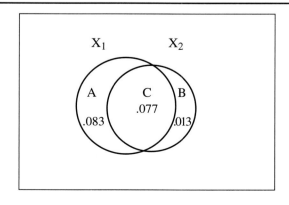

nator because r_{12} is being squared). This phenomenon, known as complementarity, will be discussed later in this section.

Completely Redundant Predictors

An interesting phenomenon occurs when the correlation between aptitude and SH reaches .75. Note what happens to the beta weights:

$$B_1 = \frac{.4 - .225}{1 - .5625} = \frac{.175}{.4375} = .4 \qquad B_2 = \frac{.3 - .3}{1 - .5625} = \frac{0}{.4375} = 0$$

The second variable does not account for any of the criterion variance that is not already accounted for by the first variable. As you can see from Figure 17.4, the overlap is complete. Knowing the number of study hours in this case does not help the prediction of HSG that we can make from knowing aptitude alone. There are even stranger combinations of correlations that can occur, which cannot be displayed on a Venn diagram, but before I mention these, I want to present another way to understand the beta weights.

Partial Regression Slopes

The beta weights we have been calculating for our standardized multiple regression equations are called *standardized partial regression coefficients*, or slopes. The slope concept gives us a way to visualize multiple regression with two predictors (I'll deal with the "partial" concept shortly). First, recall that in the one-predictor equation, $Y = bx + a$; b is the slope of the regression line with respect to the x axis, and in the standardized version of the equation, $z_{y'} = rz_x$, the slope is r. To visualize the case of two predictors imagine a square garden with soft earth but nothing planted yet. One corner is labeled the origin—the zero point for both variables. As you move away from the origin along one border, imagine that values for aptitude go up. If you move along the other border from the origin (at right angles to the first), the values of SH get larger. The corner diagonally opposite from the origin represents the highest values for both aptitude and SH (see Figure 17.5). For any student in our study, we can find some spot in the garden that represents his or her values for both aptitude and SH. Then we push a stick into the ground at that point so that the height of the stick represents that student's value for HSG. If we imagine a marble placed on top of the stick, the marble is a point in three-dimensional space that represents all three variables at the same time. In general, the marbles will be higher off the ground as we move away from the origin and closer to the

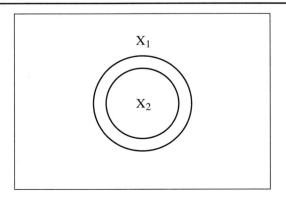

Figure 17.4

Completely Redundant Predictors

Figure 17.5

Picturing a Regression
Plane in Three
Dimensions

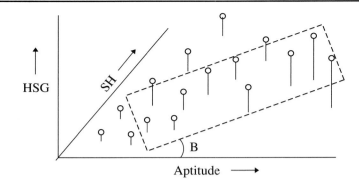

diagonally opposite corner (assuming positive validities, as usual). The
problem of multiple regression with two predictors is to find a plane that
comes as close as possible to all the marbles (i.e., points).

Imagine a large plate of glass that is very thin and permeable, so that
the sticks and marbles can pass through it without breaking it. We are not
going to be able to place the glass so that all the marbles are simultaneously
right at the surface of the glass, unless R equals 1.0. Instead, we try to minimize the distances (actually the squared distances) of the marbles from the
glass (the distance from marble to glass is measured along the stick—either
the amount of stick poking above the glass or the amount of additional stick
length needed to reach the glass). When we find the best possible position
for the plate of glass (our geometric plane), it will make some angle with
the ground, but that angle can be measured from different perspectives.

If we erect a wall along the aptitude border, the glass will intersect with
the wall, forming a line. The angle of that line with the aptitude border is the
beta that corresponds to aptitude. If we are dealing with the case of total
overlap shown in Figure 17.4, beta would be zero for SH. That's like our
plate of glass being hinged along the SH border. As the plate is raised more
and more from the ground, the angle with aptitude increases, but there is no
angle with SH. In the more usual case, the plane will pass through the origin
(remember that we are dealing with standardized scores) but will not be
hinged at either border—it will create different angles with each border.
Each angle captures just part of the slope of the plane, which is why each
angle, beta, is called a partial slope (as in Figure 17.5). If aptitude has much
more influence on HSG than SH, the angle will be fairly steep with the aptitude border and fairly shallow with the SH border.

The geometric model will get too abstract to help us when there are
more than two predictors, but there is another way to understand the betas
that will work for any number of predictors. The beta weight of a predictor
is the number of standard deviations by which the criterion changes when
that predictor is increased by one standard deviation, *provided that the
other predictors remain constant*. The way predictors can be very highly
correlated, there are situations where it is not possible to make a considerable change in one predictor without changing some others (imagine making a large change in a person's job satisfaction without changing his or her
stress level). This is one reason why some statisticians do not like to focus
on beta weights when trying to assess the relative importance of different
predictors. But in my simple example of two predictors with a .5 correlation,
the interpretation is straightforward. According to the beta we found for
aptitude, a person who is one standard deviation higher in aptitude than
another person is predicted to have about one third of a standard deviation

higher high-school grades if both spend the same amount of time studying. The beta for SH tells us that we should predict an increase of only .133 SDs in HSG if one person's study hours is a standard deviation above another's but both have the same aptitude. Just as the slope in simple linear regression is related to r when scores are standardized (actually, it is identical to r), so the partial regression slopes are closely related to (though not identical to) correlations known as semipartial correlations.

Degrees of Freedom

In the geometric model involving two predictors we have a regression plane instead of a regression line and two slopes instead of just one (we still have only one intercept—the value of the criterion when both predictors are simultaneously zero—but this is zero when dealing with standardized scores, as we have been doing all along). Just as two points can always be made to fall on the same straight line, and therefore lead to perfect correlation, any three points can always be put on the surface of a single plane and so lead to $R = 1.0$. This relationship will help explain the concept of degrees of freedom for multiple correlation. When you sample only two subjects, you will always get perfect linear correlation ($r = +1$ or -1) regardless of the two variables being measured. So, in the case of $N = 2$, there are no degrees of freedom at all. In the one-predictor case, df $= N - 2$. Similarly, if you select three subjects, and measure two predictors, along with the criterion, you will get $R = 1.0$, because in this case you lose three degrees of freedom. For the two-predictor case, df $= N - 3$. In general, df $= N - P - 1$, where P is the number of predictors; you lose one df for each slope in the regression equation (i.e., each predictor) and one for the intercept. It will be important to keep track of the df when we test our multiple regression for significance in Section B.

Semipartial Correlations

In an experiment you could assign a different number of study hours to each subject or subgroup of subjects. By randomly assigning subjects you could be reasonably sure that subjects with different numbers of study hours would not be systematically different in aptitude. However, if your research is observational, and you are measuring both number of study hours (chosen by the subjects themselves), and aptitude, it is certainly possible that SH and aptitude will be correlated.

The SH/aptitude correlation implies that if aptitude has some correlation with a variable of interest (like HSG), SH can be expected automatically to have some correlation with the same variable (but less) just because it goes along with aptitude—"riding on its coattails," so to speak. For instance, if the SH/aptitude correlation is .5, and aptitude correlates .4 with a test of spatial ability (e.g., mental rotation), SH would be expected to have a .2 correlation with that test (.5 × .4 = .2), even though there's no direct connection between how much time one studies each week in high school and one's ability to mentally rotate objects. If the correlation between SH and spatial ability does turn out to be .2, we know that all of this correlation is "spurious"—that is, due entirely to the correlation of aptitude and spatial ability, with SH just riding along. One way you can see this is by calculating the betas; you'll find that the beta for SH is zero, when the correlation between SH and spatial ability is .2. The zero beta means that SH is completely redundant with aptitude in this case; it has none of its own direct relation with spatial aptitude.

If **SH** has more than a .2 correlation with spatial ability, we know that it has some of its *own* correlation with spatial ability beyond that which is attributable to aptitude (if it has a correlation with spatial ability that is considerably *less* than .2, that represents another interesting situation that I will get to soon). That extra amount of correlation is called the *semipartial correlation* or *part correlation* (what makes the semipartial correlation "semi" will be explained shortly), and it can be pictured in a Venn diagram. Looking again at Figure 17.3, we see that the variance accounted for by SH is $.3^2 = .09$. But part of that variance overlaps with aptitude. The area labeled B, which equals .013, is the HSG variance accounted for by SH *beyond* that which is accounted for by aptitude. This is the variance accounted for by changing the number of study hours among subjects with the *same* aptitude. Area B represents the square of a correlation; if you take the square root of Area B (in this case, $\sqrt{.0133} = .115$), you get the semi-partial correlation of SH with HSG. This correlation tells us how strongly related SH is with HSG when aptitude is *not* allowed to vary. Similarly, if you want to know the part of the correlation between aptitude and HSG that is *not* related to changes in study hours, take the square root of area A ($\sqrt{.0833} = .289$). Note that the semipartial correlations are lower than the validities. This is usually the case, due to the overlap of partially redundant predictors. However, there are notable exceptions that I will get to shortly.

Calculating the Semipartial Correlation

There is a way to find the semipartial correlation directly from the raw data that is instructive, although it would be unnecessarily tedious to perform these steps in real life. Suppose that you are interested in the relation between obesity and cholesterol level (CL), and you have already calculated the correlation between a subject's weight and her CL. But that's not exactly what you want. Taller people tend to weigh more, but height should not be related to CL. What you really want is a measure of how overweight or underweight the subject is relative to the ideal weight for her height (for simplicity I'll ignore other factors such as frame size and body fat percentage). One way to get this measure is to use height to predict weight in a simple linear regression. Then subtract the predictions from subjects' actual weights to get the residuals. A positive residual indicates how overweight a subject is, and a negative residual indicates how underweight. Finally, find the correlation between the residuals (the weights adjusted for height) and CL. This correlation is actually the semipartial correlation between weight and CL partialling out height (from weight). It is a *semi*partial correlation because you are using height to get residuals for weight, but you are not using height to predict CL. If you were to use height to find residuals for both CL and weight, you could obtain what is called a *partial* correlation, but that will be discussed later in this section.

The easier way to calculate a semipartial correlation is to use a formula very similar to the one we used for B. If you have already calculated the three correlations involved, the semipartial correlation is calculated as follows:

$$r_{y(1.2)} = \frac{r_{1y} - r_{2y}r_{12}}{\sqrt{1 - r_{12}^2}}$$

Formula 17.7

In the symbol $r_{y(1.2)}$, the variable number after the period indicates which variable is being partialled out. The parentheses indicate that the partialling only occurs with respect to the variables in the parentheses; that the variable Y is not in the parentheses tells us that this is a semipartial correlation. To find $r_{y(2.1)}$ just reverse r_{1y} and r_{2y} in Formula 17.7. Remember that if you have calculated R^2 for two predictors already, the squared semipartial correlation for one predictor is just R^2 minus the r^2 of the other predictor.

Suppressor Variables

Once you have calculated the semipartial correlation, it is easy to calculate the beta weight for that variable according to the following relation:

$$B_1 = \frac{r_{y(1.2)}}{\sqrt{1 - r_{12}^2}}$$ **Formula 17.8**

For B_2, just use $r_{y(2.1)}$ in this formula, instead of $r_{y(1.2)}$. For the CL example, let us suppose that the weight/height correlation (r_{12}) is .5, the weight/CL correlation (r_{1y}) is .3, and the height/CL correlation (r_{2y}) is 0. The semipartial correlation of weight and CL, with height partialled out, is:

$$r_{y(1.2)} = \frac{r_{1y} - r_{2y}r_{12}}{\sqrt{1 - r_{12}^2}} = \frac{.3 - 0(.5)}{\sqrt{1 - .5^2}} = \frac{.3}{.866} = .346$$

In this case the semipartial correlation is *higher* than the validity. This result is indicative of a phenomenon called *suppression*; some of the variance in one of our predictors that is not related to the criterion (e.g., some of the variability in weight is due to height, and that part should not be related to CL) is being "suppressed" by partialling it out. The suppressor variable serves to "clean up" or adjust one of the predictors by eliminating extraneous variance—that is, variance not related to the criterion. In this example, the weight residuals have less variance than the original weights, but these residuals are just as closely related to CL (the variance that has been removed is that related to height, which has no connection with CL). In the general case, a suppressor variable has a moderate to high correlation with one of the predictors but very little correlation with the criterion. (*Note*: There is no way to display these relationships in a Venn diagram.)

The usual way to spot a suppressor variable is by looking at the beta weights in a multiple regression equation; although the suppressor may have a slight positive validity, its beta weight will be negative. Let's look at the beta weights for our CL example. For weight, we can find B in terms of the corresponding semipartial correlation (already calculated above), using Formula 17.8:

$$B_1 = \frac{r_{y(1.2)}}{\sqrt{1 - r_{12}^2}} = \frac{.346}{\sqrt{1 - .5^2}} = \frac{.346}{.866} = .4$$

The B for height can be found the same way, of course, or you can skip the calculation of the semipartial correlation and find B directly from the correlations using Formula 17.3. In this example, height is B_2 so according to Formula 17.3:

$$B_2 = \frac{r_{2y} - r_{1y}r_{12}}{1 - r_{12}^2} = \frac{0 - .3(.5)}{1 - .5^2} = \frac{-.15}{.75} = -.2$$

Therefore, the regression equation is simply $z_{y'} = .4z_{x_1} - .2z_{x_2}$. This equation tells us that, for a given weight, the *lower* the value for height (more negative z score), the more we *add* to our prediction of CL (we are subtracting a more negative number). This makes sense because, for a given weight, shorter people are more obese and should have a higher CL, if indeed CL is related to obesity.

Another example of suppression can be found in a variation of the aptitude/grades example. Suppose you have high school grades for students about to take the SAT, and you want to predict SAT scores from grades. If the number of hours spent studying (SH) for high school exams does not correlate highly with SAT scores, SH can be used as a suppressor variable to improve the grades/SAT correlation. Basically, a student's good grade average would count more highly as a predictor of aptitude if he or she got those grades with little studying. Conversely, if a student got those grades with a great deal of studying, a greater amount would be subtracted from his/her grade average; the new adjusted-down grade average would be a better predictor of this overachieving student's true aptitude.

You might think that an hour spent studying by a high-aptitude student would improve grades more than an hour spent by a low-aptitude student. However, that situation cannot be represented by our simple model, in which a student gets a certain boost in grades for each hour studied *added* to a boost in grades for each additional unit of aptitude (e.g., the boost per hour is the same, regardless of the student's aptitude, and is represented by the slope for SH). A change in the effectiveness of an hour of studying that depends on aptitude means that there is an interaction between aptitude and SH, in the same sense that two IVs can interact in a two-way ANOVA. The test for possible interactions between predictors is a complication I will consider in Section C.

In practice, suppressor variables are often incorporated into the predictors before the multiple regression is performed. For instance, height may be combined with weight to create an obesity index; when comparing crime rates among cities, researchers divide by the population to find the per capita rate. Negative beta weights come up fairly often in multiple regression by accident, especially when dealing with fairly redundant predictors, but to discover a reliable, useful suppressor variable that was not obvious from the beginning is quite rare. One form of statistical analysis in which suppressor variables are used very deliberately is the analysis of covariance—the main topic of the next chapter.

Complementary Variables

Related to suppression is another phenomenon in which the semipartial correlation can be higher than the validity; it is called *complementarity*. Returning to the aptitude/SH example, imagine a population in which smarter students study less, so r_{12} is negative even though the validities are still positive. You can imagine that in this hypothetical population there is some tendency for smart students to be lazy (knowing they can get by with little studying) and for lower-aptitude students to compensate by studying harder; given these tendencies, r_{12} could come out to $-.2$. The negative correlation means that I cannot show this situation in a Venn diagram—you can't show a negative overlap. But the effect of the negative overlap is to increase R^2 squared above $r_{1y}^2 + r_{2y}^2$ (which equals .25 for the aptitude/SH example). You can use Formula 17.6 to show that if r_{12} is $-.2$, $R^2 = .31$. The semipartial rs are also increased compared to their validities.

A common situation in which complementarity arises in psychological research occurs when both speed and accuracy are measured in the performance of some task. In many situations there is a "speed/accuracy trade-off" leading to a negative correlation between the two measures (this can happen when subjects try to type on a word processor very fast). Depending on how overall performance is measured, both speed and accuracy can have positive correlations with performance even though the two predictors are negatively correlated with each other. In that case, if one scores highly on speed, a high score on accuracy does not represent redundant information; in fact, it is the opposite of what is expected and therefore adds information. Scoring highly on both speed and accuracy is especially impressive and predictive of good performance, because neither speed nor accuracy came at the expense of the other.

It is possible for neither of the complementary variables to have much of a correlation with the criterion by itself, but taken together they do. For instance, if the criterion is a measure of leadership quality, friendliness may not correlate very highly with it, nor aggressiveness. But if aggressiveness and friendliness are highly negatively correlated, the combination of the two may correlate quite well with leadership. The relatively rare person who is both aggressive and friendly may appear to be a particularly good leader, whereas someone who is passive and unfriendly would probably score especially low on the leadership scale.

The Raw-Score Prediction Formula

Semipartial correlations are useful for answering theoretical questions, like what is the relationship between aptitude and high school grades if amount of studying is controlled. However, if your focus is on making predictions—for example, predicting college grades from a combination of high school grades and aptitude—you will be more interested in the multiple regression equation itself. Unfortunately, the equation we have been dealing with so far, based on the beta weights, is not in a form that is convenient for predictions. To use Formula 17.4, you have to convert the value on each predictor to a z score, and then the prediction you get, which is also in the form of a z score, must be converted back to a raw score to be meaningful for practical purposes. When predictions are your goal, you will probably want to work with a raw-score formula. Fortunately, it is easy to convert Formula 17.4 into a raw-score formula.

Recall that in simple regression, the raw-score slope, b, equals $r(s_y/s_x)$. Similarly, the unstandardized partial regression slope, symbolized as b_i because there is more than one, equals $B_i(s_y/s_{x_i})$, where B_i is the corresponding beta weight, s_y is the standard deviation of the criterion, and s_{x_i} is the standard deviation of the predictor we're dealing with (some authors use uppercase B for raw scores and lowercase b for standardized scores, but because you became accustomed to using b for raw scores in Chapter 10, I decided to retain that notation). The raw-score formula for two predictors is $Y' = b_0 + b_1X_1 + b_2X_2$. The constant b_0 is the Y intercept (symbolized as a in simple regression); it is the value of the criterion when all the predictors are at zero. In the standardized equation, b_0 drops out because it is always zero. The formula for b_0 is analogous to Formula 10.4B for a in simple regression; for two predictors it is $b_0 = \overline{Y} - b_1\overline{X}_1 - b_2\overline{X}_2$.

To show how this formula can apply to our HSG example, I need to specify means and standard deviations for each variable. Suppose that aptitude is measured as a T score with $\overline{X}_1 = 50$ and $s_1 = 10$. For study hours, the mean ($\overline{X}_2$) could be 10 hours/week, with $s_2 = 5$; for HSG, the mean ($\overline{Y}$) could be 85 with $s_y = 8$. Using the betas from our first example (in which $r_{12} = .2$), $b_1 = .35(8/10) = .28$, $b_2 = .23(8/5) = .37$, and $b_0 = 85 - .28(50) - .37(10) = 85 - 14 - 3.7 = 67.3$. So the raw-score equation is $Y' = 67.3 + .28X_1 + .37X_2$. A student who has an above-average aptitude of 60 and studies an ambitious 20 hours a week would be predicted to have a grade average equal to $67.3 + .28(60) + .37(20) = 67.3 + 16.8 + 7.4 = 91.5$.

Note that the unstandardized weights tell us nothing about the relative importance of the two predictors because they can be strongly influenced by the standard deviations of the variables, which can change dramatically just by changing the units of our measurements. In this example, b_2 is greater than b_1, even though the first predictor (aptitude) has the higher validity and the higher semipartial correlation. In the example involving height and weight, changing from inches and pounds to centimeters and kilograms would drastically change the unstandardized weights but would not affect the beta weights and semipartial correlations.

Partial Correlation

As you would guess from the name, the semipartial correlation is closely related to the partial correlation, especially in terms of calculation. However, these two types of correlation tend to be used for very different purposes. If you are trying to understand the factors that affect some interesting target variable, such as self-esteem, and you want to know the relative importance of various factors, such as salary, social support, and job satisfaction, you will most likely want to take a good look at the semipartial correlations for different predictor variables. However, if your focus is on the relationship between two particular variables, such as coffee consumption (CC) and cholesterol level (CL), but there are a number of variables that affect both of them, you will probably want to calculate a partial correlation.

The problem with observing a fairly high correlation between CC and CL is that it could be due to some third variable, such as stress. Under high levels of stress, especially at work, some people may drink more coffee. It is also possible that stress directly affects CL. If these two statements are both true, CC and CL will have some spurious correlation with each other just because each is related to stress. Any two variables related to stress will be correlated, even if there is no direct connection between them. If we are interested in the direct effect of CC on CL, we would want to hold stress at a constant level. This is generally not feasible to do experimentally, but we can use some independent measure of stress to predict both CC and CL and get residuals for both variables. The correlation between these two sets of residuals is the partial correlation of CC and CL with stress partialled out of both variables.

As you have probably guessed, it is not necessary to calculate residuals for either variable to find a partial correlation; you can use a formula similar to the one for the semipartial correlation:

$$r_{y1.2} = \frac{r_{1y} - r_{2y}r_{12}}{\sqrt{(1 - r_{2y}^2)(1 - r_{12}^2)}} \qquad \textbf{Formula 17.9}$$

The symbol $r_{y1.2}$ tells us that we are looking at the correlation of the first predictor with the criterion, partialling out the variable that follows the period—in this case, predictor 2. Notice there are no parentheses in this symbol because the partialled variable is partialled out of *both* of the variables represented in front of the period. Simple correlations between two variables that involve no partialling in either variable are called *zero-order* correlations in this context (the more partialled variables the higher the order of the correlation). Validities are zero-order correlations.

Like the semipartial correlation, the partial correlation can be related to a Venn diagram (provided that you are not dealing with complementary relationships or suppressor variables). If the correlation between coffee consumption (CC) and cholesterol level (CL) is .5, the correlation between stress and CL is .6, the correlation between stress and CC is .4, and CL plays the role of the criterion, the areas will be as shown in Figure 17.6. Area A is the square of the semipartial correlation between CC and CL partialling stress out of CC only; $r_{y(1.2)} = \sqrt{\text{Area A}} = \sqrt{.08} = .28$. However, the square of the partial correlation between CC and CL, partialling stress out of both,

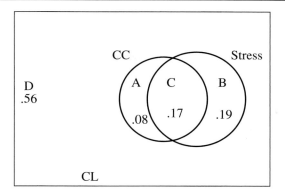

Figure 17.6

Areas Needed to Find
a Partial Correlation

equals Area A/(Area A + Area D). (Note that Area D equals $1 - R^2$.) The squared partial correlation of CC and CL answers the question How does the proportion of CL variance explained by CC, but not stress (i.e., Area A), compare to the total proportion of CL variance not explained by stress (i.e., Area A + Area D)? Of course, Area A plus Area D will sum to less than 1, so the partial correlation can be expected to be *larger* than the semipartial correlation (it can be equal but never smaller). In this example, $r^2_{y1.2} = .08/(.08 + .56) = .08/.64 = .125$, so $r_{y1.2} = \sqrt{.125} = .354$, which agrees (within rounding error) with the result you would get by using Formula 17.9, and is considerably larger than the semipartial correlation found for the same example.

The variable partialled out (like stress in the previous example) is called a *covariate* and is usually thought of as a nuisance variable with respect to the problem at hand. For instance, if we are studying the relation between social skills and memory in elderly subjects, age can be a nuisance variable. Social skills and memory can be correlated simply because both decline with age in the elderly. If you want to see that within a given age there is still a correlation between memory and social skills, you would partial out age. In the common case, where the covariate is correlated highly with each of the two variables of interest, the partial correlation will be lower than the original correlation. However, the phenomena of suppression and complementarity apply to this type of correlation as they do for the semipartial, so the partial correlation can be larger than the original.

There is no limit to the number of variables that can be partialled out at the same time. For instance, in the CC/CL example you could decide to partial out an obesity measure, in addition to stress. To see whether there is a direct connection between CC and CL, you would want to partial out all the extraneous variables that could make it look like there's a connection between CC and CL when there isn't one. Before computers, the calculation involved in partialling out more than one variable was daunting, but that is no longer a consideration. However, there is a price to pay for partialling out additional variables. You lose one degree of freedom in your significance test for each variable partialled out, as will be shown shortly. This is less of a problem for large samples but is a consideration when the sample size minus the number of partialled variables starts to fall below 30.

An even more serious problem involves the accuracy with which covariates are measured. This is generally not a problem for physical attributes like age and weight, but a variable such as stress is difficult to measure precisely or reliably. Inaccurately measured covariates can produce very misleading and biased partial correlations. Therefore, the reliability of potential covariates should be made as high as possible. Moreover, if you fail to include an important covariate among the variables you are partialling out, you could again end up with a misleading result.

A partial correlation is tested for significance like any other Pearson's r, except for the degrees of freedom. If a t test is used, the formula is:

$$ t = \frac{pr\sqrt{N - V - 2}}{\sqrt{1 - pr^2}} \qquad \textbf{Formula 17.10} $$

where N is the sample size, V is the number of variables partialled out, and pr is the partial correlation being tested. The critical t is based on $N - V - 2$ degrees of freedom. Note that when no variables are being partialled out, $V = 0$, and Formula 17.11 becomes the same as Formula 9.6, for testing Pearson's r in the simple case. As V gets larger, however, part of the numerator of the t test gets smaller, which is one reason why you wouldn't want to partial out a variable that doesn't account for much variance in your variables of interest.

Partial correlations are more often used as alternatives to multiple regression than as part of the multiple regression process. I will revisit the partial correlation concept in the context of the analysis of covariance in the next chapter. In the meantime, it is the semipartial correlation that will play an intimate role in the multiple regression procedures described in this chapter—especially those that involve theoretical models.

Finding the Best Prediction Equation

When you combine two predictors, the multiple R will always be at least as large as the larger of the two validities and usually larger than that. With partially redundant predictors, R^2 will be less than the sum of the two squared validities but, again, not smaller than the larger of the two. And R^2 is larger than $r_{1y}^2 + r_{2y}^2$ when there is suppression or complementarity between the predictors. So, are two predictors always better than one, except in the unusual case of total redundancy? The answer is actually No in many common cases. I'll refer to a predictor as valid if its correlation with the criterion, on its own, is statistically significant. If you have two valid predictors that you would like to use, the one with the larger validity will be used for certain. Whether the smaller (in terms of validity) predictor should be added to create a multiple regression equation depends on how much explained variance is added when the smaller predictor is included— that is, how much bigger is R^2 for both variables compared to r^2 for the larger predictor alone (r_{large}^2)? You may recall that $R^2 - r_{large}^2$ is equal to the square of the semipartial correlation of the smaller predictor. Thus, testing the statistical significance of the increase in R^2 produced by a particular variable is equivalent to testing its semipartial correlation (sr) for significance. An sr can be tested for significance by means of a t formula very similar to the one used for testing a pr.

$$ t = \frac{sr\sqrt{N - P - 1}}{\sqrt{1 - R^2}} \qquad \textbf{Formula 17.11A} $$

where N is the sample size, P is the total number of predictors being used and R^2 is the total amount of variance being accounted for when the variable being tested is included. The df for the critical t is, of course, $N - P - 1$.

If it turns out that the predictor with the smaller validity is not adding a significant amount of criterion variance accounted for, it should not be included. The more highly correlated (i.e., redundant) the two predictors are, the less likely it is that the smaller one will make a significant contribution (unless it is small enough to act as a suppressor variable). On the other hand, one advantage of multiple regression is that even when neither validity is significant, the combination of two predictors can be. The multiple R can be tested for significance in much the same way as the individual validities (as you will see in Section B), and it can attain significance even when the validities do not. This can happen when the two predictors are independent ($r_{12} = 0$), but it is more likely to occur when they are complementary ($r_{12} < 0$). In such a case, one and often both predictors will have significant semipartial correlations, and it is reasonable to combine them into one multiple regression equation. Similarly, you can have one valid predictor, and one that is not valid, but if the smaller one is acting as a suppressor or a complement it can add significantly to the larger one, which would justify including both of them in the regression equation.

It is important to remember that just because R is significant for the combination of two predictors does not mean you should use both. If the two predictors are partially redundant, one of them may not be adding significantly to the other, as described in the preceding paragraph. Also, the usual cautions about sample size and power apply. If you are using small samples, you may have a good chance of failing to reach statistical significance even when the correlations are fairly high. On the other hand, with sufficiently large samples even tiny, unimportant correlations will be statistically significant. You should always be conscious of the magnitude of your correlations and their importance in terms of your actual variables, rather than paying all your attention to significance levels.

Hierarchical (Theory-Based) Regression

I have spent a good deal of space on the practical problem of predicting a variable not because psychologists do this often, but because it provides a simple, mechanical framework with which to explain the workings of multiple regression. However, even though psychologists rarely need to make predictions, there are many areas of research in which the variables of interest cannot be (or should not be) controlled in the laboratory—such as self-esteem, degree of abuse in childhood, and attitude toward school—and in which most of the variables affect most of the others. In such cases, multiple regression can be used to help tease out the true relationships between the variables and decide which affect the criterion most directly and which seem to affect the criterion by means of their effects on other, mediating, variables. The effects of uninteresting (nuisance) variables can be separated from the effects of the variables of interest.

For instance, both study hours and parental income are likely to be correlated with success in high school, but whereas we would expect study hours to affect grades rather directly, we would expect the effects of income to be indirect (unless students are literally buying their grades), with higher income leading to a better study environment, private tutors, and so forth. In fact, one indirect effect may involve study hours; students with less household income may have to get part-time jobs and have less time available for studying. Therefore, the correlation between study hours and grades may be contaminated (i.e., partially confounded) by income level. If your research question is "What is the correlation between study hours and grades when income is controlled?" (i.e., what would the correlation be in a group of students all at the same level of parental income?) the answer is given by the semipartial correlation that you would get by first entering income in the equation to predict grades, and then taking the square root of the explained variance added by study hours. Given this particular theoretical question, you could enter income into the equation, even if study hours had the larger validity and even if the validity for income were not statistically significant. Occasionally, nuisance variables act as suppressors (e.g., forcing height into the equation before looking at the relation between weight and cholesterol level), but more often, as in the case of income in this example, controlling for a nuisance variable lowers the correlation of interest.

In the preceding model, your theory of how the variables affect each other and your question of interest dictate that income be a *forced* variable and study hours a *free* variable. Forced variables are entered into the regression equation automatically, and then the free variables are evaluated. When you have several predictors, you may have more than one forced variable, and then you can specify the order in which they are to enter the equation. This ordering creates a hierarchy, so when the ordering of variables is based on a theoretical model, the procedure is usually called *hierarchical regression*. This form of regression uses many of the tools developed for prediction but involves a greater exercise of judgment guided by theory and knowledge of the area of research. Of course, there isn't much room for complications when you have only two predictors; the real distinction between regression used for prediction and regression used for theoretical purposes will become apparent when we deal with more than two predictors in the next section.

SUMMARY

1. If two predictors are each correlated with a *criterion*, but not with each other, the variance they account for together is equal to the sum of their squared *validities* (i.e., correlations with the criterion). The *coefficient of multiple correlation* (R) is equal to the square root of that sum, and the weights by which you multiply the z score of each predictor in the standardized regression equation are the validities.

2. When two predictors with highly positive validities have a moderately positive correlation with each other, they will be partially redundant. The R^2 for the two predictors will be less than the sum of the two squared validities, and the beta weights (i.e., *partial regression slopes*) of the standardized regression equation will be less than the corresponding validities. As the correlation of the two predictors increases, R^2 decreases until it equals the larger of the two r^2s and one predictor is *completely redundant* with the other.

3. If the correlation between the predictors is high, and one predictor has a very low validity, it may act as a *suppressor variable*, subtracting nuisance variance from the other predictor and causing one of the beta weights to be negative. Suppression is one situation that can cause R^2 to be greater than the sum of the two squared validities. This phenomenon also occurs if the two predictors have a negative correlation with each other, in which case they are said to be *complementary*.

4. Just as regression with one predictor involves finding the straight line that minimizes the squared errors of the predictions, regression with two predictors amounts to finding the best fitting (two-dimensional) regression *plane*. The plane has two partial regression slopes, and the degrees of freedom are reduced by the number of partial slopes, plus one for the intercept (df $= N - P - 1$).

5. The *semipartial* correlation (sr) of a predictor with the criterion is the correlation between the criterion and the residuals of that predictor (found by using the other predictor to predict the first one). The squared sr equals R^2 (for the two predictors) minus the r^2 for the other predictor.

6. The *partial* correlation (*pr*) between two variables is their correlation after holding a third variable (usually a nuisance variable, called a *covariate*) constant. The covariate (or a combination of several covariates) is used to predict both of the variables of interest, and then the two sets of residuals are correlated to find pr. The pr can be higher or lower than the original correlation.

7. Semipartial correlations and beta weights can be used to assess the relative importance of two predictors. However, for actually making and using predictions, the raw-score form of the regression equation is more convenient. But because the raw-score slopes involve the standard deviations of the criterion and the predictors (which depend on the measurement scale), these slopes are not interpretable in the same way as the standardized slopes.

8. *R* should be statistically significant to justify using both predictors, and the predictor with the smaller validity should be adding a significant amount of explained variance. *R* can be significant even when neither validity is, but in that case the two predictors should be included only if they have a suppressor or complementary relationship that makes sense.

9. Rather than making predictions, psychological researchers usually use multiple regression to control for nuisance variables and/or to determine which variables have the more direct effect on the criterion. When the order in which variables are to be entered is specified in advance, usually based on a theory or model of how the variables interact, the regression is said to be *hierarchical*.

EXERCISES

*1. If a rating based on reference letters from previous employers has a .2 correlation with future job performance, a rating based on a job interview has a .4 correlation, and these two predictors have no correlation with each other,
   a. What is the multiple correlation coefficient, and what proportion of the variance in job performance is accounted for by a combination of the predictors?
   b. Answer the questions in part a again if the two validities are both .6.
   c. Can you answer the same questions if the two validities are .7 and .8? Why not?
   d. If the correlation between the job interview and job performance is .8, what is the largest correlation the reference letter rating can have with job performance if there is no correlation between the two predictors?

2. Three predictors have correlations with a criterion variable that are .28, −.43, and .61, respectively. If the correlation between any two of the predictors is zero,
   a. Write the multiple regression equation.
   b. How much variance in the criterion is accounted for by the predictions made by

the regression equation you wrote for part a?
   c. How large is the correlation between the predictions from your regression equation and the actual values of the criterion?

*3. Suppose that the "concreteness" of a word has a .4 correlation with the probability of recall and that the "imageability" of a word has a .5 correlation with the same criterion. Write the standardized regression equation, calculate *R*, and find the semipartial correlation of each predictor with respect to the other if the correlation between concreteness and imageability is
   a. .3
   b. .6
   c. .8
   d. .9
   e. What can you say about the situation represented in part c?
   f. What can you say about the situation represented in part d?

4. Suppose that your dependent variable is the likeability (*L*) of your 12 subjects, as measured by their coworkers. You find that self-confidence (SC) has a .35 correlation with *L*

and that humility (H) has a .25 correlation with L. If the correlation between self-confidence and humility is $-.6$,

a. Write the standardized equation for predicting L from both SC and H.

b. What proportion of the variance in L is accounted for by the combination of SC and H? What is the multiple R between L and the predictions from your equation in part a?

c. What is the semipartial correlation for each predictor? Test each sr for significance at the .01 level.

d. How much greater is R^2 for the combination of the two predictors than the sum of the two squared validities? Does this relation make sense? Explain.

*5. The correlation between IQ and recall for a list of words is .18, and between interest in the task and recall, $r = .34$. If the correlation between interest in the task and IQ is .12,

a. Calculate R directly from the three correlations given.

b. How much larger is R^2 than the proportion of variance accounted for by IQ alone? By interest alone?

c. Use your answers to part b to find the semipartial correlation for each predictor, assuming that there are 50 subjects. Test each sr for significance at the .01 level.

d. Use the semipartial correlations you found in part c to calculate the two beta weights. Write the standardized regression equation for predicting recall from IQ and interest.

*6. Suppose that the standardized regression equation for predicting college grades (CG) from both high school grades (HSG) and IQ is CG = .512 HSG + .276 IQ. If the validities are .60 and .44 for HSG and IQ, respectively, for a group of 15 students,

a. Find the proportion of variance in CV accounted for by the combination of HSG and IQ.

b. Use a Venn diagram to find the semipartial correlation for each predictor. Test each sr for significance at the .05 level.

c. Use the Venn diagram you drew for part b to find the partial correlation between CG and HSG, partialling out the effects of IQ. Is this partial correlation statistically significant at the .05 level if you have only 10 subjects?

7. For Exercise 6 assume that the means for college grades, high school grades, and IQ

are 2.8, 83, and 106, respectively. The corresponding SDs are .6, 12, and 15.

a. Write the unstandardized regression equation.

b. What value for CG is predicted for a student with an IQ of 112 and a high school grade average of 87?

c. What value for CG is predicted when IQ equals 95 and HSG equals 79?

d. What value for CG is predicted when IQ equals 121 and HSG equals 75?

*8. The average resting heart rate ($\overline{Y}$) for the members of a large running club is 62 bpm with $s_y = 5$. The average number of minutes ($\overline{X}_1$) spent running each day is 72 with $s_1 = 15$, and the average number of miles ($\overline{X}_2$) run per day is 9.5 with $s_2 = 2.5$. If both running time and running distance have correlations of $-.3$ with resting heart rate, and the two predictors have a correlation of $+.7$ with each other,

a. Write the standardized multiple regression equation for predicting heart rate.

b. How much unique explained variance does each predictor add to the other?

c. Write the unstandardized multiple regression equation.

d. If a club member runs 5 miles in 50 minutes each day, what heart rate would be predicted for that member?

e. If a club member runs 9.5 miles in 72 minutes each day, what would be the predicted heart rate? Why did you not need the regression equation to make this prediction?

*9. In a study of 20 of her own patients, a psychoanalyst found a correlation of .6 between a measure of "insight" and a measure of symptom improvement (relative to the start of therapy). She realized, however, that both symptom improvement and insight tend to increase with more years of therapy, so she decided to calculate the partial correlation between insight and symptom improvement, controlling for the number of therapy years. First, she found that the correlation is .8 between symptom improvement and number of therapy years.

a. How high would the partial correlation be if the correlation between insight and number of therapy years were .5? Would the partial correlation be statistically significant at the .05 level?

b. How high would the partial correlation be if the correlation between insight and number of therapy years were .7? Would the partial correlation be statistically significant at the .05 level?

c. Explain why the partial correlation between insight and symptom improvement is lower than the original (i.e., zero-order) correlation in parts a and b.

10. Suppose that a study of 12 students produces a correlation of .7 between a measure of eye-hand coordination (EHC) and performance on a tracking task (TT). Later it is discovered that a personality measure called

need for achievement (NA) has a correlation of .5 with TT, but only .1 with EHC.

a. Calculate the partial correlation between EHC and TT, partialling out NA.

b. What is the relation between the original EHC/TT correlation and the partial correlation? Explain what happened.

c. Is the partial correlation statistically significant at the .05 level?

When you have more than two predictors that you would like to consider for your regression equation, the decision concerning which predictors to include can get quite complicated. The methods used for deciding which predictors to include differ depending on whether you are being guided by theoretical questions or trying to make the best predictions. The procedures for prediction are simpler in that they are more mechanical, so I will start with those. Normally, I would follow the presentation of such procedures with an explanation of appropriate significance tests, but in the case of multiple regression, significance tests are often part of the process, so I will begin with the most important one.

B

BASIC STATISTICAL PROCEDURES

The Significance Test for Multiple *R*

Probably the most basic test in multiple regression is the test of the null hypothesis that multiple *R* in the population is zero. (Just as a lowercase rho represents *r* in the population, an uppercase rho is sometimes used to represent the population value of multiple *R*, but uppercase rho resembles an uppercase *P* when written as a symbol, and that can be confusing. Fortunately, this symbol isn't called for much, so I will avoid using it and just express the concept in words when necessary.) This test employs the *F* ratio and therefore may not seem related to the *t* test for a simple Pearson's *r* unless you square both sides of Formula 9.6 to obtain:

$$t^2 = \frac{r^2(N-2)}{1-r^2}$$

Because t^2 follows the *F* distribution with $df_{num} = 1$ and $df_{denom} = N - 2$, we can rewrite the test of a simple *r* as:

$$F = \frac{(N-2)r^2}{1-r^2}$$

Now compare that formula to the test for multiple *R*:

$$F = \frac{(N-P-1)R^2}{P(1-R^2)} \qquad \qquad \textbf{Formula 17.12}$$

where *P* equals the number of predictors (critical *F* is found with $df_{num} = P$ and $df_{denom} = N - P - 1$). As you can see, when $P = 1$, the formula reduces to the test of a simple *r*. Now you can also see the cost of adding predictors to your multiple regression equation. Although R^2 will almost always increase (and never decrease) when a predictor is added, the degrees of freedom for the error term (i.e., $N - P - 1$) decrease, which reduces the numerator of the

F ratio. At the same time, as P increases, so does the denominator, which also causes a decrease in F. Unless a predictor is adding a considerable amount of explained variance, its inclusion will actually reduce F and decrease the likelihood of obtaining a significant R.

The F test of R is closely related to the F test in ANOVA. In multiple regression, the SS for the criterion (SS_y) is divided into two pieces: the portion that can be accounted for by the predictors ($R^2 SS_y$), which is called SS_{model} and appears in the numerator, and a portion $(1 - R^2)SS_y$, known as SS_{error} or $SS_{residual}$, which appears in the denominator. Each SS is divided by its own degrees of freedom, which leads to a ratio of variances, as shown in the following formula (which is equivalent to Formula 17.12):

$$F = \frac{\dfrac{R^2 SS_y}{P}}{\dfrac{(1 - R^2)SS_y}{N - P - 1}}$$

If you think of SS_{model} as $SS_{between}$, and $SS_{residual}$ as SS_{within}, you can see that the preceding formula looks like the formula for a one-way ANOVA. (The number of predictors you need when multiple regression is used to perform a one-way ANOVA is $P = k - 1$, so SS_{model} is being divided by $k - 1$, and $N - P - 1 = N - (k - 1) - 1 = N - k + 1 - 1 = N - k$, so $SS_{residual}$ is being divided by $N - k$, which can be written as $N_T - k$.) Because the term SS_y appears in both the numerator and denominator, it can be eliminated from the formula. Algebraically rearranging terms to avoid having fractions in both the numerator and denominator leads to Formula 17.12. Conversely, the ordinary one-way ANOVA is equivalent to testing eta squared (the proportion of variance accounted for by the independent variable) for significance using Formula 17.12. The connection between ANOVA and multiple regression will be explored further in the next chapter.

Tests for the Significance of Individual Predictors

A little terminology should be introduced here. In multiple regression, the predictors are often referred to as independent variables (IVs), even though they are usually not manipulated by the experimenter, and the criterion is considered the dependent variable (DV). These terms make sense if you think of changes in the predictors producing changes in the criterion. The actual pattern of causal relationships can be hard to determine, but it is usually convenient to think in terms of IVs and a DV.

The simplest form of multiple regression, sometimes called the standard or conventional procedure (unfortunately, the terminology for referring to different forms of multiple regression is not universal), involves entering all of your predictors into the equation at once without testing them along the way to eliminate some of them. Once all of your predictors have been entered, you can test the unique contribution of each individual predictor to the total amount of criterion variance accounted for by the whole set. This can be done by testing the semipartial correlation of each predictor with Formula 17.11A, as presented in Section A. To highlight the similarity between the test of a single sr and the test for R^2 as a whole, I will square Formula 17.11A to create an F ratio for testing the null hypothesis that an $sr = 0$.

$$F = \frac{(N - P - 1)\, sr^2}{(1 - R^2)} \qquad\qquad \textbf{Formula 17.11B}$$

The degrees of freedom for the critical F are 1 and $N - P - 1$. (The df for the numerator is 1 because you are testing the significance of just one predictor rather than the whole set.)

When your goal is making predictions, it is common to focus on the unstandardized regression equation and test the significance of each variable's raw-score partial regression slope (b_i) to see if it is significantly different from zero. Rejecting the null hypothesis for a particular b_i implies that that variable makes a statistically significant contribution to the prediction equation. A b_i is tested simply by dividing it by its standard error to create a one-sample t test (the standard error for each partial regression slope is routinely given by most statistical software packages). However, the t for a particular b will be the same as the t for testing the beta weight or the corresponding sr, so there is no need to perform all of these tests. (In addition, for each semipartial correlation, there is a corresponding partial correlation that you would get if you partialled out all of the predictors—except, of course, for the one you are focusing on—from the criterion. The t test for the corresponding pr, which was shown in Section A, gives the same F when it is squared as you get from testing sr with the preceding formula. So there is no need to test both. Normally, you are not interested in both and want to focus on either the sr or the pr according to your research question.)

As I mentioned in Section A, it is important to keep in mind that a variable that contributes significantly to the prediction equation in a statistical sense may not be contributing enough to be important in any practical sense. With large sample sizes, even tiny contributions can become statistically significant, so it is important to keep in mind the actual magnitudes of your srs. On the other hand, if your sample sizes are too small, you are likely to fail to find significance even for variables that make an important contribution to your equation.

It is also very important to keep in mind that the unique contribution for a particular predictor (i.e., its sr^2 and therefore its sr) depends on which other predictors are included in the set. Unlike the predictor's validity, its sr will usually be different if an additional predictor is added or one is taken away. A predictor with a significant sr in one set of predictors may fail to reach significance if the set is changed. Thus the decision about which predictors to include is critical and not always easy to make, unless, of course, you have only two predictors. As mentioned in Section A, it is customary to start with the predictor that has the larger validity (assuming that it is statistically significant) and add the second predictor only if it contributes significantly (i.e., its b, sr, or pr differs from zero at the .05 level). However, if neither predictor has a significant validity, it is still possible that the combination of the two will attain significance. If the multiple R is significant, and neither predictor has a significant validity by itself, it would make sense to include both.

Forward Selection

It would be easy to extend the preceding procedure to the case in which there are more than two predictors. The second predictor added is the one that would have the highest sr in a two-predictor model with the first, provided that that sr is statistically significant. The third predictor chosen is the one with the highest sr given that the first two predictors are in the equation. Again, the sr must be statistically significant. The procedure stops when the largest sr among the remaining predictors is not statistically significant. The preceding procedure is called *forward selection*, and it is the simplest, but certainly not the most common, method for deciding which predictors to include. It has a couple of major problems, one of which is noticeable even when there are only two predictors. With forward selection it is easy to miss both suppressor and complementary relationships. If one of these conditions applies in a two-predictor situation, it is quite possible that neither predictor will have a significant validity on its own, so forward selection would stop before even beginning.

The other major problem arises only when you have more than two predictors, and it is a situation that becomes increasingly complex as the number of predictors increases. I'll illustrate this problem with three predictors. Imagine that you wish to predict a student's future college GPA from her high school grade average (HSG), the total of his or her SAT scores (TSAT), and some assessment of the student's study habits (SH). The intercorrelations among these variables is most easily displayed in a correlation matrix, as shown in Table 17.1.

Table 17.1			
	GPA	**HSG**	**TSAT**
HSG	.6		
TSAT	.5	.46	
SH	.5	.43	.00

Inspecting the correlation matrix is often the first step in a multiple regression analysis. Most statistical software packages when asked to intercorrelate these variables would produce a full matrix with each correlation repeated twice and a diagonal consisting of each variable correlated with itself ($r = 1.0$ for each, of course). I'm presenting a streamlined version of the matrix that is easier to read and contains all the necessary information. For this example, assume that 37 college students were selected at random and measured on the four variables at the end of their first year. With this sample size all of the correlations in the table are statistically significant, except, of course, for the TSAT/SH correlation, which is zero.

The first column of Table 17.1 contains the validities. In this example, high school grades would be the best single predictor of college grades. It is realistic that TSAT and SH would have substantial correlations with both high school and college grades, whereas there is no correlation between SH and TSAT (I'm using the SAT to represent some ideal measure of a natural aptitude for learning). With forward selection HSG is entered into the regression equation at the first step. At the second step SH is entered; even though it has the same validity as TSAT, it is slightly less redundant with HSG. R goes up from .6 (HSG alone) to .657; the F for testing R, however, goes down from 19.7 to 12.9 (still easily significant). F goes down because even though R^2 has gone up a bit, the numerator of Formula 17.11 is now being divided by 2 rather than 1.

The semipartial correlation of SH with GPA, given that HSG was entered, is .268. Entering this value into Formula 17.11B, we get:

$$F = \frac{(37 - 2 - 1).268^2}{(1 - .657^2)} = \frac{(34).0718}{.5684} = 4.293$$

With df = 1, 34, this F is significant at the .05 level (just barely), which is why SH could be added to the regression equation. A significance test of the beta weight for SH (which, using Formula 17.8, equals .297) would, of course, yield the same F value, as would a test of the raw-score slope, b.

Finally, the semipartial correlation of TSAT is computed with both HSG and SH already in the regression equation. The sr for TSAT is .329, which yields a significant F of 7.74. With all three predictors in the equation the R rises to .735, and the F testing this R remains virtually the same. The most dramatic change due to the addition of the third predictor, however, involves the srs of the two other predictors: The sr for HSG drops from a significant .426 with only SH added to the equation to a nonsignificant .200 when both SH and TSAT are added. Meanwhile, the sr for SH rises from a barely significant .268 to an easily significant .341. The beta weights are shown in the final regression equation: $z_{y'} = .257z_{HSG} + .390z_{SH} + .382z_{TSAT}$.

If we use forward selection to construct our multiple regression equation, we will end up including a predictor, HSG, that does not make a significant contribution to the three-predictor equation. The fact that this can happen with forward selection tells us both that forward selection is far from perfect as a procedure for constructing multiple regression equations and that multiple regression can get quite complicated with more than two predictors.

The paradoxical results in this problem are due to the fact that although HSG looks like a good predictor from its validity, it is not a very helpful predictor when SH and TSAT are added. HSG is not very helpful in the *full* model (all predictors are included) because it can be predicted quite well by a combination of SH and TSAT. This is not immediately obvious from the correlation matrix because HSG is not highly correlated with either of the other predictors (.43 and .46). However, the semipartial correlation of HSG with GPA in the full model is found by using SH and TSAT together to predict HSG, and then the residuals are correlated with GPA. Because SH and TSAT are not correlated with each other, when combined they have a correlation with HSG that is equal to $\sqrt{(.46^2 + .43^2)} = .3965 = .63$, which is quite high. Consequently, HSG explains little of the criterion variance ($.2^2 = .04$, or 4%) that is not explained by a combination of SH and TSAT. If SH and TSAT were both entered on the first step, HSG would not be added at all, but forward selection always begins with the highest validity and never looks back.

Backward Elimination

One solution to the drawbacks of forward selection is the opposite procedure known as *backward elimination* or *backward deletion*. This method begins by entering all of the predictors in which one is interested. Then the predictor with the smallest sr is found; if its sr is not statistically significant, that predictor is removed. This is the same as finding the predictor whose removal produces the smallest reduction in R^2 (recall that sr^2 for a particular predictor is equal to R^2 with that predictor in the equation minus R^2 after that predictor has been removed). Then the regression equation is recalculated with the remaining predictors and again the smallest contributor is deleted if its regression coefficient does not differ significantly from zero. This method continues until none of the predictors in the model has a nonsignificant sr. Applied to our example, backward elimination begins where forward selection ended (although this will certainly not always be the case)—with all three predictors included in the equation. The second step is to find the predictor with the smallest sr, and test it for significance.

In our example the sr for HSG is only .2, which leads to an F of 2.86, corresponding to a p just slightly *greater than* .10. Therefore, on the second step HSG is removed from the equation. R goes down from .735 to .707 (R^2 goes down from .54 to .50; the difference is equal to sr^2 for HSG), and the F for testing R actually goes up from 12.9 to 17.0. The equation now contains only SH and TSAT, each of which contributes equally and significantly, so the procedure stops. This is a leaner, and in many ways better, model than the three-predictor model produced by forward selection. But the advantage of backward elimination would be even more obvious if our sample contained only 34 instead of 37 students.

With the smaller sample, forward selection would stop after adding only HSG to the equation; the sr for SH would fall just short of significance at the .05 level. However, with this slightly reduced sample size, backward elimination would still retain SH and TSAT in the model—their srs being easily significant at the .05 level (even with HSG included). In this case, the model produced by backward elimination is clearly superior to the one found by forward selection; for instance, $R = .707$ for the combination of SH and TSAT, but only .6 for HSG alone. For three predictors, backward elimination clearly seems the way to go, but as the number of predictors increases, there is a growing problem with backward elimination.

The problem with backward elimination is that by putting all the predictors in the mix at once, you are maximizing the chance that some complicated and accidental suppressor or complementary relationship will be found among some of the predictors. If you are not guided by theory and are just trying to find the best prediction equation, it can be too easy for backward elimination to find a highly predictive equation that makes no sense and is not likely to be replicated in a new, random sample.

However, the technique that is most likely to capitalize on accidental relationships among predictors in a particular sample is not backward elimination but rather another procedure known variously as *setwise regression*, *the method of all possible regressions*, or *all subsets regression*. As these names imply, every possible combination of predictors is tried and tested (for our example there are seven possible subsets: each predictor by itself, each possible pair, and all three predictors). Sometimes this procedure is tried for exploratory purposes, but rarely would it be relied upon to find the best prediction equation for future situations.

Stepwise Regression

With elements of both forward selection and backward elimination a method generally known as *stepwise regression* seems to be the most popular. All three of these methods are sometimes known collectively as stepwise regression procedures because all involve a series of steps at which the model is evaluated and a predictor is added or deleted, but when the term is used to refer to a specific method, it is the one that will be described next. The method begins like forward selection, adding predictors only if they make a significant contribution. However, after each predictor is added, there is a backward elimination step. Predictors that had been added on previous steps are reevaluated, and if one is no longer contributing significantly, it is deleted. Deleted predictors can be added again on future steps if they become significant in the context of predictors added subsequently. However, to make this back and forth procedure more stable, it is common to set a more liberal alpha for deleting a predictor than for adding one (e.g., delete if alpha $>.1$, but add if alpha $<.05$).

Applying stepwise regression to our example, with $N = 37$, HSG, SH and TSAT would be added in that order. However, after all three have been added, the method finds that HSG is no longer making a significant contribution and deletes it. The final result is the same as backward elimination—for this example, a model containing only SH and TSAT. Unfortunately, if $N = 34$, stepwise regression is no better than forward selection for this example, adding HSG and then stopping. In general, stepwise regression has the advantage of rechecking previous predictors and cleaning up the equation by getting rid of dead wood, but it does not run the more extreme risk of capitalizing on chance that comes with backward elimination. For these reasons, stepwise regression has become quite popular as a method for finding a reliable equation for making predictions.

The Misuse of Stepwise Regression

It is usually considered reasonable to use some form of stepwise regression when your goal is a practical one (e.g., you want to predict longevity for the purposes of maximizing the profits of an insurance company). From a practical standpoint, it doesn't matter if you have picked the most sensible or meaningful set of variables to predict longevity, or whether these variables have relatively direct or indirect effects, as long as they do a good job of minimizing error. If you want to feel even more comfortable about your set of predictors, you might try several different selection methods to see if you end up with a similar equation in each case. There is a good deal of controversy, however, concerning the use of any stepwise regression procedures for theoretical purposes. For instance, in an attempt to understand the factors that contribute to self-esteem, a researcher may measure a variety of variables (e.g., income, body image, depression), find their correlations with each other and with self-esteem, and submit the data to a stepwise regression procedure. The relative size of the beta weights for the various predictors could be used to make statements about the relative importance of these predictors in producing a person's level of self-esteem (e.g., if number of social contacts turned out to have the largest beta weight in the final prediction equation, that variable would be considered the most important contributor to self-esteem). There are debates about whether the beta weight, sr^2, or some other measure best reflects the importance of a predictor, but a growing consensus of researchers feels that any form of stepwise regression is a seriously flawed procedure for discovering or confirming a theoretical model (e.g., Thompson, 1995) and that such procedures should never be used in an attempt to understand the underlying structure of some psychological phenomenon. You might use an "all-subsets" regression to get some ideas about the relations among your predictors and the criterion, but to confirm a theoretical model some form of sequential regression is highly recommended, as discussed near the end of this section (see "Multiple Regression as a Research Tool"). The most significant problems affecting the use of stepwise regression are discussed next. Note that most of these problems are relevant even if you are using stepwise regression only for the most practical purposes.

Problems Associated with Having Many Predictors

As the number of predictors increases, there are problems that get more serious, regardless of the selection procedure used to enter the predictors into a model. (*Note*: The more general term *regressor* is often used instead of predictor, especially when prediction is not the main goal, as is often the case. However, I will continue to use the term *predictor* because this term should already sound familiar to you.) First, there is the issue of multiple significance tests. When you are shopping around for the next predictor to add (or eliminate) among many alternatives, using an ordinary significance test at the .05 level in each case is like performing many t tests in a multigroup experiment without protection against the accumulation of Type I errors (see Chapter 13). It has been suggested that a Bonferroni adjustment be used based on the number of predictors available at a particular step. Alternatively, if you enter all your predictors and obtain a significant R value, you could be less concerned about adjusting your critical value at each step of your stepwise procedure. However, even if your F is significant with all potential predictors included, this will probably not be the most useful, efficient, or reliable model to use. A stepwise procedure can be helpful in trimming your model.

It is important to realize that if you are testing a multiple R for a regression equation that was derived as the result of a stepwise regression procedure, and you are using Formula 17.12 with the df based on the number of predictors in the final model, the alpha you use to look up the critical

F is not the true alpha. The fact that you had many predictors from which to choose, but chose a particularly good subset of them, means that you have increased the chance of making a Type I error. Under some restricted circumstances, it is possible to determine corrected critical values that keep alpha to your originally selected level (usually .05); for example, see Wilkinson (1979).

Multicollinearity

The chief problem that increases with the number of predictors is *multicollinearity*. Strictly speaking, two variables are collinear only if they are perfectly correlated, but the term is very often used to refer to two variables that are highly correlated. However, when there are many predictors, multicollinearity can occur even if no pair of variables is highly correlated; it can occur when one predictor is itself predicted very well by a combination of other predictors. This occurred to some extent in the previous example, in which a combination of SH and TSAT was highly correlated with HSG. The correlation between one predictor and a combination of other predictors is often called a *cross correlation*. If you square the cross correlation and subtract it from 1.0, you get a measure called *tolerance*, which is commonly used to determine whether a particular predictor is likely to be a strong contributor to a regression model.

$$\text{Tolerance} = 1 - R_C^2 \hspace{3cm} \textbf{Formula 17.13}$$

where R_C is the cross correlation.

The cross correlation between HSG and the combination of SH and TSAT was previously found (at the end of the subsection on Forward Selection) to be .63. The tolerance for HSG in this model is therefore $1 - .63^2 = 1 - .40 = .60$—not extremely low, but considerably lower than either of the other two predictors. The higher the cross correlation, the lower the tolerance, so low tolerance indicates that a predictor is rather redundant when the other predictors are included and thus is not likely to have a large *sr*. The more predictors you have in your set, the more likely it becomes (and the harder it can be to see from the intercorrelation matrix) that at least one of them will have low tolerance with respect to the others—especially if all of them have reasonably high validities.

A high value for tolerance means that the predictor in question is relatively independent of the other predictors; a tolerance of 1.0 means that the predictor in question is not correlated at all with any other predictor or any combination of other predictors. On the other hand, a tolerance of zero means that the predictor is perfectly correlated with some other predictor or combination of predictors. When a predictor has a zero tolerance, the beta weights for the regression equation cannot be found; the matrix that is used to solve for the beta weights (with more than two or three predictors it would be extremely tedious to write formulas for finding beta weights without the use of matrix algebra) is said to be *singular* (with two predictors you can see that the beta formulas in Section A involve dividing by zero when the two predictors are perfectly correlated). Regression weights cannot be found until a predictor is removed from the set.

Zero tolerance is extremely unlikely with real variables, except for special cases. For instance, if a questionnaire has several scales that add up to a total score, the total score will have zero tolerance if it is included with all of the subscales that add up to it (each subscale will have zero tolerance, as well). A similar problem occurs if all of the subscales must always add up to the same fixed number, like 100. If you include all of the subscales, any one of them will be perfectly predicted by the others. Eliminating any one of the subscales will relieve the problem.

As an alternative to tolerance you can use its reciprocal known as the *Variance Inflation Factor* (VIF). For our example, the VIF of HSG = $1/.6$ = 1.67. If a predictor has a high VIF (or low tolerance), its beta weight will be associated with a good deal of variability and therefore can easily change radically in a new random sample. As a simple, extreme example, imagine that two depression measures are correlated .9 with each other and have similar validities—for one, $r_{1y} = .42$ and for the other, $r_{2y} = .40$. The beta weights come out to be .32, and .12, respectively (you should check this for yourself). Notice that even though the validities are very close, the beta weights are quite different. In a new sample it would hardly be surprising if the validities of the two variables were reversed, with their intercorrelation remaining virtually the same. However, reversing the validities would also reverse the beta weights, so a slight change in the correlations with the criterion translates to a very large change in the beta weights. Both variables in this case have low tolerance (.19), which leads to their beta weights being unstable (i.e., easily changed by minor sampling fluctuations). Thus, if your goal is to have stable beta weights, it makes sense to avoid predictors that have very low tolerance. One way to do this is to combine similar variables that are highly intercorrelated into a single index, such as combining income, education level, and job status into a single index of socioeconomic status. Groups of predictor variables that are highly interrelated can be identified with a procedure known as factor analysis.

Adjusted R and R^2

There is another problem that becomes increasingly serious as you add predictors without increasing the sample size. This problem is related to the loss in degrees of freedom ($N - P - 1$). In Chapter 9, I pointed out that the expected correlation in a sample is not zero when the population correlation is zero, but rather $\sqrt{1/(N - 1)}$. When you have more than one predictor, but R is zero in the population, the expected R for your sample is $\sqrt{P/(N - 1)}$. That means that R cannot be calculated if you don't have more subjects than you have predictors and that R is always a perfect 1.0 if $N = P + 1$. Suppose you have 26 subjects with five predictors—you can expect a sample R of $\sqrt{.2} = .447$ even if the population R is 0. Thus, your sample R tends to overestimate population R. You can use the following formula to adjust your sample R so that it gives a reasonable (and nearly unbiased) estimate of the R in the population:

$$\text{Adjusted } R = \sqrt{R^2 - \frac{P(1 - R^2)}{N - P - 1}} \qquad \textbf{Formula 17.14A}$$

Of course, adjusted R^2 is found using the same formula without the square root. Note that when N is very large compared to P, the term being subtracted in the preceding formula is small, and the adjusted R is just slightly less than the square root of R^2, which, of course, is R. On the other hand, when the number of predictors is high compared to the number of degrees of freedom, the correction can be quite severe.

Formula 17.14A is instructive in showing how the adjustment of R depends on the relation between P and N. However, this formula does not make it obvious that it is not unusual for the adjustment to be so severe that the value under the square root sign becomes negative, in which case the formula cannot be used. For example, imagine that R is a respectable looking .5, but was obtained by using five predictors with a sample of only 20 subjects. The value under the square root sign of Formula 17.14A would be: $.25 - (5 \times .75)/14 = .25 - .268$, which is slightly *less* than zero. An algebraically equivalent formula, related to Formula 12.23 (which was described in Section C of Chapter 12), makes this problem more obvious. Because it is probably more common to adjust R^2 than R itself, I will show the squared version of the formula here; just remember that the square root of the following formula will always yield the same value as Formula 17.14A.

$$\text{Adjusted } R^2 = R^2\left(1 - \frac{1}{F}\right) \qquad \textbf{Formula 17.14B}$$

where F is the value you would use to test the significance of your R. Returning to my example of $R = .5$ obtained with $P = 5$ and $N = 20$, the F for testing this R, obtained from Formula 17.12, would be $14 \times .25/5 \times .75 = 3.5/3.75 = .933$. However, a glance at Formula 17.14B will show you that any F less than 1.0 will cause a number larger than 1 to be subtracted from 1 thus making the adjusted R^2 negative. In such cases, the adjusted R^2, and therefore the adjusted R as well, are considered to be zero. An F less than one offers no evidence that R in the population is greater than zero. On the other hand, you can also see that as F gets larger, the correction factor multiplying R^2 increases, eventually heading for its maximum value of 1.0 (i.e., no adjustment at all) as F becomes infinitely large.

For theoretical purposes adjusted R is a better estimate than our sample R of how well the criterion can be accounted for by the variables on which we are focusing. However, if you are interested in making predictions, a more practical problem than estimating R in the population is estimating R in a new random sample of the same size. Unfortunately, this cannot be accomplished with a simple formula; this problem is discussed next.

Shrinkage

If we really need to make useful predictions in the future based on a regression formula found for our initial sample, we will want to know how well our regression formula will work in a new sample. Adjusted R is not a very good indication of this. Adjusted R does not tell us what R to expect if we applied the beta weights from our sample to predictions in the entire population; it tells us what R to expect if the entire regression analysis were redone in the population, which includes finding the beta weights that work best for the population (i.e., the true population beta weights). Of course, the population beta weights are going to work better for predictions in the population than the beta weights for our sample, and this discrepancy is greater for smaller samples. If we knew how high an R to expect in the population when using our sample beta weights, we would have our best guess for R in the next sample; but there is no easy way to estimate this. The expected R for the next sample is what Darlington (1990) calls "shrunken" $R(R_s)$, and the problem of getting a smaller R when you use the beta weights from one sample to make predictions in another sample is called *shrinkage*. The surest thing we can say about R_s is that it is very unlikely to be as high as adjusted R, and it is usually considerably lower.

The problem of shrinkage points out the need for replication to see how well your beta weights really work on new samples. To get an idea of this (i.e., to estimate R_s) from your initial sample, the simplest thing to do is to randomly divide your sample in half (this is useful, of course, only if you have a large sample), calculate the beta weights in one of the halves, and then use these beta weights to predict scores in the other half. The correlation between the predictions and actual scores in the other half is your estimate of R_s. This method is called *cross validation*. A variation of this method, known as *double cross validation*, involves finding beta weights for both halves, then finding R_s in the opposite half in each case, and averaging the two R_ss.

When your sample is not large enough for cross validation, there is a reasonable alternative that is based on something called the **PRESS** statistic (*Pred*icted *Re*sidual *SS*). This method has the advantage of using your entire sample to estimate R_s, rather than splitting it in half. One at a time, subjects are left out of the data set and predicted by a regression based on the $N - 1$ remaining subjects. In each case the predicted score is subtracted from the actual score, and then all of these residuals are squared and summed to create the **PRESS** statistic, which can then be used to produce an estimate of R_s. Such a tedious technique could not have become popular without modern, high-speed computers. With these computers, the **PRESS** statistic can be calculated on samples large enough for cross-validation, but some statisticians argue in favor of the simpler cross- or double cross validation methods (which are applicable to a wider range of regression methods), when the sample size permits.

Too Few Predictors

The problems that occur with many predictors are often easy to see; the problems associated with too few predictors can be more subtle and more threatening to the validity of the analysis. For your regression weights to be unbiased estimates of their corresponding population weights, it is important to include in your set of predictors all of the related variables that affect the criterion. It does not bias your results to leave out a predictor that is independent of all the predictors in the included set, but, of course, if such a variable has a reasonable validity, you will very much want to include it. On the other hand, if you have included income but have left out educational level in predicting something like self-esteem, the regression weight for income will likely be misleadingly high. Unless multicollinearity is extreme, the problem of leaving out relevant predictors can be a worse one (Myers & Well, 1995). Therefore, once you have chosen some predictors, it is important to include all other predictors that are reasonably related to the ones initially chosen. Researchers generally include all of the obviously relevant predictors and let their regression methods decide which ones to throw out. Of course, if your sample size is too small, it will be too easy for important predictors to get thrown out for failing to reach statistical significance, which brings me to my next point.

Minimal Sample Size

As with all the parametric statistics described in this text, larger sample sizes generally yield greater power, more reliability, and fewer difficulties with respect to the assumptions of the test. Sample size is a special concern in multiple regression, however, in part because using many predictors eats up degrees of freedom in a way that is not always immediately noticeable. Everyone agrees that more is better, but there is much debate about the minimal sample size below which the reliability of the regression model rapidly declines. For meeting the usual normal distribution assumptions, it is desirable that the degrees of freedom, $N - P - 1$, be at least 40, so you would want to add at least 41 to the number of predictors to have a reasonable sample size. In the past, it was suggested that a sample size that is 10 times the number of predictors would yield adequate power, but this is reasonable only if you are dealing with at least five predictors. Another version of this rule that is suitable for any number of predictors is that the sample size should be at least 50 plus 8 times the number of predictors to have a reasonable amount of power to detect medium effect sizes. However, if you are using one of the stepwise regression procedures, it has been recommended that you use at least 30 or 40 subjects for each predictor to minimize the problems associated with capitalizing on chance.

Basic Assumptions of Multiple Regression

The multiple regression formulas in this chapter are all based on the assumption that there is a linear relationship between each predictor and the other predictors, as well as the criterion, in the population. If this is not true, increasing the sample size won't help, and this assumption is too important to be ignored. However, you don't necessarily have to get rid of the nonlinear variables; Section C describes some alternative solutions that may allow you to keep these variables.

The usual assumption of random sampling applies, and as usual in psychological research, it is frequently violated in studies analyzed by multiple regression. Unfortunately, when using multiple regression, we cannot rely on the random assignment of subjects to groups or conditions, so we must be cautious about avoiding any obvious biases in the selection of subjects and about generalizing our results to populations not adequately represented in our sample. Nonrepresentative sampling can also reduce the power of your statistics and result in underestimation of true correlations if it results in a restricted range for some of the variables.

Multivariate Outliers

Just as bivariate normality was assumed in testing simple correlations for significance, multivariate normality is assumed when testing multiple R and related statistics. As in the bivariate case, adequate sample sizes render violations of this assumption harmless. Homoscedasticity (i.e., equal error variance at all points along the regression surface) is also assumed, but again violations are less serious as sample size increases. Nonetheless, it makes good sense to look at the distribution of each variable measured in

Figure 17.7

Bivariate Points with Large Values for Either Leverage (A), or Residual (B), or Both (C)

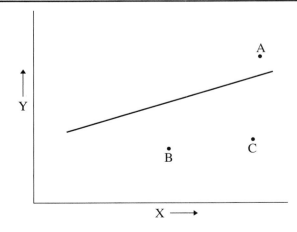

your sample and to inspect scatterplots for all pairs of variables. In addition to spotting curvilinear relationships and heteroscedasticity, you may find bivariate outliers, as described in Chapter 9. Harder to spot, but potentially just as distorting, are *multivariate outliers*, combinations of values on three or more variables that are very unusual and may indicate measurement errors or rare (possibly interesting) psychological phenomena. I cannot graph multivariate outliers for you, but I will use a bivariate graph to illustrate some of the tools you can use to spot multivariate outliers.

Measuring Leverage and Residuals

In Figure 17.7, it is easy to spot that point C is a bivariate outlier and could greatly affect the slope of the regression line, but if we learn to quantify the deviance of point C, we will be better prepared for the multivariate case. First, we notice that point C is an extreme value for the predictor variable, X; the further a score is from the mean of X, the greater its potential to slant the regression line in its direction. If we think of the regression line as a lever with its fulcrum at the mean of X, extreme scores on X can be said to have more *leverage* because they more easily rotate the regression line. If we have only one predictor, we can express the leverage of a point in terms of its z score on X. If we have many predictors, leverage is a joint product of a subject's deviations on all the predictors, so a multivariate measure of deviation is required. The measure most often used is called *hat diag* (symbolized h_i) for reasons that can only be described in terms of matrix algebra. What you need to know is that points for which hat diag is more than 3 times its mean (the mean of hat diag for any set of predictors is $(P + 1)/N$) can be considered outliers that have the potential to unduly influence your regression. In Figure 17.7, point A has as much leverage as point C but is much less problematic, which leads us to a second important diagnostic measure of a point's deviance, its residual.

Whether you have just one or many predictors, a point's residual is its value on the criterion minus its predicted value. To make the residual a meaningful measure of distance, you can standardize it. What is normally done is to divide the residual by the appropriate standard error to convert it to a t test; the result is called the studentized residual (after student's t distribution). Large residuals indicate points that are far from the regression surface and could have a large influence on it. In Figure 17.7, both points B and C have large residuals; however, point B is near the mean of X and is, therefore, not in a position to rotate the line. It could pull the whole line down and change its intercept, but that is usually not very important. Point C, on the other hand, has both leverage and a large residual; the combination of these two factors is known as *influence*, and point C clearly has a large value for influence.

Measuring Influence

Even just one point with high influence can have a noticeable effect on the regression line (or surface), so it is useful to quantify the influence of each point in your data set. The most common measure of influence is *Cook's D* (for distance), which is the sum of the squared changes in the raw-score regression weights that result from removing a particular point and recalculating the regression equation. The magnitude of Cook's D for any point is a product of both the leverage and the residual associated with that point; the way Cook's D is calculated, a value greater than 1.0 is often considered large. If your statistical software package gives you Cook's D for each of your data points, you can quickly scan for points that may be having an undue influence on your regression analysis. Be aware, however, that controversy surrounds the size of Cook's D that should be considered large, and some statisticians argue for concentrating on both leverage and residual measures, instead. The topic of *regression diagnostics* is the object of considerable investigation at present, and new recommendations are sure to emerge (see Kleinbaum et al., 1998 for a fairly recent discussion of this topic).

So what should you do if you find a point that has an unequivocally large Cook's D or can be shown in some other way to be a serious outlier? The steps are the same as those described for dealing with bivariate outliers. First make sure that the outlier is not the result of a data transcription error. Then check for circumstances that might have distorted one of your data measurements (e.g., an instrument that was not reset or a subject who didn't read the instructions, entered responses in the wrong places, or fell asleep). Of course, it is acceptable to delete an outlier that is the result of any kind of error, but if a subject is merely unusual and represents some small subset of your population, deleting that subject can distort the levels of your significance tests. On the other hand, if it seems that the probability is very low that a similar outlier will be selected in the next sample of the same size, from a practical standpoint, deleting the outlier may result in a more useful prediction equation. Finally, if you are dealing with a fairly large number of outliers, you should consider one of the methods for robust regression.

Regression with Dichotomous Predictors

In Chapter 10 (Section C), I pointed out that you can calculate a Pearson correlation coefficient even when one of the variables is dichotomous (i.e., has only two values, such as male and female); the coefficient is then called a point-biserial r (r_{pb}). In fact, it is possible for some, or even all, of the predictors in a multiple regression to be dichotomous. It is certainly not uncommon to add dichotomous predictors, such as gender, marital status, employment status, and so forth, to a multiple regression model that contains several quantitative predictors (e.g., age, income, IQ). To explain how

Figure 17.8

Regression with a
Dichotomous Predictor

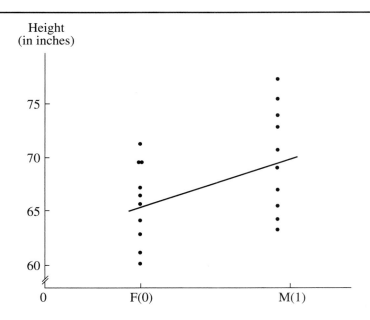

dichotomous predictors work, I'll begin with a very simple case—predicting adult height for a newborn child using gender only. The scatterplot and the regression line are shown in Figure 17.8.

The standardized regression slope is r_{pb} (for these data, .62), but the raw-score slope (b) is just the difference between the two sample means (4 inches). If females are coded as 0, and males as 1, the raw-score regression formula for predicting height in inches from gender is $Y' = 4X + 65$. For all females, the predicted height is $4(0) + 65 = 65$ inches, and for males it is $4(1) + 65 = 69$ inches. To the dichotomous predictor, gender, you can add a quantitative predictor like the baby's weight at birth. The intercorrelation between the baby's gender and birth weight is another point-biserial r, but it is used like any intercorrelation of predictors in multiple regression.

If a categorical predictor is not dichotomous, but rather has more than two levels (e.g., religious affiliation), it must be converted to a series of dichotomous predictors to use the methods of this chapter (this will be described in Chapter 18). If all of the predictors are categorical, the analysis may be viewed as a form of analysis of variance. The connection between regression and ANOVA will be dealt with in the next chapter. If the criterion is dichotomous, the multiple regression methods of this chapter are not the optimal approach to making useful predictions. A more appropriate method is briefly described in Section C.

Multiple Regression as a Research Tool

The stepwise methods described in the preceding represent objective ways of deciding which of many potential predictors should be retained and combined into your regression equation to make the best predictions. The beta weights of your final equation give you some idea of which variables seem to be having more influence on the criterion, and which less. However, the beta weights depend on which predictors are chosen. Add or delete one predictor and you can change all the beta weights radically. On the other hand, once you have chosen a particular set of predictors, the beta weights are determined. It doesn't matter what method you used or in what order the predictors were entered; for a given set of predictors, you would get the same result by entering all of them simultaneously. It is for this reason that all of the various stepwise procedures are sometimes referred to as *simultaneous regression* procedures in contrast to the sequential procedures I will describe next.

Although there are debates about which of the stepwise procedures most often yields the most accurate and reliable prediction equation, there seems to be widespread agreement that stepwise procedures should not be used to guide theoretical models in psychological research or to answer theoretical questions. Knowledge of a particular area in psychology and reasonable theories for how variables interact with each other in that domain can help a researcher place constraints on his or her regression model or place the focus on some variables over others. The most important considerations are: Which variables *cause* changes in other variables and which variables can be controlled or manipulated in a practical way? The answer to these questions can help a researcher place his or her variables into a hierarchy that can be used to control a regression analysis.

Entering Predictors by Blocks

It is possible to put all of your predictors in an exact order to create a *complete hierarchical regression analysis*, but when you have a large number of predictors, it is probably more common to arrange them in sets called *blocks*. I'll deal with blocks of predictors first and then discuss individual predictors in the next subsection. For an example involving blocks, let's go back to the problem of predicting longevity that was raised in Chapter 10. If your goal were to make the most accurate predictions possible, you would use family history, known genetic factors, obesity, diet, exercise, smoking, other health habits, psychological stress, and a number of other factors that could include geography (local pollution, toxic waste, etc.), and family income. To find the best prediction equation, it is acceptable to throw all of these different types of variables indiscriminately into a stepwise procedure. But although an insurance company may just want to make predictions and look at all of these IVs equally, a psychologist will usually have different purposes. A psychologist will see that people can control their health habits but not their family history. She will also see that smoking has a fairly direct (negative) effect on longevity but income has an indirect effect. Winning the lottery doesn't automatically improve your health, but it gives you the opportunity to receive better health care, may reduce stress, lead to a better diet, and so forth. A researcher interested in health habits would probably consider both genetic and income factors as nuisance variables to be controlled. This researcher might divide the previous variables into two sets: health habits in one and all of the others in another.

When a block is entered into a regression model, all of its variables are entered simultaneously. To determine the amount of longevity variance directly attributable to health habits, the block of nuisance variables is entered first. R^2 is calculated for the block. Then the block of health habit IVs is entered and R^2 is recalculated. The increase in R^2 that results from adding the health habits block tells us something about the amount of longevity variance that is under personal control. To test the significance of the increase in R^2, you can use a formula very similar to the one for testing a semipartial correlation. Let us use P to represent the number of predictors in both blocks combined and K to represent the number of predictors in the health habits block (i.e., the block to be added). R^2_P will represent the R^2 with all predictors from both blocks in the model, and R^2_{P-K} will represent R^2 with just the nuisance variables—without the health habits. Then the F ratio for testing the increase in R^2 due to the addition of the health block is:

$$F = \frac{(N - P - 1)(R^2_P - R^2_{P-K})}{K(1 - R^2_P)}$$ **Formula 17.15**

This F is tested against a critical F with K and $N - P - 1$ degrees of freedom. You can think of the test for a semipartial correlation as a special case of the preceding formula in which only one predictor is being added (i.e., $K = 1$).

It is not likely that a stepwise regression treating all the predictors individually would just happen to enter all the nuisance variables before any of the health habits. When the order is constrained by theory before the analysis is performed, the method is called *sequential* (or hierarchical) *regression*. When the predictors are entered into the regression in a predetermined order, whether individually or as blocks, and the increment in R^2 is determined for each predictor or block, the method can be called a *hierarchical variance decomposition*, or an incremental partitioning of variance (multiplying the increment in R^2 for a block by the total criterion variance tells you how much explained variance is associated with that block). When dealing with blocks, it is not very meaningful to come up with a beta weight for each block, so the results are presented in terms of sums of squares or variance estimates rather than a regression equation.

It is also considered legitimate to combine sequential and simultaneous regression. For instance, the entire block of nuisance variables can be forced into the model, but the health habits can be entered as *free* variables—in which case a stepwise procedure can be used to determine which individual health habit variables should be included after the nuisance block has been entered. The beta weights from the health habits regression can then be used as estimates of the relative importance of each health habit, but Darlington (1990) argues that the semipartial correlations make for more meaningful estimates. The individual predictors are ranked in the same order by both measures, but the *sr*s have the useful property that when one predictor has $sr_1 = .2$ and another has $sr_2 = .4$, it is meaningful to say that the second predictor is twice as important in the regression equation as the first.

Blocks of predictors can be used to answer practical questions, as in the following general case. There is a set of predictors that has been used traditionally in your field, but the predictors can be divided into two mutually exclusive sets: one that is easily and cheaply measured and another that is difficult and costly. By entering the cheaper set first, you can determine whether the costly measures add a statistically significant, as well as useful, amount of explained variance. Hierarchical decompositions are more commonly performed for theoretical reasons, but unfortunately, they are also commonly misinterpreted (Pedhazur, 1982). When there are more than two or three blocks, the proper ordering of the blocks may not be obvious and can be controversial. Comparing the variance increments due to different blocks entered at different stages can be misleading because they vary in how direct their effects are on the criterion. The problem of determining the order of entry for predictors is more easily described in terms of individual predictors; that is the approach I will take in the next subsection.

Ordering the Predictors

In a complete hierarchical analysis the exact order of entry is specified for all the predictors. Placing the IVs in a causal sequence is the most logical way to determine a hierarchy, as shown in the following example. Imagine that researchers have found the school grades of adopted children to correlate positively with the IQ of their adoptive parents (APIQ), even when the adoption has occurred too late to affect the child's IQ. APIQ cannot have a direct effect on a child's grades, but consider the following causal chain: Higher APIQ causes more encouragement for education (EE), which in turn causes the student to spend more hours studying (SH), which then results in higher grades (GPA). This causal path is depicted in Figure 17.9. The direction of causation is clear, and, for each pair of variables, proceeds in only one direction: APIQ and EE are measured at the beginning of a semester, SH during, and GPA at the end. This establishes a complete hierarchy—an exact ordering of the IVs for entry into a regression model.

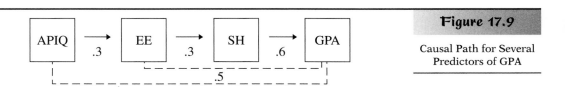

Figure 17.9

Causal Path for Several Predictors of GPA

Figure 17.10

Three Overlapping Predictors (for Illustrating a Hierarchical Regression Analysis)

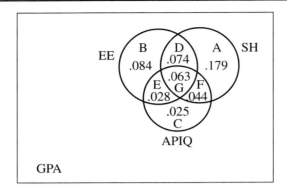

By filling in correlations in this model, I can show you how a complete hierarchical variance decomposition works with several predictors. Suppose that the correlations with GPA increase as the IV gets closer in the causal chain: .4 for APIQ, .5 for EE, and .6 for SH. The intercorrelations of the predictors I used to get the values for the Venn diagram in Figure 17.10 are not consistent, but they make the diagram easier to look at. In a simultaneous analysis (e.g., any stepwise procedure), only the unique contributions of each predictor are counted; the sr^2 terms would be given by areas A, B, and C for SH, EE, and APIQ, respectively (provided all three were entered). However, in a sequential analysis all of the overlapping areas are assigned to one IV or another according to the order in which they are entered.

The most indirect IV, APIQ, is entered first. Because no other IVs are in the model at this point, $sr^2 = r^2 = .4^2 = .16$, which corresponds to areas C + E + F + G. Then EE is added. Given that SH is not yet in the picture, EE adds areas B and D, so its $sr^2_{EE} = .084 + .074 = .158$. Adding SH at this point adds only area A, so $sr^2_{SH} = .179$. Note that variables earlier in the causal chain get more credit; the correlation between SH and GPA in this example is partly due to the indirect effects of APIQ and EE, whereas APIQ is not being affected by any other variables in the model. The three sr^2s sum to .16 + .158 + .179 = .497, which equals R^2 for the three predictors. The sr^2s always add up to R^2 in a hierarchical variance decomposition, whereas this will only happen in simultaneous regression if all the predictors are mutually independent.

When dealing with ordered individual predictors, as in the preceding example, you have the option of presenting your results as an ordinary hierarchical regression rather than a variance decomposition; the sr^2s would be converted to beta weights and put in the form of a regression equation. However, this approach has been replaced in recent years by one form or another of causal analysis. Rather than use a causal model to assess the relative importance of the predictor variables, researchers more often use the correlations in their data to test their causal models and determine both the direct and indirect effects of their variables on the criterion. The oldest, and probably simplest, form of causal analysis is path analysis, which will be described briefly in Section C.

Publishing the Results of Multiple Regression

Multiple regressions reported in the psychological literature are usually used to confirm some hypothesis and are therefore more likely to be hierarchical and concerned with changes in R^2 than simultaneous and linked to a prediction equation. The following example fits into the former category. Lee and Robbins (1998) hypothesized that women reporting higher levels of social connectedness (a feeling of belonging socially) would also report lower levels of trait anxiety (a proneness to feelings of anxiety). Moreover, they hypothesized that "social connectedness would uniquely contribute to lower trait anxiety, above and beyond the effects of collective self-esteem and perceived social support" (p. 339). The latter two variables were viewed as temporary sources of belonging and not as fundamental in determining feelings of anxiety. In addition, the subject's age and length of residency were considered uninteresting variables that should be controlled for. Accordingly, "a hierarchical multiple regression analysis was conducted to test for significance the contributions of age and length of residence (Step 1), social support and collective self-esteem (Step 2), and social connectedness (Step 3). . . . As predicted by the first hypothesis, social connectedness was uniquely related to lower scores on trait anxiety ($R^2_{\text{change}} = 16\%$)" (p. 340). In their results, age and length of residency turned out to have a negligible effect; R^2 was rounded off to zero in Step 1. With the Step 2 variables added to Step 1, the R^2 was .29. Finally, the addition of social connectedness in Step 3 produced an R^2 equal to .45, which is .16 more than the R^2 for Step 2. Although the Step 2 variables added significant variance accounted for when social connectedness was not in the model, they explained very little additional variance when social connectedness was added—confirming the authors' hypothesis concerning the primary importance of social connectedness in moderating feelings of anxiety.

B

SUMMARY

1. Multiple R is tested in a way that is similar to a simple correlation, except that for a given R, a larger number of predictors results in a smaller F ratio. An alternative to ANOVA is to find eta squared and test it for significance like multiple R^2.
2. The significance test for a semipartial correlation is equivalent to testing the beta weight or raw-score slope for that predictor. However, the *sr* or partial slope for a predictor depends on which other predictors are included in the regression equation.
3. There are several mechanical ways of deciding which predictors to include in a regression equation. The simplest is *forward selection*. The predictor with the highest validity is added first (if significant), and then the predictor with the highest *sr* relative to the first is added (if significant). This process continues until none of the remaining predictors has a significant *sr* when added to the predictors already included. This method has the disadvantage that the final equation can include one or more predictors that no longer have a significant *sr*. *Stepwise regression* corrects this problem by retesting predictors at every step and deleting those with nonsignificant *sr*'s. This latter method is now the most popular.
4. *Backward elimination* begins with all of the predictors included and then drops the predictor with the smallest *sr* if it is not significant. This process continues until all the remaining predictors have significant *sr*'s. The main disadvantage of this method is that it raises the probability that accidental suppressor and complementary relationships will be found. This problem is even worse, however, with *all subsets regression*, which tests every possible number and combination of predictors.
5. When predictors are highly correlated with each other, their beta weights can easily change a great deal from sample to sample. Having many predictors increases the chances of *multicollinearity*—one predic-

tor may be predicted almost perfectly by a combination of other predictors (this is measured by *tolerance* or its reciprocal, the *Variance Inflation Factor*). If a predictor can be perfectly predicted by other predictors, the correlation matrix is said to be *singular*, and a regression equation cannot be found.

6. The multiple R found for a sample is an overestimate of the R that would be found if the regression were recomputed for the population or another sample. The *adjusted R* (based on a simple formula) is a good estimate for the population R. However, the expected R for a new random sample ("shrunken" R), using the beta weights from a previous sample, will usually be less than adjusted R. One way to estimate this R is by *cross validation*, using the beta weights from one random half of the original sample to make predictions in the other.

7. One serious problem in multiple regression is leaving out an important predictor variable that is correlated with other predictors that *are* included. Because this can produce seriously misleading results, it is better to include all relevant variables from the beginning and to decide later which ones need to be deleted. However, if a large number of predictors are included in the final model, a sample size that is at least 10 times the number of predictors is recommended for adequate power and consistency with multivariate normality assumptions.

8. Multivariate outliers can greatly influence the beta weights of your regression equation. The *influence* of an outlier can be measured by *Cook's D*, which is a function of both a point's leverage (multivariate distance from the predictors' means) and its residual (distance from the regression line/surface). Outliers found to be due to errors or accidents can be deleted, but deleting a legitimate outlier requires modifying your statistical tests.

9. In *simultaneous regression* (including stepwise procedures) each predictor is represented by its unique contribution to the variance of the criterion (sr^2), and these contributions will usually not add up to R^2. On the other hand, in *sequential regression*, portions of the criterion variance that are shared by more than one predictor are assigned to particular predictors according to the order in which they are entered, and these hierarchical components do add up to R^2.

10. The sequence for entering predictors may be determined beforehand by a theoretical (causal) model; often nuisance variables are entered first. It is common to group the predictors into blocks, with only the order of the blocks being determined. The contribution of a particular predictor or block of predictors can be tested for significance with respect to the predictors already entered.

EXERCISES

*1. Assuming that 50 subjects were tested in Exercise 5 of the previous section,
   a. Test R for significance at the .05 level using Formula 17.11B.
   b. Test both semipartial correlations at the .05 level using Formula 17.11B.
   c. Based on the results of part a and b, which predictor(s), if any, would you rec-ommend including in the final prediction equation?

2. Assuming that 15 subjects were tested in Exercise 6 of the previous section,
   a. Test R for significance at the .05 level.
   b. Test both semipartial correlations at the .05 level using Formula 17.11B.

c. Test the significance of the partial correlation between CG and HSG in terms of an F ratio (square the result of Formula 17.10). How does the F for this significance test compare to the F you used to test the significance of the semipartial correlation of HSG in part b? Explain why this relation makes sense.

*3. If eta squared is .22 in a one-way ANOVA that consists of four groups of 28 subjects each, is the F ratio for the ANOVA significant at the .05 level?

4. Five predictor variables have been measured on 60 subjects.
   a. What is the tolerance and VIF of the fifth predictor if its cross correlation with the other four predictors is .1? .4? .8?
   b. How large is the adjusted R if R between a combination of the five predictors and the criterion is .1? .4? .8?

*5. Suppose that life expectancy (LE) is being predicted from three variables measured on 30 subjects: longevity of parents (LPAR), amount of exercise (AMEX), and amount of cigarette smoking (NCIG).
   a. If the correlation between LPAR and LE is .3, and the correlation between AMEX and LE is .2, what is the multiple R if you predict LE from just LPAR and AMEX, and the correlation between LPAR and AMEX is −.3?
   b. Is R significant at the .05 level?
   c. Test the significance of the semipartial correlation for each predictor.
   d. If R^2 is .26 when all three of the predictors are used to predict LE, are these predictors explaining a statistically significant amount of the variance in LE?
   e. If a fourth predictor were added to the model, and R^2 was increased to .28, would the regression model be statistically significant? What is the magnitude of the semipartial correlation associated with the fourth predictor?
   f. Given the four-predictor model mentioned in part e, what is your estimate for R in the population (i.e., the adjusted R)?

6. Given the information in Exercise 5,
   a. If NCIG has a −.3 correlation with AMEX, and a −.1 correlation with LPAR, what are the tolerance and VIF for each of the three predictors?
   b. If a hierarchical regression is performed such that LPAR is entered first, and the two predictors that are under a subject's

control are then added as a block, do the two latter predictors add significantly to the regression?

*7. In a study of longevity (criterion), it was found that a genetic factor had a validity of .46, while a health habits factor had a validity of .34. There were 30 participants in the study, and the correlation between the two factors was .18.
   a. Calculate multiple R and test it for significance at the .05 level.
   b. Calculate the semipartial correlation for each factor and test each for significance.
   c. Suppose that what I'm calling the genetic factor is actually a block of two predictors and the health habits factor is a block of three predictors. Does the health habits block add a significant amount of variance accounted for to the block of genetic predictors?
   d. What is the best estimate you could make of the multiple correlation that would be obtained by performing the study on the entire population (remember that R in part a really comes from 5 different predictors)?

8. In a study of 25 subjects, the correlation between income and happiness was .35 and the correlation between optimism and happiness was .43. If happiness is considered the criterion, and the correlation between income and optimism is .21,
   a. How much variance is accounted for by each predictor, given that income is to be entered first in a hierarchical regression analysis?
   b. If three variables involving satisfaction with friends, family, and a significant other are added as a block to the predictors in part a, and R^2 increases by .17 as a result, are these new variables adding a significant amount of variance at the .05 level?

*9. Suppose that you are trying to predict a student's success in college (SSC), and you have already measured eight potential predictors: IQ, parents' support for higher education (PSUP), total SAT score, student's need for achievement (SNACH), parents' level of education (PLED), number of hours student spends studying per week (STHRS), student's motivation for higher education (SMHE), and student's writing ability (SWA). The correlations among all eight predictors and the criterion are shown in the following table. Assume that the total sample size is 80. (*Note*: The following correlations were devised to make the relations among the variables clear and the exercises relatively simple; I'm not implying that such a matrix of correlations is even mathematically possible.)

	SSC	IQ	PSUP	SAT	SNACH	PLED	STHRS	SMHE
IQ	.52							
PSUP	.32	.05						
SAT	.47	.89	.15					
SNACH	.31	.03	−.04	.06				
PLED	.25	.11	.92	.17	.08			
STHRS	.41	.04	−.02	.12	.88	.07		
SMHE	.38	.09	.14	.14	.86	.13	.79	
SWA	.29	.77	.07	.84	.01	.10	.02	−.03

a. If R between SSC and a combination of all eight predictors were .8, how large would adjusted R be?

b. Given that R for all eight predictors is .8, if removing IQ and SAT reduced R to .77, would those two predictors as a block be adding a significant amount of variance when added to the other six predictors?

c. If you measure multivariate leverage for this problem with hat diag, what mean would you expect for this measure? What value for hat diag would be large enough to be considered worrisome?

d. In a forward selection procedure, which predictor would be entered first?

e. If you were planning a hierarchical regression, how might you group these predictors into blocks? Explain your choices.

f. If you were going to build a multiple regression model with only three of the predictors, and you wanted to predict as much of the variance in the criterion as possible, which three predictors would you select? Explain how you arrived at your decision.

g. If a stepwise regression produced the model you selected in part f, would it be legitimate to test R using Formula 17.12,

and a critical F based on an alpha of .05? Explain.

10. Suppose that you are trying to predict longevity (L) from seven potential predictors: income (INC), number of cigarettes smoked per week (CIG), level of stress (LOS), amount of alcohol consumed per week (ALC), longevity of parents (LP), amount of fat eaten per week (FAT), and an overall index of constitutional health (CONH). The correlations among all seven predictors and the criterion are shown in the following table. Assume that the total sample size is 50. (The note from the previous exercise applies here, as well.)

a. If R between L and a combination of all seven predictors were .6, how large would adjusted R be?

b. Given that R with all seven predictors included is .6, if removing LP and CONH reduced R to .57, would those two predictors as a block be adding a significant amount of variance when added to the other five predictors?

c. In a forward selection procedure, which predictor would be entered first?

d. If you were planning a hierarchical regression, how might you group these predictors into blocks? Explain your choices.

e. If you were going to build a multiple regression model with only three of the predictors, and you wanted to predict as much of the variance in the criterion as possible, which three predictors would you select? Explain how you arrived at your decision.

	L	INC	CIG	LOS	ALC	LP	FAT
INC	.22						
CIG	−.42	−.02					
LOS	−.41	−.69	.15				
ALC	−.28	−.07	.54	.06			
LP	.55	.10	.12	.01	−.08		
FAT	−.31	− 04	.46	.12	.81	.07	
CONH	.48	.09	−.14	−.04	.06	.73	−.09

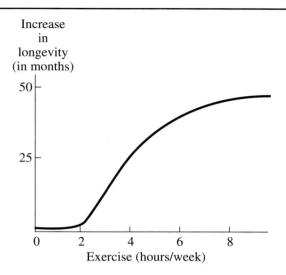

Increase
in
longevity
(in months)

Exercise (hours/week)

Figure 17.11

Predictor That Has a
Curvilinear Relation with
the Criterion

Dealing with Curvilinear Relationships

Data Transformations

In Chapter 9, I stressed that Pearson's *r* is not an appropriate measure of how strongly two variables are related if they have a curvilinear relationship. In a similar vein, I have pointed out that the multiple regression procedures in this chapter assume that each predictor (IV) has a linear relationship with the criterion (DV). Unfortunately, curvilinear relationships among important variables are not uncommon and may have to be dealt with in any multiple regression analysis. The health habits/longevity regression problem can give us several examples of curvilinear relationships. For instance, the relationship between exercise and longevity probably resembles the graph drawn in Figure 17.11. Below some threshold exercise does not help at all, and for those exercising quite a bit already, a large increase may not be very helpful. Of course, there must be some point beyond which increasing exercise actually decreases health, and the line on the graph begins to go back down, but for our purposes such extreme exercisers are rare enough to ignore. If we also exclude from our study people who exercise very little (below 2 hours a week in Figure 17.11), the graph looks like a logarithmic function—that is, longevity appears to be proportional to the log of the number of weekly exercise hours. If we take the log of exercise and plot it against longevity, the graph will be close to a straight line. This means that Pearson's *r* becomes a good description of the relationship, and log exercise can be used as a predictor in the multiple regression. (If you have forgotten what the log function is, recall that the numbers 10, 100, and 1000 can be rewritten as powers of 10: 10^1, 10^2, 10^3. The *common log* of these numbers is just the exponent part, so log (10) = 1, log (100) = 2, and log (1000) = 3. Any positive number can be written as a power of 10 and thus be transformed to a common log. The *natural* log uses a different base known as *e* but has the same effect on the shape of the graph.)

Depending on the shape of the graph, some other transformation of the IV (e.g., square root) may be better at straightening the line on the graph, but the log works well for many psychological variables for which the proportion of the increase in the IV determines the change in the criterion rather than the absolute amount. For instance, a 10% increase in income

**OPTIONAL
MATERIAL**

may have the same effect on job satisfaction regardless of the initial income level (leading to a log function), but it is safe to say that the same dollar amount (e.g., a $2,000 annual raise) will not produce the same increase in satisfaction for high- as for low-income workers (i.e., the function will not be linear). Although experienced researchers may be able to tell from looking at the graph which transformation is most likely to make it linear, it is considered acceptable to try many possible transformations and to pick the one that gives the most linear result. Of course, it is preferable if the transformation makes some sense (see Chapter 3, Section C) for the particular variable in question.

Polynomial Regression

Many variables related to health have some optimal level below *and* above which health is harmed. For instance, whereas most people in the United States eat too much fat in their diet, it is possible to eat too little. Fat plays an important function in the body's metabolism, and overzealous dieters can restrict fat so much as to harm their health. Surprisingly, recent studies have shown that there is an optimal level for alcohol ingestion and that moderate levels of consumption improve longevity more than abstaining entirely and more than, of course, overindulging. Suppose that the function relating alcohol consumption to longevity were the one graphed in Figure 17.12.

A graph that has a U shape usually indicates that the criterion is proportional to the square of the predictor (X^2), but an inverted U (as in Figure 17.12) suggests that there is a minus sign in front of the quadratic (i.e., X^2) term. Moreover, unless the peak of the curve is at $X = 0$ (or the minimum if the quadratic term is positive), there is going to be a linear component to the function, as well. The curve in Figure 17.12 follows the equation $Y = -.8X^2 + 8X - 10$. So, no simple transformation of X will straighten the graph; you have to apply the function just mentioned to X to make the graph perfectly linear. But how are you going to guess the coefficients for X

Figure 17.12

Curvilinear Relation with One Reversal (Suggests a Quadratic Function)

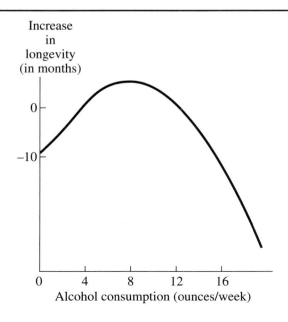

and X^2? The answer is to turn a simple linear regression into a multiple regression problem. If you use both X and X^2 as predictors of the criterion and find the raw-score regression equation, it will come out to $Y' = -.8X^2 + 8X - 10$. Normally, the points on the graph won't fall exactly on the function but will be scattered around it, so R will be less than 1.0. But with a function like this, the R you get using both X and X^2 will be higher than Pearson's r (which just uses X). In this context, R can be thought of as a coefficient of curvilinear correlation.

In Figure 17.12, the line goes up and then down; there is one reversal. If there were two reversals, you would need a cubic (i.e., X^3) component to fit the function; three reversals require X^4, and so forth. These components correspond, of course, to the trend components you learned about in Chapter 13, Section C. The overall process of finding the best weighted combination of powers of X to fit a function is called *polynomial regression*. In psychology it is very rare to have functions with more than one reversal, so researchers almost never use powers higher than X^2 as potential predictors. Also, given the usual distributions of continuous variables in psychological research, and the usual sample sizes, the power for detecting quadratic effects tends to be small, and for cubic and higher effects, much smaller still.

When powers of X are used, even if just X^2, it is preferable to perform a hierarchical regression entering the linear component (plain X) first, then X^2, and so forth. This allows each predictor to be "centered" (usually the sample mean is subtracted from each subject's value, so the mean of the transformed values is zero) without affecting the contributions of each component. Centering avoids the problems that arise from the inevitably high correlation between X and the powers of X (e.g., its squared value). That is why the trend coefficients in Table A.12 have been centered.

So far I have described why you might need a multiple regression approach to deal with a single curvilinear predictor. You may be wondering what to do if you have many predictors and several may have curvilinear relationships with the criterion. It is not unreasonable to add the squared versions of all your predictors to see what happens. However, this doubles the number of predictors (with all the problems that can cause), so you may want to look at a scatterplot of each predictor with the criterion to see if the X^2 component seems to be needed in each case. Actually, performing a linear regression first and then looking at a scatterplot of the residuals versus the original predictor can make it easier to see if a curvilinear relationship is present. In the absence of a curvilinear correlation, the residuals should be randomly scattered.

Moderator Variables

Let us return to the example in which high school grades (HSG) are being predicted by a measure of aptitude, such as a standard IQ test, and the number of weekly study hours (SH). The standard regression model resembles a two-way ANOVA with two main effects but no interaction. The predicted HSG is increased by a certain number of points for each IQ point (b_{IQ}) and a certain number of points for each study hour (b_{SH}), but the increase for each study hour is the same regardless of IQ—one study hour by a low-IQ student produces as much increase in HSG as one study hour by a high-IQ student, according to the standard model. If there is an interaction between IQ and SH such that combining high levels in both give HSG a boost beyond simply adding the increases, you would expect to get from each predictor alone, the simple additive model will not make the best predictions possible. To include an interaction term in this multiple regression model, we need only add a new variable that is formed by multiplying IQ by SH. This third predictor allows us to test the presence of an interaction, and if the interaction is significant, we are justified in including this third predictor to improve our predictions and our resulting R. A significant interaction means that we can't say that the slope for a predictor (b_i) is some particular number. For instance, if SH and IQ interact, b_{SH} keeps changing as IQ changes; b_{SH} may get higher, indicating a steeper slope (i.e., more effective studying) as IQ gets higher.

When interaction terms are added to multiple regression, the method is often called *moderated multiple regression*, and the interaction effects that are tested are called moderator effects (produced by moderator variables). In the preceding example, we can say that the relation between study hours and HSG is moderated by IQ if the SH/HSG relation changes as the value of IQ changes. Of course, under the same conditions, we can also say that SH moderates the IQ/HSG relation. If the IQ/HSG relation is your primary focus, you may, in fact, prefer to view SH as a moderating variable. Alternatively, you may simply view SH and IQ as interacting with each other in their effects on HSG.

It is a good idea to look for curvilinear relationships before looking for moderator effects or interactions because two highly correlated curvilinear predictors will often appear to interact when they really don't. Suppose that along with IQ and SH, you included SAT scores to predict HSG. It is easy to imagine that IQ and SAT will be highly correlated. You should also imagine that both have a curvilinear relation with HSG such that a given increase in aptitude has a bigger effect when aptitude is low than when it is high (see Figure 17.13). The interaction predictor, IQ $\times$ SAT, may prove statistically significant because it is so similar to IQ $\times$ IQ (i.e., IQ^2), which tests the curvilinearity of IQ. In this example, it could appear that increases in SAT are less effective when IQ is high, as though the two variables interact, when in reality increases in SAT are less effective when *SAT* levels are high (and IQ levels happen to be high at the same time because of the high SAT/IQ correlation). If any of your predictors are curvilinear, they should be transformed before being added to the regression model, or make sure to add the appropriate squared term to see if there is any real interaction beyond the

Figure 17.13
Two Curvilinear Predictors (They Can Appear to Interact)

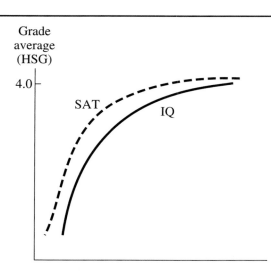

spurious interaction produced by a pair of correlated curvilinear predictors (Darlington, 1990). As with the inclusion of curvilinear predictors, it is a good idea to center your variables before including interaction terms; this will make it easier to interpret your basic effects if there is an interaction.

To test for all possible two-way interactions in a set of predictors, you can include all the possible cross-product terms (products of two predictors at a time), but this will increase the total number of predictors even more quickly than in the case of adding squared terms to test for curvilinear functions. For a set of six predictors, there are $(6 \times 5)/2 = 15$ possible two-way predictors to add, so the total set $(6 + 15)$ is more than 3 times the original, and the problem accelerates with more predictors. Therefore, it is a good idea to look for interactions only when there is a good theoretical reason or evidence from a previous study. Fortunately, researchers rarely need to consider testing for three-way or higher-order interactions and generally ignore such possibilities, especially if none of the two-way interactions are significant. Of course, if a two-way interaction term is not significant, it should be dropped from the model; otherwise, it is likely to reduce the power for testing the two predictors whose cross-product is being included.

Unfortunately, it is surprisingly difficult to attain statistical significance for an interaction in an observational study, even when you have only two predictors and the underlying variables really do interact with each other. The power for detecting an interaction is much higher in an experimental study, because the experimenter can assign equal numbers of subjects to various combinations of extreme conditions (e.g., some of the subjects with the highest IQs can be assigned to a cell of the design in which they are allowed very little time to study for a test). In an observational study, however, most subjects will have middle values on both of the predictor variables, and it will be relatively rare to find subjects with extreme values on both. But it is this combination of extremes that gives the most power to detect interactions, which is why there is a great advantage to being able to assign subjects to extreme combinations. It can take 20 times as many subjects in an observational study to have as much power to detect an interaction as a corresponding experimental study (McClelland & Judd, 1993). This problem becomes significantly worse for three-way and higher interactions, for which observational studies normally have so little power it is pointless to even test for them.

Multiple Regression with a Dichotomous Criterion

So far we have not dealt with the case in which the criterion is dichotomous. If it were rare in psychological research to deal with criteria that are dichotomous, this kind of regression would not be so important, but in fact the criterion of interest is frequently a dichotomy. For example, some psychologists would like to predict which teenagers will graduate high school and which will drop out, which marriages will end in divorce and which will not, which schizophrenics will respond well to psychotherapy and which can only be reached by drugs, and so forth. If you have a dichotomous criterion and all of your predictors are categorical, you can use the nonparametric procedures of Chapter 20. If your criterion is dichotomous (i.e., it consists of two distinct groups), and some, if not all, of your predictors are quantitative, you can use the multiple regression methods of this chapter to find a regression equation to predict which of the two groups a subject will be in (the two groups are given arbitrary numbers, like 1 and 2).

I will illustrate how R^2 can be found when you have two predictors and a dichotomous criterion. Suppose that you are predicting whether or not someone will go to college from his or her family's income and his or her high school grade average (HSG). Next, suppose that you have a group of 20 students who went to college and another 20 who did not. For these students the r_{pb} between income and group membership is .45; for HSG, r_{pb} is .35 (the point-biserial r is easily obtained from the two-group t value by Formula 10.13). If income and HSG were uncorrelated, you could square and sum the point-biserial rs to obtain R^2; if r_{12} is not zero, you can obtain R^2 from Formula 17.6. Assuming that $r_{12} = .3$:

$$R^2 = \frac{.45^2 + .35^2 - 2(.45)(.35)(.3)}{1 - .3^2} = \frac{.2025 + .1225 - .0945}{1 - .04}$$

$$= \frac{.2305}{.96} = .24$$

Testing R^2 by Formula 17.12 yields an F ratio equal to 5.84, which exceeds the critical F for 2, 38 df at the .01 level.

If the two DVs had a higher positive correlation, R^2 would be lower; if the two DVs were complementary (negatively related), R^2 would be greater than the sum of the two squared point-biserial rs. The multiple regression equation you would find for predicting a dichotomous criterion would be closely related to the discriminant function that would best weigh the predictors in a *discriminant analysis* (see Chapter 18, Section C). In fact, the R^2 you would calculate for quantitative predictors and a dichotomous criterion is easily transformed into a statistic, Hotelling's T^2, which is used to test the difference between two groups when several dependent variables are combined into a multivariate analysis of variance (MANOVA). Moreover, many of the difficulties that can be encountered in discriminant analysis and MANOVA (e.g., highly correlated predictors; outlier observations) have their counterparts in multiple regression and are dealt with in ways similar to those described in this chapter, as you will see if you read Section C of the next chapter.

If your criterion is categorical, but consists of more than two groups, and some, if not all, of your predictors are quantitative, you cannot use the multiple regression techniques of this chapter; you could probably perform some form of discriminant analysis instead. However, even though it is possible to perform an ordinary multiple regression with a dichotomous criterion, this is not the best statistical tool for the job. A more appropriate

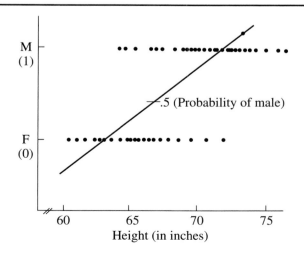

Figure 17.14

Regression with a Dichotomous Criterion

version of multiple regression has been devised for this case and has become quite popular in recent years. It is called *logistic regression*.

Logistic Regression

Logistic regression is a rather sophisticated procedure that I will only briefly introduce. To understand the need for such a complex technique, let us look at the shortcomings of ordinary regression with a dichotomous criterion. I'll begin with the example near the end of Section B, in which height is being predicted from gender. Like any two-variable regression problem, this one can be turned around so that I am regressing (i.e., predicting) gender (Y) on (i.e., from) height, as in Figure 17.14. (You can imagine that an archaeologist has discovered early human skeletons and wishes to infer their genders from their heights.) When you reverse the predictor and the criterion, the regression equation changes as well; in this case, $Y' = .095X - 5.9$. Instead of passing through the mean heights of men and women, the regression line in Figure 17.14 has a less obvious task. Ideally, for each value of height along the X axis, the regression line should pass through a point that represents the proportion of subjects at that height who are men (or whichever gender is coded with the higher number). If we use the regression equation to predict these proportions, middle values of height yield predicted proportions near .5, but insert 73 inches into the equation (not an unusual height for a man), and the prediction is an impossible 1.035 (you would be predicting that more than 100% of the people with that height are men). For 62 inches the prediction is an equally impossible −.01.

There are other problems with ordinary regression when the criterion is dichotomous. The scattering of points around the regression line at any value for X could hardly be further from a normal distribution in Figure 17.14, and there's little chance of having homoscedasticity. The recommended solution is to forget about trying to use a straight line for predictions in this case. Because you are trying to predict the probability of one gender or the other from height, the regression line should start with zero and end with 1.0. It also makes sense for the line to be a curve, like the one shown in Figure 17.15. Notice that for a height change from 58 to 60 inches or from 73 to 75 inches there is little change in the odds of being one gen-

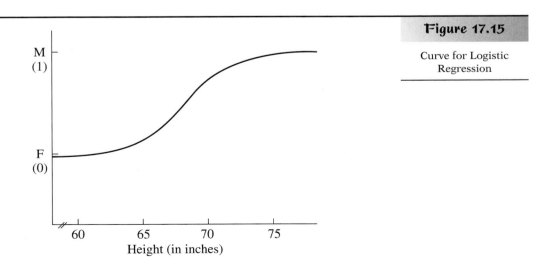

Figure 17.15

Curve for Logistic
Regression

Figure 17.16

Logistic Curves with
Shallow (Left) and Steep
(Right) Slopes

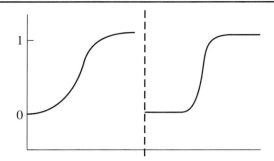

der or the other. (*Note*: In logistic regression it is common to deal with odds and ratios of odds. For instance, if the probability of being male at a certain height is .75, the odds are 3 to 1 in favor of being male at that height.) But from 66 to 68 inches there is a much larger change, which is perfectly reasonable. The curve is based on a logarithmic transformation of the ratio of the two numbers representing the odds in each case; this new ratio is called the *logistic transformation* or *logit*, for short. This is not the only transformation or curve that is reasonable—sometimes a *probit* transformation, based on the cumulative normal distribution, is used instead—but it is much better than a straight line for most purposes and avoids the problems with assumptions that you would get trying to use ordinary regression with a dichotomous criterion.

In the one-predictor case, the logistic curve has two parameters corresponding to the slope and intercept of a straight line, respectively. In logistic regression the intercept corresponds to where the curve is centered (for the gender/ height problem, you'd expect the center to be midway between the average height of women and the average height of men), and the slope corresponds to how sharply the curve bends (Figure 17.16 depicts curves with relatively shallow and sharp slopes). If gender and height were very highly correlated (i.e., r_{pb} is close to 1.0), there would be little overlap between the genders, and the curve would rise sharply, as in the right half of Figure 17.16.

With two predictors, you would use a logistic surface that bends appropriately rather than the simple plane of ordinary regression. Logistic regression would produce two beta weights or two partial slopes just as in ordinary regression. Interpreting the partial slopes in logistic regression is a bit more complicated than in ordinary regression; in fact, the whole topic of logistic regression requires at least a large chapter of its own. A brief, but helpful, introduction to this topic can be found in Darlington (1990). Despite the complexities of interpreting a logistic regression, the ease with which such analyses can be performed by statistical software, and their usefulness in dealing with dichotomous outcomes, have led to their popularity, which is likely only to grow in the foreseeable future.

Path Analysis

As you know by now, a high correlation between two variables does not prove they are causally related. However, the experiments required to demonstrate causality can be very expensive or even unethical (imagine assigning teenagers to a smoking group). Therefore, it can be very desirable to glean whatever causal information we can from observed correlations,

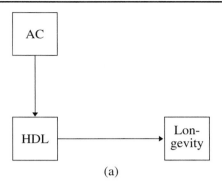

Figure 17.17a

Path Analysis Model
That Is Overidentified
(Can Be Tested)

(a)

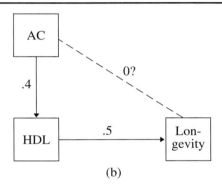

Figure 17.17b

Path Analysis Model That
Is Just-Identified (and
therefore Untestable) If
AC/Longevity Path Is
Included

(b)

using some kind of causal analysis. Although causal analysis is not used to find or to prove particular causal connections, a theoretical model of how a set of variables is causally connected can be tested using causal analysis; the model can be found to be consistent with the observed correlations, or those correlations can make a particular model very unlikely. The oldest form of causal analysis is *path analysis*. I will illustrate the basic concepts of path analysis in terms of the simplest type of causal model.

Consider the controversial finding that moderate drinkers of alcohol live longer than abstainers. If we sample only people who drink somewhere between zero and the optimal amount, the correlation between alcohol consumption (AC) and longevity will be positive. One possible explanation for this relation is the finding that, up to the optimal level, AC correlates positively with the level of HDL cholesterol in the blood (the so-called good type of cholesterol), which in turn correlates positively with longevity. This explanation can be expressed as a causal model, as shown in Figure 17.17a.

The arrows in Figure 17.17a indicate the proposed direction of causation for each pair of variables. This model is called *recursive* because all of the causal relations are unidirectional. AC is called an *exogenous* variable because it is not influenced by any variable in the model; HDL and longevity are *endogenous*. Each arrow in the model represents a *path* by which one variable can affect another, and these paths can be associated with path coefficients. Finding these coefficients is the essence of path analysis. Note that there is no path that goes directly from AC to longevity; that path coefficient is therefore automatically zero. The absence of that path represents our theory that AC does not have a direct affect on longevity—that it affects longevity only indirectly through the HDL variable.

The path coefficients for the two paths shown in the model can be found from the correlations among the variables. Let us suppose that the correlation between AC and HDL is .4 and between HDL and longevity it is .5. In this simple model, the path coefficients would correspond to the ordinary correlations for the two direct effects (see Figure 17.17b). Suppose also that the correlation between AC and longevity is .2. Are the path coefficients in Figure 17.17b consistent with the observed data (and the absence of a path between AC and longevity)? The answer is Yes in this case because a correlation of .2 is just what you would expect from the indirect effect of a .5 and .4 correlation (.5 × .4 = .2). If we were to calculate the path coefficient from AC to longevity, it would be like calculating a beta weight in multiple regression, where longevity is the criterion, and AC and HDL are variables 1 and 2, respectively:

$$B_1 = \frac{r_{1y} - r_{2y}r_{12}}{1 - r_{12}^2} = \frac{.2 - (.5)(.4)}{1 - .4^2} = \frac{.2 - .2}{.84} = 0$$

The zero beta weight shows that there is no direct effect between AC and longevity (i.e., the AC/longevity correlation is due entirely to the correlation of each of these variables with HDL), and therefore their path coefficient is zero. The more the observed AC/longevity correlation deviates from .2 (given that the other correlations remain .5 and .4), the greater the inconsistency between the data and the model. If a significance test demonstrates a significant discrepancy between the data and the model, we must conclude that the model depicted in Figure 17.17a is not applicable to these variables. Adding an arrow to indicate a direct effect of AC on longevity would be a good alternative, but, in general, rejecting one model as wrong does not tell you which model is right, and it is not considered valid to test every possible model just to see which one is not rejected by the data (this practice could inflate your Type I error rate).

The model shown in Figure 17.17a is considered "overidentified"; three correlations are observed, but only two path coefficients are free to vary (one is fixed at zero). This allows the model to be tested and, possibly, rejected. The path coefficient hypothesized to be zero is like our null hypothesis, except in this case rejecting the null hypothesis is usually not what we want to happen. Consequently, with path analysis low power becomes more of a problem than with traditional null hypothesis testing; low power can mislead us into believing that our model is correct, when a larger sample size would have led to the opposite conclusion. Bear in mind that, as usual, accepting the null does not provide strong evidence that the null hypothesis (in this case, our causal model) is true, but rejecting the null (regardless of the sample size) is considered reasonable evidence that it is false.

Had I added an arrow from AC to longevity in Figure 17.17a, the model would have been "just-identified" and could not have been tested. Once a direct effect from AC to longevity is allowed, any pattern of observed correlations will work; a just-identified model can always account for the data.

Figure 17.18	
A (Somewhat) Complex Path Analysis Model	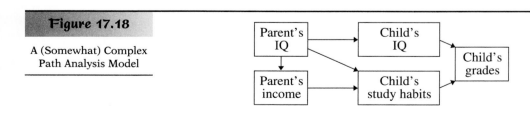

Of course, a just-identified model may be the correct one, but it is preferable when an overidentified model fits the data because the latter type of model is more parsimonious and informative, and is vulnerable to being rejected by the data. Ordinary multiple regression is like a just-identified causal model in which the criterion is an endogenous variable and all of the predictors are exogenous. No causal testing can be done.

Path analyses can be quite complicated, with variables being affected by several direct and indirect pathways at the same time (see Figure 17.18 for an example) and with several multiple regressions required to find the path coefficients. Direct effects can be separated from indirect effects, or total effect coefficients can be found that combine direct and indirect effects but ignore spurious and ambiguous effects; these effect coefficients can be used to compare meaningfully the impacts of two (or more) variables on a third. However, as complicated as the models can get, traditional path analysis has many restrictions that keep its calculations relatively simple. Curvilinear relations and interactions are assumed not to exist. It is also assumed that the variables are measured without error and that all the relevant variables have been included in the model. Finally, for each pair of variables causation can occur in only one direction.

Unfortunately, real-world research problems rarely fit these restrictions. To perform causal modeling in more complex and realistic situations, more sophisticated procedures have been developed in recent years, using high-speed computers to do the calculations. The modern method is called *structural equation modeling*, and it can deal with various complications, such as causation flowing in both directions between a pair of variables, and the effects of *latent* variables that cannot be accurately captured by any one measure. Of course, this more advanced procedure can also handle simple path analyses, as previously described, so it is becoming very popular, indeed.

SUMMARY

1. Many important psychological variables have a curvilinear relationship with some criterion. For instance, if the increase of the DV becomes less at higher values of the IV (the curve seems to level off), the two variables may have a logarithmic relation. In that case, the log of the IV may have the desired linear relation with the criterion and can be used as a predictor in multiple linear regression. In other cases, a number of possible transformations (e.g., square root of IV) may be explored before finding the one that yields the best linear relation.

2. As the IV increases, the criterion may go up, level off, and then go down (or the opposite). Such a reversal indicates that you should add a predictor that is the square of your IV so that you are including both X and X^2. Additional reversals of the criterion require higher powers of the IV as predictors, but this situation rarely arises. You can add X^2 predictors for all your IVs, but it makes more sense to inspect the appropriate scatterplots to see if a curvilinear relation looks likely in each case.

3. If the effect of one predictor depends on the level of another predictor, the two predictors are said to interact. To account for an interaction's effect on the criterion, you should add a predictor that is the product of the two predictors involved. However, you should first check that the interaction is not an artifact of having highly correlated, curvilinear predictors. Also, because the power for detecting interactions among observed (rather than manipulated) variables is surprisingly low, it is usually best to test for interactions only when they are theoretically interesting and likely to be fairly large.

4. Combining dichotomous and continuous predictors in multiple regression is straightforward and produces no problems. However, when the criterion is dichotomous, standard regression methods can result in strange predictions that translate to probabilities less than zero or greater than 1.0. The preferred method for predicting a dichotomous criterion is *logistic regression,* in which a symmetrical logarithmic curve that starts at probability zero and ends at 1.0 replaces the straight line of simple regression. The method generalizes to any number of predictors (e.g., a logistic "surface" is found if there are two predictors).

5. Path analysis begins with a model that specifies which of your variables have direct causal relations, and in which directions. The observed correlations for pairs of variables are used to test the model and thus determine whether there may be causal connections not proposed by the model. Traditional path analysis can only be used with certain restrictions, such as requiring that the model be *recursive* (causation can only flow in one direction between a pair of variables). More modern methods, particularly structural equation modeling, can handle more complex forms of causal analysis.

EXERCISES

*1. Suppose a subject eats 10 small snacks that vary in calorie content and rates each on a satisfaction scale (0 = not satisfying at all; 10 = completely satisfying). The data are as follows:

Snack No.	1	2	3	4	5	6	7	8	9	10
Calories	40	5	10	2	0	30	100	35	15	75
Rating	6	3	4	2	1	5	9	7	5	8

a. Calculate the Pearson correlation between calories and satisfaction ratings (the correlations in this and the following exercise will be very high, so retain at least four digits past the decimal point in your intermediate calculations to obtain a sufficiently accurate result). Plot the data on a scatter graph. What pattern do you see in the scatter graph that suggests that the correlation could be even higher?

b. Transform the calories variable by taking the square root of each value. Calculate Pearson's *r* between the ratings and the transformed variable. Plot the scatter graph for this correlation. Based on an inspection of this scatter graph and the one in part a, explain the difference in the magnitudes of the corresponding correlations.

2. This exercise is based on the data in Exercise 1.

a. Center the ratings variable (i.e., subtract the mean of the ratings from each one). Square the values of the centered ratings variable to create a second new variable. Calculate the correlation between the centered ratings and the squared centered ratings.

b. Calculate the correlations between the calories variable and both the centered and squared centered ratings variable. Calculate the multiple *R* for predicting calories from both the centered and squared centered ratings. Explain what you accomplished with this multiple regression.

*3. For a set of eight predictor variables, what is the total number of predictors you would end up with if, along with the original set, you included all the squared terms and all of the possible two-way interactions?

4. The equation $Y = 2X^5 - 5X^3 + 3X + 24$ can describe a function with, at most, how many reversals?

*5. Two groups of 10 subjects each perform a series of clerical tasks. One group is exposed to several distracting conditions, and the other group serves as a control. Both the speed and the accuracy with which they perform the tasks are measured. If the *t* tests between the two groups yield 1.5 for speed and 1.7 for accuracy (the control group scoring more highly on both measures, of course), and the correlation between the speed and accuracy measures is $-.5$,

a. What is the value of R^2 for predicting group membership?

b. Test R^2 for significance at the .05 level.

6. An educator is testing the claim that students at a new progressive high school are obtaining higher SAT scores than students at the old traditional school. Verbal and math scores for students from the two schools are as follows:

TRADITIONAL		PROGRESSIVE	
Verbal	Math	Verbal	Math
460	540	620	630
550	590	520	410
570	520	680	500
470	450	520	490
510	450	600	530
390	480	500	440
500	520	650	510
500	550	580	500

a. What is the value of R^2 for predicting group membership?

b. Test R^2 for significance at the .05 level.

c. Create a total SAT score for each student, and perform a t test between the two groups on the total score.

d. Compare the square of your t value in part c to the F ratio you used in part b. What does this comparison tell you about the advantage of multiple regression over simpler methods (e.g., using equal weights to form a sum of scores)?

*7. Suppose that a measure of academic motivation has a .6 correlation with the number of hours studied per week and that the latter variable has a correlation of .3 with GPA. What correlation between motivation and GPA would be consistent with a causal model that posits no direct effect of motivation on GPA?

8. Given that the correlation is .7 between the IQ of a sample of mothers and the IQ of their daughters, and that there is a .4 correlation between the IQ of the daughters and their high school grade averages (HSG), is a .5 correlation between the mothers' IQ and their daughters' HSG consistent with a causal model that posits no direct effect between the IQ of a mother and the HSG of her daughter? Explain.

9. List a pair of psychological variables for which you would expect a strong interaction, and describe the nature of that interaction.

10. Describe a realistic criterion and set of predictor variables, for which logistic regression would be the appropriate analysis.

The standardized multiple regression equation for two uncorrelated predictors:

$$z_{y'} = r_{yx_1}z_{x_1} + r_{yx_2}z_{x_2}$$

Formula 17.1

The multiple correlation coefficient for two uncorrelated predictors:

$$R = \sqrt{r_{yx_1}^2 + r_{yx_2}^2}$$

Formula 17.2

The standardized partial regression coefficients (beta weights) for two correlated predictors:

$$B_1 = \frac{r_{1y} - r_{2y}r_{12}}{1 - r_{12}^2} \qquad B_2 = \frac{r_{2y} - r_{1y}r_{12}}{1 - r_{12}^2}$$

Formula 17.3

The standardized multiple regression equation for two correlated predictors:

$$z_{y'} = B_1 z_{x_1} + B_2 z_{x_2}$$

Formula 17.4

The multiple correlation coefficient for two correlated predictors (in terms of the beta weights):

$$R = \sqrt{B_1 r_{1y} + B_2 r_{2y}}$$

Formula 17.5

The squared multiple correlation coefficient for two correlated predictors (in terms of the pairwise correlation coefficients):

$$R^2 = \frac{r_{1y}^2 + r_{2y}^2 - 2r_{1y}r_{2y}r_{12}}{1 - r_{12}^2}$$

Formula 17.6

The semipartial correlation (sr) between X_1 and the criterion, partialling out X_2 from X_1 (but not from the criterion):

$$r_{y(1.2)} = \frac{r_{1y} - r_{2y}r_{12}}{\sqrt{1 - r_{12}^2}}$$

Formula 17.7

The beta weight associated with X_1 in terms of its semipartial correlation:

$$B_1 = \frac{r_{y(1.2)}}{\sqrt{1 - r_{12}^2}}$$

Formula 17.8

The partial correlation coefficient between X_1 and Y, partialling out X_2 from both (in terms of the pairwise correlation coefficients):

$$r_{y1.2} = \frac{r_{1y} - r_{2y}r_{12}}{\sqrt{(1 - r_{2y}^2)(1 - r_{12}^2)}}$$

Formula 17.9

The t value for testing the statistical significance of a partial correlation:

$$t = \frac{pr\sqrt{N - V - 2}}{\sqrt{1 - pr^2}}$$

Formula 17.10

The t value for testing the statistical significance of a semipartial correlation:

$$t = \frac{sr\sqrt{N - P - 1}}{\sqrt{1 - R^2}}$$

Formula 17.11A

The F ratio for testing the statistical significance of a semipartial correlation:

$$F = \frac{(N - P - 1)sr^2}{(1 - R^2)}$$

Formula 17.11B

The F ratio for testing the statistical significance of multiple R:

$$F = \frac{(N - P - 1)R^2}{P(1 - R^2)}$$

Formula 17.12

Tolerance, in terms of the coefficient of cross correlation:

$$\text{Tolerance} = 1 - R_C^2$$

Formula 17.13

Estimate of R in the population based on the R found for a sample:

$$\text{Adjusted } R = \sqrt{R^2 - \frac{P(1 - R^2)}{N - P - 1}}$$

Formula 17.14A

Estimate of R^2 in the population based on the R^2 found in the sample, and the F ratio used to test its significance:

$$\text{Adjusted } R^2 = R^2\left(1 - \frac{1}{F}\right)$$

Formula 17.14B

The F ratio for testing the statistical significance of the change in R^2 when adding a block of K predictors to a model already containing $P - K$ predictors:

$$F = \frac{(N - P - 1)(R_P^2 - R_{P-K}^2)}{K(1 - R_P^2)}$$

Formula 17.15

THE REGRESSION APPROACH TO ANOVA

We will need to use the following from previous chapters:

Symbols:
s: Unbiased standard deviation of a sample
SS: Sum of squared deviations from the mean
r: Pearson's product-moment correlation coefficient
b: (Raw-score) slope of a regression line
η^2: Proportion of variance accounted for in ANOVA

Formulas:
Formula 10.5: The raw-score regression line
Formula 10.7B: The variance of residuals
Formula 17.10: The t test for pr
Formula 17.11B: The F ratio for testing sr
Formula 17.12: The F ratio for testing R
Formula 17.15: F ratio for testing the added variance of a block of predictors

Concepts:
The limitations of linear correlation
The assumptions of linear regression
The breakdown of SS in a one-way ANOVA

Chapter

$\mathcal{A}$

CONCEPTUAL FOUNDATION

As I hope you recall from Chapter 10, when you have two distinct groups, and you are comparing them with respect to some quantitative variable, you have a choice. You can perform a t test for independent samples, or you can calculate Pearson's correlation coefficient (which in this case is called r_{pb}). To perform the correlation you have to assign arbitrary numbers to the two groups, but this is not difficult. For several reasons, the easiest choice is to use 0 and 1. Now suppose you are dealing with three distinct groups, such as normals, neurotics, and schizophrenics, and you are comparing them on some psychological measure. You can test for differences with a one-way ANOVA (Chapter 12), but can you still find Pearson's correlation by assigning arbitrary numbers to the three groups, perhaps 0, 1, and 2? Of course you can, but the answer in this case won't be meaningful. The problem is that the size of r will depend on the order in which you assign the arbitrary numbers.

This is not the case with two groups. With two groups, r_{pb} is the same size regardless of which group is assigned 0; the *sign* of r_{pb} depends on which group is 0, but that is why we usually ignore the sign of r_{pb}. However, if we get one r when 0 = normal, 1 = neurotic, and 2 = schizophrenic and a different r when 0 = normal, 1 = schizophrenic, and 2 = neurotic, neither one can be correct. Perhaps, you think that the first of these orderings makes more sense, but what if the three groups were three different types of neurotics, such as obsessives, phobics, and depressives? More likely you are thinking Why should I care about performing a correlation at all when I can just perform a one-way ANOVA, if necessary? The answer is not a practical one; learning how to view the analysis of distinct groups in terms of correlations will deepen your understanding of both ANOVA and correlation and will prepare you for some more advanced topics, such as dealing with unbalanced factorial ANOVAs.

If your IV were quantitative, say three dosage levels of a drug or number of sessions per week of psychotherapy, the numbers associated with your different groups would be quantitatively meaningful, and you could use ordinary regression, or the trend analysis I described in Section C of Chapter 13. However, if the three levels of your IV are just different categories, like three different types of neuroses, you have the same situation I depicted in Chapter 12, Section C, in which I drew a broken regression line to connect the sample means of the three groups. The broken regression line was just a convenient device I used to show that the predicted score for any subject was the sample mean for that subject's group. Now I can show you how the same predictions can be made using the multiple regression methods of the previous chapter.

Dummy Coding

Assigning the numbers 0 and 1 (or any two convenient numbers) to two groups is most often called *dummy coding* (*dummy* because the numbers have no meaning and *coding* because numbers are being assigned to represent categories). As previously mentioned, when only two levels of dummy coding are used with a qualitative variable, the correlation is legitimate because the ordering of the codes does not affect the magnitude of the correlation. The trick for dealing with more than two groups is to use more than one dichotomous variable to make your predictions.

For instance, with three groups, two dichotomous IVs are needed. If we have normals, neurotics, and psychotics, one predictor (X_1) could be "presence of neurosis" and the second (X_2) could be "presence of psychosis." A normal subject would be coded as 0 on both, whereas neurotics and psychotics would be 1 on one of the predictors, and 0 on the other, as appropriate. No subject would be coded 1 for both predictors. Suppose the DV were willingness to seek psychotherapy (on a scale from 0 to 10) with the sample means as follows: normals = 3, neurotics = 7, and psychotics = 1. If both dichotomous predictors are entered into a multiple regression, the raw-score prediction equation would come out to be $Y' = 4X_1 - 2X_2 + 3$ (plug in 0 and 0, 0 and 1, and 1 and 0 for X_1 and X_2 to see that the predictions come out to the sample means). If we added affective or personality disorders as a fourth group, we would need to add a third dichotomous predictor. The number of predictors is always one less than the number of groups or categories (i.e., $k - 1$), which is equal to df_{bet} in a one-way ANOVA (this is not an accident).

The Regression Plane

In the case of three groups, we use two predictors, so instead of using a straight (or broken) line to predict the criterion, we are using a plane. Just as the scatterplot for a point-biserial correlation looks strange (points are arranged along two vertical lines), the scatterplot involving two dichotomous predictors is strange in three dimensions. Going back to the garden analogy in Section A of the previous chapter, imagine planting three sticks such that a right angle is formed on the ground (see Figure 18.1). All of the subjects' values on the criterion would be placed at different heights along the three sticks (one stick for each group). The regression plane would be angled so that it would pass through the mean of the points along each stick. That way the prediction for each subject is his or her own group mean, but we didn't need to use a broken line. Of course, adding a fourth group, and therefore a third predictor, means we would need four dimen-

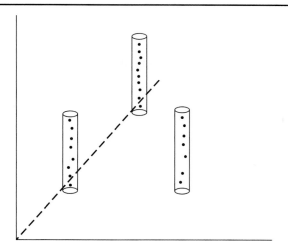

Figure 18.1

The Scatterplot
for the Data from
Three Groups

sions to picture the regression, but fortunately, picturing the regression surface is not essential for understanding the basic concept.

Effect Coding

In the preceding three-group example it was natural to code the "normal" category as zero on both predictors, which results in the sample mean of normal being the intercept of (i.e., the constant added to) the prediction equation. This system would also be reasonable any time you have a control group and several different treatment groups. The group that is coded 0 on all predictors is called the *base*, or *reference* group, and the slopes in the raw-score prediction equation are the differences in means between each group and the base group (in the equation $Y' = 4X_1 - 2X_2 + 3$, 4 is the difference between neurotics and normals, -2 is the difference between psychotics and normals, and 3 is the mean for normals). However, this coding system wouldn't make much sense for a study of obsessives, phobics, and depressives. Which group should be used as the basis for comparison?

A very useful form of coding, especially when none of the groups provides a basis for comparison, is *effect coding*. One group is chosen arbitrarily as the base and each of the other groups is coded 1 for its own variable and 0 for all of the others. Instead of being coded 0 on all variables, the base is coded as -1 for all variables. For instance, if you are comparing four groups of neurotics—obsessives, phobics, depressives, and hysterics—one possible effect coding scheme is shown in Table 18.1.

In this case the hysterics are the base group, but this was chosen arbitrarily. The advantage of effect coding is that the raw-score regression equation it produces when all the predictors are entered has two very convenient properties. First, the intercept of the equation turns out to be the grand mean of all the subjects (assuming equal-sized groups) rather than the mean

	X_1	X_2	X_3	**Table 18.1**
Obsessives	1	0	0	
Phobics	0	1	0	
Depressives	0	0	1	
Hysterics	-1	-1	-1	

of the base group. Second, the slope for each predictor equals the mean of the group it represents (i.e., the group coded 1 on that predictor) minus the grand mean. The base group, as in simple dummy coding, doesn't have its own predictor or slope in the equation, but its mean automatically equals the grand mean minus the slopes of the other predictors. For instance, if the means were obsessives = 7, phobics = 4, depressives = 8, and hysterics = 5, the regression equation would be $Y' = X_1 - 2X_2 + 2X_3 + 6$. (To get the mean of the obsessives, enter 1, 0, and 0 for X_1, X_2, and X_3, respectively, which yields $1 - 0 + 0 + 6 = 7$. Enter the codes for the other groups in Table 18.1, and you'll see that you get the correct means.)

When you use effect coding, the regression equation for four equal-sized groups always has the following form: $Y' = b_1X_1 + b_2X_2 + b_3X_3 + \mu$, where μ equals the grand mean. If we define the "effect" of each group as its mean minus the grand mean, each b in the preceding equation is the effect of the corresponding group. We can symbolize the effect of each group in the population as α_i and simplify the preceding equation to $Y' = \mu + \alpha_i$. This means that the predicted score for any subject is the grand mean plus the effect of the group he or she is in, which added together equal the mean for the group that the subject is in (although the base group does not have a corresponding b, it does have an alpha, which is equal to 0 minus the three other effects—together all four effects must sum to zero).

The General Linear Model

If we want an equation that represents the actual score of each subject rather than the score predicted for that subject, we have to add an error term (or residual) that is unique to each subject; it is the amount by which that subject differs from his or her predicted score. The equation becomes:

$$Y = \mu + \alpha_i + \epsilon_{ij} \qquad \qquad \textbf{Formula 18.1}$$

where i is the subject's group number and j is some arbitrary number that distinguishes subjects within each group. The reason I put the equation for subjects' scores into the form of Formula 18.1 is that this is the form most often used to represent the *general linear model* (GLM) when applied to ANOVA. The GLM is a way of expressing the theoretical model that is tested by analysis of variance. The null hypothesis of a one-way ANOVA can be stated as $\mu_1 = \mu_2 = \mu_3$, and so forth, or as $\alpha_1 = \alpha_2 = \alpha_3 = \cdots = 0$. The GLM makes it easy to see that the analysis of variance is just a special case of multiple (linear) regression. In other words, we don't need a separate system of statistical analysis to deal with groups; we don't need t tests or ANOVAs. Dummy or effect coding can convert any t test or ANOVA into a linear or multiple regression. The GLM may not seem terribly profound or useful in the case of a one-way ANOVA, but it becomes an increasingly helpful model as your ANOVA design grows more complex, especially if you are mixing categorical and quantitative predictors.

Equivalence of Testing ANOVA and R^2

Another way to see the connection between ANOVA and multiple regression is to look at the test for multiple R and compare it to the test for a one-way ANOVA, as I did in the previous chapter. The connection will be easier to see if I rewrite the F test for R^2 (Formula 17.12) in terms of η^2 (the proportion of variance in the DV accounted for by the IV in an ANOVA).

$$F = \frac{(N_T - k)\eta^2}{(k-1)(1-\eta^2)}$$

Formula 18.2

As you may recall, testing a one-way ANOVA for significance is equivalent to testing η^2 for significance. Had we performed a multiple regression with dummy-coded predictors, the multiple R^2 we would have obtained would be exactly equal to η^2, and you can see that Formula 17.12 yields exactly the same F ratio as Formula 18.2 when you realize that, for a one-way ANOVA, the number of dummy-coded predictors (P) is always equal to $k-1$ and N_T is just N in a multiple regression ($k = P + 1$, so $N_T - k = N - P - 1$).

Two-Way ANOVA as Regression

As soon as you add a second factor to your ANOVA design, you can begin to appreciate the simplicity of the ANOVA approach as compared with multiple regression. However, looking at a two-way ANOVA as a special case of multiple regression will deepen your understanding of both procedures and provide a framework for dealing with unbalanced designs. I'll begin with the simplest case: the 2×2 ANOVA. Imagine that you want to predict adult height for a 1-year-old infant based on two factors: gender and whether or not the infant has been breast-fed continuously since birth. If we are planning to analyze our results with a two-way ANOVA, we would probably try to select an equal number of babies in each of the four cells of our design.

Now imagine that you have equal cell sizes and the adult height of all of your subjects. If your emphasis is on finding out whether or not breast-feeding affects height, and whether it does so equally for both genders, the two-way ANOVA is the logical analysis. If your emphasis is on actual height predictions for each subject, multiple regression seems more appropriate. But it really doesn't matter because, using effect coding, the predictions from multiple regression will turn out to be the cell means of the ANOVA, and the significance tests for each predictor will correspond to the significance tests of the ANOVA.

With dichotomous variables, the effect codes are always -1 and $+1$. If we are coding gender (X) and our outcome is adult height (Y), it makes sense to code female as -1 and male as $+1$. With only gender, the prediction equation would be something like: $Y' = \mu + \alpha = 67 + 2X$, so $Y' = 67 - 2 = 65$ for females and $Y' = 67 + 2 = 69$ for males (the raw-score slope, b, is 2, and the grand mean, or the mean height for all humans, is 67). Anticipating that breast-feeding will increase height, I will code "not breast-fed for the first year of life" (nBF) as -1 and "breast-fed" (BF) as $+1$. If the multiple regression returns a slope that is negative, I will know that I was wrong about the direction I picked.

The Interaction Term

If we use only these two dichotomous predictors, however, we will not be able to detect an interaction; the sum of squares for the interaction will end up as part of the error sum of squares. To capture the sum of squares due to any interaction between the two factors, we need a third predictor. As mentioned in Section C of the previous chapter, the appropriate predictor for the interaction is just the product of the two factors: gender $\times$ feed (I will call the breast-feeding factor "feed," for short). Each subject has values on three predictors, which are the appropriate codes for his or her cell.

The codes for the interaction predictor are found by simply multiplying the codes for the other two predictors, as shown in Table 18.2. (*Note:* The advantage of effect coding is that the correlation between each factor and the interaction is zero—for a balanced design—which is not the case for other coding schemes.)

Table 18.2

	X_1	X_2	X_1X_2
Female/nBF[a]	−1	−1	+1
Female/BF[b]	−1	+1	−1
Male/nBF	+1	−1	−1
Male/BF	+1	+1	+1

[a]nBF = not breast-fed
[b]BF = breast-fed

By entering all three predictors along with each subject's actual adult height (as the criterion) into a multiple regression procedure, you will be able to get the same results you would get from performing the corresponding two-way ANOVA. Suppose the cell means for the actual heights turned out to be those shown in Table 18.3. (Note that I built in a small amount of interaction such that breast-feeding has a greater effect on females than males.)

Table 18.3

	Female	Male	Mean
nBF	65	69	67
BF	67	70	68.5
Mean	66	69.5	$\overline{X}_G = 67.75$

The GLM for the Two-Way ANOVA

The multiple regression equation would be $Y' = 67.75 + 1.75X_1 + .75X_2 - .25X_1X_2$. If you substitute the codes in Table 18.2, the Y' for each cell will be its cell mean as given in Table 18.3. The general (GLM) equation for any two-way ANOVA is:

$$Y = \mu + \alpha_i + \beta_j + \alpha_i\beta_j + \epsilon_{ijk}$$

Formula 18.3

where i is the level of the first factor, j is the level of the second factor, and k is an arbitrarily assigned number representing the kth subject in one of the cells. The βs, of course, are the effects of the different levels of the second factor, and the $\alpha\beta$s are the interaction effects of each cell.

Getting back to the equation for our example, each slope can be tested for statistical significance. The t value you get for testing 1.75 in the preceding equation is equivalent to testing the main effect of gender in the two-way ANOVA; just square the t value to get the F ratio for the gender main effect. Similarly, testing .75 in the preceding equation is equivalent to testing the main effect of the feed factor, and testing the $-.25$ slope tells you whether the interaction is statistically significant.

Two-Way ANOVA with Multiple Levels

Unfortunately, this simple correspondence between the slopes of predictors and the effects of the two-way ANOVA breaks down when there are more levels of each factor. As in the one-way ANOVA, more levels means more predictors. A 4×2 ANOVA (e.g., four diagnostic groups by gender) requires only one predictor for the two-level factor but three predictors for the main effect of the four-level factor, and three more predictors for the interaction, for a total of seven predictors (e.g., the gender variable would be multiplied by each of the three diagnostic group predictors to form the three interaction predictors). Testing the slope of the gender predictor would corres`ond to a test of the main effect of gender, but the main effect of diagnostic group would be spread over three predictors, as would be the interaction. So to test the interaction you would have to rerun the multiple regression without the three interaction predictors and find the change in R^2 between the two regressions (i.e., with all the predictors included versus all predictors except for the interaction predictors). The F ratio you would use to test the change in R^2 (from Formula 17.15) would be identical to the F ratio for the interaction in the corresponding two-way ANOVA. Similarly, the main effect of diagnostic group would be tested by testing the change in R^2 with and without the three diagnostic group predictors in the regression model.

A 3×5 ANOVA would require 14 predictors: two for the first factor, four for the second factor, and eight (2×4) for the interaction; the number of predictors required to test an effect is the number of degrees of freedom for that effect. In general, a $J \times K$ ANOVA requires a total of $J \times K - 1$ predictors (one less than the number of cells), which is the overall df_{bet} for a two-way ANOVA. Adding factors can add complexity very quickly. A $3 \times 3 \times 3$ ANOVA requires $3 \times 3 \times 3 - 1 = 26$ predictors, so you can see why the ANOVA model can be so much more convenient than multiple regression when all the predictors are categorical.

The GLM for Higher-Order ANOVA

The full general linear model for an ANOVA with three factors is $Y = \mu + \alpha_i + \beta_j + \gamma_k + \alpha_i\beta_j + \alpha_i\gamma_k + \beta_j\gamma_k + \alpha_i\beta_j\gamma_k + \epsilon_{ijkl}$. As mentioned, the number of predictors needed to represent this model can get quite large. However, as I have already shown for the two-way case, when all the factors are dichotomous, the multiple regression model is no more complicated than the corresponding three-way ANOVA model. When all of the factors are dichotomous, each can be represented by a single dummy variable that serves as a predictor. If you code each predictor as -1 and $+1$, and form the appropriate product of predictors to represent each two-way and the three-way interaction, you can replace the $2 \times 2 \times 2$ ANOVA with seven predictors in a multiple regression, with the test of each predictor corresponding to one of the F ratios of the ANOVA. In general, an ANOVA consisting of N factors, all of which are dichotomous, requires $2^N - 1$ predictors. The correspondence between ANOVA and regression is simple, however, only when the design is balanced—that is, when there are the same number of scores in every cell of the design. When the design is not balanced, there is more than one way to perform the ANOVA, and these different ways are best understood in terms of their corresponding regression models. To discuss the analysis of unbalanced factorial designs, I will deal only with the simplest case—the two-way ANOVA.

Analyzing Unbalanced Designs

The first step in analyzing a two-way design is to divide SS_{total} into two components: SS_{bet} (as though all the cells of the two-way were different groups in a one-way analysis) and SS_W (the average of all the cell variances). The second step is to divide SS_{bet} into three components corresponding to the two main effects and the interaction. In Chapter 14 there was no need to

Figure 18.2

Orthogonal Effects for a
Balanced Two-Way
ANOVA

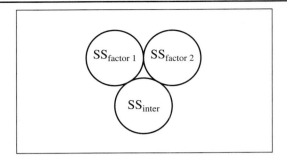

calculate SS_{inter} separately; it could be found by subtracting the SS components for the two main effects from SS_{bet}. We could do that because we were dealing with balanced designs. When a two-way design is balanced, the three components of SS_{bet} are mutually orthogonal. Now that you know a little bit about multiple regression I can make that statement more meaningful. One way to understand this orthogonality is to consider a 2×2 ANOVA with effect coding. Each subject will have values for the three predictors (-1s and/or $+1$s, according to which cell he or she is in) and a value for the DV. If you calculate the correlation coefficient for each possible pair of predictors over all of the subjects, it will be zero—but only if the cell sizes are either all equal or proportional (this latter possibility is a little tricky, so I will postpone its discussion just a bit). Because the predictors are mutually orthogonal (i.e., uncorrelated), we can add their squared validities (r_{pb}^2, in each case) to attain the R^2 for the set. Each r^2 multiplied by SS_{total} corresponds to an SS component of the ANOVA, and their total equals $R^2 SS_{total}$, which equals SS_{bet} for balanced or proportional designs. This mutual orthogonality can be shown in the form of a Venn diagram, like those in the previous chapter, but this time I will multiply all of the r^2s by SS_{total} so that the diagram shows the SS components of a two-way ANOVA. Designs that are balanced or proportional yield diagrams with no overlap of the circles, as shown in Figure 18.2.

There are three main reasons why a psychologist may end up with an unbalanced design. The first, and by far the most common, reason is that an experiment was planned with equal-sized cells, but subjects were lost at a point when it was too late to replace them (e.g., a subject may have dropped out just before the end of a long-term study, or during the data analysis stage, it is discovered that a subject's data are unusable for some reason). In the past, researchers would try to replace the missing data with estimated data or randomly drop other subjects just to balance the design. In recent years, however, statistical software has made it easy to analyze unbalanced designs, so there is less motivation to drop subjects or replace missing ones. The second reason for unbalanced designs is convenience. You may be dealing with natural groupings of subjects (e.g., students in classes or patients at a mental hospital who fall into different diagnostic categories), and you may want to use all of the subjects available in each group (dropping subjects at random to equalize cell sizes reduces the power of your statistical test).

Proportional Coding

The third reason is the least common: Unequal cell sizes are planned to reflect differences in the population. This reason does not apply to IVs that involve experimental manipulations but rather to IVs that deal with existing differences between people. If one of your factors is handedness, it can make sense to sample 5 times as many right-handers as left-handers to reflect the population. If your other factor is gender and all of your subjects are selected for having dyslexia, you may want to sample 3 or 4 times as many boys as girls. If you're going to plan unequal cell sizes in a two-way design, there is an advantage to making the cell sizes proportional. Table 18.4 shows proportional cell sizes for a gender/handedness study of dyslexia.

	Girls	Boys	Total
Left-handed	5	15	20
Right-handed	25	75	100
Total	30	90	120

Table 18.4

Notice that the left-handed/right-handed ratio of cell sizes is 1 to 5 for *both* girls and boys. Also, the girl/boy ratio is the same (1 to 3) for *both* left- and right-handers. This equality of proportions across rows and columns is the definition of a proportional design. With such a design, the main effects are not correlated; one main effect cannot dribble over into the other, as it did in the unbalanced design of Table 14.10. However, if you use ordinary effect coding, the main effects will generally be correlated with the interaction as depicted in Figure 18.3. (Bear in mind that the areas of overlap can be negative—although this is impossible to draw—in which case SS_{bet} could actually be larger than the sum of its parts.) The advantage of planning proportional cell sizes is that the overlap shown in Figure 18.3 can be avoided by simply using proportional coding instead of effect coding. For the design shown in Table 18.4 you would start with effect coding and then multiply the codes for all the girls by 3 and the codes for all the left-handers by 5. Now the codes for each predictor will add up to zero. As usual, the interaction predictor would be the product of these codes. Proportional designs and coding can be extended to more complex ANOVA designs, but this is rarely done.

The Regression Approach

In the more usual case of the unbalanced design, the cell sizes are not proportional, and the main effects are correlated with each other as well as with the interaction, as shown in Figure 18.4. (Again, note that any of the overlaps can be negative due to complementary relationships.) There is no form of coding that will eliminate these overlaps. The problem is how to assign the overlap areas. For example, if a main effect overlaps with the interaction, which gets custody of the overlap area? The simplest solution is to ignore all of the overlapping areas and count only the parts of an effect

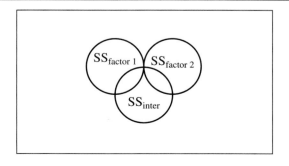

Figure 18.3

The Possible Overlap of Effects in a Proportional Design If Proportional Coding Is Not Used

Figure 18.4

The Possible Areas of
Overlap in an
Unbalanced Two-Way
ANOVA

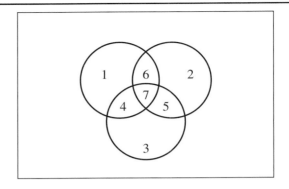

that are unique—i.e., not shared (areas 1, 2, and 3 in Figure 18.4). You may
recognize this solution as a form of simultaneous regression, which was
described in the previous chapter. For a 2×2 design you simply enter all
three predictors into the model and test the partial slope or semipartial correlation of each. These significance tests will exactly match the results you
would get from the unweighted-means analysis described in Chapter 14,
Section C. For a multilevel two-way design you would have to run the
regression with and without each set of predictors and test the change in R^2
to test the significance of each effect. Conveniently, the simultaneous regression approach (usually referred to just as *the regression approach*) is the
default for SPSS and other statistical software when you use the general
linear model (GLM) to perform the ANOVA (it is called Type III sum of
squares in both SPSS and SAS).

The simultaneous regression approach treats each cell as though it had
the same number of subjects, and therefore gives it the same weight as
all other cells. This approach makes the most sense when equal cell sizes
were planned, but data have been lost randomly. It should be noted, however, that if data have been lost for reasons related to your experimental
treatments (e.g., subjects drop out of the placebo group because they are
not getting relief), you can lose the random assignment that is critical to
drawing conclusions from your experiment (e.g., the sickest subjects drop
out of the placebo group but not the drug group). There may not be any
coding or statistical trick that can give you valid results.

Method II Approach

If your different-sized cells are chosen by convenience, you can use the simultaneous regression method (designated Method I by Overall and Spiegel,
1969), but more often than not the size of convenient samples reflects underlying differences in the population. Some coding system that weights the cells
according to the subpopulations they represent may be more appropriate,
but in practice such weights are rarely used. The chief alternative to simultaneous regression for unbalanced designs is some form of sequential or hierarchical regression. The regression designated as Method II by Overall and
Spiegel (1969), and Type II sums of squares by SPSS, ignores the overlap
between the two main effects, but assigns to each main effect its overlap with
the interaction. In terms of Figure 18.4 the main effects are assigned areas
1 and 4, and 2 and 5, respectively, and areas 6 and 7 are ignored. This method
is not controversial when applied to the proportional case, because the main
effects don't overlap, or when applied to a case in which there is little interac-

tion, but for many unbalanced designs the results of Method II are not easy to interpret. Consequently, this method is rarely used.

The Hierarchical Approach

Method III of Overall and Spiegel (which is paradoxically labeled Type I sums of squares in both the SPSS and SAS statistical packages) entails a complete hierarchical regression. The order in which the two main effects are entered must be specified, but the interaction is entered last, as it is in Method II. If main effect A is entered first, its sums of squares will consist of areas 1, 4, 6, and 7, and main effect B will consist of areas 2 and 5. If effect B is entered first, it will grab areas 6 and 7 for itself, leaving only areas 1 and 4 for A. In all three methods the interaction always comes out to be area 3 in Figure 18.4.

If the main effects are uncorrelated, as in a proportional design, Methods II and III produce the same results. Method III has the convenient feature that the sums of squares for the three effects add up exactly to SS_{bet}, even when the design is not balanced or proportional. The disadvantage of this method is that you must come up with some rationale for deciding which main effect to enter first. Consequently, this method almost never makes sense when both factors involve experimental manipulations. If, however, one of the factors is based on individual differences that are considered a nuisance, it may make sense to add that effect first. If both factors are based on individual differences, one may take precedence based on the chain of causality. Suppose you are studying the effects of birth order (first-born/later born) and introversion (introvert/extrovert) on a child's moral development. An association between the two factors can make it difficult to balance the four cells. In a hierarchical analysis of an unbalanced design it would make sense to enter birth order first; some of the variance across levels of introversion may actually be the result of birth order, but obviously the reverse is not possible.

Methods for Controlling Variance

In ANOVA, as in multiple regression, there is a single dependent variable on which subjects vary. The goal is to explain or account for as much of this variance as possible. Some of this accounted-for variance may be *produced* by one of your independent variables (e.g., giving drugs to some subjects but not others), and some is just explained (e.g., gender), but in any case, the leftover variance is considered error, and we want it to be as small as possible. Any variable that affects the magnitude of your dependent variable, but is otherwise not interesting to you (e.g., reaction time is affected by age, but you may be interested only in different characteristics of the stimulus), can be considered a nuisance variable. Nuisance variables are not so bad once you have identified and measured them because you can then take the variance due to the nuisance variable (I'll call this nuisance variance) out of the error pile and put it into the explained pile, or just throw it away. There are several ways to do this in the context of ANOVA.

Perhaps the simplest way to remove nuisance variance from your error term is to screen your subjects to control for the nuisance variable. If age affects your DV, you can use only subjects that fall in a narrow age range. If smoking affects your DV, use only nonsmokers. The obvious disadvantage of this method is that you cannot generalize your results beyond the narrow range of subjects you allowed in your study. Probably the best way to remove nuisance variance in general is not to restrict the range of your subjects but to use the same subject in all conditions; unfortunately, there are many practical situations in which this is not possible, such as comparing methods for treating patients for neurotic symptoms or teaching students a particular skill. Note that even when repeated measures are possible, the traditional RM ANOVA may not be the most powerful way to analyze the data (I will return to this point shortly).

Matching subjects is often the next best method. You can usually match subjects on several nuisance variables at once, although the precision of the matching is likely to be less for each nuisance variable as you add more. Unfortunately, this method becomes increasingly cumbersome as the number of levels of the IV increases; matching four or five subjects together at a time is not trivial. The biggest problem with this method, however, is that you may not know who all of your subjects will be at the beginning of the study. If your subjects are to be the next 30 patients to enter a clinic, matching subjects as you go along is not feasible. Matching retroactively at the end of the study is awkward and rarely attempted.

A reasonable alternative to matching subjects is to create "blocks" of subjects who are similar on some nuisance variable and then treat the blocks as different levels of an additional factor in your ANOVA. If the nuisance variable is categorical, it can be simply added as a factor to the ANOVA, just as I added gender to create a two-way ANOVA in Section A of Chapter 14 (recall that the error term of the F ratio for the other factor was reduced as a result).

If the nuisance variable is quantitative, it can be broken somewhat arbitrarily into ranges to create blocks of subjects. If the blocks are specified at the beginning of the experiment, subjects can be sampled so that all blocks have the same number of subjects and so that the number of subjects in a block is a multiple of the number of levels of the factor of interest. That way, subjects can be randomly assigned to the conditions, with an equal number of subjects in each condition from each block. Treatment by block designs are very useful and powerful, but with quantitative nuisance variables you are sometimes throwing away valuable information by reducing your quantitative measure to a few arbitrary categories. Although blocks can be created after an experiment has been completed, in such cases the design is not likely to come out balanced.

Quantitative nuisance variables are usually called *covariates*. If a covariate has a strong linear relationship with the dependent variable, and the factor of interest is categorical (e.g., treatment groups), it is likely that the most powerful statistical procedure you can use is the *analysis of covariance*. For some designs, this method can also be used as a more powerful alternative to the traditional repeated measures analysis. This popular method of statistical analysis is the focus of Section B.

SUMMARY

1. An independent variable with k qualitative levels can be represented by $k - 1$ dichotomous predictors by the use of *dummy coding*. Each level of the IV is assigned 0 on all but one of the predictors, and 1 on its own predictor. The exception is that one group, usually a control group, has no predictor of its own; it is assigned 0 on all of the predictors.

2. *Effect coding* is similar to dummy coding, except that one group is selected arbitrarily to be scored -1 on all predictors. Effect coding makes more sense when there is no natural control or reference group and has the added advantage (when the groups are all the same size) that the raw-score slope associated with each predictor represents the "effect" of the group corresponding to that predictor (the effect of a

group is the difference between that group's mean and the grand mean). The *general linear model* for the one-way ANOVA expresses each subject's score as the sum of the population grand mean, the effect of that subject's group in the population, and an error term unique to that subject.

3. Performing a multiple regression on the DV using all $k - 1$ predictors is equivalent to performing the ordinary one-way ANOVA. The R^2 from the multiple regression will always be exactly the same as eta squared from the corresponding one-way ANOVA. Therefore, testing R^2 for significance yields the same F ratio as testing eta squared for significance, which yields the same F as the ordinary ANOVA.

4. Performing a 2×2 ANOVA with multiple regression requires three dichotomous predictors to account for both main effects *and* the interaction (for N dichotomous factors, $2^N - 1$ predictors are required). The interaction predictor is found by simply multiplying the predictor values for the two factors. If either factor has multiple levels, the interaction will require multiple predictors (e.g., in a 3×5 ANOVA, the interaction is represented by $2 \times 4 = 8$ predictors), which can be entered as a block to find the change in R^2. In general, a $J \times K$ ANOVA requires $J \times K - 1$ predictor variables to test all three F ratios.

5. The general linear model for a three-way ANOVA consists of the sum of the grand mean, the three main effects, the three two-way interaction effects, the three-way interaction effect, and an error term unique to the individual subject.

6. When a two-way ANOVA is balanced, the variance explained by any one effect will not overlap with the variance explained by any other effect. Alternatively, if the cell sizes of the two-way design follow the same proportions across each row and down each column (so as to be proportional to corresponding segments of the larger population), proportional coding can be used to make the effects orthogonal (i.e., not overlapping).

7. Unfortunately, designs often become unbalanced through the random loss of subjects or the use of convenient samples, in which case there are three choices for analyzing the results.

   a. *Method I* (Type III *SS*) ignores the areas of overlap and assigns to each effect its unique portion of the total *SS*. This is equivalent to a simultaneous regression entering all three effects and is therefore most often referred to as the *regression approach* (or, sometimes, the *analysis of unique sources*).

   b. *Method II* (Type II *S*) ignores the overlap between the two main effects but assigns to each main effect its overlap with the interaction. This method has a straightforward interpretation for proportional designs but can be difficult to interpret for other unbalanced designs and is therefore rarely used.

   c. *Method III* (Type I *SS*) is hierarchical; areas of overlap are assigned according to the order in which the main effects are entered, with the interaction always entered last. This method makes the most sense when there is a rationale for entering one main effect before the other, as when one effect is a nuisance variable or takes precedence causally.

8. The power of an ANOVA can be increased without increasing the sample size if the error term can be reduced. This reduction can be accomplished by restricting the sample (which reduces generalizability), matching subjects beforehand (when possible), or grouping subjects into blocks on some nuisance variable. However, if the nuisance variable has a strong linear correlation with the DV, a more powerful method than blocking is the analysis of covariance.

1. Subjects' responses in a motor learning experiment are measured under one or another of three conditions: rewards for correct responses, punishments for incorrect responses, and neither rewards nor punishments (control).
   a. Write out the dummy codes for a set of predictors to analyze the results by multiple regression.
   b. Write out the regression equation, given that the group means are as follows: rewards = 45, punishments = 40, control = 28.

*2. a. An experiment is designed to test four different antidepressant drugs (e.g., Elavil, Prozac, Zoloft, and Celexa) against each other and against a placebo. Using dummy coding, write out the codes for a set of variables that could be used to test the means of the five groups by multiple regression.
   b. Recent immigrants from five different countries (e.g., Japan, Greece, India, Mexico, and Egypt) are being compared in terms of blood chemistry changes after being placed on a typical U.S. diet for 6 months. Given that effect coding will be used to analyze the data, write out the codes for one possible set of dummy predictors to analyze this experiment by multiple regression.
   c. If the means for the experiment in part b turned out to be $\overline{X}_{Japan} = 16$, $\overline{X}_{Greece} = 9$, $\overline{X}_{India} = 7$, $\overline{X}_{Mexico} = 12$, and $\overline{X}_{Egypt} = 11$, write out the regression equation that fits with your effect coding from part b.

3. a. Find η^2 for the ANOVA in Exercise 12B4. Test that η^2 for significance using Formula 18.2, and compare the F ratio for testing η^2 with the F ratio you found for Exercise 12B4.
   b. Repeat part a for the ANOVA in Exercise 12B8.

*4. a. Consider an experiment in which political attitudes are being measured for liberals, conservatives, and middle-of-the-roaders from four regions of the United States (Northwest, Southwest, Northeast, and Southeast). Write out the effect codes corresponding to a balanced two-way ANOVA for this experiment that does *not* include the interaction.
   b. Imagine that the experiment in part a involves only two types of regions: urban/suburban and rural. Write out the effect codes for analyzing this 3×2 ANOVA, including the interaction.

5. Suppose the cell means for heart rate from a 2×2 study are as follows: nonsmoker/nonexerciser = 70, smoker/nonexerciser = 80, nonsmoker/exerciser = 60, smoker/exerciser = 66. Write out the effect codes for this design, and write out the regression equation based on those codes and the cell means given.

*6. a. To analyze a $2 \times 2 \times 2 \times 2$ ANOVA, and test for all possible interactions, how many dummy predictors would be required? How many such predictors would be required for a $2 \times 3 \times 4 \times 5$ ANOVA?
   b. Write out the general linear model for a four-factor ANOVA.

7. Suppose that at a large university there are 500 psychology majors: 50 freshmen, 100 sophomores, 150 juniors, and 200 seniors. Sixty percent of the psychology majors are female, and 40% are male.
   a. If you are going to sample 100 of these students for a two-way ANOVA, in which attitudes toward psychology as a career are measured as a function of class year and gender, how many subjects should be in each cell of the design so that proportional coding can be used?
   b. Suppose that the subjects are grouped into lower- (freshmen and sophomores) and upper-year (juniors and seniors) students. How many subjects should be sampled in each cell of the resulting 2×2 design if the total number of subjects is to be 50?
   c. What codes could you use for the cells of the design in part b if you want your coding to be proportional?

*8. Imagine that a two-way unbalanced ANOVA design in which the factors are self-esteem (low or high) and task difficulty (easy, hard, impossible) results in overlapping sums of squares as shown in the Venn diagram in Figure 18.5. Assume that the error term for all effects is 50.
   a. Find the F ratio for each of the three effects using Method I, the (simultaneous) regression approach.
   b. Find the three F ratios using Method II.
   c. Find the three F ratios using Method III (hierarchical regression) if self-esteem is entered first.

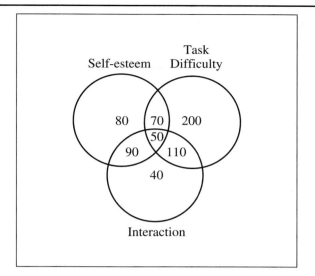

Figure 18.5

Overlapping Areas for
Exercise 18A6

The analysis of covariance (ANCOVA) is a general method that can be applied to virtually any ANOVA design, but in this section I will present in detail only the simple case: one factor, one covariate, and no repeated measures. As in any ANOVA, the factor can be a truly independent one with subjects randomly assigned to the different levels (or treatments) or a quasi-independent (observational) one on which subjects have already been selected either by themselves (e.g., vegetarians vs. nonvegetarians) or by nature (e.g., gender). The statistical procedures are the same for both types of factors, but, as you will see, the second case presents a number of difficulties for interpreting the results.

There are a number of ways to describe an ANCOVA, but the most common and traditional way is as a mixture of an ordinary ANOVA and simple linear regression. The traditional description is a good one in that it provides a context for understanding the various assumptions that underlie ANCOVA and for viewing the mechanics of the process. I will get to that shortly. However, now that you have learned the basic concepts of multiple regression, there is an easier way for me to explain ANCOVA to you.

B

**BASIC
STATISTICAL
PROCEDURES**

Simple ANCOVA as Multiple Regression

In the simplest possible ANCOVA, there are only two groups (i.e., one IV with two levels), one DV and one covariate. For example, a researcher may be testing a new drug that is supposed to improve short-term memory in subjects over 70 years old. The IV (X_1) is the grouping variable: drug group versus placebo group. The DV could be the number of words recalled from a test list (Y), and the covariate, age (X_2). Of course, subjects could have been matched for age at the outset of the study, but one advantage of ANCOVA is that it doesn't have to be planned. It is possible that our researcher was obtaining subjects a few at a time over the course of a year but was wise enough to collect demographic information on each subject, including age. Let's suppose that the correlation between age and memory (r_{2y}) is $-.5$. If the drug works fairly well, the correlation between the IV and the DV (r_{1y}) could come out to $+.3$ (note that this is a point-biserial correlation and we can make it come out positive by the way we assign dummy codes to the two groups). There is one more correlation to consider: the point-biserial correlation between the IV and the covariate (r_{12}). If we are careful to assign our subjects randomly, there is no reason to expect one group to be older on average than the other, but, of course, some small age difference is likely by chance. To keep this example as simple as possible, I will assume that r_{12} is so tiny that it rounds off to zero. Thus, there is no overlap between age and the grouping variable in predicting the DV.

As you should recall from the previous chapter, the multiple R obtained from predicting the DV from both the IV *and* the covariate, given the correlations I just described, will be higher than the correlation of just the IV with the DV. Of greater relevance to ANCOVA is the fact that the semipartial correlation between the IV and DV (with the covariate included), even if no larger than the validity of the IV alone, can have a greater chance of attaining statistical significance. To show how this can happen, I will begin by testing r_{1y} alone for statistical significance. Without the covariate we would test r_{1y} like any other correlation. I will use the F form of the test to be consistent with the ANCOVA approach. If there are 18 subjects in each group for a total N of 36, the test comes out as follows:

$$F = \frac{(N-2)r^2}{(1-r^2)} = \frac{34(.09)}{.91} = \frac{3.06}{.91} = 3.36$$

The critical F for 1 and 34 df is about 4.13 at the .05 level, so the group/memory correlation of .3 is not significant by itself. (This is the same result you would get by performing a pooled-variance t test between the two groups on the memory scores and squaring to get an F ratio.) However, when we add the covariate, it is appropriate to test r_{1y} as a semipartial correlation using Formula 17.11B. Because there is no overlap between the IV and the covariate, the appropriate sr is the same as r_{1y}, and therefore sr^2 = .09. Also, because there is no overlap, R^2 for predicting the DV from both group and age equals $(+.3)^2 + (-.5)^2 = .09 + .25 = .34$. The total number of predictors (P) is 2. The F ratio for testing the semipartial group/memory correlation therefore is:

$$F = \frac{(N-P-1)sr^2}{(1-R^2)} = \frac{(36-2-1).09}{1-.34} = \frac{2.97}{.66} = 4.5$$

The critical F for 1 and 33 df is about 4.14 at the .05 level, so this time the correlation (and therefore the difference of the groups) is statistically significant.

ANCOVA Reduces the Error Term

The group/memory correlation did not change with the addition of the covariate, so why did its significance level effectively change? The answer can be seen by comparing the test for r with the test for sr. If there is no IV/covariate correlation, r_{1y} and its corresponding sr will be the same, so the two numerators will be almost the same—except for losing one extra degree of freedom for the covariate in the sr test. This tends to make the F ratio *smaller* for the sr formula. Fortunately, as N gets very large, the loss of one df has little effect, but this comparison points out that there is a price to pay for adding covariates, especially when the sample size is small. On the other hand, the denominator of the sr formula contains R^2 instead of r^2 (or sr^2), so to the extent that the covariate is correlated with the DV, a greater portion will be subtracted from the denominator for the sr than the r test, tending to make the F ratio for sr (i.e., the F ratio for ANCOVA, in this simple case) larger than the F ratio for r_{1y} (i.e., the F ratio for the ordinary ANOVA). With a reasonable DV/covariate correlation and a reasonable sample size, the increase in F due to R^2 being larger than r^2 will easily outweigh the decrease due to losing a degree of freedom. For experimental studies this is the most important advantage of ANCOVA.

The decrease in the denominator of the test for sr relative to the test for r occurs because your covariate is allowing you to take some of the error variance in the dependent variable (i.e., variance not accounted for by the IV) and remove it from the error term of your ANOVA—this portion is now accounted for. In terms of our example, knowing the subjects' ages allows you to adjust your memory measure accordingly; the same number of words recalled becomes a higher score for an older subject. Once adjusted, subjects' scores won't vary because of memory changes related to age, and therefore memory changes due to the drug can more easily attain significance. (Age acts in a way that is analogous to a suppressor variable in multiple regression.)

ANCOVA Adjusts the Group Means

In real applications there will usually be some overlap between the IV and the covariate. This overlap can be positive or negative, so sr may come out smaller or larger than r_{1y}, which can hurt or help the F ratio. But even if the covariate seems to be helping us, we really don't want this kind of help. We want the difference in means for the dependent variable to be a reflection of our IV only, neither helped nor hurt by the covariate. To make it clear how the covariate can help in a way that is quite misleading, I'll use a simple example. Suppose a random sample of men is found to have an average foot length of 12 inches, and a random sample of women averages 10 inches. Most of this 2-inch difference can be accounted for by the fact that men are taller, but perhaps not all. If you want to know if men have bigger feet beyond the height difference—that is, do men tend to have larger feet than women who are the same height?—you can use height as a covariate when comparing foot length. One result of performing an ANCOVA is an adjustment of the group means. If it turns out that the 2-inch foot-length difference between the genders can be accounted for entirely by height, the adjusted foot-length means (controlling for height) would be identical, and the F ratio for the ANCOVA would be zero. ANCOVA is thus telling us that the 2-inch foot-length difference has nothing to do with the size of men's and women's feet in proportion to their body size.

It could also turn out that the man usually has the larger foot when he is compared to a woman of the same height, but that the difference between the adjusted means is only half an inch. In that case, the true (i.e., proportional) foot-size difference is being helped by the covariate (i.e., height difference), and we don't want that kind of help. One reason we do ANCOVAs is that we want to find the true group differences for our variable of interest—not confounded by other variables—and we want our F ratio to reflect that true difference when testing it for statistical significance. Unfortunately, the lower the correlation between the DV and the covariate, the less accurate is the adjustment of the means. This inaccuracy gets worse as the groups differ more on the covariate. One of the reasons it can be problematic to deal with intact rather than random groups (as is the case with gender) is that there can easily be large differences on the covariate, and therefore misleading adjustments, when the DV/covariate correlation is not very high. I will deal with the problems raised by intact groups at the end of this section.

Of course, there are also examples in which differences on the covariate are tending to decrease the mean differences for your dependent variable. Imagine that two regular math classes from the same high school have been chosen to participate in an experiment; one class is taught basic probability in the traditional manner, and the other is taught in terms of some new method. At the end of the experiment, the mean test scores for the new method are somewhat higher, but the results are not significant. Then it is discovered that, just by chance, the class given the traditional method had a higher average on last year's math grades and that there is a strong correlation between last year's grades and the probability test scores. Thus, initial differences between the two groups are acting against the new method. An ANCOVA would raise the F (or t) for this experiment in two ways. First, the mean test scores would be adjusted: down for the traditional method group (to compensate for their apparently higher math ability) and up for the new method group. This adjustment would increase the difference between the two group means. Second, the error term would be reduced by removing variance attributable to last year's math grades. The ANCOVA can be viewed as a type of partial correlation; the nuisance variable or covariate (e.g., age, initial math ability) is correlated with both the dependent variable and the independent variable (to the extent that the groups differ on the covariate). I will come back to this view near the end of this section.

The Linear Regression Approach to ANCOVA

The multiple regression approach can certainly get a bit tedious when the IV has many levels. On the other hand, analyzing an ANCOVA design in terms of ordinary ANOVA combined with linear regression is both educational and useful. The next example will involve an IV with three levels, and I will demonstrate the traditional approach to ANCOVA (except that I will not use the traditional computational formulas, opting instead for versions that are more instructive).

I will illustrate the use of ANCOVA in terms of an experiment in which two types of therapy are compared with a control group for the treatment of acrophobia (fear of heights). Subjects are randomly assigned to psychodynamic therapy, behavioral therapy, or a control condition in which subjects read inspirational stories about mountain climbing, hot-air ballooning, and so forth. At the end of 6 months, each subject's phobia is measured in terms of how many rungs up a tall, outdoor ladder he or she is willing to climb, so higher numbers indicate less phobia and presumably greater progress due to the therapy (or control) condition. It would be reasonable to match the subjects in threes on phobic intensity at the beginning of the experiment, but we will assume that the subjects were run as they entered the phobia clinic. As an alternative to matching, various covariates could be measured (e.g., degree of neuroticism, subjective rating of phobic intensity), but arguably the best covariate is an initial measure of phobic intensity on the same scale as the dependent variable (i.e., number of rungs climbed on the ladder). Due to a few random drop-outs from our hypothetical experiment, the sample sizes are slightly unequal. The data are shown in Table 18.5.

The initial phobia scores serve as the covariate, and the final scores are the dependent variable; the analysis begins with the calculation of the ordinary ANOVA for each of these measures. Because there is nothing new about the calculation of these two ANOVAs, I will give you the results; you can use the methods of Chapter 12 to check that you have not forgotten

| | PSYCHODYNAMIC | | BEHAVIORAL | | CONTROL | | **Table 18.5** |
	Initial	Final	Initial	Final	Initial	Final	
	1	3	1	2	2	2	
	2	4	2	6	3	4	
	2	3	2	5	3	3	
	3	5	3	8	4	5	
	3	7	3	6	4	3	
	4	7	4	9	5	7	
	4	6	5	8	5	5	
	5	9	6	6	6	7	
	5	6			6	4	
	6	8			7	9	
					7	8	
					7	6	
Means	3.5	5.8	3.25	6.25	4.92	5.25	
SDs	1.58	2.04	1.67	2.19	1.73	2.18	

how to obtain them yourself. The next step in the traditional analysis is to form the cross-products of the DV and covariate (as you would to calculate a correlation) and then perform an ANOVA on these cross-products, as well. The purpose of dealing with cross products is to save computational steps; this is a holdover from the days before modern handheld calculators. Because electronic calculators now make it easy to obtain correlation coefficients, and because these coefficients are more meaningful than sums of cross products, I will demonstrate the calculation of the ANCOVA in terms of the appropriate correlations. To simplify the description, I will break the process into discrete steps.

Step 1. Compute the ANOVA for the Dependent Variable

First, calculate the means and standard deviations of the DV for each group. Looking at the means in Table 18.5 we see that both therapy groups performed better than the control mean of 5.25, but behavioral therapy led to an increase of only one rung, and psychodynamic therapy, only about half a rung, on the average. Not surprisingly, the ordinary one-way ANOVA is not statistically significant. In fact, F is less than 1.0, as you can see in Table 18.6.

Source	SS	df	MS	F	p	**Table 18.6**
Between	4.95	2	2.475	.542	>.05	
Within	123.35	27	4.57			
Total	128.3	29				

Step 2. Compute the ANOVA for the Covariate

To distinguish the sums of squares for the covariate from those for the DV, I will use SSX for the former and SSY for the latter (it is traditional to label the DV as Y and the covariate as X). The SS components from an ANOVA of the covariate are $SSX_{total} = 92.0$, $SSX_{bet} = 17.08$, and $SSX_{error} = 74.92$. The F ratio equals 3.078; you will need this result for post hoc comparisons. The F ratio falls short of significance, which is likely if the samples are random (or randomly assigned) and the covariate is measured before the experiment begins. Significant differences on the covariate would suggest that you should look carefully at the randomness of your samples (or group assignments), but bear in mind that significant results can be expected 5% of the time even with perfect randomness (assuming you use alpha = .05). Note that the covariate means run opposite to the DV means for this example, with the control mean highest, and the behavioral mean lowest (see Table 18.5).

Step 3. Compute the Correlation between the DV and the Covariate across (i.e., Ignoring) the Groups and Find the Adjusted SS_{total}

You can use the methods of Chapter 9 (or a statistical calculator) to calculate the correlation between the DV and the covariate for all 30 subjects. For these data, $r_{xy} = .607$. The DV is then adjusted for the covariate by using the latter to predict the former and then finding the residual for each subject. In this example, the residual is that part of each subject's final phobia score that is not predictable from his or her initial phobia score. The SS for these residuals is the total SS for the ANCOVA. Because it will be smaller than the total SS for the DV (SSY_{total}), it is known as the adjusted total SS. I will use the symbol SSA for adjusted sums of squares; it is only the DV that gets adjusted, so there should not be any confusion. Also, when I mention residuals, I will always be referring to the DV adjusted for the covariate.

Fortunately, there is a simple relation between any correlation and the SS of the corresponding residuals, so you don't actually have to calculate any residuals. Recall Formula 10.7B: $\sigma^2_{est\ y} = \sigma^2_y(1 - r^2)$. The variance of the estimate is the variance of the residuals. If you multiply both sides of Formula 10.7B by N, you obtain the following general formula: $SS_{residuals} = SS(1 - r^2)$. Applied to the total SS of ANCOVA, the formula can be written as:

$$SSA_{total} = SSY_{total}(1 - r^2_{xy})$$ **Formula 18.4**

For this problem, $SSA_{total} = 128.3(1 - .607^2) = 128.3(.632) = 81.08$.

Step 4. Compute the DV/Covariate Correlation Separately for Each Group and Find the Pooled Within-Group Correlation

You might think that you could perform an ordinary ANOVA on the residuals mentioned in the previous step and that would be your ANCOVA. That would work only if the groups (on average) did not differ at all on the covariate, in which case you would have reduced your error term without affecting the numerator. However, there will almost always be at least small group differences on the covariate, and these must be taken into account. One way to appreciate the impact of covariate differences is to consider their influence on the overall r_{xy} in relation to the r_{xy}s within each group. In Figure 18.6 you can see that the correlation between the DV and the covariate is high within each group, but the way the groups are arranged diminishes the overall correlation, r_{xy}. This is similar to the situation in the present example, in which the covariate means are negatively correlated with the DV means. In other cases, the covariate means will be positively correlated with the DV means, as shown in Figure 18.7. In these cases, the overall correlation is enhanced—that is, r_{xy} will generally be larger than the correlations within groups. For the data in Table 18.5, the rs within the three groups are .826 (control), .860 (Psychodynamic), and .606 (Behavioral).

A critical element of the ANCOVA involves the discrepancy between the overall r_{xy} and a correlation based on pooling together the within-group correlations. It might seem reasonable to just average the three correlations for the three groups in our example to get the pooled r (r_p), but even if the groups were the same size, it would not be so simple. A weighted average must be taken that adjusts not only for sample size, but for the standard deviations of both the DV and covariate in each group. As the following formula shows, the r for each group must be multiplied by the two SDs and the df for that group, and these products must be summed over all the groups (this sum is an important intermediate statistic called the "within-group sum of cross products," which I will abbreviate to SCPW);

$$SCPW = \sum(n_i - 1)s_{x_i}s_{y_i}r_i \qquad \textbf{Formula 18.5}$$

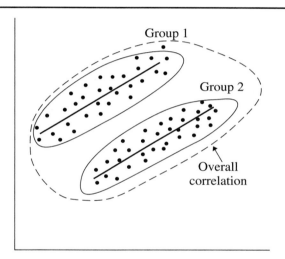

Figure 18.6

Two-Group Case in Which the Means on the Covariate Are Negatively Related to the Means on the DV (Tends to Decrease r_{xy} with Respect to r_p)

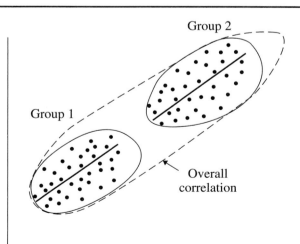

Figure 18.7

Two-Group Case in Which the Means on the Covariate Are Positively Related to the Means on the DV (Tends to Increase r_{xy} with Respect to r_p)

For this problem, SCPW = $11(1.73)(2.18).826 + 9(1.58)(2.04).86 + 7(1.67)$ $(2.19).606 = 34.27 + 24.94 + 15.52 = 74.73$. (*Note:* If all the groups are the same size, $n_i - 1$ becomes just $n - 1$, and this term can be placed in front of the summation sign to reduce the amount of calculation.)

To find r_p divide SCPW by the square root of the product of the within-group error terms for both the DV and the covariate, as shown in Formula 18.6:

$$r_p = \frac{\text{SCPW}}{\sqrt{\text{SSX}_\text{error}\text{SSY}_\text{error}}} \qquad\qquad \textbf{Formula 18.6}$$

In this case:

$$r_p = \frac{74.73}{\sqrt{(74.92)(123.35)}} = \frac{74.73}{96.13} = .777$$

Note that this value is close to the simple average of the three rs, which will usually be the case, and that r_p is between the highest and lowest of the three rs, which will always be the case. Also, note that r_{xy} is lower than r_p due to the negative relation between the group means on the DV and the group means on the covariate (just such a negative relation is depicted in Figure 18.6).

Step 5. Find the Adjusted SS$_\text{error}$

The appropriate error term for the ANCOVA is based on the correlations within the groups. It should make sense that you can make more accurate predictions of the DV using these correlations than the overall r_{xy}. The error term is adjusted for r_p using a formula analogous to the one in step 3:

$$SSA_\text{error} = SSY_\text{error}(1 - r_p^2) \qquad\qquad \textbf{Formula 18.7}$$

Therefore, $SSA_\text{error} = 123.35(1 - .777^2) = 123.35(.396) = 48.85$.

Step 6. Find the Adjusted SS$_\text{between}$ and Complete the Summary Table for the ANCOVA

As in an ordinary ANOVA, $SSA_\text{total} = SSA_\text{between} + SSA_\text{error}$, so the SSA_between can be obtained by subtraction:

$$SSA_\text{between} = SSA_\text{total} - SSA_\text{error} \qquad\qquad \textbf{Formula 18.8}$$

For our problem, $SSA_\text{between} = 81.08 - 48.85 = 32.23$. The completed ANCOVA is shown in Table 18.7.

As you can see, the SS_between has been increased, and the SS_error has been reduced relative to the ordinary ANOVA, leading to a much larger, and in this case, statistically significant, F ratio (totals are not included; because of the covariate, the totals would not be relevant). Note that the critical value of F is based on $k - 1$ df for the numerator, as usual, but on $N - k - 1$ df for the denominator; one additional df has been lost in the error term to account for the covariate (if more than one covariate were

Table 18.7	**Source**	**adj.SS**	**df**	**adj.MS**	**F**	**p**
	Between	32.23	2	16.12	8.57	<.005
	Within	48.85	26	1.88		

used, you would lose one df for each). As with any significant one-way ANOVA, you can reject the null hypothesis that all of the populations have the same mean, but if there are more than two conditions, you would need to perform follow-up tests to determine which pairs of groups differ significantly (e.g., Are both therapies significantly better than the control group? Do the two therapies differ significantly from each other?). These pairwise comparisons require the next step.

Step 7. Calculate the Pooled Regression Slope and Use It to Find the Adjusted Group Means

Linear regression is used within the groups to reduce the error term, and it is also used to adjust the group means to compensate for differences on the covariate. To accomplish the latter function, the regression slopes within the groups must be pooled in the same way as was done for the correlation coefficients. In fact, one way to find the pooled within-group regression slope, b_p, is to divide SCPW, as found in step 4, by SSX_{error}. Alternatively, b_p can be found from r_p by making use of the fact that the slope of a regression line is equal to the correlation coefficient times the appropriate ratio of standard deviations, or, in terms of SSs, predicting Y from X leads to the following formula for the slope: $b_y = r_{xy}[\sqrt{(SSY/SSX)}]$. Changing the subscripts to suit the present purpose yields Formula 18.9:

$$b_p = r_p \sqrt{\frac{SSY_{error}}{SSX_{error}}}$$

Formula 18.9

In this case, $b_p = .777 \sqrt{(123.35/74.92)} = .777(1.283) = .997$.

Part of the advantage of ANCOVA is that we recognize that a group mean may be higher or lower than it should be because of its value on the covariate. For instance, in the present example, the control group accidentally consists of subjects with less than average initial phobia (more rungs climbed), and, given the high correlation between initial and final scores, the control group can be expected to have a higher mean for the DV than it would if it had an average score for the covariate. The pooled slope tells us how many extra units of the DV to expect for each extra unit of the covariate. By subtracting the extra units, the group's mean on the DV is adjusted to what it would be if the group had been average on the covariate. The adjustment is embodied in Formula 18.10:

$$\text{Adj. } \overline{Y}_i = \overline{Y}_i - b_p(\overline{X}_i - \overline{X}_G)$$

Formula 18.10

where $\overline{X}_i$ is that group's mean on the covariate, and $\overline{X}_G$ is the mean of the covariate across all subjects (i.e., the grand mean).

For the three groups in our example, the adjusted means are:

Control: $\quad$ adj. $\overline{Y} = 5.25 - .997(4.92 - 4.0) = 5.25 - .92 = 4.33$.

Psychodynamic: adj. $\overline{Y} = 5.8 - .997(3.5 - 4.0) = 5.8 + .5 = 6.3$.

Behavioral: $\quad$ adj. $\overline{Y} = 6.25 - .997(3.25 - 4.0) = 6.25 + .75 = 7.0$.

Note that the control group mean has been adjusted downward, whereas the therapy group means have increased. The therapy results look more impressive, and the separation of the group means is greater. We are entitled to report the adjusted means because we have corrected for a bias due to an extraneous factor (initial phobia levels). Moreover, these are the means you would use to conduct post hoc tests to determine which pairs of conditions differ significantly.

The increased spread of the means explains why SS_{bet} for the ANCOVA is larger than it was for the original ANOVA. This increase can be expected when groups with a larger mean on the covariate tend to have a lower mean on the DV (I'm assuming that the DV/covariate correlation is positive within the groups). This situation can easily lead to r_{xy} being considerably lower than r_p, as in this example. If groups with larger means on the covariate tend to have larger means on the DV, the adjusted means are likely to end up closer to each other than the original means, thus lowering SS_{bet}; in this case, r_{xy} could become even larger than r_p.

Step 8. Test for Homogeneity of the Regression Slopes

Pooling the within-group regression slopes to create b_p rests on the assumption of *homogeneity of regression*: In each population represented by a sample in our study the relation between the DV and the covariate is the same, and any differences in the bs of the different samples are accidental, each b being an imperfect reflection of the true population slope. This, of course, is perfectly analogous to assuming homogeneity of variance to pool sample variances together, and like homogeneity of variance, homogeneity of regression can be tested. In fact, it should be tested before proceeding with an ANCOVA because you might not want to proceed without homogeneity of regression appearing reasonable and, at least, would proceed differently if homogeneity of regression were rejected. I have waited until this point to discuss the test for homogeneity of regression because the discussion will be clearer now that you have seen the basic analysis, and I can use some of the results from that analysis. Strategically, if not computationally, this step should be considered step 1.

Adding a covariate to a one-way ANOVA is, in some ways, like adding a second factor. In a two-way ANOVA there is the possibility that the two factors will interact in their effect on the DV, and similarly the covariate can interact with the IV. The interaction can be displayed just as for a two-way ANOVA; plot the DV versus the covariate for each group on the same graph, as I have done in Figure 18.6. Parallel lines indicate that the regression slopes are the same from group to group; the more the lines diverge from being parallel, the more evidence there is against the assumption of homogeneity of regression. However, in the standard analysis of covariance, as just described, there is no attempt to separate the SS for any IV/covariate interaction, so SS_{inter} remains part of the adjusted error term for the ANCOVA. The adjusted error term (SSA_{error}) is based on using b_p to find residuals within each group, but clearly a smaller error term would be found by using each group's b to find residuals for that group. I'll call this smaller error term the "within-group regression error term" (SSR_{error}). The more the individual group bs vary around b_p, the greater will be the discrepancy between SSA_{error} and SSR_{error}. This discrepancy, $SSA_{error} - SSR_{error}$, is the between-groups SS for regression (SSR_{bet}), and the larger it is, the more likely it is that homogeneity of regression will be rejected.

The actual test for homogeneity of regression is a one-way ANOVA using the sums of squares just described. Because SSA_{error} has already been calculated, and SSR_{bet} can be found by subtraction, only one new SS has to be calculated, SSR_{error}. To find this term, we need to find the SS for the residuals separately for each group. We can apply Formula 18.1 to each group, using the correlation for that group, and then add across groups.

The *SSY* for each group is found by multiplying the variance of that group on the DV by $n - 1$ for that group. The calculations for the present example are as follows:

Control: $SSR_{error} = SSY(1 - r^2) = 52.276(.318) = 16.62$

Psychodynamic: $37.454(1 - .86^2) = 37.454(.2604) = 9.75$

Behavioral: $33.573(1 - .606^2) = 33.573(.633) = 21.25$

The total $SSR_{error} = 16.62 + 9.75 + 21.25 = 47.62$.

$SSR_{bet} = SSA_{error} - SSR_{error} = 48.85 - 47.62 = 1.23$.

The complete analysis is shown in Table 18.8 (note that calculating correlations separately for each group entails losing two degrees of freedom for each group in the error term).

Source	SS	df	MS	F	p
Between	1.23	2	.62	.31	>.05
Within	47.62	24	1.98		
Total	48.85	26			

Table 18.8

What I have really done in Table 18.8 is remove the DV/covariate interaction $SS(SSR_{bet})$ from the ANCOVA error term and then test the interaction against the new smaller error term. With the *F* ratio nowhere near significance, homogeneity of regression is a reasonable assumption for these data. Do bear in mind, however, that if you are dealing with small sample sizes, the homogeneity of regression test won't have much power. If your within-group slopes differ considerably and the homogeneity test falls just short of significance, there is no reason to feel confident about homogeneity of regression in the population. On the other hand, with very large samples, even small differences in regression slope can lead to significance, but this would not necessarily make the ANCOVA results entirely misleading and useless.

Post Hoc Comparisons

As with any significant one-way ANOVA, you will probably want to follow a significant ANCOVA with pairwise or complex comparisons to localize the significant differences. If you are going to perform Fisher's protected *t* tests, as would be reasonable for the preceding ANCOVA, it is appropriate to use the adjusted means in the numerator. The adjusted error term from the ANCOVA would be used in the denominator, but a correction factor is needed that depends on how much your groups vary on the covariate. Assuming that the subjects have been randomly assigned, the correction for MSA_{error} ($CMSA_{error}$) is given by the following formula:

$$CMSA_{error} = MSA_{error}\left(\frac{1 + F_X}{df_{error}}\right)$$

Formula 18.11

where F_X is the *F* ratio that tests for differences on the covariate (from step 2); df_{error} corresponds to that *F* ratio. For this problem, $CMSA_{error} = 1.88(1 + 3.078/27) = 1.88(1.114) = 2.094$. The *t* test for comparing the control group with the behavioral therapy group, using Formula 13.3, is as follows:

$$t = \frac{\overline{X}_i - \overline{X}_j}{\sqrt{CMSA_{error}\left(\frac{1}{n_i} + \frac{1}{n_j}\right)}} = \frac{7.0 - 4.33}{\sqrt{2.094\left(\frac{1}{8} + \frac{1}{12}\right)}} = \frac{2.67}{.66} = 4.04$$

The comparisons for the two other pairs follow.

Control group vs. psychodynamic therapy:

$$t = \frac{6.3 - 4.33}{\sqrt{2.094\left(\frac{1}{10} + \frac{1}{12}\right)}} = \frac{1.97}{.62} = 3.18$$

Behavioral therapy vs. psychodynamic therapy:

$$t = \frac{7.0 - 6.3}{\sqrt{2.094\left(\frac{1}{8} + \frac{1}{10}\right)}} = \frac{.7}{.686} = 1.02$$

The critical t for 26 df at the .05 level is 2.056; therefore, both therapy groups differ significantly from the control group, but the two therapy groups are not significantly different from each other.

Looking at Formula 18.11, you can see that the correction will be small whenever F is small relative to the sample size, which is likely for experimental studies. (In a typical study with randomized groups, F will be close to 1.0; if about 100 subjects are involved, the corrected error term will be only 1% higher than the ANCOVA error term.) However, if you are using intact rather than randomly assigned groups, a different correction for your error term should be applied before calculating your post hoc comparisons (Bryant & Paulson, 1976). Although the ordinary critical t is quite accurate for following up a significant ANCOVA with three groups and only one covariate (as used earlier), with more than three groups and/or more than one covariate, a modification of Tukey's procedure that requires its own tables (see Stevens, 1999) is recommended. Of course, you don't need special tables to apply the Bonferroni or Scheffé methods, but the modified Tukey test will give you adequate Type I error protection, along with greater power for pairwise comparisons that were not planned.

Performing ANCOVA by Multiple Regression

At the beginning of this section, I described how ANCOVA can be accomplished by multiple regression. If you create two dummy variables to represent the three groups in the preceding example and combine those with the covariate to predict the DV, you will find that the multiple R^2 is .620. The r^2 relating the DV to the covariate alone is $.607^2 = .3684$; therefore, the change in R^2 due to adding the IV is $.620 - .3684 = .2516$. The test for this increase in R^2 is based on Formula 17.15. For this example:

$$F = \frac{26(.2516)}{2(1 - .62)} = \frac{6.542}{.76} = 8.61$$

which agrees, except for the error due to rounding off the correlations, with the results of the ANCOVA found by the traditional method.

I can test for homogeneity of regression by adding two cross-product predictors to test for an interaction between the IV and the covariate.

The addition of these two predictors raises R^2 to .629. The increase in R^2, $.629 - .620 = .009$, is again tested with Formula 17.15:

$$F = \frac{24(.009)}{2(1 - .629)} = \frac{.216}{.742} = .29$$

This result is within rounding error of the result in step 8.

I also mentioned that ANCOVA can be performed as a partial correlation. In the first step, create residuals by predicting the DV from the covariate; in the second step use the two IV predictors and the covariate to predict the residuals. In the second step, the covariate will serve as a suppressor (it is not correlated with its own residuals), thus adjusting the IV for the covariate, as well. The resulting R^2 equals .398, corresponding to a partial r of .631. The meaning of this partial r will be shown in the next subsection.

Power and Effect Size

The effect size in an ANCOVA is estimated the same way as in an ordinary ANOVA. Eta squared is calculated using the adjusted sums of squares: $\eta^2 = SSA_{bet}/SSA_{total}$. For our example, $\eta^2 = 32.23/81.08 = .398$. As you can see, this agrees with the square of the partial correlation just found. The unbiased estimate of omega squared is found by Formula 12.20, also using the adjusted SS components. The increased η^2 for the ANCOVA (as compared to the meager .039 that would have been obtained from the ordinary ANOVA) translates to an increase in power that did not require an increase in sample size. In most studies with randomly assigned subjects, the increased power of ANCOVA comes from the correlation of the covariate with the DV, which leads to a reduction in the error term. In the preceding example, there is also a considerable increase in the separation of the adjusted means as compared to the ordinary means, leading to a further increase in power. But this contribution can go in either direction and is usually small compared to the effect of the reduced error term in experimental studies.

The Assumptions of ANCOVA

The analysis outlined in this section requires the usual assumptions of ANOVA (normality, independence, homogeneity of variance) as applied to the adjusted DV (i.e., the residuals). Because the residuals are obtained through *linear* regression, the following assumption must be added:

1. *The relation between the covariate and the DV in the population is linear.* If the relation is actually curvilinear, it is not likely that ANCOVA will be very helpful, and the results could be misleading. Sometimes the covariate can be transformed to make the relation linear. Otherwise, you can either use polynomial regression within the ANCOVA (see Section C of previous chapter), or create a blocking variable, instead. The latter solution will be discussed further under "Alternatives to ANCOVA." Also, as in linear regression, homoscedasticity and bivariate normality are usually assumed as well (the latter can be violated considerably if the samples are large).

   Because the within-group correlations are averaged to adjust the error term, the next assumption is required.
2. *Homogeneity of regression.* That is, the group regression coefficients or slopes are all equal in the population. This is the assumption tested in

step 8. If the variances of the DV and covariate are the same for all groups, this assumption implies that the DV/covariate correlation is the same in each group. This assumption is also equivalent to stating that there is no interaction between the covariate and the IV. Imagine that in the age/memory example the drug did not merely improve memory at every age but virtually eliminated the effect of age on memory. In that case the regression slope would be much flatter in the drug group, and averaging its slope with that of the control group would not make sense. Of course, this interaction would be an interesting finding in itself. But like an interaction in a two-way ANOVA, it can obscure the main effects. If the memory drug flattens the effect of age, the drug-control memory difference (which is what $SS_{between}$ is based on) will not be constant but will increase with age. Such an interaction can be accounted for within the context of ANCOVA by a method devised by Johnson and Neyman (1936), but this complex procedure goes beyond the scope of this text (see Huitema, 1980). A simpler, albeit less powerful, alternative is the randomized-blocks design, which will be discussed shortly.

As with any partial correlation, it is critical that the covariate accurately measure the variable that you are using it to measure. Hence, the third important assumption.

3. *The covariate is measured without error*. The more random error of measurement that is added to the covariate, the lower will be the DV/covariate correlation, which in turn lowers the power of the ANCOVA. A more serious problem, however, is that random error in the covariate can make it look like two groups differ on the DV when they really don't. This problem is less serious with randomly assigned groups because differences on the covariate are not expected to be large and are not expected to occur more often in one direction than the other. However, when dealing with intact groups, a consistent difference on the covariate may be unavoidable and lead to a bias in the results of the DV. Returning to the example of gender and foot length, even if the height difference between men and women accounts completely for the difference in foot length, a less than perfect correlation between height and foot length can make it appear that there is a foot length difference anyway. In Figure 18.8, you can see that the intercepts are different for men and women, implying that the adjusted DV means will be as well. You can also see that this problem occurs because the slope is reduced when the correlation is less than perfect; the problem disappears completely only when the correlation is perfect (see the "perfect correlation" lines in Figure 18.8). This is just one of several reasons why it can be problematic to draw conclusions from an ANCOVA involving preexisting groups. I will discuss other reasons shortly.

Additional Considerations

Removing variance associated with the covariate makes sense in that the covariate is considered an irrelevant nuisance variable. If the covariate, however, were influenced by the independent variable, removing covariate variance could also result in removing some of the variance produced by your treatment conditions. Although the following is not a formal statistical assumption of ANCOVA, it is an important consideration in designing any experiment from which you want to draw valid conclusions after performing an ANCOVA.

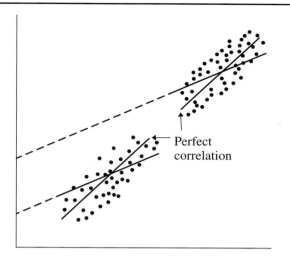

The IV and the covariate are independent. You can be certain that this condition has been met with randomly assigned subjects if the covariate is measured before the experimental treatment is administered or the covariate is some aspect of the individual that cannot be changed by any of the treatments (e.g., age or IQ). On the other hand, if the covariate is measured at the end of the experiment and could have been affected by the experiment, partialling out the covariate can inadvertently reduce (or enhance) the apparent impact of your IV. For example, imagine you have two groups in which generosity is measured (e.g., the DV could be willingness to volunteer for more experiments) after a manipulation (e.g., getting the subject to look in a mirror has been shown to increase generosity) or a control condition. Suppose you measure generosity with a questionnaire (I'll call this measure QGEN) at the end of the study, so as not to alert subjects to your interest in generosity. If you use QGEN as a covariate to reduce the error term (which it is likely to do), it will probably reduce the difference of your group means and may therefore hurt more than help compared to an ordinary ANOVA. The manipulation that affects generosity as measured by the DV could also affect the subjects' responses to QGEN in the same direction. Adjusting the means for differences in the covariate will then reduce group differences on the DV. The ANCOVA is penalizing you as though your group differences are in part due to initial generosity differences, when in fact the covariate differences are most likely due to the treatment.

Factorial ANCOVA

The two-way factorial ANCOVA is a fairly straightforward extension of the one-way ANCOVA, especially if the two-way design is balanced. The error term for all three effects is based on first calculating the DV/covariate correlation separately for each cell of the design and then taking a weighted average over all the cells as in step 4 of the procedure in this section. The pooled within-cell correlation is then used to adjust the error term as in step 5. As you would expect, the ordinary two-way ANCOVA requires the assumption that the regression slopes are the same for all of the populations represented by the cells in the design. The pooled within-cell regression slope is used to adjust the cell means using Formula 18.10. In a balanced design, the row and column marginal means can be found by simply averaging the appropriate cell means. Higher-order factorial ANCOVAs are analyzed in an analogous fashion.

However, the calculation of the SS for each of the three effects in a two-way ANCOVA is considerably more complicated than in the one-way design; these three SS components will generally not add up to the overall $SS_{between}$ of the two-way design (i.e., the $SS_{between}$ you would get by treating all the cells of the two-way ANOVA as different groups in a one-way design), even when the design is balanced. Nonetheless, the two-way ANCOVA can be analyzed by multiple regression just as in the one-way case. The covariate is entered into the equation with all of the predictors for the main effects and interaction. Each effect can be evaluated by finding the change in R^2 due to removing that effect from the full model. Fortunately, the multiple regression approach is rarely necessary; many statistical software packages, including SPSS, make it easy to add one or more covariates to a two-way or higher-order factorial ANOVA.

Using Two or More Covariates

As you might guess from studying the material on mustiple regression, if your goal is to predict the DV as well as possible, two covariates relatively unrelated to each other can do the job much better than just one. If you are comparing the political attitudes of retired couples in different regions of the country or in areas of different population densities (e.g., urban, suburban, rural), you might want to control for both age and income. On the other hand, if you have two possible covariates that are highly redundant, you would probably use only the better of the two. You lose a degree of freedom in your error term for each covariate (SS_{error} is divided by a smaller number, yielding a larger error term), so you pay a price for each covariate added, which is negligible only when your sample size is very large.

Just as you would use linear regression to predict the DV from one covariate, you use multiple regression to predict the DV when there are two or more covariates. The assumptions of ANCOVA, which are fairly restrictive even with only one covariate, can get rather difficult to satisfy with several covariates. For instance, homogeneity of regression is extended to require not only that the slope relating each covariate to the DV be the same for each group (or cell) but that the covariances for each pair of covariates be the same for each group (e.g., in the retired-couples example, it would be a problem if age and income were moderately related in urban areas but not at all in rural areas). Also, in most cases it is necessary to assume a joint multivariate normal distribution for all of your covariates and the DV to test the ANCOVA for significance (a large sample size will usually help with this assumption).

Finally, the more covariates you use, the greater the likelihood of problems with multicollinearity. Unlike an exploratory multiple regression, in which you try not to leave out any relevant variable and therefore freely include redundant variables for some selection procedure to sort out, an ANCOVA generally works best with randomly assigned groups and no more than two or three covariates that are strongly (and linearly) related to the DV but not so much to each other. The temptation to use several covariates is likely to be greater when performing an ANCOVA on preexisting groups, but the analysis in that case has its own problems, which will be discussed shortly.

Alternatives to ANCOVA

When subjects are randomly assigned to groups, and neither matched nor repeated-measures designs are feasible, and a covariate that is linearly related to the DV and is unaffected by the IV can be measured accurately, it makes sense to perform an ANCOVA. For instance, if you were studying motivational factors affecting problem-solving, you might use an IQ measure as a covariate. If you were comparing different methods for teaching long division to fourth-graders, you might get more specific and use their math grades from the previous semester or their scores on a specially designed arithmetic pretest. However, even when a repeated-measures design is possible, it is often better to perform an ANCOVA instead.

Mixed-Design ANOVA

If you have only two measures of the DV for each subject, one taken before and one taken after some experimental manipulation or condition, you can either perform a mixed-design ANOVA or use the initial measure (sometimes called a baseline measure) as your covariate in a one-way ANCOVA on the after scores. That is what I did in the acrophobia example in this section. Had I performed a mixed-design ANOVA on the same data, I would have ignored the main effects and focused on the interaction. The F for the interaction would be the same as performing a one-way ANOVA comparing the three therapy groups on the before-after (covariate-DV) difference scores (see Chapter 16). In this way, the before-after ANOVA can be seen as a crude form of ANCOVA. When you treat before and after scores as repeated measures, you adjust the after score by subtracting the before score from it. In ANCOVA, you use the before score to predict the after score and adjust the after score by subtracting the prediction. In the former case, you get a simple difference score, and in the latter case, the residual from linear regression. Which is better? If the regression slope is 1.0 (which is very close to being the case for the acrophobia example), the prediction is the same as the before score, and the two methods are virtually identical (the before-after design has the edge because you lose a degree of freedom with ANCOVA). The more the slope differs from 1.0, the more the residuals will vary less than the difference scores, and the greater the advantage of ANCOVA. Whenever you have before and after scores for two or more groups of subjects, you should consider performing an ANCOVA.

Randomized-Blocks Design

When the relationship between the covariate and the DV is curvilinear, or the relationship is linear with so much scatter that Pearson's r is not more than about .4, an alternative to ANCOVA, the randomized-blocks (RB) or treatment × blocks design (introduced in Chapter 15) should be considered. The covariate is turned into a second factor in the ANOVA by dividing it into categories. For instance, if drugs for memory enhancement are being studied in the elderly with age as a covariate, subjects can be divided into blocks by decades (sixties, seventies, eighties, etc.). The original factor is drug group, but adding the age factor creates a two-way (treatment × blocks) ANOVA. The error term becomes the average variance within each *cell* (each decade/drug combination), which is smaller than the error term for the one-way ANOVA to the extent that decade has an effect on memory scores. This is similar to the error reduction that was created in Chapter 14, Section A by adding gender as a factor. The only difference is that gender is already categorical, whereas age has to be artificially divided into categories. This design works best when you know the covariate values before running the experiment, you assign subjects to conditions randomly within each block, and the size of each block is a multiple of the number of conditions.

The most precise form of the RB design is to create blocks (based on similarity of covariate values) such that the number of subjects in each block is equal to the number of conditions. This matching of subjects allows you to perform a matched *t* test or RM ANOVA, but unless the DV/covariate correlation is quite low, an ANCOVA will be more powerful. This combination of matching subjects in blocks and performing the ANCOVA is usually more powerful than an ANCOVA without matching. If covariate values were not used to match or assign subjects, forming blocks post hoc (after the experiment has been performed) is not usually recommended; the ANCOVA is a better choice in nearly every case. However, a carefully planned RB design can be the best choice when the ANCOVA assumptions are not likely to be met—especially when the correlation between the DV and the covariate is not very linear or there is an interaction between the IV and the covariate (in the latter case, the RB analysis is a simpler alternative to the Johnson-Neyman method for analyzing an ANCOVA when homogeneity of regression appears violated). Bear in mind that when planning an RB design, you can use the expected DV/covariate correlation, sample size, and number of treatment conditions to determine the optimal number of blocks (see Feldt, 1958).

Using ANCOVA with Intact Groups

As you know by now, it is difficult to draw any causal conclusions from observed differences between self-selected (intact, preexisting) groups. For instance, if you observe that vegetarians have lower cholesterol levels (CL) than nonvegetarians, you cannot be sure that this is the result of dietary differences; it may be that people who choose to be vegetarians are more likely to exercise or are less likely to smoke or have a lower percentage of body fat. It is not uncommon to use ANCOVA to remove the effects of some or all of these nuisance variables to rule out alternative explanations and to see if there is still a difference on the grouping variable. Although ANCOVA can be useful in such situations, statisticians caution that ANCOVA can never entirely solve the problem of inferring causation with intact groups, and in the absence of careful interpretation, the use of ANCOVA in these situations can result in misleading conclusions.

One of the problems that arises with the comparison of intact groups is that you may not have identified all of the important variables that need to be controlled. In the vegetarian example, suppose you sample only non-smokers and use percentage of body fat and hours per week of exercise as covariates. Any remaining difference in CL between the two groups can still be due to some covariate you have not measured; those who choose to become vegetarians may differ on various psychological variables (some of which may affect CL), such as optimism or the ability to deal with stress. On the other hand, suppose you add a third covariate, education level, and the cholesterol difference between the two groups nearly vanishes. Does that prove a role for education? Not necessarily. By controlling for education level, you are also controlling, to some extent, for other variables closely correlated with education. It is possible that people who have more education can deal with stress more easily, so if vegetarians tend to have more education, they may have lower CL because they have less stress.

These problems are not unique to this type of ANCOVA design, of course; they are the problems usually associated with multiple regression, which rarely involves the random assignment of subjects to groups. However, given the frequent association of ANOVA and ANCOVA with randomized experimental designs, the use of ANCOVA with intact groups may increase the temptation to make erroneous causal conclusions. In addition to these problems of interpretation, the ANCOVA with intact groups is more susceptible to potential violations of the assumptions described previously. I will turn to these next.

A lack of homogeneity of regression becomes more likely with intact groups. If you are studying salary differences between men and women, you would want to use the number of years at the present company as a covariate. But to do that you must assume that the slope relating salary to years is the same for men as it is for women. Given the current socio-economic situation, that is not a reasonable assumption to make. Also, as already mentioned, mean differences between groups on a covariate can produce spurious adjustments of the means that make it look like there are DV differences when there are not. Intact groups are more likely than random groups to differ on the covariate and are therefore more likely to produce biased results that become more misleading as the measurement of the covariate becomes less precise (any covariate differences shown by random groups will not be consistent from one replication to another, but intact groups may well show a consistent bias). Some statisticians consider a significant difference on the covariate among intact groups a sign that the ANCOVA should not be performed. However, when interpreted cautiously, ANCOVA can be a useful tool in trying to locate the true causes of group differences when the random assignment of subjects is impossible, unethical, or simply not feasible.

1. In the simplest form of ANCOVA, the IV has only two levels, and there is only one covariate. If the IV and covariate are uncorrelated, the ANCOVA can be expressed simply as a test of the semipartial correlation between the IV and DV. Although the sr in this case is the same as the r_{pb} between the IV and the DV, the test of sr benefits from subtracting the variance accounted for by the covariate from the error term, thus increasing the power of the test relative to testing r_{pb}.

2. When the correlation between the IV and the covariate is not zero (i.e., when the groups differ on the covariate), the ANCOVA procedure can be used to adjust the group means. The adjusted means reflect how each group would be expected to score on the DV if all the groups had the same average score for the covariate. Depending on the IV/covariate relationship, the adjustment could bring the means closer together, tending to reduce the F ratio, or spread them further apart, increasing the power of the test. In either case, the adjusted means provide a more accurate picture of the effects of the IV, unconfounded by the covariate.

3. A one-way ANCOVA can be performed as a multiple regression in which the covariate is entered into the model, and then the additional (DV) variance accounted for by the IV is assessed. Alternatively, ANCOVA can be viewed as a mixture of traditional ANOVA and linear regression. From the latter perspective an ANCOVA can be performed by the following series of steps:

Step 1. Compute the ordinary ANOVA for the DV and find the means and SDs for each group.

B

SUMMARY

Step 2. Repeat Step 1 for the covariate.

Step 3. Calculate Pearson's r between the DV and covariate across all the subjects in the study. The adjusted total SS is the original SS_{total} multiplied by $1 - r_{all}^2$.

Step 4. Calculate the DV/covariate correlation separately for each group and then pool these rs to obtain r_p. The pooling involves a weighted average that is based on the sample sizes and the SDs of both the DV and covariate in each group.

Step 5. The adjusted error SS is the original SS_{error} multiplied by $1 - r_p^2$.

Step 6. The adjusted $SS_{between}$ can be found by subtraction, and then the ANCOVA summary table can be completed using the adjusted SS components.

Step 7. The pooled regression slope, b_p, can be found by multiplying r_p by the square root of the ratio of the original SS_{error} for the DV to the SS_{error} for the covariate. The adjustment to a group's mean depends on the difference between that group's mean on the covariate and the grand mean of the covariate. That difference is multiplied by b_p and then subtracted from the original group mean.

Step 8. The test for homogeneity of regression is a test to see whether there is an interaction between the IV and the covariate. If this step finds a statistically significant result, the other steps should not be performed. SS_{error} is calculated separately for each group and added to form a new, smaller error term. The difference between this new error term and the adjusted SS_{error} from the ANCOVA (i.e., SSR_{bet}) is tested for significance by its own ANOVA.

4. If appropriate, post hoc comparisons are performed on the adjusted group means. The error term from the ANCOVA is used after being multiplied by a correction factor that is related to how much the groups differ on the covariate.

5. In addition to the usual ANOVA assumptions, the following must be assumed for ANCOVA:
   a. The relationship between the covariate and the DV is linear in the population.
   b. The regression slope is the same for all of the populations represented by levels of the IV (i.e., there is no IV/covariate interaction in the population).
   c. The covariate is measured without error. Measurement error can lead to lowered power when dealing with randomly assigned groups and to serious bias when dealing with preexisting groups.

6. To draw valid causal conclusions from your ANCOVA results, the IV and the covariate must be independent; that is, the IV manipulation must not affect the values on the covariate. If you are dealing with intact groups, it is not unlikely that the groups will differ significantly on the covariate; in fact, the covariate may have been chosen because it is a confounding variable whose effects you want to remove. You should be particularly cautious when interpretating the results in these cases in part because of the danger of not including an important covariate or including a covariate that is not measuring exactly what you think it is measuring. Also, be aware that the use of ANCOVA with intact groups is usually associated with a greater risk of violating the assumptions of ANCOVA.

7. Adding a second or third covariate can improve the ANCOVA if the covariates are not highly correlated with each other. However, as the number of covariates increases, there is a loss of degrees of freedom, the assumptions of ANCOVA can be harder to satisfy, and problems

can arise due to multicollinearity. ANCOVA can also be extended to two-way and higher-order factorial designs.

8. The ideal situation for ANCOVA is when subjects have been randomly assigned to experimental conditions, and you have a precisely measured covariate that has a strong linear correlation with your dependent variable. If the covariate is a measure on the same scale as the DV, but taken before the experimental manipulation, the ANCOVA is usually more powerful than performing a repeated measures analysis. However, if the relation between the covariate and the DV is curvilinear or involves an interaction, forming randomized blocks based on the covariate should be considered as an alternative procedure.

EXERCISES

1. For a five-group one-way ANOVA with 15 subjects per group, $SS_{bet} = 440$ and $SS_w = 2800$. Suppose a covariate is found that has an average within-group correlation (r_p) with the dependent variable of .4. Calculate the F ratio for the ANCOVA when the correlation across groups (r_{xy}) is equal to: a. .1, b. .3, c. .45, d. .5
   e. For a given value of r_p, what happens to the ANCOVA F as r_{xy} increases? Why should this be the case? (*Hint*: Consider what must be happening to the separation of the adjusted group means relative to the original means, in each case.)
*2. Three different methods were used to teach geometry at a school; a different class of 30 students was assigned to each method. To account for possible differences in math ability among the classes (and reduce error vari-

   a. Perform a one-way ANCOVA on the data.
   b. Test for homogeneity of regression. Are the data for this problem consistent with that assumption?
   c. Find the adjusted mean for each group and conduct Fisher's protected t tests using the corrected error term.
3. A researcher has tested four methods for raising HDL ("good") cholesterol levels: diet, exercise, drugs, and vitamins. Ten subjects were randomly assigned to each group. For each subject a baseline measure was taken and then CL was measured again after 6 months of treatment. The means for each experimental treatment are given (in arbitrary units) in the following table (standard deviations are in parentheses). Pearson's r between the baseline and treatment scores is also given for each group.

	Exercise	Diet	Vitamins	Drugs
Baseline	56.3 (3.06)	55.3 (3.46)	52.9 (2.60)	55.1 (2.02)
Treatment	58.1 (4.48)	56.3 (4.03)	53.0 (4.19)	57.9 (4.70)
Pearson's r	.589	.454	.530	.737

ance), each student's performance on a recent math aptitude test was used as a covariate. The means and standard deviations for the geometry grades and the aptitude measure for each group are given below. The total sum of squares was 20,642.4 for the geometry grades and 5,652.3 for the aptitude scores. The correlation between geometry and covariate scores over all subjects was (r_{xy}) = .45.

a. Given that the total sums of squares are 367.6 for the baseline and 822.78 for the treatment, complete the ordinary ANOVA for both the baseline and treatment scores, testing both F ratios for significance at the .05 level.
b. Given that the baseline/treatment correlation over all the subjects (r_{xy}) is .63, perform a one-way ANCOVA on the data.

	METHOD I		METHOD II		METHOD III	
	Aptitude	Geometry	Aptitude	Geometry	Aptitude	Geometry
Means	40.2	83.8	45.1	79.5	38.4	87.2
SDs	7.5	12.4	6.7	20.1	8.3	11.1
Within-group r	.53		.65		.48	

c. Test for homogeneity of regression. Are the data for this problem consistent with that assumption?

d. Find the adjusted mean for each group and conduct Tukey's HSD tests using the corrected error term. (*Note*: Special tables are required to find q when dealing with an ANCOVA, but using the ordinary q is a rough approximation, and makes for a useful exercise.)

e. Explain the relation between the ANOVA F for the treatment and the ANCOVA F in terms of the relation between the original treatment means and the adjusted means.

*4. In this study the time (in minutes) to solve a logic problem is recorded for subjects given a concrete hint, an abstract hint, or no hint at all. The covariate is the time required to solve a somewhat related problem presented the previous day without a hint. The solution times are given in the following table.

Covar	Concrete	Covar	Abstract	Covar	No Hint
14	10	21	17	15	20
25	12	15	13	13	25
21	15	19	16	20	14
8	11	14	12	19	17
17	5	9	7	7	12
19	7	12	8	10	18
7	2	8	3	9	7

a. Conduct an ANCOVA on the preceding data and compare to the ANOVA results without the covariate.

b. Test for homogeneity of regression. Are the data for this problem consistent with that assumption?

c. Find the adjusted mean for each group and conduct Fisher's protected t tests using the corrected error term.

d. How did initial differences on the covariate affect the ANCOVA as compared to the ordinary ANOVA?

5. In Exercise 12B7, job satisfaction was compared among three major academic areas of a university. The one-way ANOVA on those data, reproduced in the next table, did not

yield significant results. However, systematic salary differences frequently exist across different areas, and salary may be related to job satisfaction (even in academia), so if the salaries of the original participants were available, it would make sense to redo the analysis using salary as a covariate. As this was a hypothetical study to begin with, it was easy to add the salary data to the original table (the salary numbers are in thousands of dollars of annual income).

a. Perform a one-way ANCOVA on the data. Can the null hypothesis be rejected?

b. What effect did the covariate have on the original analysis? Explain why it had this effect.

c. Find the adjusted mean for each group.

d. Test the homogeneity of regression assumption.

e. Describe two potential threats to the validity of the conclusion you drew in part a.

*6. In a cognitive experiment, subjects are randomly assigned to one of four training sessions that differ in terms of the method subjects learn for memorizing a list of words. The methods are: clustering (grouping the words by common category), chaining (associating each word with the next one on the list), visual imagery (creating bizarre images that include several words), and rote rehearsal (this is a control group, as subjects are expected to use this method without training). Subjects are tested for recall both before and after training. The number of words correctly recalled from a list of 40 is shown for each of the 10 subjects in each condition in the following table:

a. Conduct an ANCOVA on these data.

b. Conduct a mixed design ANOVA on the data and test the interaction for significance. (*Hint*: it will be sufficient to conduct a one-way ANOVA on the before-after difference scores.) Which statistic from the ANCOVA helps to explain why the results of the two analyses are similar?

	SOCIAL SCIENCES		NATURAL SCIENCES		HUMANITIES	
	Salary	Satisfaction	Salary	Satisfaction	Salary	Satisfaction
	72	6	77	8	70	7
	75	7	84	10	62	5
	77	10	80	7	68	9
	80	8	75	8	60	4
	65	8				
	74	9				

CLUSTERING		CHAINING		VISUAL IMAGERY		CONTROL	
Before	After	Before	After	Before	After	Before	After
22	24	17	19	18	23	21	20
18	19	10	13	13	20	17	14
25	28	12	11	24	29	19	20
11	15	22	21	12	16	13	11
10	16	14	20	10	13	16	19
15	13	9	14	17	21	10	11
20	22	21	21	15	18	14	13
17	18	15	18	19	25	23	20
13	17	19	22	21	27	20	25
23	21	16	16	14	18	15	16

    c. Find the adjusted mean for each group and conduct Tukey's HSD tests using the corrected error term (the note for Exercise 3 applies here as well).

    d. Which statistic calculated in part c explains why the results of parts a and b are so similar?

7. a. Conduct an ANCOVA using only Visual Imagery and Control groups from Exercise 6.

    b. Calculate the point-biserial correlations between group membership and both the before and after scores (*Hint*: calculate the two-group t test in each case and then convert to r_{pb}) and the before/after corre-lation across both groups. Use these three correlations to calculate the partial correlation between group membership and the after scores, partialling out the before scores from both.

    c. Show that the F ratio for testing the partial correlation in part b for significance is the same as the F ratio for the ANCOVA in part a. Explain the connection.

*8. Perform an ANCOVA on the After scores in Exercise 16B7, using the Before scores as the covariate. Compare your ANCOVA F to the F for the interaction in Exercise 16B7. Explain why these two F ratios are comparable.

Multivariate Analysis of Variance

Multifactor experiments have become very popular in recent years, in part because they allow for the testing of complex interactions, but also because they can be an efficient (not to mention economical) way to test several hypotheses in one experiment, with one set of subjects. This need for efficiency is driven to some extent by the ever-increasing demand to publish, as well as the scarcity of funding. Given the current situation, it is not surprising that researchers rarely measure only one dependent variable. Once you have invested the resources to conduct an elaborate experiment, the cost is usually not increased very much by measuring additional variables; it makes sense to squeeze in as many extra measures or tasks as you can without exhausting the subjects and without one task interfering with another. Having gathered measures on several DVs, you can then test each DV separately with the appropriate ANOVA design. However, if each DV is tested at the .05 level, you are increasing the risk of making a Type I error in the overall study—that is, you are increasing the experimentwise alpha. You can use the Bonferroni adjustment to reduce the alpha for each test, but there is an alternative that is potentially more powerful. This method, in which all of the DVs are tested simultaneously, is called the multivariate analysis of variance (MANOVA); the term *multivariate* refers to the incorporation of more than one DV in the test (all of the ANOVA techniques you have learned thus far are known as univariate tests).

    Although it seems clear to me that the most common use of MANOVA at present is the control of experimentwise alpha, it is certainly not the most interesting use. I think it is safe to say that the most interesting use of

**OPTIONAL
MATERIAL**

MANOVA is to find a combination of the DVs that distinguishes your groups better than any of the individual DVs separately. In fact, the MANOVA can attain significance even when none of the DVs does by itself. This is the type of situation I will use to introduce MANOVA. The choice of my first example is also dictated by the fact that MANOVA is much simpler when it is performed on only two groups.

The Two-Group Case: Hotelling's T^2

Imagine that a sample of high school students is divided into two groups depending on their parents' scores on a questionnaire measuring parental attitudes toward education. One group of students has parents who place a high value on education, and the other group has parents who place relatively little value on education. Each student is measured on two variables: scholastic aptitude (for simplicity I'll use IQ) and an average of grades for the previous semester. The results are shown in Figure 18.9. Notice that almost all of the students from "high value" (HV) homes (the filled circles) have grades that are relatively high for their IQs, whereas nearly all the students from "low value" (LV) homes show the opposite pattern. However, if you performed a t test between the two groups for IQ alone, it would be nearly zero, and although the HV students have somewhat higher grades on average, a t test on grades alone is not likely to be significant, either. But you can see that the two groups are fairly well separated on the graph, so it should come as no surprise that there is a way to combine the two DVs into a quantity that will distinguish the groups significantly.

A simple difference score, IQ − grades (i.e., IQ minus grades), would separate the groups rather well, with the LV students clearly having the higher scores. (This difference score is essentially an *under*achievement score; in this hypothetical example students whose parents do not value education do not get grades as high as their IQs suggest they could, whereas their HV counterparts tend to be overachievers). However, the MANOVA procedure can almost always improve on a simple difference score by finding the weighted combination of the two variables that produces the largest t value possible. (If you used GPA on a four-point scale to replace grades, it would have to be multiplied by an appropriate constant before it would be reasonable to take a difference score, but even if you transform both variables to z scores, the MANOVA procedure will find a weighted combination that is better than just a simple difference. In many cases, a sum works better than a difference score, in which case MANOVA finds the best weighted

Figure 18.9

Scatterplot in which Two Groups of Students Differ Distinctly on Two Variables

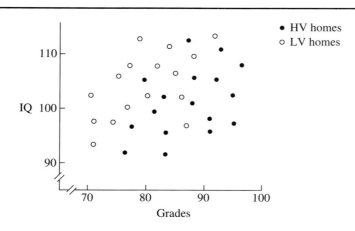

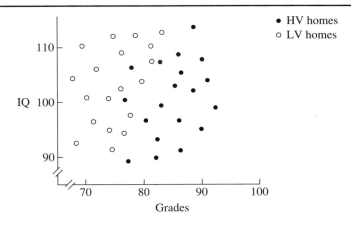

Figure 18.10

Scatterplot in which Two Groups of Students Differ Strongly on One Variable and Weakly on a Second Variable

sum.) Given the variables in this problem, the *discriminant function*, which creates the new variable to be tested, can be written as $W_1 \times IQ + W_2 \times$ Grades + Constant (the constant is not relevant to the present discussion). For the data in Figure 18.9, the weights would come out close to $W_1 = 1$ and $W_2 = -1$, leading to something resembling a simple difference score. However, for the data in Figure 18.10, the weights would be quite different.

Looking at the data in Figure 18.10, you can see that once again the two groups are well separated, but this time the grades variable is doing most of the discrimination, with IQ contributing little. The weights for the discriminant function would reflect that; the weight multiplying the z score for grades would be considerably larger than the weight for the IQ z score.

It is not a major complication to use three, four, or even more variables to discriminate between the two groups of students. The raw-score (i.e., unstandardized) discriminant function for four variables would be written as $W_1X_1 + W_2X_2 + W_3X_3 + W_4X_4 +$ Constant. This equation can, of course, be expanded to accommodate any number of variables. Adding more variables nearly always improves your ability to discriminate between the groups, but you pay a price in terms of losing degrees of freedom, as you will see when I discuss testing the discriminant function for statistical significance. Unless a variable is improving your discrimination considerably, adding it can actually reduce your power and hurt your significance test.

Going back to the two-variable case, the weights of the discriminant function become increasingly unreliable (in terms of changing if you repeat the experiment with a new random sample) as the correlation of the two variables increases. It is not a good idea to use two variables that are nearly redundant (e.g., both SAT scores and IQ). The likelihood of two of your variables being highly correlated increases as you add more variables, which is another reason not to add variables casually. The weights of your discriminant function depend on the discriminating power of each variable individually (its r_{pb} with the grouping variable) and the intercorrelations among the variables. When your variables have fairly high intercorrelations, the *discriminant loading* of each variable can be a more stable indicator of its contribution to the discriminant function. A DV's discriminant loading is its ordinary Pearson correlation with the scores from the discriminant function.

High positive correlations among your DVs reduce the power of MANOVA when all the DVs vary in the same direction across your groups (Cole, Maxwell, Arvey, & Salas, 1994), which is probably the most common case. In fact, it has been suggested that one can obtain more power by running separate univariate ANOVAs for each of the highly correlated DVs and adjusting the alpha for each test according to the Bonferroni inequality. On the other hand, MANOVA becomes particularly interesting and powerful when some of the intercorrelations among the DVs are negative or when some of the DVs vary little from group to group but correlate highly (either positively or negatively) with other DVs. A DV that fits the latter description is acting like a suppressor variable in multiple regression. The advantage of that type of relation was discussed in the previous chapter.

In the two-group case the discriminant weights will be closely related to the beta weights of multiple regression when you use your variables to predict the grouping variable (which can just be coded arbitrarily as 1 for one group, and 2 for the other). This was touched upon in Chapter 17, section C, under "Multiple Regression with a Dichotomous Criterion." Because discriminant functions are not used nearly as often as multiple regression equations, I will not go into much detail on that topic. The way that discriminant functions are most often used is as the basis of the MANOVA procedure, and when performing MANOVA, there is usually no need to look at the underlying discriminant function. We are often only interested in testing its significance by methods I will turn to next.

Testing T^2 for Significance

It is not easy to calculate a discriminant function, even when you have only two groups to discriminate (this requires matrix algebra and is best left to statistical software), but it is fairly easy to understand how it works. The discriminant function creates a new score for each subject by taking a weighted combination of that subject's scores on the various dependent variables. Then, a t test is performed on the two groups using the new scores. There are an infinite number of possible discriminant functions that could be tested, but the one that *is* tested is the one that creates the highest possible t value. Because you are creating the best possible combination of two or more variables to obtain your t value, it is not fair to compare it to the usual critical t. When combining two or more variables, you have a greater chance of getting a high t value by accident. The last step of MANOVA involves finding the appropriate null hypothesis distribution.

One problem in testing our new t value is that the t distribution cannot adjust for the different number of variables that can go into our combination. We will have to square the t value so that it follows an F distribution. To indicate that our new t value has not only been squared but that it is based on a combination of variables, it is customary to refer to it as T^2—and in particular, Hotelling's T^2—after the mathematician who determined its distribution under the null hypothesis. T^2 follows an F distribution, but it first must be reduced by a factor that is related to the number of variables, P, that were used to create it. The formula is:

$$F = \frac{n_1 + n_2 - P - 1}{P(n_1 + n_2 - 2)} T^2 \qquad \textbf{Formula 18.12}$$

where, n_1 and n_2 are the sizes of the two groups. The critical F is found with P and $n_1 + n_2 - P - 1$ degrees of freedom. Notice that when the sample sizes are fairly large compared to P, T^2 is multiplied by approximately $1/P$. Of course, when $P = 1$, there is no adjustment at all.

The one case in which it is quite easy to calculate T^2 is when there are only two groups and two DVs. Suppose you have equal-sized groups of left- and right-handers and have calculated t tests for both a verbal test and a spatial test. If across all your subjects the two DVs have a zero correlation, you can find the square of the point-biserial correlation corresponding to each t test (use Formula 10.13 without taking the square root) and add the two together. The resulting r_{pb}^2 can be converted back to a t value by using Formula 10.12 (for testing the significance of r_{pb}). However, if you use the square of that formula to get t^2 instead, what you are really getting is T^2 for the combination of the two DVs. T^2 can then be tested with Formula 18.12. If you have any number of DVs, and each possible pair has a zero correlation over all your subjects, you can add all the squared point-biserial rs and convert to T^2, as just described. However, if you have two DVs and they have a nonzero correlation with each other, you can use multiple regression to combine the point-biserial rs and find R^2. The F ratio used in multiple regression to test R^2 would give the same result as the F for testing T^2 in this case. In other words, the significance test of a MANOVA with two groups is the same as the significance test for a multiple regression to predict group membership from your set of dependent variables. An example with two correlated DVs and two groups will help me to make the connection between MANOVA and multiple regression clear.

Consider the data displayed in Figure 18.9, and imagine that a t test between 18 LV homes and 18 HV homes produces a t value of 1.5 for IQ and 1.7 for grades. Converting these t values to $r_{pb}s$ using Formula 10.13 yields $r_{1y} = .249$, and $r_{2y} = .280$, respectively. If we assume that IQ and grades actually have a slightly negative correlation with each other—i.e., $r_{12} = -.20$—we are ready to plug the three correlations into Formula 17.6 to find R^2:

$$R^2 = \frac{.249^2 + .280^2 - 2(.249)(.280)(-.20)}{1 - (-.20)^2} = \frac{.062 + .0784 - (-.0279)}{.96}$$

$$= \frac{.168}{.96} = .175$$

Now we can convert R^2 to T^2 by using the square of Formula 10.12 (with a change in notation):

$$T^2 = \frac{R^2}{\frac{1 - R^2}{df}} = \frac{.175}{\frac{.825}{34}} = \frac{.175}{.024} = 7.2$$

Finally, we can test T^2 for significance with Formula 18.12:

$$F = \frac{18 + 18 - 2 - 1}{2(18 + 18 - 2)}(7.2) = \frac{33}{68}(7.2) = .4583(7.2) = 3.50$$

Because 3.5 is larger than $F_{.05}$ (2, 34) = 3.28, T^2, and therefore the MANOVA, is statistically significant. Note that neither IQ nor grades were associated with a significant t value, but when combined they yielded a significant F for MANOVA. In this example, the MANOVA took advantage of the complementary relationship of the two DVs, in the same way that multiple regression would do in predicting the LV/HV grouping variable as the criterion. This example demonstrates a particularly good situation for the use of MANOVA.

Effect Size in the Two-Group Case

If you divide an ordinary t value by the square root of $n/2$ (if the groups are not the same size, n has to be replaced by the harmonic mean of the two sample sizes), you get g, a sample estimate of the effect size in the population. If you divide T^2 by $n/2$ (again, you need the harmonic mean if the ns are unequal) you get MD^2, where MD is a multivariate version of g, called the *Mahalanobis distance*.

In Figure 18.11 I have reproduced Figure 18.9, but added a diagonal measure of distance. The means of the LV and HV groups are not far apart

Figure 18.11

Scatterplot of Two
Groups of Students
Measured on Two
Different Variables
Including the
Discriminant Function
and the Group Centroids

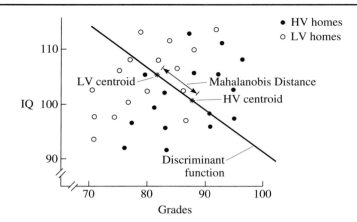

Figure 18.11

Scatterplot of Two
Groups of Students
Measured on Two
Different Variables
Including the
Discriminant Function
and the Group Centroids

on either IQ or grades separately, but if you create a new axis from the discriminant function that optimally combines the two variables, you can see that the groups are well separated on the new axis. Each group has a mean (called a *centroid*) in the two-dimensional space formed by the two variables. The *MD* is the standardized distance between the centroids, taking into account the correlation between the two variables. If you had three discriminator variables, you could draw the points of the two groups in three-dimensional space, but you would still have two centroids and one distance between them. The *MD* can be found, of course, if you have even more discriminator variables, but unfortunately I can't draw such a case. Because $T^2 = (n/2)MD^2$, even a tiny *MD* can attain statistical significance with a large enough sample size. That is why it is useful to know *MD* in addition to T^2, so you can evaluate whether the groups are separated enough to be easily discriminable. I will return to this notion when I briefly discuss discriminant analysis.

The Multigroup Case: MANOVA

The relation between Hotelling's T^2 and MANOVA is analogous to the relation between the ordinary *t* test and the univariate ANOVA. Consistent with this analogy, the Hotelling's T^2 statistic cannot be applied when you have more than two groups, but its principles do apply. A more flexible statistic, which will work for any number of groups and any number of variables, is the one known as Wilk's lambda; it is symbolized as Λ (an uppercase Greek letter, corresponding to the Roman L), and calculated simply as: $\Lambda = SS_W/SS_{total}$. This statistic should remind you of η^2 (eta squared). In fact, $\Lambda = 1 - \eta^2$. Just as the significance of η^2 can be tested with an *F* ratio, so can Λ. In the simple case of only two groups and *P* variables, Wilks' Λ can be tested in an exact way with the following *F* ratio:

$$F = \frac{1 - \Lambda}{\Lambda} \left(\frac{n_1 + n_2 - P - 1}{P} \right)$$ **Formula 18.13**

The critical *F* is based on *P* and $n_1 + n_2 - P - 1$ degrees of freedom. The ratio of $1 - \Lambda$ to Λ is equal to SS_{bet} / SS_W; and when this ratio is multiplied by the ratio of dfs, as it is in Formula 18.13, the result is the familiar ratio, MS_{bet}/MS_W, which you know from the one-way ANOVA, and it produces the same value as Formula 18.12. (In the two-group case, $\Lambda = df/(T^2 + df)$, where $df = n_1 + n_2 - 2$.)

The problem that you encounter as soon as you have more than two groups (and more than one discriminator variable) is that more than one discriminant function can be found. If you insist that the scores from each of the discriminant functions you find be completely uncorrelated with those from each and every one of the others (and we always do), there is, fortunately, a limit to the number of discriminant functions you can find for any given MANOVA problem. The maximum number of discriminant functions, s, cannot be more than P or $k - 1$ (where k is the number of groups), whichever is *smaller*. We can write this as $s = \min(k - 1, P)$. Unfortunately, there is no universal agreement on how to test these discriminant functions for statistical significance.

Consider the case of three groups and two variables. The first discriminant function that is found is the combination of the two variables that yields the largest possible F ratio in an ordinary one-way ANOVA. This combination of variables provides what is called the largest or *greatest characteristic root* (gcr). However, it is possible to create a second discriminant function whose scores are not correlated with the scores from the first function. (It is not possible to create a third function with scores uncorrelated with the first two; we know this because, for this case, $s = 2$). Each function corresponds to its own lowercase lambda (λ), which can be tested for significance.

Having more than one little lambda to test is analogous to having several pairs of means to test in a one-way ANOVA—there is more than one way to go about it while trying to keep Type I errors down and to maximize power at the same time. The most common approach is to form an overall Wilk's Λ by pooling (through multiplication) the little lambdas and then to test Λ with an F ratio (the F ratio will follow an F distribution only approximately if you have more than three groups *and* more than two DVs). Pillai's trace (sometimes called the Pillai-Bartlett statistic) is another way of pooling the little lambdas, and it leads to a statistic that appears to be more robust than Wilk's Λ with respect to violations of the assumptions of MANOVA (see the subsection on Assumptions). Therefore, statisticians tend to prefer Pillai's trace especially when sample sizes are small or unequal. A third way to pool the little lambdas, Hotelling's trace (sometimes called the Hotelling-Lawley trace), is reported and tested by the common statistical software packages but is rarely used. All three of the statistics just described lead to similar F ratios in most cases, and it is not very common to attain significance with one but not the others. All of these statistics (including the one to be described next) can be calculated when there are only two groups, but in that case they all lead to exactly the same F ratio.

A different approach to testing a multigroup MANOVA for significance is to test only the gcr for significance, usually with Roy's largest root test. Unfortunately, it is possible to attain significance with one of the "multiple-root" tests previously described, even though the gcr is not significant. In such a case, it is quite difficult to pinpoint the source of your multivariate group differences, which is why some authors of statistical texts (notably, Harris, 1985) strongly prefer gcr tests. The gcr test is a reasonable alternative when its assumptions are met and when the largest root (corresponding to the best discriminant function) is considerably larger than any of the others. But consider the following situation. The three groups are normals, neurotics, and psychotics, and the two variables are degree of orientation to reality and inclination to seek psychotherapy. The best discriminant function might consist primarily of the reality variable, with normals and neurotics being similar but very different from psychotics. The second function might be almost as discriminating as the first, but if it were weighted most heavily on the psychotherapy variable, it would be discriminating the neurotics from the other two groups. When several discriminant functions are about equally good, a multiple-root test, like Wilk's lambda or Pillai's trace, is likely to be more powerful than a gcr test.

The multiple-root test should be followed by a gcr test if you are interested in understanding why your groups differ; as already mentioned, a significant multiple-root test does not guarantee a significant gcr test (Harris, 1985). A significant gcr test, whether or not it is preceded by a multiple root test, can be followed by a separate test of the next largest discriminant function, and so on, until you reach a root that is not significant. (Alternatively, you can recalculate Wilk's Λ without the largest root, test it for significance, and then drop the second largest and retest until Wilk's Λ is no longer significant.) Each significant discriminant function (if standardized) can be understood in terms of the weight each of the variables has in that function (or the discriminant loading of each variable). I'll say a bit more about this in the section on discriminant analysis.

Follow-Up Tests

Any ANOVA design can be performed as a MANOVA; in fact, factorial MANOVAs are quite common. A different set of discriminant functions is found for each main effect, as well as for the interactions. A significant two-way interaction might be followed by separate one-way MANOVAs (i.e., simple main effects), and a significant main effect in the absence of interaction might be followed by pairwise comparisons. However, even with a (relatively) simple one-way multigroup MANOVA, follow-up tests can get quite difficult to interpret when several discriminant functions are significant.

For instance, if you redid the MANOVA for each pair of groups in the normals/neurotics/psychotics example, you would get a very different discriminant function in each case. The situation can get even more complex and difficult to interpret as the number of groups and variables increases. However, the most common way of following up a significant one-way MANOVA is with separate univariate ANOVAs for each DV. Then any significant univariate ANOVA can be followed up as described in Chapter 13. Of course, this method of following up a MANOVA is appropriate when you are not interested in multivariate relations and are simply trying to control Type I errors. With respect to this last point, bear in mind that following a significant MANOVA with separate tests for each DV involves a danger analogous to following a significant ANOVA with protected t tests. Just as adding one control or other kind of group that is radically different from the others (i.e., the complete null is not true) destroys the protection afforded by a significant ANOVA, adding one DV that clearly differentiates the groups (e.g., a manipulation check) can make the MANOVA significant and thus give you permission to test a series of DVs that may be essentially unaffected by the independent variable.

Assumptions

All of the assumptions of MANOVA are analogous to assumptions with which you should already be familiar. First, the DVs should each be normally distributed in the population and together follow a multivariate normal distribution. For instance, if there are only two DVs, they should follow a bivariate normal distribution as described in Chapter 9 as the basis of the significance test for linear correlation. It is generally believed that a situation analogous to the central limit theorem for the univariate case applies to the multivariate case, so violations of multivariate normality are not considered serious when the sample size is fairly large. Unfortunately, as in the case of bivariate outliers, multivariate outliers can distort your results. Multivariate outliers can be found with the same models described for multiple regression (see Chapter 17, section B).

Second, the DVs should have the same variance in every population being sampled (i.e., homogeneity of variance), and, in addition, the covariance of any pair of DVs should be the same in every population being sampled. The last part of this assumption is essentially the same as the requirement, described in Chapter 16, that pairs of RM levels in a mixed design have the same covariance at each level of the between-groups factor (i.e., multigroup sphericity). In both cases, this assumption can be tested with Box's M test but is generally not a problem when all of the groups are the same size. It is also assumed that no pair of DVs exhibits a curvilinear relation. These assumptions are also the basis of the procedure to be discussed next, discriminant analysis.

Discriminant Analysis

When a MANOVA is performed, the underlying discriminant functions are tested for significance, but the discriminant functions themselves are often ignored. Sometimes the standardized weight or the discriminant loading of each variable is inspected to characterize a discriminant function and better understand how the groups can be differentiated. Occasionally, it is appropriate to go a step further and use a discriminant function to predict an individual's group membership; that procedure is called *discriminant analysis*. Discriminant analysis (DA) is to MANOVA as linear regression and prediction is to merely testing the significance of the linear relation between two variables. As in the case of MANOVA, DA is much simpler in the two-group case, so that is where I'll begin.

In a typical two-group MANOVA situation you might be comparing right- and left-handers on a battery of cognitive tasks, especially those tasks known to be lateralized in the brain, to see if the two groups are significantly different. You want to see if handedness has an impact on cognitive functioning. The emphasis of discriminant analysis is different. In discriminant analysis you want to find a set of variables that differentiates the two groups as well as possible. You might start with a set of variables that seem likely to differ between the two groups and perform a *stepwise discriminant analysis* in which variables that contribute well to the discrimination (based on statistical tests) are retained and those which contribute little are dropped (this procedure is similar to stepwise regression, as described in Chapter 17). The weights of the resulting standardized (i.e., variables are converted to z scores) discriminant function (also called a *canonical variate* because discriminant analysis is a special case of canonical correlation), if significant, can be compared to get an idea of which variables are doing the best job of differentiating the two groups. (The absolute sizes of the weights, but not their relation to one other, are arbitrary and are usually normalized so that the squared weights sum to 1.0). Unfortunately, highly correlated DVs can lead to unreliable and misleading relative weights, so an effort is generally made to combine similar variables or delete redundant ones.

Using MANOVA to Test Repeated Measures

There is another common application of MANOVA with which you will be confronted if you use SPSS to perform an RM ANOVA: MANOVA can be used as an alternative to the univariate one-way RM ANOVA. To understand how this is done, it will help to recall the direct-difference method for the matched *t* test. By creating difference scores, a two-sample test is converted to a one-sample test against the null hypothesis that the mean of the difference scores is zero in the population. Now suppose that your RM factor has three levels (e.g., before, during, and after some treatment). You can create two sets of difference scores, such as before-during (BD), and during-after (DA). (The third difference score, before-after, would be exactly the same as the sum of the other two—because there are only two df, there can only be two sets of nonredundant difference scores.) Even though you now have two dependent variables, you can still perform a one-sample test to determine whether your difference scores differ significantly from zero. This can be accomplished by performing a one-sample MANOVA. The MANOVA procedure will find the weighted combination of BD and DA that produces a mean score as far from zero as possible.

Finding the best weighted average of the difference scores sounds like an advantage over the ordinary RM ANOVA, which just deals with ordinary averages, and it can be—but you pay a price for the customized combinations of MANOVA. The price is a considerable loss of degrees of freedom in the error term. For a one-way RM ANOVA, df_{error} equals $(n - 1)(P - 1)$, where n is the number of different subjects (or matched blocks), and P is the number of levels of the RM factor. If you perform the analysis as a one-sample MANOVA on $P - 1$ difference scores, df_{error} drops to $n - P + 1$ (try a few values for n and P, and you will notice the differences). In fact, you cannot use the MANOVA approach to RM analysis when the number of subjects is less than the number of RM levels (i.e., $n < P$); your error term won't have any degrees of freedom. And when n is only slightly greater than P, the power of the MANOVA approach is usually less than the RM ANOVA.

So, why is the MANOVA alternative strongly encouraged by many statisticians? Because MANOVA does not take a simple average of the variances of the possible difference scores and therefore does not assume that these variances are all the same (the sphericity assumption), the MANOVA approach is not vulnerable to the Type I error inflation that occurs with RM ANOVA when sphericity does not exist in the population. Of course, there are adjustments you can make to RM ANOVA, as you learned in Chapter 15, but when your sample is fairly large a MANOVA is likely to have more power than an adjusted RM ANOVA. Because it is not an easy matter to determine which approach has greater power for various combinations of sample sizes and departures from sphericity (Davidson, 1972), several rules of thumb have been proposed. One recommendation is that when you are dealing with eight RM levels or less, and epsilon (ϵ) is no more than .85, use the MANOVA if your sample size is at least 30 more than the number of RM levels, and use the RM ANOVA (with ϵ-adjusted df) if the sample size is less than that (Algina & Keselman, 1997). As ϵ approaches 1.0, sphericity becomes a reasonable assumption, and even with large sample sizes, it is difficult to know which procedure will have the greater power. Fortunately, modern software provides an easy way to avoid deciding which procedure is more appropriate for your data. Simply plan in all cases to perform both a MANOVA and an ϵ-adjusted ANOVA on your data (SPSS performs both as the default); if only one of the procedures yields significant results, that's the procedure you would report. However, this is a reasonable approach with respect to controlling Type I errors only if you use half of your alpha for each test (usually .025 for each).

Complex MANOVA

The MANOVA approach can be used with designs more complicated than the one-way RM ANOVA. For instance, in a two-way mixed design, MANOVA can be used to test the main effect of the RM factor, just as described for the one-way RM ANOVA. In addition, the interaction of the mixed design can be tested by forming the appropriate difference scores separately for each group of subjects and then using a two- or multigroup (i.e., one-way) MANOVA. A significant one-way MANOVA indicates that the groups differ in their level-to-level RM differences, which demonstrates a group by RM interaction. The MANOVA approach can also be extended to factorial RM ANOVAs (as described in Section C of Chapter 15) and designs that are called *doubly* multivariate. The latter design is one in which a set of DVs is measured at several points in time or under several different conditions within the same subjects.

1. In the simplest form of multivariate analysis of variance (MANOVA) there are two independent groups of subjects measured on two dependent variables. The MANOVA procedure finds the weighted combination of the two DVs that yields the largest possible t value for comparing the two groups. The weighted combination of (two or more) DVs is called the discriminant function, and the t value that is based on it, when squared, is called Hotelling's T^2.

SUMMARY

2. T^2 can be converted into an F ratio for significance testing; the larger the number of DVs that contributed to T^2, the more the F ratio is reduced before testing. The T^2 value is a product of (half) the sample size, and an effect size measure like g (called the Mahalanobis distance), which measures the multivariate separation between the two groups.

3. When there are more than two independent groups, the T^2 statistic is usually replaced by Wilks' Λ, the ratio of error variability (i.e., SS_W) to total variability (you generally want Λ to be small). However, when there are more than two groups and more than one DV, there is more than one discriminant function that can be found (the maximum number is one less than the number of groups or the number of DVs, whichever is smaller) and therefore more than one lambda to calculate. The most common way to test a MANOVA for significance is to pool the lambdas from all the possible discriminant functions and test the pooled Wilks' Λ with an approximate F ratio (Pillai's trace is a way of combining the lambdas that is more robust when the assumptions of MANOVA are violated).

4. When there are more than one discriminant function that can be found, the first one calculated is the one that produces the largest F ratio; this one is called the *greatest characteristic root* (gcr). An alternative to testing the pooled lambdas is to test only the gcr (usually with Roy's test). The gcr test has an advantage when one of the discriminant functions is much larger than all of the others. If, in addition to finding out whether the groups differ significantly, you want to explore and interpret the discriminant functions, you can follow a significant gcr test by testing successively smaller discriminant functions until you come to one that is not significant. Alternatively, you can follow a significant test of the pooled lambdas by dropping the largest discriminant function, retesting, and continuing the process until the pooled lambda is no longer significant.

5. The most common use of MANOVA is to control Type I errors when testing several DVs in the same experiment; a significant MANOVA is then followed by univariate tests of each DV. However, if the DVs are virtually uncorrelated, or one of the DVs very obviously differs among the groups, it may be more legitimate (and powerful) to skip the MANOVA test and test all of the DVs separately, using the Bonferroni

adjustment. Another use for MANOVA is to find combinations of DVs that discriminate the groups more efficiently than any one DV. In this case you want to avoid using DVs that are highly correlated because this will lead to unreliable discriminant functions.

6. If you want to use a set of DVs to predict which of several groups a particular subject is likely to belong to, you want to use a procedure called discriminant analysis (DA). DA finds discriminant functions, as in MANOVA, and then uses these functions to predict which of several possible groups each subject will belong to. DA can also be used as a theoretical tool to understand how groups differ in complex ways involving several DVs simultaneously.

7. It may not be efficient or helpful to use all of the DVs you have available for a particular discriminant analysis. There are procedures, such as stepwise discriminant analysis, that help you systematically to find the subset of your DVs that does the best job of discriminating among your groups. These stepwise procedures are similar to the procedures for stepwise multiple regression.

8. The MANOVA procedure can be used as a substitute for the one-way RM ANOVA by forming difference scores (between pairs of RM levels) and then finding the weighted combination of these difference scores that best discriminates them from zero (the usual expectation under the null hypothesis). Because MANOVA does not require the sphericity assumption, the df does not have to be conservatively adjusted. However, in the process of customizing the combination of difference scores, MANOVA has fewer degrees of freedom available than the corresponding RM ANOVA ($n - P + 1$ for MANOVA, but $(n - 1)(P - 1)$ for RM ANOVA). MANOVA cannot be used in place of RM ANOVA when there are fewer subjects than treatment levels, and MANOVA is not recommended when the number of subjects is only slightly larger than the number of treatments. However, when the sample size is relatively large, MANOVA is likely to have more power than RM ANOVA, especially if the sphericity assumption does not seem to apply to your data.

EXERCISES

*1. In a two-group experiment, three dependent variables were combined to give a maximum t value of 3.8.
  a. What is the value of T^2?
  b. Assuming both groups contain 12 subjects each, test T^2 for significance.
  c. Find the Mahalanobis distance between these two groups.
  d. Recalculate parts b and c if the sizes of the two groups are 10 and 20.

2. In a two-group experiment, four dependent variables were combined to maximize the separation of the groups. $SS_{bet} = 55$ and $SS_W = 200$.
  a. What is the value of Λ?
  b. Assuming one group contains 20 subjects and the other 25 subjects, test Λ for significance.
  c. What is the value of T^2?

d. Find the Mahalanobis distance between these two groups.

*3. Nine men and nine women are tested on two different variables. In each case, the t test falls short of significance; $t = 1.9$ for the first DV, and 1.8 for the second. The correlation between the two DVs over all subjects is zero.
  a. What is the value of T^2?
  b. Find the Mahalanobis distance between these two groups.
  c. What is the value of Wilks' Λ?
  d. Test T^2 for significance. Explain the advantage of using two variables rather than one to discriminate the two groups of subjects.

4. Redo the previous exercise, assuming that the correlation between the two DVs is:
  a. +.7
  b. −.3

*5. Suppose you have planned an experiment in which each of your 12 subjects is measured under six different conditions.
   a. What is the df for the error term if you perform a one-way RM ANOVA on your data?
   b. What is the df for the error term if you perform a MANOVA on your data?

6. Suppose you have planned an experiment in which each of your 20 subjects is measured under four different conditions.
   a. What is the df for the error term if you perform a one-way RM ANOVA on your data?
   b. What is the df for the error term if you perform a MANOVA on your data?

The general linear model for the one-factor case:

$$Y = \mu + \alpha_j + \epsilon_{ij}$$

Formula 18.1

KEY FORMULAS

The significance test for eta squared (equivalent to a one-way ANOVA):

$$F = \frac{\eta^2(N_T - k)}{(1 - \eta^2)(k - 1)}$$

Formula 18.2

The GLM for the two-factor case:

$$Y = \mu + \alpha_i + \beta_j + \alpha_i\beta_j + \epsilon_{ijk}$$

Formula 18.3

The total SS for a one-way ANCOVA:

$$SSA_{total} = SSY_{total}(1 - r_{xy}^2)$$

Formula 18.4

The within-group sum of cross products (based on the DV/covariate correlation within each group):

$$SCPW = \sum (n_i - 1)s_{x_i}s_{y_i}r_i$$

Formula 18.5

The pooled correlation coefficient (based on a weighted average of the within-group DV/covariate correlations):

$$r_p = \frac{SCPW}{\sqrt{SSX_{error}SSY_{error}}}$$

Formula 18.6

The error SS for a one-way ANCOVA:

$$SSA_{error} = SSY_{error}(1 - r_p^2)$$

Formula 18.7

The between-group SS for a one-way ANCOVA (by subtraction):

$$SSA_{between} = SSA_{total} - SSA_{error}$$

Formula 18.8

The pooled regression slope (equivalent to a weighted average of the within-group regression slopes):

$$b_p = r_p\sqrt{\frac{SSY_{error}}{SSX_{error}}}$$

Formula 18.9

The adjusted means for a one-way ANCOVA (adjusting for group differences on the covariate):

$$\text{Adj. } \overline{Y}_i = \overline{Y}_i - b_p(\overline{X}_i - \overline{X}_G)$$

Formula 18.10

The corrected within-group error term for post hoc comparisons following an ANCOVA with randomly assigned groups:

$$CMSA_{error} = MSA_{error}\left(1 + \frac{F_X}{df_{error}}\right)$$ **Formula 18.11**

The F ratio for testing the significance of T^2 calculated for P dependent variables and two independent groups:

$$F = \frac{n_1 + n_2 - P - 1}{P(n_1 + n_2 - 2)}\, T^2$$ **Formula 18.12**

The F ratio for testing the significance of Wilks' lambda calculated for P dependent variables and two independent groups:

$$F = \frac{1 - \Lambda}{\Lambda}\left(\frac{n_1 + n_2 - P - 1}{P}\right)$$ **Formula 18.13**

THE BINOMIAL DISTRIBUTION

Part Seven
Nonparametric
Statistics

19

Chapter

You will need to use the following from previous chapters:

Symbols:
μ: Mean of a population
$\overline{X}$: Mean of a sample
σ: Standard deviation of a population

Formula:
Formula 4.1: The z score

Concepts:
Null hypothesis distribution
Normal distribution
One- and two-tailed hypothesis tests
Addition rule for mutually exclusive events
Multiplication rule for independent events

A

**CONCEPTUAL
FOUNDATION**

I will begin the discussion of nonparametric statistics by describing a situation in which you could not use any of the statistical procedures already presented in this text. For this example I need to bring back our psychic friend from Chapter 5. This time, instead of predicting math aptitude, the psychic claims that he can predict the gender of a child soon after its conception. To test his claim we find four women who very recently became pregnant and ask the psychic to make a prediction about each child. (Perhaps he places his hand on each woman's abdomen and feels "vibrations.") Then we wait, of course, to find out if the psychic is correct in each instance. Suppose that the psychic is correct in all four instances. Should we believe that the psychic has some special ability? (By the way, I am not using this example because I am certain that psychic powers don't exist, but rather because it is easy to believe that there are at least some people claiming psychic powers who have no special abilities at all.)

By now you must know what Dr. Null would say about this claim. Dr. Null would say that the psychic was just lucky this time and that he has no psychic ability. Before we challenge Dr. Null's assessment, we would like to know the probability of Dr. Null's making us look foolish. Dr. Null cannot beat the psychic in this case, but if Dr. Null also makes four correct predictions in a row, our psychic will no longer seem so impressive because Dr. Null makes his predictions at random—perhaps by flipping a coin: heads, it's a boy, tails, it's a girl. In this simple case it is not difficult to calculate the probability of making four correct predictions in a row by chance. First, note that the probability of being correct about any one gender prediction by chance is .5. (Although the number of boys that are born is not exactly equal to the number of girls, I will assume they are equal for the examples in this text.) To find the probability of being correct four times in a row, we need to use the multiplication rule for independent events (see Chapter 4, Section C). We multiply .5 by itself four times: $p = .5 \times .5 \times .5 \times .5 = .0625$.

Now we know that Dr. Null has a .0625 chance of predicting just as well as the psychic. This is not a very large probability (1 out of 16), but if we had set alpha at .05, we could not reject Dr. Null's hypothesis that the psychic has no ability. Fortunately, it was very easy in this example to calculate Dr. Null's probability of duplicating the psychic's performance, but you will not always be quite this fortunate. Suppose that the psychic makes 10 predictions and is right in eight of the cases. This level of performance sounds pretty good, but we need to find Dr. Null's chance of being right eight, nine, or all ten times, and this involves more than a simple application of the multiplication rule.

The Origin of the Binomial Distribution

The approach to null hypothesis testing that I have taken in previous chapters is to find the null hypothesis distribution (NHD) and then to locate particular experimental results on that distribution. Constructing the NHD when you are dealing with a total of only four dichotomous predictions is not difficult, but it will require the application of some simple probability rules. Constructing the NHD when there are 10 predictions to be made is considerably more tedious, but it uses the same principles. In both cases, the NHD is a form of the *binomial distribution*; there is a different binomial distribution for each number of predictions to be made. Of course, in real life you would not have to construct the binomial distribution yourself. You would either find a table of the distribution, use an approximation of the distribution if appropriate, or analyze your data by computer. However, to understand this distribution and its use, it will help to see how the distribution arises.

A binomial distribution may arise whenever events or observations can be classified into one (and only one) of two categories, each with some probability of occurrence (e.g., male or female, right or wrong); such events are called *dichotomous events*. The probabilities corresponding to the two categories are usually symbolized as P and Q. Because P and Q must add up to 1.0, knowing P automatically tells you what Q is; in fact, the two probabilities are often referred to as P and $1 - P$. (Note that I am using an uppercase P for the probability of a single event, whereas I will continue to use a lowercase p to represent a p value for testing a null hypothesis.) The simplest case is when $P = Q = .5$, so this is the case I will consider first. Fortunately, this case frequently represents the null hypothesis, such as when you are flipping a coin to see if it is fair (i.e., the probability of heads equals the probability of tails).

To get a binomial (or Bernouilli) distribution you need to have a sequence of *Bernouilli trials* (after Jacques Bernouilli, 1654–1705, a Swiss mathematician). Bernouilli trials are dichotomous events that are independent of each other and for which P and Q do not change as more and more trials occur. The total number of trials is usually symbolized as N; the binomial distribution is a function of both P and N. The reason you have a distribution at all is that whenever there are N trials, some of them will fall into one category and some will fall into the other, and this division into categories can change for each new set of N trials. The number of trials that fall into the first category (the one with probability P) is usually called X; this is the variable that is distributed. For instance, P can stand for the proportion of women eligible for jury duty, and Q, for the proportion of men. Assuming that each jury selected consists of 12 people ($N = 12$), X (the number of women on a particular jury) can vary from 0 to 12. If you look at thousands of juries, you will see that some values of X are more common than others; you are not likely to get a uniform distribution. If juries were selected at random with respect to gender, an infinite number of juries would give you a perfect binomial distribution, as shown in Figure 19.1. To show you how this distribution is constructed, and why the middle values of X turn out to be the most frequent, I will begin with a case that has a small N.

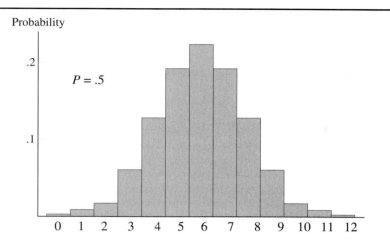

Probability

P = .5

0 1 2 3 4 5 6 7 8 9 10 11 12

Figure 19.1

Binomial Distribution for
$N = 12, P = .5$

The Binomial Distribution with $N = 4$

A very simple experiment consists of flipping the same coin four times to see if it is fair. With only four flips it is not difficult to write out every possible sequence of heads and tails that can occur. Because there are two possible outcomes for each toss (H or T) and four tosses, there are $2 \times 2 \times 2 \times 2 = 16$ different sequences that can occur. They are shown in Table 19.1.

HHHH	HTHH	THHH	TTHH
HHHT	HTHT	THHT	TTHT
HHTH	HTTH	THTH	TTTH
HHTT	HTTT	THTT	TTTT

Table 19.1

If our focus is on the number of heads (X) in each sequence, some of the sequences shown in Table 19.1 can be lumped together. For instance, the following four sequences are alike because each contains just one head: HTTT, THTT, TTHT, TTTH. For each possible number of heads (zero to four) we can count the number of sequences as follows: 0H: one sequence; 1H: four sequences; 2H: six sequences; 3H: four sequences; 4H: one sequence. If the probability of heads is P and the probability of tails is Q, the probability of any particular sequence is found by multiplying the appropriate Ps and Qs. For example, the probability of flipping HTTH would be $P \times Q \times Q \times P$. Matters are greatly simplified if you assume that the coin is fair (i.e., H_0 is true). In this case, $P = Q = .5$, so every sequence has the same probability, $.5 \times .5 \times .5 \times .5 = .0625$. Note that we could have arrived at the same answer by observing that there are 16 possible sequences, and if each is equally likely (as is the case when $P = Q$), the probability of any one sequence occurring is 1 out of 16, which equals 1/16 = .0625.

Now, if we want to know the probability of flipping just one head in four flips, we have to add the probabilities of the four sequences that contain only one head each (using the addition rule of probability; see Chapter 4, Section C). So, the probability of one head = .0625 + .0625 + .0625 + .0625 = .25. Again, we could have arrived at the same conclusion by noting that 4 out of the 16 equally likely sequences shown in Table 19.1 contain one head, and 4/16 = .25.

Figure 19.2

Binomial Distribution for
$N = 4, P = .5$

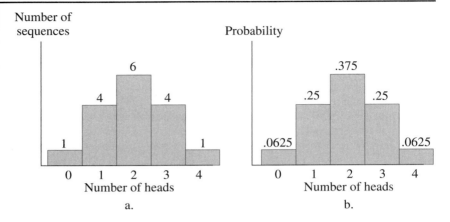

a.   b.

By finding the number of sequences with zero, one, two, three, and four heads, we obtain the frequency distribution shown in Figure 19.2a. Dividing each number of sequences by the total number of sequences (16), we can convert the frequency distribution into a probability distribution: the binomial distribution with $P = .5$, $N = 4$ (see Figure 19.2b). Looking at the figures, the first thing you might notice about these distributions is that they are symmetrical; this is because $P = Q$. If X were the number of blond children two brunette parents might have, P would be less than Q, and three blondes in a family of four siblings would be much less likely than one blond in four. The distribution would be positively skewed. The next thing you might notice about Figure 19.2 is that the symmetrical binomial distribution looks a little like the normal distribution. This resemblance increases as N gets larger; for example, take another look at Figure 19.1, in which $P = .5$ and $N = 12$.

The Binomial Distribution with $N = 12$

The binomial distribution with $N = 12$ can be constructed using the same method we employed for $N = 4$. However, with $N = 12$ there are a total of 4,096 sequences to write out (2 multiplied by itself 12 times, or 2^{12}). Then we would have to count how many contained one head, two heads, and so on. There is a mathematical shortcut, but it is based on combinations and permutations, which I will not discuss until Section C. Table A.13 shows the probabilities for all the binomial distributions from $N = 1$ to $N = 15$ (assuming $P = .5$). To show how we can use these probabilities to test statistical hypotheses, I will return to the example involving juries ($N = 12$), in which X represents the number of women.

For the sake of the example, assume that each jury is selected at random from the list of registered voters and that this list contains an equal number of women and men. Suppose that the docket includes a case involving a divorced couple, and the husband is claiming that the jury—nine women and three men—has been unfairly stacked against him. If the null hypothesis is that the jury is just a random selection from the voter list (i.e., $P = Q$), we can use Table A.13 to find the p value that corresponds to a nine-woman–three-man jury. Remember that the p value can be viewed as Dr. Null's chance of beating your experimental result, and a tie goes to Dr. Null. In this example Dr. Null has to randomly select a jury that is just as unbalanced or even more unbalanced than the one in question. First, you see from the table that when $N = 12$ and $X = 9$, $p = .0537$. That is,

Dr. Null has a .0537 chance of selecting such a jury at random. But we must also include the chance that Dr. Null will select a jury of 10, 11, or 12 women, because he beats us in each of those cases. Summing the probabilities for $X = 9$, 10, 11, and 12, we have $.0537 + .0161 + .0029 + .0002 = .0729$. This is not a large probability, but if we use an alpha of .05 to make our decision about fairness, $p > .05$, and we cannot reject the null hypothesis that this jury was selected at random (from a pool of an equal number of women and men).

In the preceding example, I performed a one-tailed test. I looked only at the chance of the jury having an excess of women. If the jury had had nine men and three women, the husband might not have complained, but from the standpoint of testing for random selection this jury would be just as unbalanced as the previous one. In most cases a two-tailed test is the more reasonable approach. To perform a two-tailed test, we need to find the two-tailed p value. All we have to do is take the one-tailed p value just found and double it: $.0729 \times 2 = .1458$. This makes the p value even larger, so if the one-tailed p was not less than .05, the two-tailed p will not be significant either. Could we reject the null hypothesis with a jury of ten women and two men (or vice versa)? In that case, $p = .0161 + .0029 + .0002 = .0192$, and the two-tailed $p = .0192 \times 2 = .0384$. Even with a two-tailed test, p is less than .05, so we can reject the null hypothesis. Of course, this doesn't mean that a jury of ten women and two men (or vice versa) could *not* have been selected at random—but assuming a human element of bias could have entered the jury selection process, finding the p value can help us decide how seriously to consider the possibility that the jury was not randomly selected.

The binomial distribution can also be applied to the evaluation of experiments, such as the gender predictions of the psychic. If the psychic makes 12 independent gender predictions (for 12 different, unrelated pregnant women) and is wrong only twice, we know from the preceding calculations that p (two-tailed) $< .05$, so we can reject the null hypothesis that the psychic has no special ability. On the other hand, if the psychic were wrong 3 out of 12 times, $p > .05$ (even for a one-tailed test), so we could not reject the null hypothesis. (Note that rejecting H_0 for one experiment would not convince many scientists that psychic abilities exist. The more controversial the research hypothesis, the more evidence would have to accumulate before the scientific community would really be comfortable in rejecting the null hypothesis.)

When the Binomial Distribution Is Not Symmetrical

So far I have been dealing only with the symmetrical binomial distribution, but there are plenty of circumstances in which P is less than or more than .5. For instance, in the jury problem, imagine a city whose citizens belong to either of two racial or ethnic groups, which I will refer to as X and Y. Forty percent of the people belong to group X (i.e., $P = .4$) and 60% belong to group Y (i.e., $Q = .6$). In this case, selecting a jury consisting of four Xs and eight Ys would be fairly likely and much more likely than selecting eight Xs and four Ys. Unfortunately, Table A.13 could not help us find p values for different jury combinations. We would need a version of Table A.13 constructed for $P = .4$.

Complete tables for various values of P (other than .5) are not common, but there are tables that will give you the critical value for X as a function of both N and P for a particular alpha level. For instance, for the preceding example ($N = 12$, $P = .4$, $\alpha = .05$), the critical value of X would be 9, which means that randomly selecting a jury with nine Xs has a probability of less than .05 (so would a jury with even more Xs, of course). Bear in mind that this kind of table is set up for a one-tailed test. When the distribution is not symmetrical, the two-tailed p is not simply twice the size of the one-tailed p. I have not included such a table in this text because these tables are rarely used. Alternative procedures for finding p values in such cases will be discussed shortly and in the next chapter.

I pointed out earlier that the symmetrical binomial distribution bears some resemblance to the normal distribution and that this resemblance increases as N gets larger. In fact, the binomial distribution becomes virtually indistinguishable from the normal distribution when N is very large, and the two distributions become truly identical when N is infinitely large. For any particular value of P, even when P is not equal to .5, the binomial distribution becomes more symmetrical and more like the normal distribution as N gets larger. However, as P gets further from .5, it takes a larger N before the distribution begins to look symmetrical. When N is large enough, the binomial distribution resembles a normal distribution that has a mean of NP and a standard deviation of $\sqrt{NPQ}$. This resemblance can be used to simplify null hypothesis testing in situations that would otherwise call for the binomial distribution. An example of the normal approximation to the binomial distribution follows.

The Normal Approximation to the Binomial Distribution

Consider a grand jury, which can contain as many as 48 individuals. Assume again that equal numbers of women and men are eligible to serve as jurors. With $N = 48$ and $P = .5$, the binomial distribution looks a lot like the normal distribution (see Figure 19.3). Indeed, this binomial distribution can be approximated, without much error, by a normal distribution with a mean of 24 ($\mu = NP = 48 \times .5$) and a standard deviation of 3.46 ($\sigma = \sqrt{48 \cdot .5 \cdot .5} = \sqrt{12}$). Assuming that we are now working with a normal distribution and we want to find the p value associated with a particular value of X, we need to convert X to a z score. According to Formula 4.1:

$$z = \frac{X - \mu}{\sigma}$$
Formula 4.1

If we substitute expressions for μ and σ in terms of N, P, and Q, we create the following new formula, Formula 19.1, for finding z scores when the normal distribution is being used to approximate a binomial distribution:

$$z = \frac{X - NP}{\sqrt{NPQ}}$$
Formula 19.1

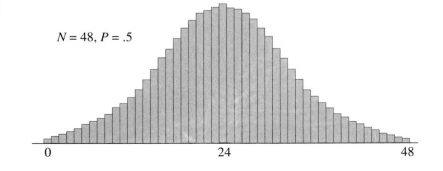

Figure 19.3

Binomial Distribution for
$N = 48$, $P = .5$

$N = 48$, $P = .5$

0   24   48

This formula can be used to determine how likely it is to select a jury with a number of women equal to X or more. For example, a particular grand jury ($N = 48$) contains 30 women. What is the likelihood of randomly selecting such a jury with this many women or more? First, we find the z score according to Formula 19.1:

$$z = \frac{X - 24}{3.46} = \frac{30 - 24}{3.46} = \frac{6}{3.46} = 1.73$$

Looking up this z score in Table A.1, we see that the p value (i.e., the area beyond z) is equal to .0418. With a one-tailed test we could reject the null hypothesis at the .05 level, but not with a two-tailed test.

Another Example of the Normal Approximation

The preceding z score formula can be used in the same way no matter what the value of P is, as long as N is sufficiently large. Returning to the example of two ethnic or racial groups in a city, assume that group X comprises 30% of the population and group Y comprises 70% (i.e., $P = .3$, $Q = .7$). A grand jury is found to have only 10 Xs and 38 Ys. What is the chance of randomly selecting only 10 Xs (or fewer) for a grand jury in this case? We begin by using Formula 19.1:

$$z = \frac{X - NP}{\sqrt{NPQ}} = \frac{10 - 48(.3)}{\sqrt{48(.3)(.7)}} = \frac{10 - 14.4}{\sqrt{10.08}} = \frac{-4.4}{3.17} = -1.38$$

The p value corresponding to this z score is .0838, which does not permit us to reject the null hypothesis at the .05 level. Selecting only 10 Xs for a jury of 48 is not extremely unusual given that the X group is only 30% of the population. ($NP = 14.4$ is the number of Xs to be expected on the average grand jury, in this example.)

I have not said how large N must be to justify using the normal distribution as an approximation of the binomial distribution. There is no exact answer, of course, but most researchers agree that when $P = .5$, N does not have to be more than 20 or 25 before the error of approximation becomes negligible. As P gets closer to 0 or 1, a larger N is needed to maintain a good approximation. For example, when P is only .1, an N of at least 100 is preferred. As a general rule, when P is not near .5, NPQ should be at least 9.

The z Test for Proportions

The results of large-scale studies, such as national surveys or polls, are usually presented in terms of percentages or proportions rather than as actual frequencies. For instance, a newspaper might report that 58% of those sampled favor candidate A, whereas 42% favor candidate B. Knowing N, we could determine the actual number of people favoring one of the candidates and use Formula 20.1 to test the hypothesis that the two candidates are equally favored in the population. On the other hand, we can modify Formula 19.1 to get a formula for testing proportions directly. All we need to do is divide both the numerator and the denominator of Formula 19.1 by N, as follows:

$$z = \frac{\frac{1}{N}(X - NP)}{\frac{1}{N}\sqrt{NPQ}} = \frac{\frac{X}{N} - \frac{NP}{N}}{\frac{\sqrt{NPQ}}{N}}$$

This leads to Formula 19.2:

$$z = \frac{p - P}{\sqrt{\dfrac{PQ}{N}}}$$ **Formula 19.2**

where p (not to be confused with a p value) is the proportion in the X category, and P, as before, is the population proportion, according to the null hypothesis. To use this formula, you must convert data given as percentages to proportions, but that only entails dividing the percentage by 100. For example, if 58% of those polled favor candidate A, $p = 58/100 = .58$. If the null hypothesis is that the two candidates are equally favored, $P = .5$. Assuming the sample consists of 200 people ($N = 200$), Formula 19.2 yields the following result:

$$z = \frac{.58 - .5}{\sqrt{\dfrac{(.5)(.5)}{200}}} = \frac{.08}{\sqrt{.00125}} = \frac{.08}{.0353} = 2.26$$

This z score is large enough to allow us to reject the null hypothesis and conclude that there is a preference for candidate A in the population. Note that we would have obtained the same z score by finding that $X = 116$ (58% of 200), $NP = 100$, $NPQ = 50$ and inserting these values in Formula 19.1. Formula 19.2, however, gives us the convenience of dealing directly with proportions.

If there are three or more candidates in an election poll, and you want to know if they are all equally favored, you cannot use the binomial distribution at all; you are dealing with a *multinomial* situation and need to use the methods described in the next chapter. In the next section of this chapter, I will focus on a particular application of the binomial distribution for null hypothesis testing: an alternative to the matched t test.

SUMMARY

1. The binomial distribution arises from a series of independent, dichotomous events. The two possible outcomes of each event have probabilities P and Q, which sum to 1.0 (i.e., $Q = 1 - P$).
2. Each simple event is commonly called a trial, and the total number of trials is symbolized by N. The number of trials falling in the category with probability P is labeled X. The distribution of X, which follows the binomial distribution, is a function of both N and P.
3. When $P = Q = .5$, the binomial distribution will be symmetrical. As N increases, this distribution more closely resembles the normal distribution. Even when $P \neq Q$, the resemblance to the normal distribution increases as N increases; as N becomes infinitely large, the binomial distribution becomes identical to the normal distribution.
4. The binomial distributions for $P = Q = .5$ and N ranging from 1 to 15 are given in Table A.13. This table can be used to test the null hypothesis that $P = .5$. For example, if three women are members of a jury of 12, look at the binomial distribution for $N = 12$ and find the probability corresponding to $X = 3$. Then also find the probabilities for more extreme values of X (in this case, $X = 2, 1,$ and 0). The sum of these probabilities is the one-tailed p value for testing whether $P = .5$.
5. To perform a two-tailed test, double the p value you find using the preceding procedures. If $P \neq Q$, the binomial distribution will not be symmetrical, and it will not be easy to perform a two-tailed test.
6. When N is sufficiently large, the binomial distribution closely resembles a normal distribution with a mean of NP and a standard devia-

tion of $\sqrt{NPQ}$. The probability that X or more trials will land in the P category can be approximated by calculating the appropriate z score and looking up the area beyond that z score.

7. If $P = .5$, the normal distribution becomes a good approximation when N reaches about 25. If P is considerably more or less than .5, a larger N is needed for a good approximation. A good rule of thumb is that when P is not near .5, NPQ should be at least 9 if you wish to use the normal approximation.

EXERCISES

1. a. Write out all the possible gender sequences for five children born into one family.
   b. Assuming that $P = Q$ for each birth, construct the binomial distribution for the genders of the five children.

*2. A particular woman has given birth to 11 children: nine boys and two girls. Assume $P = .5$ and use Table A.13 to answer the following questions:
   a. What is the probability of having nine or more boys?
   b. What is the probability of having nine or more children of the same gender?
   c. Would you reject the null hypothesis (i.e., $P = .5$) at the .05 level with a one-tailed test? With a two-tailed test?

3. Fourteen infants are simultaneously shown a picture of a human face and a colorful ball of the same size (relative positions of the pictures are varied).
   a. If 10 of the infants spend more time looking at the face than the ball, can the null hypothesis (no preference between the face and the ball) be rejected at the .05 level (two-tailed test)?
   b. In a two-tailed test, how many infants must spend more time looking at the face than the ball to allow the researcher to reject the null hypothesis at the .05 level? At the .01 level?

*4. He didn't get a chance to study, so Johnny guessed on all 100 questions of his true-or-false history test.
   a. If Johnny scored 58 correct, can we conclude ($\alpha = .05$, two-tailed) that he actually knew some of the answers and wasn't guessing (i.e., can we reject $P = .5$)?
   b. How many questions would Johnny have to get correct for us to conclude that he was not just guessing randomly?

5. Jane's history test consisted of 50 multiple-choice questions (four choices for each question).
   a. If she gets 20 correct, can we conclude that she wasn't merely guessing?
   b. How many questions would Jane have to get correct for us to conclude that she was not just guessing?

*6. Suppose that 85% of the population is right-handed (Q) and 15% are left-handed (P).
   a. If out of 120 randomly selected civil engineers, 27 are found to be left-handed, what is the z score for testing the null hypothesis? Can we reject the null hypothesis that $P = .15$ for this profession?
   b. If 480 civil engineers are sampled and 108 are found to be left-handed, what is the z score for testing the null hypothesis?
   c. How does the z score in part a compare to the z score in part b? Can you determine the general rule that is being illustrated?

7. Fifty women involved in abusive marriages filled out a questionnaire. The results indicated that 30 of these women had been abused as children. If you know that 20% of all women were abused as children, test the null hypothesis that the women in the study are a random selection from the general population (use $\alpha = .05$, two-tailed). What can you conclude about the likelihood that women abused as children will end up in abusive marriages?

*8. In the town of Springfield, 70% of the voters are registered as Republicans and 30% as Democrats. If 37% of 80 voters polled at random say that they plan to vote for Bert Jones, a Democrat, for mayor, can we conclude ($\alpha = .05$, two-tailed) that the people of Springfield are not going to vote strictly along party lines?

B

BASIC STATISTICAL PROCEDURES

One important application of the binomial distribution arises when two stimuli are being compared but the comparison cannot be easily quantified. For instance, a music teacher can listen to two students play the same piece of music and be quite sure which student is superior without being able to quantify the difference—especially if the difference is slight. This kind of comparison can form the basis of an experiment.

Suppose that music students are closely matched in pairs so that the two students in each pair are virtually indistinguishable in their ability to play a particular piece of music. Then one member of each pair is chosen at random to participate in a mental practice session in which the subject is guided in using visual, auditory, and kinesthetic imagery to rehearse the piece of music selected for the study. The other member of each pair is exposed to a control procedure (perhaps some relaxation exercises) for the same period of time. Then each pair of subjects comes before a panel of well-trained judges who do not know which subject received imagery training and which did not. One at a time, the two subjects play the same piece of music, and the judges must decide which subject had mastered the piece more successfully. (The order in which the subjects play is determined randomly for each pair.) In each pair, either the experimental subject or the control subject is rated superior (ties are avoided by having an odd number of judges on the panel).

If the difference between the members of each pair could be reliably quantified (e.g., if each member of the pair could be given a rating and then one rating could be subtracted from the other), a matched t test could be performed on the difference scores. If it is only possible to judge which member of each pair is better, an alternative to the matched t test, called the *sign test*, can be performed. The six-step procedure I have been using to test null hypotheses based on parametric statistics can also be applied to nonparametric statistics, as I will now demonstrate.

Step 1. State the Hypotheses

The null hypothesis is that subjects given imagery training will play no better (or worse) than the control subjects. If P represents the probability of the experimental subject of the pair being rated superior, the null hypothesis can be stated symbolically as H_0: $P = .5$. The two-tailed alternative hypothesis would be stated as H_A: $P \neq .5$, whereas a one-tailed H_A could be stated as $P > .5$ or $P < .5$.

Step 2. Select the Statistical Test and the Significance Level

Because the difference in each pair of subjects will not be measured precisely but only categorized in terms of direction, the appropriate test is the sign test. The same considerations concerning the selection of alpha for parametric statistics apply to nonparametric procedures as well. Therefore, we will stay with the convention of setting alpha at .05.

Step 3. Select the Samples and Collect the Data

It may not be feasible to obtain subjects through random selection from the entire population of interest. However, as usual, it is critical that the assignment of subjects within each pair be random. To use the normal distribution as an approximation of the null hypothesis distribution, you need to have at least about 20 pairs. Imagine that you have 20 pairs of subjects who are to be tested. Each pair is given a plus (+) if the experimental subject is judged superior and a minus (−) if the control subject is judged superior. The total number of + signs is referred to as X. It is because we are only considering the direction, or sign, of the difference in each pair that the test is called the sign test. For this example, we will assume that there are 15 pluses and 5 minuses (and no ties), so $X = 15$.

Step 4. Find the Region of Rejection

Because we are using the normal distribution and have set alpha equal to .05 (two-tailed), the region of rejection is the portion of the distribution above $z = +1.96$ or below $z = -1.96$.

Step 5. Calculate the Test Statistic

The appropriate statistic, if we are using the normal approximation, is z as calculated by Formula 19.1. However, using the normal curve, which is smooth and continuous, to approximate the binomial distribution, which is discrete and steplike, becomes quite crude when N is fairly small. The biggest problem is that for the binomial distribution the probability of some value, say 3, is equivalent to the area of the rectangular bar that extends from 2 to 3 (the first bar goes from 0 to 1, the second from 1 to 2, etc.), whereas for the normal distribution the probability of 3 corresponds to the area enclosed by a range from 2.5 to 3.5. This discrepancy is reduced by subtracting half of a unit in the numerator of Formula 19.1. This is called the *correction for continuity* (Yates, 1934). Because sometimes the numerator will be negative (when $X < NP$), we need to take the absolute value of the numerator before subtracting .5; otherwise we would actually be making the numerator larger whenever it was negative. Including the correction factor in Formula 19.1 yields Formula 19.3:

$$z = \frac{|X - NP| - .5}{\sqrt{NPQ}}$$ **Formula 19.3**

As N increases, the steps of the binomial distribution get smaller, and the discrepancy between the normal and binomial distributions becomes smaller as well. Therefore, for large N the continuity correction makes very little difference and is usually ignored. As usual, there is some disagreement about how large N should be before the continuity correction is not needed. It is safe to say that when N is greater than 100, the continuity correction makes too little difference to worry about. For the example in this section, N is small enough that the continuity correction makes a noticeable difference, so we will use Formula 19.3 to find our z score:

$$z = \frac{|X - NP| - .5}{\sqrt{NPQ}} = \frac{|15 - 20(.5)| - .5}{\sqrt{20(.5)(.5)}} = \frac{|15 - 10| - .5}{\sqrt{5}}$$

$$= \frac{5 - .5}{2.24} = \frac{4.5}{2.24} = 2.01$$

Step 6. Make the Statistical Decision

Because the calculated z falls in the region of rejection ($2.01 > 1.96$), we can reject the null hypothesis that $P = .5$. Having rejected the null hypothesis, we can conclude that the imagery training has had more of an effect than the control procedure. Too many experimental subjects have performed better than their control subject counterparts for us to conclude that this is merely an accident of sampling.

Interpreting the Results

You might argue that the subjects in this hypothetical study are not representative of the entire population of interest and that this limits our conclusion to saying that the imagery training only works on subjects with some musical background or with other characteristics that resemble the subjects in our study. That argument has a good deal of merit, but at least the random assignment of subjects to type of training, the random order of playing within each pair, and the blindness of the judges to the subjects' assignment all ensure that the results are internally valid (that is, within the sample of subjects that was tested).

Assumptions of the Sign Test

Dichotomous Events

It is assumed that the outcome of each simple event or trial must fall into one of two possible categories and that these two categories are mutually exclusive and exhaustive. That is, the event cannot fall into both categories simultaneously, and there is no third category. Therefore, if the probabilities for the two categories are P and Q, $P + Q$ will always equal 1.0. In terms of the preceding example, we are assuming that one member of each pair must be superior; a tie is not possible. In reality, the sign test is sometimes performed by discarding any trials that result in a tie, thus reducing the sample size (e.g., if there were two ties out of the 20 pairs, N would equal 18) and therefore the power of the test. Bear in mind, however, that if more than a few ties occur, the validity of the sign test is weakened.

Independent Events

It is assumed that the outcome of one trial in no way influences the outcome of any other trial. In the preceding example, each decision about a pair of subjects constitutes a trial. The assumption holds for our example because there is no connection between the decision for one pair of subjects and the decision for any other pair.

Stationary Process

It is also assumed that the probability of each category remains the same throughout the entire experiment. Whatever the value is for P on one trial, it is assumed to be the same for all trials. There is no reason for P to change over the course of trials in the experiment used for the example in this section. It is possible, however, to imagine a situation in which P changes over time, even if the trials are independent. For instance, if the trials are mayoral elections in a small city, and P is the probability of the mayor's belonging to a particular ethnic group, P can change between elections because of shifts in the relative percentages of various ethnic groups in that city.

Normal Approximation

To use the normal distribution to represent the NHD, as we did in our example, we need to assume that there is a negligible amount of error involved in this approximation. If $P = .5$ (as stated by H_0) and N is at least 20, the normal distribution is a reasonably good approximation. If N is less than 16 and $P = .5$, you can use Table A.13 to obtain the p value for your sign test exactly.

Distribution-Free Tests

Note that we do *not* have to assume that the dependent variable is normally distributed. (In fact, we don't know what the distribution of the differences between pairs of subjects would look like if we were able to measure them accurately rather than just assigning them a $+$ or a $-$.) That is why the sign test is sometimes called a *distribution-free test*. On the other hand, to perform a matched t test you need to assume that the difference scores from all the pairs of subjects are normally distributed. For the sign test, we do need to know that the binomial distribution will represent the null hypothesis distribution (NHD), but this is guaranteed by the first three assumptions just described.

The Gambler's Fallacy

The assumption of independent events will be true in a wide range of situations, including the mechanics underlying games of chance. For instance, successive flips of the same coin can be assumed to be independent, except in rather strange circumstances, such as the following. A sticky coin that is flipped can accumulate some dirt on the side that lands face down, and the coin can therefore develop an increasing bias; the outcome of a trial would then depend to some extent on the outcome of the trial before it (e.g., a string of heads leads to a buildup of dirt on the tail side, making heads even more likely). On the other hand, even when random trials are completely independent, it can appear to some people as though they are not independent. This is called the *gambler's fallacy* (as mentioned in Chapter 4, Section C). In one version of the fallacy, it seems that after a run of trials that fall in one category, the probability of future trials falling in the other category increases, so there will tend to be equal numbers of trials in each category. It "feels" as though there is some force of nature trying to even out the two categories (if $P = .5$)—so after a string of heads, for example, you expect that the chance of a tail has increased. However, if the trials are truly independent (as they usually are in such situations), this will not happen; after a "lucky" streak of 10 heads in a row, the probability of a tail is still equal to .5. Although there tends to be an equal number of heads and tails after a coin has been flipped many times, there is no process that compensates for an unusual run of trials in the same category—the coin has no memory of its past flips.

When to Use the Binomial Distribution for Null Hypothesis Testing

The Sign Test

The sign test is an appropriate alternative to the matched t test when only the direction, but not the magnitude, of the difference between two subjects in a pair (or two measurements of the same subject) can be determined. The sign test can also be applied in cases where a matched t test had been originally planned but the difference scores obtained are so far from resembling a normal distribution that this assumption of the matched t test appears to be severely violated. This violation of the normality assumption can threaten the accuracy of a t test, especially if the number of paired scores is less than about 20. If you are worried that the assumptions of the matched t test will not be met, you can ignore the magnitudes of the difference scores and add up only their signs. On the other hand, the sign test will usually have considerably less power than the matched t test in a situation where both tests could apply. Therefore, if each of the assumptions of the matched t test are fairly reasonable for your case, the matched t test is to be preferred over the sign test. A third alternative, which is usually intermediate in power between the two tests just mentioned (but closer to the t test), is the Wilcoxon signed-rank test. The Wilcoxon test can only be applied, however, when the difference scores can be rank ordered; this approach will be discussed further in Chapter 21.

Observational Research

There are many grouping variables that consist of two, and only two, mutually exclusive categories (e.g., male or female, married or not married). If the proportion of each category in the population is fairly well known, it is possible to test a subpopulation to see if it differs from the general population with respect to those categories. For instance, imagine that you test 20 leading architects for handedness and find that eight are left-handed. If the instance of left-handedness is only 10% in the population (i.e., $P = .1$), the binomial distribution with $N = 20$, $P = .1$ can be used to represent the null hypothesis for the population of architects. You would need to use a table for this particular binomial distribution (or you could construct one with the methods described in Section C); the normal approximation would not be sufficiently accurate in this case. Whichever method you used, you would find that H_0 could be rejected, so you could conclude that the proportion of left-handers is higher among architects than in the general population. However, you could *not* determine whether left-handedness contributes to becoming an architect based on this finding. You have only observed an association between handedness and being an architect.

Experimental Research

In some cases, a subject's response to an experimental task or condition can fall into only one or the other of two categories (e.g., the subject solved the problem or not; used a particular strategy for solution or not). Consider a study of seating preferences in a theater. One subject at a time enters a small theater (ostensibly to see a short film); the theater contains only two seats. Both seats are at the same distance from the screen, but one is a bit to the left of center, whereas the other is equally to the right of center. The variable of interest is which seat is chosen, and this variable is dichotomous, having only two values (i.e., left and right). The null hypothesis is that the left and right seats have an equal probability of being chosen. If one side is chosen by more subjects than the other, the binomial distribution can be used to decide whether the null hypothesis can be rejected. (Assume that only right-handers are included in the study.) If the experiment involved two different groups of subjects—such as left-handers and right-handers—and you observed seating preferences for both groups, you would have a two-factor experiment (handedness vs. seating preference) in which both factors are dichotomous. To analyze the results of such an experiment, you would need the methods I will discuss in the next chapter.

1. The sign test can be used in place of a matched t test when the amount of difference between the members of a matched pair cannot be determined, but the direction of that difference can be. Because the direction of each difference must fall into one of only two categories (such as $+$ or $-$), the binomial distribution can be used to determine whether a given imbalance between those categories is likely to occur by chance. For most experimental manipulations, an equal number of pluses and minuses would be expected under the null hypothesis.

2. Unless N is very large, using the normal distribution to approximate the binomial distribution introduces a small systematic error that can be compensated for with a *correction for continuity*—subtracting half of a unit from the absolute value of the difference in the numerator of the z score.

3. The valid application of the sign test requires the following assumptions:

   a. *Dichotomous events*. Each simple event or trial can fall into only one or the other of two categories—not both categories simultaneously or some third category. The probabilities of the two categories, P and Q, must sum to 1.0.

   b. *Independent events*. The outcome of one trial does not influence the outcome of any other trial.

   c. *Stationary process*. The probabilities of each category (i.e., P and Q) remain the same for all trials in the experiment.

4. The most common gambler's fallacy is that the probability of truly independent events will change to compensate for a string of events in one category—to ensure a proper balance between the categories (e.g., after a string of heads is flipped, a tail is now overdue—that is, more likely than usual). The illusion is produced by the fact that the numbers of events in different categories *tend* to balance out over the long run (according to their initial probabilities), but there is no "force of nature" operating on any one particular event to change its probability and balance things out, regardless of which events have occurred in the past (i.e., if the events are truly independent, which they are in many games of chance and psychological experiments).

5. The binomial distribution has the following (fairly) common uses in psychological research:

   a. *The sign test*. The binomial distribution can be used as an alternative to the matched or repeated measures t test when it is possible to determine the direction of the difference between paired observations but not the amount of that difference. The sign test can be planned (as when you make no attempt to measure the amount of difference but assess only its direction) or unplanned (as when you plan a matched t test but the sample size is fairly small and the difference scores are very far from following a normal distribution).

   b. *Observational research*. The binomial distribution applies when this kind of research involves counting the number of individuals in each of two categories within a specified group (e.g., counting the number of people who are heavy smokers [or not] among patients with heart disease). The values of P and Q are based on estimates of the proportion of each category in the general population (e.g., if it is estimated that 30% of the population are heavy smokers, $P = .3$ and $Q = .7$).

   c. *Experimental research*. The binomial distribution is appropriate when the dependent variable is not quantifiable but can be categorized as one of two alternatives. (For example, given a choice between two toys—identical except that one is painted red and the other blue—do equal numbers of babies choose each one?)

B

SUMMARY

*1. Perform the sign test on the data from Exercise 11B5 using the same alpha level and number of tails. Did you reach the same conclusion with the sign test as with the matched *t* test? If not, explain the discrepancy.

2. Perform the sign test on the data from Exercise 11B6 using the same alpha level and number of tails. Did you reach the same conclusion with the sign test as with the matched *t* test? If not, explain the discrepancy.

*3. Six students create two paintings each. Each student creates one painting while listening to music and the other while listening to white noise in the background. If for five of the six students, the painting produced with music is judged to be more creative than the other, can you reject the null hypothesis ($\alpha = .05$, two-tailed) that music makes no difference? (Use Table A.13.)

4. Imagine that the experiment described in the previous exercise involved 60, instead of 6, subjects and that for 50 of the students the painting produced with music was rated more highly.
   a. Use the normal approximation to test the null hypothesis.
   b. How does the experiment in this exercise compare with the one in Exercise 3 in terms of the proportion of "music" paintings judged more highly? Why are the conclusions different?

*5. Does the mental condition of a chronic schizophrenic tend to deteriorate over time spent in a mental institution? To answer this question, nine patients were assessed clinically after 2 years on a ward and again 1 year later. These clinical ratings appear in the following table:

Patient No.	1	2	3	4	5	6	7	8	9
Time 1	5	7	4	2	5	3	5	6	4
Time 2	3	6	5	2	4	4	6	5	3

Assume that these clinical ratings are so crude that it would be misleading to calculate the difference score for each subject and perform a matched *t* test. However, the direction (i.e., sign) of each difference is considered meaningful, so the sign test can be performed. Test the null hypothesis ($\alpha = .05$, two-tailed) that there is no difference over time for such patients.

6. One hundred and fifty schizophrenics have been taking an experimental drug for the last 4 months. Eighty of the patients exhibited some improvement; the rest did not. Assuming that half the patients would show some improvement over 4 months without the drug, use the sign test ($\alpha = .01$) to determine whether you can conclude that the new drug has some effectiveness.

*7. In a simulated personnel selection study, each male subject interviews two candidates for the same job: an average-looking man and a male model. Two equally strong sets of credentials are devised, and the candidates are randomly paired with the credentials for each interview. Thirty-two subjects choose the male model for the job, and 18 choose the average-looking man. Can you reject the null hypothesis (at the .05 level) that attractiveness does not influence hiring decisions with a one-tailed test? With a two-tailed test?

8. Suppose that in recent years 30 experiments have been performed to determine whether violent cartoons increase or decrease aggressive behavior in children. Twenty-two studies demonstrated increased aggression and 8 produced results in the opposite direction. Based on this collection of studies, can you reject the null hypothesis (at the .05 level) that violent cartoons have no effect on aggressive behavior?

The Classical Approach to Probability

C

OPTIONAL MATERIAL

Chapter 4 discussed probability in terms of the normal distribution. Events were defined as some range of values (e.g., over 6 feet tall; IQ between 90 and 110); the probability of any very precise value (e.g., exactly 6 feet tall) was assumed to be virtually zero. The probability of an event was defined in terms of the relative amount of area under the distribution that is enclosed by the event (e.g., if 20% of the distribution represents people above 6 feet tall, the probability of randomly selecting someone that tall is .2). When you are dealing with discrete events (e.g., getting either heads or tails in a coin toss), probability can be defined in a different way. Specific events can be counted and compared to the total number of possible events. However, the counting can get quite complicated, as you will soon see.

You have already seen how simple events (trials), such as individual coin tosses, can pile up in different ways to form the binomial distribution. Another example that was mentioned is the birth of a child. If a family has one child, only two gender outcomes are possible: boy or girl. However, if the family has four children, quite a few outcomes are possible. Each possible sequence of genders represents a different outcome; having a boy first followed by three girls (BGGG) is a different outcome from having a boy *after* three girls (GGGB). As I pointed out in Section A (in the context of tossing a coin four times), the total number of different outcomes would be $2^4 = 2 \times 2 \times 2 \times 2 = 16$.

Complex events can be defined in terms of the number of different outcomes they include. For instance, if a complex event is defined as a family with four children, only one of which is a boy, there are four different outcomes (simple events) that can be included (BGGG, GBGG, GGBG, GGGB). If all of the specific outcomes you are dealing with are equally likely, the probability of a complex event can be found by counting the outcomes included in the event and counting the total number of possible outcomes. This is known as the *classical approach to probability*. The probability of an event A (i.e., $p(A)$) is defined as a proportion, or ratio, as follows:

$$p(A) = \frac{\text{number of outcomes included in } A}{\text{total number of possible outcomes}} \qquad \textbf{Formula 19.4}$$

Using Formula 19.4, the probability that a family with four children will contain exactly one boy is 4/16 = .25. The probability of having two boys and two girls is 6/16 = .375. (You can use Table 19.1 to count the number of sequences that contain two of each category.) This counting method is sufficient as long as we can assume that all of the particular outcomes or sequences are equally likely, and this assumption will be true if $P = Q$. However, there are interesting cases in which P and Q are clearly different, and these cases require more complex calculations (to be described later in this section).

The Rules of Probability Applied to Discrete Variables

Let's review the addition and multiplication rules of probability (first presented in Chapter 4, Section C) as they apply to discrete events. One of the easiest examples for illustrating these rules involves selections from an ordinary deck of 52 playing cards, 13 cards in each of four suits (hearts, diamonds, clubs, spades). Ten of the 13 cards in each suit are numbered (assuming that the ace counts as 1), and the other three are picture cards (jack, queen, and king). The probability of selecting a heart on the first draw is, according to the classical approach to probability, 13 (the number of simple events that are classified as a heart) divided by 52 (the total number of simple events), which equals .25.

The Addition Rule

We can use the addition rule to find the answer to the more complex question: What is the probability of picking either an ace or a picture card on the first draw? The probability of picking an ace is 4/52 = .077, and the probability of choosing a picture card is 12/52 (three picture cards in each of four suits) = .231. Because these two events are mutually exclusive (one card cannot be both an ace and a picture card), we can use Formula 4.7:

$$p(A \text{ or } B) = p(A) + p(B) = .077 + .231 = .308$$

On the other hand, if we want to know the probability that the first selection will be either a club ($p = .25$) or a picture card ($p = .231$), the preceding addition rule is not valid. These two events are not mutually exclusive: a card can be a club *and* a picture card at the same time (there are three such cards). The addition rule modified for overlapping events (Formula 4.8) must be used instead:

$$p(A \text{ or } B) = p(A) + p(B) - p(A \text{ and } B) = .25 + .231 - .058 = .423$$

The Multiplication Rule

If our question concerns more than one selection from the deck of cards, we will need to use some form of the multiplication rule. If we want to know the probability of drawing two picture cards in a row, and we *replace* the first card before drawing the second, we can use the multiplication rule for independent events (Formula 4.9):

$$p(A \text{ and } B) = p(A)p(B) = .231 \cdot .231 = .0533$$

If we draw two cards in succession *without* replacing the first card, the probability of the second card will be altered according to which card is selected first. If the first card drawn is a picture card ($p = .231$), and it is not replaced, there will be only 11 picture cards and a total of only 51 cards left in the deck for the second draw. The probability of drawing a picture card on the second pick is 11/51 = .216. The probability of drawing two picture cards in a row without replacement is given by the multiplication rule for dependent events (Formula 4.10), expressed in terms of conditional probability:

$$p(A \text{ and } B) = p(A)p(B \mid A) = .231 \cdot .216 = .050$$

Permutations and Combinations

In Section A, I mentioned that the problem with counting is that when you get to as many as 12 flips of a coin, there are $2^{12} = 4,096$ possible outcomes to write out. To understand how to do the counting mathematically (without writing out all the sequences), you need to know something about permutations and combinations. A *permutation* is just a particular ordering of items. If you have four different items, for instance, they can be placed in a variety of orders, or permutations. Suppose you are organizing a symposium with four speakers, and you have to decide in what order they should speak. How many orders do you have to choose from? You have four choices for your opening speaker, but once you have made that choice, you have only three choices left for the second presenter. Then there are only two choices left for the third slot; after that choice is made, the remaining speaker is automatically placed in the fourth slot. Because *each* of the four choices for opening speaker can be paired with any one of the three remaining speakers for the second slot, we multiply 4 by 3. Then we have to multiply by 2 for the third slot choice, and finally by 1 for the fourth slot (of course, this last step does not change the product). The number of orders (permutations) of four distinct items is $4 \times 3 \times 2 \times 1 = 24$. The general rule is that you take the number of items, multiply by that number minus 1, subtract 1 again and multiply, and continue this process until you are down to the number 1. This multiplication sequence arises often enough in mathematics that its product is given a name: *factorial*. The symbol for a factorial is the exclamation point. For example, 5! (pronounced "five factorial") stands for $5 \times 4 \times 3 \times 2 \times 1$, which is equal to 120. If you had five different speakers, 120 orders are possible.

You may wonder how permutations can apply to 12 flips of a coin; in that case you do not have 12 different items. In fact, you have only two different categories into which the 12 flips fall. This is where *combinations* enter the picture. I'll begin with an easier example consisting of six flips of a coin. The kind of question we are usually interested in (especially if we are constructing a binomial distribution) is: How many of the different possible sequences created by the six flips contain exactly X heads? If we used permutations to decide how many different orderings there were, the answer would be $6! = 720$. But we know this is too many. If $X = 2$, two of the flips will be in the same category: heads. It doesn't matter if you reverse the two heads, so 6! is twice as large as it should be. But if there are two heads, there are four tails, and the order of the four tails is also irrelevant, so the 6! contains 4! that should also be removed. The solution is to divide the 6! by both 2! and 4!, which gives us:

$$\frac{6!}{2!4!} = \frac{720}{(2)(24)} = \frac{720}{48} = 15$$

Fifteen of the possible sequences created when you flip a coin six times will have exactly two heads. Because the total number of possible outcomes is $2^6 = 64$, the probability of getting exactly two heads is $15/64 = .23$.

As a general rule, when you have N items and X of them fall into one category and the remainder $(N - X)$ fall into a second category, the number of different sequences is given by Formula 19.5:

$$\binom{N}{X} = \frac{N!}{X!(N - X)!} \qquad \textbf{Formula 19.5}$$

The term on the left is often referred to as a *binomial coefficient*, which can be symbolized as either $\binom{N}{X}$ or $_NC_X$ and is expressed as "N taken X at a time." Note that $_NC_X$ will always yield the same value as $_NC_{N-X}$. In the preceding example, the number of sequences with exactly two tails (and four heads) is the same as the number of sequences with exactly two heads (and four tails).

Combinations have many uses. I used combinations, without mentioning the term, in Chapter 13, when I showed how many different t tests could be performed with a particular number of group means. Because X is always equal to 2 when dealing with t tests, the combination formula can be simplified to:

$$\frac{N!}{2!(N - 2)!}$$

Now note that $N! = N \times (N - 1)! = N \times (N - 1) \times (N - 2)!$ and so on (e.g., $5! = 5 \times 4! = 5 \times 4 \times 3!$). So the preceding formula can be rewritten as:

$$\frac{N \cdot (N - 1) \cdot (N - 2)!}{2! \cdot (N - 2)!}$$

Canceling out the $(N - 2)!$ and noting that $2! = 2$, we have reduced the formula to $N(N - 1)/2$, which is equivalent to Formula 13.1.

Constructing the Binomial Distribution

Combinations can be used to construct a binomial distribution. If $N = 6$, we find $_NC_0$, $_NC_1$, $_NC_2$, $_NC_3$, $_NC_4$, $_NC_5$, and $_NC_6$ and divide each by the total number of outcomes (which, in the case of six tosses of a coin, is 64). ($_NC_4$, $_NC_5$, and $_NC_6$ will be the same as $_NC_2$, $_NC_1$ and $_NC_0$, respectively.) To find $_NC_0$ you must know that 0! is defined as equal to 1, so $_NC_0$ always equals 1. Counting the outcomes for a particular X and dividing by the total number of outcomes is appropriate only when all the outcomes are equally likely, and this is only true when $P = Q = .5$. When P and Q are not equal, a bit more work is required to find the binomial distribution.

Suppose we are interested in the eye color, rather than the gender, of the children in a family. Assuming that each parent has one gene for brown eyes and one gene for blue eyes, elementary genetics tells us that the chance of having a blue-eyed child is .25 (because blue eyes are recessive, you need two blue-eyed genes to have blue eyes). If the parents have four children, the probability that exactly one child will have blue eyes is *not* the same as the probability that exactly one child will *not* have blue eyes, even though there are four possible outcomes in each case. Using the multiplication rule for independent events, the probability that the first child will have blue eyes and the next three will not is $.25 \times .75 \times .75 \times .75 = .105$. All four sequences that contain one blue-eyed child will have the same probability, so the probability of having exactly one blue-eyed child is $.105 \times .4 = .42$. On the other hand, the probability of having three blue-eyed children in a row followed by one who does not have blue eyes is $.25 \times .25 \times .25 \times .75 = .012$. Again, there are four (equally likely) sequences with three blue-eyed children, so the probability of having exactly one child who is *not* blue-eyed is $.012 \times 4 = .048$. Note the asymmetry. Having only one blue-eyed child is much more likely ($p = .42$) than having only one child who does not have blue eyes ($p = .048$). When $P \neq Q$, all outcomes are not equally likely; the probability of an outcome depends on the value of X. Only outcomes that share the same value of X will have the same probability. Formula 19.6 is the general formula for constructing the binomial distribution:

$$p(x) = \binom{N}{X} P^X Q^{N-X}$$

Formula 19.6

We can use this formula to find the probability corresponding to any value of X when P is not equal to Q. (Note that when $P = Q$, the second part of the formula, $P^X Q^{N-X}$, always comes out to $.5^N$, or $1/2^N$.) For instance, if we want to find the probability of obtaining two blue-eyed children out of six ($P = .25$, as in the preceding example), we can use Formula 19.6:

$$\binom{6}{2} P^2 Q^4 = \frac{6!}{2!4!}(.25)^2(.75)^4 = \frac{6 \cdot 5 \cdot 4!}{2!4!}(.0625)(.32) = \frac{6 \cdot 5}{2 \cdot 1}.02$$

$$= 15(.02) = .3$$

If we continue this procedure for all possible values of X (in this case, from 0 to 6), we can use the resulting values to graph the binomial distribution for $N = 6$, $P = .25$, as shown in Figure 19.4.

The Empirical Approach to Probability

The classical approach to probability is particularly appropriate when dealing with games of chance based on mechanical devices (e.g., a roulette wheel) or a countable set of similar objects (e.g., a deck of playing cards).

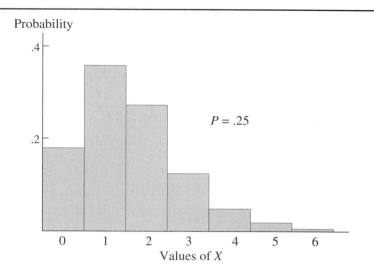

Figure 19.4

Binomial Distribution for
$N = 6, P = .25$

Probability

Values of X

$P = .25$

On the other hand, when I mentioned at the end of Section B that the probability of selecting a left-hander from the population might be .1, I was alluding to a case in which it is very unlikely that the classical approach to probability would be employed. To use this approach you would have to know the exact number of people in the population and the exact number of left-handers. Considering that it would be difficult to obtain such exact information for real populations, it is more likely that you would use the *empirical approach to probability*. The empirical approach is based on sampling the population to estimate the proportion that falls into some category, such as left-handers. This estimated proportion is then used as the probability that a random selection from the population will fall in the category of interest. Some sort of estimate would most likely be used, as well, to determine the probability corresponding to each ethnic group in the grand jury selection examples described in Section A. However, once you have estimated P and Q empirically, you can use the preceding formulas to answer such questions as: In a random sample of 10 people, what is the chance that none (or exactly two) will be left-handed?

SUMMARY

1. Given a series of N trials with only two possible outcomes for each, a total of 2^N different sequences can occur (e.g., for $N = 6$ the number of sequences is $2^6 = 64$).
2. A complex event can be defined in such a way that several different sequences or outcomes are included. For instance, if a coin is flipped six times and the event is defined as obtaining exactly two heads, there are 15 different sequences included in that event.
3. If all the sequences of two simple events are equally likely (i.e., $P = .5$), the *classical approach* to probability can be used to determine the probability of a complex event by counting. The probability of the event is a proportion: the number of outcomes (or sequences) included in that event divided by the total number of possible outcomes. Using this approach, the probability of obtaining exactly two heads out of six flips of a fair coin is $15/64 = .23$.
4. A *permutation* is an ordering of distinct items. The number of permutations that are possible for N items is $N!$ (read as "N factorial"), which

equals $N \times (N - 1) \times (N - 2) \times (N - 3)$, and so on, down to the number 1 (e.g., $6! = 6 \times 5 \times 4 \times 3 \times 2 \times 1 = 720$). Thus six people in a line can be arranged into 720 different orders.

5. If out of N items, X fall into one category and the remainder (i.e., $N - X$) fall into a second category, *combinations* (symbolized as $\binom{N}{X}$ or $_NC_X$) can be used to determine the number of different possible sequences. The number of possible sequences is given by $N!$ divided by the product of $X!$ and $[(N - X)!]$. Thus $_NC_X$ will always yield the same value as $_NC_{N-X}$. (Note that $0! = 1$ by definition.)

6. When $P = .5$, combinations are all that is needed to find the binomial distribution. First, find the total number of possible sequences. Then find $_NC_X$ for each value of X from 0 to N. Finally, divide each $_NC_X$ by the total number of sequences to find the probability corresponding to each value of X.

7. When $P \neq .5$, all sequences are not equally likely. The probability of a sequence depends on the value of X; only sequences that share the same value for X will have the same probability. To find the probability corresponding to a value of X, use the binomial coefficient to determine the number of sequences and then multiply P and Q the appropriate number of times each to calculate the probability of each sequence (see Formula 19.6).

8. The *empirical approach* to probability uses sampling to estimate proportions corresponding to different categories in a population, rather than exhaustive counting, as in the classical approach.

EXERCISES

*1. On your first draw from a deck of 52 playing cards,
   a. What is the probability of selecting a numbered card higher than 5?
   b. What is the probability of selecting either a red card (a heart or a diamond) or a spade?
   c. What is the probability of selecting either a red card or a numbered card higher than 5?

2. If patients with schizophrenia make up 40% of the psychiatric population,
   a. What is the probability that the next three patients selected at random from the psychiatric population will have schizophrenia?
   b. What is the probability that none of the next four patients selected at random from the psychiatric population will have schizophrenia?

*3. On your first two draws from a deck of 52 playing cards (without replacement),
   a. What is the probability of selecting two hearts?
   b. What is the probability of selecting a heart and a spade?

   c. What is the probability of selecting two numbered cards higher than 5?

4. If the host of a radio show has eight songs to play in the next half hour, in how many different orders can he or she play the songs?

*5. In a seminar class of 15 students, three must be chosen to make a joint presentation in the next class.
   a. How many different three-person groups can be formed?
   b. If five students are to present jointly, how many different groups can be formed?

6. If a family has sixteen children,
   a. What is the probability that all of them will be of the same gender (assuming that boys and girls are equally likely to be born)?
   b. What is the probability that there will be an equal number of each gender?

*7. If a quiz consists of 10 multiple-choice questions with five possible answers for each, and a student guesses on all 10 questions,
   a. What is the probability that the student will obtain a score of 8?
   b. What is the probability that the student will obtain a score of 8 or more?

8. If it is estimated that 15% of all fifth-grade boys have some form of dyslexia,
   a. What is the probability that there will be exactly three dyslexics in a random group of eight fifth-grade boys?
   b. What is the probability that there will be no dyslexics in a random group of eight fifth-grade boys?

z score when the normal distribution is used to approximate the binomial distribution. Because this formula does not contain a correction for continuity, it is most appropriate when N is large:

$$z = \frac{X - NP}{\sqrt{NPQ}}$$

Formula 19.1

z score for testing a proportion (this is the same as Formula 19.1, except that all terms have been divided by N, so the formula is expressed in terms of proportions instead of frequencies):

$$z = \frac{p - P}{\sqrt{\dfrac{PQ}{N}}}$$

Formula 19.2

z score for the binomial test (this is the same as Formula 19.1, except that it includes a correction for continuity):

$$z = \frac{|X - NP| - .5}{\sqrt{NPQ}}$$

Formula 19.3

The classical approach to determining the probability of an event:

$$p(A) = \frac{\text{number of outcomes included in } A}{\text{total number of possible outcomes}}$$

Formula 19.4

The number of combinations when X out of N items fall into one category and the remainder fall into a second category:

$$\binom{N}{X} = \frac{N!}{X!(N - X)!}$$

Formula 19.5

The probability of X for a binomial distribution for any values of P and N:

$$p(x) = \binom{N}{X} P^X Q^{N-X}$$

Formula 19.6

CHI-SQUARE TESTS

You will need to use the following from previous chapters:

Symbols:
N: Sample size
X: Frequency of first category
P: Probability corresponding to first category
Q: Probability corresponding to second category

Formula:
Formula 19.1: Normal approximation to the binomial distribution

Concepts:
Binomial distribution
Classical and empirical approaches to probability

20

Chapter

CONCEPTUAL FOUNDATION

In the previous chapter I showed how the binomial distribution can be used to determine the probability of randomly selecting a jury that misrepresents the population to various degrees. This probability estimate could help you decide whether a particular jury is so lopsided that you should doubt that the selection was truly random. I pointed out that if there were more than two subpopulations to be considered in the same problem, you could not use the binomial distribution, and the problem would become more complicated. In this chapter I will show how to consider any number of subpopulations with just a small increase in the complexity of the statistical procedure compared to the two-group case.

The Multinomial Distribution

Imagine that the population of a city is made up of three ethnic groups, which I will label A, B, and C. The proportion of the total population belonging to each group is designated P, Q, and R, respectively ($P + Q + R = 1.0$, so $R = 1 - P - Q$). A grand jury ($N = 48$) is randomly selected from the population; the jury consists of X members from group A, Y members from group B, and Z members from group C ($X + Y + Z = N$, so $Z = N - X - Y$). Suppose that $P = .5$, $Q = .33$, and $R = .17$ and that $X = 28$, $Y = 18$, and $Z = 2$. A spokesperson for group C might suggest that the overrepresentation of group A ($X = 28$) and group B ($Y = 18$) at the expense of group C ($Z = 2$) is the result of deliberate bias. Certainly, X and Y are larger than expected based on the proportions P and Q (expected $X = PN = .5 \times 48 = 24$; expected $Y = QN = .33 \times 48 = 16$), and Z is less than expected ($RN = .17 \times 48 = 8$), but is this jury composition very unlikely to occur by chance?

To calculate the probability of selecting a jury with the preceding composition, you could use an extension of the procedures described in the previous chapter for dealing with the binomial distribution. You could test the null hypothesis that $P = .5$ and $Q = .33$, but you would not be dealing with a binomial distribution. Because there are more than two categories in which the jurors can fall, you would be calculating values of a *multinomial distribution*. Just as the binomial distribution depends on the values of N and P, the multinomial distribution in the case of three categories depends on N, P, and Q. As the number of categories increases, so does the complexity of the multinomial distribution. In fact, even with only three categories, the calculations associated with the multinomial distribution are so tedious that they are universally avoided. The easy way to avoid these calculations is to use an approximation—just as we used the normal distribution to approximate the binomial distribution in the last chapter.

The Chi-Square Distribution

To get from a binomial approximation to a multinomial approximation, I will employ the same trick that I used in Chapter 12 to get from the t test to ANOVA: I will take the formula for the two-group case and square it. Squaring Formula 19.1 yields the following expression:

$$z^2 = \frac{(X - NP)^2}{NPQ}$$

Assuming that the z scores follow a normal distribution before squaring, the squared z scores follow a different mathematical distribution, called the *chi-square distribution*. The symbol for chi square is the Greek letter chi (pronounced "kie" to rhyme with "pie") squared (χ^2). As with the t distribution, there is a whole family of chi-square distributions; the shape of each chi-square distribution depends on its number of degrees of freedom. However, unless df is very large, all chi-square distributions tend to be positively skewed because the value of χ^2 cannot fall below zero (as df becomes extremely large, the mean of the distribution also becomes extremely large, and the skewing becomes negligible—with infinite df the chi-square distribution becomes identical to the normal distribution). The distribution of squared z scores just referred to is a chi-square distribution with one degree of freedom [$\chi^2(1)$]. A more typical chi-square distribution (df = 4) is shown in Figure 20.1.

If the preceding formula for z^2 is used instead of Formula 19.1 to test an hypothesis concerning a binomial distribution, instead of finding the critical value from the normal table (Table A.1), we need to look in Table A.14 under df = 1. For instance, for a .05 two-tailed test, we would use ±1.96 as our critical z, but for $\chi^2(1)$ the corresponding critical value would be 3.84. Note that $3.84 = 1.96^2$. Squaring the test statistic requires squaring the corresponding critical value as well.

Expected and Observed Frequencies

When there are only two categories, it doesn't matter whether you use Formula 19.1 or its squared version, as long as you use the correct critical value and perform a two-tailed test. (Note that if the correction for continuity is appropriate, you would make the correction before squaring.) The advantage of squaring is that it readily leads to an expanded formula that can accommodate any number of categories. However, to accommo-

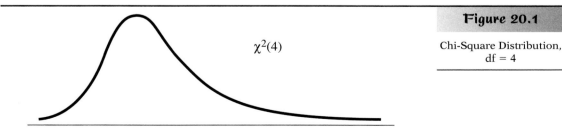

$\chi^2(4)$

Figure 20.1

Chi-Square Distribution, df = 4

date more categories, it is convenient to adopt some new symbols. Consider the numerator of Formula 19.1: $X - NP$. The term NP is the number of trials we *expected* to fall in the category with probability P, and X is the number of trials that actually did. In the terminology of the chi-square test, NP is the expected frequency (f_e), and X is the frequency that was actually obtained, or observed (f_o). I will apply this terminology to the example I posed earlier of three ethnic groups being represented on a grand jury. I mentioned that the expected frequencies for groups A, B, and C were 24, 16, and 8, respectively. The obtained frequencies were 28, 18, and 2. It is customary to put both the obtained and expected frequencies in a table, as shown in Table 20.1.

Table 20.1		
Group A	Group B	Group C
f_o 28	18	2
f_e 24	16	8

The Chi-Square Statistic

Now we need a formula (similar to Formula 19.1 squared, but with new symbols) to measure the discrepancy between the obtained and expected frequencies. The appropriate formula is Formula 20.1:

$$\chi^2 = \sum \frac{(f_o - f_e)^2}{f_e}$$

Formula 20.1

I left out the indexes on the summation sign, but the sum goes from 1 to k, where k is the number of groups ($k = 3$ in this example). Thus the formula is applied k times, once for each pairing of observed and expected frequencies. The test statistic produced by the formula is called the chi-square statistic because it follows the chi-square distribution (approximately) when the null hypothesis is true (and certain assumptions, to be discussed later, are met). It is also referred to as *Pearson's chi-square statistic*, after Karl Pearson, who devised the formula (as well as the correlation coefficient described in Chapter 9). Whereas the numerator of this formula is the same as Formula 19.1 after squaring and substituting new symbols, the denominator does not seem to correspond. Despite appearances, however, this formula always gives the same answer as the square of Formula 19.1 in the two-group case. Applying Formula 20.1 to our example, we get:

$$\chi^2 = \frac{(28 - 24)^2}{24} + \frac{(18 - 16)^2}{16} + \frac{(2 - 8)^2}{8} = \frac{16}{24} + \frac{4}{16} + \frac{36}{8} = 5.42$$

If you are dealing with only two categories and N is not large, a more accurate value for χ^2 is given by adding Yates' continuity correction to Formula 20.1 to create Formula 20.2 (this is the same correction I added to Formula 19.1 to create Formula 19.3):

$$\chi^2 = \sum \frac{(|f_o - f_e| - .5)^2}{f_e}$$

Formula 20.2

Critical Values of Chi-Square

Before we can make any decision about the null hypothesis, we need to look up the appropriate critical value of the chi-square distribution. To do this we need to know the number of degrees of freedom. In general, if there are k categories, df $= k - 1$ because once $k - 1$ of the f_e have been fixed, the last one has to be whatever number is needed to make all of the f_es add up to N. (The expected frequencies must always add up to the same number as the obtained frequencies.) In this example, df $= 3 - 1 = 2$. Looking in Table A.14 (or Table 20.2, which is a portion of Table A.14), we find that χ^2_{crit} for df $= 2$ and alpha $= .05$ is 5.99. As you can see from Figure 20.2, the

		AREA IN THE UPPER TAIL			
df	.10	.05	.025	.01	.005
1	2.71	3.84	5.02	6.63	7.88
2	4.61	5.99	7.38	9.21	10.60
3	6.25	7.81	9.35	11.34	12.84
4	7.78	9.49	11.14	13.28	14.86
5	9.24	11.07	12.83	15.09	16.75

Table 20.2

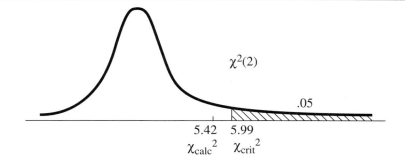

Figure 20.2

Chi-Square Distribution, df = 2 (Area beyond Critical Value)

calculated χ^2 (5.42), being less than the critical χ^2, does not fall in the region of rejection. Although it may seem that group C is underrepresented on the jury, we cannot reject the null hypothesis that this jury is a random selection from the population at the .05 level (given that $P = .5$, $Q = .33$, and $R = .17$, as stated in the example).

If you look across the rows of Table 20.2, you will see that the critical values increase as alpha is reduced, just as they do for the other statistics in this text. However, unlike t or F, the critical value of χ^2 gets larger as the degrees of freedom increase. (The df play a different role for χ^2, being based only on the number of categories rather than on the number of subjects.) You may also notice that the χ^2 table does not give you a choice between one- and two-tailed tests. This is the same situation we encountered in using the tables of the F distributions (and for essentially the same reason); I will deal with this issue next.

Tails of the Chi-Square Distribution

Note that the rejection zone appears in only one tail of the distribution, the positive tail. Recall that the positive tail of the F distribution represents ANOVAs in which the group means are unusually spread out, and the negative tail represents cases in which the group means are unusually close together. Similarly, unusually large discrepancies between expected and obtained frequencies lead to chi-square values in the positive tail, whereas studies in which the obtained frequencies are unusually close to their expected values lead to values in the negative tail (near zero). The rejection zone is placed entirely in the positive tail of the chi-square distribution because only large discrepancies between expected and obtained values are inconsistent with the null hypothesis; extremely small discrepancies, no matter how unusual, serve only to support the null hypothesis. Thus in one sense, all chi-square tests are one-tailed.

In another sense chi-square tests are multitailed because there are several ways to get large discrepancies. Imagine Dr. Null trying to duplicate our somewhat lopsided grand jury. He would take a truly random selection from the population (with $P = .5$, $Q = .33$, and $R = .17$) and then calculate the χ^2 corresponding to his jury. If his χ^2 exceeded ours, Dr. Null would win. But it doesn't matter what kind of imbalance led to Dr. Null's large χ^2—he can win by selecting a jury that *over*represents group C or even one in which group C's representation is exactly as expected, but there is an unexpected imbalance between groups A and B. A significant χ^2 statistic tells you that there is some mismatch between your observed and expected frequencies, but you would have to conduct additional analyses on pairs of groups (using a Bonferroni adjustment to your α) to make more specific conclusions.

Expected Frequencies Based on No Preference

In the preceding example, the expected frequencies were based on the known proportions of the different subgroups in a population. This is not the most common situation in which chi-square tests are used. More often, especially in an experimental context, the expected frequencies are based on more abstract theoretical considerations. For instance, imagine that a developmental psychologist is studying color preference in toddlers. Each child is told that he or she can take one toy out of four that are offered. All four toys are identical except for color: red, blue, yellow, or green. Forty children are run in the experiment, and their color preferences are as follows: red, 13; blue, 9; yellow, 15; and green, 3. These are the obtained frequencies. The expected frequencies depend on the null hypothesis. If the null hypothesis is that toddlers in general have no preference for color, we would expect the choices of colors to be equally divided among the entire population of toddlers. Hence, the expected frequencies would be 10 for each color (the f_e must add up to N). The f_o and f_e are shown in Table 20.3.

Table 20.3

	Red	Blue	Yellow	Green
f_o	13	9	15	3
f_e	10	10	10	10

Applying Formula 20.1 to the data in the table, we get:

$$\chi^2 = \frac{(13-10)^2}{10} + \frac{(9-10)^2}{10} + \frac{(15-10)^2}{10} + \frac{(3-10)^2}{10}$$

$$= \frac{9}{10} + \frac{1}{10} + \frac{25}{10} + \frac{49}{10} = 8.4$$

The number of degrees of freedom is one less than the number of categories, so df = 3 for this example. Therefore, the critical value for a .05 test (from Table A.14) is 7.82. Because our calculated χ^2 (8.4) exceeds the critical value, we can reject the null hypothesis that the population of toddlers is equally divided with respect to color preference. We can conclude that toddlers, in general, have preferences among the four primary colors used in the study. (A toy company might be interested in this conclusion.)

The Varieties of One-Way Chi-Square Tests

The chi-square tests described thus far can be considered one-way tests in the same sense that an ANOVA can be one-way: All of the categories are considered levels of the same independent (or quasi-independent) variable. For instance, a one-way chi-square test can involve any number of different religions or any number of different political parties. However, if the design of the study included several different religions *and* several different political parties in a completely crossed factorial design (for example, you wish to determine for each religion the number of subjects in each political party), you would be dealing with a "two-way" chi-square test, which is the subject of Section B.

The one-way chi-square test is often referred to as a *goodness-of-fit test*; this label is especially apt when the emphasis of the test is on measuring the fit between observed frequencies and frequencies predicted by a theory or an hypothesized population distribution.

Population Proportions Are Known

It is rare that proportions of the population are known exactly, but for many categorical variables we have excellent estimates of the population proportions. For instance, from voter registration data we could find out the proportion of citizens belonging to each political party within a particular locale. Then, by sampling the readers of a particular newspaper, for instance, and performing a chi-square test, we could decide if the politics of these readers represented a random selection of the electorate. The preceding example concerning three ethnic groups and their representation on a grand jury is of this type.

A related use of the chi-square test is to compare the proportions in one population with those in another. For instance, knowing the proportions of certain mental illnesses in industrial nations, you could test a sample from the population of a less-developed nation to see if this particular country exhibits significantly different proportions of the same mental illnesses.

Expected Frequencies Are Hypothesized to Be Equal

In conducting tests involving games of chance or experimental research, researchers commonly hypothesize equal expected frequencies. If you were throwing a die 90 times to see if it were "loaded" (i.e., biased), you would expect each side to turn up 15 times (90/6 sides = 15 per side) if the die were fair (i.e., if the null hypothesis were true). In a social psychology experiment, descriptions of hypothetical job applicants could be paired with names representing different ethnic groups. If subjects are selecting the best applicant from each set without prejudice, the various names should be selected equally often (assuming that the name-description pairings are properly counterbalanced).

Of course, games of chance do not always involve equal frequencies. For example, if a pair of dice is thrown, the number 7 is much more likely to come up than the numbers 2 or 12. The expected frequencies can be determined by the kind of counting techniques described in the previous chapter. Similarly, the choices in an experimental design might lead to predicted frequencies that are based on a theoretical model and are not all equal.

The Shape of a Distribution Is Being Tested

One of the original applications of the χ^2 goodness-of-fit test was to test the shape of a population distribution for some continuous variable based on the distribution of that variable within a sample. The expected frequencies depend on the distribution shape you wish to test. For instance, if you want to see if body weight has a normal distribution in the population, you could generate expected frequencies based on areas from the normal table. Out of a random sample of 100 subjects from a normal distribution, you would expect about 34 people to fall between $z = 0$ and $z = +1$ (and 34 people to fall between $z = 0$ and $z = -1$), about 13.5 (or 14) people to fall between $z = +1$ and $z = +2$ (and the same number between -1 and -2), and about 2.5 (or 3) people to fall beyond $z = +2$ (or -2). These are your expected frequencies. Then you would convert the 100 body weights to z scores and sort them into the same categories to find the obtained frequencies. For a more sensitive test you would break the distribution into somewhat more than six categories.

The chi-square statistic can be used to measure how well the frequencies in your sample fit the normal distribution. However, for an interval/ratio variable like body weight, you would probably use a more appropriate procedure, such as the Kolmogorov–Smirnov test (see Hays, 1994). Rejecting the null hypothesis implies that your sample distribution is inconsistent with a normally distributed population. If you know your sample was truly random, this would be evidence that the population is not normally distributed (this could be important information if you had planned to use a parametric statistic that assumes your variable *is* normally distributed in the population). On the other hand, if you know that the population has a normal distribution, rejecting the null hypothesis can be taken as evidence that the sample is not random or that it is not from the hypothesized population. Depending on the situation, this could be a case when you are hoping that the null hypothesis is *not* rejected.

A Theoretical Model Is Being Tested

Sometimes a well-formulated theory can make quantitative predictions. For instance, a theory of memory might predict that the probability of recalling a word from a list is related to the frequency with which that word occurs in written language. If a particular list consists of words from four frequency categories, specific predictions might be made about the number of words that will be recalled from each category. These are expected frequencies that can be tested against the frequencies observed in an experiment to measure the goodness-of-fit. This is a case in which you definitely do *not* want to reject the null hypothesis; you want your calculated χ^2 to be as close to zero as possible. Because your expected frequencies represent your research hypothesis, and not some null hypothesis, you are hoping for a good fit between the expected and observed frequencies. If you have a χ^2 table with values in the left tail of the distribution (the shorter tail near zero), you may be able to demonstrate that your calculated value of χ^2 is unusually small (of course, you are more likely to conduct the test with statistical software and thereby obtain an exact p value).

SUMMARY

1. When dealing with more than two categories, the probability that a sample of a certain size will break down into those categories in a particular way depends on the appropriate *multinomial distribution*.
2. Because it is tedious to find exact multinomial probabilities, the *chi-square* (χ^2) *distribution* is used as an approximation. The number of degrees of freedom for a one-variable chi-square test is one less than the number of categories. Knowing df and α, you can find the critical value of χ^2 in Table A.14.

3. *Pearson's chi-square statistic* is based on finding the difference between the expected and observed frequencies for each category, squaring the difference, and dividing by the expected frequency. The sum over all categories is the χ^2 statistic, which follows the chi-square distribution (approximately) when the null hypothesis is true. Either the normal approximation to the binomial distribution, or the chi-square test, can be used when there are only two categories. In both cases, the correction for continuity should be used, unless N is large.

4. Large discrepancies between expected and observed frequencies produce large values of χ^2, which fall in the positive (right) tail of the distribution and, if large enough, can lead to rejection of the null hypothesis. Unusually small discrepancies can produce χ^2 values near zero, which fall in the negative (left) tail but, regardless of how small, never lead to rejection of the null hypothesis.

5. When you are testing to see whether subjects have any preference among the categories available, the null hypothesis is usually that the categories will be selected equally often (i.e., subjects will have no preference). The expected frequencies are found in this case by dividing the sample size by the number of categories. Expected frequencies may also be based on the proportions of various groups in the population (e.g., if 10% of the population are vegetarians, the null hypothesis would predict that 10% of the subjects in each of your samples would be vegetarian).

6. The one-way chi-square test is often called a *goodness-of-fit* test, especially when the frequencies of values in a sample are tested to see how well they fit a hypothesized population distribution, or when specific predictions are made from a theory and you are testing the agreement between your data and the predictions. In the latter case especially, you would want your χ^2 value to be unusually *small*, leading to a tiny p value in the *left* tail of the chi-square distribution, indicating a good match between expected and observed frequencies.

EXERCISES

*1. A one-way chi-square test involves eight different categories.
   a. How many degrees of freedom are associated with the test?
   b. What is the critical value for χ^2 at $\alpha = .05$? At $\alpha = .01$?

2. Rework part a of Exercise 19A6 using the chi-square statistic instead of the normal approximation to the binomial distribution. How does the value for χ^2 compare with the z score you calculated for that exercise?

*3. A soft drink manufacturer is conducting a blind taste test to compare its best-selling product (X) with two leading competitors (Y and Z). Each subject tastes all three and selects the one that tastes best to him or her.
   a. What is the appropriate null hypothesis for this study?

   b. If 27 subjects prefer product X, 15 prefer product Y, and 24 prefer product Z, can you reject the null hypothesis at the .05 level?

4. Suppose that the taste test in Exercise 3 was conducted with twice as many subjects, but the proportion choosing each brand did not change (i.e., the number of subjects selecting each brand also doubled).
   a. Recalculate the chi-square statistic with twice the number of subjects.
   b. Compare the value of chi-square in part a with the value you found for Exercise 3.
   c. What general rule is being illustrated here?

*5. A gambler suspects that a pair of dice he has been playing with are loaded. He rolls one of the dice 120 times and observes the following frequencies: one, 30; two, 17; three, 27; four, 14; five, 13; six, 19. Can he reject the null hypothesis that the die is fair at the .05 level? At the .01 level?

6. A famous logic problem has four possible answers (A, B, C, and D). Extensive study has demonstrated that 40% of the population choose the correct answer, A, 26% choose B, 20% choose C, and 14% choose D. A new study has been conducted with 50 subjects to determine whether presenting the problem in concrete terms changes the way subjects solve the problem. In the new study, 24 subjects choose A, 8 subjects choose B, 16 subjects choose C, and only 2 subjects choose D. Can you reject the null hypothesis that the concrete presentation does not alter the way subjects respond to the problem?

*7. It has been suggested that admissions to psychiatric hospitals may vary by season. One

hypothetical hospital admitted 100 patients last year: 30 in the spring; 40 in the summer; 20 in the fall; and 10 in the winter. Use the chi-square test to evaluate the hypothesis that mental illness emergencies are evenly distributed throughout the year.

8. Of the 100 psychiatric patients referred to in the previous exercise, 60 were diagnosed as schizophrenic, 30 were severely depressed, and 10 had a bipolar disorder. Assuming that the national percentages for psychiatric admissions are 55% schizophrenic, 39% depressive, and 6% bipolar, use the chi-square test to evaluate the null hypothesis that this particular hospital is receiving a random selection of psychiatric patients from the national population.

B

BASIC STATISTICAL PROCEDURES

Two-Variable Contingency Tables

Most of the interesting questions in psychological research involve the relationship between (at least) two variables rather than the distribution of only one variable. That is why, in psychological research, the one-way chi-square test is not used nearly as often as the two-way chi-square. When one of two variables is categorical (e.g., psychiatric diagnoses; different sets of experimental instructions) and the other is continuous, the parametric tests already discussed in this text may be appropriate. It is when both of the variables can only be measured categorically that the two-variable chi-square test is needed.

For instance, suppose a researcher believes that adults whose parents were divorced are more likely to get divorced themselves compared to adults whose parents never divorced. We cannot quantify the degree of divorce; a couple either gets divorced or they do not. To simplify the problem we will focus only on the female members of each couple. Suppose that the researcher has interviewed 30 women who have been married: 10 whose parents were divorced before the subject was 18 years old and 20 whose parents were married until the subject was at least 18. Half of the 30 women in this hypothetical study have gone through their own divorce; the other half are still married for the first time. To know whether the divorce of a person's parents makes the person more likely to divorce, we need to see the breakdown in each category—that is, how many currently divorced women come from "broken" homes and how many do not, and similarly for those still married. These frequency data are generally presented in a *contingency* (or cross-classification) *table*, in which each combination of levels from the two variables (e.g., parental divorce and subject's own divorce) is represented as a cell in the table. The preceding example involves two levels for each variable and therefore can be represented by a 2 × 2 contingency table, such as the one in Table 20.4. The data in such a table are often referred to as *cross-classified categorical data*.

Pearson's Chi-Square Test of Association

You can see at once from Table 20.4 that among the subjects whose parents were divorced, about twice as many have been divorced as not; the reverse

	Parents Divorced	Parents Married	Row Sums	**Table 20.4**
Self divorced	7	8	15	
Self married	3	12	15	
Column sums	10	20	30	

trend is evident for those whose parents were not divorced. There seems to be some association between parental divorce and one's own divorce. Of course, Dr. Null would claim that this association is accidental and that the results are just as likely to come out in the opposite direction for the next 30 subjects. *Pearson's chi-square test of association* would allow us to decide whether to reject the null hypothesis in this case (i.e., that parental divorce is *not* associated with the likelihood of one's own divorce), but first we have to determine the expected frequency for each cell. The naive approach would be to use the same logic as the one-way chi-square and divide the total N by 4 to get the same expected frequency in each cell (i.e., 30/4 = 7.5). However, the marginal sums in Table 20.4 remind us that twice as many subjects did *not* have divorced parents as did, and this relationship should hold within each row of the table if the two variables are not associated. Given the marginal sums in Table 20.4, Dr. Null expects the cell frequencies to be as shown in Table 20.5. Note that whereas Dr. Null expects the same marginal sums as were actually obtained, he expects that the ratios of cell frequencies within any row or column will reflect the ratios of the corresponding marginal sums, as shown in Table 20.5 (e.g., the ratio of the column sums is 1 to 2, so that is the ratio of the f_es in each row of Table 20.5). The next example will illustrate a simple procedure for finding the appropriate expected frequencies.

	Parents Divorced	Parents Married	Row Sums	**Table 20.5**
Self divorced	5	10	15	
Self married	5	10	15	
Column sums	10	20	30	

An Example of Hypothesis Testing with Categorical Data

The preceding example was correlational; the experimenter noted which subjects had been through divorce, but of course had no control over who got divorced. In the next example, I will present a design in which one of the variables is actually manipulated by the experimenter. Also, I will extend the chi-square test of association to a table larger than 2 × 2.

Imagine that a psychiatrist has been frustrated in her attempts to help chronic schizophrenics. She designs an experiment to test four therapeutic approaches to see if any treatment is better than the others for improving the lives of her patients. The four treatments are intense individual psychodynamic therapy, constant unconditional positive regard and Rogerian therapy, extensive group therapy and social skills training, and a token economy system. The dependent variable is the patient's improvement over a 6-month period, measured in terms of three categories: became less schizophrenic (improved), became more schizophrenic (got worse), or showed no change. The different proportions of improvement among these four treatments can be tested for significance with the usual six-step procedure.

Step 1. State the Hypotheses

In the case of the two-variable chi-square test, the null hypothesis is that there is no association or correlation between the two variables—that is, the way that one of the variables is distributed into categories does not change at different levels of the second variable. Stated yet another way, the null hypothesis asserts that the two variables are independent of each other. Hence this test is often called the *chi-square test for independence*. For this example, the null hypothesis is H_0: method of treatment and degree of improvement are independent. As usual, the alternative hypothesis is the negation of H_0; H_A: method of treatment and degree of improvement are *not* independent.

Step 2. Select the Statistical Test and the Significance Level

The data consist of the frequencies in categories arranged along two dimensions, so the two-way chi-square test is appropriate. As usual we will set alpha = .05.

Step 3. Select the Samples and Collect the Data

Eighty schizophrenics meeting certain criteria (not responsive to previous treatment, more than a certain number of years on the ward, etc.) are selected and then assigned at random to the four treatments, with the constraint that 20 are assigned to each group. After 6 months of treatment, each patient is rated as having improved, having gotten worse, or having remained the same. The data are displayed in Table 20.6, a 3×4 contingency table.

Table 20.6

Observed Frequencies	Psychodynamic Therapy	Rogerian Therapy	Group Therapy	Token Economy	Row Sums
Improved	6	4	8	12	30
No change	6	14	3	5	28
Got worse	8	2	9	3	22
Column sums	20	20	20	20	$N = 80$

Step 4. Find the Region of Rejection

We will be using a chi-square distribution to represent the null hypothesis distribution. But to know which chi-square distribution is appropriate, we need to know the number of degrees of freedom with which we are dealing. As in the case of the one-variable chi-square test, the df depend on the number of categories rather than on the number of subjects. However, when there are two variables, the number of categories for each must be considered. The two-variable case can always be represented by an $R \times C$ contingency table, where R stands for the number of rows and C stands for the number of columns. Formula 20.3 for df can then be stated as follows:

$$df = (R - 1)(C - 1) \qquad \textbf{Formula 20.3}$$

For the preceding 3×4 table, df = $(3 - 1)(4 - 1) = (2)(3) = 6$. Looking in Table A.14 for df = 6 and alpha = .05, we find that the critical value for χ^2 is 12.59. The region of rejection is that area of the $\chi^2(6)$ distribution above 12.59. Chi-square tests of independence are always one-tailed, in that large χ^2 values can lead to statistical significance, but there is no value of χ^2 so small that it would lead to the rejection of the null hypothesis.

Step 5. Calculate the Test Statistic

For chi-square tests, this step begins with finding the expected frequencies. In the one-variable case this usually involves either dividing the total sample size (N) by the number of categories or multiplying N by the population proportion corresponding to each category. In the test for independence of two variables, however, finding the f_e is a bit more complicated. Fortunately, there is a mathematical trick that is easy to apply. The expected frequency for a particular cell is found by multiplying the two marginal sums to which it contributes and then dividing by the total N. This trick can be expressed as Formula 20.4:

$$f_e = \frac{(\text{row sum})(\text{column sum})}{N} \qquad \textbf{Formula 20.4}$$

Each f_e can be calculated using this formula, but fortunately, not all 12 cells need to be calculated in this way. The number of degrees of freedom associated with Table 20.6 tells us how many f_es must be calculated. Because df = 6, we know that only six of the f_e are free to vary; the remaining cells can be found by subtraction (within each row and column, the f_es must add up to the same number as the f_os). However, if you want to save yourself some calculation effort in this way, you will have to choose an appropriate set of six cells to calculate; one possibility is shown in Table 20.7.

Expected Frequencies	Psychodynamic Therapy	Rogerian Therapy	Group Therapy	Token Economy	Row Sums
Improved	7.5	7.5	7.5		30
No change	7	7	7		28
Got worse					22
Column sums	20	20	20	20	80

Table 20.7

To illustrate how the f_es in Table 20.7 were calculated, I will use Formula 20.4 to find the f_e for the "Rogerian Therapy, Did Not Change" cell.

$$f_e = \frac{(\text{row sum})(\text{column sum})}{N} = \frac{(28)(20)}{80} = \frac{560}{80} = 7$$

After finding the six f_es as shown, we can find the remaining f_es by subtraction. You can find the f_e for the "Got worse" cells in the first three columns (subtract the other two f_e from the column sum, 20) and then each f_e in the "Token Economy" column by subtracting the other three treatments from each row sum. (The column sums will not always be equal to each other, but this is more likely to occur for a condition that is experimentally assigned.) The 3 × 4 contingency table with the observed frequencies and all of the expected frequencies (in parentheses) is shown as Table 20.8.

Note that if you calculate each of the f_es separately (12, in this case), you can then check your calculation by seeing whether the f_es add up to the same marginal sums as the f_os. You can also check that the f_es in each column follow the same relative proportions as the row sums; for instance if all of the column sums are the same, as in this case, all of the f_es in a given row will be

Table 20.8	Psychodynamic Therapy	Rogerian Therapy	Group Therapy	Token Economy	Row Sums
Improved	6 (7.5)	4 (7.5)	8 (7.5)	12 (7.5)	30
No change	6 (7)	14 (7)	3 (7)	5 (7)	28
Got worse	8 (5.5)	2 (5.5)	9 (5.5)	3 (5.5)	22
Column sums	20	20	20	20	80

the same. We can now apply Formula 20.1, wherein the summation sign tells us to find the value for every cell and then add up all of those values:

$$\chi^2 = \sum \frac{(f_o - f_e)^2}{f_e}$$

$$= \frac{(6 - 7.5)^2}{7.5} + \frac{(4 - 7.5)^2}{7.5} + \frac{(8 - 7.5)^2}{7.5} + \frac{(12 - 7.5)^2}{7.5}$$

$$+ \frac{(6 - 7)^2}{7} + \frac{(14 - 7)^2}{7} + \frac{(3 - 7)^2}{7} + \frac{(5 - 7)^2}{7}$$

$$+ \frac{(8 - 5.5)^2}{5.5} + \frac{(2 - 5.5)^2}{5.5} + \frac{(9 - 5.5)^2}{5.5} + \frac{(3 - 5.5)^2}{5.5}$$

$$= .3 + 1.63 + .03 + 2.7 + .143 + 7 + 2.29 + .57 + 1.14$$

$$+ 2.23 + 2.23 + 1.14 = 21.4$$

Step 6. Make the Statistical Decision

The calculated value of χ^2 (21.4) is larger than the critical value (12.59), and therefore it lands in the region of rejection. We can reject the null hypothesis and conclude that the tendency toward improvement is not independent of the type of treatment; that is, the various treatments differ in the proportion of the population that would experience improvement, show no change, or become worse. However, just because the two variables are significantly related, we cannot conclude that the relation is a strong one. A tiny association between the two variables can lead to a statistically significant result if the sample size is sufficient. Methods for assessing the degree of association between two categorical variables are described in Section C.

The Simplest Case: 2 × 2 Tables

Table 20.9		
a	b
c	d

In the case of a 2 × 2 table, there is only one degree of freedom and one f_e that must be calculated. For this case only, the calculation of the f_es can be combined with the calculation of χ^2 to create a simplified formula. Assume that the observed frequencies are assigned letters as in Table 20.9.

A simplified formula for χ^2 can then be written in terms of those letters as follows:

$$\chi^2 = \frac{N(ad - bc)^2}{(a + b)(c + d)(a + c)(b + d)} \qquad \textbf{Formula 20.5}$$

To illustrate the use of this formula, I will apply it to the data of Table 20.4 (the divorce example):

$$\chi^2 = \frac{30(84 - 24)^2}{(15)(15)(10)(20)} = \frac{30(3600)}{45,000} = \frac{108,000}{45,000} = 2.4$$

The value for χ^2 obtained from Formula 20.5 is the same, of course, as the value that you would obtain from Formula 20.1 (it would be a useful exercise at this point to demonstrate this for yourself). As you might guess, this shortcut is no longer very important now that most data analysis is performed by computer. However, 2×2 contingency tables are quite common, and, unless your table is only part of a larger, more complex study, it just might be easier to use Formula 20.5 than to enter the data into the computer.

It has been suggested that Formula 20.2, which includes the continuity correction proposed by Yates (1934), be used in the two-way case when there is only one degree of freedom—that is, for the 2×2 table. However, because the Yates' correction is only helpful for a special case of the 2×2 table that rarely occurs (see "Fisher's Exact Test" in Section C), and there are good reasons not to use the correction for other 2×2 tables (Conover, 1974; Overall, 1980), I have not included it in this section. Other corrections have been recommended (e.g., Greenwood & Nikulin, 1996), but they are more difficult to explain and not commonly used, so I have not included them, either.

Assumptions of the Chi-Square Test

The chi-square statistic will not follow the chi-square distribution exactly unless the sample size is infinitely large. The chi-square distribution can be a reasonably good approximation, however, provided that the following assumptions have been met. These assumptions apply to both goodness-of-fit tests and tests of association.

Mutually Exclusive and Exhaustive Categories

The categories in the chi-square test must be exhaustive so that every observation falls into one category or another (this may require an "other" category in which to place subjects not easily classified). The categories must also be mutually exclusive so that no observation can simultaneously fall in more than one category.

Independence of Observations

We assume that observations are independent, just as we did with respect to random samples for parametric tests. This assumption is usually violated when the same subject is categorized more than once. For instance, suppose that men and women are judging whether they liked or disliked various romantic movies shown to them, and you are testing to see if the two genders like the same proportion of romantic movies. If five movies are being judged by a total of 10 men and 10 women, and each subject judges all five movies, there will be a total of 100 observations (20 subjects $\times$ 5 movies). However, these 100 observations will not be mutually independent (e.g., a subject who hates romantic movies may tend to judge all five movies harshly). Because a violation of independence seriously undermines the validity of the test, it is safest to ensure that each subject contributes only one categorical observation (i.e., that the total number of observations equals the number of different subjects).

Size of Expected Frequencies

For the chi-square distribution to be a reasonably accurate approximation of the distribution of the chi-square statistic, cautious statisticians recommend that the expected frequency for each cell should be at least 5 when df is greater than one, and 10 when df equals one. Haber (1980) favors the more relaxed criterion that the average of all the f_es be at least 5. A reasonable compromise is to require that no f_e is less than 1, and not more than 20% of the f_es are less than 5, although quite a few other rules have been proposed (Greenwood & Nikulin, 1996).

If you have quite a few categories, and small frequencies in most of them, you should consider combining some categories if that makes sense for your study. However, the danger is that this restructuring will be performed in an arbitrary way that capitalizes on chance and leads to more Type I errors than your alpha level would suggest. Therefore, when collapsing categories, you should avoid the temptation to shop around for the scheme that gives you the highest χ^2. (When and how to collapse categories is another topic about which there is a good deal of statistical literature.) For a special type of 2 × 2 table with small f_e, the chi-square approximation can be avoided by using an exact multinomial test. This option is discussed further in Section C.

Some Uses for the Chi-Square Test for Independence

The chi-square test for independence can be planned when you want to see the relationship between two variables, both of which are being measured on a categorical scale (e.g., psychiatric diagnosis and season of birth). The two-variable chi-square test can also be used for a study originally designed to be analyzed by a t test or a one-way ANOVA, but for which the distribution of the dependent variable is very far from normal. For instance, if a floor or ceiling effect causes most of the scores to have the same low or high value (e.g., you are counting the number of times children initiate fights in the playground, and most children don't initiate any), it may be reasonable to convert the original multivalued scale to a few broad categories. Bear in mind that this usually results in a considerable loss of power, so you should only change to a categorical scale when the distribution of values is too far from the normal distribution (and your sample size is too small) to permit the use of parametric statistics. As usual, studies can differ with respect to whether the variables are manipulated or merely observed.

Observational Research

In observational research, both of the categorical variables would be grouping (or measured) variables (i.e., not experimentally manipulated). In the example involving divorce, frequencies were based on preexisting differences—the researcher had no influence on who would get divorced. (Thus the researcher could draw no conclusions about parental divorce *causing* the later divorce of the children.) For another example involving two grouping variables, imagine testing the hypothesis that agoraphobia (fear of leaving home, of being in a crowd, etc.) is associated with the early loss of a loved one. A group of agoraphobics and a comparison group of persons with some other form of neurosis (preferably from the same mental health clinic) are obtained. Each patient is then categorized as having lost a loved one or not, thus forming a 2 × 2 contingency table.

Experimental Research

In experimental research, one of the two categorical variables must be a truly independent variable (i.e., manipulated by the experimenter); the other variable is a dependent variable consisting of distinct categories. The example in this section involving the treatment of schizophrenics falls into this category.

Publishing the Results of a Chi-Square Test

The results of the treatment for schizophrenia experiment could be reported in the following manner: "The 4×3 contingency table revealed a statistically significant association between the method of treatment and the direction of clinical improvement, χ^2 (6, $N = 80$) = 21.4, $p < .05$." The first number in the parentheses following χ^2 is the number of degrees of freedom associated with the chi-square statistic (for this example, df = 6). The second number is the sample size. The number of degrees of freedom, together with α, determines the critical value needed for statistical significance; the sample size allows the reader to compute a measure of the strength of association between the two variables. The latter measure provides important information that is separate from your decision about the null hypothesis. Measures of strength of association in a contingency table are discussed in Section C.

An Excerpt from the Psychological Literature

The two-way chi-square test appears very frequently in the psychological literature; the example that follows shows how an experimental design that might have led to a t test produced data for which a chi-square test was more appropriate. Schwartz, Slater, and Birchler (1994) randomly assigned half of a group of 34 chronic back pain patients to a stressful interview condition and the other half to a neutral conversation condition. After the stressful or neutral condition, all subjects were asked to pedal an exercise bicycle at a steady rate for 20 minutes but were instructed to stop if they experienced considerable back pain. Subjects who had been in the stressful condition were expected to stop after fewer minutes than subjects who had been in the control group. If only a few subjects had pedaled the full 20 minutes, a t test could have been performed on the number of minutes pedaled by subjects in the two groups. However, the results did not lend themselves to such an analysis, as described in the following excerpt.

> Examination of the bicycling data revealed a skewed distribution that was due to a ceiling effect (i.e., 53% of the subjects persisted in the bicycle task for the maximum amount of time). The persistence data were dichotomized, and a contingency table analysis was performed to determine whether there were significant differences in the proportion of patients in each interview group who persisted in the physical activity task. A significantly greater proportion of patients in the stress interview condition terminated the bicycling task prematurely ($n = 11$), compared with patients in the neutral talking condition ($n = 5$), $\chi^2 (1, N = 34) = 4.25, p < .05$.

SUMMARY

1. When subjects can be classified according to the categories of two variables at the same time, a two-way contingency table can be formed. The usual null hypothesis is that the two variables are independent of each other, so a subject's classification for one variable does not affect his or her classification on the other variable.
2. Pearson's chi-square test of association (or independence) begins with the determination of the expected frequency for each cell of the two-way table. The f_es in each row or column must yield the same sum as the corresponding f_os, and the proportions within each column or row

of the table should correspond to the proportions of the column or row sums, respectively.

3. The expected frequency for each cell is the product of its row sum and column sum divided by the total N. However, the degrees of freedom for the table determines how many f_es must be calculated in this way (the remainder can be found by subtraction). The df $= (R - 1)(C - 1)$. The same chi-square formula used for the one-way test is then applied to each cell of the contingency table and then these values are summed.

4. There is a shortcut formula that avoids the calculation of expected frequencies, but it applies only to the 2×2 table. It involves the squared difference of the two diagonal cross products multiplied by N and then divided by the product of all four marginal sums.

5. **The Assumptions of the Chi-Square Test**
   a. *Mutually exclusive and exhaustive categories*. Each observation falls into one, and only one, category.
   b. *Independence of observations*. Usually this assumption is satisfied by having each frequency count represent a different subject.
   c. *Size of expected frequencies*. The conservative rule of thumb is that no f_e should be less than 5 (when df $= 1$, each f_e should be at least 10). A more relaxed rule states that f_e must never be less than 1 and no more than 20% of the f_es can be less than 5.

6. **When to Use the Chi-Square Test for Independence**
   Two-way chi-square tests are generally used with one of three types of study designs:
   a. *Two grouping variables* (e.g., proportions of left- and right-handed persons in various professions).
   b. *One grouping variable and one measured variable* (e.g., proportions of babies from different cultures expressing one of several possible emotional reactions to an experimental condition, such as removing a toy).
   c. *One experimentally manipulated variable and one measured variable* (e.g., exposing children to violent, neutral, or peaceful cartoons and categorizing their subsequent behavior as aggressive, neutral, or cooperative).

EXERCISES

1. Is one's personality related to one's choice of college major? To address one aspect of this question, 25 physics majors, 35 literature majors, and 45 psychology majors were tested for introversion/extraversion. The results were as follows for each major: physics, 15 introverts and 10 extroverts; literature, 17 introverts and 18 extroverts; psychology, 19 introverts and 26 extroverts. Display these results in a 3×2 contingency table and perform the appropriate chi-square test.

*2. Does a diet high in sugar cause hyperactivity in young children? Ten fourth-graders whose parents admit to providing a high-sugar diet and 20 fourth-graders on a more normal diet are rated for hyperactivity in the classroom. The data are shown in the following contingency table.

	Normal Diet	High-Sugar Diet
Hyperactive	2	4
Not hyperactive	18	6

a. Perform the chi-square test (use Formula 20.1) and decide whether the null hypothesis (i.e., hyperactivity is not associated with dietary sugar) can be rejected at the .05 level.
b. Recalculate the value of χ^2 using Formula 20.5, and check that the value is the same.
c. If the null hypothesis were rejected, could you conclude that a large amount

of sugar can *cause* hyperactivity? Why or why not?

*3. In spite of what may seem obvious, medical research has shown that being cold is not related to catching a cold. Imagine that in one such study, 40 subjects are exposed to cold viruses. Then half of the subjects are randomly assigned to remain in a cold room for several hours, and the other half remain in a room kept at ordinary room temperature. At the end of the experiment, the numbers of subjects who catch a cold are as shown in the following table:

	Cold Room	Normal Room
Caught cold	14	12
No cold	6	8

a. Test the null hypothesis ($\alpha = .05$) that being cold is unrelated to catching a cold.
b. Given your statistical decision in part a, which type of error (Type I or Type II) could you be making?

4. A social psychologist is studying whether people are more likely to help a poor person or a rich person who they find lying on the floor. The three conditions all involve an elderly woman who falls down in a shopping mall (when only one person at a time is nearby). The independent variable concerns the apparent wealth of the woman; she is dressed to appear either poor, wealthy, or middle class. The reaction of each bystander is classified in one of three ways: ignoring her, asking if she is all right, and helping her to her feet. The data appear in the contingency table below. Test the null hypothesis at the .01 level. Is there evidence for an association between the apparent wealth of the victim and the amount of help provided by a bystander?

	Poor	Middle Class	Wealthy
Ignores	16	10	7
Talks to her	8	6	5
Helps her up	6	14	18

*5. Is there a connection between a person's position on the death penalty and his or her opinion about gun control? Fifty people are polled about their views on both issues. The data appear in the following table. Calculate χ^2 for these data. Can you conclude that opinions on the two issues are not independent?

	Favors No Gun Control	Favors Some Gun Control	Favors Strict Gun Control
For the death penalty	8	12	7
Against the death penalty	4	10	9

6. In Exercise 20A6, the responses of 50 subjects to a logic problem were compared to expected frequencies from extensive prior research. Suppose instead that the experiment were conducted with two groups of 50 subjects each and that one group is presented with the problem in the traditional (abstract) way, whereas the other group receives the more concrete presentation. Suppose that the data for the concrete subjects are the same as given in Exercise 20A6 and that the abstract subjects perform exactly according to the expectations stated in that exercise, as shown in the following table:

	A	B	C	D
Abstract	20	13	10	7
Concrete	24	8	16	2

a. Calculate χ^2 for the data. Can you reject the null hypothesis that type of presentation and type of response to the problem are independent?
b. Compare the value for χ^2 in this problem with the value you calculated for Exercise 20A6. Can you explain the advantage of having expectations based on a population instead of another group of the same size?

*7. The use of the polygraph for lie detection remains controversial. In a typical laboratory experiment to test the technique, half the subjects are told to commit a mock crime (to remove a camera from a cabinet); the other half of the subjects remain "innocent." The polygrapher, blind to each subject's condition, must make a judgment of guilt or innocence based on a subject's physiological responses to a series of questions. The hypothetical data for one such study are shown in the following table. Test the null hypothesis ($\alpha = .05$) that the poly-grapher's judgments are unrelated to the subject's guilt (i.e., that this form of lie detection is totally ineffective).

	Innocent	Guilty
Judged innocent	10	3
Judged guilty	5	12

8. In Exercise 12B4, the dependent variable was the amount of time a subject listened to tape-recorded cries for help from the next room before getting up to do something. If some subjects never respond within the time allotted for the experiment, the validity of using parametric statistical techniques could be questioned. As an alternative, subjects could be classified as fast or slow responders (and possibly, nonresponders). The data from Exercise 12B4 were used to classify subjects as fast responders (less than 12 seconds to respond) or slow responders (12 seconds or more). The resulting contingency table is shown in the following table:

	Child's Voice	Adult Female	Adult Male
Fast responder	5	3	1
Slow responder	2	4	6

   a. Test the null hypothesis ($\alpha = .05$) that speed of response is independent of type of voice heard.
   b. How does your conclusion in part a compare with the conclusion you drew in Exercise 12B4? Categorizing the dependent variable throws away information; how do you think that loss of information affects power?

*9. The director of a mental health clinic, which trains all of its therapists in a standardized way, is interested in knowing whether some of its therapists are more effective than others. Twenty patients (carefully matched for severity of problem, age, and other relevant

factors) are assigned to each of five therapists for a 6-month period, at the end of which each patient is categorized as "improved," "no change," or "worse." The data appear in the following table. Test the null hypothesis ($\alpha = .01$) that the assignment of a particular therapist is unrelated to the patient's progress.

	Dr. A	Dr. B	Dr. C	Dr. D	Dr. E
Improved	15	11	16	13	10
No change	5	3	0	4	6
Worse	0	6	4	3	4

10. A magazine publisher is testing four different covers for her anniversary issue and wants to know if each one appeals equally to people of both genders. Fifty subjects are brought in, and each selects the cover that would most draw him or her to buy the magazine. The number of people of each gender who chose each cover are as follows:

Gender of Subject	Cover I	Cover II	Cover III	Cover IV
Female	12	10	5	3
Male	5	9	1	5

   a. Test the null hypothesis ($\alpha = .05$) that cover preference is independent of gender.
   b. If you rejected the null hypothesis in part a, test the preferences of each gender separately (i.e., perform two one-way chi-square tests) to see if the preferences are equal within each gender. If you did not reject the null hypothesis in part a, ignore gender (i.e., sum the frequencies for each cover for the two genders) and perform a one-way chi-square test to see if the four covers are equally preferred.

OPTIONAL MATERIAL

Measuring Strength of Association

When you perform a chi-square test, you want the value for χ^2 to be high—at least high enough to be statistically significant. But what does it mean if χ^2 is extremely high? A very high χ^2 value tells you that your p value is very small, which implies that your results would be significant even if you chose an unusually small alpha level. However, as with the t and F statistics, the size of the chi-square statistic from a two-way test tells you nothing about the strength of the association between the two variables. A very weak dependency between two variables can lead to a large χ^2 value if the sample size (N) is sufficiently large. In fact, if you multiply each cell frequency in a contingency table by the same constant, the chi-square statistic will be multiplied by that constant. This principle also applies to the one-way test. For instance, if flipping a fair coin 10 times produces seven heads and three tails, χ^2 will equal a nonsignificant 1.6 (even less with Yates's correction). However, if you obtain the same proportions of heads and tails in 100 flips (i.e., 70H, 30T), the χ^2 value will be 16 ($p < .001$). It would be helpful to supplement the χ^2 value with a measure of strength of association (or an estimate of effect size) that is not affected by sample size. A number of such measures have been devised; the most widely used of these are described in the following.

Phi Coefficient

In the case of a 2 × 2 contingency table, there is a very natural choice for a measure of association: Pearson's r. Recall that in Chapter 10, Section C, I described how to calculate a Pearson correlation coefficient when you have two distinct groups of scores (i.e., when one variable consists of only two categories and the other variable is continuous). The two levels of the grouping variable can be assigned any two values (0 and 1 are the most convenient), and the ordinary correlation formula can then be applied. The result is often called the point-biserial correlation (r_{pb}) as a reminder that one of the variables has only two values and that the sign of the correlation is probably irrelevant. A similar procedure can be applied to a 2 × 2 contingency table. In this case, both variables are dichotomous, so both can be assigned arbitrary values such as 0 and 1. Then, any one of the Pearson correlation formulas can be applied. The result is called the *phi* (φ) *coefficient*, or sometimes, the *fourfold point correlation*.

If you have already found the χ^2 value, you do not have to calculate φ from scratch. The phi coefficient is a simple function of both the chi-square statistic and the total sample size, as given by Formula 20.6:

$$\phi = \sqrt{\frac{\chi^2}{N}}$$ **Formula 20.6**

This formula should remind you of Formula 10.13 for finding r_{pb} from a two-group t value. Note that multiplying every cell frequency by the same constant results in N being multiplied by that constant, as well as χ^2. In Formula 20.6 both the numerator and the denominator would be multiplied by the same constant, leaving the overall value unchanged. Thus, increasing the sample size without changing the relative proportions does not affect the phi coefficient. Also note that in a 2 × 2 table, χ^2 can never exceed N (otherwise φ would exceed 1.0); χ^2 will equal N only if all of the frequencies fall in two diagonal cells and the other two cells are empty.

Applying Formula 20.6 to the divorce example in Section B (Table 20.4), we find:

$$\phi = \sqrt{\frac{2.4}{30}} = \sqrt{.08} = .283$$

The general practice is to take the positive square root in Formula 20.6, so phi ranges between 0 and +1.0. However, it is possible in some cases to assign 0 and 1 to each variable in a meaningful way so that when calculated directly from the raw data, the sign of the phi coefficient is meaningful. For instance, in the divorce example, you could assign 1 for "divorced" and 0 for "not divorced" for both "parents" and "self." In that case, a positive correlation would indicate a tendency for greater divorce among those whose parents are divorced; a negative correlation would indicate the opposite tendency. The sign of the phi coefficient becomes irrelevant, though, when strength of association is measured by squaring φ. You may recall that squaring a correlation coefficient is useful because it tells you the proportion of variance accounted for. Squaring φ gives you an analogous measure;

squaring both sides of Formula 20.6 results in Formula 20.7 for ϕ^2, as follows:

$$\phi^2 = \frac{\chi^2}{N}$$ **Formula 20.7**

Cramér's Phi and the Contingency Coefficient

The value of ϕ^2 provides useful information to accompany χ^2; the latter allows you to determine statistical significance, whereas the former tells you the degree of association between the two variables in your present sample. It would be nice to have this supplemental information for chi-square problems that do not involve a 2×2 contingency table. Fortunately, only a slight modification of Formula 20.6 is needed to create a phi coefficient that is appropriate for any contingency table. Actually, no modification of that formula is required if one of the variables has only two levels, but if both variables have more than two levels, Cramér (1946) suggested using the following statistic:

$$\phi_C = \sqrt{\frac{\chi^2}{N(L - 1)}}$$ **Formula 20.8**

where L is either the number of rows or the number of columns, whichever is smaller. This statistic is often called *Cramér's phi*, and like ordinary phi, it ranges from 0 to 1.0. For the main example of Section B, which involves a 4×3 table, $L = 3$ and the value for Cramér's phi is:

$$\phi_C = \sqrt{\frac{21.4}{80(3 - 1)}} = \sqrt{.134} = .366$$

A related statistic, called the *contingency coefficient*, is given by Formula 20.9:

$$C = \sqrt{\frac{\chi^2}{\chi^2 + N}}$$ **Formula 20.9**

However, because C does not generally range from 0 to 1, statisticians find that Cramér's phi is preferable for descriptive purposes.

The Cross-Product Ratio (or Odds Ratio)

A strength of association measure for the 2×2 table that is commonly employed in biological and medical research is the *cross-product ratio*, also called the *odds ratio*. This is a very simple measure that consists of the ratio between the products of the diagonally opposite cells in the 2×2 table. In terms of the letter designations in Table 20.9, Formula 20.10 for the cross-product ratio is:

$$\text{Cross-product ratio} = \frac{ad}{bc}$$ **Formula 20.10**

For the divorce example in Section B (Table 20.4), the cross-product ratio is:

$$\frac{ad}{bc} = \frac{(7)(12)}{(8)(3)} = \frac{84}{24} = 3.5$$

A ratio of 3.5 indicates a moderately high degree of association between the two variables (if N had not been so low for this example, the chi-square value would have been significant). The ratio can vary from zero to an infinitely high number. Numbers much larger or smaller than 1.0 represent a strong degree of association; a ratio of 1.0 indicates a total lack of association between the variables. As applied to the divorce example, the cross-product ratio compares the odds of getting divorced given that your parents were divorced (7/3) with the odds of getting divorced given your parents did not get divorced (8/12)—hence the alternative term for this measure, the odds ratio (divide 7/3 by 8/12 and you will get 3.5). In medical applications, you might want to know the odds of getting positive test results if you have the disease being tested for, compared to the odds of getting false positive results. Unless the odds ratio is very high, it might not be worthwhile to take the test.

Power as a Function of Strength of Association

As in the *t* test or ANOVA, the power of a two-way chi-square test depends on both a measure of effect size (or the strength of the association between the variables) and the size of the sample. For a given strength of association (and α), the sample size required to achieve a particular level of power can be estimated using tables prepared for that purpose (Cohen, 1988).

Measuring Interrater Agreement When Using Nominal Scales

This chapter deals with the situation in which all of your variables are categorical. Such variables require that a subject (which needn't be a person, of course, but could be a rat, a city, or even a painting) be placed into one or another of a set of mutually exclusive and exhaustive categories. Sometimes this is easy, as when the subject tells you what category he or she is in (e.g., with respect to religion, political party, etc.). Sometimes it is a matter of simple observation—e.g., the child picks up the yellow toy first, or the blue one, etc. And, of course, it is easy when you start out with an interval/ratio scale and create distinct categories later (e.g., the child picks up no toys, just one toy, more than one toy). On the other hand, the classification of subjects can sometimes be rather difficult. Consider the problem of classifying facial expressions of emotion.

It is fairly easy to distinguish happiness from sadness by simple observation of subjects' faces, but distinguishing anger from disgust is not so easy, especially when the subjects experiencing the emotions are not experiencing them very strongly. Classifying these emotions from the faces of people from a different culture than your own can be even more difficult. It is important for the evaluation of some theories of emotion (e.g., that emotions are expressed similarly across cultures; Ekman, 1982) to determine how reliably emotional expressions can be categorized. A method is needed to determine the extent to which different raters agree on their categorizations.

I have referred to this concept before with respect to continuous variables when I mentioned *interrater reliability* in Chapter 9. For instance, if two raters were judging the amount of happiness exhibited by the faces of several subjects (assigning numbers from 0 to 10), Pearson's *r* could be used to quantify the amount of agreement between the two raters. If the raters merely ranked the faces in order from least to most happiness shown, Pearson's *r* could still be calculated, though the correlation would be treated as a *Spearman rank-order correlation coefficient* for inferential purposes (this will be discussed further in Section B of the next chapter). However, if facial expressions are to be categorized by two different raters into one of four emotional categories (e.g., anger, fear, disgust, and contempt), an ordinary correlation coefficient cannot be found. Fortunately, the contingency tables you have learned about in this chapter can be used to quantify the amount of agreement between the two raters. A two-way table is set up with the same categories along each dimension, the rows representing one rater and the columns the other.

The Contingency Table for Agreement

To illustrate this type of contingency table, I have set up the following example. Imagine that 32 photographs of faces expressing emotion are categorized by two well-trained raters. The columns in Table 20.10 represent the judgments of the first rater and the rows represent the judgments of the second rater.

   The upper left cell of the table shows, for instance, that there were six photos that both raters classified as exhibiting anger. The first rater classified a total of seven photos as exhibiting contempt, while the second rater placed a total of eight photos in that category. Two of the photos classified as disgust by the first rater were classified as fear by the second rater. The cells along the diagonal from upper left to lower right represent the instances in which the two raters agreed. Adding those cells (6 + 4 + 5 + 4) yields the total number of agreements, 19. This number is meaningful, of course, only as a proportion of the total number of judgments: 19/32 = .59. Expressed as a percentage, this amount of agreement (59%) does not represent an overwhelming amount of agreement; this percentage suggests that the categorization called for is a difficult one and not extremely reliable.

Table 20.10		Anger	Fear	Disgust	Contempt	Sum
	Anger	6	0	1	2	9
	Fear	0	4	2	0	6
	Disgust	2	1	5	1	9
	Contempt	1	1	2	4	8
	Sum	9	6	10	7	32

Kappa as a Measure of Agreement

Unfortunately, the situation is even worse than the proportion just calculated would suggest. If the two raters were classifying the faces each in their own arbitrary way with no criteria in common, we would still expect some agreement by accident. To get a clearer measure of agreement, we need to determine the likely number of agreements due to chance and subtract that number from the original number of agreements. To find the expected number of chance agreements, you need to calculate the expected frequencies of the diagonal cells just as you would for the two-way chi-square test. For instance, the expected frequency for anger is its column total (9) times its row total (9) divided by the total N (32). Thus, $f_e = 81/32 = 2.53$. Similarly, the expected frequencies for fear, disgust, and contempt are 1.125, 2.81, and 1.75, respectively. So the total number of expected agreements is 8.22, more than a quarter of the total number of judgments. The number of agreements corrected for chance is $19 - 8.22 = 10.78$. To be fair, the total number of judgments should be corrected by this same factor before we calculate a chance-corrected proportion. The new total is $32 - 8.22 = 23.78$ and the new proportion is $10.78/23.78 = .45$. This adjusted proportion, corrected for chance agreements, is called *kappa*, and was devised by Jacob Cohen (1960), the same psychologist/statistician who formalized power analysis for psychological research. The formula for kappa follows:

$$\kappa = \frac{\sum f_o - \sum f_e}{N - \sum f_e}$$ **Formula 20.11**

Note that f_o and f_e refer to the observed and expected frequencies in the diagonal cells only and N is the total number of items being classified.

There are ways to test kappa for statistical significance, but these are only relevant if you are wondering whether there is any basis for two raters to agree in their classifications on a particular dimension. More often, we expect at least some agreement between raters, but need to know if the agreement is high enough to be useful. For example, in the case of the emotion classification study just described if the proportion of agreement were considered to be high enough, we might feel confident in using those photographs upon which the raters agreed as stimuli in a new experiment.

Fisher's Exact Test

It is important to remember that, in normal practice, the chi-square statistic does not follow the chi-square distribution exactly, and the use of the chi-square distribution as an approximation is not recommended when one or more of the f_e is near zero. A similar problem occurs with using the normal approximation to the binomial distribution. For small sample sizes (especially less than 16), it is recommended that you use the exact binomial distribution (Table A.13). Calculating exact probabilities is also a possibility for a 2×2 contingency table when one or more of the f_e is less than 5. These calculations involve determining probabilities of the multinomial distribution directly rather than approximating them, and therefore they are legitimate no matter how small the frequencies are. The procedure is known as *Fisher's exact test* (named for R. A. Fisher, who first devised the test). As you could guess, Fisher's exact test would be very tedious to calculate without a computer, but given the present availability of computers, the hand calculation of the test is no longer a consideration. The more serious problem concerning this test is that it assumes that the marginal sums are fixed, which is not usually the case when chi-square tests are performed. I will give an example of a test with fixed marginals next.

Fixed versus Free Marginal Sums

In the treatment of schizophrenia example, the sums for each treatment were fixed because we assigned the same number of patients to each condition. On the other hand, the sums for each improvement level were not fixed; they were free to vary depending on how well the treatments worked, in general. However, those sums could have been fixed, as well. For instance, the improvement variable could have been based on a median split of all 80 patients involved in the study. One level of improvement (one row in the table) could have consisted of patients in the top half of "improvement" for all patients (in a study where less than half the patients improve at all, a patient could be in the top half of improvement for not getting too much worse); the other level would be the bottom half. In such a case, the row sums would also be fixed (for this example, at 40 and 40). Of course, the cell frequencies would be free to vary within those constraints (e.g., in a very extreme case, two therapies could have 20 patients in the top half of improvement and none in the bottom half, with the other two therapies exhibiting the reverse pattern).

Studies that lead to fixed marginals (i.e., fixed for both columns and rows) are not common. In fact, in a typical observational study, in which subjects are not assigned to treatments, both the column and row marginal sums are free to vary. For instance, a sample of subjects can be given a set of questionnaires to determine, among other things, whether they are introverts or extraverts and whether they are "day" or "night" people. If we make a 2 × 2 contingency table to look at the frequencies of day and night people among both introverts and extraverts, we have no control over the column or row sums. One reason why I did not recommend Yates' continuity correction for a 2 × 2 table is that it works well only when dealing with fixed marginals, which is rarely the case in psychological research.

Contingency Tables Involving More Than Two Variables

If you can only look at two variables at a time, the research questions you can ask are quite limited. You saw in Chapter 14, for instance, how two independent variables can interact in their influence on a third, dependent variable. A similar situation can arise with three categorical variables. For example, the dependent variable of interest might be whether a subject returns for a second session of an experiment; this variable is categorical and has only two levels: returns and does not return. If we wanted to know the effect of the experimenter's behavior on the subject's return, we could have the experimenter treat half the subjects rudely and half the subjects politely. (Subjects would be run individually, of course.) The results would fit in a 2 × 2 contingency table.

A more interesting question, however, would explore the role of self-esteem in the relationship between experimenter behavior and subject return. Research in social psychology has suggested that subjects with low self-esteem are more helpful to a *rude* experimenter, whereas the reverse is true for subjects with high self-esteem (e.g., Lyon & Greenberg, 1991). Repeating the 2 × 2 contingency table once for a group of low self-esteem subjects and once again for a group of high self-esteem subjects produces a three-variable contingency table. Rather than trying to represent a cube on a flat page, it is easier to display the 2 × 2 × 2 contingency table as two 2 × 2 tables side by side, as depicted in Table 20.11.

The major problem in applying the chi-square test to a contingency table with three or more variables is finding the appropriate expected frequencies. The trick we used in Section B is not valid when there are more than two variables. The solution that has become very popular in recent years is to use a *log-linear model*. A discussion of the log-linear model is beyond the scope of this text; I will just point out that once the expected frequencies have been estimated with such a model, the ordinary chi-square test (or the increasingly popular *likelihood ratio test*) can then be applied. Although log-linear analyses are now performed by most of the major computerized statistical packages, an advanced understanding of statistics is required to use these techniques properly.

Table 20.11	LOW SELF-ESTEEM		HIGH SELF-ESTEEM	
	Polite	Rude	Polite	Rude
Return				
Don't return				

1. The size of the chi-square statistic in a two-variable design does not tell you the degree to which the two variables are related. A very weak association can produce a large value for χ^2 if the sample size is very large.

2. For a 2×2 contingency table, the strength of association can be measured by assigning arbitrary numbers (usually 0 and 1) to the levels of each variable and then applying the Pearson correlation formula. The result is called the *phi coefficient* or the *fourfold point correlation*.

3. Like other correlation coefficients, ϕ ranges from -1 to 0 to $+1$; however, the sign of ϕ is not always meaningful. If you are not concerned about the sign, you can find ϕ easily by taking the positive square root of χ^2 divided by N (the sample size). An alternative measure of association for 2×2 tables, often used in medical research, is the *cross-product ratio* or *odds ratio*. It is the product of the frequencies in one set of diagonally opposite cells divided by the product of the other diagonal cells.

4. For contingency tables larger than 2×2, a modified ϕ coefficient, known as *Cramér's phi*, is recommended as a measure of association. Cramér's phi is preferred to a related statistic, the *contingency coefficient*, because the former always ranges from 0 to 1, whereas the latter need not.

5. Placing subjects (or faces, paintings, etc.) into categories is sometimes a difficult task, and it can be important to measure the degree to which two independent judges agree on these categorizations. The number of items classified identically by two judges can be divided by the total number of items classified to create a proportion, but it is more useful to correct this proportion for chance by subtracting the expected number of agreements from both the numerator and denominator of the proportion. The resulting index of agreement is called *Cohen's kappa*.

6. For 2×2 tables in which the expected frequencies are too small to justify the chi-square approximation, the probability of the null hypothesis producing the observed results (or more deviant results) can be calculated exactly using *Fisher's exact test*. Strictly speaking, however, this test is only valid if all of the marginal sums are fixed. For tables larger than 2×2 with small f_e, collapsing categories can solve the problem, but in so doing you must be cautious not to capitalize on chance.

7. If more than two categorical variables are included in a contingency table, determining the expected frequencies is not straightforward. The f_e can be estimated, however, using the *log-linear model*. The analysis is completed by using the ordinary chi-square test or the *likelihood ratio test*.

SUMMARY

EXERCISES

*1. a. Calculate the ϕ coefficient for the data in Exercise 20B2 using the χ^2 value that you found for that exercise.

   b. What can you say about the strength of the relationship between dietary sugar and hyperactivity in the data of Exercise 20B2?

2. a. Calculate the ϕ coefficient for the data in Exercise 20B3, using the χ^2 value that you found for that exercise.

   b. Calculate the cross-product ratio for the same data.

   c. What can you say about the strength of the relationship between being cold and catching cold in the data of Exercise 20B3?

*3. a. Calculate the ϕ coefficient for the data in Exercise 20B7.

   b. Calculate the cross-product ratio for the same data.

c. What do these measures tell you about the accuracy of the lie detector test in Exercise 20B7?

4. a. Calculate both Cramér's phi and the contingency coefficient for the data in Exercise 20B4.

b. What can you say about the degree of relationship between the apparent wealth of the person in need and the amount of help given in that hypothetical experiment?

*5. a. Calculate both Cramér's phi and the contingency coefficient for the data in Exercise 20B9.

b. What can you say about the degree of relationship between the therapist assigned and the amount of improvement in the patient?

6. In a study of a new treatment for schizophrenia, each patient must be rated as improved, unchanged, or worse. Suppose that two psychologists are asked to make these judgments for the same set of 30 patients. The two-way contingency table for their judgments is as follows:

	Improved	Unchanged	Worse	Sum
Improved	14	4	2	20
Unchanged	1	5	2	8
Worse	1	1	0	2
Sum	16	10	4	30

a. Calculate Cohen's kappa for these data.

b. Based on the kappa you calculated, do you trust the ratings of these psychologists? Explain.

*7. Suppose that two raters are classifying dream reports into one or another of five categories (A through E). Of the 100 reports classified, the two raters agree on 70 of them. One of the judges classifies 20, 10, 30, 15, and 25 into categories A, B, C, D, E, respectively, whereas the other judge classifies 30, 25, 15, 20, and 10 into these same categories. Calculate Cohen's kappa for these data.

8. Devise an experiment whose results could be represented by a three-way contingency table.

KEY FORMULAS

Pearson chi-square statistic for df > 1:

$$\chi^2 = \sum \frac{(f_o - f_e)^2}{f_e}$$

Formula 20.1

Pearson chi-square statistic for df $= 1$:

$$\chi^2 = \sum \frac{(|f_o - f_e| - .5)^2}{f_e}$$

Formula 20.2

Degrees of freedom for a two-variable contingency table:

$$df = (R - 1)(C - 1)$$

Formula 20.3

Expected frequencies in a two-variable contingency table:

$$f_e = \frac{(\text{row sum})(\text{column sum})}{N}$$

Formula 20.4

Shortcut formula for the chi-square statistic for a 2×2 table:

$$\chi^2 = \frac{N(ad - bc)^2}{(a + b)(c + d)(a + c)(b + d)}$$

Formula 20.5

Phi coefficient for a 2×2 table as a function of the chi-square statistic:

$$\phi = \sqrt{\frac{\chi^2}{N}}$$

Formula 20.6

Phi coefficient squared (measures proportion of variance accounted for):

$$\phi^2 = \frac{\chi^2}{N}$$

Formula 20.7

Cramér's phi coefficient for contingency tables larger than 2×2 (varies between 0 and 1):

$$\phi_C = \sqrt{\frac{\chi^2}{N(L-1)}}$$ **Formula 20.8**

Contingency coefficient for tables larger than 2×2 (maximum value is usually less than 1):

$$C = \sqrt{\frac{\chi^2}{\chi^2 + N}}$$ **Formula 20.9**

Cross-product (or odds) ratio:

$$\text{Cross-product ratio} = \frac{ad}{bc}$$ **Formula 20.10**

Cohen's kappa:

$$\kappa = \frac{\sum f_o - \sum f_e}{N - \sum f_e}$$ **Formula 20.11**

STATISTICAL TESTS FOR ORDINAL DATA

Chapter 21

CONCEPTUAL FOUNDATION

You will need to use the following from previous chapters:

Symbol:
Σ: Summation sign

Formula:
Formula 9.1: Pearson correlation coefficient

Concepts:
Ordinal scales
Median
Correlation

Remember the Martian from Chapter 8 who wanted to test whether men and women differ in height? Imagine that the Martian has rounded up 10 men and 12 women but does not have any kind of ruler to measure their heights. Can he, she, or it still perform a statistical test to decide whether the populations of men and women are identical with respect to height? The answer is Yes. Perhaps you're thinking that a crude ruler could be made on the spot to serve the Martian's purpose, but I have a different solution in mind—one that will introduce the concepts of this chapter.

Ranking Data

One simple solution to the Martian's dilemma is to line up all 22 of the humans in order of height. This is easy to do without any kind of ruler. Of course, it will occasionally be difficult to determine which of two people is taller, but assuming such ties are rare, the statistical procedures for ordinal data are still valid. I will discuss what to do about tied scores later in this section; for now I will assume that all 22 people have been placed in height order.

The next step is to assign *ranks* to the people standing in line. It is arbitrary whether you assign the rank of 1 to the tallest or the shortest person in line, but it is common to assign the lowest rank to the one with the most of that variable—so I will give the rank of 1 to the tallest person, 2 to the next tallest, and so on, assigning the rank of 22 to the shortest. If this group of adult humans is typical, we would expect the women to be clustered toward one end of the line and the men to be concentrated near the other end. Nonetheless, we expect a fair amount of mixing of the genders near the middle of the line. In terms of ranks, we expect the men to have lower ranks than the women, for the most part.

Comparing the Ranks from Two Separate Groups

Suppose that the Martian finds the sum of ranks for men to be considerably smaller than that for women. Can it conclude that the population of men differs from the population of women in height? As you know, Dr. Null would claim that this difference in the sums of ranks is the result of accidentally drawing unusual samples of men and women—that typical samples would not exhibit such a difference. What we need is a test statistic based

728

on the sum of ranks that follows a known mathematical distribution when the null hypothesis is true. The statistic most commonly used is called U, and the statistical test based on U is called the *Mann-Whitney U-test*, after the two statisticians who devised this approach (Mann & Whitney, 1947). Wilcoxon (1949) simplified the test somewhat by working out tables of probabilities that are expressed directly in terms of sums of ranks (eliminating the necessity to calculate the U statistic); therefore his version of the test is sometimes referred to as the *Wilcoxon rank-sum test*. Because U is still commonly used, I will discuss that approach briefly. But because Wilcoxon's version is simpler, it is the one that I will follow in this chapter. Regardless of whose version is used to find the critical value or p value, the test is sometimes called the *Wilcoxon-Mann-Whitney test*. However, I will use the more common designation: *the Mann-Whitney test*. This name is less confusing because another common ordinal test, based on matched samples (described in Section B), is most often called the *Wilcoxon test*. If the U statistic is being used rather than Wilcoxon's approach, I will refer to the test specifically as the Mann-Whitney U test.

The Sum of Ranks

To make it easier to illustrate the mathematics of dealing with ranks, I will use an example with very small samples: four men and five women (i.e., $n_m = 4$ and $n_w = 5$). Again, we line up all of the men and women by height and assign them ranks from 1 to 9. One quantity that will prove useful is the sum of the ranks (S_R) for all people in the line (i.e., $\sum_{i=1}^{N} R_i$). If the total number of people ranked is N, the sum of the ranks will be:

$$S_R = \frac{N(N + 1)}{2}$$

Formula 21.1

For $N = 9$, $S_R = 9(10)/2 = 90/2 = 45$. If there is no overlap between the men and women in the line (e.g., the men are ranked 1 to 4 and the women 5 to 9), S_R for the men can be found by using Formula 21.1:

$$S_R = \frac{n_m(n_m + 1)}{2} = \frac{4(5)}{2} = \frac{20}{2} = 10$$

The sum of ranks for the women can be found by subtracting the sum for the men from the total sum: $45 - 10 = 35$. No matter how the men and women are arranged in line, adding the sum of ranks for men to the sum of ranks for women will always give you the sum of ranks for the whole group (i.e., 1 to N), as found by inserting N into Formula 21.1.

The U Statistic

Another way to measure the discrepancy in height order between the men and the women standing in our hypothetical line is to note how many women are shorter than each man (or conversely, how many men are shorter than each woman). I will illustrate this measure by taking the simple case in which men are ranked 1 to 4 and women are ranked 5 to 9. The man ranked 1 is taller than all five women, so he gets 5 "points." In fact, in this example, all four men receive 5 points each, so the men receive a total of 20 points. Because none of the women are taller than any of the men, the women are assigned 0 points. Both of these values, 0 and 20, are possible values for the U statistic. The convention is to define U as the smaller of the two possible values, so for the example above, $U = 0$. As you might guess, $U = 0$ is the lowest possible value, and it occurs whenever there is no overlap in ranks for the two subgroups.

Determining the U value directly can be tedious when the two ns are fairly large and there is a good deal of mixing between the two subgroups. Fortunately, there is a simple formula that allows you to calculate U in terms of the sum of ranks for two groups labeled A and B, respectively:

$$U_A = n_A n_B + \frac{n_A(n_A + 1)}{2} - \sum R_A \qquad \text{Formula 21.2}$$

where $\sum R_A$ represents the sum of ranks for members of group A.

You may notice that the middle term in Formula 21.2 is the sum of ranks from 1 to n_A. This happens to be the lowest value that $\sum R_A$ can take on. If the A group has all the low ranks (i.e., there is no overlap between groups), $\sum R_A$ will equal the middle term, and U_A will equal $n_A n_B$. (This means $U_B = 0$ because the sum of U_A and U_B will always equal $n_A n_B$.) Generally, $\sum R_A$ will be higher than the middle term, so some amount will be subtracted from U_A. However, $\sum R_A$ (or $\sum R_B$, as appropriate) can be used directly to complete the test, as I will show in Section B.

Dealing with Tied Scores

The statistical procedures presented in this chapter rest on the assumption that the variables involved are actually continuous, even though they are being measured on an ordinal scale. For instance, just because we have no precise way to measure beauty does not mean beauty exists at only particular levels with no gradations in between. Beauty varies continuously. Theoretically, with any two paintings that have a similar degree of beauty, we should be able to find (or create) a third painting whose beauty falls somewhere between the first two. The implication of this principle is that when we rank objects according to beauty (or some other continuous variable that is hard to measure), we should not encounter two objects that have exactly the same amount of beauty—that is, there should not be any ties.

In practice, our ability to discern differences is limited, so some ties are inevitable—but ties should be fairly rare. If ties are common in our data, either the variable is not continuous, or our ability to discern differences is too crude. In either case the validity of the tests in this chapter would be questionable. In fact, if there are many ties (e.g., most boys observed in a playground for a week initiate either zero or one fight, and a few initiate more than one), it may be necessary to divide your scale into a few broad categories and use statistics for qualitative data (e.g., chi-square tests). If you decide to assign ranks in spite of having several ties, you should use the following method before applying any of the ordinal statistical procedures described in this chapter.

To illustrate ranking with tied scores, I will rank order a set of 10 subjects with the following IQ scores: 109, 112, 120, 100, 95, 112, 104, 100, 112, 115. The first step is to place these scores in numerical order and assign a unique rank to each score, even though the rank seems arbitrary in the case of tied scores; see Table 21.1.

Table 21.1	Score	120	115	112	112	112	109	104	100	100	95
	Rank	1	2	3	4	5	6	7	8	9	10

The next step is to calculate the average rank for each group of tied scores in Table 21.1. For instance, the score 112 has been given the ranks 3, 4, and 5. The average of these three ranks (3 + 4 + 5)/3 is 4; therefore, each score of 112 is given the same rank: 4. The score 100 has ranks 8 and 9. The average of 8 and 9 is 8.5, so each score of 100 receives the rank of 8.5. Making these changes gives the final ranks, as shown in Table 21.2. It is important to note that the sum of the ranks is the same in Tables 21.1 and 21.2.

Score	120	115	112	112	112	109	104	100	100	95
Rank	1	2	4	4	4	6	7	8.5	8.5	10

Table 21.2

For Table 21.1, S_R is given by Formula 21.1:

$$S_R = \frac{N(N+1)}{2} = \frac{10(10+1)}{2} = \frac{110}{2} = 55$$

You can check for yourself that the ranks in Table 21.2 also add up to 55. This is why the first step—ranking all the scores and ignoring ties—is important. Otherwise there is a strong tendency to lose track of the total number of ranks and assign ranks that result in a different sum. (For example, you might tend to give the rank of 3 to all three of the 112s and then the rank of 4 to 109 or a rank of 4 to all of the 112s but a rank of only 5 to 109.)

In some cases, the subjects in a study will not produce any scores but rather will have to be ranked directly (e.g., students in a music class who are ranked based on their ability to sing). The same two-step procedure is applied. First, you assign a unique rank to each subject (even if two students with indistinguishable singing ability must be ordered arbitrarily for the moment) and then calculate and assign average ranks for each set of tied subjects.

When to Use the Mann-Whitney Test

The Mann-Whitney test is generally used to answer the same types of research questions that are analyzed by a t test of two independent groups. You may wish to compare the means of two preexisting populations (e.g., smokers, and nonsmokers) on some continuous dependent variable, or to assign subjects randomly to two different treatment groups and then compare the means on some continuous dependent variable that you expect to be altered differentially by the treatments.

However, there are two very different types of data situations for which you would want to substitute the Mann-Whitney test for the t test described in Chapter 7. One of these situations has already been introduced: The variable of interest cannot be measured precisely, but subjects can be placed in a meaningful order. This situation can occur when a precise scale exists but is not immediately available (as when the Martian cannot find a ruler). More commonly, a precise scale for the variable has not yet been devised. The other situation occurs when the interval/ratio data collected do not meet the distributional assumptions of the t test. Both of these situations will be discussed in greater detail shortly. When the assumptions of the t test have been met, it will have more power than the Mann-Whitney test, but it is important to note that in most common situations the power of the Mann-Whitney test is nearly as high as that of a t test on the same data.

Variables That Are Not Measured Precisely

Many variables do not lend themselves to precise measurement; beauty has already been mentioned as an example. For another example, consider the problem of quantifying aggressiveness in children. One possibility is to measure aggressiveness in a playground in terms of the number of fights initiated by each child. The chief problem with this scale is the "floor effect"—too many children will score zero. The scale is not sensitive enough to distinguish between two children (neither of whom starts fights directly): one who teases others and one who is very shy and passive. Although it may be difficult to devise a precise aggressiveness scale for observed playground behavior, it is quite possible that observers could rank order a group of the boys from most to least aggressive (just as ethologists rank order animals in a colony based on position in the dominance hierarchy). Then the ranks could be summed separately for those boys who for instance, come from homes where there has been a divorce and those who do not, to perform the Mann-Whitney test. The rank order method would be more sensitive and have greater power to detect population differences than merely categorizing the boys as initiators or noninitiators of fights.

Some variables are measured by psychologists using self-report questionnaires. For example, anxiety is often measured by asking the subject a series of questions about different anxiety symptoms. The subject assigns a rating to each symptom (e.g., indicating how often or how intensely she or he experiences that symptom), and then all of these ratings are added to yield a single anxiety score. These scores are usually treated as though they were derived from an interval scale. (That is, the psychologist assumes that two subjects scoring 20 and 30, respectively, are as different in anxiety level as two subjects scoring 30 and 40.) Means and standard deviations are calculated and parametric statistics are performed. However, it can be argued that subjective ratings represent an ordinal scale at best. (The subject scoring 30 may be more anxious than the subject scoring 20, and the subject scoring 40 may be more anxious than the subject scoring 30, but we cannot compare the differences.) From this point of view, the calculation of means, standard deviations, and parametric statistics is not justified. This is a controversial issue in psychological research. I am not going to attempt to solve the controversy in this text. But I can tell you that if you are worried that the scale you are using does not have the interval property or does not even come close, you can rank order your scores and use the methods for ordinal data, such as the Mann-Whitney test.

Variables That Do Not Meet Parametric Assumptions

The other common situation in which the Mann-Whitney test may be appropriate occurs when a t test for two independent samples has been planned but the data do not fit the assumptions of a parametric test. I will assume that an interval or ratio scale of measurement is being used, as required for the t test, but that the sample data indicate a distribution that is not consistent with the assumption of a normal distribution in the population. For instance, two groups of subjects try to solve a problem that requires insight (e.g., they must see that some ordinary object can be used as a tool); subjects in one group get a hint and subjects in the other do not. The amount of time it takes to find a solution is recorded for each subject.

In this situation you may find that most subjects in the "hint" group solve the problem quickly, with a few subjects taking quite a long time, and perhaps one or two not solving the problem at all. In the "no hint" group there would probably be more subjects taking a long time or not solving the problem at all. Given such data, the assumption that solution times are normally distributed in the population seems unreasonable. Fortunately, the Central Limit Theorem implies that with sufficiently large sample sizes the sampling distribution of the mean will be approximately normal, in spite of the population distribution. But what if your samples are fairly small? You can combine the data from the two groups, rank order all the subjects, sum the ranks separately for each group, and perform the Mann-Whitney test. Even subjects who never solved the problem could be included; they would be tied at the lowest rank. Just remember that too many ties threaten the validity of the test.

Repeated Measures or Matched Samples

If each subject has been measured on an interval/ratio scale twice (or subjects have been matched in pairs), but the difference scores present a distribution that is inconsistent with a normal distribution in the population (and the sample is relatively small), the Sign test described in Chapter 19 can be performed as a substitute for the matched t test. Unfortunately, the Sign test throws away all of the quantitative information, retaining only the direction for each pair, and therefore has considerably less power than the matched t test for a given sample size. A more powerful alternative to the Sign test, and one that is also distribution-free, is the *Wilcoxon matched-pairs signed-ranks test*.

For an example of an experiment that may require this test, let us return to the problem-solving experiment. However, this time we will imagine that each subject serves in both conditions; two similar problems are solved, one with a hint and one without (assume orders and problems are properly counterbalanced). Thus, two amounts of time are recorded and the "no hint" − "hint" difference score is calculated for each subject. Suppose that for a few subjects the hint greatly reduces solution time, leading to a few large positive difference scores. However, for the majority of subjects the hint makes little difference, either because they solve both problems quickly or because they have difficulty even with the hint. This leads to several small difference scores, including a few negative ones. With two clusters of difference scores, small and large, and very few moderate scores in between, the assumption that a normal distribution is being represented may be untenable. The normality assumption can be avoided by assigning ranks to the difference scores. First, all of the difference scores are placed in order of magnitude regardless of sign (e.g., −1 is smaller than −2 or +2). Then the ranks are summed separately for the negative and positive differences; the smaller of the two sums is equal to a statistic called Wilcoxon's T. Therefore, the test is sometimes referred to as *Wilcoxon's T test*. This test can also be used when an interval/ratio scale has not been used but it is possible to rank the magnitude of the differences directly or when the scale used to measure the dependent variable is ordinal to begin with. An example for the latter case will be presented in the next section.

1. To test whether two populations differ on some continuous variable without measuring the variable precisely (i.e., on an interval or ratio scale), you can take samples of both populations and put them in a single order. If you assign ranks to the entire group and then sum the ranks separately for members of each population, you can determine whether there is a large discrepancy between the two sums.

SUMMARY

2. The discrepancy in the sums of ranks for any two groups ranked together on some variable can be tested for significance by calculating the *Mann-Whitney U statistic*, or by using Wilcoxon's table (the Wilcoxon rank-sum test) that gives critical values for sums of ranks. In either case, the procedure is often called the *Mann-Whitney test*.

3. The sum of ranks (S_R) for any N items ranked 1 to N is equal to $N(N + 1)/2$.

4. Most ordinal tests rest on the assumption that the variable being measured is actually continuous, so true ties are virtually impossible and apparent ties are due to the crudeness of the measurement. In assigning ranks, tied scores are initially given ranks as though they were different, and then these ranks are replaced by the average of the ranks for the tied scores.

5. The Mann-Whitney test is usually used when a variable cannot be measured precisely, but it is possible to determine which of two subjects has more of that variable (i.e., subjects can be ranked without too many ties) or when a variable has been measured precisely, but its distribution does not meet the assumptions for a two-group *t* test.

6. If subjects are paired across two conditions or measured twice, and their difference scores do not meet the requirements of the matched *t* test, but the differences can be ranked, it should be possible to test for a significant difference between the two conditions by using the Wilcoxon matched-pairs signed-ranks test. All of the differences are rank-ordered for magnitude, ignoring the signs of the differences, and then the ranks are summed separately for positive and negative differences. A large discrepancy in these two sums could lead to a significant difference, as will be shown in Section B.

EXERCISES

1. Describe two variables that would be difficult to measure on an interval/ratio scale but on which subjects could be rank ordered.

*2. If the top 16 tennis players in the world are ranked from 1 to 16, what is the sum of the ranks?

3. If the top seven tennis players in the previous exercise are from the United States,
    a. What is the sum of ranks for the U.S. players?
    b. What is the sum of ranks for the rest of the players?

*4. Suppose that ten women and nine men are rank ordered on some variable, and the result is that the two genders alternate, with women occupying the highest and lowest ranks. If men are labeled A and women B,
    a. What are the values of ΣR_A and ΣR_B?
    b. What are the values of U_A, U_B, and U?

5. Rank order the following sets of numbers, assigning the average rank to each set of ties. For each set, sum the ranks to check that you have assigned the ranks correctly.
    a. 102, 99, 101, 98, 104, 99, 100, 99, 101, 100, 102, 101, 99, 103
    b. 72, 68, 72, 75, 69, 68, 73, 70, 72, 70, 76, 68, 71, 74, 72

*6. Imagine that 10 patients have rated their fear of public speaking on a 10-point scale before and after a new controversial form of phobia treatment, as shown in the following table:

Before	9	8	10	9	7	8	6	7	5	9
After	8	5	9	6	8	4	9	7	7	10

   a. Rank the after-before differences in terms of absolute magnitude (ignoring the signs), but include the sign in parentheses next to each difference.
   b. Sum the ranks separately for positive and negative differences. Does it look like the new treatment reduced the phobia?

Testing for a Difference in Ranks between Two Independent Groups: The Mann-Whitney Test

B

BASIC STATISTICAL PROCEDURES

As an example in which two groups of subjects are measured for some variable on an ordinal scale, imagine testing one aspect of Freud's theory of personality. Freud suggested that toddlers subjected to a strict form of toilet training will later exhibit an anal retentive personality, including stubbornness, stinginess, and neatness. An English teacher decides that she will test the prediction concerning neatness by rank ordering her 21 students according to the neatness and organization of their assignments. She then consults the parents of her students to find out which students had been strictly toilet trained and which had not. The difference in neatness between the "strict" group and the "lenient" group can be tested with the usual six-step method.

Step 1. State the Hypotheses

The hypotheses for inference with ordinal data usually involve statements about the entire population distributions rather than just measures of central tendency. For the present example the appropriate null hypothesis would be that the distribution of the strictly trained population (with respect to neatness) is *identical* to the distribution of the leniently trained population. The alternative hypothesis is that the two population distributions differ. However, there are many ways in which the distributions can differ. If the null hypothesis is rejected, it could mean that the distributions differ only in central tendency, but it could also mean that the distributions have the same central tendency but differ in variability and/or shape. Only if there is some basis for assuming that the two population distributions are similar in shape and variability can the rejection of the null hypothesis be taken to imply a difference in central tendency. For ordinal data a difference in central tendency would be expressed as a difference in the *medians* of the two distributions because it is not proper to calculate means for ordinal data. For this example, any difference in the two populations would suggest that further experimental investigation of this aspect of Freudian theory is warranted.

Step 2. Select the Statistical Test and the Significance Level

Because the dependent variable (neatness) has been measured on an ordinal scale and the independent variable (strict versus lenient toilet training) is categorical, the appropriate procedure for testing the difference between the two groups is the Mann-Whitney test. I will use Wilcoxon's approach, as you will see in steps 4 and 5. As usual, alpha will be set to .05.

Step 3. Select the Samples and Collect the Data

In this example, the two samples are obviously based on convenience. Truly random samples from the general population would increase the generalizability of the results. However, in testing a controversial hypothesis it is not uncommon to use a convenient sample for an exploratory study before launching a more comprehensive investigation.

Assume that in ranking the 21 students the teacher found that the two neatest were indistinguishable and therefore tied for the lowest rank. Also assume that the three sloppiest students were tied for last place. Students who were then found to have been strictly trained were assigned the

Table 21.3

Subject No.	Rank	Group
1	1.5	A
2	1.5	A
3	3	B
4	4	A
5	5	B
6	6	A
7	7	A
8	8	A
9	9	B
10	10	B
11	11	A
12	12	A
13	13	B
14	14	B
15	15	B
16	16	A
17	17	B
18	18	B
19	20	B
20	20	B
21	20	B

label "group A" because these students represent the smaller subgroup. The remainder (those in the larger subgroup) were labeled "group B," as shown in Table 21.3.

Step 4. Find the Region of Rejection

The critical values for the Mann-Whitney (or *rank-sum*) test can be found in Table A.15. First, we need to count the subjects in group A; we label this total n_A. Similarly, the number of subjects in group B equals n_B. In this example, $n_A = 9$ and $n_B = 12$. (If n_A were equal to n_B, of course it would not matter which subgroup was marked A and which was marked B.) In Table A.15 we look at the panel labeled $n_A = 9$ and the row labeled $n_B = 12$. We look for our values under the column .025 because we are performing a two-tailed test with alpha = .05. The corresponding entry in the table is 71–127. This means that if our observed value is anywhere between 71 and 127, the null hypothesis *cannot* be rejected. Thus there are two regions of rejection: less than or equal to 71 and greater than or equal to 127. (Large samples cannot be found in Table A.15; for large samples you would have to use the normal approximation described shortly.)

Step 5. Calculate the Test Statistic

According to Wilcoxon's approach, all we have to do is sum the ranks for the smaller of the two groups (i.e., ΣR_A; if $n_A = n_B$, either sum will do). Summing the ranks marked A in Table 21.3, we get $\Sigma R_A = 1.5 + 1.5 + 4 + 6 + 7 + 8 + 11 + 12 + 16 = 67$. (Note: If we were using a table of critical values for the U statistic, we could now calculate U with Formula 21.2:

$$U_A = n_A n_B + \frac{n_A(n_A + 1)}{2} - \Sigma R_A = 9 \cdot 12 + \frac{9(10)}{2} - 67$$

$$= 108 + 45 - 67 = 86$$

$$U_B = n_A n_B - U_A = 108 - 86 = 22$$

Because U_B is less than U_A, U_B (22) would be taken as the value of U.)

To check that you have assigned tied ranks correctly, find ΣR_B and add this sum to ΣR_A. The total of all ranks should equal S_R, as given by Formula 21.1,

$$S_R = \frac{N(N + 1)}{2} = \frac{21(22)}{2} = \frac{462}{2} = 231$$

where $N = n_A + n_B$. Using the ranks for group B in Table 21.3, we find that $\Sigma R_B = 164$. Then we see that $\Sigma R_A + \Sigma R_B = 67 + 164 = 231$, which, fortunately, equals the S_R found in the equation.

Step 6. Make the Statistical Decision

The value for ΣR_A (67) is less than 71, so it is in the region of rejection. The null hypothesis—that the two population distributions are identical—can be rejected. To the extent that we can assume that the two distributions are similar in shape and variability, we can conclude that those who were strictly toilet trained have neater work habits. In any case, we are entitled to say that the two populations differ in some way and that further investigation may be warranted to explore the basis for this difference. Note that we *cannot* conclude that strict toilet training *causes* neatness in later life because the teacher had no control over the way the parents trained their children. It is possible, for instance, that neatness is hereditary and that neat people tend to be strict about toilet training. Only by randomly assigning toddlers to toilet training conditions can we test whether strict toilet training leads to neatness later in life. Of course, the Mann-Whitney test is applied in the same way when the groups are based on random assignment as when preexisting populations are sampled; it is only the causal conclusions that would differ.

The Normal Approximation to the Mann-Whitney Test

If you look at Table A.15, you will see that it cannot be used when n_B is larger than 15. With fairly large samples, a normal approximation can be used to perform the Mann-Whitney test. Fortunately, when N (i.e., $n_A + n_B$) is greater than about 20, the distribution of ΣR_A begins to resemble the normal distribution with a mean of $.5(n_A)(N + 1)$ and a standard deviation of:

$$\sigma = \sqrt{\frac{n_A n_B(N + 1)}{12}}$$

Therefore, when the samples are too large to use Table A.15, the Mann-Whitney test can be performed by calculating the following test statistic:

$$z = \frac{\Sigma R_A - .5(n_A)(N + 1)}{\sqrt{\frac{n_A n_B(N + 1)}{12}}} \qquad \text{**Formula 21.3**}$$

The z score found by this formula should follow the standard normal distribution (approximately) when the null hypothesis is true, so you can find the region of rejection from Table A.1 (e.g., for $\alpha = .05$, two-tailed, critical $z = \pm 1.96$). For the neatness example, the z score would be:

$$z = \frac{67 - .5(9)(22)}{\sqrt{\frac{(9)(12)(22)}{12}}} = \frac{67 - 99}{\sqrt{198}} = \frac{-32}{14.1} = -2.274$$

This z score would be significant at the .05 level (two-tailed), in agreement with our previous conclusion. The minus sign on the z score reminds us that group A has the smaller ranks (more neatness). Although Table A.15 is more accurate for the sample sizes in this example, the normal approximation becomes quite reasonable when n_B is larger than 10.

Effect Size of the Mann-Whitney (Rank-Sum) Test

The size of the z statistic for a Mann-Whitney test tells you no more about the size of the effect than does the t value from a two-group t test. The ranks of the subjects in two samples can overlap a great deal, and yet still yield a large z-score, if the samples are very large. If you want to know the extent to which the ranks do *not* overlap, in terms of a measure that is independent of sample size, you can calculate the *rank biserial correlation coefficient* devised by Glass (1965). Very simply, this correlation coefficient, symbolized as r_G, relates the difference in average rank between the two samples to the total size of the two samples combined, as follows:

$$r_G = \frac{2(\overline{R}_A - \overline{R}_B)}{N_T} \qquad \text{**Formula 21.4**}$$

For the preceding example, $\bar{R}_A = \Sigma R_A/n_a = 67/9 = 7.444$, and $\bar{R}_B = 164/12 = 13.667$, so the Glass rank biserial correlation is:

$$r_G = \frac{2(7.444 - 13.667)}{21} = \frac{-12.445}{21} = -.593$$

Although r_G is not equivalent to a true Pearson correlation coefficient, it does range from -1 to $+1$, so the value of $-.593$ just calculated represents a strong relationship between rank and group membership. A magnitude of 1.0 is only attained when there is no overlap in the rankings of the two groups. The negative sign in this case merely indicates that it is the smaller group that has the lower ranks, and therefore the sign of r_G will always match the sign of z. For this example, the minus sign tells us that the results are in the direction predicted by Freud's theory: strict toilet training produces a neater child. (If the two samples were the same size, a negative z or r_G would indicate that the group you arbitrarily designated as group A has the lower-numbered ranks, which is usually associated with larger amounts of whatever the dependent variable is measuring.)

Assumptions of the Mann-Whitney Test

Independent random sampling. The observations within each set of scores should be independent, and the two sets of scores should be independent of each other. As with the *t* test, random assignment often substitutes for randomly sampling from the population of interest.
The dependent variable is continuous. This assumption implies that ties will be rare.

Publishing the Results of the Mann-Whitney Test

The Mann-Whitney test is most likely to be used in research areas that involve fairly small samples that are likely to differ in size and variability—for instance, neuropsychology. The following study illustrates the use of the Mann-Whitney test in neuropsychological research and the way the results of this test would be reported in a journal.

Tranel and Damasio (1994) conducted a study of skin conductance responses (SCRs) in brain-damaged subjects. Psychological SCRs were elicited by presenting emotion-evoking slides. Subjects were divided into groups based on the location of their brain damage; for example, all the subjects in Group 2 had damage in the inferior parietal area. Each group of brain-damaged subjects was compared with a group of 20 control subjects who were free of neurological and psychiatric disease. Qualitative observations of Group 2 subjects suggested that damage to the right cerebral hemisphere had a much greater effect on SCRs than damage on the left side. This led to the following statistical test:

To check the reliability of our conclusion regarding Group 2, we conducted three statistical comparisons (Mann-Whitney U tests based on psychological SCRs). The comparison of the four right parietal subjects with controls was significant ($U = 0$, $p < .001$); by contrast, the comparison of the six left parietal subjects with controls was nonsignificant ($U = 44$, $p > .05$). A comparison of the four right parietal subjects with the six left parietal subjects was significant ($U = 4$, $p < .05$) (pp. 432–433).

The authors state that their use of nonparametric techniques was motivated by their small sample sizes and a lack of homogeneity of variance. Note that the U of zero for one of the tests indicates that all four of the right parietal subjects had smaller SCRs than any of the control subjects.

Ranking the Differences between Paired Scores: The Wilcoxon Signed-Ranks Test

Although the most common use, by far, for the *Wilcoxon matched-pairs signed-ranks test* is to deal with a fairly small set of interval/ratio difference scores whose distribution is very far from the normal curve, the test is also needed when the dependent variable is measured on an ordinal scale that lacks precise intervals. It is true that psychologists frequently use parametric statistics with apparently little concern about the precision of their measurement scales, but for this example I'll imagine a cautious researcher who is using the Wilcoxon test because her measurements involve an ordinal scale. Suppose that moderately phobic patients are being treated for their fear of public speaking by means of a controversial form of therapy (implosive therapy—the patient is forced to speak in front of a large group of people). Each patient's phobia is rated on a 10-point intensity scale both before and after five sessions of this treatment. The before-after difference scores can be tested with the usual six-step procedure.

Step 1. State the Hypotheses

The null hypothesis for this kind of test has the same form as for the Mann-Whitney test. H_0 states that the population distribution of phobics after treatment is identical to the population distribution before treatment. H_A states only that the two population distributions differ in some way.

Step 2. Select the Statistical Test and the Significance Level

In our hypothetical example the researcher feels that the phobic intensity scale is precise enough that the difference scores can be rank ordered (i.e., a difference of 3 on one part of the scale is larger than a difference of 2 on any other part of the scale) but not precise enough to perform the matched t test. The appropriate alternative is the Wilcoxon signed-ranks test. As usual, alpha will be set to .05.

Step 3. Select the Sample and Collect the Data

A sample of 14 moderate phobics is selected from a mental health clinic, and the intensity of each subject's phobia is rated on the 10-point scale. After the treatment each subject is rated for phobic intensity again, and the before-after difference score is calculated. The data appear in Table 21.4.

Step 4. Find the Region of Rejection

The critical values for the Wilcoxon T statistic are given in Table A.16. We need to know only alpha and the sample size (i.e., the number of *pairs* of scores) to find critical T, but there is one complication concerning the size of our sample. Three of the subjects obtained the same rating before and after the treatment; their difference scores were zero. Because the T statistic depends on summing ranks separately for positive and negative difference scores, zero difference scores present a problem. Some statisticians recommend dropping the subjects with the zero difference scores and

Table 21.4

Before	After	Difference
4	6	−2
5	1	+4
7	4	+3
6	5	+1
5	5	0
7	2	+5
7	8	−1
4	1	+3
6	3	+3
6	7	−1
5	0	+5
7	7	0
5	4	+1
5	5	0

	Difference Score	Initial Rank	Final Rank
Table 21.5	0	1	(+)1.5
	0	2	(−)1.5
	−1	3	(−)4.5
	−1	4	(−)4.5
	+1	5	(+)4.5
	+1	6	(+)4.5
	−2	7	(−)7
	+3	8	(+)9
	+3	9	(+)9
	+3	10	(+)9
	+4	11	(+)11
	+5	12	(+)12.5
	+5	13	(+)12.5

reducing the sample size (N) accordingly. However, the more conservative approach seems to be retaining the zeroes and arbitrarily labeling half of them as positive and half as negative. But even this approach has a problem if there is an odd number of zeroes, and therefore it is recommended to drop one of the zeroes in such a case, and reduce N by 1. Following this approach and noting that there are three zero differences in our example, we will drop the third one (the last subject) and reduce our N to 13. Now we can look in Table A.16 under $N = 13$ and $\alpha = .05$ (two-tailed) to find that $T_{crit} = 17$. Because small values of T are inconsistent with the null hypothesis, the region of rejection consists of T values *equal to or less than 17*.

Step 5. Calculate the Test Statistic

First, the difference scores must be placed in order of magnitude, *ignoring sign*. Basically, we are treating the positive and negative difference scores like the two groups in a Mann-Whitney test. The scores are all ranked together by magnitude, but we keep track of which group (in this case, sign) is associated with each score. If ties occur, they are given averaged ranks as described in Section A. In Table 21.5, the difference scores from Table 21.4 have been placed in order and assigned ranks.

Notice that next to each final rank in Table 21.5, I placed a plus or a minus sign in parentheses as a reminder of whether that rank represents a positive or negative difference score (plus and minus signs have also been placed arbitrarily on the ranks corresponding to the two remaining zero difference scores). This will help us to sum the ranks separately for the negative and positive differences. For the negative differences, $\Sigma R_{minus} = 1.5 + 4.5 + 4.5 + 7 = 17.5$. The value of T is either ΣR_{minus} or ΣR_{plus}, whichever is smaller. Because it is obvious that ΣR_{plus} will be larger than 17.5, we can stop at this point and state that $T = 17.5$.

To check your work find both sums ($\Sigma R_{plus} = 73.5$, in this example), and make sure that $\Sigma R_{minus} + \Sigma R_{plus} = S_R$, where S_R is given by Formula 20.1. For this example,

$$S_R = \frac{N(N + 1)}{2} = \frac{13(14)}{2} = \frac{182}{2} = 91$$

and $\Sigma R_{minus} + \Sigma R_{plus} = 17.5 + 73.5 = 91$, so it appears that our calculations are accurate (recall that N was reduced to 13).

Step 6. Make the Statistical Decision

Because our T value (17.5) is *not* less than or equal to the critical T (17), we cannot reject the null hypothesis that the "before" and "after" populations are identical with respect to phobic intensity. Our T value is too large for us to ignore chance factors as an explanation for our results. This is disappointing because the results looked promising, but we must bear in mind that our test lacked power for two reasons: the sample size is quite small and the Wilcoxon test generally has less power than a matched t test (some quantitative information in the data is being lost when converting to ranks). In fact, a matched t calculated for the data in Table 20.4 comes out to 2.46, which is easily significant at the .05 level, two-tailed.

The Power of the Wilcoxon Test

The Wilcoxon test usually has somewhat less power than a matched t test on the same data, so the matched t test is preferred whenever its assumptions are met. However, given that you have decided on an ordinal test, unless the matching is poor indeed, the signed-rank test just described will have more power than a Mann-Whitney test on the same data (just as a matched t test will generally beat an independent-samples t test on the same data). The degree of matching can be measured by means of the Spearman correlation coefficient, r_s, which will be described shortly.

It is also important to note that the power of the Wilcoxon test is often considerably greater than the Sign test for the same data. For instance, in the present example our Wilcoxon T would have been significant at the .05 level with a one-tailed test (although this would hardly be justified with a controversial treatment). By comparison, the Sign test does not come close. To perform the Sign test on the data of Table 21.4, we throw away the three zero differences and count the number of plus and minus signs. With $N = 11$, and $X = 3$ (or 8, if you prefer), the p value adds up to about .113 for a one-tailed test (see Chapter 19). The Sign test throws away even more information than the Wilcoxon test; for instance, the Sign test is not equipped to take advantage of the fact that in the preceding example all four of the negative scores are fairly small.

The Normal Approximation to the Wilcoxon Test

Although Table A.16 contains critical values for sample sizes as large as 25 pairs (more than is usually used in a Wilcoxon test), the distribution of the T statistic begins to resemble the normal distribution when there are as few as 10 pairs. To be on the safe side, I would recommend using a table of critical values, if one is handy, unless you have more than 20 pairs. The normal distribution that is approached has a mean (μ_T) of $.25N(N + 1)$ (this is midway between the lowest possible sum of ranks, which is zero, and the highest, which is S_R), and a variance of:

$$\sigma_T^2 = \frac{N(N + 1)(2N + 1)}{24}$$

Thus, a particular T value can be converted to a z score by Formula 21.5:

$$z = \frac{T - .25N(N + 1)}{\sqrt{\frac{N(N + 1)(2N + 1)}{24}}}$$ **Formula 21.5**

For the present example:

$$z = \frac{17.5 - .25(13)(14)}{\sqrt{\dfrac{13(14)(27)}{24}}} = \frac{17.5 - 45.5}{\sqrt{204.75}} = \frac{-28}{14.31} = -1.957$$

Notice that this z score just barely misses statistical significance at the .05 (two-tailed) level, which is consistent with the fact that our observed T was just slightly above the critical T (17) in Table A.16. Be aware that the sign of the z score can be misleading; compare ΣR_{minus} and ΣR_{plus} to determine the direction of your results.

Effect Size of the Wilcoxon (Signed Ranks) Test

Analogous to the rank biserial correlation coefficient we calculated in order to express the effect size for the Mann-Whitney test is the *matched-pairs rank biserial correlation coefficient*, which serves the same function for the Wilcoxon test. Like r_G, the matched-pairs version, symbolized as r_C, is not a true Pearson r, but it ranges from -1 to $+1$, reaching its maximum magnitude when all of the differences share the same sign. This correlation is a function of the difference between the sums of ranks for positive and negative differences, related to the total sample size, as follows:

$$r_C = \frac{2(\Sigma R_{plus} - \Sigma R_{minus})}{N(N+1)} \qquad \textbf{Formula 21.6}$$

For the preceding example, $\Sigma R_{plus} = 73.5$, and $\Sigma R_{minus} = 17.5$, so the matched-pairs biserial correlation is:

$$r_C = \frac{2(73.5 - 17.5)}{13(14)} = \frac{112}{182} = +.615$$

The positive sign in this case indicates that it is the positive differences that have the higher ranks. (The sign of r_C will not always match the sign of z, but the latter sign is really not meaningful and should be ignored.) For this example, because the after scores are being subtracted from the before scores, the positive sign tells us that the after scores are generally *smaller* than the before scores, demonstrating the expected reduction in phobia ratings following the treatment. The value of .615 just calculated implies a rather large effect size for the (implosive therapy) treatment.

Assumptions of the Wilcoxon Signed-Ranks Test

The assumptions for the Wilcoxon test are essentially the same as for the Mann-Whitney test.

1. *Independent random sampling.* The observations within each set of scores should be independent. However, the two sets of scores are not independent of each other; for each score in one set there is a related score (either through the matching of subjects or repeated measures) in the other set.

2. *The dependent variable is continuous.* This assumption implies that ties will be rare. I have already described how to deal with a tie involving two scores in the same pair (i.e., a zero difference score). When ties occur among nonzero difference scores, the average rank can be given to each of the tied scores (as in the example above) with little effect on the T statistic. However, if there are too many ties, the critical values

from Table A.16 will not be valid. If using the normal approximation, Formula 21.5 can be corrected for the number of ties, but I will not present that correction here (see Siegel & Castellan, 1988).

When to Use the Wilcoxon Test

Like the matched t test, the Wilcoxon test can be performed when the scores are from matched pairs of subjects or when each pair of scores comes from the same subject. With respect to the dependent variable, there are three situations that could lead to the use of the Wilcoxon test.

1. *The differences between pairs of observations are ranked directly.* Without actually trying to measure the degree of schizophrenia exhibited by any of your patients, you might try to rank order them in terms of their amount of change (improvement or decline) in response to a new form of therapy.
2. *The difference scores are derived from ordinal data but can be accurately ranked.* This is the situation described in the previous example. Normally, the differences of pairs of ordinal scores would not be considered rankable, so this situation would require an elaborate justification and therefore rarely arises. If the differences cannot be ranked, the direction of the differences would still be meaningful, so the Sign test could be applied.
3. *The difference scores are derived from interval/ratio data but do not meet distributional assumptions.* The matched t test assumes that the difference scores follow a normal distribution in the population. If the difference scores in your data seem incompatible with that assumption, they can be rank ordered and submitted to the Wilcoxon test. This is probably the most common use of the Wilcoxon test.

Correlation with Ordinal Data:
The Spearman Correlation Coefficient

Before concluding this section, I want to introduce another statistical procedure that is used with ordinal data, because it is easy to explain and rather commonly used. For an example of the need for this statistic,

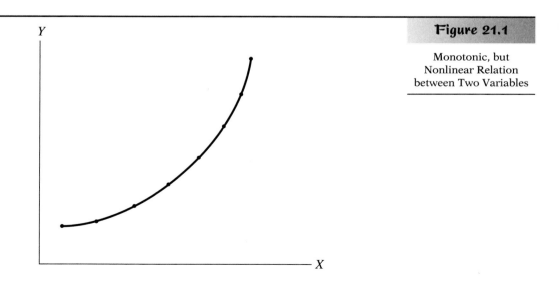

Figure 21.1

Monotonic, but Nonlinear Relation between Two Variables

imagine that your dependent variable is the result of rank ordering boys in a playground according to the amount of aggressiveness displayed by each one (the boys can then be divided into two groups, such as first-born versus later-born children). Ranking the boys may be easier than trying to measure their aggressiveness precisely, but whose judgment can be trusted in ranking the boys correctly? Even experts can disagree dramatically in making such judgments. We can train an observer to use specific criteria in ranking the boys, but how can we know that the criteria are being applied in a reliable fashion? The simplest solution is to have two raters, trained the same way, rank the boys independently (i.e., each rater is unaware of the other's rankings). Ideally, the two raters will rank each boy in the same position. Usually, however, there will be some discrepancy between the two sets of rankings. We need to measure the amount of discrepancy or, conversely, the amount of agreement, between the raters so we can decide whether any observer's set of rankings can be trusted. Also, if the agreement between the two raters is high, we can feel justified in averaging the two sets of rankings.

The amount of agreement between two sets of rankings can be determined by calculating the Pearson correlation coefficient, just as we would for any two variables measured on an interval or ratio scale. When the two variables are measured on an ordinal scale, the correlation coefficient that results from applying Pearson's formula is often called the *Spearman correlation* (r_s), or the *rank correlation coefficient*. As with the point-biserial r (r_{rb}) and the phi coefficient (ϕ), the special symbol, r_s, reminds us that although this is a Pearson correlation coefficient, it is calculated for data involving at least one variable that was not measured on an interval or ratio scale. As with r_{pb} and ϕ, there is a shortcut formula for calculating r_s; it is based on the difference of ranks on the two variables for each subject. However, this shortcut formula is rendered inaccurate by tied scores, which are not uncommon when dealing with ordinal data, so you are better off just calculating Pearson's correlation coefficient on the ranks using your favorite Pearson's r formula. Consequently, there is no need for me to work out a numerical example of Spearman's rank correlation coefficient.

It is important to note that there are cases in which both variables are measured on interval/ratio scales and yet it is preferable to rank order both variables and find r_s rather than calculate r directly for the original scores. For instance, a few extreme outliers can have a devastating effect on r, but their impact can be reduced by converting the data to ranks. When you do this, the highest score on one variable can be very far from the next highest, but in terms of ranking it is just one rank higher. Also, recall that Pearson's r measures only the degree of *linear* relationship. If the relationship between two variables follows a curve (as in Figure 21.1), even a curve that keeps rising, Pearson's r can be deceptively low. A curvilinear correlation coefficient can be calculated if that is what you are interested in, or the data can be transformed to create a fairly linear relationship. However, ranking the scores may lead to the answer you want. The correlation of the ranks will give the relationship credit for being *monotonic* (whenever X goes up, Y goes up, and vice versa), even if it's far from linear. In fact, if the relationship is perfectly monotonic (as in Figure 21.1), r_s will equal 1.0. Therefore, if it is the degree of monotonicity, rather than the linearity of the relationship that you wish to measure, r_s will serve you better than Pearson's r.

Determining Statistical Significance for Spearman's *r*

For a Pearson correlation the null hypothesis is that $\rho = 0$. For normally distributed variables, $\rho = 0$ implies that the two variables are independent.

This implication does not hold, however, if we make no assumptions about the shape of the distributions of the variables, as is the case with the Spearman correlation. So rather than $H_0: \rho_s = 0$, the null hypothesis that we are really interested in with ordinal data is H_0: The two variables are independent in the population. The two-tailed alternative hypothesis is H_A: The two variables are *not* independent in the population.

Although r_s can be found by applying the ordinary Pearson correlation formula to ranked data, the distribution of r_s is not the same as the distribution of r that is calculated for interval/ratio data unless the sample size is extremely large (theoretically, infinite). You cannot use Table A.5 to find the critical values for a Spearman r, nor can you use the t value from Formula 9.6, unless your sample size is rather large. Special tables or approximating formulas must be used to determine the significance of Spearman's r for small sample sizes. However, now that Spearman correlation is available from the major statistical software packages, I will not bother to include a special table in this text (see Siegel & Castellan, 1988). In general, it is easier to obtain a high correlation with ranks than it is with interval/ratio data, so the critical values for r_s are higher than the corresponding critical values for r. Perfect Spearman correlation requires only that ordinal positions match across two variables, but perfect Pearson correlation requires, in addition, that the variables have a perfect *linear* relationship.

When to Use the Spearman Correlation

The situations calling for the use of r_s can be divided into three cases, according to the measurement scale applied to each variable:

Both variables have been measured on an ordinal scale. The situation in which you wish to measure interrater reliability for ranked data is a common example, as previously mentioned.

One variable has been measured on an ordinal scale and the other on an interval or ratio scale. If one of the two variables consists of ranks, or otherwise lacks precise intervals, and the other is measured on an interval/ratio scale, the latter variable must be converted to ranks. If you were to correlate the two sets of numbers without ranking the interval/ratio data, you would be "comparing apples to oranges," and you would not know what distribution could be used to represent the null hypothesis or where to obtain critical values.

Both variables have been measured on an interval/ratio scale. In this case, you would only convert to ranks to calculate r_s instead of r if the distributional assumptions of r were severely violated and the sample was relatively small. Another reason to calculate r_s instead of r with interval/ratio data is if the relationship is far from linear and you just want to assess the degree to which the relationship is monotonic (i.e., whenever X goes up, Y goes up, and vice versa).

B

SUMMARY

1. The null hypothesis for the Mann-Whitney test is that the two populations being sampled not only have the same central tendency but are identically distributed. If the larger of your two samples is not more than 15, the sum of the ranks for the smaller group can be compared to the critical values in Table A.15. If you are conducting a two-tailed test, and this sum is not between the two critical values (or equal to either one of them), the null hypothesis can be rejected.
2. If the larger of your two samples is more than 15, and the smaller more than 5, there is a formula based on the sum of the ranks in the smaller

group (and the sample sizes) whose results will follow (approximately) a normal distribution when the null hypothesis is true. The usual critical values of z can be used to test for a significant difference in the sums of the ranks. There is also a *rank biserial correlation coefficient, r_G,* that takes into account the sample sizes to provide a measure of effect size for the Mann-Whitney test.

3. In the Wilcoxon Signed-Ranks Test, the difference scores are ranked for magnitude ignoring their signs (smaller differences must be given higher ranks, i.e., smaller rank numbers), and then summed separately for each sign. The smaller of the two sums is labeled T and can be compared to a critical value in Table A.16. If you have more than about 20 pairs of scores, T can be used in a formula that follows the normal distribution quite well.

4. The Wilcoxon test is most often used as a substitute for the matched t test when a fairly small sample of difference scores seem to follow a very strange distribution. The Wilcoxon test is also appropriate when differences can be ranked but cannot be measured precisely. The Wilcoxon test is not as powerful as the matched t test when the latter's assumptions are met, but it is usually more powerful than the Sign test. The effect size of the Wilcoxon test can be measured by the *matched-pairs rank biserial correlation coefficient, r_C,* which relates the difference in positive and negative ranks to the total sample size.

5. When you have two measures for each subject, and the measures are in the form of ranks, you can nonetheless calculate the ordinary Pearson's r for the two variables. This r is called the Spearman correlation coefficient (r_s) as a reminder that its distribution differs from an ordinary Pearson's r. The critical values for r_s are higher than for r, although the difference becomes negligible for very large sample sizes.

6. The Spearman correlation is called for when one or both variables have been measured on an ordinal scale to begin with or when one or both variables have been measured on an interval/ratio scale but has a distribution very far from the normal distribution and the sample is not very large. Also, if you are more concerned about the relation of the two variables being monotonic than linear, the correlation based on ranks may be more suited to your purposes.

7. The tests based on the sums of ranks for two groups (Mann-Whitney), the ranks of differences (Wilcoxon test), or the correlation of ranks (Spearman) are distribution-free (e.g. none of these requires the variable being measured to follow a normal distribution). The only assumptions required by these tests are
   a. *Independent random sampling.* Random assignment may be used for the Mann-Whitney test, but (within that constraint) all of the scores should be independent of each other both within and across groups. Each pair of scores in the Wilcoxon test, or a Spearman correlation, should be independent of any other.
   b. *The dependent variable is continuous.* This implies that tied ranks will be rare. If there are more than a few ties, you may need to use correction factors, or switch to categorical tests (see Chapter 20).

EXERCISES

*1. Imagine that judges for the last Mr. Universe body-building contest discover that the second, third, fifth, sixth, and eighth ranked finalists out of the top 20 broke the rules by using steroid drugs. Use these data to test the null hypothesis that steroid drugs offer no advantage in a body-building contest at
a. The .05 level.
b. The .01 level.

2. Suppose that sixth-grade boys are ranked for their activity level in a classroom. Boys labeled A regularly eat a large amount of sweets, whereas the boys labeled B are allowed only a limited amount of sweets. Activity decreases from left to right:

A A A B A A B A A B B A B B A B B A B B B B

a. Test the null hypothesis (α = .05) that activity levels are unrelated to the amount of sweets eaten.
b. Calculate r_G for these data. Does the effect size appear to be large?

*3. The following data are the number of trials required to learn a maze for patients with left- versus right-hemisphere brain damage:

Left Damage	Right Damage
5	9
3	13
8	8
6	7
	11
	6

a. Perform a t test for two independent samples on these data, and calculate r_{pb}.
b. Using the same data, perform the Mann-Whitney test, and calculate r_G.
c. Are the results of the two significance tests comparable? Are the effect sizes similar?

4. In Exercise 7B9, the results of an industrial psychology experiment were given in terms of the number of clerical tasks successfully completed by the subjects in two experimental groups. The data are

Individual Motivation: 11, 17, 14, 10, 11, 15, 10, 8, 12, 15.
Group Motivation: 10, 15, 14, 8, 9, 14, 6, 7, 11, 13.

a. Perform the Mann-Whitney test on these data α = .05, (one-tailed).

b. Using the normal approximation formula, find the z score corresponding to the Mann-Whitney test.
c. Compare the z score you found in part b with the t value that you calculated for Exercise 7B9. Are the two tests comparable in this case?

*5. A behavioral psychologist is studying the disruptive effects of a drug on maze-learning in rats. Five rats are injected with the drug before being placed in the maze, and five rats receive a control (i.e., saline) injection. The amount of time in minutes each rat takes to complete the maze follows:

Drug: 18, 47, 15, 10, 16
Control: 8, 3, 5, 12, 9

a. Test the null hypothesis (α = .05) that the drug has no effect (i.e., it has the same effect as saline) by performing the t test for two independent groups. Also, calculate the effect size, g.
b. Test the same hypothesis using the Mann-Whitney test, and calculate r_G.
c. Are your conclusions in parts a and b different? What problem in the data for this exercise is eliminated by ranking?
d. Are both effect size measures large for these data? Explain.

6. Does partying the night before a very difficult math test improve performance? The following are the exam scores for four students who partied all night before the test and five students who spent the evening studying:

Students Who Partied: 20, 50, 10, 60
Students Who Studied: 80, 55, 90, 45, 85

a. Test the null hypothesis (α = .05) that partying and studying are equally helpful for math performance by conducting the t test for two independent groups.
b. Test the same hypothesis using the Mann-Whitney test.
c. Are your conclusions in parts a and b different? Can you see why the Mann-Whitney test might have less power in this case?

*7. Twelve subjects completed a course in memory training. Three subjects actually exhibited *poorer* memory performance after the training, but these difference scores were the smallest in magnitude of all the subjects. The remaining subjects exhibited various amounts of improvement.

a. Use the table for Wilcoxon's signed-ranks test to decide whether the null hypothesis—that the memory training makes no difference—can be rejected at the .05 level, two-tailed.

b. Calculate an appropriate effect-size measure for these data. Does the effect size appear to be large?

8. In Exercise 19B5 you were asked to treat the clinical ratings in the following table as so crude that the amount of difference in ratings from Time 1 to Time 2 could not be used. For this exercise, assume that the differences of the clinical ratings can be meaningfully ranked.

Patient No.	1	2	3	4	5	6	7	8	9
Time 1	5	7	4	2	5	3	5	6	4
Time 2	3	6	5	2	4	4	6	5	3

a. Perform Wilcoxon's signed-ranks test on these data, and test the results at the .05 level (two-tailed) using Table A.16.

b. Calculate r_C for these data.

*9. The following data are from Exercise 11B6, in which you were asked to perform a matched t test on the IQ scores of teenaged couples.

Boy	110	100	120	90	108	115	122	110	127	118
Girl	105	108	110	95	105	125	118	116	118	126

a. Perform a one-tailed Wilcoxon test at the .05 level, using Table A.16 to determine whether there is evidence supporting the hypothesis that boys tend to date girls who have a lower IQ.

b. Repeat the test in part a, using Formula 21.4. Compare the z value you just calculated with the t value calculated in Exercise 11B6.

c. Rank the boys and girls separately and calculate the Spearman correlation coeffi-

cient between the two sets of ranks. Would the value for r_s be significant as a Pearson r?

10. In Exercise 11B7, subjects had to quickly recognize objects placed in either their right or left hands. The data from that exercise follow:

Subject No.	Left	Right
1	8	10
2	5	9
3	11	14
4	9	7
5	7	10
6	8	5
7	10	15
8	7	7
9	12	11
10	6	12
11	11	11
12	9	10

a. Perform a one-tailed Wilcoxon test at the .05 level, using Table A.16 to determine whether there is evidence supporting the hypothesis that objects are recognized more quickly in the right hand by right-handers.

b. Repeat the test in part a, using Formula 21.4. Compare your result to the result of the matched t test, and explain any discrepancy.

c. Calculate the Spearman correlation coefficient between the left- and right-hand scores. Explain the relation between the size of r_s and the power of the Wilcoxon test.

*11. Calculate r_s for the data in Exercise 9B9, and compare it to the Pearson r already calculated for those data.

12. Calculate r_s for the data in Exercise 9A5, and compare it to the Pearson r calculated for those data in Exercise 9B3. Can you explain why the magnitude of r_s is so much smaller than the magnitude of r for these data?

Testing for Differences in Ranks among Several Groups: The Kruskal-Wallis Test

Children raised in different cultures may differ in the intensity with which they express emotion during an upsetting experience. To compare two cultures, the emotional expressions of children from the two groups can be rank ordered together, and the Mann-Whitney test can be performed. But what if you wish to compare children from three or more cultures in the same study? Fortunately, it is easy to extend the Mann-Whitney test so that it can accommodate any number of groups.

The extended test is called the *Kruskal-Wallis one-way analysis of variance by ranks*, or more commonly, the Kruskal-Wallis test. This test begins exactly like the Mann-Whitney test: You combine all of the subjects from all of the subgroups (e.g., cultures) into one large group and rank order them on the dependent variable (e.g., intensity of emotional expression); ties are given average ranks, as usual. Then, as you might expect, you sum the ranks separately for each subgroup in the study. The only new thing to learn about the Kruskal-Wallis test is how to derive a convenient test statistic from the several sums of ranks.

The test statistic devised by Kruskal and Wallis (1952) is referred to as H; hence, the test is sometimes called the *Kruskal-Wallis H test*. To simplify the notation in the formula for H, I will use the symbol T_i to represent the sum of the ranks in one of the subgroups (i.e., the ith group). (You may recall that this notation parallels the use of T_i to represent the sum of scores in each group in the one-way ANOVA, as in Formula 12.8.) I will use n_i to represent the number of subjects in one of the subgroups, k to represent the number of subgroups, and N to represent the total number of subjects in the study (i.e., $N = \sum_{i=1}^{k} n_i$). Using this notation, Formula 21.7 for H is:

$$H = \frac{12}{N(N+1)} \sum_{i=1}^{k} \left(\frac{T_i^2}{n_i} \right) - 3(N+1)$$ **Formula 21.7**

For example, if the sum of ranks for four people from culture 1 were 40, the sum for five people from culture 2 were 50, and the sum for six people from culture 3 were 30, H would be:

$$H = \frac{12}{15(16)} \left(\frac{40^2}{4} + \frac{50^2}{5} + \frac{30^2}{6} \right) - 3(16)$$

$$= \frac{1}{20}(400 + 500 + 150) - 48 = 52.5 - 48 = 4.5$$

H falls to its minimum value (near zero) when the ranks are equally divided among the groups. Usually, the researcher wants the sums of ranks to be as disparate as possible, leading to a large and (it is hoped) significant H. But we need to know the distribution of H when the null hypothesis is true to decide when H is large enough to be statistically significant. Conveniently, when all of the populations involved are identical and the samples are not too tiny, H follows a distribution that resembles the chi-square distribution with df = $k - 1$. Therefore, we can use Table A.14 to find the critical value for H. For the preceding example, df = $3 - 1 = 2$, so critical H for $\alpha = .05$ is 5.99. Note that we are using only the upper tail of the chi-square distribution, as we did with the F distribution for analysis of variance (and for similar reasons).

Our observed H (4.5), being less than the critical H, does not allow us to reject the null hypothesis. Incidentally, if none of our subgroups were larger than 5, the chi-square distribution would not be considered a good enough approximation, and we would be obliged to use a special table to find the critical value for H (see Siegel & Castellan, 1988). Had our observed H for this example been greater than 5.99, we could have concluded that the cultural populations sampled are not all identical. If, in addition, we had made the common assumption that all the population distributions are similar in shape and variability, we could have concluded that the median intensity of emotional expression was not the same for all of the populations.

As you would guess, the Kruskal-Wallis test requires the same assumptions as the Mann-Whitney test and is used in the same circumstances. Moreover, ties are handled in a similar fashion, and there is a correction factor for H that should be used if more than about 25% of the observations result in ties (Siegel & Castellan, 1988).

Follow-up Tests and Effect Size for the Kruskal-Wallis (H) Test

If our H statistic were significant in the previous example, we could only declare that the three populations are not identical, or at best that the three population medians are not all the same. But, as in any multi-group study in which the levels are considered fixed (i.e., not chosen at randon), we would want to know more specifically which pairs of cultures can be considered different—is culture 2 different from culture 3? Is culture 1 different from culture 2 (or 3)? The simplest way to address this issue would be to conduct ordinary Mann-Whitney rank-sum tests (as described in Section B) for each pair of cultures. With three groups, there are only three pairs to test, so when H is significant, it seems reasonable to conduct each follow-up test at the .05 level. However, when dealing with more than three groups, it would be a good idea to use some form of Bonferroni correction to maintain control over the possible rate of Type I errors, or to use a more specialized procedure designed for following up nonparametric tests (Siegel & Castellan, 1988).

The Kruskal-Wallis (K-W) test is a good alternative to the one-way independent-groups ANOVA when your samples are fairly small and your dependent variable is quantitative, but oddly distributed—and, especially when there are outliers and/or the distribution appears to be different from one level of your factor to another. The power of the K-W test generally compares favorably with an ANOVA conducted on the same data; the K-W test is usually estimated to have at least 90% of the power of its parametric counterpart. Just as for a one-way ANOVA, there is a corresponding value for eta squared that can be calculated for the K-W test, and used as an estimate of the overall effect size. I'll use the subscript H to remind you that this version of η^2 is based on the H statistic, adjusted for the size of the samples and the number of groups.

$$\eta_H^2 = \frac{H - k + 1}{N - k}$$ **Formula 21.8**

For the preceding example, H was 4.5 and the 3 groups contained a total of 15 subjects, so:

$$\eta_H^2 = \frac{4.5 - 3 + 1}{15 - 3} = \frac{2.5}{12} = .2083$$

Although .2083 is a fairly large proportion of variance to be accounted for, it should not be surprising that the data in this example did not attain statistical significance, given the very small sample sizes, and the somewhat reduced power available to the K-W test.

Testing for Differences in Ranks among Matched Subjects: The Friedman Test

Although the power of the Kruskal-Wallis test compares reasonably well with ANOVA in most common situations, suppose you want (or need) the added power of matching subjects (or repeated measures) even though you have more than two treatment conditions and your data are measured on an ordinal scale. In that case, you would need a substitute for the repeated-measures ANOVA, such as the *Friedman test*.

The Friedman test is particularly easy to apply because rather than having to rank order all of the subjects (or difference scores) with respect to each other, you only order the subjects (or repeated observations) within each matched set. For instance, if there are only three different treatment conditions (e.g., three methods for teaching children to play piano), you would only rank order three subjects (or repeated observations) at a time. Suppose that the subject taught by method I is always the best piano player of the three matched subjects and that method II always produces the second best, and method III, the third. This ideal situation is depicted in Table 21.6.

Subject	Method I	Method II	Method III	
A	1	2	3	**Table 21.6**
B	1	2	3	
C	1	2	3	
D	1	2	3	
E	1	2	3	
F	1	2	3	
ΣR	6	12	18	

The data shown in Table 21.6 produce the largest possible separation in the sums of ranks for each method and the best chance of rejecting the null hypothesis (that the three methods produce identical populations). On the other hand, when the three teaching methods produce inconsistent results from subject to subject, the sums of ranks are more similar. The worst case—in which all methods produce the same sum of ranks—is shown in Table 21.7.

Subject	Method I	Method II	Method III	
A	3	1	2	**Table 21.7**
B	2	3	1	
C	3	2	1	
D	1	2	3	
E	2	1	3	
F	1	3	2	
ΣR	12	12	12	

What we need is a test statistic that can measure the degree to which these sums of ranks differ. The statistic devised by Milton Friedman (1937), who much later won the Nobel prize for economics—I will call it F_r—is found by Formula 21.9:

$$F_r = \frac{12}{Nc(c+1)} \sum_{i=1}^{c} T_i^2 - 3N(c+1) \qquad \textbf{Formula 21.9}$$

where c is the number of (repeated) treatments, N is the number of matched sets of observations (e.g., the number of rows in Table 21.6), and T_i is the sum of ranks for the ith treatment.

The resemblance between the Friedman test and the Kruskal-Wallis H statistic should be obvious. Like H, F_r follows a distribution that approaches the chi-square distribution with df $= c - 1$, when the null hypothesis is true and the total number of observations is at least about eight. (Tables are available to find critical values of F_r for small samples; see Siegel & Castellan, 1988.)

Applying Formula 21.9 to the data in Table 21.6, we get

$$F_r = \frac{12}{(6)(3)(4)}(6^2 + 12^2 + 18^2) - (3)(6)(4)$$

$$= .167(504) - 72 = 84 - 72 = 12$$

This is the highest F_r can get when $N = 6$ and $c = 3$. The critical value for F_r is a chi-square value that can be found in Table A.14; with df $= 3 - 1 = 2$, and $\alpha = .05$, the critical $F_r = 5.99$. Therefore, we can reject the null hypothesis for the data in Table 21.6 and conclude that the three teaching methods do *not* produce identical populations. To decide which pairs of methods differ significantly, you would have to make post hoc comparisons, as described by Siegel and Castellan (1988).

The Friedman test is sometimes described as an extension of the Wilcoxon matched-pairs test, but in some ways it is more an extension of the Sign test. If the Friedman test is performed with just $c = 2$ conditions, each score in a pair of scores is ranked 1 or 2. This really is no different from assigning a plus or minus to each pair according to which member of the pair rates more highly on the dependent variable. In fact, when dealing with only two conditions, the Friedman test will always provide the same p value as the Sign test (provided that you use the normal approximation for the binomial distribution).

On the other hand, the Friedman test resembles the RM ANOVA in that the size of the test statistic reflects subject-to-subject consistency with respect to conditions. To the extent that subjects tend to agree about the different conditions (i.e., all subjects tend to respond best to the same treatment, second best to some other treatment, and so on), the data will resemble Table 21.6 producing sums of ranks that differ considerably, which, in turn, leads to a large value for F_r. Of course, in a RM ANOVA subjects can be quite consistent with each other, and the difference in treatment means can be quite small or rather large. In the Friedman test, however, the mean differences automatically increase as subject-to-subject consistency increases.

Like the RM ANOVA, the Friedman test can be used with repeated measures from the same subject or measures from sets of subjects that have been matched as closely as possible (the size of each set must equal c). Also, the Friedman test can be based either on ranking each set of observations directly (e.g., rank ordering the piano playing of each set of three matched subjects taught by different methods) or on interval/ratio data that has been converted to ranks because parametric assumptions have been severely violated. Like other procedures for ordinal data, the Friedman test is affected by the presence of tied ranks, so a correction factor should be applied to Formula 21.9 when there are more than just a few ties (see Siegel & Castellan, 1988).

Kendall's Coefficient of Concordance

As I just mentioned in comparing it to the RM ANOVA, the Friedman test is really a test of the consistency from subject to subject across the various conditions. This aspect of the test can be exploited for another purpose. Instead of three experimental conditions (as represented by the three methods in Table 21.6), suppose we had eight paintings to be judged for creativity. And, instead of six different subjects, imagine that we had only three, who each rank order the eight paintings according to the creativity they exhibit. A value for F_r can be calculated by Formula 21.9 (with $c = 8$, and

$N = 3$) and tested for significance as in the usual Friedman test. However, the result would be interpreted differently in this case. We would not be directly interested in the fact that the paintings led to significantly different ratings for creativity; rather, a significant result would tell us that the three subjects were consistent with each other (beyond what could easily be attributed to chance) in giving their ratings. If we were worried that creativity is so subjective and difficult to define that subjects were not likely to agree, a significant result would be reassuring. A significant result might tell us that the creativity ratings of the three subjects (i.e., raters or judges) could be averaged together (or more properly a median taken) and used for another part of the experiment (e.g., to see if the creativity of an artist's painting is related to the stress reduction it produces in a viewer). (Bear in mind, however, that with so few raters serving as subjects, the use of the chi-square distribution for significance testing would not be justified; you would need a special table of critical values.)

If we were using the Friedman test in this way to assess interrater reliabilty, we would be interested not only in the statistical significance of the reliability but also in its magnitude. Reliability that barely reaches statistical significance is usually not good enough. It would be helpful to quantify the subject-to-subject consistency in addition to testing its significance. In the case of only two raters, this could be done by calculating r_s, the Spearman rank correlation coefficient (e.g., over the eight paintings being rated). With more than two raters, we could calculate r_s for each possible pair of raters and then average these correlation coefficients. Fortunately, there is an easier way to arrive at this average. First, transform your value for F_r into a value for W, Kendall's (1970) symbol for his coefficient of concordance, according to the following simple formula:

$$W = \frac{F_r}{N(c - 1)}$$ **Formula 21.10**

For perfect consistency, as in Table 21.6, W equals 1. You can confirm this with Formula 21.10 applied to the data from Table 21.6. For that table, F_r was equal to 12, so $W = 12/6(3 - 1) = 12/12 = 1$. For a total lack of consistency, as in Table 21.7, F_r equals 0, and therefore W equals 0, as well. W can be a useful way to quantify the degree to which several raters mutually agree. However, a more descriptive and familiar measure is the average r_s over all possible pairs of raters. This is found easily from W by the following formula (Kendall, 1970):

$$\bar{r}_s = \frac{NW - 1}{N - 1}$$ **Formula 21.11**

Not surprisingly, when W equals 1 the formula reduces to $(N - 1)/(N - 1)$, so the average Spearman correlation coefficient also equals 1.

1. *The Kruskal-Wallis test* is a straightforward extension of the Mann-Whitney test that is used when there are more than two independent groups. A statistic called H is calculated from the sums of ranks (and sample sizes) for the subgroups; H approximately follows the chi-square distribution with df $= k - 1$ (where k is the number of groups) when the null hypothesis is true. For very small samples a special table of critical values must be used. When H is statistically significant, post hoc pairwise comparisons are usually performed to determine which pairs of groups differ significantly. An effect-size measure, analogous to eta squared for an ANOVA, can also be calculated.

SUMMARY

2. *The Friedman test* is a good substitute for the repeated-measures ANOVA when the dependent variable is measured on an ordinal scale or the data are not consistent with the assumptions of parametric statistics. The observations are ranked separately for each matched set (there are c observations per set, where c is the number of different treatment conditions), and then the ranks are summed separately for each treatment. Consistent treatment rankings from subject to subject lead to relatively large values of a test statistic, F_r, which varies like the chi-square distribution (with df $= c - 1$) when the null hypothesis is true. Special tables are needed for critical values of F_r when the sample is very small (less than eight). As with H, significant values of F_r need to be followed by pairwise comparisons to localize the source of the effect.

3. Instead of an experimenter ranking each subject across several conditions, the subjects could be serving as judges or raters who rank order the several conditions on some dimension (which of several possible magazine covers each subject likes most, likes second most, etc.). In that case, the Friedman test determines the statistical significance of the consistency among the several raters. To quantify the amount of interrater reliability on a scale from 0 to 1, F_r can be transformed into W, Kendall's coefficient of concordance. W can then be transformed into a related but more descriptive measure, the average Spearman correlation coefficient for all the possible pairs of raters.

EXERCISES

*1. Suppose that you are comparing three architects, three musicians, and three lawyers on a spatial ability test, and you find that the three architects occupy the three highest ranks and the three lawyers occupy the three lowest ranks.
   a. Find the value of H.
   b. Even though the sample sizes are too small to justify it, use the chi-square approximation to test the significance of H at the .05 level.

2. a. Perform the Kruskal-Wallis test for the data in Exercise 12B4, and calculate η_H^2.
   b. Does your statistical conclusion in part a differ from your conclusion for Exercise 12B4? How does the power of the Kruskal-Wallis test seem to compare with the power of the one-way ANOVA?
   c. Regardless of the statistical decision in part a, perform all of the possible pairwise comparisons using the Mann-Whitney test.

*3. Perform the Kruskal-Wallis test for the data in Exercise 12B7, and calculate eta squared. Compare η_H^2 to η^2 for the ANOVA on the same data. What does this comparison suggest about the loss of information/precision

involved with ordinal as compared to parametric tests?

4. a. Given the F ratio you calculated for Exercise 15B3, what statistical conclusion would you have reached if you had set alpha to .01?
   b. Perform the Friedman test for the data in Exercise 15B3, with alpha = .01.
   c. Does your statistical conclusion in part a differ from your conclusion in part b? How does the power of the Friedman test seem to compare with the power of the one-way RM ANOVA?

*5. Perform the Friedman test for the data in Exercise 15B7, with alpha = .05. As a follow-up, perform the Wilcoxon test for each pair of conditions that are adjacent in time.

6. For the data in Exercise 15B4,
   a. Perform the Friedman test, with $\alpha = .01$.
   b. Imagine that the data in Exercise 15B4 represent the ratings of subjects, who are assessing the four texts for readability on a scale that goes from 1 to 20. Calculate the coefficient of concordance for these sets of ratings. What is the average Spearman correlation among the possible pairs of raters?

The sum of the ranks from 1 to N:

$$S_R = \frac{N(N + 1)}{2}$$

Formula 21.1

The Mann-Whitney U statistic:

$$U_A = n_A n_B + \frac{n_A(n_A + 1)}{2} - \sum R_A$$

Formula 21.2

Normal approximation to the Mann-Whitney test:

$$z = \frac{\sum R_A - .5(n_A)(N + 1)}{\sqrt{\frac{n_A n_B(N + 1)}{12}}}$$

Formula 21.3

Glass's rank biserial correlation coefficient (measures effect size for a Mann-Whitney rank-sum test):

$$r_G = \frac{2(\overline{R}_A - \overline{R}_B)}{N}$$

Formula 21.4

Normal approximation to the Wilcoxon test:

$$z = \frac{T - .25N(N + 1)}{\sqrt{\frac{N(N + 1)(2N + 1)}{24}}}$$

Formula 21.5

The matched-pairs rank biserial correlation coefficient (measures effect size for a Wilcoxon signed-ranks test):

$$r_C = \frac{2(\sum R_{\text{plus}} - \sum R_{\text{minus}})}{N(N + 1)}$$

Formula 21.6

Kruskal-Wallis H statistic:

$$H = \frac{12}{N(N + 1)} \sum_{i=1}^{k} \left(\frac{T_i^2}{n_i} \right) - 3(N + 1)$$

Formula 21.7

Eta squared (a measure of effect size) for the Kruskal-Wallis H test:

$$\eta_H^2 = \frac{H - k + 1}{N - k}$$

Formula 21.8

F_r statistic for Friedman test:

$$F_r = \frac{12}{Nc(c + 1)} \sum_{i=1}^{c} T_i^2 - 3N(c + 1)$$

Formula 21.9

Kendall's coefficient of concordance, as a function of F_r from the Friedman test:

$$W = \frac{F_r}{N(c - 1)}$$

Formula 21.10

The average Spearman rank correlation coefficient over all possible pairs of raters, as a function of Kendall's W and the number of raters:

$$\overline{r}_s = \frac{NW - 1}{N - 1}$$

Formula 21.11

Note: All of the entries in the following tables were computed by the author, except where otherwise indicated.

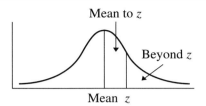

Mean to *z*

Beyond *z*

Mean *z*

z	Mean to z	Beyond z	z	Mean to z	Beyond z
.00	.0000	.5000	.43	.1664	.3336
.01	.0040	.4960	.44	.1700	.3300
.02	.0080	.4920	.45	.1736	.3264
.03	.0120	.4880	.46	.1772	.3228
.04	.0160	.4840	.47	.1808	.3192
.05	.0199	.4801	.48	.1844	.3156
.06	.0239	.4761	.49	.1879	.3121
.07	.0279	.4721	.50	.1915	.3085
.08	.0319	.4681	.51	.1950	.3050
.09	.0359	.4641	.52	.1985	.3015
.10	.0398	.4602	.53	.2019	.2981
.11	.0438	.4562	.54	.2054	.2946
.12	.0478	.4522	.55	.2088	.2912
.13	.0517	.4483	.56	.2123	.2877
.14	.0557	.4443	.57	.2157	.2843
.15	.0596	.4404	.58	.2190	.2810
.16	.0636	.4364	.59	.2224	.2776
.17	.0675	.4325	.60	.2257	.2743
.18	.0714	.4286	.61	.2291	.2709
.19	.0753	.4247	.62	.2324	.2676
.20	.0793	.4207	.63	.2357	.2643
.21	.0832	.4168	.64	.2389	.2611
.22	.0871	.4129	.65	.2422	.2578
.23	.0910	.4090	.66	.2454	.2546
.24	.0948	.4052	.67	.2486	.2514
.25	.0987	.4013	.68	.2517	.2483
.26	.1026	.3974	.69	.2549	.2451
.27	.1064	.3936	.70	.2580	.2420
.28	.1103	.3897	.71	.2611	.2389
.29	.1141	.3859	.72	.2642	.2358
.30	.1179	.3821	.73	.2673	.2327
.31	.1217	.3783	.74	.2704	.2296
.32	.1255	.3745	.75	.2734	.2266
.33	.1293	.3707	.76	.2764	.2236
.34	.1331	.3669	.77	.2794	.2206
.35	.1368	.3632	.78	.2823	.2177
.36	.1406	.3594	.79	.2852	.2148
.37	.1443	.3557	.80	.2881	.2119
.38	.1480	.3520	.81	.2910	.2090
.39	.1517	.3483	.82	.2939	.2061
.40	.1554	.3446	.83	.2967	.2033
.41	.1591	.3409	.84	.2995	.2005
.42	.1628	.3372	.85	.3023	.1977

Table A.1 (continued)

Areas under the Standard Normal Distribution

z	Mean to z	Beyond z	z	Mean to z	Beyond z
.86	.3051	.1949	1.42	.4222	.0778
.87	.3078	.1922	1.43	.4236	.0764
.88	.3106	.1894	1.44	.4251	.0749
.89	.3133	.1867	1.45	.4265	.0735
.90	.3159	.1841	1.46	.4279	.0721
.91	.3186	.1814	1.47	.4292	.0708
.92	.3212	.1788	1.48	.4306	.0694
.93	.3238	.1762	1.49	.4319	.0681
.94	.3264	.1736	1.50	.4332	.0668
.95	.3289	.1711	1.51	.4345	.0655
.96	.3315	.1685	1.52	.4357	.0643
.97	.3340	.1660	1.53	.4370	.0630
.98	.3365	.1635	1.54	.4382	.0618
.99	.3389	.1611	1.55	.4394	.0606
1.00	.3413	.1587	1.56	.4406	.0594
1.01	.3438	.1562	1.57	.4418	.0582
1.02	.3461	.1539	1.58	.4429	.0571
1.03	.3485	.1515	1.59	.4441	.0559
1.04	.3508	.1492	1.60	.4452	.0548
1.05	.3531	.1469	1.61	.4463	.0537
1.06	.3554	.1446	1.62	.4474	.0526
1.07	.3577	.1423	1.63	.4484	.0516
1.08	.3599	.1401	1.64	.4495	.0505
1.09	.3621	.1379	1.65	.4505	.0495
1.10	.3643	.1357	1.66	.4515	.0485
1.11	.3665	.1335	1.67	.4525	.0475
1.12	.3686	.1314	1.68	.4535	.0465
1.13	.3708	.1292	1.69	.4545	.0455
1.14	.3729	.1271	1.70	.4554	.0446
1.15	.3749	.1251	1.71	.4564	.0436
1.16	.3770	.1230	1.72	.4573	.0427
1.17	.3790	.1210	1.73	.4582	.0418
1.18	.3810	.1190	1.74	.4591	.0409
1.19	.3830	.1170	1.75	.4599	.0401
1.20	.3849	.1151	1.76	.4608	.0392
1.21	.3869	.1131	1.77	.4616	.0384
1.22	.3888	.1112	1.78	.4625	.0375
1.23	.3907	.1093	1.79	.4633	.0367
1.24	.3925	.1075	1.80	.4641	.0359
1.25	.3944	.1056	1.81	.4649	.0351
1.26	.3962	.1038	1.82	.4656	.0344
1.27	.3980	.1020	1.83	.4664	.0336
1.28	.3997	.1003	1.84	.4671	.0329
1.29	.4015	.0985	1.85	.4678	.0322
1.30	.4032	.0968	1.86	.4686	.0314
1.31	.4049	.0951	1.87	.4693	.0307
1.32	.4066	.0934	1.88	.4699	.0301
1.33	.4082	.0918	1.89	.4706	.0294
1.34	.4099	.0901	1.90	.4713	.0287
1.35	.4115	.0885	1.91	.4719	.0281
1.36	.4131	.0869	1.92	.4726	.0274
1.37	.4147	.0853	1.93	.4732	.0268
1.38	.4162	.0838	1.94	.4738	.0262
1.39	.4177	.0823	1.95	.4744	.0256
1.40	.4192	.0808	1.96	.4750	.0250
1.41	.4207	.0793	1.97	.4756	.0244

z	Mean to z	Beyond z	z	Mean to z	Beyond z
1.98	.4761	.0239	2.54	.4945	.0055
1.99	.4767	.0233	2.55	.4946	.0054
2.00	.4772	.0228	2.56	.4948	.0052
2.01	.4778	.0222	2.57	.4949	.0051
2.02	.4783	.0217	2.58	.4951	.0049
2.03	.4788	.0212	2.59	.4952	.0048
2.04	.4793	.0207	2.60	.4953	.0047
2.05	.4798	.0202	2.61	.4955	.0045
2.06	.4803	.0197	2.62	.4956	.0044
2.07	.4808	.0192	2.63	.4957	.0043
2.08	.4812	.0188	2.64	.4959	.0041
2.09	.4817	.0183	2.65	.4960	.0040
2.10	.4821	.0179	2.66	.4961	.0039
2.11	.4826	.0174	2.67	.4962	.0038
2.12	.4830	.0170	2.68	.4963	.0037
2.13	.4834	.0166	2.69	.4964	.0036
2.14	.4838	.0162	2.70	.4965	.0035
2.15	.4842	.0158	2.71	.4966	.0034
2.16	.4846	.0154	2.72	.4967	.0033
2.17	.4850	.0150	2.73	.4968	.0032
2.18	.4854	.0146	2.74	.4969	.0031
2.19	.4857	.0143	2.75	.4970	.0030
2.20	.4861	.0139	2.76	.4971	.0029
2.21	.4864	.0136	2.77	.4972	.0028
2.22	.4868	.0132	2.78	.4973	.0027
2.23	.4871	.0129	2.79	.4974	.0026
2.24	.4875	.0125	2.80	.4974	.0026
2.25	.4878	.0122	2.81	.4975	.0025
2.26	.4881	.0119	2.82	.4976	.0024
2.27	.4884	.0116	2.83	.4977	.0023
2.28	.4887	.0113	2.84	.4977	.0023
2.29	.4890	.0110	2.85	.4978	.0022
2.30	.4893	.0107	2.86	.4979	.0021
2.31	.4896	.0104	2.87	.4979	.0021
2.32	.4898	.0102	2.88	.4980	.0020
2.33	.4901	.0099	2.89	.4981	.0019
2.34	.4904	.0096	2.90	.4981	.0019
2.35	.4906	.0094	2.91	.4982	.0018
2.36	.4909	.0091	2.92	.4982	.0018
2.37	.4911	.0089	2.93	.4983	.0017
2.38	.4913	.0087	2.94	.4984	.0016
2.39	.4916	.0084	2.95	.4984	.0016
2.40	.4918	.0082	2.96	.4985	.0015
2.41	.4920	.0080	2.97	.4985	.0015
2.42	.4922	.0078	2.98	.4986	.0014
2.43	.4925	.0075	2.99	.4986	.0014
2.44	.4927	.0073	3.00	.4987	.0013
2.45	.4929	.0071	3.20	.4993	.0007
2.46	.4931	.0069			
2.47	.4932	.0068	3.40	.4997	.0003
2.48	.4934	.0066			
2.49	.4936	.0064	3.60	.4998	.0002
2.50	.4938	.0062			
2.51	.4940	.0060	3.80	.4999	.0001
2.52	.4941	.0059			
2.53	.4943	.0057	4.00	.49997	.00003

Table A.1
(*continued*)

Areas under
the Standard Normal
Distribution

Table A.2
Critical Values of the
t Distribution

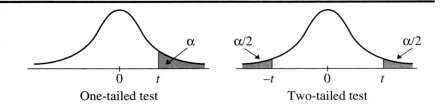

One-tailed test   Two-tailed test

	LEVEL OF SIGNIFICANCE FOR ONE-TAILED TEST					
	.10	.05	.025	.01	.005	.0005
	LEVEL OF SIGNIFICANCE FOR TWO-TAILED TEST					
df	.20	.10	.05	.02	.01	.001
1	3.078	6.314	12.706	31.821	63.657	636.620
2	1.886	2.920	4.303	6.965	9.925	31.599
3	1.638	2.353	3.182	4.541	5.841	12.924
4	1.533	2.132	2.776	3.747	4.604	8.610
5	1.476	2.015	2.571	3.365	4.032	6.869
6	1.440	1.943	2.447	3.143	3.707	5.959
7	1.415	1.895	2.365	2.998	3.499	5.408
8	1.397	1.860	2.306	2.896	3.355	5.041
9	1.383	1.833	2.262	2.821	3.250	4.781
10	1.372	1.812	2.228	2.764	3.169	4.587
11	1.363	1.796	2.201	2.718	3.106	4.437
12	1.356	1.782	2.179	2.681	3.055	4.318
13	1.350	1.771	2.160	2.650	3.012	4.221
14	1.345	1.761	2.145	2.624	2.977	4.140
15	1.341	1.753	2.131	2.602	2.947	4.073
16	1.337	1.746	2.120	2.583	2.921	4.015
17	1.333	1.740	2.110	2.567	2.898	3.965
18	1.330	1.734	2.101	2.552	2.878	3.922
19	1.328	1.729	2.093	2.539	2.861	3.883
20	1.325	1.725	2.086	2.528	2.845	3.850
21	1.323	1.721	2.080	2.518	2.831	3.819
22	1.321	1.717	2.074	2.508	2.819	3.792
23	1.319	1.714	2.069	2.500	2.807	3.768
24	1.318	1.711	2.064	2.492	2.797	3.745
25	1.316	1.708	2.060	2.485	2.787	3.725
26	1.315	1.706	2.056	2.479	2.779	3.707
27	1.314	1.703	2.052	2.473	2.771	3.690
28	1.313	1.701	2.048	2.467	2.763	3.674
29	1.311	1.699	2.045	2.462	2.756	3.659
30	1.310	1.697	2.042	2.457	2.750	3.646
40	1.303	1.684	2.021	2.423	2.704	3.551
60	1.296	1.671	2.000	2.390	2.660	3.460
120	1.289	1.658	1.980	2.358	2.617	3.373
∞	1.282	1.645	1.960	2.326	2.576	3.291

	ONE-TAILED TEST (α)				Table A.3
	.05	.025	.01	.005	Power as a Function of δ and Significance Criterion (α)
	TWO-TAILED TEST (α)				
δ	.10	.05	.02	.01	
0.5	.14	.08	.03	.02	
0.6	.16	.09	.04	.02	
0.7	.18	.11	.05	.03	
0.8	.21	.13	.06	.04	
0.9	.23	.15	.08	.05	
1.0	.26	.17	.09	.06	
1.1	.29	.20	.11	.07	
1.2	.33	.22	.13	.08	
1.3	.37	.26	.15	.10	
1.4	.40	.29	.18	.12	
1.5	.44	.32	.20	.14	
1.6	.48	.36	.23	.16	
1.7	.52	.40	.27	.19	
1.8	.56	.44	.30	.22	
1.9	.60	.48	.33	.25	
2.0	.64	.52	.37	.28	
2.1	.68	.56	.41	.32	
2.2	.71	.60	.45	.35	
2.3	.74	.63	.49	.39	
2.4	.77	.67	.53	.43	
2.5	.80	.71	.57	.47	
2.6	.83	.74	.61	.51	
2.7	.85	.77	.65	.55	
2.8	.88	.80	.68	.59	
2.9	.90	.83	.72	.63	
3.0	.91	.85	.75	.66	
3.1	.93	.87	.78	.70	
3.2	.94	.89	.81	.73	
3.3	.95	.91	.84	.77	
3.4	.96	.93	.86	.80	
3.5	.97	.94	.88	.82	
3.6	.97	.95	.90	.85	
3.7	.98	.96	.92	.87	
3.8	.98	.97	.93	.89	
3.9	.99	.97	.94	.91	
4.0	.99	.98	.95	.92	
4.1	.99	.98	.96	.94	
4.2	.99	.99	.97	.95	
4.3	*	.99	.98	.96	
4.4		.99	.98	.97	
4.5		.99	.99	.97	
4.6		*	.99	.98	
4.7			.99	.98	
4.8			.99	.99	
4.9			*	.99	
5.0				.99	

* The power at and below this point is greater than .995.

	\delta as a Function of Significance Criterion (\alpha) and Power				

Table A.4 δ as a Function of Significance Criterion (α) and Power

	ONE-TAILED TEST (α)			
	.05	**.025**	**.01**	**.005**
	TWO-TAILED TEST (α)			
Power	**.10**	**.05**	**.02**	**.01**
.25	0.97	1.29	1.65	1.90
.50	1.64	1.96	2.33	2.58
.60	1.90	2.21	2.58	2.83
.67	2.08	2.39	2.76	3.01
.70	2.17	2.48	2.85	3.10
.75	2.32	2.63	3.00	3.25
.80	2.49	2.80	3.17	3.42
.85	2.68	3.00	3.36	3.61
.90	2.93	3.24	3.61	3.86
.95	3.29	3.60	3.97	4.22
.99	3.97	4.29	4.65	4.90
.999	4.37	5.05	5.42	5.67

	LEVELS OF SIGNIFICANCE FOR A ONE-TAILED TEST			
	.05	.025	.01	.005
	LEVELS OF SIGNIFICANCE FOR A TWO-TAILED TEST			
df	.10	.05	.02	.01
2	.900	.950	.980	.990
3	.805	.878	.934	.959
4	.729	.811	.882	.917
5	.669	.755	.833	.875
6	.622	.707	.789	.834
7	.582	.666	.750	.798
8	.549	.632	.716	.765
9	.521	.602	.685	.735
10	.498	.576	.658	.708
11	.476	.553	.634	.684
12	.458	.533	.612	.661
13	.441	.514	.592	.641
14	.426	.497	.574	.623
15	.412	.482	.558	.606
16	.400	.468	.542	.590
17	.389	.456	.529	.575
18	.379	.444	.516	.562
19	.369	.433	.503	.549
20	.360	.423	.492	.537
21	.351	.413	.482	.526
22	.344	.404	.472	.515
23	.337	.396	.462	.505
24	.330	.388	.453	.496
25	.323	.381	.445	.487
26	.317	.374	.437	.479
27	.311	.367	.430	.471
28	.306	.361	.423	.463
29	.301	.355	.416	.456
30	.296	.349	.409	.449
35	.275	.325	.381	.418
40	.257	.304	.358	.393
45	.243	.288	.338	.372
50	.231	.273	.322	.354
55	.220	.261	.307	.339
60	.211	.250	.295	.325
70	.195	.232	.274	.302
80	.183	.217	.256	.283
90	.173	.205	.242	.267
100	.164	.195	.230	.254
120	.150	.178	.210	.232
150	.134	.159	.189	.208
200	.116	.138	.164	.181
300	.095	.113	.134	.148
400	.082	.098	.116	.128
500	.073	.088	.104	.115
1000	.052	.062	.073	.081

Table A.5
Critical Values of Pearson's r (df $= N - 2$)

	Table A.6							

Table of Fisher's Transformation of r to Z

r	Z	r	Z	r	Z	r	Z
0.000	0.000	0.250	0.255	0.500	0.549	0.750	0.973
0.005	0.005	0.255	0.261	0.505	0.556	0.755	0.984
0.010	0.010	0.260	0.266	0.510	0.563	0.760	0.996
0.015	0.015	0.265	0.271	0.515	0.570	0.765	1.008
0.020	0.020	0.270	0.277	0.520	0.576	0.770	1.020
0.025	0.025	0.275	0.282	0.525	0.583	0.775	1.033
0.030	0.030	0.280	0.288	0.530	0.590	0.780	1.045
0.035	0.035	0.285	0.293	0.535	0.597	0.785	1.058
0.040	0.040	0.290	0.299	0.540	0.604	0.790	1.071
0.045	0.045	0.295	0.304	0.545	0.611	0.795	1.085
0.050	0.050	0.300	0.310	0.550	0.618	0.800	1.099
0.055	0.055	0.305	0.315	0.555	0.626	0.805	1.113
0.060	0.060	0.310	0.320	0.560	0.633	0.810	1.127
0.065	0.065	0.315	0.326	0.565	0.640	0.815	1.142
0.070	0.070	0.320	0.332	0.570	0.648	0.820	1.157
0.075	0.075	0.325	0.337	0.575	0.655	0.825	1.172
0.080	0.080	0.330	0.343	0.580	0.662	0.830	1.188
0.085	0.085	0.335	0.348	0.585	0.670	0.835	1.204
0.090	0.090	0.340	0.354	0.590	0.678	0.840	1.221
0.095	0.095	0.345	0.360	0.595	0.685	0.845	1.238
0.100	0.100	0.350	0.365	0.600	0.693	0.850	1.256
0.105	0.105	0.355	0.371	0.605	0.701	0.855	1.274
0.110	0.110	0.360	0.377	0.610	0.709	0.860	1.293
0.115	0.116	0.365	0.383	0.615	0.717	0.865	1.313
0.120	0.121	0.370	0.388	0.620	0.725	0.870	1.333
0.125	0.126	0.375	0.394	0.625	0.733	0.875	1.354
0.130	0.131	0.380	0.400	0.630	0.741	0.880	1.376
0.135	0.136	0.385	0.406	0.635	0.750	0.885	1.398
0.140	0.141	0.390	0.412	0.640	0.758	0.890	1.422
0.145	0.146	0.395	0.418	0.645	0.767	0.895	1.447
0.150	0.151	0.400	0.424	0.650	0.775	0.900	1.472
0.155	0.156	0.405	0.430	0.655	0.784	0.905	1.499
0.160	0.161	0.410	0.436	0.660	0.793	0.910	1.528
0.165	0.167	0.415	0.442	0.665	0.802	0.915	1.557
0.170	0.172	0.420	0.448	0.670	0.811	0.920	1.589
0.175	0.177	0.425	0.454	0.675	0.820	0.925	1.623
0.180	0.182	0.430	0.460	0.680	0.829	0.930	1.658
0.185	0.187	0.435	0.466	0.685	0.838	0.935	1.697
0.190	0.192	0.440	0.472	0.690	0.848	0.940	1.738
0.195	0.198	0.445	0.478	0.695	0.858	0.945	1.783
0.200	0.203	0.450	0.485	0.700	0.867	0.950	1.832
0.205	0.208	0.455	0.491	0.705	0.877	0.955	1.886
0.210	0.213	0.460	0.497	0.710	0.887	0.960	1.946
0.215	0.218	0.465	0.504	0.715	0.897	0.965	2.014
0.220	0.224	0.470	0.510	0.720	0.908	0.970	2.092
0.225	0.229	0.475	0.517	0.725	0.918	0.975	2.185
0.230	0.234	0.480	0.523	0.730	0.929	0.980	2.298
0.235	0.239	0.485	0.530	0.735	0.940	0.985	2.443
0.240	0.245	0.490	0.536	0.740	0.950	0.990	2.647
0.245	0.250	0.495	0.543	0.745	0.962	0.995	2.994

Table A.7

Critical Values of the F Distribution for α = .05

df NUMERATOR

df Denominator	1	2	3	4	5	6	7	8	9	10	12	15	20	25	30	40	60	120	∞
3	10.13	9.55	9.28	9.12	9.01	8.94	8.89	8.85	8.81	8.79	8.74	8.70	8.66	8.63	8.62	8.59	8.57	8.55	8.53
4	7.71	6.94	6.59	6.39	6.26	6.16	6.09	6.04	6.00	5.96	5.91	5.86	5.80	5.77	5.75	5.72	5.69	5.66	5.63
5	6.61	5.79	5.41	5.19	5.05	4.95	4.88	4.82	4.77	4.74	4.68	4.62	4.56	4.52	4.50	4.46	4.43	4.40	4.36
6	5.99	5.14	4.76	4.53	4.39	4.28	4.21	4.15	4.10	4.06	4.00	3.94	3.87	3.83	3.81	3.77	3.74	3.70	3.67
7	5.59	4.74	4.35	4.12	3.97	3.87	3.79	3.73	3.68	3.64	3.57	3.51	3.44	3.40	3.38	3.34	3.30	3.27	3.23
8	5.32	4.46	4.07	3.84	3.69	3.58	3.50	3.44	3.39	3.35	3.28	3.22	3.15	3.11	3.08	3.04	3.01	2.97	2.93
9	5.12	4.26	3.86	3.63	3.48	3.37	3.29	3.23	3.18	3.14	3.07	3.01	2.94	2.89	2.86	2.83	2.79	2.75	2.71
10	4.96	4.10	3.71	3.48	3.33	3.22	3.14	3.07	3.02	2.98	2.91	2.85	2.77	2.73	2.70	2.66	2.62	2.58	2.54
11	4.84	3.98	3.59	3.36	3.20	3.09	3.01	2.95	2.90	2.85	2.79	2.72	2.65	2.60	2.57	2.53	2.49	2.45	2.40
12	4.75	3.89	3.49	3.26	3.11	3.00	2.91	2.85	2.80	2.75	2.69	2.62	2.54	2.50	2.47	2.43	2.38	2.34	2.30
13	4.67	3.81	3.41	3.18	3.03	2.92	2.83	2.77	2.71	2.67	2.60	2.53	2.46	2.41	2.38	2.34	2.30	2.25	2.21
14	4.60	3.74	3.34	3.11	2.96	2.85	2.76	2.70	2.65	2.60	2.53	2.46	2.39	2.34	2.31	2.27	2.22	2.18	2.13
15	4.54	3.68	3.29	3.06	2.90	2.79	2.71	2.64	2.59	2.54	2.48	2.40	2.33	2.28	2.25	2.20	2.16	2.11	2.07
16	4.49	3.63	3.24	3.01	2.85	2.74	2.66	2.59	2.54	2.49	2.42	2.35	2.28	2.23	2.19	2.15	2.11	2.06	2.01
17	4.45	3.59	3.20	2.96	2.81	2.70	2.61	2.55	2.49	2.45	2.38	2.31	2.23	2.18	2.15	2.10	2.06	2.01	1.96
18	4.41	3.55	3.16	2.93	2.77	2.66	2.58	2.51	2.46	2.41	2.34	2.27	2.19	2.14	2.11	2.06	2.02	1.97	1.92
19	4.38	3.52	3.13	2.90	2.74	2.63	2.54	2.48	2.42	2.38	2.31	2.23	2.16	2.11	2.07	2.03	1.98	1.93	1.88
20	4.35	3.49	3.10	2.87	2.71	2.60	2.51	2.45	2.39	2.35	2.28	2.20	2.12	2.07	2.04	1.99	1.95	1.90	1.84
21	4.32	3.47	3.07	2.84	2.68	2.57	2.49	2.42	2.37	2.32	2.25	2.18	2.10	2.04	2.01	1.96	1.92	1.87	1.81
22	4.30	3.44	3.05	2.82	2.66	2.55	2.46	2.40	2.34	2.30	2.23	2.15	2.07	2.02	1.98	1.94	1.89	1.84	1.78
23	4.28	3.42	3.03	2.80	2.64	2.53	2.44	2.37	2.32	2.27	2.20	2.13	2.05	2.00	1.96	1.91	1.86	1.81	1.76
24	4.26	3.40	3.01	2.78	2.62	2.51	2.42	2.36	2.30	2.25	2.18	2.11	2.03	1.97	1.94	1.89	1.84	1.79	1.73
25	4.24	3.39	2.99	2.76	2.60	2.49	2.40	2.34	2.28	2.24	2.16	2.09	2.01	1.95	1.92	1.87	1.82	1.77	1.71
26	4.23	3.37	2.98	2.74	2.59	2.47	2.39	2.32	2.27	2.22	2.15	2.07	1.99	1.94	1.90	1.85	1.80	1.75	1.69
27	4.21	3.35	2.96	2.73	2.57	2.46	2.37	2.31	2.25	2.20	2.13	2.06	1.97	1.92	1.88	1.84	1.79	1.73	1.67
28	4.20	3.34	2.95	2.71	2.56	2.45	2.36	2.29	2.24	2.19	2.12	2.04	1.96	1.91	1.87	1.82	1.77	1.71	1.65
29	4.18	3.33	2.93	2.70	2.55	2.43	2.35	2.28	2.22	2.18	2.10	2.03	1.94	1.90	1.85	1.81	1.75	1.70	1.64
30	4.17	3.32	2.92	2.69	2.53	2.42	2.33	2.27	2.21	2.16	2.09	2.01	1.93	1.88	1.84	1.79	1.74	1.68	1.62
40	4.08	3.23	2.84	2.61	2.45	2.34	2.25	2.18	2.12	2.08	2.00	1.92	1.84	1.78	1.74	1.69	1.64	1.58	1.51
60	4.00	3.15	2.76	2.53	2.37	2.25	2.17	2.10	2.04	1.99	1.92	1.84	1.75	1.69	1.65	1.59	1.53	1.47	1.39
120	3.92	3.07	2.68	2.45	2.29	2.17	2.09	2.02	1.96	1.91	1.83	1.75	1.66	1.60	1.55	1.50	1.43	1.35	1.25
∞	3.84	3.00	2.60	2.37	2.21	2.10	2.01	1.94	1.88	1.83	1.75	1.67	1.57	1.51	1.46	1.39	1.32	1.22	1.00

α = .05

Table A.8
Critical Values of the F Distribution for $\alpha = .025$

$\alpha = .025$

df NUMERATOR

df Denominator	1	2	3	4	5	6	7	8	9	10	12	15	20	25	30	40	60	120	∞
3	17.44	16.04	15.44	15.10	14.88	14.73	14.62	14.54	14.47	14.42	14.34	14.25	14.17	14.12	14.08	14.04	13.99	13.95	13.90
4	12.22	10.65	9.98	9.60	9.36	9.20	9.07	8.98	8.90	8.84	8.75	8.66	8.56	8.50	8.46	8.41	8.36	8.31	8.26
5	10.01	8.43	7.76	7.39	7.15	6.98	6.85	6.76	6.68	6.62	6.52	6.43	6.33	6.27	6.23	6.18	6.12	6.07	6.02
6	8.81	7.26	6.60	6.23	5.99	5.82	5.70	5.60	5.52	5.46	5.37	5.27	5.17	5.11	5.07	5.01	4.96	4.90	4.85
7	8.07	6.54	5.89	5.52	5.29	5.12	4.99	4.90	4.82	4.76	4.67	4.57	4.47	4.40	4.36	4.31	4.25	4.20	4.14
8	7.57	6.06	5.42	5.05	4.82	4.65	4.53	4.43	4.36	4.30	4.20	4.10	4.00	3.94	3.89	3.84	3.78	3.73	3.67
9	7.21	5.71	5.08	4.72	4.48	4.32	4.20	4.10	4.03	3.96	3.87	3.77	3.67	3.60	3.56	3.51	3.45	3.39	3.33
10	6.94	5.46	4.83	4.47	4.24	4.07	3.95	3.85	3.78	3.72	3.62	3.52	3.42	3.35	3.31	3.26	3.20	3.14	3.08
11	6.72	5.26	4.63	4.28	4.04	3.88	3.76	3.66	3.59	3.53	3.43	3.33	3.23	3.16	3.12	3.06	3.00	2.94	2.88
12	6.55	5.10	4.47	4.12	3.89	3.73	3.61	3.51	3.44	3.37	3.28	3.18	3.07	3.01	2.96	2.91	2.85	2.79	2.72
13	6.41	4.97	4.35	4.00	3.77	3.60	3.48	3.39	3.31	3.25	3.15	3.05	2.95	2.88	2.84	2.78	2.72	2.66	2.60
14	6.30	4.86	4.24	3.89	3.66	3.50	3.38	3.29	3.21	3.15	3.05	2.95	2.84	2.78	2.73	2.67	2.61	2.55	2.49
15	6.20	4.77	4.15	3.80	3.58	3.41	3.29	3.20	3.12	3.06	2.96	2.86	2.76	2.69	2.64	2.59	2.52	2.46	2.40
16	6.12	4.69	4.08	3.73	3.50	3.34	3.22	3.12	3.05	2.99	2.89	2.79	2.68	2.61	2.57	2.51	2.45	2.38	2.32
17	6.04	4.62	4.01	3.66	3.44	3.28	3.16	3.06	2.98	2.92	2.82	2.72	2.62	2.55	2.50	2.44	2.38	2.32	2.25
18	5.98	4.56	3.95	3.61	3.38	3.22	3.10	3.01	2.93	2.87	2.77	2.67	2.56	2.49	2.44	2.38	2.32	2.26	2.19
19	5.92	4.51	3.90	3.56	3.33	3.17	3.05	2.96	2.88	2.82	2.72	2.62	2.51	2.44	2.39	2.33	2.27	2.20	2.13
20	5.87	4.46	3.86	3.51	3.29	3.13	3.01	2.91	2.84	2.77	2.68	2.57	2.46	2.40	2.35	2.29	2.22	2.16	2.09
21	5.83	4.42	3.82	3.48	3.25	3.09	2.97	2.87	2.80	2.73	2.64	2.53	2.42	2.36	2.31	2.25	2.18	2.11	2.04
22	5.79	4.38	3.78	3.44	3.22	3.05	2.93	2.84	2.76	2.70	2.60	2.50	2.39	2.32	2.27	2.21	2.14	2.08	2.00
23	5.75	4.35	3.75	3.41	3.18	3.02	2.90	2.81	2.73	2.67	2.57	2.47	2.36	2.29	2.24	2.18	2.11	2.04	1.97
24	5.72	4.32	3.72	3.38	3.15	2.99	2.87	2.78	2.70	2.64	2.54	2.44	2.33	2.26	2.21	2.15	2.08	2.01	1.94
25	5.69	4.29	3.69	3.35	3.13	2.97	2.85	2.75	2.68	2.61	2.51	2.41	2.30	2.23	2.18	2.12	2.05	1.98	1.91
26	5.66	4.27	3.67	3.33	3.10	2.94	2.82	2.73	2.65	2.59	2.49	2.39	2.28	2.21	2.16	2.09	2.03	1.95	1.88
27	5.63	4.24	3.65	3.31	3.08	2.92	2.80	2.71	2.63	2.57	2.47	2.36	2.25	2.18	2.13	2.07	2.00	1.93	1.85
28	5.61	4.22	3.63	3.29	3.06	2.90	2.78	2.69	2.61	2.55	2.45	2.34	2.23	2.16	2.11	2.05	1.98	1.91	1.83
29	5.59	4.20	3.61	3.27	3.04	2.88	2.76	2.67	2.59	2.53	2.43	2.32	2.21	2.14	2.09	2.03	1.96	1.89	1.81
30	5.57	4.18	3.59	3.25	3.03	2.87	2.75	2.65	2.57	2.51	2.41	2.31	2.20	2.12	2.09	2.01	1.94	1.87	1.79
40	5.42	4.05	3.46	3.13	2.90	2.74	2.62	2.53	2.45	2.39	2.29	2.18	2.07	1.99	1.94	1.88	1.80	1.72	1.64
60	5.29	3.93	3.34	3.01	2.79	2.63	2.51	2.41	2.33	2.27	2.17	2.06	1.94	1.87	1.82	1.74	1.67	1.58	1.48
120	5.15	3.80	3.23	2.89	2.67	2.52	2.39	2.30	2.22	2.16	2.05	1.94	1.82	1.75	1.69	1.61	1.53	1.43	1.31
∞	5.02	3.69	3.12	2.79	2.57	2.41	2.29	2.19	2.11	2.05	1.94	1.83	1.71	1.63	1.57	1.48	1.39	1.27	1.00

Table A.9

Critical Values of the F Distribution for α = .01

df NUMERATOR

df Denominator	1	2	3	4	5	6	7	8	9	10	12	15	20	25	30	40	60	120	∞
3	34.12	30.82	29.46	28.71	28.24	27.91	27.67	27.49	27.35	27.23	27.05	26.87	26.69	26.58	26.50	26.41	26.32	26.22	26.13
4	21.20	18.00	16.69	15.98	15.52	15.21	14.98	14.80	14.66	14.55	14.37	14.20	14.02	13.91	13.84	13.75	13.65	13.56	13.46
5	16.26	13.27	12.06	11.39	10.97	10.67	10.46	10.29	10.16	10.05	9.89	9.72	9.55	9.45	9.38	9.29	9.20	9.11	9.02
6	13.75	10.92	9.78	9.15	8.75	8.47	8.26	8.10	7.98	7.87	7.72	7.56	7.40	7.30	7.23	7.14	7.06	6.97	6.88
7	12.25	9.55	8.45	7.85	7.46	7.19	6.99	6.84	6.72	6.62	6.47	6.31	6.16	6.06	5.99	5.91	5.82	5.74	5.65
8	11.26	8.65	7.59	7.01	6.63	6.37	6.18	6.03	5.91	5.81	5.67	5.52	5.36	5.26	5.20	5.12	5.03	4.95	4.86
9	10.56	8.02	6.99	6.42	6.06	5.80	5.61	5.47	5.35	5.26	5.11	4.96	4.81	4.71	4.65	4.57	4.48	4.40	4.31
10	10.04	7.56	6.55	5.99	5.64	5.39	5.20	5.06	4.94	4.85	4.71	4.56	4.41	4.31	4.25	4.17	4.08	4.00	3.91
11	9.65	7.21	6.22	5.67	5.32	5.07	4.89	4.74	4.63	4.54	4.40	4.25	4.10	4.01	3.94	3.86	3.78	3.69	3.60
12	9.33	6.93	5.95	5.41	5.06	4.82	4.64	4.50	4.39	4.30	4.16	4.01	3.86	3.76	3.70	3.62	3.54	3.45	3.36
13	9.07	6.70	5.74	5.21	4.86	4.62	4.44	4.30	4.19	4.10	3.96	3.82	3.66	3.57	3.51	3.43	3.34	3.25	3.17
14	8.86	6.51	5.56	5.04	4.69	4.46	4.28	4.14	4.03	3.94	3.80	3.66	3.51	3.41	3.35	3.27	3.18	3.09	3.00
15	8.68	6.36	5.42	4.89	4.56	4.32	4.14	4.00	3.89	3.80	3.67	3.52	3.37	3.28	3.21	3.13	3.05	2.96	2.87
16	8.53	6.23	5.29	4.77	4.44	4.20	4.03	3.89	3.78	3.69	3.55	3.41	3.26	3.16	3.10	3.02	2.93	2.84	2.75
17	8.40	6.11	5.18	4.67	4.34	4.10	3.93	3.79	3.68	3.59	3.46	3.31	3.16	3.087	3.00	2.92	2.83	2.75	2.65
18	8.29	6.01	5.09	4.58	4.25	4.01	3.84	3.71	3.60	3.51	3.37	3.23	3.08	3.098	2.92	2.84	2.75	2.66	2.57
19	8.18	5.93	5.01	4.50	4.17	3.94	3.77	3.63	3.52	3.43	3.30	3.15	3.00	2.91	2.84	2.76	2.67	2.58	2.49
20	8.10	5.85	4.94	4.43	4.10	3.87	3.70	3.56	3.46	3.37	3.23	3.09	2.94	2.84	2.78	2.69	2.61	2.52	2.42
21	8.02	5.78	4.87	4.37	4.04	3.81	3.64	3.51	3.40	3.31	3.17	3.03	2.88	2.78	2.72	2.64	2.55	2.46	2.36
22	7.95	5.72	4.82	4.31	3.99	3.76	3.59	3.45	3.35	3.26	3.12	2.98	2.83	2.753	2.67	2.58	2.50	2.40	2.31
23	7.88	5.66	4.76	4.26	3.94	3.71	3.54	3.41	3.30	3.21	3.07	2.93	2.78	2.68	2.62	2.54	2.45	2.35	2.26
24	7.82	5.61	4.72	4.22	3.90	3.67	3.50	3.36	3.26	3.17	3.03	2.89	2.74	2.64	2.58	2.49	2.40	2.31	2.21
25	7.77	5.57	4.68	4.18	3.85	3.63	3.46	3.32	3.22	3.13	2.99	2.85	2.70	2.60	2.54	2.45	2.36	2.27	2.17
26	7.72	5.53	4.64	4.14	3.82	3.59	3.42	3.29	3.18	3.09	2.96	2.81	2.66	2.57	2.50	2.42	2.33	2.23	2.13
27	7.68	5.49	4.60	4.11	3.78	3.56	3.39	3.26	3.15	3.06	2.93	2.78	2.63	2.53	2.47	2.38	2.29	2.20	2.10
28	7.64	5.45	4.57	4.07	3.75	3.53	3.36	3.23	3.12	3.03	2.90	2.75	2.60	2.51	2.44	2.35	2.26	2.17	2.06
29	7.60	5.42	4.54	4.04	3.73	3.50	3.33	3.20	3.09	3.00	2.87	2.73	2.57	2.47	2.41	2.33	2.23	2.14	2.03
30	7.56	5.39	4.51	4.02	3.70	3.47	3.30	3.17	3.07	2.98	2.84	2.70	2.55	2.45	2.39	2.30	2.21	2.11	2.01
40	7.31	5.18	4.31	3.83	3.51	3.29	3.12	2.99	2.89	2.80	2.66	2.52	2.37	2.27	2.20	2.11	2.02	1.92	1.80
60	7.08	4.98	4.13	3.65	3.34	3.12	2.95	2.82	2.72	2.63	2.50	2.35	2.20	2.10	2.03	1.94	1.84	1.73	1.60
120	6.85	4.79	3.95	3.48	3.17	2.96	2.79	2.66	2.56	2.47	2.34	2.19	2.03	1.93	1.86	1.76	1.66	1.53	1.38
∞	6.63	4.61	3.78	3.32	3.02	2.80	2.64	2.51	2.41	2.32	2.18	2.04	1.88	1.77	1.70	1.59	1.47	1.32	1.00

Table A.10
Power of ANOVA
($\alpha = .05$)

$k = 2$ (ϕ)

df_w	1.0	1.2	1.4	1.6	1.8	2.0	2.2	2.6	3.0
4	.20	.26	.33	.41	.49	.57	.65	.78	.88
8	.24	.32	.41	.51	.61	.70	.78	.89	.96
12	.26	.35	.44	.55	.65	.74	.81	.92	.97
16	.26	.36	.46	.57	.67	.76	.83	.93	.98
20	.27	.37	.47	.58	.68	.77	.84	.94	.98
30	.28	.38	.48	.59	.69	.78	.85	.94	.98
60	.29	.39	.50	.61	.71	.79	.86	.95	.99
∞	.29	.40	.51	.62	.72	.81	.88	.96	.99

$k = 3$ (ϕ)

df_w	1.0	1.2	1.4	1.6	1.8	2.0	2.2	2.6	3.0
4	.18	.23	.30	.38	.46	.54	.62	.76	.86
8	.23	.32	.42	.52	.63	.72	.80	.92	.97
12	.26	.36	.47	.58	.69	.78	.86	.95	.99
16	.27	.38	.49	.61	.72	.81	.88	.96	.99
20	.28	.39	.51	.63	.74	.83	.89	.97	.99
30	.29	.41	.53	.65	.76	.85	.91	.98	*
60	.31	.43	.55	.68	.78	.87	.92	.98	*
∞	.32	.44	.57	.70	.80	.88	.94	.99	*

$k = 4$ (ϕ)

df_w	1.0	1.2	1.4	1.6	1.8	2.0	2.2	2.6	3.0
4	.17	.23	.29	.37	.45	.53	.61	.75	.86
8	.24	.33	.43	.54	.65	.75	.83	.94	.98
12	.27	.38	.50	.62	.73	.82	.89	.97	.99
16	.29	.40	.53	.66	.77	.86	.92	.98	*
20	.30	.42	.55	.68	.79	.87	.93	.99	*
30	.32	.45	.58	.71	.82	.90	.95	.99	*
60	.34	.47	.61	.74	.84	.92	.96	.99	*
∞	.36	.50	.64	.77	.87	.93	.97	*	*

$k = 5$ (ϕ)

df_w	1.0	1.2	1.4	1.6	1.8	2.0	2.2	2.6	3.0
4	.17	.22	.29	.36	.45	.53	.61	.75	.86
8	.24	.34	.45	.56	.67	.77	.85	.95	.99
12	.28	.39	.52	.65	.76	.85	.92	.98	*
16	.30	.43	.56	.69	.81	.89	.94	.99	*
20	.32	.45	.59	.72	.83	.91	.96	.99	*
30	.34	.48	.63	.76	.86	.93	.97	*	*
60	.37	.52	.67	.80	.89	.95	.98	*	*
∞	.40	.55	.71	.83	.92	.96	.99	*	*

*Power ≥ .995.

Source: Adapted from "Tables of the Power of the *F* Test," by M.L. Tiku, *Journal of the American Statistical Association*, vol. 62, 1967, pp. 525–539. Copyright © 1967 by American Statistical Association. Reprinted by permission of the American Statistical Association via the Copyright Clearance Center.

Table A.11

Critical Values of the Studentized Range Statistic (q) for $\alpha = .05$

NUMBER OF GROUPS (OR NUMBER OF STEPS BETWEEN ORDERED MEANS)

df for Error Term	2	3	4	5	6	7	8	9	10	11	12	13	14	15	16	17	18	19	20
1	17.97	26.98	32.82	37.08	40.41	43.12	45.40	47.36	49.07	50.59	51.96	53.20	54.33	55.36	56.32	57.22	58.04	58.83	59.56
2	6.08	8.33	9.80	10.88	11.74	12.44	13.03	13.54	13.99	14.39	14.75	15.08	15.38	15.65	15.91	16.14	16.37	16.57	16.77
3	4.50	5.91	6.82	7.50	8.04	8.48	8.85	9.18	9.46	9.72	9.95	10.15	10.35	10.52	10.69	10.84	10.98	11.11	11.24
4	3.93	5.04	5.76	6.29	6.71	7.05	7.35	7.60	7.83	8.03	8.21	8.37	8.52	8.66	8.79	8.91	9.03	9.13	9.23
5	3.64	4.60	5.22	5.67	6.03	6.33	6.58	6.80	6.99	7.17	7.32	7.47	7.60	7.72	7.83	7.93	8.03	8.12	8.21
6	3.46	4.34	4.90	5.30	5.63	5.90	6.12	6.32	6.49	6.65	6.79	6.92	7.03	7.14	7.24	7.34	7.43	7.51	7.59
7	3.34	4.16	4.68	5.06	5.36	5.61	5.82	6.00	6.16	6.30	6.43	6.55	6.66	6.76	6.85	6.94	7.02	7.10	7.17
8	3.26	4.04	4.53	4.89	5.17	5.40	5.60	5.77	5.92	6.05	6.18	6.29	6.39	6.48	6.57	6.65	6.73	6.80	6.87
9	3.20	3.95	4.41	4.76	5.02	5.24	5.43	5.59	5.74	5.87	5.98	6.09	6.19	6.28	6.36	6.44	6.51	6.58	6.64
10	3.15	3.88	4.33	4.65	4.91	5.12	5.30	5.46	5.60	5.72	5.83	5.93	6.03	6.11	6.19	6.27	6.34	6.40	6.47
11	3.11	3.82	4.26	4.57	4.82	5.03	5.20	5.35	5.49	5.61	5.71	5.81	5.90	5.98	6.06	6.13	6.20	6.27	6.33
12	3.08	3.77	4.20	4.51	4.75	4.95	5.12	5.27	5.39	5.51	5.61	5.71	5.80	5.88	5.95	6.02	6.09	6.15	6.21
13	3.06	3.73	4.15	4.45	4.69	4.88	5.05	5.19	5.32	5.43	5.53	5.63	5.71	5.79	5.86	5.93	5.99	6.05	6.11
14	3.03	3.70	4.11	4.41	4.64	4.83	4.99	5.13	5.25	5.36	5.46	5.55	5.64	5.71	5.79	5.85	5.91	5.97	6.03
15	3.01	3.67	4.08	4.37	4.59	4.78	4.94	5.08	5.20	5.31	5.40	5.49	5.57	5.65	5.72	5.78	5.85	5.90	5.96
16	3.00	3.65	4.05	4.33	4.56	4.74	4.90	5.03	5.15	5.26	5.35	5.44	5.52	5.59	5.66	5.73	5.79	5.84	5.90
17	2.98	3.63	4.02	4.30	4.52	4.70	4.86	4.99	5.11	5.21	5.31	5.39	5.47	5.54	5.61	5.67	5.73	5.79	5.84
18	2.97	3.61	4.00	4.28	4.49	4.67	4.82	4.96	5.07	5.17	5.27	5.35	5.43	5.50	5.57	5.63	5.69	5.74	5.79
19	2.96	3.59	3.98	4.25	4.47	4.65	4.79	4.92	5.04	5.14	5.23	5.31	5.39	5.46	5.53	5.59	5.65	5.70	5.75
20	2.95	3.58	3.96	4.23	4.45	4.62	4.77	4.90	5.01	5.11	5.20	5.28	5.36	5.43	5.49	5.55	5.61	5.66	5.71
24	2.92	3.53	3.90	4.17	4.37	4.54	4.68	4.81	4.92	5.01	5.10	5.18	5.25	5.32	5.38	5.44	5.49	5.55	5.59
30	2.89	3.49	3.85	4.10	4.30	4.46	4.60	4.72	4.82	4.92	5.00	5.08	5.15	5.21	5.27	5.33	5.38	5.43	5.47
40	2.86	3.44	3.79	4.04	4.23	4.39	4.52	4.63	4.73	4.82	4.90	4.98	5.04	5.11	5.16	5.22	5.27	5.31	5.36
60	2.83	3.40	3.74	3.98	4.16	4.31	4.44	4.55	4.65	4.73	4.81	4.88	4.94	5.00	5.06	5.11	5.15	5.20	5.24
120	2.80	3.36	3.68	3.92	4.10	4.24	4.36	4.47	4.56	4.64	4.71	4.78	4.84	4.90	4.95	5.00	5.04	5.09	5.13
∞	2.77	3.31	3.63	3.86	4.03	4.17	4.29	4.39	4.47	4.55	4.62	4.68	4.74	4.80	4.85	4.89	4.93	4.97	5.01

Source: Adapted from *Biometrika Tables for Statisticians*, Vol 1, 3rd ed., by E. Pearson & H. Hartley, Table 29. Copyright © 1966 University Press. Used with the permission of the Biometrika Trustees.

Table A.12
Orthogonal Polynomial Trend Coefficients

k	Trend	1	2	3	4	5	6	7	8	9	10	$\Sigma\, C_i^2$
3	Linear	−1	0	1								2
	Quadratic	1	−2	1								6
4	Linear	−3	−1	1	3							20
	Quadratic	1	−1	−1	1							4
	Cubic	−1	3	−3	1							20
5	Linear	−2	−1	0	1	2						10
	Quadratic	2	−1	−2	−1	2						14
	Cubic	−1	2	0	−2	1						10
	Quartic	1	−4	6	−4	1						70
6	Linear	−5	−3	−1	1	3	5					70
	Quadratic	5	−1	−4	−4	−1	5					84
	Cubic	−5	7	4	−4	−7	5					180
	Quartic	1	−3	2	2	−3	1					28
7	Linear	−3	−2	−1	0	1	2	3				28
	Quadratic	5	0	−3	−4	−3	0	5				84
	Cubic	−1	1	1	0	−1	−1	1				6
	Quartic	3	−7	1	6	1	−7	3				154
8	Linear	−7	−5	−3	−1	1	3	5	7			168
	Quadratic	7	1	−3	−5	−5	−3	1	7			168
	Cubic	−7	5	7	3	−3	−7	−5	7			264
	Quartic	7	−13	−3	9	9	−3	−13	7			616
	Quintic	−7	23	−17	−15	15	17	−23	7			2184
9	Linear	−4	−3	−2	−1	0	1	2	3	4		60
	Quadratic	28	7	−8	−17	−20	−17	−8	7	28		2772
	Cubic	−14	7	13	9	0	−9	−13	−7	14		990
	Quartic	14	−21	−11	9	18	9	−11	−21	14		2002
	Quintic	−4	11	−4	−9	0	9	4	−11	4		468
10	Linear	−9	−7	−5	−3	−1	1	3	5	7	9	330
	Quadratic	6	2	−1	−3	−4	−4	−3	−1	2	6	132
	Cubic	−42	14	35	31	12	−12	−31	−35	−14	42	8580
	Quartic	18	−22	−17	3	18	18	3	−17	−22	18	2860
	Quintic	−6	14	−1	−11	−6	6	11	1	−14	6	780

n	X	p	n	X	p	n	X	p
1	0	.5000		1	.0176	13	0	.0001
	1	.5000		2	.0703		1	.0016
2	0	.2500		3	.1641		2	.0095
	1	.5000		4	.2461		3	.0349
	2	.2500		5	.2461		4	.0873
3	0	.1250		6	.1641		5	.1571
	1	.3750		7	.0703		6	.2095
	2	.3750		8	.0176		7	.2095
	3	.1250		9	.0020		8	.1571
4	0	.0625	10	0	.0010		9	.0873
	1	.2500		1	.0098		10	.0349
	2	.3750		2	.0439		11	.0095
	3	.2500		3	.1172		12	.0016
	4	.0625		4	.2051		13	.0001
5	0	.0312		5	.2461	14	0	.0001
	1	.1562		6	.2051		1	.0009
	2	.3125		7	.1172		2	.0056
	3	.3125		8	.0439		3	.0222
	4	.1562		9	.0098		4	.0611
	5	.0312		10	.0010		5	.1222
6	0	.0156	11	0	.0005		6	.1833
	1	.0938		1	.0054		7	.2095
	2	.2344		2	.0269		8	.1833
	3	.3125		3	.0806		9	.1222
	4	.2344		4	.1611		10	.0611
	5	.0938		5	.2256		11	.0222
	6	.0156		6	.2256		12	.0056
7	0	.0078		7	.1611		13	.0009
	1	.0547		8	.0806		14	.0001
	2	.1641		9	.0269	15	0	.0000
	3	.2734		10	.0054		1	.0005
	4	.2734		11	.0005		2	.0032
	5	.1641	12	0	.0002		3	.0139
	6	.0547		1	.0029		4	.0417
	7	.0078		2	.0161		5	.0916
8	0	.0039		3	.0537		6	.1527
	1	.0312		4	.1208		7	.1964
	2	.1094		5	.1934		8	.1964
	3	.2188		6	.2256		9	.1527
	4	.2734		7	.1934		10	.0916
	5	.2188		8	.1208		11	.0417
	6	.1094		9	.0537		12	.0139
	7	.0312		10	.0161		13	.0032
	8	.0039		11	.0029		14	.0005
9	0	.0020		12	.0002		15	.0000

Table A.13
Probabilities of the Binomial Distribution for $P = .5$

Table A.14
Critical Values of the
χ^2 Distribution

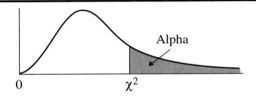

df	ALPHA (AREA IN THE UPPER TAIL)				
	.10	.05	.025	.01	.005
1	2.71	3.84	5.02	6.63	7.88
2	4.61	5.99	7.38	9.21	10.60
3	6.25	7.81	9.35	11.35	12.84
4	7.78	9.49	11.14	13.28	14.86
5	9.24	11.07	12.83	15.09	16.75
6	10.64	12.59	14.45	16.81	18.55
7	12.02	14.07	16.01	18.48	20.28
8	13.36	15.51	17.54	20.09	21.96
9	14.68	16.92	19.02	21.67	23.59
10	15.99	18.31	20.48	23.21	25.19
11	17.28	19.68	21.92	24.72	26.75
12	18.55	21.03	23.34	26.22	28.30
13	19.81	22.36	24.74	27.69	29.82
14	21.06	23.69	26.12	29.14	31.32
15	22.31	25.00	27.49	30.58	32.80
16	23.54	26.30	28.85	32.00	34.27
17	24.77	27.59	30.19	33.41	35.72
18	25.99	28.87	31.53	34.81	37.15
19	27.20	30.14	32.85	36.19	38.58
20	28.41	31.41	34.17	37.56	40.00
21	29.62	32.67	35.48	38.93	41.40
22	30.81	33.92	36.78	40.29	42.80
23	32.01	35.17	38.08	41.64	44.18
24	33.20	36.42	39.37	42.98	45.56
25	34.38	37.65	40.65	44.31	46.93
26	35.56	38.89	41.92	45.64	48.29
27	36.74	40.11	43.19	46.96	49.64
28	37.92	41.34	44.46	48.28	50.99
29	39.09	42.56	45.72	49.59	52.34
30	40.26	43.77	46.98	50.89	53.67
40	51.80	55.76	59.34	63.69	66.78
50	63.16	67.50	71.42	76.16	79.50
60	74.40	79.08	83.30	88.39	91.96
70	85.53	90.53	95.03	100.43	104.23
80	96.58	101.88	106.63	112.34	116.33
90	107.56	113.14	118.14	124.12	128.31
100	118.50	124.34	129.56	135.81	140.18

Table A.15
Critical Values of ΣR_A for the Mann-Whitney (Rank-Sum) Test

$n_A = 3$
LEVEL OF SIGNIFICANCE FOR ONE-TAILED TEST

n_B	.05	.025	.01	.005
3	6–15			
4	6–18			
5	7–20	6–21		
6	8–22	7–23		
7	8–25	7–26	6–27	
8	9–27	8–28	6–30	
9	10–29	8–31	7–32	6–33
10	10–32	9–33	7–35	6–36
11	11–34	9–36	7–38	6–39
12	11–37	10–38	8–40	7–41
13	12–39	10–41	8–43	7–44
14	13–41	11–43	8–46	7–47
15	13–44	11–46	9–48	8–49

$n_A = 4$
LEVEL OF SIGNIFICANCE FOR ONE-TAILED TEST

n_B	.05	.025	.01	.005
4	11–25	10–26		
5	12–28	11–29	10–30	
6	13–31	12–32	11–33	10–34
7	14–34	13–35	11–37	10–38
8	15–37	14–48	12–40	11–41
9	16–40	14–42	13–43	11–45
10	17–43	15–45	13–47	12–48
11	18–46	16–48	14–50	12–52
12	19–49	17–51	15–53	13–55
13	20–52	18–54	15–57	13–59
14	21–55	19–57	16–60	14–62
15	22–58	20–60	17–63	15–65

$n_A = 5$
LEVEL OF SIGNIFICANCE FOR ONE-TAILED TEST

n_B	.05	.025	.01	.005
5	19–36	17–38	16–39	15–40
6	20–40	18–42	17–43	16–44
7	21–44	20–45	18–47	16–49
8	23–47	21–49	19–51	17–53
9	24–51	22–53	20–55	18–57
10	26–54	23–57	21–59	19–61
11	27–58	24–61	22–63	20–65
12	28–62	26–64	23–67	21–69
13	30–65	27–68	24–71	22–73
14	31–69	28–72	25–75	22–78
15	33–72	29–76	26–79	23–82

$n_A = 6$
LEVEL OF SIGNIFICANCE FOR ONE-TAILED TEST

n_B	.05	.025	.01	.005
6	28–50	26–52	24–54	23–55
7	29–55	27–57	25–59	24–60
8	31–59	29–61	27–63	25–65
9	33–63	31–65	28–68	26–70
10	35–67	32–70	29–73	27–75
11	37–71	34–74	30–78	28–80
12	38–76	35–79	32–82	30–84
13	40–80	37–83	33–87	31–89
14	42–84	38–88	34–92	32–94
15	44–88	40–92	36–96	33–99

$n_A = 7$
LEVEL OF SIGNIFICANCE FOR ONE-TAILED TEST

n_B	.05	.025	.01	.005
7	39–66	36–69	34–71	32–73
8	41–71	38–74	35–77	34–78
9	43–76	40–79	37–82	35–84
10	45–81	42–84	39–87	37–89
11	47–86	44–89	40–93	38–95
12	49–91	46–94	42–98	40–100
13	52–95	48–99	44–103	41–106
14	54–100	50–104	45–109	43–111
15	56–105	52–109	47–114	44–117

$n_A = 8$
LEVEL OF SIGNIFICANCE FOR ONE-TAILED TEST

n_B	.05	.025	.01	.005
8	51–85	49–87	45–91	43–93
9	54–90	51–93	47–97	45–99
10	56–96	53–99	49–103	47–105
11	59–101	55–105	51–109	49–111
12	62–106	58–110	53–115	51–117
13	64–112	60–116	56–120	53–123
14	67–117	62–122	58–126	54–130
15	69–123	65–127	60–132	56–136

Table A.15 (continued)
Critical Values of ΣR_A for the Mann-Whitney (Rank-Sum) Test

$n_A = 9$
LEVEL OF SIGNIFICANCE FOR ONE-TAILED TEST

n_B	.05	.025	.01	.005
9	66–105	62–109	59–112	56–115
10	69–111	65–115	61–119	58–122
11	72–117	68–121	63–126	61–128
12	75–123	71–127	66–132	63–135
13	78–129	73–134	68–139	65–142
14	81–135	76–140	71–145	67–149
15	84–141	79–146	73–152	69–156

$n_A = 10$
LEVEL OF SIGNIFICANCE FOR ONE-TAILED TEST

n_B	.05	.025	.01	.005
10	82–128	78–132	74–136	71–139
11	86–134	81–139	77–143	73–147
12	89–141	84–146	79–151	76–154
13	92–148	88–152	82–158	79–161
14	96–154	91–159	85–165	81–169
15	99–161	94–166	88–172	84–176

$n_A = 11$
LEVEL OF SIGNIFICANCE FOR ONE-TAILED TEST

n_B	.05	.025	.01	.005
11	100–153	96–157	91–162	87–166
12	104–160	99–165	94–170	90–174
13	108–167	103–172	97–178	93–182
14	112–174	106–180	100–186	96–190
15	116–181	110–187	103–194	99–198

$n_A = 12$
LEVEL OF SIGNIFICANCE FOR ONE-TAILED TEST

n_B	.05	.025	.01	.005
12	120–180	115–185	109–191	105–195
13	125–187	119–193	113–199	109–203
14	129–195	123–201	116–208	112–212
15	133–203	127–209	120–216	115–221

$n_A = 13$
LEVEL OF SIGNIFICANCE FOR ONE-TAILED TEST

n_B	.05	.025	.01	.005
13	142–209	136–215	130–221	125–226
14	147–217	141–223	134–230	129–235
15	152–225	145–232	138–239	133–244

$n_A = 14$
LEVEL OF SIGNIFICANCE FOR ONE-TAILED TEST

n_B	.05	.025	.01	.005
14	166–240	160–246	152–254	147–259
15	171–249	164–256	156–264	151–269

$n_A = 15$
LEVEL OF SIGNIFICANCE FOR ONE-TAILED TEST

n_B	.05	.025	.01	.005
15	192–273	184–281	176–289	171–294

Source: An adaptation of Table 1 from "Extended Tables of Critical Values for Wilcoxon's Test Statistic," by L. R. Verdooren, *Biometrika*, Vol. 50, 1963, pp. 177–186. Used with permission of the Biometrika Trustees.

	LEVEL OF SIGNIFICANCE FOR ONE-TAILED TEST*				
	.05	.025	.01	.005	
	LEVEL OF SIGNIFICANCE FOR TWO-TAILED TEST*				
n	.10	.05	.02	.01	
5	0	†	†	†	
6	2	0	†	†	
7	3	2	0	†	
8	5	3	1	0	
9	8	5	3	1	
10	10	8	5	3	
11	13	10	7	5	
12	17	13	9	7	
13	21	17	12	9	
14	25	21	15	12	
15	30	25	19	15	
16	35	29	23	19	
17	41	34	27	23	
18	47	40	32	27	
19	53	46	37	32	
20	60	52	43	37	
21	67	58	49	42	
22	75	65	55	48	
23	83	73	62	54	
24	91	81	69	61	
25	100	89	76	68	

Table A.16
Critical Values of *T* for the Wilcoxon Signed-Ranks Test

*To be significant, the obtained T must be *equal to* or *less than* the critical value.
†No decision is possible for this combination of sample size and alpha.

CHAPTER 1

Section A

2. *a*) ratio *c*) nominal *e*) ordinal (but often treated as interval) *g*) nominal *i*) ratio

3. *a*) discrete *c*) discrete *e*) continuous

5. *a*) size of reward *b*) number of words recalled *c*) ratio

7. *a*) correlational *b*) correlational *c*) experimental *d*) experimental

Section B

1. *a*) $4 + 6 + 8 + 10 = 28$ *c*) $10 + 20 + 30 + 40 + 50 = 150$ *e*) $4 + 16 + 36 + 64 + 100 = 220$ *g*) $3^2 + 5^2 + 7^2 + 9^2 + 11^2 = 285$

2. *a*) $5 + 9 + 13 + 17 + 21 = 65$ *c*) $(30)(35) = 1050$ *e*) $(-1) + (-1) + (-1) + (-1) + (-1) = -5$ *g*) $9 + 11 + 13 + 15 + 17 = 65$

4. *a*) $9N$ *c*) $3 \Sigma D$ *e*) $\Sigma Z^2 + 4N$

6. *a*) 144.01 *c*) 99.71 *e*) 7.35 *g*) 6.00

8. *a*) 55.6 *c*) 99.0 *e*) 1.4

Section C

2. $\sum\limits_{j=1}^{12}\sum\limits_{i=1}^{9}X_{ij}$

4. $\sum\limits_{j=1}^{N}\sum\limits_{i=1}^{6}X_{ij}$

CHAPTER 2

Section A

1. *a*)

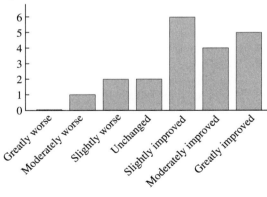

Number of patients

Progress	f	cf	rf	crf	cpf
Greatly improved	5	20	.25	1.00	100
Moderately improved	4	15	.20	.75	75
Slightly improved	6	11	.30	.55	55
Unchanged	2	5	.10	.25	25
Slightly worse	2	3	.10	.15	15
Moderately worse	1	1	.05	.05	5
Greatly worse	0	0	0	0	0

c) five patients; 25%
e) moderately improved; unchanged

2.

No. of Words	f	cf	rf	crf	cpf
29	1	25	.04	1.00	100
28	2	24	.08	.96	96
27	0	22	0	.88	88
26	3	22	.12	.88	88
25	5	19	.20	.76	76
24	7	14	.28	.56	56
23	4	7	.16	.28	28
22	0	3	0	.12	12
21	1	3	.04	.12	12
20	0	2	0	.08	8
19	2	2	.08	.08	8

a) 28% *c*) 76; 88

5.

Score	f	cf	rf	crf	cpf
10	2	20	.10	1.00	100
9	2	18	.10	.90	90
8	4	16	.20	.80	80
7	3	12	.15	.60	60
6	2	9	.10	.45	45
5	2	7	.10	.35	35
4	1	5	.05	.25	25
3	2	4	.10	.20	20
2	1	2	.05	.10	10
1	1	1	.05	.05	5

a) two; 10% *c*) 35; 90

7.

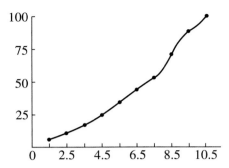

a) 6.9 *b*) 8.3 *c*) 22.5 *d*) 52.5

Section B

1. *a*)

Interval	f	cf	rf	crf	cpf
145–149	1	50	.02	1.00	100
140–144	0	49	0	.98	98
135–139	3	49	.06	.98	98
130–134	3	46	.06	.92	92
125–129	4	43	.08	.86	86
120–124	7	39	.14	.78	78
115–119	9	32	.18	.64	64
110–114	10	23	.20	.46	46
105–109	6	13	.12	.26	26
100–104	4	7	.08	.14	14
95–99	3	3	.06	.06	6

b)

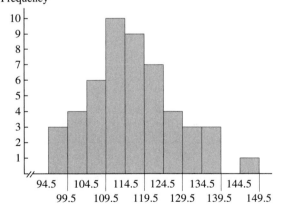

c) first quartile = 104.5 + 4.58 = 109.08;
third quartile = 119.5 + 3.928 = 123.43
d) 40th percentile = 109.5 + 3.5 = 113;
60th percentile 114.5 + 3.88 = 118.38
e) PR for 125 = 78 + .8 = 78.8
f) PR for 108 = 14 + 8.4 = 22.4

2. *a*)

Test Scores	f	cf	rf	crf	cpf
95–99	4	60	.066	1.00	100
90–94	8	56	.133	.933	93
85–89	12	48	.20	.800	80
80–84	13	36	.216	.600	60
75–79	10	23	.166	.383	38.3
70–74	5	13	.083	.216	21.6
65–69	3	8	.05	.133	13.3
60–64	3	5	.05	.083	8.3
55–59	2	2	.033	.033	3.3

b)

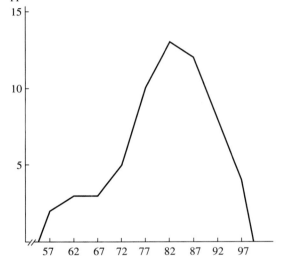

c) 84.5 + 5 = 89.5
d) 75th percentile = 84.5 + 3.75 = 88.25;
60th percentile = 79.5 + 5 = 84.5
e) PR for 88 = 60 + 14 = 74, 100 − 74 = 26%
f) PR for 81 = 38.3 + 6.48 = 44.78

4. *a*)

Speed	f	cf	rf	crf	cpf
85–89	1	25	.04	1.00	100
80–84	1	24	.04	.96	96
75–79	3	23	.12	.92	92
70–74	2	20	.08	.80	80
65–69	3	18	.12	.72	72
60–64	3	15	.12	.60	60
55–59	5	12	.20	.48	48
50–54	3	7	.12	.28	28
45–49	4	4	.16	.16	16

b) about 70% *c*) about 61
d) 40th percentile = 54.5 + 3 = 57.5

e) first quartile = 49.5 + 3.75 = 53.25;
third quartile = 69.5 + 1.88 = 71.38
f) PR for 62 = 48 + 6 = 54

5. *a*)

Number of Dreams	f	cf	rf	crf	cpf
35–39	4	40	.10	1.00	100
30–34	3	36	.075	.90	90
25–29	6	33	.15	.825	82.5
20–24	8	27	.20	.675	67.5
15–19	4	19	.10	.475	47.5
10–14	7	15	.175	.375	37.5
5–9	2	8	.05	.20	20
0–4	6	6	.15	.15	15

b)

Subjects

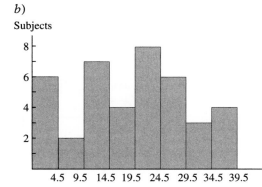

4.5 9.5 14.5 19.5 24.5 29.5 34.5 39.5

c) 7 subjects = 17.5%;
PR = (33 + 3)2.5 = 83.25 *d*) about 35
e) PR = (8 + .7)2.5 = 21.75
f) PR = (19 + .8)2.5 = 49.5

7.

Quiz score	f	cpf%
20–21	1	100
18–19	1	98
16–17	8	96
14–15	7	80
12–13	14	66
10–11	9	38
8–9	6	20
6–7	1	8
4–5	1	6
2–3	1	4
0–1	1	2

a)

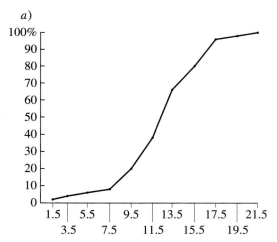

b) 25%tile = 10.3; 50%tile = 12.4; 75%tile = 14.8
c) 10%tile = 7.8; 90%tile = 16.8
d) PR (8) = 11
e) PR (18) = 96.5

Section C

2.

Stem	Leaf
5*	
5.	5 6
6*	1 3 4
6.	5 8 9
7*	0 1 2 3 4
7.	5 5 6 6 6 7 7 8 9 9
8*	0 0 0 1 1 2 2 2 3 3 4 4 4
8.	5 5 5 5 6 6 6 7 7 7 9 9
9*	0 0 1 2 2 2 4 4
9.	5 6 7 9

4.

Stem	Leaf
0*	0 0 0 1 3 4
0.	7 8
1*	1 1 2 3 3 3 4
1.	5 6 8 9
2*	0 1 2 3 3 3 4 4
2.	5 6 6 6 7 8
3*	1 2 3
3.	6 7 8 8

CHAPTER 3

Section A

1. a) mode b) mean c) mode d) median
   e) mean

3. negatively skewed

6. a) mean = 11.75; SS = 79.5; σ^2 = 9.9375
   b) MD = 2.75; σ = 3.152

7. df = 8 − 1 = 7; s^2 = 79.5/7 = 11.357;
   s = 3.37

9. mean = 27.4375; mode = 26; median = 27.5;
   range = 36 − 17 = 19; SIQ range = (31.5 −
   23.75)/2 = 3.875; MD = 67/16 = 4.1875;
   s = $\sqrt{(453.9375)/15}$ = 5.501

Section B

1. μ = 3750/35 = 107.14

3. missing score = 6

6. σ = $\sqrt{(784.125 − 749.39)}$ = 5.89;
   s = $\sqrt{.143(6273 − 5995.125)}$ = 6.30

7. a) s = $\sqrt{.143(1709 − 1431.125)}$ = 6.30.
   The answer is the same as the answer in
   Exercise 6; if you subtract a constant from
   every number in a set, the standard
   deviation will not change.
   b) s = $\sqrt{.143(697 − 666.125)}$ = 2.1. The
   answer is one-third the answer in Exercise
   6; if you divide every number in a set by a
   constant, the standard deviation will be
   divided by that constant.

9. a) s = $\sqrt{(5/4)}$ (4.5) = 5.031 b) 4.617
   c) 4.523

Section C

1.

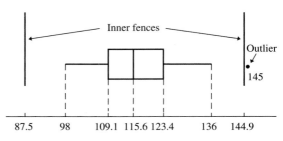

3.

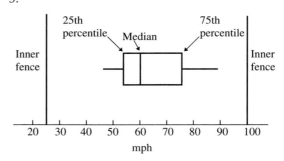

4. σ = 9.615; population skewness =
   11,856/(9 × 888.9) = 1.48

7. σ = 1.287; skewness = 12.8246/(9 × 2.132)
   = .668. The transformation reduces
   skewness.

9. kurtosis = .024

CHAPTER 4

Section A

1. a) (60 − 65)/3 = −1.667; (65 − 65)/3 = 0
   c) −3(3) + 65 = 56 inches; 61 inches;
   64.1 inches; 58.7 inches

3. a) 100(−0.2) + 500 = 480 b) 100(1.3) +
   500 = 630 c) 190 d) 690

5. a) (15 − 18)/3 = −1; SAT = 100(−1) + 500 =
   400 b) (10 − 18)/3 = −2.667; SAT = 233.3
   c) 566.7 d) 800

7. a) .0714 c) .3106 e) .4826

10. a) 17.5/$\sqrt{5}$ = 7.83 b) 17.5/$\sqrt{25}$ = 3.5
    c) 1.565 d) 0.7

12. N = $(32.2/4.6)^2$ = 49

Section B

1. a) 95.25% b) 28.77% c) 4.75% d) 98.26%

3. a) .1056; .5987 b) .6480; .0030 c) .1587 −
   .0516 = .1071; .4332 − .0987 = .3345;
   .4834 + .1480 = .6314

5. Between 336 and 664

7. *a)* $z = -2.81$ *b)* .0025

9. *a)* $z = -4.04$ *b)* $z = -8.08$ *c)* If sample size is multiplied by C, z is multiplied by $\sqrt{C}$

11. *a)* $z = 3.16$ *b)* $z = 3.16$ *c)* z scores are unaffected by (linear) changes in scale; to evaluate a z score you don't have to know what measurement scale was used.

Section C

1. *b*

3. *d*

4. *a)* $.0401 + .1056 = .1457$
   b) $.2357 + .148 + .1587 + .0668 = .6092$
   c) $.5091 + .4013 - .1354 = .775$

6. *a)* $.2643 + .3783 = .6426$
   b) $.3574 + .3783 - .114 = .6217$
   c) $(.7357)(.7357) = .541$
   d) $2 \times (.2514)(.1587) = .08$

7. *a)* $(.25)(.25) = .0625$
   b) $P(H) \times P(H|H) = (.25)(12/51) = .0577$
   c) $P(H) \times P(S|H) = (.25)(13/51) = .0637$

CHAPTER 5

Section A

1. *a)* .0885; .177 *b)* .2420; .484
   c) .0139; .0278

3. *a)* 1.16 *b)* .75

5. $z = 2.4$, two-tailed $p = .0164$

8. *d*

would be investigated further even though it is worthless. *d)* Others might ignore the drug even though it has some value *e)* If overlooking a valuable drug is very much worse than raising expectations about what is actually a worthless drug, it could justify making alpha larger.

10. $N \geq 198$

Section B

1. no; $z_{calc} = -1.18 < z_{crit} = 1.645$

2. no; $z_{calc} = 1.6 < z_{crit} = 1.96$

4. *a)* reject null; $z_{calc} = 2.05 > z_{crit} = 1.96$
   b) accept null; $z_{calc} < z_{crit} = 2.575$; as alpha becomes smaller, the chance of attaining statistical significance decreases.

6. no; $z_{calc} = (470 - 500)/(100/\sqrt{9}) = -.9$

8. *a)* accept null; $z_{calc} = 1.83 < z_{crit} = 1.96$
   b) reject null; $z_{calc} > z_{crit} = 1.645$ *c)* The drug

Section C

1. *b*

3. Anywhere between 0 and 20 (there is no way to estimate the number of Type I errors in this case); none of the 20 significant experiments can be a Type II error.

5. *a)* $.05/.45 = .111 = 11.1\%$
   b) $.095/.135 = .703 = 70.3\%$
   c) $.02/.66 = .0303 = 3.03\%$

CHAPTER 6

Section A

1. *a)* .524 *b)* .175 (one-third as large as the answer in part *a*)

3. *a)* df $= 9$ *b)* ± 2.262; ± 3.250 *c)* 1.833; 2.821

5. *a)* $t_{calc} = 6.20$; $t_{crit} = 2.08$ *b)* $t_{calc} = 4.39$; t_{calc} in part *a* divided by $\sqrt{2}$ equals t_{calc} in part *b* (because there are half as many subjects in part *b*)

7. $t = -6.1875/3.356 = -1.84$; $1.84 < t_{.05}(15) = 2.131$, so cannot reject the null.

9. The *p* value will be smaller for the experiment with the larger sample size, because as N increases, the tails of the t distribution become thinner, and there is less area beyond a given t value.

Section B

1. *a)* accept null; $t_{calc} = 2.044 < t_{crit} = 2.132$
   b) accept null; could be making a Type II error

3. *a)* $\mu_{\text{lower}} = 4.78$; $\mu_{\text{upper}} = 5.62$
   b) $\mu_{\text{lower}} = 4.99$; $\mu_{\text{upper}} = 5.41$
   c) width of CI in part *a* = .84, width of CI in part *b* = .42; if you multiply the sample size by *C*, you divide the width by $\sqrt{C}$. (This relationship does not hold for small samples because changing the sample size also changes the critical values of the *t* distribution.)

4. *a)* $\mu_{\text{lower}} = 2.32$, $\mu_{\text{upper}} = 2.88$ *b)* Because $\mu = 3$ does not fall within the 95% CI for the American couples, we can say the American couples differ significantly from the Europeans at the .05 level.

7. *a)* $t_{\text{calc}} = 2.87$; reject null at .05 level but not at .01 level *b)* $\mu_{\text{lower}} = 8.41$, $\mu_{\text{upper}} = 12.99$

9. CI goes from 20.66 to 34.96; yes, 34 is in the 95% CI, so it would not be significant at the .05 level

Section C

1. 10% trimmed $t = 2.95/1.15 = 2.57$; CI = $10.75 \pm 2.365(1.15)$, goes from 8.03 to 13.47; yes.

3. 20% trimmed $t = 6.5/1.435 = 4.53$; $4.53 > t_{.05}(9) = 2.262$, so reject null; CI = $27.5 \pm 2.262(1.435)$, goes from 24.25 to 30.75; the original sample had heavy tails, and the trimming eliminated them, thus increasing the *t* value.

CHAPTER 7

Section A

1. *a)* $z = (47.5 - 45.8)/\sqrt{5.5^2/150 + 42/100} = 2.83$ *b)* .0023

3. $s_p^2 = [100(12) + 50(8)]/150 = 10.67$

5. $s_p^2 = (120 + 180)/2 = 150$

7. *a)* $t = (27.2 - 34.4)/\sqrt{4^2/15 + 14^2/15} = -1.92$ *b)* $t = \dfrac{27.2 - 34.4}{\sqrt{106[(1/15) + (1/15)]}} = -1.92$

   Note: The answers to parts *a* and *b* should be the same. When the two samples are the same size, the pooled-variances *t* equals the separate-variances *t*.

Section B

1. $t_{\text{calc}} = (52 - 44)/\sqrt{130.2(1/7 + 1/10)} = 1.42 < t_{\text{crit}} = 2.132$; accept null

3. $t_{\text{calc}} = (87.2 - 82.9)/\sqrt{(28.09 + 19.36/1/12)} = 2.16 > t_{\text{crit}} = 2.074$; reject null (false expectations *can* affect student performance)

4. *a)* $4.4 \pm (1.65)(1.98)$; $\mu_{\text{lower}} = 1.13$, $\mu_{\text{upper}} = 7.67$ *b)* $4.4 \pm (1.65)(2.617)$; $\mu_{\text{lower}} = .07$, $\mu_{\text{upper}} = 8.73$ *c)* Because zero is not contained in either the 95% CI or the 99%

CI, the null hypothesis can be rejected at both the .05 and the .01 levels.

7. *a)* accept null; $t_{\text{calc}} = (21.11 - 17.14)/\sqrt{67.55(1/9 + 1/7)} = .96$ *b)* $\mu_{\text{lower}} = -4.91$, $\mu_{\text{upper}} = 12.85$

8. *a)* reject null; $t_{\text{calc}} = (21.11 - 13.83)/\sqrt{37.36(1/9 + 1/6)} = 2.26 > t_{\text{crit}} = 2.16$; the value of *t* changed from .96 to 2.26 with the removal of a single outlier, so the *t* test seems very susceptible to outliers. *b)* $\bar{X}_1 - \bar{X}_2 = 7.28$ minutes; $\mu_{\text{lower}} = .325$, $\mu_{\text{upper}} = 14.24$ *c)* Yes, the separate-variances test would be recommended because the sample sizes are small and different, and the variances are quite different.

9. *a)* accept null; $t_{\text{calc}} = 1.18 < t_{\text{crit}} = 2.101$ *b)* no, because the sample sizes are equal *c)* $t_{.05}(9) = 2.262$; no, because the *t* test was not significant with a smaller critical value

Section C

1. $t_{\text{sep}} = 1.40$; rounded $df_{\text{Welch}} = 12$; $t_{.05}(12) = 2.179 > 1.40$, so cannot reject null

3. *a)* $t_{\text{pool}} = 1.36$ *b)* $t_{\text{sep}} = 1.90$; the separate-variances *t* value is larger

5. $t'_{s-\nu} = .732$; $df'_{\text{Welch}} = 9.7$

CHAPTER 8

Section A

2. **d** = .3

4. δ = .3$\sqrt{28/2}$ = 1.12; δ is less than the critical value required for significance, so the results are expected to be significant less than half of the time

6. **d** = 1.5; δ = 4.74

8. *a)* *g* = .70 *b)* *g* = 1.70

10. *d; b* and *c* reduce Type II errors without affecting Type I errors, and *a* reduces Type II errors by allowing Type I errors to increase.

Section B

1. *a)* β = .68, power = .32; β = .29, power = .71
   b) β = .86, power = .14; β = .53, power = .47; reducing alpha increases beta
   c) required delta = 1.7; required delta = 3.25

3. *a)* 2 × (3.1/.7)² = 39.2; 40 participants are required in each group *b)* 54 participants per group

5. *a)* *N* = 138 *b)* *N* = 224

7. *a)* harmonic mean = 13.33; *b)* power = .54 (approximately); the experiment is probably not worth doing

9. *a)* 16 participants per group *b)* **d** = 1.32

11. *a)* δ = 2.23; power = .60 (approximately) *b)* **d** must be at least .91

Section C

1. *a)* less-biased *g* = .665 *b)* less-biased *g* = 1.657

3. *a)* about .46 *b)* about .58

5. *a)* The approx. CI for **d** goes from .077 to .964.
   b) The approx. CI for **d** goes from .7 to 2.7

CHAPTER 9

Section A

3.

Annual salary

a) The general trend is toward a positive correlation. *b)* Misleading because of the outlier at (4, 50).

4.

Number of details recalled

a) Low; the points do not come close to fitting on one straight line. *b)* Somewhat

misleading; the degree of linear correlation is low, but there appears to be a meaningful (curvilinear) relationship between anxiety and number of details recalled.

5.

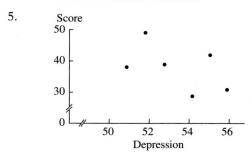

Score

a) Negative; points slope down as you move to the right in the scatterplot. *b)* Moderate; the trend is noticeable, but there is considerable scatter with respect to any straight line.

7. *a*

9. *a)* −1.0 *b)* −.45

Section B

1. *a)* *r* = (122,000/20 − 75 × 80)/(10 × 12) = .83 *b)* Yes, $r_{calc} > r_{crit}$ = .5614 *c)* There is a high degree of linear relationship.

4. *a)* $r_{calc} = 1/8(2029 - 9 \times 6 \times 34.56)/(2.74 \times 11.39) = .653 < r_{crit} = .666$; accept null
   b) $r_{calc} = .957 > r_{crit} = .707$; reject null
   c) You find out that the person with four years at the company is related to the president of the company.

5. *a)* $r_{calc} = 1/9(1267 - 10 \times 6.7 \times 20.4)/(3.23 \times 5.13) = -.669$
   b) reject null; $|r_{calc}| > r_{crit} = .632$

6. *a)* $r_{calc} = .845$ *b)* reject null; $r_{calc} > r_{crit} = .805$; no, $r_{calc} < r_{crit} = .878$

8. *a)* $r_{calc} = 1/5(140 - 6 \times 3.33 \times 4.83)/(3.33 \times 2.86) = .91$ *b)* reject null; $r_{calc} > r_{crit} = .882$

9. *a)* $r_{calc} = -.718$ *b)* reject null; $|r_{calc}| > r_{crit} = .707$ *c)* Because subjects were randomly assigned to exercise durations, we can conclude that an increase in exercise duration (within the limits of the study) *causes* a lowering of serum cholesterol level.

11. $r = -.709$; it is significant at the .05 level, and just barely at the .01 level.

Section C

1. ρ has to be at least .905 in magnitude

3. *a)* .05 *c)* .31 *e)* .867 *g)* 1.832 *i)* .245 *k)* .74 *m)* .922 *o)* .987

5. *a)* $\delta = .5\sqrt{9} = 1.5$; power = .32
   b) $\delta = .669\sqrt{9} = 2.0$; power = .52
   c) $N = (3.24/.669)^2 + 1 = 24.5$; 25 schizophrenics

7. accept null; $z_{calc} = (.81 - .203)/\sqrt{1/7 + 1/12} = 1.28 < z_{crit} = 1.96$

9. *a)* no, $z_{calc} = (.549 - .424)/\sqrt{1/77} = 1.10 < z_{crit} = 1.96$
   b) no, $z_{calc} = (.549 - .31)/\sqrt{1/77 + 1/117} = 1.63 < z_{crit} = 1.96$

CHAPTER 10

Section A

2. *a)* $z_{Y'} = .4 \times 1.5 = .6$ *b)* $z_{Y'} = .4(-.9) = -.36$

4. $z'_{exam1} = -.709 z_{phobia}$; $r^2 = .503$ of the variance is accounted for

7. *a)* $b_{YX} = .6(4/3) = .8$ *b)* $a_{YX} = 32 - .8(69) = -23.2$ *c)* No, it is the predicted waist size of someone whose height is zero.
   d) $Y' = .8X - 23.2$

8. *a)* 34.4 *b)* 26.4 *c)* 71.5 inches

10. $r = \sqrt{.5} = .707$

12. *b*

Section B

1. *a)* $Y' = 3.56X + 10.4$ *b)* $Y' = 24.6$ *c)* about 14 years

3. *a)* $r^2 = -.669^2 = .45$ *b)* $Y' = -1.06X + 27.51$
   c) 27.51; it is the expected orientation score of someone who just entered the hospital.
   d) about 16.5 years

5. *a)* $Y' = -1.7X + 95.5$ *b)* 78.5 *c)* -2.65; no

7. *a)* $Y' = -4.02X + 225.4$ *d)* 169.12
   e) $169.12 \pm (3.707)(15.27) \times \sqrt{1 + .125 + (14 - 6)2/[7(13.2)]}$;
   $Y'_{lower} = 92.7$; $Y'_{upper} = 245.53$

9. *a)* $Y' = .425X + 2.95$ *b)* $Y' = .763X - 2.75$

10. *a)* $r = .87$ *b)* $r = .239$ *c)* the r in part b is much smaller because the effect of age has been removed.

Section C

1. *a)* $r_{pb} = \dfrac{\dfrac{190}{16} - (.5625)(19.375)}{(.496)(7.936)}$

   $= \dfrac{11.875 - 10.9}{3.936} = .248$

   b) $t = \dfrac{\sqrt{16 - 2}\,(.248)}{1\sqrt{-.248^2}} = \dfrac{3.74(.248)}{\sqrt{.9385}}$

   $= \dfrac{.9275}{.9688} = .96$

3. *a)* $r^2_{pb} = 10^2/(10^2 + 38) = .725$ *b)* $r^2_{pb} = .20$
   c) estimated $\omega^2 = (10^2 - 1)/(10^2 + 38 + 1) = .712$; estimated $\omega^2 = .198$

5. *a)* $\omega^2 = .8^2/(.8^2 + 4) = .138$ *b)* $\omega^2 = .059$
   c) $\mathbf{d} = 2$

CHAPTER 11

Section A

2. *a)* $t = (35 - 39)/\sqrt{(7^2 + 5^2)/10} = -1.47$
   b) No $(t_{crit} = 2.101)$

3. *a)* $t = \dfrac{35 - 39}{\sqrt{(7^2 + 5^2)/10 - 2(.1)(7)(5)/10}} = -.155$
   c) -1.74

5. *a)* $\overline{X} = 0$; $s = 4.62$ *b)* $\overline{X} = 1.44$; $s = 6.15$

8. *d*

Section B

2. *a)* reject null; $t_{calc} = 1.6/(1.075/\sqrt{10}) =$
   $4.71 > t_{crit} = 2.262$ *b)* The matched t is
   much larger than the independent groups t
   (1.18) from Exercise 7B9 because of the
   close matching.

4. *a)* $t = 8.5/(10.46/\sqrt{8}) = 2.30$;
   no, $t_{calc} < t_{crit} = 3.499$ *b)* $\mu_D = 8.5 \pm$
   3.499×3.70; $\mu_{lower} = -4.44$, $\mu_{upper} = 21.44$

5. $t_{calc} = \dfrac{4.83 - 3.33}{\sqrt{\dfrac{(11.07 + 8.17)}{6} - \dfrac{2(.91)(3.33)(2.858)}{6}}}$
   $= 2.645$

6. accept null; $t = .6/(7.575/\sqrt{10}) = .25$

8. reject null; $t = 1.67/(2.345/\sqrt{9}) = 2.13 >$
   $t_{crit} = 1.860$

Section C

1. *a)* $\delta = (.4)\sqrt{1/(1 - .5)}\sqrt{25/2} = 2$;
   power $= .52$ *b)* $\delta = 2.58$; power $=$ about .74
   c) power $=$ about .51

3. *a)* $\delta = 2.63 = .3\sqrt{1/(1 - .6)}\sqrt{N/2}$; $N = 62$
   b) $\delta = 3.25$; $N = 94$

5. *d*

CHAPTER 12

Section A

1. $MS_w = (100 + 225 + 144 + 121 + 100)/5 = 138$

3. Five groups with 17 subjects in each group

5. $F = 80 \times 18.67/150.6 = 9.92$

7. *a)* $F_{calc} = 8 \times 3.33/10.11 = 2.64$ *b)* $F_{crit} = 2.95$ *c)* The null hypothesis cannot be rejected.

9. *d*

Section B

1. $F_{calc} = 229.24/42.88 = 5.346$; $F_{.05}(4, 63) = 2.52$ (approximately); reject the null hypothesis.

3. *a)* $F = 0/365 = 0$ *c)* Because all three sample means are the same, you would have known that MS_{bet} must be zero, and that therefore F must be zero.

5. *a)* $F_{calc} = 5.06/6.37 = .79$ *b)* $F_{.01} = 6.36$
   c) accept null

d)

Source	SS	df	MS	F	p
Between groups	10.11	2	5.06	.794	>.05
Within groups	95.5	15	6.37		

8. *c)*

Source	SS	df	MS	F	p
Between groups	64.2	3	21.4	9.11	<.01
Within groups	37.6	16	2.35		

9. *a)* $F_{calc} = 35.51/9.74 = 3.64$ *b)* $F_{crit} = 2.45$
   c) reject null

11. *a)* $k = 5$, df $= 30$, $\phi = 2$ (approximately);
    therefore, power $= .93$
    b) $\phi = 1.6$ (approximately); therefore,
    power $= .76$ *c)* $\phi = 1.0$ (approximately);
    therefore, power $= .34$

13. *a)* $k = 2$; $n = 30$; $\phi = 2.46$; approximating
    df_W as 60 and interpolating between $\phi = 2.2$
    and 2.6 in the table, power is about .92

b) $k = 2$; min. $f = .15$; $n = 178$ yields $\phi = 2.0$, which yields power close to .8; so no more than about 180 participants should be used per group

15. a) reject homogeneity of variance assumption; $F_{calc} = .49/.16 = 3.063 > F_{.025}(11, 19) = 2.77$ (interpolated) b) no

Section C

1. a) $\eta^2 = 2(5)/[2(5) + 27] = .27; .15; .10$

3. a) $\eta^2 = 10.11/(10.11 + 95.5) = .096$; you cannot estimate ω^2 using Formula 12.14, because F is less than 1.

5. a) .0099 b) .059 c) .39 d) .5

CHAPTER 13

Section A

1. a) $5(5 - 1)/2 = 10$ c) 45

3. a) front vs. middle: $t = (34.3 - 28.7)/\sqrt{2 \times 150.6/80} = 2.886$; middle vs. back: $t = 1.49$; front vs. back: $t = 4.38$ b) $t_{crit} = 1.96$; front vs. middle and front vs. back exceed the critical t

5. a) front vs. middle: $t = 1.443$; middle vs. back: $t = .747$; front vs. back: $t = 2.19$ b) The t value is divided by 2. (In general, if n is divided by C, t will be divided by $\sqrt{C}$.)

8. a) $\alpha_{pc} = .00625$ b) $\alpha_{pc} = .0033$ c) $\alpha_{pc} = .00714$

10. b

Section B

2. LSD = 5.78; HSD = 7.02; a) only child vs. adult male b) same as part a c) LSD; e.g., female vs. male comes much closer to significance with LSD than with HSD d) no; the F ratio was not significant, so the follow-up t tests would not be protected

3. a) LSD = 3.26 b) HSD = $3.87\sqrt{10.11/8} = 4.35$ c) Each CI = $\bar{X}_i - \bar{X}_j \pm 4.35$; so for Marij/Amph, -3.35 to 5.35; for Marij/Val, -2.35 to 6.35; for Marij/Alc, -1.35 to 7.35; for Amph/Val, -1.35 to 7.35; for Amph/Alc, $-.35$ to 8.35; for Val/Alc, -3.35 to 5.35

5. a) athletes vs. controls: reject null, $t_{calc} = (14 - 11.57)/\sqrt{1.176(1/7 + 1/6)} = 4.03 > t_{crit} = 2.145$; athletes vs. musicians: reject null, $t_{calc} = 2.5 > t_{crit} = 2.145$; musicians vs.

controls: accept null, $t_{calc} = 1.0$ b) Instead of merely concluding that all three populations do not have the same mean, you can conclude that the athletes differ from both the musicians and controls, but that the latter two groups do not differ (significantly) from each other.

7. a) $\alpha_{pc} = .05/5 = .01$; $\bar{X}_3$ vs. $\bar{X}_6$: $t = (17.8 - 12)/\sqrt{9.74(1/8 + 1/7)} = 3.59 > t_{.01}(40) = 2.704$; only $\bar{X}_3$ b) harmonic $n = 7.48$; modLSD = 4.61; only $\bar{X}_3$ c) The Bonferroni t (i.e., $t_{.01}$) when multiplied by the square root of 2 is still considerably less than the q_{crit} used to get modLSD.

9. a) The complex comparison that makes the most sense is to compare the average of the animal phobias to the average of the nonanimal phobias: $L = 1/4\bar{X}_{rat} + 1/4\bar{X}_{dog} + 1/4\bar{X}_{spider} + 1/4\bar{X}_{snake} - 1/4\bar{X}_{party} - 1/4\bar{X}_{speak} - 1/4\bar{X}_{claus} - 1/4\bar{X}_{acro}$ b) $L = 9.5$; $SS_{contrast} = 16(9.5)^2/.5 = 2,888$; $F = 2,888/18.7 = 154.44$; $F_S = 7(2.09) = 14.63 < 154.44$ c) HSD = 4.714; the F for Scheffé for each pair = $8D^2/18.7$, where D is the difference of the two means being compared ($F_S = 14.63$). d) $8L^2/18.7 = 14.63$, $L = 5.85$. This value for L is larger than HSD (for a pairwise comparison, L is just the difference of the two means). Scheffé's test is more conservative, hence less powerful for pairwise comparisons.

11. One logical set of orthogonal contrasts consists of: "Girls in Training" (i.e., the average of Athletes and Musicians) vs.

Controls; and Athletes vs. Musicians. For Girls in Training vs. Controls: $L = 1/2(14) + 1/2(12.25) - 11.57 = 1.555$; so

$$SS_{contrast} = \frac{L^2}{\sum \frac{c_i^2}{n_i}} = \frac{-1.555^2}{\frac{1^2}{7} + \frac{.5^2}{6} + \frac{.5^2}{4}}$$

$$= \frac{2.418}{.143 + .0417 + .0625} = \frac{2.418}{.2472}$$

$$= 9.78$$

$F_{calc} = 9.78/1.176 = 8.32$; For Athletes vs. Musicians: $L = 1.75$; so

$$SS_{contrast} = \frac{1.75^2}{\frac{1^2}{6} + \frac{-1^2}{4}} = \frac{3.0625}{.167 + .25}$$

$$= \frac{3.0625}{.417} = 7.344$$

$F_{calc} = 7.344/1.176 = 6.24$. For planned comparisons, the critical $F = F_{.05}(1, 14) = 4.6$, so both of these contrasts would be significant.

Section C

1. a) $F_{ANOVA} = 5.58/4.0 = 1.39$, ns
   b) $F_{linear} = 19.44/4.0 = 4.86$, $p < .05$; $F_{quad} = 2.74/4.0 = .685$, ns c) The F for the linear

component is much larger than the F for the ANOVA; testing trend components usually leads to a more powerful test than the ordinary ANOVA for quantitative levels. d) The curve rises sharply and then levels off; largest number of reversals = 3.

3. a) $F_{regress} = 1102.1/6.47 = 170.3$
   b) $F_{ANOVA} = 87.2$; the F for regression is nearly twice as large as the F for the ANOVA. c) $SS_{linear} = 1102.1$ d) $SS_{quad} = 8.03$; $SS_{linear} + SS_{quad} = 1110.13$; this sum is the same as SS_{bet} (within rounding error); with only three groups, the linear and quadratic components are the only ones possible.

5. a) $F_{regress} = 116.58/.465 = 250.7$ b) $F_{linear} = 116.58/.28 = 415$; $F_{quad} = 1.43/.28 = 5.1$, $p < .05$; $F_{cubic} = 4.067/.28 = 14.5$, $p < .01$; $F_{quart} = .788/.28 = 2.8$, ns; the cubic is the highest trend component that is significant; the quintic is the highest component that could be tested. c) The curve rises sharply in the middle, but flattens out on both ends, suggesting a cubic component (even though there are no actual reversals, the curvature changes twice). d) the transformed IV produces the same SS as the quadratic component in part b.

CHAPTER 14

Section A

1. a) 10; 50 c) 18; 90

3. a)

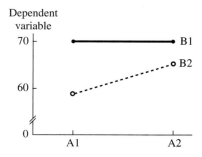

The marginal means are $A_1 = 65$, $A_2 = 67.5$, $B_1 = 70$, $B_2 = 62.5$ b) All three effects could be significant.

5. a)

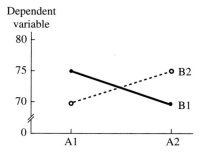

The marginal means are $A_1 = A_2 = B_1 = B_2 = 72.5$ b) The interaction might be significant, but neither of the main effects can be significant.

7. d) The lines on a graph of cell means will not be parallel.

Section B

2. *a)*

Source	SS	df	MS	F	p
Imagery group	2.45	1	2.45	1.53	>.05
Word type	22.05	1	22.05	13.8	<.01
Interaction	14.45	1	14.45	9.03	<.01
Within-cells	25.6	16	1.6		

b)

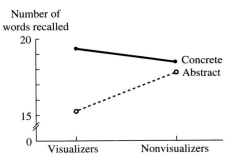

Number of words recalled

Whereas the recall of concrete words is slightly higher for visualizers than for nonvisualizers, the recall of abstract words is considerably higher for nonvisualizers than for visualizers. *c)* There was no significant difference between the two imagery groups, $F(1, 16) = 1.53, p > .05$, but recall was significantly higher for the concrete than the abstract condition, $F(1, 16) = 13.8, p < .01, \eta^2 = .34$. The latter main effect was qualified by a significant interaction, $F(1, 16) = 9.03, p < .01$.
d) η_p^2 for imagery group $= 2.45/28.05 = .087$; η_p^2 for word type $= 22.05/64.55 = .34$.

4. *a)*

Source	SS	df	MS	F	p
Difficulty	100	4	25	2.5	<.05
Reward	150	2	75	7.5	<.01
Interaction	40	8	5	.5	>.05
Within-cells	900	90	10		
Total	1190	104			

b) 7

c) η_p^2 for difficulty $= 10/100 = .1$; η_p^2 for reward $= 15/105 = .143$; η_p^2 for interaction $= 4/94 = .043$.

6. *a)*

Source	SS	df	MS	F	p
Type of therapy	119.7	2	59.85	43.7	<.01
Presence of drug	68.27	1	68.27	49.8	<.01
Interaction	13.63	2	6.82	4.97	<.05
Within-cells	74.0	54	1.37		

b) SS_{cont} for Psy/Group $= 10 (1.5^2)/4$
$= 5.625$; $F = 5.625/1.37 = 4.11 > F_{.05}(1, 54)$
$= 4.01$, so this 2×2 contrast is significant.
For Psy/Behav: $F = 1.6$, n.s.; for Group/Behav: $F = 9.65, p < .05$.
c) For therapy type:
$\omega_p^2 = [119.7 - 2(1.37)]/[119.7 + 58(1.37)]$
$= .587$; for presence of drug: ω_p^2
$= 66.9/149.1 = .45$.

8. *a)*

Source	SS	df	MS	F	p
Agreement	20.0	1	20.0	8.51	<.05
Intent	24.2	1	24.2	10.3	<.01
Interaction	20.0	1	20.0	8.51	<.05
Within-cells	37.6	16	2.35		

10. *a)*

Source	SS	df	MS	F	p
Between-Cells	127.275	9			
Therapy	79.935	1	79.935	19.98	<.01
Sessions	14.10	4	3.525	0.88	>.05
Interaction	33.24	4	8.31	2.08	<.01
Within-Cells	560.0	140	4.0		
Total	687.275	149			

b) $\eta_p^2 = 79.935/(79.935 + 560) = .125$; g_p
$= (7.78 - 6.32)/2 = .73$; $\omega_p^2 = (79.935 - 4)/$
$[79.935 + 149(4)] = .112$
c) largest 2×2 $L = (8.0 - 5.3)$
$- (6.8 - 7.0) = 2.9$; $SS_{cont} = 15(2.9)^2/4$
$= 31.5375$; $F_{cont} = 31.5375/4.0 = 7.88$
$> F_{.05}(1, 140) = 3.9$, so the largest 2×2
interaction contrast is significant at the .05
level; g or $d_c = \sqrt{(2 \times 7.88/15)} = 1.025$
d) No, because $7.88 < F_S = 4 \times 2.44 = 9.76$;
yes, because the interaction for the entire
two-way ANOVA was not significant, no
part of it will be significant by Scheffé's test.

Section C

3. *a)* $L_{ex+diet} = 661 - 525 = 141$; $L_{ex-only}$
$= 603 - 552 = 51$; $L_{inter} = 141 - 51 = 90$;

$SS_{cont} = 20(90)^2/140 = 1157.14; F_{cont}$
$= 1157.14/350 = 3.31 < F_{.05}(1,240) = 3.88$, so
not significant; $d_c = \sqrt{(2 \times 3.31/20)} = .575$.
b) $SS_{quad} = 20(-57)^2/84 = 773.57; SS_{cubic}$
$= 20(-59)^2/180 = 386.78$
c) $SS_{quad} = 20(-82)^2/84 = 1600.95; F_{quad}$
$= 4.57, p < .05; SS_{cubic} = 20(-54)^2/180$
$= 324; F_{cubic} = .93, p > .05$

5. a)

Source	SS	df	MS	F	p
Age	264.9	1	264.9	38.1	<.01
Languages	116.8	1	116.8	16.8	<.01
Interaction	1.51	1	1.51	.22	>.05
Within-cells	159.8	23	6.95		

b) The older children committed
significantly fewer errors than the younger
children, $F(1, 23) = 38.1, p < .01$, est. ω^2
$= .47$, and bilinguals committed significantly fewer errors than monolinguals,
$F(1, 23) = 16.8, p < .01$, est. $\omega^2 = .20$. The
interaction was very small, and did not
approach significance.

6. a) $F_{emotion} = 74.37/14.3 = 5.2, p < .01; F_{relax}$
$= 64.4/14.3 = 4.5, p < .05; F_{dark} = 31.6/14.3$
$= 2.21$, n.s.; $F_{emo \times rel} = 55.77/14.3 = 3.9$,
$p < .05; F_{emo \times dark} = 17.17/14.3 = 1.2$, n.s.;
$F_{rel \times dark} = 127.3/14.3 = 8.9, p < .01$;
$F_{emo \times rel \times dark} = 25.73/14.3 = 1.8$, n.s.
b) For emotion: $\eta_p^2 = .122$; for relaxation:
$\eta_p^2 = .039$; for darkness: $\eta_p^2 = .019$. Assuming
that a moderate effect size is about .06 (or
6%), only the main effect of emotion is
larger than moderate in size.

8. a)

Source	SS	df	MS	F	p
Drug	496.8	3	165.6	60.65	<.001
Therapy	32.28	2	16.14	5.91	<.01
Depression	36.55	1	6.55	13.4	<.01
Drug × Therapy	384.15	6	64.03	23.45	<.001
Therapy × Depression	31.89	2	15.95	5.84	<.05
Drug × Depression	20.26	3	6.75	2.47	>.05
Drug × Therapy × Depression	10.2	6	1.7	.62	>.05
Within-groups	131	48	2.73		

b) In both graphs, most of the interaction
involves the Antianxiety and placebo
conditions, and the amount of interaction
for the depressed patients is not very
different when compared to the
nondepressed patients. This similarity
suggests that the three-way interaction is
not large, which is consistent with the F
ratio being less than 1.0 for the three-way
interaction in this example.
c) You could begin by exploring the large
drug by therapy interaction, perhaps by
looking at the simple effect of therapy for
each drug. Then you could explore the
therapy by depression interaction, perhaps
by looking at the simple effect of depression
for each type of therapy.
d) $L = [(11.5 - 8.7) - (11 - 14)] - [(19 -$
$14.5) - (12 - 10)] = [2.8 - (-3)] - [4.5 -$
$2] = 5.8 - 2.5 = 3.3; SS_{contrast} = nL^2/\Sigma c^2 =$
$3(3.3)^2/8 = 32.67/8 = 4.08375; F_{contrast} =$
$4.08375/2.73 = 1.5$ (not significant, but
better than the overall three-way
interaction).

10. $2^5 - 1 = 32 - 1 = 31; 5(5 - 1)/2 = 20/2 = 10$.

CHAPTER 15

Section A

2.

Longest string
recalled

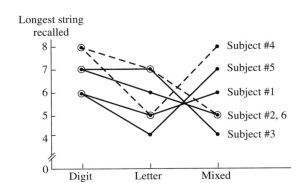

There is a fairly small amount of interaction between digit and letter, and a considerably larger amount of interaction between letter and mixed.

3. *a)* $SS_{inter} = 375 - 40 - 185 = 150$; $MS_{RM} = 40/5 = 8$; $MS_{inter} = 150/95 = 1.58$; $F = 8/1.58 = 5.06$ *b)* $SS_W = 185 + 150 = 335$; $MS_W = 335/114 = 2.94$; $F = 8/2.94 = 2.72$

6. *b)* $F = 36.56/15.206 = 2.404$; no, because $2.404 < F_{.05}(4, 28) = 2.701$.

8. c

10. b

Section B

1. *a)*

Source	SS	df	MS	F	p
Type of music	83.73	2	41.87	8.32	<.05
Residual	40.27	8	5.03		

b) Yes, because $8.32 > F_{.05}(1, 4) = 7.71$

3. *a)* The size of the imagined audience had a significant effect on systolic blood pressure, as shown by a one-way repeated measures ANOVA, $F(2, 22) = 7.07, p < .05$. *b)* Allow time for blood pressure to return to baseline level before presenting the next condition. *c)* $\eta^2_{RM} = 2(7.07)/[2(7.07) + 22] = .39$; yes, this effect appears to be quite large; could be misleading if the correlations among conditions are not as high in future experiments. *d)* LSD = 4,13; large differs significantly from both one and twenty

4. *a)* Yes

Source	SS	df	MS	F	p
Text	76.75	3	25.58	22.3	<.01
Residual	27.50	24	1.15		

b) Type I error *c)* Yes, because 22.3 > $F_{.01}(1, 8) = 11.26$. *d)* HSD = 1.39; text C differs significantly from each of the others

6. *a)* Accept null; $F_{calc} = 3.17/1.77 = 1.79 < F_{crit} = 4.10$ *b)* No, not assuming sphericity would only make it harder to reach significance. *c)* No; it looks like sphericity is unlikely to exist in the population. *d)* Digit/letter: $t = 3.16, p < .05$; digit/mixed: $t = 1.66$, ns; letter/mixed: $t < 1$

7. *a)*

Source	SS	df	MS	F	p
Time	178.1	3	59.36	30.7	<.01
Residual	40.66	21	1.94		

b) Est. $\omega^2 = [178.1 - 3(8.129)]/(405.72 + 8.129) = 153.7/413.8 = .3714$; yes, even without the boost of repeated measures, a great deal of variance is accounted for. *c)* Before/prior: $t = 2.76, p < .05$; prior/after: $t = 7.74, p < .01$; after/day after: $t = 2.16, p > .05$; Bonferroni $\alpha = .05/3 = .0167$— with this stricter criterion, only prior/after is significant at the .05 level.

9. *a)* $f = .327$; $f_{RM} = 1.363$; $\rho = .9425$ (this is an estimate of the average intercorrelation among the pairs of levels). *b)* It is very high.

Section C

1. *a)* Three, unless you wanted the orders to be digram-balanced, in which case you would have to use all six orders. *c)* Four subjects per order.

3. $t_{lin} = 30.83/7.91 = 3.9, p < .05$; $t_{quad} = 5.83/1.64 = 3.56, p < .05$; $t_{cubic} = .167/15.2 = .011$, n.s.

5. *a)* $F_{diet} = 201.55/29.57 = 6.82, p < .05$; $F_{time} = 105.6/11.01 = 9.1, p < .01$; $F_{diet \times time} = 8.67/7.67 = 1.13$, n.s. *b)* Adj. F for all tests = $F_{.05}(1, 5) = 6.61$; no conclusions will change, but F_{time} will no longer be significant at the .01 level.

7. *a)* $F_{sleep} = 90.75/7.0 = 12.96, p < .05$; $F_{dosage} = 153.44/17.5 = 8.77, p < .01$; $F_{sleep \times dosage} = 24.3/7.1 = 3.43, p < .05$; worst-

case $F = F_{.05}(1, 5) = 6.61$; the two main effects would still be significant (though this is not an issue for the sleep-condition factor, which has only two levels), but the interaction would not be. b) $t_{linear} =$

$25.67/6.74 = 3.81, p < .05; t_{quad} = 4.0/2.28 = 1.75$, n.s.; $t_{cubic} = 5.33/4.01 = 1.33$, n.s. c) $t_{sleep \times linear} = 10.33/8.62 = 1.20$, n.s.; $t_{sleep \times quad} = 3.8/2.08 = 1.82$, n.s.; $t_{sleep \times cubic} = 17.67/6.87 = 2.57, p = .05$.

CHAPTER 16

Section A

2.

Source	SS	df	MS	F	p
Groups	88	1	88	.64	>.05
Within-groups	1380	10	138		
Time	550	1	550	41.04	<.01
Group X time	2	1	2	.15	>.05
Subject X time	134	10	13.4		

4.

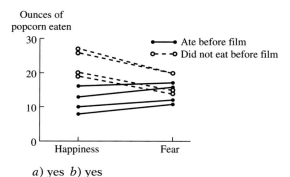

Ounces of popcorn eaten

— Ate before film
o - - o Did not eat before film

Happiness   Fear

a) yes b) yes

Section B

1. a)

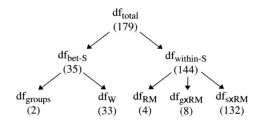

df_{total} (179)

df_{bet-S} (35)

$df_{within-S}$ (144)

df_{groups} (2) df_W (33) df_{RM} (4) df_{gXRM} (8) df_{sXRM} (132)

b) main effect of groups: $F_{.05}(2, 33) = 3.29$ (approximately); main effect of time: $F_{.05}(4, 132) = 2.45$ (approximately); interaction of group and time: $F_{.05}(8, 132) = 2.02$ (approximately).

3. a) $F_{load} = 210.25/23.46 = 8.96, p < .05$; $F_{emotion} = 12.25/.46 = 26.63, p < .01$; $F_{inter} = 64.0/.46 = 139.13, p < .01$

b)

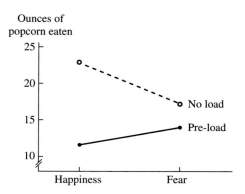

Ounces of popcorn eaten

No load
Pre-load

Happiness   Fear

The main effect of groups is due to the larger amounts eaten by the no-load group during both films. The main effect of emotion is due to a larger amount eaten during the happy film compared to the fear-evoking film. However, this main effect is qualified by a very large disordinal interaction. Happiness leads to more consumption only for the no-load group; the effect reverses for the pre-load group.

c) $t = [5.75 - (-2.25)]/\sqrt{(.917 + .917)/4} = 11.817; 11.817^2 = 139.64$, which is the F ratio for the interaction of the two factors.

5. b)

Source	SS	df	MS	F	p
Condition	2590.7	2	1295.4	71.85	<.01
Within-groups	108.2	6	18.03		
Difficulty level	315.8	3	105.3	124.9	<.01
Condition × difficulty	5.06	6	.84	1.0	>.05
Subject × difficulty	15.17	18	.84		

No. The F ratio for difficulty level is much larger than the adjusted critical F, and the interaction would remain not significant

c) For the feedback conditions: LSD = $2.447\sqrt{2(18.03)/12} = 4.24$; all three pairs of feedback conditions differ significantly; LSD is appropriate in this instance because the main effect was significant, and it has only three levels. For the difficulty levels: HSD = $4.0\sqrt{(.84)/9} = 1.22$; all possible pairs of difficulty levels differ significantly; HSD is appropriate when a main effect has more than three levels, whether or not the main effect is significant.

7. *a)* $F_{type} = .36/4.16 = .09$, n.s.; $F_{time} = 17.36/.69 = 25.0, p < .01$; $F_{inter} = .36/.69 = .52$, n.s. *b)* No. Time has only two levels, and therefore does not require the sphericity assumption. The interaction is not significant with the conventional df, so reducing the df would not change the conclusion in that case, either. *c)* The one-way ANOVA on the before-after difference scores should produce the same F ratio as the interaction in part a. The interaction term in a $2 \times J$ mixed ANOVA tests the difference among the J differences between the levels of the two-level factor.

9. *a)* $F_{sleep} = 90.75/17.63 = 5.15, p < .05$; $F_{dosage} = 153.44/12.29 = 12.48, p < .01$; $F_{sleep \times dosage} = 24.3/12.29 = 1.98$, n.s. *b)* The F ratio for the sleep condition main effect is much less in the mixed design, because its error term now includes subject-to-subject variability. However, because of the way the mixed design lumps together error terms from the two-way RM ANOVA, the F ratio for the dosage main effect is now larger, whereas the F for the interaction is smaller (both Fs now use the same error term).

Section C

2. *a)* $t_{sleep \times linear} = 10.33/12.39 = .834$, n.s.; $t_{sleep \times quad} = 3.8/3.5 = 1.09$, n.s.; $t_{sleep \times cubic} = 17.67/3.16 = 5.59, p < .05$. *b)* The numerators of the t tests in part a are the Ls for the interactions of the sleep factor with

the different trend components of the dosage factor; the corresponding SS's (found by using Formula 13.12) are $SS_{linear} = 16.017, SS_{quad} = 10.083$, and $SS_{cubic} = 46.817$. The sum of these three (orthogonal) SS components equals SS_{inter} from the two-way (RM or mixed) ANOVA.

4. *a)* $F_{RM} = 168.53/23.25 = 7.25, p < .01$; $F_{order} = 571.53/175.25 = 3.26$, ns; $F_{inter} = 26.44/23.25 = 1.14$, ns; F_{RM} is slightly higher than the $F(7.07)$ before adding order as a factor. *b)* Yes, there seems to be a good deal of position by treatment interaction. *c)* $SS_P = 69.56$; $SS_{P \times T} = 36.22$; for the main effect of position, $F_P = 1.5$, ns; for the position by treatment interaction, $F_{P \times T} = .78$, ns.

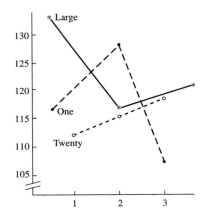

6. *a)* RE = $(23.83/23.25)(19/21)(25/23) = 1.008$ (a very slight increase in efficiency due to separating order effects). *b)* RE = $(1.77/1.92)(9/11)(13/11) = .89$ (in this case, the order effect is so small that the reduction in df outweighs the advantage of removing order effects resulting in a small decrease in efficiency relative to the ordinary RM ANOVA.

CHAPTER 17

Section A

1. *a)* $R = \sqrt{r_{yx_1}^2 + r_{yx_2}^2} = \sqrt{.2^2 + .4^2} = .45$; proportion of variance accounted for equals $R^2 = .45^2 = .2$

b) $R = \sqrt{.6^2 + .6^2} = \sqrt{.72} = .85$,

proportion of variance accounted for equals $R^2 = .85^2 = .72$ *c)* No, because $.7^2 + .8^2 = 1.13$, which represents an impossible correlation. *d)* .6, because $.8^2 + .6^2 = 1.0$

3. *a)* $z_{y'} = .275z_{x1} + .418z_{x2}$;

$R = \sqrt{.275(.4) + .418(.5)} = .565;$

$r_{y(1.2)} = .275\sqrt{1 - .3^2} = .262; r_{y(2.1)} = .40$

c) $z_{y'} = .5z_{x2}; R = .5; r_{y(1.2)} = 0;$

$r_{y(2.1)} = .5\sqrt{1 - .8^2} = .3$ e) The first predictor adds nothing to the variance explained by the second and is therefore not included in the multiple regression equation.

5. a) $R = \sqrt{\dfrac{.18^2 + .34^2 - 2(.18)(.34)(.12)}{1 - .12^2}} =$

$\sqrt{.135} =$ .3678 b) $.135 - .0324 = .1026; .135 - .1156 = .0194$ c) sr for IQ $= \sqrt{.0194} = .139; t = .95/.93 = 1.02$, n.s.; sr for interest $= \sqrt{.1026} = .32; t = 2.2/.93 = 2.36, p < .05$ but $p > .01$. d) $z_{y'} = .14z_{IQ} + .322z_{interest}$

6. a) $R^2 = .512(.6) + .276(.44) = .4286$ b) sr for HSG $= \sqrt{.235} = .485; t = 1.68/.756 = 2.22, p < .05; sr$ for IQ $= \sqrt{.0686} = .262; t = .908/.756 = 1.2$, n.s. c) partial $r^2 = .235/(.235 + .571) = .2914$; partial $r = \sqrt{.2914} = .54$ $t = .54\sqrt{(10 - 1 - 2)}/\sqrt{1 - .2914} = 1.43/.842 = 1.70$, ns

8. a) $z_{Y'} = -.176z_{min} - .176z_{miles}$ b) .016 c) $Y' = 69.6 - .059X_{Min} - .353X_{Miles}$ d) 64.9 bpm e) 62 bpm; because the club member is running the mean distance for the mean amount of time, the prediction would be the mean heart rate of the group.

9. a) $r_{1y} = .6, r_{2y} = .8, r_{12} = .5;$ $r_{y1.2} = (r_{1y} - r_{2y}r_{12})/\sqrt{(1 - r_{2y}^2)(1 - r_{12}^2)} =$ $(.6 - .4)/\sqrt{(.36)(.75)} = .2/.52 = .385;$

$t = .385\sqrt{(20 - 1 - 2)}/\sqrt{1 - .1482} = 1.587/.9229 = 1.72$ (not significant) b) $r_{y1.2} = .093; t = .383/.9956 = .385$ (not significant) c) Part of the correlation between insight and symptom improvement is due to the correlation of each of those with therapy years; finding the partial correlation removes that part of the original correlation.

Section B

1. a) $F = 47(.1353)/2(1 - .1353) = 6.359/1.729 = 3.68; F_{.05}(2,47) = 3.20$, so R is significant b) for IQ: $F = 47(.0194)/.8647 = 1.05$; for interest: $F = 47(.1026)/.8647 = 5.58$; $F_{.05}(1, 47) = 4.05$, so sr for interest is significant, but not for IQ c) IQ is not adding significantly to the prediction

equation, so it would be preferable to use interest alone.

3. Use Formula 17.12 to change R^2 to eta squared: $F = 108(.22)/[3(.78)] = 23.76/2.34 = 10.2$ (note: $P = k - 1$); $F_{.05}(3, 108)$ is about 2.69, so the one-way ANOVA is significant at the .05 level.

5. a) $R = [.3^2 + .2^2 - 2(.3)(.2)(-.3)]/[1 - (-.3)^2] = \sqrt{.166/.91} = .427$ $F = (N - P - 1)R^2/[P(1 - R_2)] =$ b) $27(.1824)/[2(.8176)] = 4.925/1.64 = 3.0;$ $F_{.05}(2, 27) = 3.35$, so R is not significant at the .05 level. c) $sr_{LPAR} = (r_{1y} - r_{2y}r_{12})/\sqrt{1 - r_{12}^2} = (.3 - .2(-.3))/\sqrt{.91} = .36/.954 = .377;$ $F = (N - P - 1)sr^2/(1 - R^2) = 27(.377^2)/.8176 = 3.837/.8176 = 4.69;$ $F_{.05}(1, 27) = 4.21$, so sr for LPAR is significant at the .05 level. Similarly, sr for AMEX $= .304$, and $F = 3.05$, so the sr for AMEX is not significant. d) $F = 26(.26)/3(.74) = 6.76/2.22 = 3.05;$ $F_{.05}(3, 26) = 2.98$, so this R^2 is significant at the .05 level. e) $F = 25(.28)/4(.72) = 7.0/2.88 = 2.43; F_{.05}(4, 25) = 2.76$, ns; sr^2 for fourth predictor is $.28 - .26 = .02$, so $sr = .1414$. Adding the fourth predictor added so little variance that it reduced the F ratio for the entire model below significance. f) adjusted $R =$ $\sqrt{R^2 - [P(1 - R^2)/(N - P - 1)]}$ $= \sqrt{.28 - [4(1 - .28)/25]}$ $= \sqrt{.28 - .1152} = .406$

7. a) $R = \sqrt{[.46^2 + .34^2 - 2(.46)(.34)(.18)]/(1 - 18^2)} = \sqrt{.2713/.9676} = .53; F = 27(.28)/2(.72) = 7.56/1.44 = 5.25; p < .05$ b) $sr_{genetic} = \sqrt{.28 - .1156} = \sqrt{.1644} = .405; sr_{health} = \sqrt{.28 - .2116} = \sqrt{.0684} = .262; F_{genetic} = 27(.1644)/(.72) = 7.165; F_{health} = 27(.0684)/(.72) = 2.565; F_{.05}(1, 27) = 4.21$, so $F_{genetic}$ contributes significantly but F_{health} does not. c) $F = (N - P - 1)(R_P^2 - R_{P-K}^2)/[K(1 - R_P^2)] = 24(.0684)/3(.72) = 1.642/2.16 = .76$ d) adjusted $R = \sqrt{.28 - [5(.72)/24]} = \sqrt{.13} = .361$

9. a) adjusted $R = \sqrt{.64 - 8(.36)/71} = \sqrt{.64 - .041} = .774$ b) $F = 71(.03)/2(.36) = 2.13/.72 = 2.96; F_{.05}(2, 71) = 3.12$ (approximately), so this block is not adding significantly to the other predictors.

c) $(8 + 1)/80 = .1125; .3375$
d) IQ—it has the highest validity
e) [IQ, SAT, SWA] [SNACH, STHRS, SMHE] [PSUP, PLED] The first group contains intelligence measures of the student; the second group contains measures of student motivation; and the third group contains measures of parental support. f) IQ, STHRS, PSUP One is chosen from each of the three groups in part e. The one chosen in each case has the highest validity, and lowest intercorrelation with the other measures. g) No, although the model would contain only three predictors, they were chosen from a larger set to give the highest possible F. The significance test should be adjusted for this.

Section C

1. a) $r = .917$; the points fall on a curve rather than a straight line. b) $r = .981$; taking the square-root transformed the scatterplot from a curve to a nearly straight line.

3. $8 + 8 + 8 * 7/2 = 16 + 28 = 44$

5. a) The point-biserial r between speed and group is: $r_{speed} = \sqrt{t^2/(t^2 + df)} = \sqrt{2.25/20.25} = .333$. Similarly, $r_{accu} = \sqrt{2.89/20.89} = .372$; $R^2 = [.111 + .138 - 2(.333)(.372)(-.5)]/[1 - (-.5)^2] = .373/.75 = .498$ b) $F = 17(.498)/[2(.502)] = 8.466/1.004 = 8.43$; $p < .01$

7. $r = .6 * .3 = .18$

CHAPTER 18

Section A

2. a)

	X_1	X_2	X_3	X_4
Elavil	1	0	0	0
Prozac	0	1	0	0
Zoloft	0	0	1	0
Celexa	0	0	0	1
Placebo	0	0	0	0

b)

	X_1	X_2	X_3	X_4
Japan	1	0	0	0
Greece	0	1	0	0
India	0	0	1	0
Mexico	0	0	0	1
Egypt	−1	−1	−1	−1

c) $Y' = 5X_1 - 2X_2 - 4X_3 + X_4 + 11$

4. a)

	X_1	X_2	X_3	X_4	X_5
Liberal/NW	1	0	1	0	0
Conservative/NW	0	1	1	0	0
MOR/NW	−1	−1	1	0	0
Liberal/SW	1	0	0	1	0
Conservative/SW	0	1	0	1	0
MOR/SW	−1	−1	0	1	0
Liberal/NE	1	0	0	0	1
Conservative/NE	0	1	0	0	1
MOR/NE	−1	−1	0	0	1
Liberal/SE	1	0	−1	−1	−1
Conservative/SE	0	1	−1	−1	−1
MOR/SE	−1	−1	−1	−1	−1

X_1 and X_2 code for political attitude, while X_3, X_4, and X_5 code for region.

b)

	X_1	X_2	X_3	X_4	X_5
Liberal/urban	1	0	1	1	0
Conservative/urban	0	1	1	0	1
MOR/urban	−1	−1	1	−1	−1
Liberal/rural	1	0	−1	−1	0
Conservative/rural	0	1	−1	0	−1
MOR/rural	−1	−1	−1	1	1

X_1 and X_2 code for political attitude, X_3 codes for region, while X_4 and X_5 code for the interaction.

6. a) $2^4 - 1 = 16 - 1 = 15$; $2 \times 3 \times 4 \times 5 - 1 = 120 - 1 = 119$ b) $Y = \mu + \alpha_i + \beta_j + \gamma_k + \delta_l + \alpha_i\beta_j + \alpha_i\gamma_k + \alpha_i\delta_l + \beta_j\gamma_k + \beta_j\delta_l + \gamma_k\delta_l + \alpha_i\beta_j\gamma_k + \alpha_i\beta_j\delta_l + \alpha_i\gamma_k\delta_l + \beta_j\gamma_k\delta_l + \alpha_i\beta_j\gamma_k\delta_l + \epsilon_{ijklm}$

8. a) $F_{SE} = 80/50 = 1.6$; $F_{diff} = 2.0$; $F_{inter} = .8$
b) $F_{SE} = 170/50 = 3.4$; $F_{diff} = 3.1$; $F_{inter} = .8$
c) $F_{SE} = 290/50 = 5.8$; $F_{diff} = 3.1$; $F_{inter} = .8$

Section B

2. a) $r_p = .532$; $F = 1151.4/162.75 = 7.07$, $p < .01$ b) $F = 718.5/151.5 = 4.74$, $p < .05$; these data are not consistent with the homogeneity of regression assumption. c) $b_p = 1.065$; adjusted means: $\overline{Y}_I = 84.9$, $\overline{Y}_{II} = 75.4$, $\overline{Y}_{III} = 90.2$; $CMSA_{error} = 174.7$; $LSD = 6.79$; methods II and II differ significantly at the .05 level.

4. *a)* $r_p = .562$; $F_{ANOVA} = 99.2/26.5 = 3.74$, $p < .05$; $F_{ANCOVA} = 129.7/19.2 = 6.76$, $p < .01$
   b) $F = 139.9/3.1 = 45.13$, $p < .01$; these data are not consistent with the homogeneity of regression assumption. *c)* $b_p = .521$; adjusted means: $\overline{Y}_{conc} = 8.09$, $\overline{Y}_{abs} = 11.05$, $\overline{Y}_{no} = 16.71$; $CMSA_{error} = 19.63$; LSD = 5.0; no hint differs significantly from both concrete and abstract at the .05 level. *d)* Initial differences on the covariate ran counter to the differences on the DV, so adjusting for the covariate increased the spread of the means and helped to increase the ANCOVA F relative to the original ANOVA F.

6. *a)* $r_p = .87$; $F_{ANCOVA} = 35.03/5.11 = 6.86$, $p < .01$ *b)* $F_{inter} = 18.0/2.64 = 6.82$, $p < .01$ *c)* $b_p = .877$; adjusted means: $\overline{Y}_{cluster} = 18.5$, $\overline{Y}_{chain} = 18.4$, $\overline{Y}_{visual} = 21.2$, $\overline{Y}_{control} = 16.6$; $CMSA_{error} = 5.16$; HSD = 2.74; the visual condition differs significantly from both the chaining and control conditions at the .05 level. *d)* b_p; because the pooled regression slope is near 1.0, the residuals from the predictions are similar to simple before-after difference scores.

8. $F_{ANCOVA} = .7/1.3 = .54$; the two F ratios are similar because the relevant regression slopes are not very far from 1.0.

Section C

1. *a)* $T^2 = 3.8^2 = 14.44$
   b) $F = [14.44 \times (24 - 3 - 1)]/[3(22)] = 4.376 > F_{.05}(3, 20) = 3.1$, so T^2 is significant at the .05 level.
   c) $MD^2 = T^2 / n/2 = 14.44/6 = 2.407$; $MD = \sqrt{2.407} = 1.55$
   d) $F = 14.44 \times (30 - 3 - 1)/3 (28) = 4.47$; harmonic mean of 10 and 20 = 13.33, $MD^2 = 14.44 / 13.33/2 = 2.167$; $MD = \sqrt{2.167} = 1.47$

3. *a)* R^2 (the sum of the two r_{pb}^2s) = $.184 + .168 = .352$; $T^2 = 16 \times [.352/(1 - .352)] = 8.69$
   b) $MD^2 = 8.69/4.5 = 1.93$; $MD = \sqrt{1.93} = 1.39$
   c) $\Lambda = 16 / (8.69 + 16) = .648$
   d) $F = (15/32) \times 8.69 = 4.07 > F_{.05}(2, 15) = 3.68$, so T^2 is significant at the .05 level. As in multiple regression with uncorrelated predictors, each *DV* captures a different part of the variance between the two groups; together the two *DV*s account for much more variance than either one alone.

5. *a)* df = $(6 - 1)(12 - 1) = 5 \times 11 = 55$
   b) df = $12 - 6 + 1 = 7$

CHAPTER 19

Section A

2. *a)* $p = .0269 + .0054 + .0005 = .0328$
   b) $.0328 \times 2 = .0656$ *c)* yes; no

4. *a)* No, $z = (58 - 50)/\sqrt{100(.5)(.5)} = 1.6$
   b) $(X - 50)/\sqrt{100(.5)(.5)} = 1.96$, $X = 59.8$, so Johnny would have to get 60 questions right

6. *a)* $z = [(27 - (120 \times .15)]/\sqrt{120(.15)(.85)} = 2.30$; yes, at the .05 level (two-tailed)
   b) $z = [(108 - (480 \times .15)]/\sqrt{480 \times .15 \times .85} = 4.60$ *c)* It is half as large. If N is multiplied by C, but the proportion obtained in the P category remains the same, z is multiplied by $\sqrt{C}$.

8. No, $z = (.37 - .30)/\sqrt{(.3 \times .7)/80} = 1.37$

Section B

1. From Table A.13 ($N = 6$), $p = (.0938 + .0156) \times 2 = .1094 \times 2 = .2188$; $p > .05$, so accept null; no, even if you used a .05, one-tailed test, you would have rejected the null hypothesis for Exercise 11B5, but not for the sign test. The sign test throws away much of the information in the data, and therefore has considerably less power than the matched t test.

3. No, $p = .2188 > .05$ (see calculation for Exercise 19B1)

5. From Table A.13 [$N = 9 - 1$ (tie) = 8], $p = (.2188 + .1094 + .0312 + .0039) \times 2 = .3633 \times 2 = .727$; accept null

7. *a)* Yes, $z = (|32 - 25| - .5)/\sqrt{50 \times .5 \times .5} = 1.88 > z_{crit} = 1.645$ *b)* no

Section C

1. a) 20/52 = .385 b) 26/52 + 13/52 = .75
   c) 26/52 + 20/52 − 10/52 = 36/52 = .69

3. a) (13/52)(12/51) = .0588
   b) (13/52)(13/51) × 2 = .127 c)
   (20/52)(19/51) = .143

5. a) $_{15}C_3$ = 15!/(3!12!) = 455 b) $_{15}C_5$ =
   15!/(5!10!) = 3,003

7. a) p(8) = $_{10}C_8 \cdot 2^8 \cdot 8^2$ = .0000737
   b) p(8) + p(9) + p(10) = .0000737
   + .000004 + .0000001 = .0000778

CHAPTER 20

Section A

1. a) df = 7 b) 14.07, 18.48

3. a) Each of the three products will be
   selected one-third of the time.
   b) no, χ^2_{calc} = $(27 − 22)^2/22 + (15 − 22)^2/22 +$
   $(24 − 22)^2/22$ = 3.55 < χ^2_{crit} = 5.99

5. χ^2_{calc} = $(30 − 20)^2/20 + (17 − 20)^2/20 + (27 −$
   $20)^2/20 + (14 − 20)^2/20 + (13 − 20)^2/20 +$
   $(19 − 20)^2/20$ = 12.2 > $\chi^2_{.05}$ = 11.07 (but not
   significant at the .01 level)

7. reject null; χ^2_{calc} = 20 > $\chi^2_{.05}$ = 7.81

Section B

2. a) accept null; χ^2_{calc} = $(2 − 4)^2/4 + (4 − 2)^2/2 +$
   $(18 − 16)^2/16 + (6 − 8)^2/8$ = 3.75 < $\chi^2_{.05}$ =
   3.84 c) No, because the experimenter had
   no control over the diets of the children.

3. a) accept null; χ^2_{calc} = .44 < $\chi^2_{.05}$ = 3.84
   b) possible Type II error

5. χ^2_{calc} = $(8 − 6.48)^2/6.48 + (12 − 11.88)^2/11.88$
   $+ (7 − 8.64)^2/8.64 + (4 − 5.52)^2/5.52 +$
   $(10 − 10.12)^2/10.12 + (9 − 7.36)^2/7.36$ = 1.45;
   no, the null hypothesis cannot be rejected.

7. reject null; χ^2_{calc} = 6.65 > $\chi^2_{.05}(1)$ = 3.84

9. accept null; χ^2_{calc} = 13.54 < $\chi^2_{.01}(8)$ = 20.09

Section C

1. a) $\phi = \sqrt{(3.75/30)}$ = .354 c) The
   relationship is moderately strong.

3. a) $\phi = \sqrt{(6.65/30)}$ = .471 b) cross-product
   ratio = (10 × 12)/(3 × 5) = 8 c) It is not
   extremely accurate, but a moderate amount
   of accuracy has been demonstrated.

5. a) $\phi_c = \sqrt{(13.54)/[100(3 − 1)]}$ = .26;
   $C = \sqrt{(13.54)/(13.54 + 100)}$ = .345
   b) The relationship is not very strong.

7. κ = (70 − 18.5)/(100 − 18.5) = .632

CHAPTER 21

Section A

2. S_R = 16(17)/2 = 136

4. a) ΣR_A = 90; ΣR_B = 100 b) U_A
   = 9(10) + 9(10)/2 − 90 = 45 = U_B = U

6. b) Sum of positives = 28; sum of negatives =
   17; yes, because the positive differences
   have the larger sum of ranks (but the
   difference is not a dramatic one).

Section B

1. If steroid users are labeled A, then n_A = 5,
   n_B = 15, and ΣR_A = 24. a) For a two-tailed

test at α = .05, the critical value from Table
A.15 is 29 (use column for .025); 24 is less
than 29, so the null hypothesis can be
rejected at this level. b) For a two-tailed test
at α = .01, the critical value is 23 (use
column for .005), so the null hypothesis
cannot be rejected at this level; however, if a
one-tailed test were justified, the critical
value would be 26, and the null hypothesis
could be rejected at this level.

3. a) t = 2.24, p < .05, one-tailed, but not two-
   tailed; r_{pb} = .621 b) If left damage is labeled
   A, then ΣR_A = 13 (assigning a rank of 1 to
   the smallest number). For a .05 two-tailed
   test the relevant critical value is 12, so the

null hypothesis cannot be rejected; but for a one-tailed test, the critical value is 13, so the null hypothesis can be rejected if a one-tailed test can be justified; $r_G = 2(3.25 - 7)/10 = -.75$. c) The results for parts a and b suggest that the two tests are very similar in this case, and the effect sizes are similar, as well (the signs attached to r_{pb} and r_G are arbitrary).

5. a) $t = (21.2 - 7.4)/\sqrt{(14.7^2 + 3.5^2)/5} = 13.8/\sqrt{45.67} = 2.04, p > .05; g = 13.8/10.7 = 1.29$. b) $n_A = n_B = 5$. If you assign lower ranks to smaller amounts of time and arbitrarily label the drug group as A, $\Sigma R_A = 5 + 7 + 8 + 9 + 10 = 39$. Because 39 is larger than the appropriate critical value (38) from Table A.15, the null hypothesis can be rejected at the .05 level (two-tailed); $r_G = 2(7.8 - 3.2)/10 = .92$. c) Yes. The drug group contains an extreme outlier, which adversely affects the t test by inflating the variance of one of the groups. Ranking minimizes the impact of the outlier, resulting in a significant Mann-Whitney test. d) Yes. r_G is close to its maximum value of 1.0, and a g that is greater than 1.2 is considered quite large.

7. a) $N = 12; T = 1 + 2 + 3 = 6 < T_{crit}$ (13), so $p < .05$, two-tailed;
b) $r_c = 2(72 - 6)/12(13) = 132/156 = .846$; yes, r_c is quite large.

9. a) $T = 24 > T_{crit}$ (10), so the null hypothesis cannot be rejected
b) $z = [24 - .25(10)(11)]/\sqrt{[10(11)(21)]/24} =$

$(24 - 27.5)/\sqrt{96.25} = -3.5/9.81 = -.36$; $t = .25$ for Exercise 11B6, which is similar to the test statistic calculated for the Wilcoxon test.
c) $r_s = .719$; yes, because $.719 > .632$.

11. $r_s = -.743$; it is a little larger in magnitude than the Pearson r $(-.718)$.

Section C

1. a) $H = 12/9(10)[36/3 + 225/3 + 576/3] - 3(10) = 7.2$ b) $7.2 > \chi^2_{.05}(2) = 5.99$, so reject null hypothesis

3. $H = 12/210[1600/6 + 1640.25/4 + 600.25/4] - 3(15) = .05714[826.8] - 45 = 47.25 - 45 = 2.25 < \chi^2_{.05}(2) = 5.99$, so cannot reject the null hypothesis. $\eta^2_H = (2.25 - 3 + 1)/14 - 3 = .25/11 = .023 << \eta^2_{ANOVA} = .252$. In this case, it looks like the loss of information involved in ranking the data greatly reduces the apparent effect size, which in turn suggests a considerable reduction in power.

5. $F_r = 12/[(8)(4)(5)] (9.5^2 + 14.5^2 + 30^2 + 26^2) - (3)(8)(5) = 20.74; 20.74 > \chi^2_{.05}(3) = 7.81$; reject null hypothesis. For Day Before vs. Prior to Film: dropping the one zero difference, $T = 2 = T_{crit}$, so the difference is (just barely) significant at the .05 level; for Prior to Film vs. After Film: $T = 0, p < .05$; for After Film vs. Day After: assigning one zero difference to be positive, and the other negative, $T = 4 > T_{crit} = 3$, so cannot reject the null hypothesis.

Abelson, R. P. (1997). On the surprising longevity of flogged horses: Why there is a case for the significance test. *Psychological Science, 8,* 12–15.

Abelson, R. P., & Prentice, D. A. (1997). Contrasts tests of interaction hypotheses. *Psychological Methods, 2,* 315–328.

Algina, J., & Keselman, H. J. (1997). Detecting repeated measures effects with univariate and multivariate statistics. *Psychological Methods, 2,* 208–218.

Algina, J., Keselman, H. J., & Penfield, R. D. (2005). An alternative to Cohen's standardized mean difference effect size: A robust parameter and confidence interval in the two independent groups case. *Psychological Methods, 10,* 317–328.

American Psychological Association (APA). (2001). *Publication manual of the American Psychological Association* (5th ed.). Washington, DC: Author.

Behrens, J. T. (1997). Principles and procedures of exploratory data analysis. *Psychological Methods, 2,* 131–160.

Berkson, J. (1938). Some difficulties of interpretation encountered in the application of the chi-square test. *Journal of the American Statistical Association, 33,* 526–542.

Blouin, D. C., & Riopelle, A. J. (2005). On confidence intervals for within-subjects designs. *Psychological Methods, 10,* 397–412.

Boik, R. J. (1979). Interactions, partial interactions, and interaction contrasts in the analysis of variance. *Psychological Bulletin, 86,* 1084–1089.

Brown, M. B., & Forsythe, A. B. (1974). The ANOVA and multiple comparisons for data with heterogeneous variances. *Biometrics, 30,* 719–724.

Bryant, J. L., & Paulson, A. S. (1976). An extension of Tukey's method of multiple comparisons to experimental designs with random concomitant variables. *Biometrika, 63,* 631–638.

Carver, R. P. (1978). The case against statistical significance testing. *Harvard Educational Review, 48,* 378–399.

Chow, S. L. (1998). Précis of statistical significance: Rationale, validity, and utility. *Behavioral and Brain Sciences, 21,* 169–239.

Cicchetti, D. V. (1972). Extension of multiple range tests to interaction tables in the analysis of variance: A rapid approximate solution. *Psychological Bulletin, 77,* 405–408.

Clinch, J. J., & Keselman, H. J. (1982). Parametric alternatives to the analysis of variance. *Journal of Educational Statistics, 7,* 207–214.

Cohen, B. H. (2002). Calculating a factorial ANOVA from means and standard deviations. *Understanding Statistics, 1,* 191–203.

Cohen, J. (1960). A coefficient of agreement for nominal scales. *Educational and Psychological Measurement, 20,* 37–46.

Cohen, J. (1973). Eta-squared and partial eta-squared in fixed factor ANOVA designs. *Educational and Psychological Measurement, 33,* 107–112.

Cohen, J. (1988). *Statistical power analysis for the behavioral sciences* (2nd ed.). Hillsdale, NJ: Erlbaum.

Cohen, J. (1994). The earth is round ($p < .05$). *American Psychologist, 49,* 997–1003.

Cole, D. A., Maxwell, S. E., Arvey, R., & Salas, E. (1994). How the power of MANOVA can both increase and decrease as a function of the intercorrelations among the dependent variables. *Psychological Bulletin, 115,* 465–474

Collins, L. M., & Sayer, A. G. (Eds.). (2001). *New methods for the analysis of change.* Washington, DC: American Psychological Association.

Conover, W. J. (1974). Some reasons for not using the Yates continuity correction on 2×2 contingency tables. *Journal of the American Statistical Association, 69,* 374–382.

Cortina, J. M., & Dunlap, W. P. (1997). On the logic and purpose of significance testing. *Psychological Methods, 2*, 161–172.

Cowles, M. (1989). *Statistics in psychology: An historical perspective.* Hillsdale, NJ: Erlbaum.

Cramér, H. (1946). *Mathematical methods of statistics.* Princeton, NJ: Princeton University Press.

Cumming, G., & Maillardet, R. (2006). Confidence intervals and replication: Where will the next mean fall? *Psychological Methods, 11*, 217–226.

Darlington, R. B. (1990). *Regression and linear models.* New York: McGraw-Hill.

Davidson, E. S., & Schenk, S. (1994). Variability in subjective responses to marijuana: Initial experiences of college students. *Addictive Behaviors, 19*, 531–538.

Davidson, M. L. (1972). Univariate versus multivariate tests in repeated measures experiments. *Psychological Bulletin, 77*, 446–452.

DeCarlo, L. T. (1997). On the meaning and use of kurtosis. *Psychological Methods, 2*, 292–307.

Denny, E. B., & Hunt, R. R. (1992). Affective valence and memory in depression: Dissociation of recall and fragment completion. *Journal of Abnormal Psychology, 101*, 575–580.

Dixon, P. (2003). The *p*-value fallacy and how to avoid it. *Canadian Journal of Experimental Psychology, 57*, 189–202.

Driskell, J. E., & Salas, E. (1991). Group decision-making under stress. *Journal of Applied Psychology, 76*, 473–478.

Dunn, O. J. (1961). Multiple comparisons among means. *Journal of the American Statistical Association, 56*, 52–64.

Dunnett, C. W. (1964). New tables for multiple comparisons with a control. *Biometrics, 20*, 482–491.

Ekman, P. (Ed.). (1982). *Emotion in the human face* (2nd ed.). London: Cambridge University Press.

Feldt, L. S. (1958). A comparison of the precision of three experimental designs employing a concomitant variable. *Psychometrika, 23*, 335–353.

Fisher, R. A. (1951). *The design of experiments* (6th ed.). Edinburgh: Oliver and Boyd.

Fisher, R. A. (1955). Statistical methods and scientific induction. *Journal of the Royal Statistical Society, Series B, 17*, 69–78.

Fisher, R. A. (1970). *Statistical methods for research workers* (14th ed.). Edinburgh: Oliver and Boyd.

Friedman, M. (1937). The use of ranks to avoid the assumption of normality implicit in the analysis of variance. *Journal of the American Statistical Association, 32*, 675–701.

Games, P. A., & Howell, J. F. (1976). Pairwise multiple comparison procedures with unequal *n*'s and/or variances: A Monte Carlo study. *Journal of Educational Statistics, 1*, 113–125.

Geisser, S., & Greenhouse, S. W. (1958). An extension of Box's results on the use of the *F* distribution in multivariate analysis. *Annals of Mathematical Statistics, 29*, 885–891.

Gillett, R. (2003). The metric comparability of meta-analytic effect-size estimators from factorial designs. *Psychological Methods, 8*, 419–433.

Glass, G. V. (1965). A ranking variable analogue of biserial correlation: Implications for short-cut item analysis. *Journal of Educational Measurement, 2*, 91–95.

Glass, G. V. (1976). Primary, secondary, and metaanalysis research. *Educational Researcher, 5*, 3–8.

Greenhouse, S. W., & Geisser, S. (1959). On methods in the analysis of profile data. *Psychometrika, 24,* 95–112.

Greenwald, A. G., Gonzalez, R., Harris, R. J., & Guthrie, D. (1996). Effect sizes and *p* values: What should be reported and what should be replicated? *Psychophysiology, 33,* 175–183.

Greenwood, P. E., & Nikulin, M. S. (1996). *A guide to chi-squared testing.* New York: Wiley.

Griggs, R. A., & Cox, J. R. (1982). The elusive thematic-materials effect in Wason's selection task. *British Journal of Psychology, 73,* 407–420.

Haber, M. (1980). A comparison of some continuity corrections for the chi-squared test on 2 × 2 tables. *Journal of the American Statistical Association, 75,* 510–515.

Harlow, L. L., Mulaik, S. A., & Steiger, J. H. (1997). *What if there were no significance tests?* Hillsdale, NJ: Erlbaum.

Harris, R. J. (1985). *A primer of multivariate statistics* (2nd ed.). Orlando, FL: Academic Press.

Harte, J. L., & Eifert, G. H. (1995). The effects of running, environment, and attentional focus on athletes' catecholamine and cortisol levels and mood. *Psychophysiology, 32,* 49–54.

Hartley, H. O. (1950). The maximum *F*-ratio as a short-cut test for heterogeneity of variance. *Biometrika, 37,* 308–312.

Hays, W. L. (1994). *Statistics* (5th ed.). New York: Harcourt Brace.

Hayter, A. J. (1986). The maximum familywise error rate of Fisher's least significant difference test. *Journal of the American Statistical Association, 81,* 1000–1004.

Hedges, L. V. (1981). Distribution theory for Glass's estimator of effect size and related estimators. *Journal of Educational Statistics, 6,* 107–128.

Hedges, L. V. (1982). Estimation of effect size from a series of independent experiments. *Psychological Bulletin, 92,* 490–499.

Hedges, L. V., Cooper, H., & Bushman, B. J. (1992). Testing the null hypothesis in meta-analysis: A comparison of combined probability and confidence interval procedures. *Psychological Bulletin, 111,* 188–194.

Hedges, L. V., & Olkin, I. (1985). *Statistical methods for meta-analysis.* San Diego, CA: Academic Press.

Hochberg, Y. (1988). A sharper Bonferroni procedure for multiple tests of significance. *Biometrika, 75,* 800–802.

Hogg, R. V., & Craig, A. T. (1995). *Introduction to mathematical statistics* (5th ed.). Englewood Cliffs, NJ: Prentice Hall.

Holm, S. (1979). A simple sequentially rejective multiple test procedure. *Scandinavian Journal of Statistics, 6,* 65–70.

Howell, D. C. (2007). Statistical methods for psychology (6th ed.). Belmont, CA: Wadsworth.

Huck, S. W., & McLean, R. A. (1975). Using a repeated measures ANOVA to analyze the data from a pretest–posttest design: A potentially confusing task. *Psychological Bulletin, 82,* 511–518.

Huitema, B. E. (1980). *The analysis of covariance and alternatives.* New York: Wiley.

Hunter, J. E. (1997). Needed: A ban on the significance test. *Psychological Science, 8,* 3–7.

Hunter, J. E., & Schmidt, F. L. (1990). *Methods for meta-analysis: Correcting error and bias in research findings.* Newbury Park, CA: Sage.

Huynh, H., & Feldt, L. S. (1976). Estimation of the Box correction for degrees of freedom from sample data in randomized block and split-plot designs. *Journal of Educational Statistics, 1,* 69–82.

Huynh, H., & Mandeville, G. K. (1979). Validity conditions in repeated measures designs. *Psychological Bulletin, 86,* 964–973.

Johnson, P. O., & Neyman, J. (1936). Tests of certain linear hypotheses and their application to some educational problems. *Statistical Research Memoirs, 1,* 57–93.

Johnson-Laird, P. N., Legrenzi, P., & Legrenzi, M. S. (1972). Reasoning and a sense of reality. *British Journal of Psychology, 63,* 395–400.

Kaye, K. L., & Bower, T. G. R. (1994). Learning and intermodal transfer of information in newborns. *Psychological Science, 5,* 286–288.

Kendall, M. G. (1970). *Rank correlation methods* (4th ed.). London: Griffin.

Keppel, G. (1991). *Design and analysis: A researcher's handbook* (3rd ed.). Englewood Cliffs, NJ: Prentice Hall.

Keren, G. (1993). A balanced approach to unbalanced designs. In G. Keren & C. Lewis (Eds.), *A handbook for data analysis in the behavioral sciences: Statistical issues* (pp. 95–127). Hillsdale, NJ: Erlbaum.

Keren, G., & Lewis, C. (1979). Partial omega squared for ANOVA designs. *Educational and Psychological Measurement, 39,* 119–128.

Killeen, P. R. (2005). An alternative to null-hypothesis significance tests. *Psychological Science, 16,* 345–353.

Kleinbaum, D. G., Kupper, L. L., Muller, K. E., & Nizam, A. (1998). *Applied regression analysis and other multivariable methods.* Pacific Grove, CA: Brooks/Cole.

Kline, R. B. (2004). *Beyond significance testing: Reforming data analysis methods in behavioral research.* Washington, DC: American Psychological Association.

Kruskal, W. H., & Wallis, W. A. (1952). Use of ranks in one-criterion variance analysis. *Journal of the American Statistical Association, 47,* 583–621.

Lee, R. M., & Robbins, S. B. (1998). The relationship between social connectedness and anxiety, self-esteem, and social identity. *Journal of Counseling Psychology, 45,* 338–345.

Levene, H. (1960). Robust tests for the equality of variances. In I. Olkin (Ed.), *Contributions to probability and statistics.* Palo Alto, CA: Stanford University Press.

Leventhal, L., & Huynh, C-L. (1996). Directional decisions for two-tailed tests: Power, error rates, and sample size. *Psychological Methods, 1,* 278–292.

Lewis, C. (1993). Analyzing means from repeated measures data. In G. Keren & C. Lewis (Eds.), *A handbook for data analysis in the behavioral sciences: Statistical issues* (pp. 73–94). Hillsdale, NJ: Erlbaum.

Loftus, G. R. (1996). Psychology will be much better science when we change the way we analyze data. *Current Directions in Psychological Science, 5,* 161–170.

Loftus, G. R., & Masson, M. E. J. (1994). Using confidence intervals in within-subject designs. *Psychonomic Bulletin & Review, 1,* 476–490.

Lykken, D. E. (1968). Statistical significance in psychological research. *Psychological Bulletin, 70,* 151–159.

Lyon, D., & Greenberg, J. (1991). Evidence of codependency in women with an alcoholic parent: Helping out Mr. Wrong. *Journal of Personality and Social Psychology, 61,* 435–439.

Mann, H. B., & Whitney, D. R. (1947). On a test of whether one of two random variables is stochastically larger than the other. *Annals of Mathematical Statistics, 18,* 50–60.

Masson, M. E. J., & Loftus, G. R. (2003). Using confidence intervals for graphically based data interpretation. *Canadian Journal of Experimental Psychology, 57,* 203–220.

Mauchly, J. W. (1940). Significance test for sphericity of a normal *n*-variate distribution. *Annals of Mathematical Statistics, 11,* 204–209.

McClelland, G. H., & Judd, C. M. (1993). Statistical difficulties of detecting interactions and moderator effects. *Psychological Bulletin, 114,* 376–390.

Meehl, P. E. (1990). Why summaries of research on psychological theories are often uninterpretable. *Psychological Reports, 66,* 195–244.

Morris, S. B., & DeShon, R. P. (2002). Combining effect size estimates in meta-analysis with repeated measures and independent-groups designs. *Psychological Methods, 7,* 105–125.

Mosteller, F. (1948). A *k*-sample slippage test for an extreme population. *Annals of Mathematical Statistics, 19,* 58–65.

Mulaik, S. A., Raju, N. S., & Harshman, R. A. (1997). There is a time and place for significance testing. In L. L. Harlow, S. A. Mulaik, & J. H. Steiger (Eds.), *What if there were no significance tests?* (pp. 65–115). Hillsdale, NJ: Erlbaum.

Myers, J. L., & Well, A. D. (1995). *Research design and statistical analysis,* Hillsdale, NJ: Erlbaum.

Neyman, J., & Pearson, E. S. (1928). On the use and interpretation of certain test criteria for purposes of statistical inference. Part I. *Biometrika, 20A,* 175–263.

Neyman, J., & Pearson, E. S. (1933). On the problem of the most efficient tests of statistical hypotheses. *Philosophical Transactions of the Royal Society of London, Series A, 231,* 289–337.

Nickerson, R. S. (2000). Null hypothesis significance testing: A review of an old and continuing controversy. *Psychological Methods, 5,* 241–301.

O'Brien, R. G. (1981). A simple test for variance effects in experimental designs. *Psychological Bulletin, 89,* 570–574.

Olejnik, S., & Algina, J. (2003). Generalized eta and omega squared statistics: Measures of effect size for some common research designs. *Psychological Methods, 8,* 434–447.

Overall, J. E. (1980). Power of chi-square tests for 2 × 2 contingency tables with small expected frequencies. *Psychological Bulletin, 87,* 132–135.

Overall, J. E., & Spiegel, D. K. (1969). Concerning least squares analysis of experimental data. *Psychological Bulletin, 72,* 311–322.

Pedhazur, E. J. (1982). *Multiple regression in behavioral research.* New York: Holt, Rinehart and Winston.

Pollard, P., & Richardson, J. T. E. (1987). On the probability of making type I errors. *Psychological Bulletin, 102,* 159–163.

Reichardt, C. S., & Gollob, H. F. (1999). Justifying the use and increasing the power of a *t* test for a randomized experiment with a convenience sample. *Psychological Methods, 4,* 117–128.

Rom, D. M. (1990). A sequentially rejective test procedure based on a modified Bonferroni inequality. *Biometrika, 77,* 663–665.

Rosenthal, R. (1979). The "file drawer problem" and tolerance for null results. *Psychological Bulletin, 86,* 638–641.

Rosenthal, R. (1991). *Meta-analytic procedures for social research* (Rev. ed.). Newbury Park, CA: Sage.

Rosenthal, R. (1993). Cumulating evidence. In G. Keren & C. Lewis (Eds.), *A handbook for data analysis in the behavioral sciences. Methodological issues* (pp. 519–559). Hillsdale, NJ: Erlbaum.

Rosenthal, R., & Rosnow, R. (1975). *The volunteer subject.* New York: Wiley.

Rosnow, R., & Rosenthal, R. (1989). Definition and interpretation of inter-action effects. *Psychological Bulletin, 105,* 143–146.

Rosnow, R., & Rosenthal, R. (1991). If you're looking at the cell means, you're not looking at only the interaction (unless all main effects are zero). *Psychological Bulletin, 110,* 574–576.

Rosnow, R., & Rosenthal, R. (1996). Contrasts and interactions redux: Five easy pieces. *Psychological Science, 7,* 253–257.

Rozeboom, W. W. (1960). The fallacy of the null hypothesis significance test. *Psychological Bulletin, 57,* 416–428.

Satterthwaite, F. E. (1946). An approximate distribution of estimates of variance components. *Biometrics Bulletin, 2,* 110–114.

Scheffé, H. (1953). A method for judging all contrasts in the analysis of variance. *Biometrika, 40,* 87–104.

Schmidt, F. L. (1996). Statistical significance testing and cumulative knowledge in psychology. *Psychological Methods, 1,* 115–129.

Schwartz, L., Slater, M. A., & Birchler, G. R. (1994). Interpersonal stress and pain behaviors in patients with chronic pain. *Journal of Consulting and Clinical Psychology, 62,* 861–864.

Seaman, M. A., Levin, J. R., & Serlin, R. C. (1991). New developments in pairwise multiple comparisons: Some powerful and practicable procedures. *Psychological Bulletin, 110,* 577–586.

Serlin, R. C., & Lapsley, D. K. (1993). Rational appraisal of psychological research and the good-enough principle. In G. Keren & C. Lewis (Eds.), *A handbook for data analysis in the behavioral sciences: Methodological issues* (pp. 199–228). Hillsdale, NJ: Erlbaum.

Shaffer, J. P. (1986). Modified sequentially rejective multiple test procedures. *Journal of the American Statistical Association, 81,* 826–831.

Sidak, Z. (1967). Rectangular confidence regions for the means of multivariate normal distributions. *Journal of the American Statistical Association, 62,* 623–633.

Siegel, S., & Castellan, N. J., Jr. (1988). *Nonparametric statistics for the behavioral sciences* (2nd ed.). New York: McGraw-Hill.

Singer, J. D., & Willett, J. B. (2003). *Applied longitudinal data analysis: Modeling change and event occurrence.* New York: Oxford University Press.

Steiger, J. H., & Fouladi, R. T. (1997). Noncentrality interval estimation and the evaluation of statistical models. In L. L. Harlow, S. A. Mulaik, & J. H. Steiger (Eds.), *What if there were no significance tests?* (pp. 221–257). Hillsdale, NJ: Erlbaum.

Sternberg, S. (1966). High-speed scanning in human memory. *Science, 153,* 652–654.

Stevens, J. (1999). *Intermediate statistics: A modern approach* (2nd ed.). Hillsdale, NJ: Erlbaum.

Thompson, B. (1993). The use of statistical significance tests in research: Bootstrap and other alternatives. *Journal of Experimental Research, 61,* 361–377.

Thompson, B. (1995). Stepwise regression and stepwise discriminant analysis need not apply here: A guidelines editorial. *Educational and Psychological Measurement, 55,* 525–534.

Tomarken, A. J., & Serlin, R. C. (1986). Comparison of ANOVA alternatives under variance heterogeneity and specific noncentrality structures. *Psychological Bulletin, 99,* 90–99.

Tranel, D., & Damasio, H. (1994). Neuroanatomical correlates of electrodermal skin conductance responses. *Psychophysiology, 31,* 427–438.

Tukey, J. W. (1977). *Exploratory data analysis.* Reading, MA: Addison-Wesley.

van Lawick-Goodall, J. (1971) *In the shadow of man*. Boston: Houghton Mifflin.

Welch, B. L. (1947). The generalization of "Student's" problem when several different population variances are involved. *Biometrika, 34*, 28–35.

Welch, B. L. (1951). On the comparison of several mean values: An alternative approach. *Biometrika, 38*, 330–336.

Wilcox, R. R. (2001). *Fundamentals of modern statistical methods: Substantially improving power and accuracy*. New York: Springer-Verlag.

Wilcox, R. R. (2003). *Applying contemporary statistical techniques*. San Diego, CA: Academic Press.

Wilcoxon, F. (1949). *Some rapid approximate statistical procedures*. Stamford, CT: American Cyanamid Company, Stamford Research Laboratories.

Wilkinson, L. (1979). Tests of significance in stepwise regression. *Psychological Bulletin, 86*, 168–174.

Wilkinson, L., & the Task Force on Statistical Inference. (1999). Statistical methods in psychology journals: Guidelines and explanations. *American Psychologist, 54*, 594–604.

Winkler, I., Kujala, T., Tiitinen, H., Sivonen, P., Alku, P., Lehtokoski, A., et al. (1999). Brain responses reveal the learning of foreign language phonemes. *Psychophysiology, 36*, 638–642.

Yates, F. (1934). Contingency tables involving small numbers and the chi-square test. *Supplement to the Journal of the Royal Statistical Society, 1* (2), 217–235.

Yuen, K. K. (1974). The two-sample trimmed t for unequal population variances. *Biometrika, 61*, 165–170.

Yuen, K. K., & Dixon, W. J. (1973). The approximate behavior and performance of the two-sample trimmed t. *Biometrika, 60*, 369–374.